11th International Conference on Aerophysics and Physical Mechanics of Classical and Quantum Systems (APhM 2017)

Journal of Physics: Conference Series Volume 1009

Moscow, Russia
21 - 24 November 2017

Editor:

Sergey T. Surzhikov

ISBN: 978-1-5108-6267-8
ISSN: 1742-6588

Printed from e-media with permission by:

Curran Associates, Inc.
57 Morehouse Lane
Red Hook, NY 12571

CURRAN ASSOCIATES INC. proceedings.com

Some format issues inherent in the e-media version may also appear in this print version.

Printed by Curran Associates, Inc. (2018)

Institute of Physics
Dirac House, Temple Back
Bristol BS1 6BE UK

Phone: 44 1 17 929 7481
Fax: 44 1 17 920 0979

techtracking@iop.org

Additional copies of this publication are available from:

Curran Associates, Inc.
57 Morehouse Lane
Red Hook, NY 12571 USA
Phone: 845-758-0400
Fax: 845-758-2633
Email: curran@proceedings.com
Web: www.proceedings.com

Table of contents

Volume 1009

11th International Conference "Aerophysics and Physical Mechanics of Classical and Quantum Systems" - APhM-2017

21–24 November 2017, Moscow, Russian Federation

Accepted papers received: 11 April 2018
Published online: 4 May 2018

Preface

Papers

Aerothermodynamics

Phisical-Chemical Kinetics

Turbulence

Low Temperature Plasma and Gas Discharges

Processes in Ramjet and Scramjet Engines

Experimental Aerothermodinamics

Radiative Gas Dynamics

IOP Conf. Series: Journal of Physics: Conf. Series **1009** (2018) 011001 doi:10.1088/1742-6596/1009/1/011001

Preface

11th International Conference "Aerophysics and Physical Mechanics of Classical and Quantum Systems" – APhM–2017 was held from 21 to 24 November 2017 at the Ishlinsky Institute for Problems in Mechanics of the Russian Academy of Sciences (IPMech RAS), Moscow, Russia.

The aim of the Conference was to bring together leading scientists and discuss experience of new research findings, databases, computer codes, information on new instruments and methods of computational and experimental aerothermodynamics, physical experiments in the field of physical and chemical kinetics in gas dynamics, gas discharge physics, and plasma physics in the processes of combustion and explosion. This year the main scientific focus of the Conference was on high speed aerothermodynamics, and, first of all, development of numerical technology of validation of author's computer codes.

The Conference work was carried out in 5 sections:

- Computational aerothermodynamics;
- Physical gasdynamics;
- Magneto- and radiative gas dynamics;
- Physical-chemical processes and heat transfer;
- Experimental aerothermodynamics.

More than one hundred delegates from seventeen universities, scientific research institutions and companies took part in the Conference.
Totally 77 oral reports devoted to the topics of a section were presented and discussed.
40 papers were submitted for publication in the JPCS. The others were published in the electronic journal «Physical-Chemical Kinetics in Gas Dynamics» by Moscow Lomonosov State University.

Most of the papers at the conference are performed under financial support from the Programs of basic researches of the Russian academy of sciences, the corresponding projects of the Russian Foundation for Basic Research, and of the Russian Science Foundation.

All papers published in this volume of *Journal of Physics: Conference Series* have been peer reviewed through processes administered by the Editors. Reviews were conducted by expert referees to the professional and scientific standards expected of a proceedings journal published by IOP Publishing.

The next APhM Conference will be held in December, 2018 at the Institute for Problems in Mechanics of the Russian Academy of Sciences, Moscow, Russia.

Founders

The Federal Agency for Scientific Organizations (FASO Russia).

IOP Conf. Series: Journal of Physics: Conf. Series **1009** (2018) 011001 doi:10.1088/1742-6596/1009/1/011001

Organisation

- *Chairmen*
  Dmitry M. Klimov
  Sergey T. Surzhikov

- *Secretary*
  Olga V. Shalashova

Organising Committee
Sergey T. Surzhikov, Michael K. Ermakov, Igor A. Kryukov, Viktor V. Kuzenov, Larisa B. Ruleva, Olga V. Shalashova, Michael A. Kotov, Roman K. Seleznev, Sergey I. Solodovnikov.

Program Committee
- Daniil Andrienko (USA),
- Yacine Babou (Spain),
- Gianpiero Colonna (Italy),
- Michael K. Ermakov (Russia),
- Dmitry M. Klimov (chairman, Russia),
- Anatoly F. Kolesnikov (Russia),
- Igor A. Kryukov (Russia),
- Joseph Shang (USA),
- Sergey T. Surzhikov (Russia),
- Valery A. Tovstonog (Russia).

Editors
- Editor: Sergey T. Surzhikov (Russia), Dr. Sci.

Invited Editors:
- Daniil Andrienko (USA), PhD,
- Yacine Babou (Spain), PhD,
- Gianpiero Colonna (Italy), PhD,
- Alexey S. Dikaljuk (Russia), PhD,
- Michael K. Ermakov (Russia), PhD,
- Igor A. Kryukov (Russia), PhD,
- Eduard V. Teodorovich (Russia), Dr. Sci.

We thank all the conference participants. We appreciate the efforts of the members of Program and Organizing Committee during the preparation and providing the Conference.

Dmitry M. Klimov
Sergey T.Surzhikov

IOP Conf. Series: Journal of Physics: Conf. Series **1009** (2018) 011002 doi:10.1088/1742-6596/1009/1/011002

Peer review statement

All papers published in this volume of *Journal of Physics: Conference Series* have been peer reviewed through processes administered by the proceedings Editors. Reviews were conducted by expert referees to the professional and scientific standards expected of a proceedings journal published by IOP Publishing.

Calculation analysis of the experimental data of HIFiRE-I using the computer code NERAT-2D

S T Surzhikov

Ishlinsky Institute for Problems in Mechanics RAS, Vernadsky prospekt 101(1), Moscow, 119526, Russia
Moscow Institute of Physics and Technology (State University), 9 Institutskiy per., Dolgoprudny, Moscow Region, 141701, Russia
Dukhov All-Russian Scientific Research Institute of Automatics, Moscow, Russia

surg@ipmnet.ru

Abstract. A computer simulation of the densities of convective heat fluxes to the surface of a blunted circular cone has been performed at hypersonic flight speeds. For comparison, the conditions implemented in the scientific program HIFiRE-I were used.
Calculation studies were performed using the author's computer code NERAT-2D and the algebraic models of turbulent mixing of Baldwin – Lomax and Prandtl. A satisfactory agreement between the calculated and experimental data is shown.

1. Introduction

The international research program HIFiRE (Hypersonic International Flight Research and Experimentation) [1– 3] was created with the aim of conducting flight experiments in the field of hypersonic technology with an emphasis on the study of aerothermodynamics and thermogasdynamics problems that are extremely difficult or impossible to solve using aerodynamic stands.

The main executors of the scientific programs were the US AFRL (United States Air Force Research Laboratory) and DSTO (Australian Defense Science and Technology Organization). Significant participation in the program was made by Boing and the University of Queensland in Australia.

With the development of hypersonic technology, it became clear that one of the main problems is the extremely high thermal and accompanying force impact on the aircraft. For the design of new devices, a detailed knowledge of these influences is necessary in conditions as close as possible to the real ones. A wide range of ground studies performed using aerodynamic tubes showed that the specific features of hypersonic aerophysics are not fully modeling on ground stands. There was an urgent problem of conducting flight experiments and filling the database on the results of hypersonic research both on ground stands and in real flight.

Obviously, such a work should be closely linked with the use of modern computing and information computer technologies. One should pay attention to the fundamental difference between the HIFiRE program and other hypersonic programs, for example, the X-41 and X-51 [4,5], the main purpose of which was to demonstrate the technological possibilities of implementing a controlled hypersonic flight. The HIFiRE program focused on the study of the fundamental scientific problems of hypersonic flight.

The main objects of research were:
1) Aerodynamics and aerothermodynamics;
2) Thermogasdynamics of scramjets;
3) Integration of the propulsion system with a high-speed glider;
4) Flight management;
5) High-temperature materials and structures;
6) Organization of measurement of high temperatures and force impacts;
7) System research in the field of organization of hypersonic flights.

Another important element of the HIFiRE program was the organization and mutual coordination of flight, bench and computer studies of the above objects.

Nine flight experiments were planned in the scientific program HIFiRE, not all of which were realized:
1) HIFiRE-0: testing of control systems, instrumentation and software for research launches;
2) HIFiRE-1 and 5: investigation of hypersonic boundary layer features near the surface of circular and elliptical blunted cones;
3) HIFiRE-2, 3 and 7: are directed to the study of the thermogasdynamics of a scramjet;
4) HIFiRE-4 and 6: research problems of control of hypersonic flow;
5) HIFiRE-8: study of integration problems of the propulsion system with a hypersonic glider.

Flight experiments were performed using rocket accelerators. Most of the experiments were performed on the descending part of the ballistic parabolic trajectory. The HIFiRE-1 experiment was implemented on March 22, 2010. The main purpose of this experiment was to study the aerodynamic heating of a blunted circular cone. This launch was preceded by intensive ground-based experiments and calculated studies using computer codes [6,7]. Despite a number of technical problems in the performance of this flight experiment, some experimental data were obtained and published in [3]. The HIFiRE-5 experiment was realized on April 23, 2012. The main task of this experiment was to study laminar and turbulent junction on an elliptical cone.

As already noted, the experiment HIFiRE-1 was preceded by bench tests and calculated studies of the flight conditions of the experimental blunted circular cone. In [6, 7] a detailed description of experiments performed on the shock tubes of the Calspan University at the Buffalo Research Center (CUBRC) under the leadership of M. Holden is given. In particular, a detailed analysis of the experimental and calculated distributions of the density of convective heat fluxes along the surface of a blunted cone with fixation of a section of a laminar-turbulent transition is given in [7]. Two experiments were thoroughly investigated: flow around the cone with the Mach number $M=6.58$ (test # 4) and at $M=7.16$ (test # 30). The results of numerical simulation using computer code DPLR [8] and different models of turbulent mixing are also given: the Baldwin-Lomax model [9], the Spalart – Allmaras model [10], and SST model [11]. In the numerical studies, the STABL code was also used [12, 13].

In this paper, we present the results of numerical simulation of convective heating of the experimental HIFiRE-I cone for the conditions at $M=6.58$ and $M=7.16$ using the author's NERAT-2D computer code. This code is based on finite difference method, the integration of a system of Reynolds-averaged Navier – Stokes equations on multi-block structured grids. Various versions of this code were used to analyze the flight and calculated data of Fire-2 [14], Apollo-4 [15], RAMC-II [16], as well as experiments [17]. For the modeling of turbulent mixing, the algebraic models of the Prandtl mixing path (PMM) [18] and the Baldwin-Lomax model (BLM) [9] were used.

2. Results of numerical simulation

The initial numerical simulation data are presented in table 1. Location of the laminar-turbulent transition was determined by input of the critical Reynolds $\mathrm{Re}_* = \rho_\infty V_\infty x / \mu_\infty$ number, where ρ_∞, μ_∞ are the density and dynamic viscosity in the oncoming stream, V_∞ is the velocity in the oncoming stream.

Table 1. Initial data for calculation of convective heating of the experimental model HIFiRE-1.

Variant	p_∞ (erg/cm^3)	ρ_∞ (g/cm^3)	T_∞ (K)	V_∞ (cm/s)	M
A	7.33×10^4	0.124×10^{-3}	214.4	1.927×10^5	6.58
B	4.62×10^4	0.720×10^{-4}	231.7	2.183×10^5	7.16

Since the experimental data are presented in [7] only for the conical and cylindrical part of the experimental blunted cone HIFiRE-1, the closing cone of the real experimental model was not taken into account.

Figure 1 shows two types of computation grids, which were used in this paper. Four blocks of the grid were used. The first type of structured nonhomogeneous computational grid (figure 1, *a*) had the following number of nodes in four blocks:

1st block (near blunted nose): $NI \times NJ = 401 \times 81$, where *NI*, *NJ* is the number of nodes along the normal to the surface and along the surface;

2nd block (along the conical and cylindrical parts of the model): $NI \times NJ = 401 \times 401$;

3rd block: $NI \times NJ = 401 \times 97$;

4th block: $NI \times NJ = 41 \times 97$.

Note that in the fourth block the index of the grid $J = 1, ..., NJ$ varies along the normal to the surface.

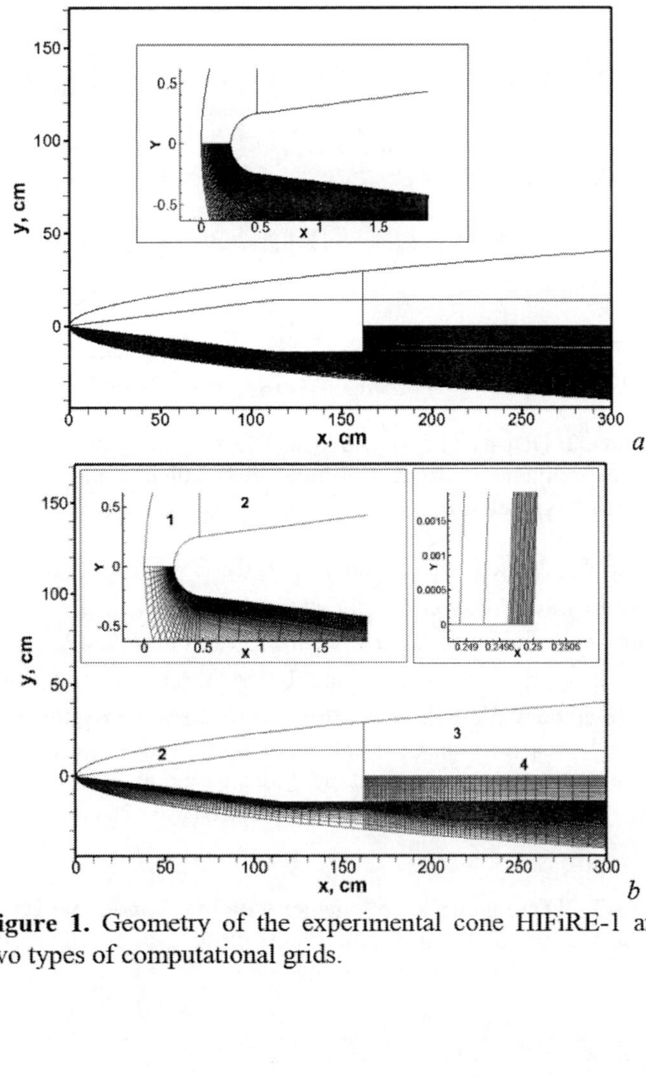

Figure 1. Geometry of the experimental cone HIFiRE-1 and two types of computational grids.

IOP Conf. Series: Journal of Physics: Conf. Series **1009** (2018) 012001 doi:10.1088/1742-6596/1009/1/012001

The second type of computation grids (figure 1, *b*) was significantly more economical:

1st block: $NI \times NJ = 101 \times 21$; 3d block: $NI \times NJ = 101 \times 55$;

2nd block: $NI \times NJ = 101 \times 201$; 4th block: $NI \times NJ = 21 \times 55$.

However, in the first cell resting on the surface, an additional grid of 20 nodes was inserted (figure1, *b*).

Figures 2 and 3 show numerical simulation results for the densities of convective heat fluxes on surfaces obtained using a grid of the 1st type. In this case, the minimum value $y^+ = y\sqrt{\frac{\rho_w}{\mu_w}\left(\frac{\partial u}{\partial y}\right)_w} = 20$, where u is the velocity along the surface at a distance y from it.

Figure 2 shows the results of calculations using the Prandtl mixing model (PMM) [18]. Here we demonstrate the effect of the critical Reynolds number and the turbulent Prandtl number on the numerical simulation results for experimental variant A (see table 1).

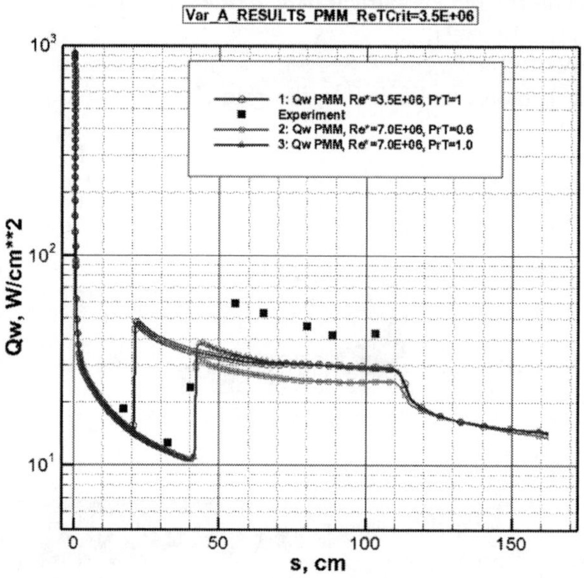

Figure 2. Density of convective heat fluxes along the surface in the experimental variant A. Model of turbulent mixing PMM. The first type of grid.

Figure 3 compares the calculated data obtained with the Baldwin – Lomax model (BLM) and the PMM model for a fixed Reynolds number Re_* for experimental variant B.

The results of similar calculations on a much coarser grid, but with a sublayer in the first cell, are shown in figure 4 (variant A) and in figure 5 (variant B). In this case, the minimum value is $y^+ = 0.5$.

Presented computational data allow us to confirm the well-known results of the simulation of heat transfer in turbulent flow about the fundamental importance of choosing a detailed calculation grid near the surface. Note that a coarse grid along the surface also leads to significant errors in determining the density of heat fluxes on a cylindrical surface ($x > 120$ cm).

3. Conclusion

Validation of the NERAT-2D computer code together with the Prandtl and Baldwin-Lomax turbulent mixing algebraic models was performed using the experimental data of HIFiRE-I.

A satisfactory description of the experimental data is shown. The importance of using a detailed grid for simulating turbulent heat transfer and good agreement with experimental data was confirmed.

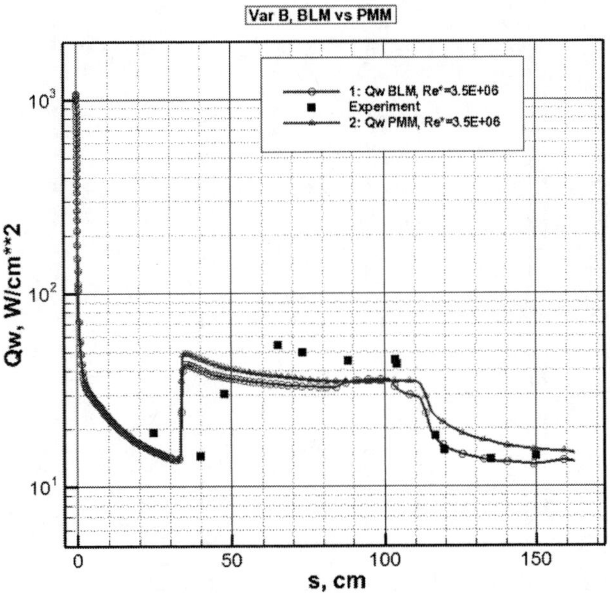

Figure 3. Density of convective heat fluxes along the surface in experimental version B. Models of turbulent mixing BLM (1) and PMM (2). The first type of grid.

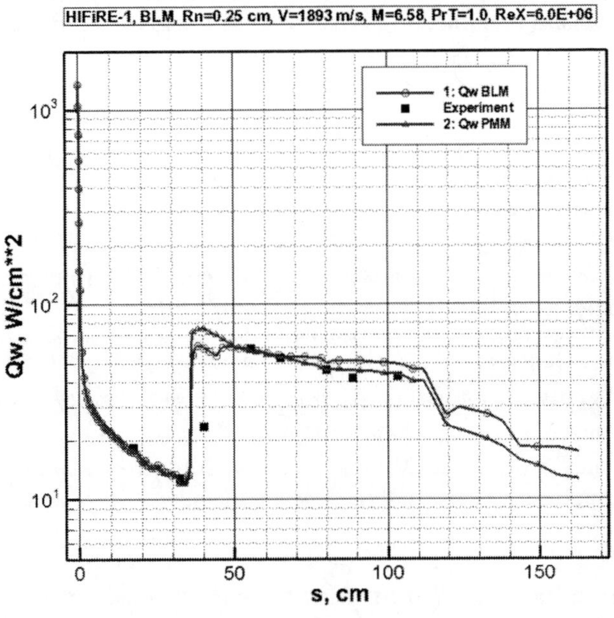

Figure 4. Density of convective heat fluxes along the surface in the experimental version A. Models of turbulent mixing BLM (1) and PMM (2). The second type of grid.

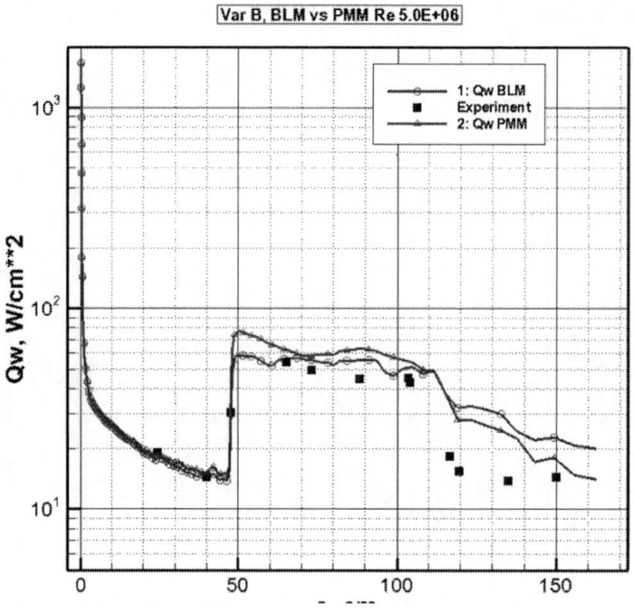

Figure 5. Density of convective heat fluxes along the surface in experimental version B. Turbulent mixing models BLM (1) and PMM (2). The second type of grid.

Acknowledgments
The work was performed within the framework of the Government program of basic research of the Russian academy of sciences (contract # AAAA–A17–117021310372–6) and partially under support of the RFBR grant # 016 – 01 – 00379.

References

[1] Kimmel R L, Adamczak D and Brisbane DSTO-AVD Team 2011 HIFiRE-1 Preliminary Aerothermodynamic Experiments *AIAA 2011-3413*
[2] Li F, Choudhari M, Chang C, Kimmel R and Adamczak D, 2011 Transition Analysis for the HIFiRE-1 Flight Experiment *AIAA 2011-3414*
[3] Stanfield S A, Kimmel R, and Adamczak D 2012 HIFiRE-1 Data Analysis: Boundary Layer Transition Experiment During Re-entry *AIAA 2012-1087*
[4] McClinton C R X-43 2006 Scramjet Power Breaks the Hypersonic Barrier Dryden Lectureship in Research for 2006 *AIAA 2006-0001*
[5] Hank J M, Murphy J S and Mutzman R C 2008 The X-51A Scramjet Engine Flight Demonstration Program *AIAA 2008-2540*
[6] Wadhams T, Mundy E, MacLean M and Holden M 2008 Ground Test Studies of the HIFiRE-1 Transition Experiment. Part 1: Experimental Results *J. of Spacecraft and Rockets* **45** (5) pp 1134– 1148
[7] MacLean M, Wadhams T, Holden M and Hohnson H 2008 Ground Test Studies of the HIFiRE–1 Transition Experiment. Part 2: Computational Analysis *J. of Spacecraft and Rockets* **45** (5) pp 1149– 1164
[8] Wright M J, Bose D and Candler G V 1998 A Data Parallel Line Relaxation Method for the Navier – Stokes Equations *AIAA J.* **38** (9) pp 1603– 1609
[9] Baldwin B S and Lomax H 1978 Thin Layer Approximation and Algebraic Model for Separated Turbulent Flows *AIAA Paper 78–0257*

IOP Conf. Series: Journal of Physics: Conf. Series **1009** (2018) 012001 doi:10.1088/1742-6596/1009/1/012001

[10] Spalart P R and Allmaras S R 1992 A One-Equation Turbulent Model for Aerodynamic Flows *AIAA 92-0439*
[11] Menter F L 1994 Two-Equation Eddy-Viscosity Turbulent Models for Engineering Applications *AIAA J.* **32** (8) pp 1598– 1605
[12] Johnson H and Candler G 2005 Hypersonic Boundary Layer Stability Analysis Using PSE–Chem *AIAA 2005-5023*
[13] Johnson H and Candler G 2006 Analysis of Laminar-Turbulent Transition in Hypersonic Flight Using PSE-Chem *AIAA 2006-33057*
[14] Surzhikov S T 2016 Radiative Gas Dynamics of the Fire-II Superorbital Space Vehicle *Technical Physics* **61**(3) pp 349–359
[15] Surzhikov S T 2016 Radiative Gas Dynamics of Apollo Command Module at Angle of Attack *AIAA paper 2016-4022*
[16] Surzhikov S T and Shang J S 2014 Three Dimensional Simulation of Shock Layer Ionization for RAMC-II Flight Tests *AIAA paper 2014-1078*
[17] Surzhikov S T 2013 Convective heating of small-radius spherical blunting for relatively low hypersonic velocities *High Temperature* **51** (2) pp 231–245
[18] Tannehill J C, Anderson D A and Pletcher R H 1997 *Computational Fluid Mechanics and Heat Transfer* Taylor & Francis 792 p

Computational study of the waverider aerothermodynamics by the UST3D computer code

D S Yatsukhno

Ishlinsky Institute for Problems in Mechanics of the Russian academy of sciences, Vernadsky prospect 101(1), Moscow, 119526, Russia

E-mail: yatsukhno-ds@rambler.ru

Abstract. The paper describes the numerical simulation results for two models of the waverider. The aerodynamic characteristics as well as three-dimensional flow fields were calculated by the UST3D code. The splitting method is applied for the Navier-Stokes system integration. The solution is obtained by the time relaxation. The governing equations approximations are applied for unstructured tetrahedron grids. The UST3D code verification was performed by the first waverider model aerodynamics. The scramjet was integrated with the second waverider model. The engine energy deposition simulation was conducted by the source term addition into the energy equation. In this particular case aerothermodynamics parameters estimation is presented.

1. Introduction

The gas dynamic construction method is, in fact, the reverse aerodynamic problem. The vehicle shape is obtained by the supersonic inviscid flow field about simple bodies or their combinations. The shock wave is used for the bottom surface flow compression and lift increasing. There is no shock detaching in the absence of leading edge blunting. The upper surface is usually parallel with regard to the freestream flow and possesses a negligible drag. This idea was made real by Nonweiler for the new vehicle called waverider [1]. The latter has a higher aerodynamic efficiency than the conventional aerospace systems. However, it should be noted that waverider configuration is suitable for the specific operation conditions which define streamlines and shock shape. The waverider aerodynamics can be improved for the off-design operation points by the optimization leading edge shape optimization [2]. The viscous effects and laminar-turbulent transition can be included in the optimization processes. The overall sizes, payload and volume efficiency are optimization criteria at the same time.

An ability of the waverider to integrate with other modern aerospace systems components is of practical interest. The scramjet-integrated vehicles can be constructed using the waverider generation principles. In particular, the above mentioned method of the high-efficiency lifting surfaces shaping can be also employed for the scramjet configuration and optimization [3, 4]. Besides, waveriders configurations for the various planet atmospheres were put forward [5, 6]. Thus, the waverider application area is fairly wide. The waverider is the reliable verification and validation model due to a considerable number of the computational and experimental data.

The first part of the numerical study contains the UST3D code verification using viscous-optimized waverider aerodynamic coefficients calculations as an example. The scramjet-integrated waverider model gas dynamic parameter evaluation under the condition of the energy source presence is given in the second part of the work.

2. Computational model

The UST3D code is based on the following governing equations [7]: the continuity equation, the Navier-Stokes equations and the energy equation. The thermal and calorific perfect gas state equations with the specific full energy relation and the Fourier law are used for the full system closing. The vector form of the system are given as

$$\frac{\partial}{\partial t}\begin{Bmatrix}\rho\\ \rho u\\ \rho v\\ \rho w\\ \rho E\end{Bmatrix}+\frac{\partial}{\partial x}\begin{Bmatrix}\rho u\\ \rho uu+p\\ \rho uv\\ \rho uw\\ \rho uE+pu\end{Bmatrix}+\frac{\partial}{\partial y}\begin{Bmatrix}\rho v\\ \rho vu\\ \rho vv+p\\ \rho vw\\ \rho vE+pv\end{Bmatrix}+\frac{\partial}{\partial z}\begin{Bmatrix}\rho w\\ \rho wu\\ \rho wv\\ \rho ww+p\\ \rho wE+pw\end{Bmatrix}=\frac{\partial}{\partial x}\begin{Bmatrix}0\\ \tau_{xx}\\ \tau_{xy}\\ \tau_{xz}\\ u\tau_{xx}+v\tau_{xy}+w\tau_{xz}\end{Bmatrix}+$$
$$+\frac{\partial}{\partial y}\begin{Bmatrix}0\\ \tau_{yx}\\ \tau_{yy}\\ \tau_{yz}\\ u\tau_{yx}+v\tau_{yy}+w\tau_{yz}\end{Bmatrix}+\frac{\partial}{\partial z}\begin{Bmatrix}0\\ \tau_{zx}\\ \tau_{zy}\\ \tau_{zz}\\ u\tau_{zx}+v\tau_{zy}+w\tau_{zz}\end{Bmatrix}+\frac{\partial}{\partial x}\begin{Bmatrix}0\\ 0\\ 0\\ 0\\ \lambda\frac{\partial T}{\partial x}\end{Bmatrix}+\frac{\partial}{\partial y}\begin{Bmatrix}0\\ 0\\ 0\\ 0\\ \lambda\frac{\partial T}{\partial y}\end{Bmatrix}+\frac{\partial}{\partial z}\begin{Bmatrix}0\\ 0\\ 0\\ 0\\ \lambda\frac{\partial T}{\partial z}\end{Bmatrix} \quad (1)$$

$$p=(\gamma-1)\rho U=(\gamma-1)\rho\left[E-0.5\left(u^2+v^2+w^2\right)\right] \quad (2)$$

$$U=c_V T \quad (3)$$

$$E=\frac{p}{(\gamma-1)\rho}+\frac{u^2+v^2+w^2}{2} \quad (4)$$

$$q=-\lambda\nabla T \quad (5)$$

Here ρ – density; u,v,w – velocity vector projections; x,y,z – Cartesian coordinates; p – pressure; E – specific full energy; λ – thermal conductivity; T – temperature; μ – dynamic viscosity; γ – ratio of specific heats; U – specific internal energy; c_V – specific heat capacity. Shear stresses is defined by the following relations

$$\tau_{xx}=\mu\left(\frac{4}{3}\frac{\partial u}{\partial x}-\frac{2}{3}\frac{\partial v}{\partial y}-\frac{2}{3}\frac{\partial w}{\partial z}\right),\ \tau_{xy}=\tau_{yx}=\mu\left(\frac{\partial v}{\partial x}+\frac{\partial u}{\partial y}\right)$$
$$\tau_{yy}=\mu\left(\frac{4}{3}\frac{\partial v}{\partial y}-\frac{2}{3}\frac{\partial u}{\partial x}-\frac{2}{3}\frac{\partial w}{\partial z}\right),\ \tau_{yz}=\tau_{zy}=\mu\left(\frac{\partial w}{\partial y}+\frac{\partial v}{\partial z}\right) \quad (6)$$
$$\tau_{zz}=\mu\left(\frac{4}{3}\frac{\partial w}{\partial z}-\frac{2}{3}\frac{\partial u}{\partial x}-\frac{2}{3}\frac{\partial v}{\partial y}\right),\ \tau_{xz}=\tau_{zx}=\mu\left(\frac{\partial u}{\partial z}+\frac{\partial w}{\partial x}\right)$$

The method for splitting into the physical processes was used for the mentioned system numerical integration. The general mathematical theory of such methods is described in [8]. There are significant distinctions between classical gas dynamics splitting methods [9] and unstructured grids splitting approach. In latter case, the averaged flow parameters derivatives values are considered instead conventional finite differences. The approximations obtained by the divergence theorem given as [7]

$$\left\langle \frac{\partial f}{\partial \alpha} \right\rangle_{V_i} = \frac{1}{V_i} \int_{V_i} \frac{\partial f}{\partial \alpha} dV = \frac{1}{V_i} \int_{V_i} \boldsymbol{fi} \cdot d\boldsymbol{S} \approx \frac{1}{V_i} \sum_{j=1}^{4} S_i^j n_{\alpha,i}^j f_i^j \qquad (7)$$

Here $\alpha = x, y, z$; i – tetrahedron index; j – surface index; $f_i^j = (u, \mathrm{v}, w, p, q_\alpha, \tau_{\alpha\beta})$ – flow parameters corresponding i – tetrahedron and j – surface; q_α – heat flux projections. The scheme for present explicit splitting method has a first approximation order by time and space. The slip boundary conditions are applied on the vehicle surface. The wall temperature is constant.

3. Numerical simulation results

3.1. Viscous-optimized waverider model calculations

The conical-derived waverider virtual model generated for the on-design Mach number of 6 is depicted in figure 1. Table 1 contains the main geometry parameters of the vehicle. There are two groups of data obtained by the various computer codes for mentioned model. The first data involves the aerodynamic coefficients which were gained by the MAXWARP code [2] and CFL3D code [10] using structured grids. The second data is the aerodynamic characteristics calculated by UST3D code which has already been applied for the validation and verification problems [11]. The code uses unstructured grids. All computations were conducted for the three flow conditions shown in table 2 at zero angle of attack. The tetrahedron grid is shown in figure 2.

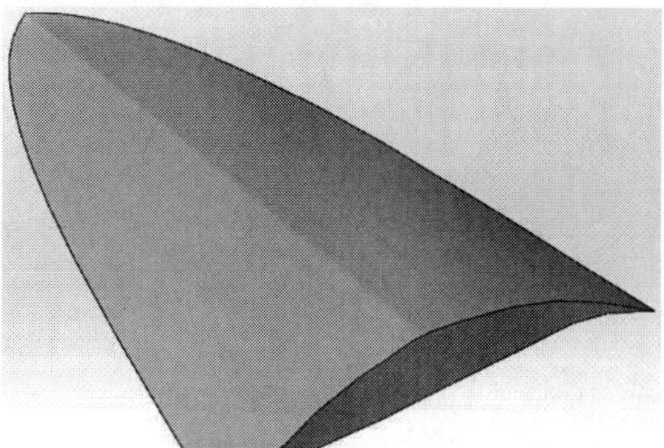

Figure 1. Waverider virtual model [10].

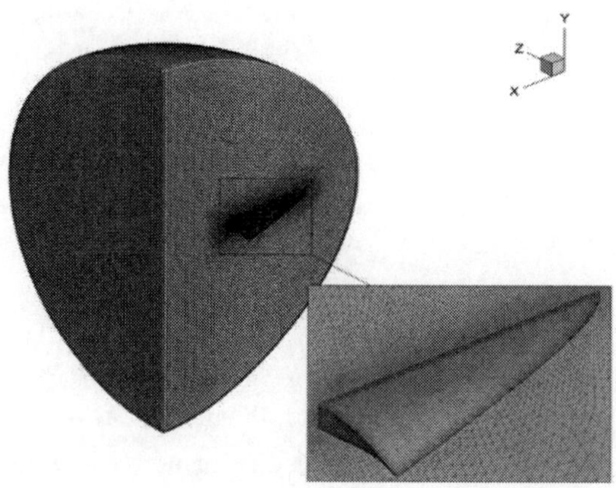

Figure 2. Unstructured grid (2515038 elements).

Table 1. Waverider reference geometry parameters.

Parameter	Value
L, cm	6000
W, cm	3725
S_{ref}, cm^2	1.423
S_{base}, cm^2	105

IOP Conf. Series: Journal of Physics: Conf. Series **1009** (2018) 012002 doi:10.1088/1742-6596/1009/1/012002

Table 2. Free-stream conditions for the various Mach number.

	$M_\infty = 4$	$M_\infty = 6$	$M_\infty = 8$
p_∞, erg/cm^3	17791	11866	8889
$\rho_\infty \cdot 10^4$, g/cm^3	0.268	0.179	0.134
T_∞, K		231.3	
$\mathrm{Re}_\infty \cdot 10^4$, cm^{-1}		2.156	

Figures 3 (a) and 3 (b) show the lift and drag coefficients which depend on the Mach number. The planform area was used as the reference area. It should be noticed that the drag coefficient contains only inviscid part which is determined by the pressure forces. The lift-to-drag ratio is presented in figure 3 (c). The values of the all parameters are decreased with Mach number increasing. The UST3D lift and drag coefficients are lower than the corresponding structured grid codes values. Nevertheless, the three various code using the different numerical methods and grid type demonstrate generally good agreement between the corresponding calculations results.

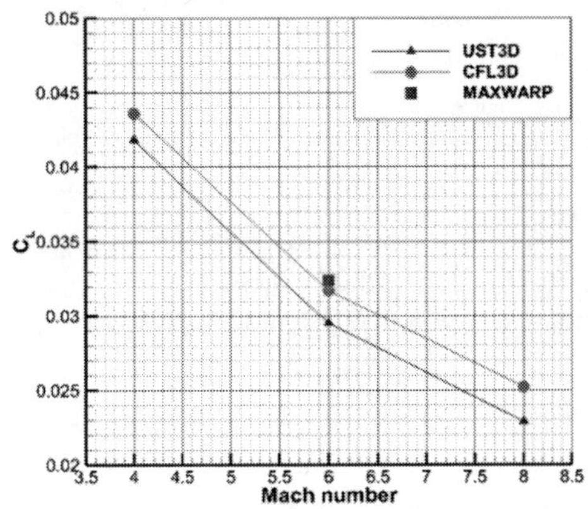

a

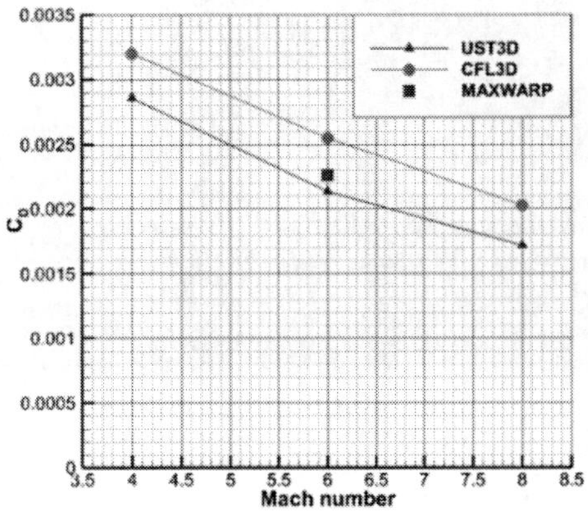

b

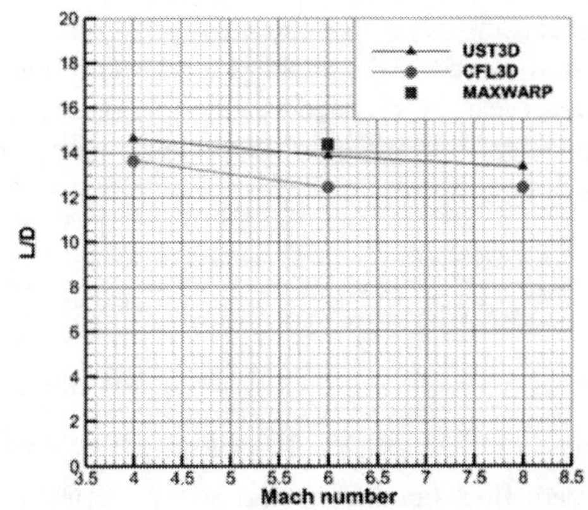

c

Figure 3. Lift (a) and drag (b) coefficients as well as lift-to-drag ratio (c) as functions of a Mach number.

The pressure distributions at the exit flow plane for the various Mach numbers are shown in figure 4. The best visual correspondence is observed between UST3D and CFL3D results [10], which were obtained for Mach 4. There are somewhat pressure contours differences for on-design Mach number and Mach 8 consisting of the different compression flow degrees about leading edge. Perhaps it is due to relatively coarse unstructured grid within the leading edge. The calculations of a decay of a discontinuity were performed by the different approaches for both codes. It partly explains obtained aerodynamic coefficients results discrepancy.

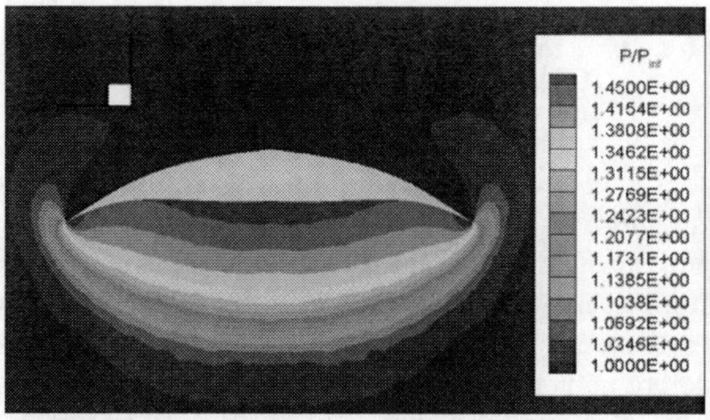

a

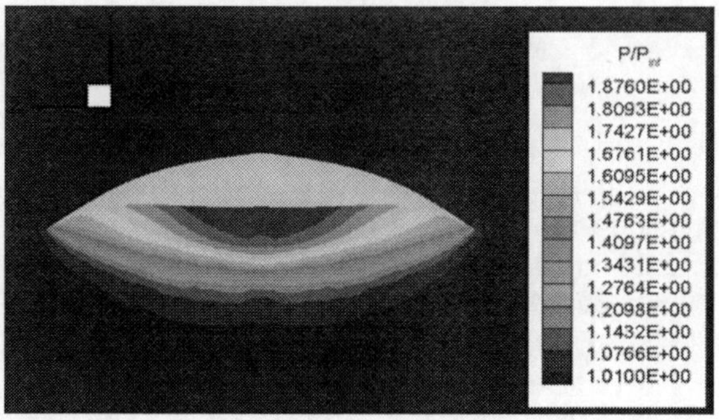

b

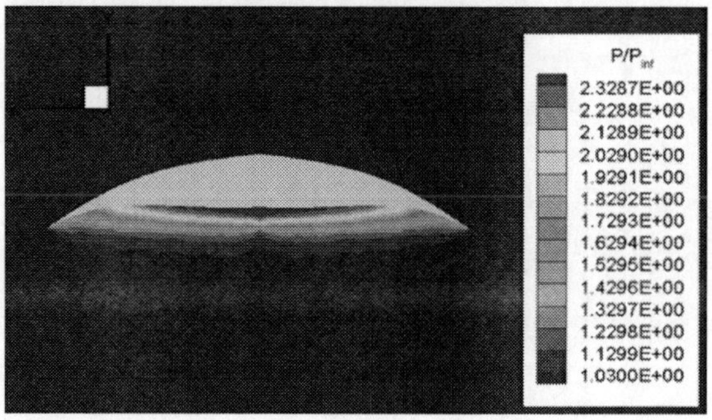

c

Figure 4. Pressures ratio contours at exit plane for Mach 4 (a), Mach 6 (b) and Mach 8 (c).

3.2. Scramjet-integrated waverider model calculations

The second task was following. The computations of the flow over a scramjet-integrated waverider were conducted at Mach 6 for two angles of attack. The waverider model and unstructured grid are presented in figures 5–7. The distinguishing feature of the work is the taking into account the source term into the energy equation. The physical processes concerning with the chemical reactions and turbulence affection on the flow parameters are outside the framework of the paper [12, 13]. The study has a following motivation: the drag reduction and lift increase can be reached by the energy impact on the flow [14, 15]. The scramjet energy deposition capacity is defined in accordance with the following relations [15]

$$q_{\max} = 0.7 \cdot \frac{I_{\infty} f_{O_2} H_u}{L V_{energy}} = 280 \ \frac{\text{W}}{\text{cm}^3} \tag{8}$$

Here I_{∞} – mass flow rate; $f_{O_2} = 0.2315$ – oxygen part; $L = 34.5$ – stoichiometric coefficient; $H_u = 119.54$ MJ/kg – fuel specific enthalpy; $V_{energy} = 3000\ \text{cm}^3$ – energy release zone volume. Energy source zone location is shown in figure 8. Free-stream flow conditions and attack angles are shown in table 3.

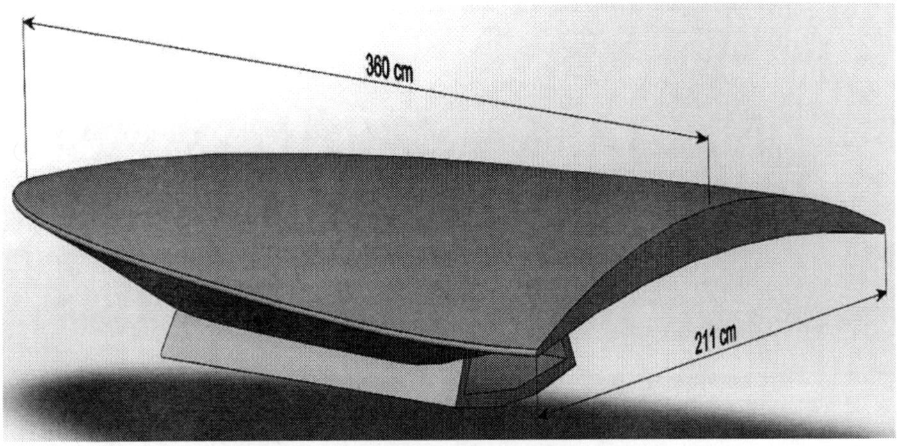

Figure 5. Scramjet-integrated model sizes.

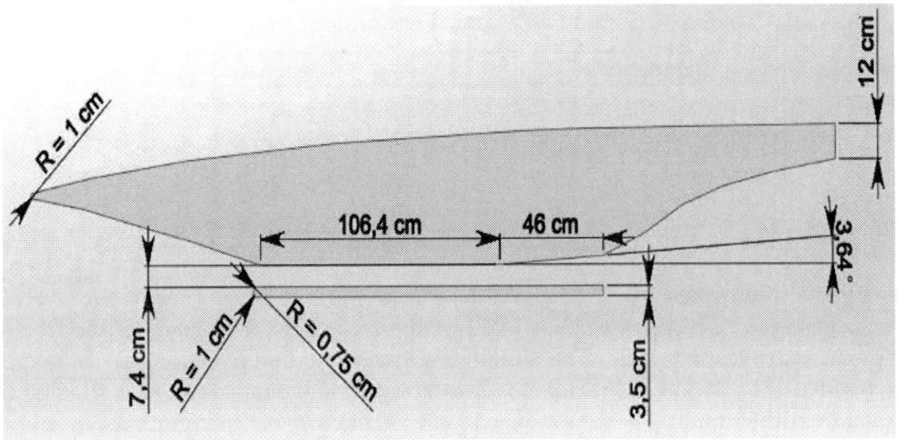

Figure 6. Other dimensions.

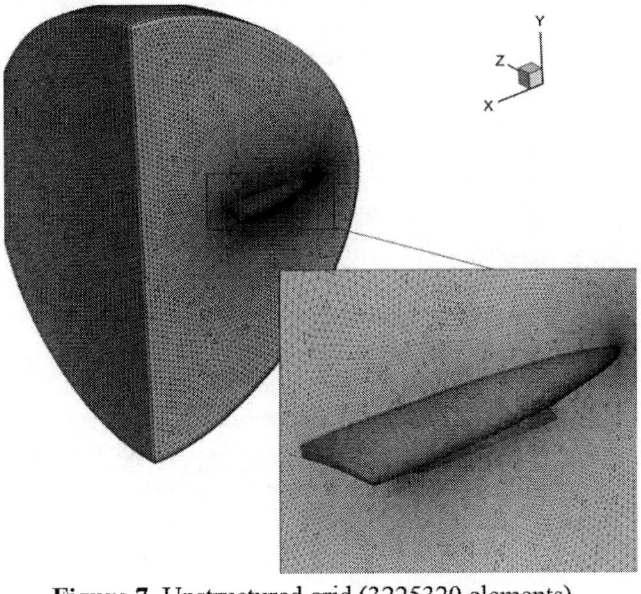

Figure 7. Unstructured grid (3225320 elements).

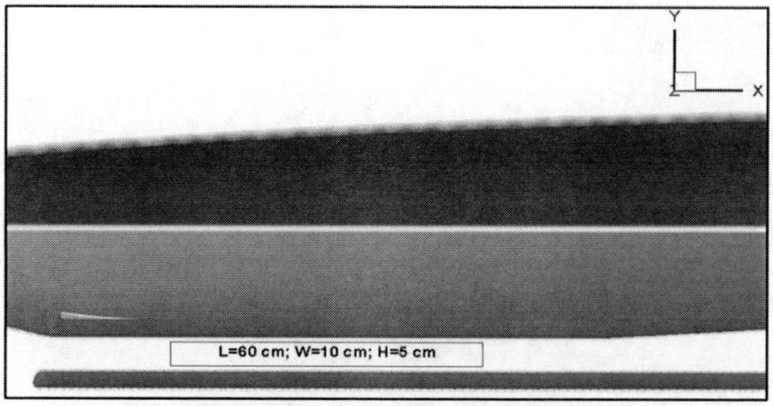

Figure 8. Energy source location.

Table 3. Scramjet-integrated waverider flow conditions.

Parameter	Value
p_∞, erg/cm^3	11900
$\rho_\infty \cdot 10^4$, g/cm^3	0.179
T_∞, K	231.3
α, °	0 and 5

Figure 9 describes Mach number distribution for various cases. There are no considerable differences in no-energy-source flows (a, b) despite the fact that various parts of the free-stream flow fall into the air intake leading edge. The additional flow stagnation is caused by the scramjet energy deposition (c). The stagnation zone is expended with respect to the no-source operation points.

The pressure distribution is shown in figure 10. The shock wave system intensity increases with the growth of angle of attack. The counter-pressure is generated by the energy source. It results in pressure increase and its redistribution behind the energy release zone (c). The strong flow compression is observed at the blunting scramjet leading edge (a, b, c).

The temperature fields in the scramjet are depicted in figure 11. The higher temperature is observed in the stagnation regions compared with the rest of the flow. The temperature influence is manifested directly at the source location and is distributed downstream. The source capacity was selected so that temperature could provide scramjet wall suitability. The maximum scramjet temperature is approximately 2200 K. The blunted nose is another configuration part which is subjected to strong heating. The maximum nose temperature is 1750 K (the calculations results are not shown here).

The upper part pressure contours at the symmetry plane are shown in Figure 12 (b). The pressure increase within the scramjet is observed in case of source deposition.

The joint information about aerodynamic characteristics is given in table 4. Another energy source with capacity $q_{max} = 400$ Wt/cm^3 was also considered. There are the following energy source features. The first is the scramjet upper wall pressure growth providing lift rise. The second is the air intake shock weakening caused by the stagnation pressure decrease. The high-capacity source provides significant lift increase and drag decrease but the scramjet wall temperature in this case is approximately 3100 K (the calculations results are not shown here).

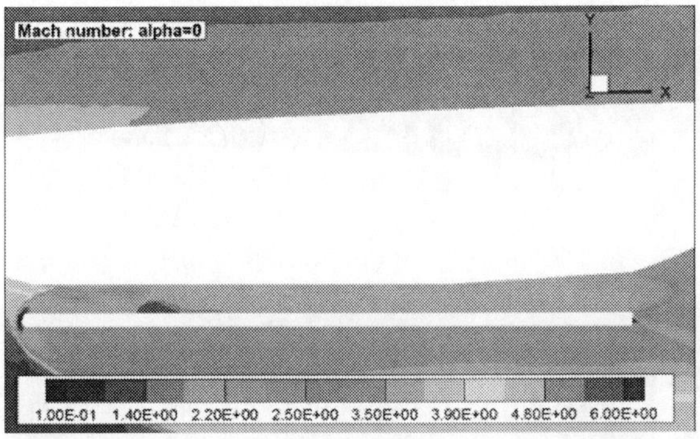

a

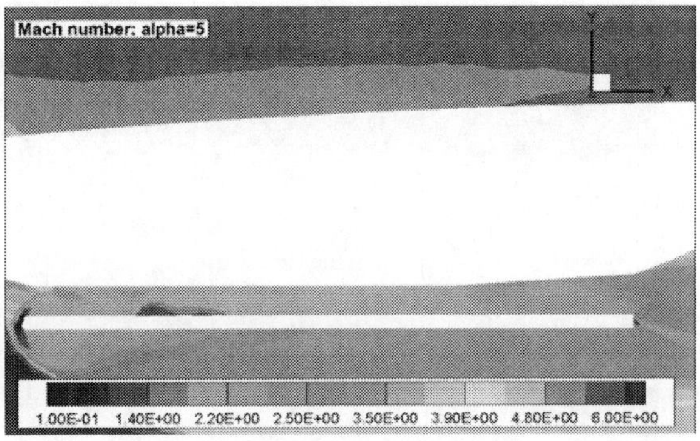

b

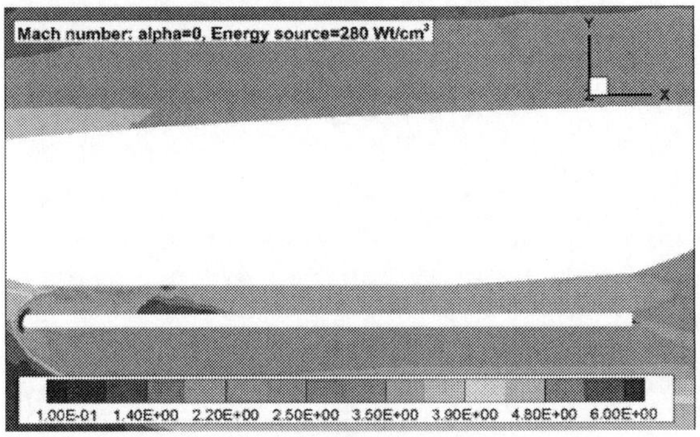

c

Figure 9. Mach number distribution: zero angle of attack (a), five degrees angle of attack (b) and zero angle of attack with energy source (c).

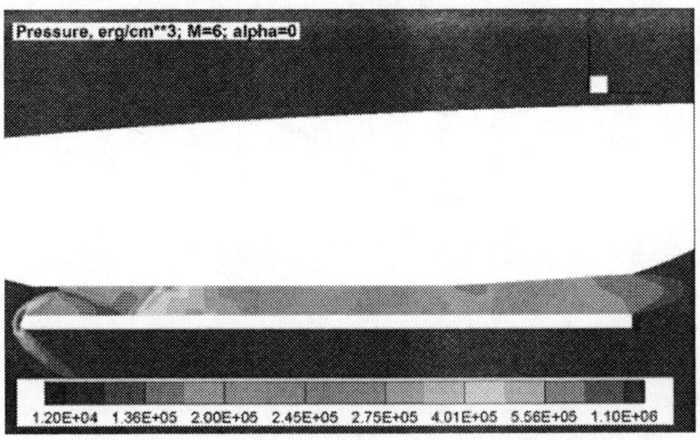

a

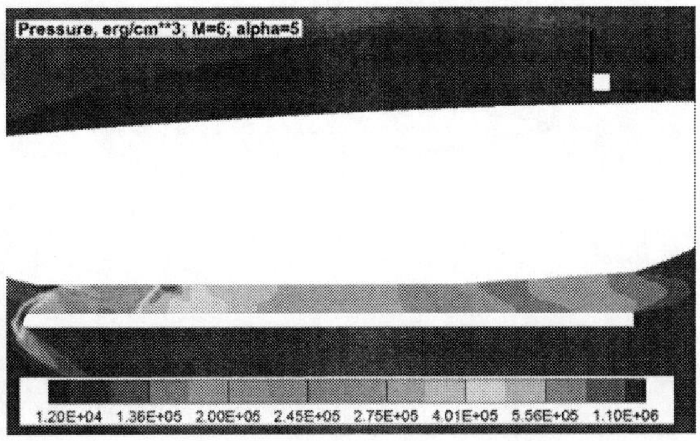

b

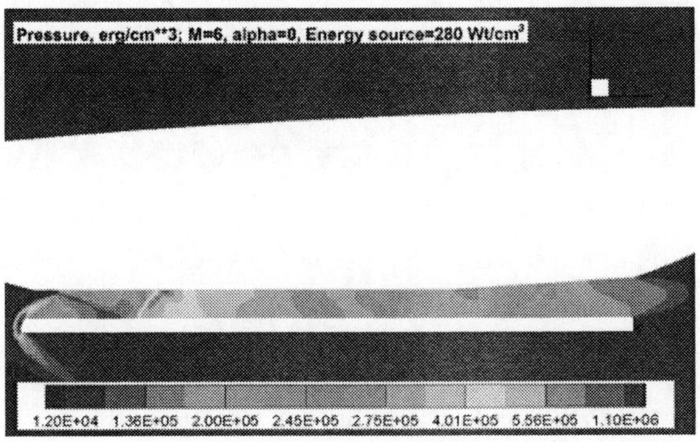

c

Figure 10. Pressure distribution: zero angle of attack (a), five degrees angle of attack (b) and zero angle of attack with energy source (c).

IOP Conf. Series: Journal of Physics: Conf. Series **1009** (2018) 012002 doi:10.1088/1742-6596/1009/1/012002

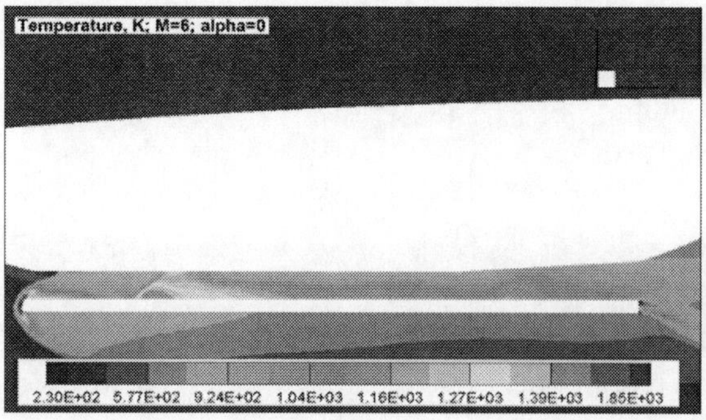

a

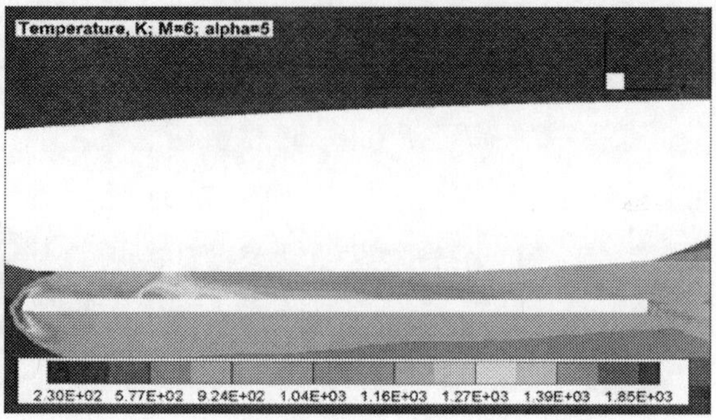

b

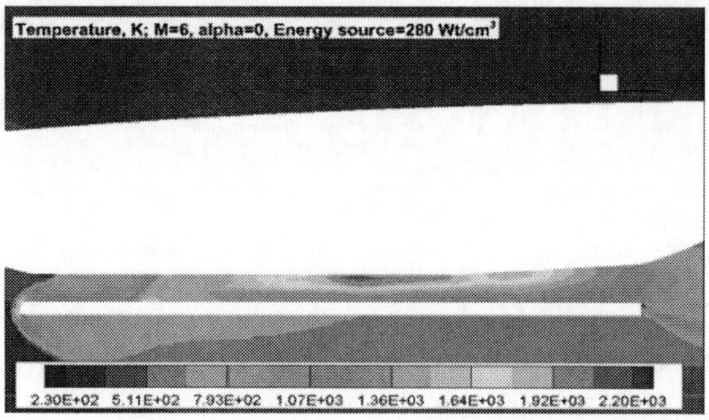

c

Figure 11. Temperature distribution: zero angle of attack (a), five degrees angle of attack (b) and zero angle of attack with energy source (c).

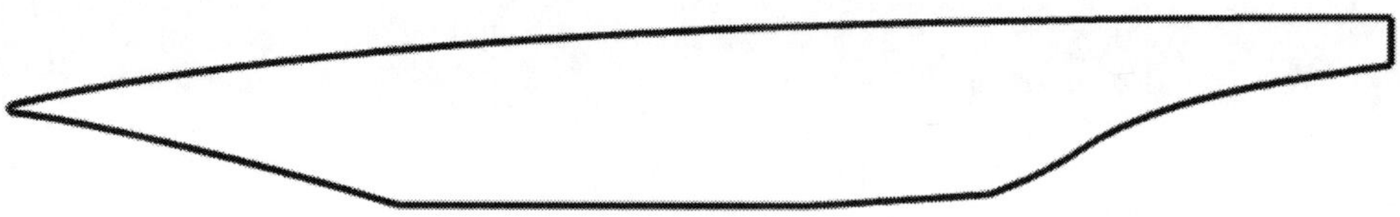

a

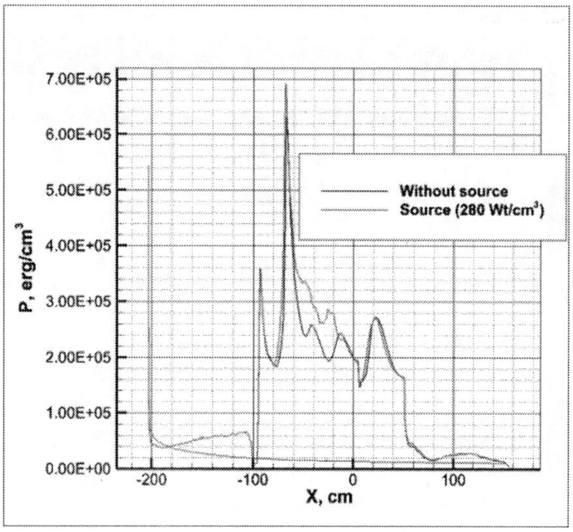

b

Figure 12. Upper surface (a) and corresponding pressure distributions (b).

IOP Conf. Series: Journal of Physics: Conf. Series **1009** (2018) 012002 doi:10.1088/1742-6596/1009/1/012002

Table 4. Aerodynamics characteristics of the scramjet-integrated vehicle.

	$q_{max}=0$		$q_{max}=280\ \frac{W}{cm^3}$	$q_{max}=400\ \frac{W}{cm^3}$
p_∞, erg/cm^3	$\alpha=0°$	$\alpha=5°$	$\alpha=0°$	
C_L	0.0217	0.429	0.0263	0.0502
C_D	0.211	0.288	0.208	0.200
L/D	0.103	1.49	0.126	0.251

4. Conclusions

The numerical study of the various hypersonic vehicles aerodynamics characteristics was conducted by the UST3D code. The good agreement between the UST3D numerical simulation results and other computer code calculations was obtained. Generally, the aerodynamic efficiency of the viscous optimized waverider obtained taking into account only pressure-based drag are not decreased considerable at the off-design operation points. The scramjet-integrated model calculations were performed with the fixed energy source and without one. The results show that the scramjet energy source allows decrease aerodynamics drag with respect to the no-source scramjet configuration. The significant increase of the lift was observed. It should be pointed that there is the additional wall heating caused by the energy source. It is necessary that energy source capacity should be selected in accordance with the scramjet chamber material thermal stability.

Acknowledgments

The work was performed within the framework of the Government program of basic research of the Russian academy of sciences (contract # AAAA-A17-117021310372-6) and partially support of the RFBR grant # 016 – 01 – 00379.

References

[1] Nonweiler T R F 1959 Aerodynamic problems of manned space vehicle *Journal of the Royal Aeronautical Society* **63** pp 521–528

[2] Corda S, Anderson J D 1988 Viscous Optimized Hypersonic Waveriders Designed from Axisymmetric Flow Fields AIAA 26th Aerospace Sciences Meeting January 11–14 Reno Nevada *AIAA Paper* 88 – 0369

[3] Ferguson F 1995 A Design Method for the Construction of Hypersonic Vehicle Configurations AIAA 6th International Aerospace Planes and Hypersonic Technologies Conference April 3–7 Chattanooga TN *AIAA Paper* 95–6009

[4] O'Neil M K L, Lewis M J 1992 Design Tradeoffs on Engine-Integrated Hypersonic Vehicles 1th Aerospace Design Conference February 3–6 Irvine CA *AIAA Paper* 92–1205

[5] Lunan D A 2015 Waverider, Revised Chronology *20th AIAA International Space Planes and Hypersonic Technologies Conference* July 6 – 9 Glasgow Scotland *AIAA Paper* 2015 – 3529

[6] Anderson J D, Lewis M J, Kothari A P and Corda S 1990 Hypersonic Waverider for Planetary Atmospheres AIAA 28th Aerospace Sciences Meeting January 8–11 Reno Nevada *AIAA Paper* 90 – 0538

[7] Zheleznyakova A L and Surzhikov S T 2013 Application of the Method of Splitting by Physical Processes for the Computation of a Hypersonic Flow over an Aircraft Model of Complex Configuration *High Temperature* **51** (6) pp 816 – 829

[8] Marchuk G I 1988 *Splitting methods* (Moscow: Science) p (in Russian)

[9] Belotserkovsky O M and Davydov Yu M 1982 *Method of large particles in gas dynamics*. (Moscow: Science) p 392 (in Russian).

[10] Takashima N and Lewis M J 1992 Navier-Stokes computations of a Viscous Optimized Waverider AIAA 30th Aerospace Sciences Meeting January 6–9 Reno Nevada

IOP Conf. Series: Journal of Physics: Conf. Series **1009** (2018) 012002 doi:10.1088/1742-6596/1009/1/012002

AIAA Paper 92–0305

[11] Surzhikov S.T. 2017 Validation of computational code UST3D by the example of experimental aerodynamic data *Journal of Physics: Conference Series* **815** 012023

[12] Seleznev R K, Surzhikov S T, Shang J A 2016 Quasi-One-Dimensional Analysis of Hydrogen-Fueled Scramjet Combustors 52th AIAA/SAE/ASEE Joint Propulsion Conference Salt Lake City *AIAA Paper* 2016 – 4569

[13] Seleznev R K 2017 Comparison of two-dimensional and quasi-one-dimensional models by the example of VAG experiment *Journal of Physics: Conference Series* **815** 012007

[14] Shang J S, Surzhikov S T, Yan H 2012 Hypersonic Nonequilibrium Flow Simulation Based on Kinetics Models *Frontiers in Aerospace Engineering* **1** (1) 12 p.

[15] Surzhikov S T and Shang J S 2013 Numerical Prediction of Convective and Radiative Heating of Scramjet Combustion Chamber with Hydrocarbon Fuels 51st Aerospace Sciences Meeting including the New Horizons Forum and Aerospace Exposition *AIAA Paper* 2013 – 1076

[16] Surzhikov S T and Shang J S 2013 Radiative Heat Exchange in a Hydrogen-Fueled Scramjet Combustion Chambers 51st Aerospace Sciences Meeting including the New Horizons Forum and Aerospace Exposition *AIAA Paper* 2013 – 0448

[17] Surzhikov S T, Zheleznyakova A L, Shang J S and Rivir R B 2013 Simulating Gasdynamic Interaction and Radiative Heating within Scramjets with Hydrocarbon Fuels 44th AIAA Thermophysics Conference *AIAA Paper* 2013 – 2642

[18] Knight D Survey of Aerodynamic Drag Reduction at High Speed by Energy Deposition *Journal of Propulsion and Power* **24** (6) pp 1153–1167

[19] Khankhasaeva Ya V, Borisov V E, Lutsky A E 2016 Energy Impact on the Flow Around Hypersonic Flying Vehicles *Physical-Chemical Kinetics in Gas Dynamics* **17**(4) (in. Russian)

IOP Conf. Series: Journal of Physics: Conf. Series **1009** (2018) 012003 doi:10.1088/1742-6596/1009/1/012003

Three-dimensional calculation of the aerothermodynamics of a double cone 25°/55° on an unstructured grid

I A Koryukov[1,2] and I A Kryukov[1]

[1]Ishlinsky Institute for Problems in Mechanics RAS, Moscow, 119526, Russia
[2]Moscow Institute of Physics and Technology (State University), Dolgoprudniy, Moscow region, 141701, Russia

E-mail: iakoryukov@mail.ru

Abstract. A three-dimensional numerical simulation of 25°/55° double-cone high-speed aerothermodynamics on an unstructured grid was carried out. For numerical simulation of the flow, a geometric model of a double cone 25°/55° was built. Numerical simulation using air and N_2 (nitrogen) was carried out. For numerical simulation, the CFD computer code UG3D developed in the Institute for Problems in Mechanics of the Russian academy of Sciences was used. This solver is based on the model of Navier-Stokes equations for a perfect gas solved on an unstructured grid. Verification and validation of the results were carried out.

1. Introduction

The high-speed flow around a double cone of 25°/55° presents considerable practical and theoretical interest. An object of interest for this flow lies at the junction of two cones, where the shock wave from the cone 25° interacts with the boundary layer of the 55° cone. Despite the simple construction of the body, in the case of numerical simulation of the flow there are some significant difficulties with exact calculation of the region of interaction of the shock wave, or the boundary layer. It makes the problem of numerical simulation of such flow a good benchmark for computer codes for high-speed flows. And as a good benchmark the problem has a lot of approved experimental data obtained by various experimental groups. One of the well-known groups is Calspan-University at Buffalo Research Center (CUBRC) [9]. CUBRC conducted many rigorous experiments, which form a significant experimental database for high-speed flows. The database is widely used for verification and validation of numerical computer codes [1-5].

In the Institute for Problems in Mechanics of the Russian academy of Sciences, there is an experimental program for the formation of database for verification and validation of computer codes for high-speed flows [6-8]. The database also contains some experimental results for a hypersonic flow near a double cone.

Experimental data for the flow near a 25°/55° double-cone obtained in CUBRC were estimated and then chosen as a good data set for validating the UG3D code. Most published results of numerical simulation of the flow [1-5] were obtained with 2D numerical codes. So, one of the goals of this work was 3D numerical simulation of the flow. We chose the laminar case with the Mach number $M = 9.59$ for code validation. Obtained results are discussed and compared with CURRC experimental data and calculated results of other authors. Some features of the obtained results are considered in more detail.

IOP Conf. Series: Journal of Physics: Conf. Series **1009** (2018) 012003 doi:10.1088/1742-6596/1009/1/012003

2. The computational flow model

For the numerical simulation of double cone flow 25°/55° aerodynamics, it is assumed that the flow is determined by compressible Navier-Stokes equations, which describe the conservation of mass, momentum, and energy in a viscous fluid:

$$\frac{\partial w}{\partial t}+\frac{\partial F^x(w)}{\partial x}+\frac{\partial F^y(w)}{\partial y}+\frac{\partial F^z(w)}{\partial z}=\frac{\partial G^x(w)}{\partial x}+\frac{\partial G^y(w)}{\partial y}+\frac{\partial G^z(w)}{\partial z}, \tag{1}$$

where F^x, F^x, F^x – are the projections of the convective flow vector (2); G^x, G^y, G^z – projections of the viscous flow vector (3); ρ - density; P - pressure; u, v, w - components of the velocity vector; E - is specific total energy of the gas; $\tau_{\alpha\beta}$ - components of the viscous stress tensor (α = x, y, z; β = x, y, z); q_x, q_y, q_z - components of the heat flux vector [13].

$$w=\begin{pmatrix}\rho\\ \rho u\\ \rho v\\ \rho w\\ \rho E\end{pmatrix} F^x=\begin{pmatrix}\rho u\\ \rho u^2+P\\ \rho uv\\ \rho uw\\ \rho uE+Pu\end{pmatrix} F^y=\begin{pmatrix}\rho v\\ \rho vu\\ \rho v^2+P\\ \rho vw\\ \rho vE+Pv\end{pmatrix} F^Z=\begin{pmatrix}\rho w\\ \rho wu\\ \rho wv\\ \rho w^2+P\\ \rho wE+Pw\end{pmatrix}, \tag{2}$$

$$G^x=\begin{pmatrix}\tau\\ \tau_{xx}\\ \tau_{yx}\\ \tau_{zx}\\ u\tau_{xx}+v\tau_{yx}+\omega\tau_{zx}-q_x\end{pmatrix} G^y=\begin{pmatrix}\tau\\ \tau_{xy}\\ \tau_{yy}\\ \tau_{zy}\\ u\tau_{xy}+v\tau_{yy}+\omega\tau_{zy}-q_y\end{pmatrix} G^z=\begin{pmatrix}\tau\\ \tau_{xz}\\ \tau_{yz}\\ \tau_{zz}\\ u\tau_{xz}+v\tau_{yz}+\omega\tau_{zz}-q_z\end{pmatrix} \tag{3}$$

A complete system of Navier-Stokes equations is used together with the perfect gas equation of state (4):

$$P=(\gamma-1)\rho U=(\gamma-1)\rho\left[E-\frac{1}{2}\left(u^2+v^2+w^2\right)\right], \tag{4}$$

where $\gamma=c_p/c_v$ - heat capacity ratio; c_p, c_v - specific heat at constant pressure and volume, respectively; $U=c_vT, E$ - specific internal and total energy of gas; T – temperature [13]. The working fluid is assumed to be a calorically perfect gas, and the ratio of specific heats is assumed to be 1.4.

The flow solver ug3D is a density-based cell-center finite-volume Navier-Stokes solver and can be considered to be a Godunov-type solver. In order to achieve second-order spatial accuracy, a piecewise linear reconstruction MUSCL approach is used [12].

3. Statement of the problem of the high-speed flow past a model

For three-dimensional numerical modeling of aerothermodynamics of the object, computer geometry of the model was created on the basis of data from open sources [10, 11] in the format *.iges in CAD (figures 1, 2).

The initial data of the oncoming stream from the Run 28, which is used in the calculations, is presented in the table 1 [11], where parameters with the index "∞" refer to the incoming stream.

4. Grid models

After the CAD geometric model of the double cone 25°/55° was built, a volume mesh was generated. For this purpose, the model was exported to the mesh generator. The CFD module of aerothermodynamics UG3D is able to perform calculations both on structured and unstructured grids.

IOP Conf. Series: Journal of Physics: Conf. Series **1009** (2018) 012003 doi:10.1088/1742-6596/1009/1/012003

However, in this case an unstructured tetrahedral volume mesh of 2.1 million cells was generated (figure 2).

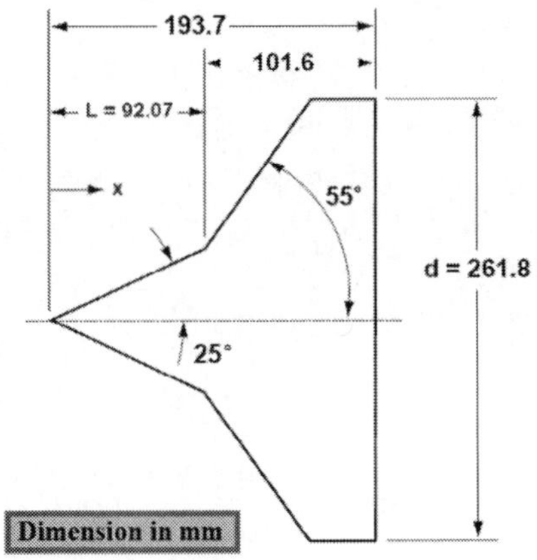

Figure 1. Schematic view of the model [11].

Figure 2. General view of the double cone model 25°/55°.

Table 1. The initial parameters.

№	V_∞, m/s	ρ_∞, kg/m^3	T_∞, K	T_W,K	M_∞	Gas	γ
Run 28	2664.0	0.0006546	185.6	293.3	9.59	N_2 (Nitrogen)	1.39
Run 28 (Air)	2664.0	6.546	185.6	293.3	9.59	Air	1.4

The unstructured type of grid was chosen because the structured grid requires a lot of computer resources to generate. In the future work, it is planned to perform a three-dimensional calculation on a structured grid. The obtained tetrahedral unstructured volume grid has regions of concentration along the entire surface of the geometric model (figures 3–6).

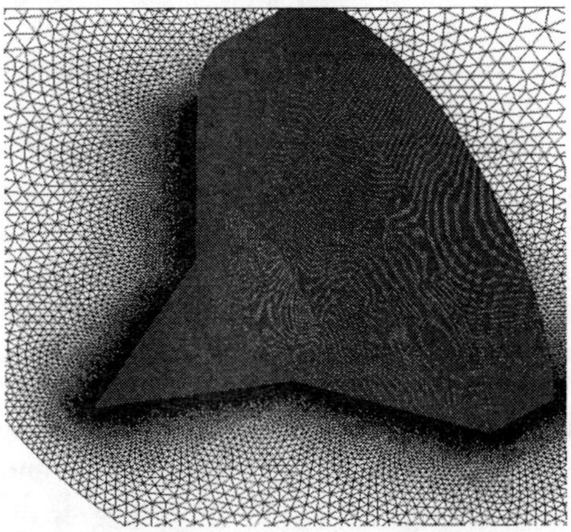

Figure 3. General view of the grid. The number of tetrahedron = 2.1 million, the number of nodes = 391135.

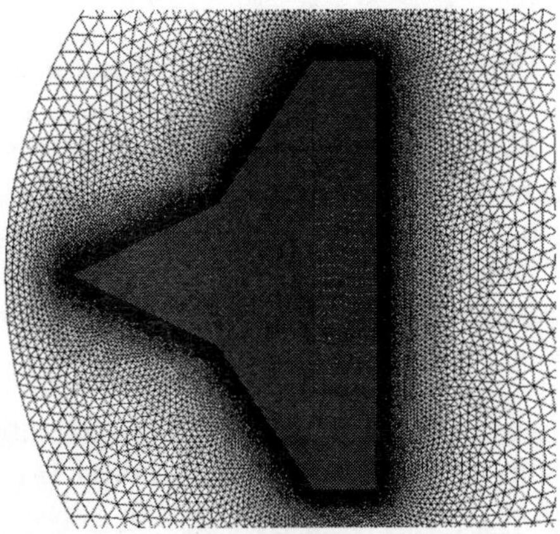

Figure 4. Grid thickening over the entire surface of the geometric model. The number of tetrahedron = 2.1 million, the number of nodes = 391135.

IOP Conf. Series: Journal of Physics: Conf. Series **1009** (2018) 012003 doi:10.1088/1742-6596/1009/1/012003

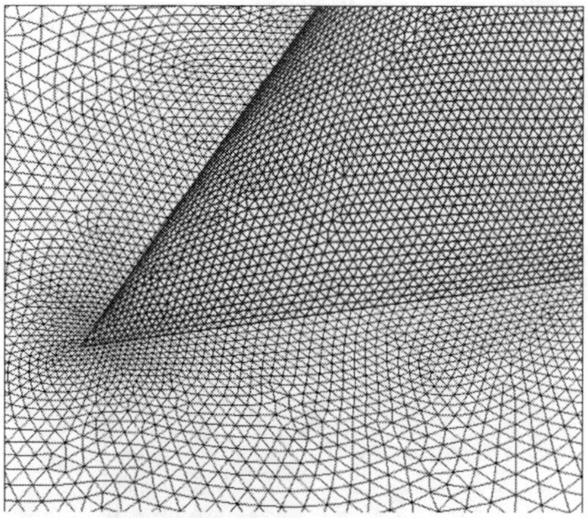

Figure 5. Grid thickening on the surface of the first cone 25°.

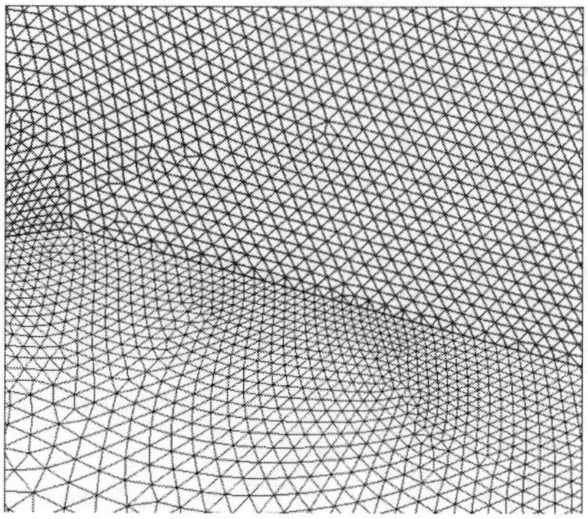

Figure 6. Grid thickening on the surface of the second cone 55°.

The quality of the grid models created was evaluated by a number of generally accepted criteria [14, 15]. The constructed grid models completely satisfy the requirements for efficient operation of the aerodynamic calculation code UG3D.

5. Results of numerical simulation

Calculations were made using the air and N_2 (nitrogen) as test gases. As an illustration of computation results, three-dimensional distributions of gas-dynamic parameters with zero angle of attack are presented (frigures 7, 8, 11-13). For example, the distribution of density, Mach number and pressure are presented below. Comparison of obtained results of the pressure coefficients with the experimental data, analytical solution in the area of 25° cone (figures 9, 10) and calculations by James N. Moss was carried out [9]. Moss uses the Monte Carlo method (DSMC) in his calculations [11]. Analytical solution of pressure coefficient on the 25° cone was described by the Taylor-McCall equations [14]:

$$\frac{\gamma-1}{2}\left[V_{max}^2 - V_r^2 - \left(\frac{dV_r}{d\theta}\right)^2\right]\left[2V_r + \frac{dV_r}{d\theta}\cot\theta + \frac{d^2V_r}{d\theta^2}\right] - \frac{dV_r}{d\theta}\left[V_2\frac{dV_r}{d\theta} + \frac{dV_r}{d\theta}\left(\frac{d^2V_r}{d\theta^2}\right)\right] = 0; \frac{dV_r}{d\theta} = V_\theta \quad (5)$$

where V_r – tangential component of the velocity vector; V_θ – normal component of the velocity vector; θ – intermediate angle of inclination of the conical generator; γ – adiabatic index; V_{max} – maximum isentropic flow rate.

All obtained results show density and pressure growth in the area of shock wave-boundary layer interaction. Using the air as a test gas the viscous and inviscid flow was computed. Firstly, an inviscid calculation was performed (figure 9). Obtained pressure coefficient has a bad agreement with experimental and Moss data in a 25–55 cone junction (experimental and Moss data show C_p growth beginning prior to junction). Calculated maximum rate of C_p is much higher than the experimental one, and also shifted by X coordinate. The second considered case was viscous calculations (figure 10). A viscous calculation shows a good agreement of pressure coefficient with Moss data in the region of cones junction as well as C_p peak values. However, the C_p maximum value is still shifted along the X coordinate.

Comparing calculated results of the pressure coefficient distribution using the N2 as a test gas with the results of Moss [11] and experiment [9], there is a great agreement of Cp peak value and the placing of Cp growth beginning with Moss' data [11] (figure 14). Comparing the results of the Cp distribution with the experimental data, one can see the difference in the peak values in the region of interaction between the shock wave and the boundary layer. Perhaps this is due to the choice of the calculation method used. Because Moss [11] uses the Monte-Carlo method and UG3D code uses the Godunov method.

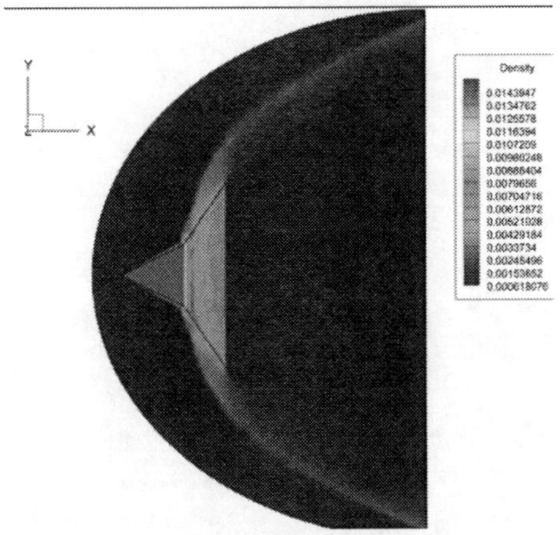

Figure 7. Distribution of density [kg/m^3] in air. $M_\infty = 9.59$.

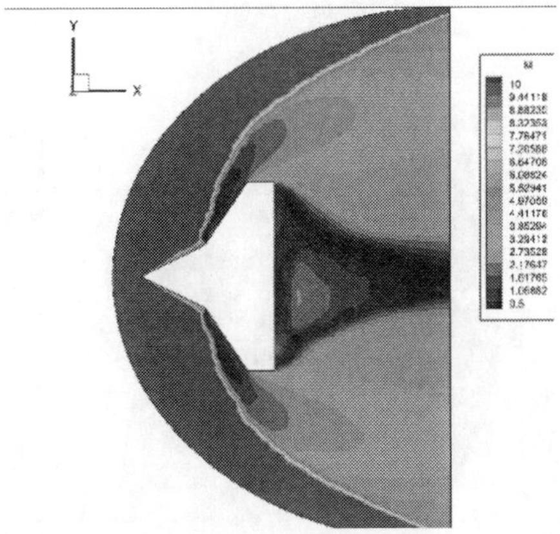

Figure 8. Distribution of the Mach number in the air. $M_\infty = 9.59$.

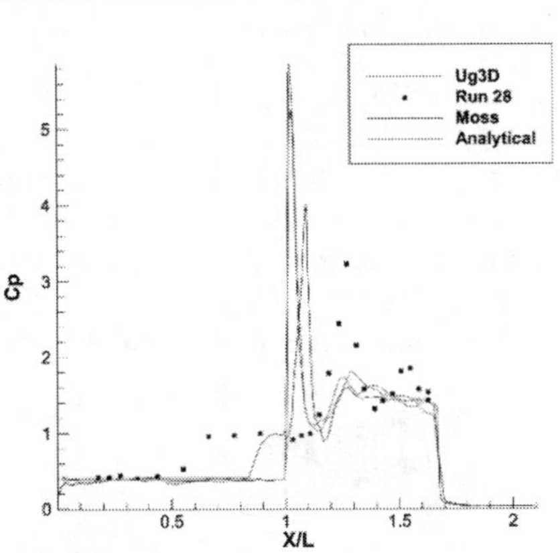

Figure 9. Graph of comparison of the pressure coefficient C_p with the experiment Moss [11] and analytical calculation. Inviscid calculation in the air.

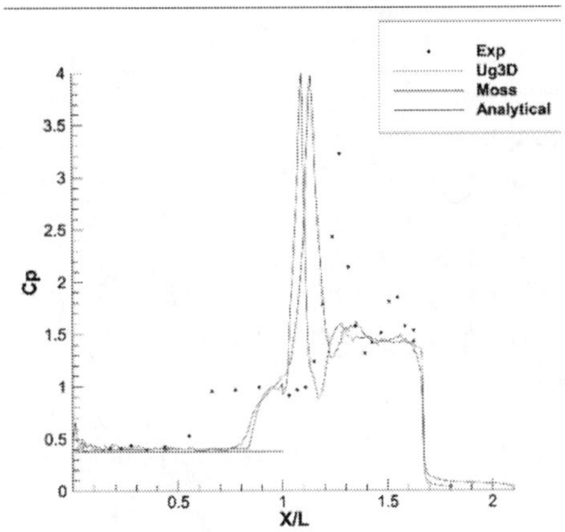

Figure 10. Graph of comparison of the pressure coefficient C_p with experiment and Moss [11] and analytical calculation. Viscous calculation in the air.

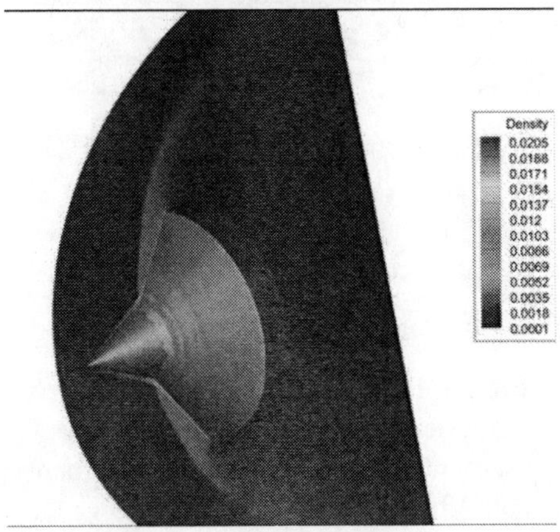

Figure 11. Distribution of density [kg/m^3] in N_2. $M_\infty = 9.59$.

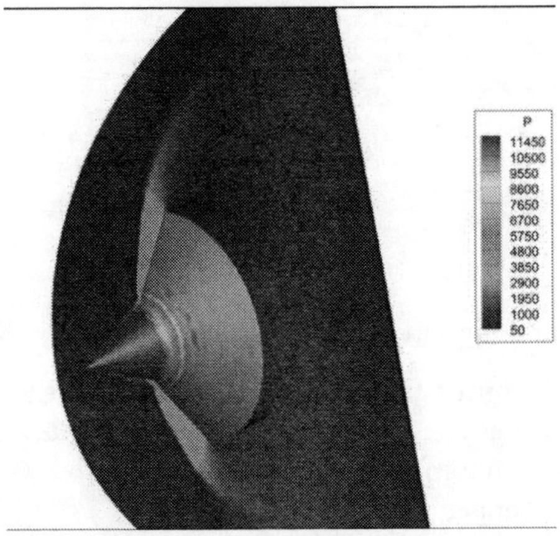

Figure 12. Pressure [Pa] distribution in N_2. $M_\infty = 9.59$.

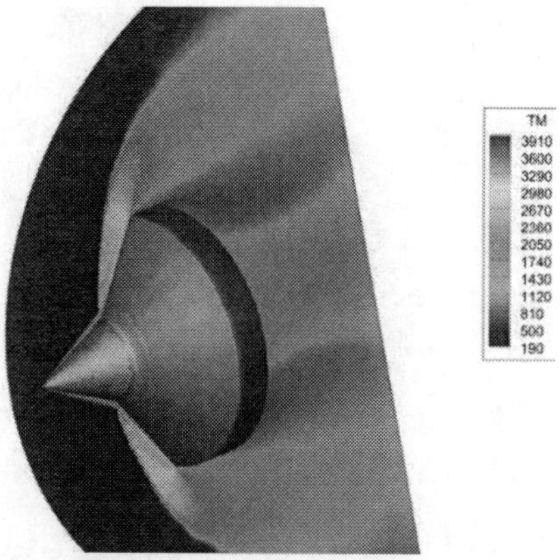

Figure 13. Temperature [K] distribution in N_2. $M_\infty = 9.59$.

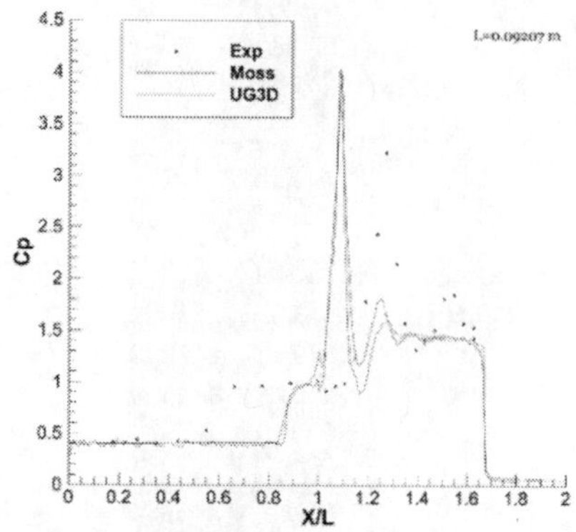

Figure 14. Graph of the comparison of the pressure coefficient C_p with the experiment and Moss (DSMC). Viscous calculation in nitrogen.

Comparison of calculated data with the results of Gnoffo calculation [10] and experimental shadowgraph are also considered. Figure 15 presents a pressure-contour comparison of Gnoffo and UG3D numerical simulations. It can be seen from the comparison that the area of maximum pressure in the shock wave – boundary layer region is shifted below, locally closer to the 25°-55° cone conjunction. It is also seen that the region of interaction of the shock wave with the boundary layer is smaller than in the results of Gnoffo. Gnoffo uses the LAURA code based on the Navier-Stokes equations [16]. An experimental shadowgraph [9] is compared to numerically calculated density gradient in figure 16.

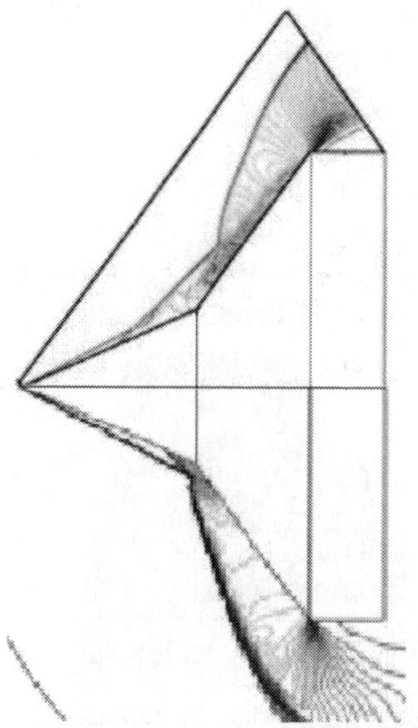

Figure 15. The result of comparing the pressure distribution on the model surface (bottom) with the results of Gnoffo [10] (upper part).

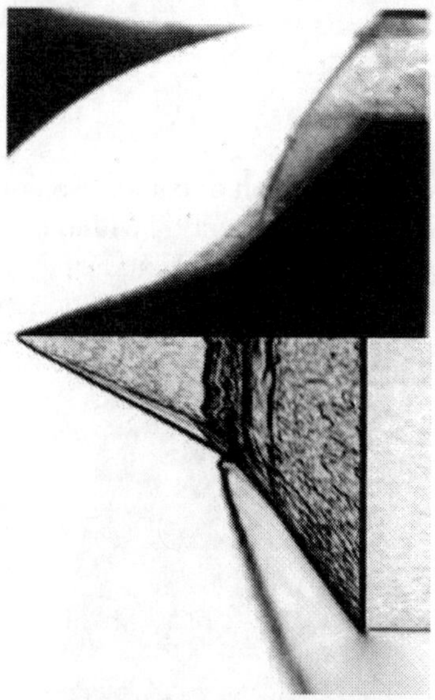

Figure 16. The result of comparing the numerical simulation of the density gradient (bottom) with the experimental data [9] (upper part).

Differences in calculated and experimental maximum rates of gas-dynamic parameters in shock wave – boundary layer interaction region may be explained by a mesh detailed insufficiently for adequate modeling of all the effects of the boundary layer in the calculation. The problem of more detailed grid with a separately constructed boundary layer generation to obtain better results is a part of the future works.

6. Conclusion

Three-dimensional numerical simulation of 25°/55° double-cone high-speed aerothermodynamics on an unstructured grid was carried out. Three-dimensional numerical simulation was carried out using the air and N_2 (nitrogen) as test gases. The CFD UG3D solver, developed in the Institute for Problems in Mechanics of the Russian academy of Sciences, was used for numerical simulation. This computer code is based on the model of Navier-Stokes equations for a perfect gas solved on an unstructured grid.

For numerical simulation of the flow, a geometric model and unstructured tetrahedral mesh of a double cone 25°/55° was generated. Verification and validation of the results were carried out. Obtained numerical results have a good agreement of pressure coefficient with the analytical solution along the 25° cone surface in all cases. A viscous calculation shows a good agreement of pressure coefficient with Moss' data in the region of cones junction as well as C_p peak values. However, the C_p maximum value is still shifted along the X coordinate. There is a great agreement of C_p peak value and the placing of C_p growth beginning with Moss' data using the N_2 as a test gas. In all cases, there is an agreement in the character of calculated pressure coefficient distribution with the experimental data but the peaks in the shock wave - boundary layer interaction region are shifted.

Further simulations of 25°/55° double-cone high-speed aerothermodynamics with a more detailed computational grid in the boundary layer area are supposed to improve the agreement of calculations with the experiment results.

The presented study was partially supported by Russian State Assignment under contract No.AAAA-A17-117021310372-6.

References

[1] Jiaao H, Jingying W and Chunhian L 2017 Numerical Simulation of High-Enthalpy Double-Cone Flows *AIAA J.* **55** 7 pp 2471–2475

[2] Xiao H, Shang Y, Wu D 2014 DSMC simulation and experimental validation of shock interaction in hypersonic low density flow, *Scientific World J.*

[3] Druguet M C, Candler G V and Nompelis I 2003 Simulations of viscous hypersonic double-cone flows: Influence of numerics. *In 16th AIAA Computational Fluid Dynamics Conf.*

[4] Candler G V, Nompelis I, Druguet M C, Holden M S, Wadhams T P, Boyd I D et al 2002 CFD validation for hypersonic flight: Hypersonic double-cone flow simulations. *In 40th AIAA Aerospace Sciences Meeting and Exhibit*

[5] Reichel S and Groll R 2013 Numerical Simulation And Experimental Validation Of A Hypersonic Flow, *For International Journal of Computational Methods and Experimental Measurements*

[6] Kotov M A, Kryukov I A, Ruleva L B , Solodovnikov S I and Surzhikov S T 2013 June 24-27 Experimental Investigations of an Aerodynamic Flow of Geometrical Models in Hypersonic Aerodynamic Shock Tube *31st AIAA Applied Aerodynamics Conference San Diego, CA, AIAA Paper* 2013–2931 15 p

[7] Kotov M A, Kryukov I A, Ruleva L B , Solodovnikov S I and Surzhikov S T 2015 Supersonic Air Flows Around Some Geometrical Primitives, *33rd AIAA Applied Aerodynamics Conference, AIAA AVIATION Forum, AIAA Paper* 2015-3012

[8] Kotov M A, Kryukov I A, Ruleva L B and Solodovnikov S I 2017 The Investigation of Shock-Wave Interaction with Aerodynamic Models, *55th AIAA Aerospace Sciences Meeting, AIAA SciTech Forum, AIAA Paper* 2017-0262

[9] Holden M S and Wadhams T P Jan. 2001 Code Validation Study of Laminar Shock/Boundary

IOP Conf. Series: Journal of Physics: Conf. Series **1009** (2018) 012003 doi:10.1088/1742-6596/1009/1/012003

Layer and Shock/Shock Interactions in Hypersonic Flow Part A: Experimental Measurements, *AIAA Paper* 2001-1031

[10] Gnoffo P A 2003 Computational Fluid Dynamics Technology For Hypersonic Applications, *NASA Langley Research Center Hampton, VA* 23681-0001

[11] James N Moss 2001 Hypersonic Flows About a 25° Sharp Cone *NASA Langley Research Center, MS 408A, Hampton, VA* 23681-2199

[12] Barth T J and Jespersen D C 1989 The design and application of upwind schemes on unstructured meshes, *AIAA Paper* 1989-0366

[13] Surzhikov S T 2017 Validation of computational code UST3D by the example of experimentl aerodynamic data *J. of Physics: Conference Since* **815** 012023

[14] Liseikin V D 1999 *Grid Generation Methods* (Berlin: Springer)

[15] Thompson J F, Soni B K, Weatherill N P 1998 *Handbook of Grid Generation* (CRC Press)

[16] Cheatwood F M and Gnoffo P A 1996 User's Manual for the Langley Aerothermodynamic Upwind Relaxation Algorithm (LAURA), *NASA TM* **4674**

Aerothermodynamics calculation of the EXPERT reentry flight vehicle

N A Kharchenko[1,2,3] **and I A Kryukov**[1,2]

[1] Ishlinsky Institute for Problems in Mechanics RAS, Vernadsky prospect 101(1), Moscow, 119526, Russia

[2] Tactical Missiles Corporation JSC, Ilicha street 7, Korolev, Moscow region, 141080, Russia

[3] Moscow Physical–Technical Institute, Institutskiy per. 9, Dolgoprudny, Moscow region, 141701, Russia

E–mail: kharchenko.na@phystech.edu

Abstract. This paper concerns the aerothermodynamics of high–speed gas flows around the EXPERT reentry flight vehicle moving at different angles of attack at an altitude of 50 km with the speed V_∞ = 5 km/s. Calculations were carried out using UST3D and UG3D computer codes developed at the Ishlinsky Institute for Problems in Mechanics RAS (IPMech RAS). The computer codes implement the numerical simulation of the flight vehicle aerothermodynamics by integrating the complete set of Navier–Stokes equations on unstructured mesh using the ideal gas model.

1. Introduction

The EXPERT reentry flight vehicle was developed by the European Space Agency (ESA) [1]. Flight tests of the vehicle were not conducted, but there is a sufficiently large amount of calculations and experimental data. Experimental data are presented this papers [2–5]. In papers [4–6] the numerical results are presented. At first glance, the geometry of this device may seem fairly simple, but the numerical simulation reveals the difficulties related to the design features of this type of vehicles.

2. Computational model

Flows around the EXPERT reentry flight vehicle were calculated using the UST3D and UG3D computational codes [7, 8]. These codes are designed to compute the aerothermodynamics of high–speed flight vehicles and can be used in a wide range of Mach numbers, altitudes and angles of attack [7–9]. The computer codes are based on the algorithms for solving the non–stationary three–dimensional set of Navier–Stokes equations (1) – (4) [10–12].

$$\frac{\partial \mathrm{w}}{\partial t}+\frac{\partial F^x(\mathrm{w})}{\partial x}+\frac{\partial F^y(\mathrm{w})}{\partial y}+\frac{\partial F^z(\mathrm{w})}{\partial z}=\frac{\partial G^x(\mathrm{w})}{\partial x}+\frac{\partial G^y(\mathrm{w})}{\partial y}+\frac{\partial G^z(\mathrm{w})}{\partial z}, \quad (1)$$

$$\mathrm{w}=\begin{pmatrix}\rho\\ \rho u\\ \rho v\\ \rho w\\ \rho E\end{pmatrix}, \quad (2)$$

$$F^x = \begin{pmatrix} \rho u \\ \rho u^2 + p \\ \rho uv \\ \rho uw \\ \rho uE + pu \end{pmatrix}, \quad F^y = \begin{pmatrix} \rho v \\ \rho uv \\ \rho v^2 + p \\ \rho vw \\ \rho vE + pv \end{pmatrix}, \quad F^z = \begin{pmatrix} \rho w \\ \rho uw \\ \rho vw \\ \rho w^2 + p \\ \rho wE + pw \end{pmatrix}, \tag{3}$$

$$G^x = \begin{pmatrix} 0 \\ \tau_{xx} \\ \tau_{yx} \\ \tau_{zx} \\ u\tau_{xx} + v\tau_{yx} + w\tau_{zx} - q_x \end{pmatrix}, \quad G^y = \begin{pmatrix} 0 \\ \tau_{xy} \\ \tau_{yy} \\ \tau_{zy} \\ u\tau_{xy} + v\tau_{yy} + w\tau_{zy} - q_y \end{pmatrix},$$

$$G^z = \begin{pmatrix} 0 \\ \tau_{xz} \\ \tau_{yz} \\ \tau_{zz} \\ u\tau_{xz} + v\tau_{yz} + w\tau_{zz} - q_z \end{pmatrix}, \tag{4}$$

The complete set of Navier–Stokes equations is used in conjunction with the equation of state for ideal gas (5).

$$p = (\gamma - 1)\rho U = (\gamma - 1)\rho\left[E - \frac{1}{2}\left(u^2 + v^2 + w^2\right)\right], \tag{5}$$

The UST3D and UG3D computer codes execute numerical integration of the system using the splitting method with respect to physical processes and the Godunov–type method [13]. The methods are explicit and use the first–order and second–order approximation in time and second–order approximation in space. The problem is solved by obtaining the convergence [7–9].

During the verification of UST3D and UG3D computational codes, the results were compared with the ones obtained by the SolidWorks Flow Simulation module for gasdynamic computations. This module is based on the algorithms for solving the non–stationary three–dimensional set of Navier–Stokes equations using the ideal gas model, as well. Codes UST3D and UG3D were validated in [7, 8].

3. Problem statement for computing flows around the EXPERT reentry flight vehicle

In order to perform a numerical simulation of the flow process, a three–dimensional surface model of the EXPERT reentry vehicle was developed (figures 1, 2).

Figure 1. Three–dimensional surface of the EXPERT reentry vehicle.

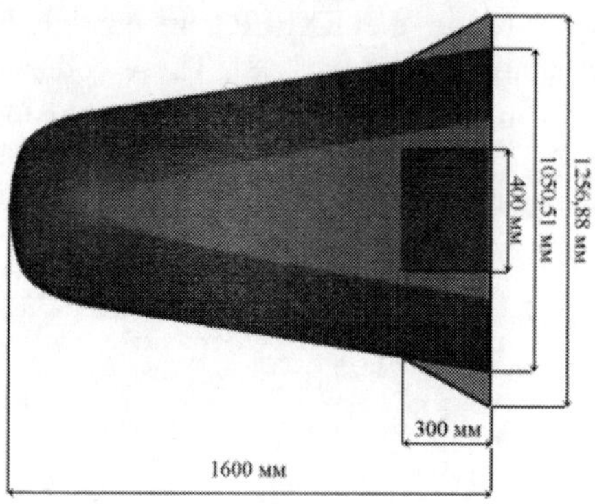

Figure 2. Dimensions of the EXPERT reentry vehicle.

IOP Conf. Series: Journal of Physics: Conf. Series **1009** (2018) 012004 doi:10.1088/1742-6596/1009/1/012004

The model was created using the SolidWorks CAD environment. The model is a complete equivalent of the vehicle and considers all its design features.

4. Computational mesh

In this paper, a three–dimensional flow field near the flight vehicle was simulated using unstructured tetrahedral meshes (figures 3–6).

Figure 3. Surface mesh of the EXPERT reentry vehicle.

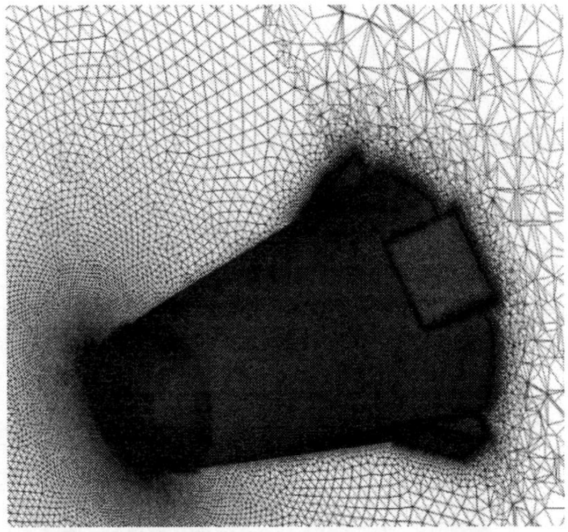

Figure 4. Volume mesh of the EXPERT reentry vehicle.

Several options of the computational mesh were generated: excluding and including the shock–wave feature of the flow structure (figures 4, 6). In the first case, the mesh was refined toward the vehicle head and toward the vehicle parts sensitive to the specific design features. The number of triangles on the surface was 98,454, and the total number of tetrahedrons in the volume was 1 024 290 cells (figures 3, 4). For an adaptive mesh, the refinement was carried out toward the head shock wave which position and structure depend on the incoming flow parameters [14]. The number of triangles on the surface was 65,210, and the total number of tetrahedrons in the volume was 5 383 515 cells (figures 5, 6).

Figure 5. Surface mesh of the EXPERT reentry vehicle.

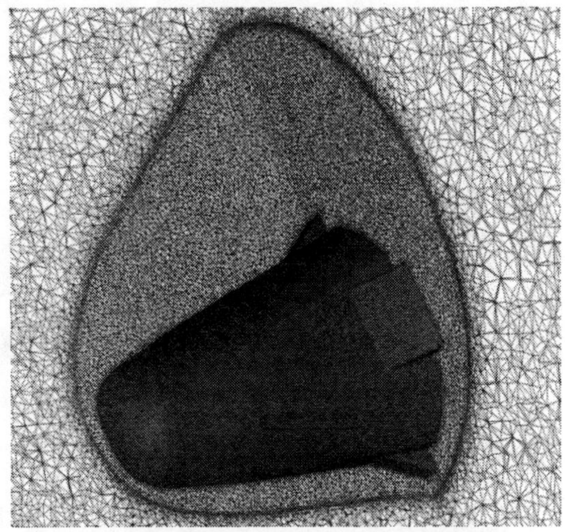

Figure 6. Volume mesh of the EXPERT reentry vehicle.

5. Results obtained

Figure 7 shows the density distribution on the surface and in the vicinity of the vehicle flying at zero angle of attack at an altitude of 50 km with the speed V_∞ = 5 km/s. Figure 8 illustrates pressure distribution on the vehicle surface. The calculation was carried out using the UST3D and UG3D computer codes and the SolidWorks Flow Simulation module. Index "a" denotes the results calculated using the adaptive mesh

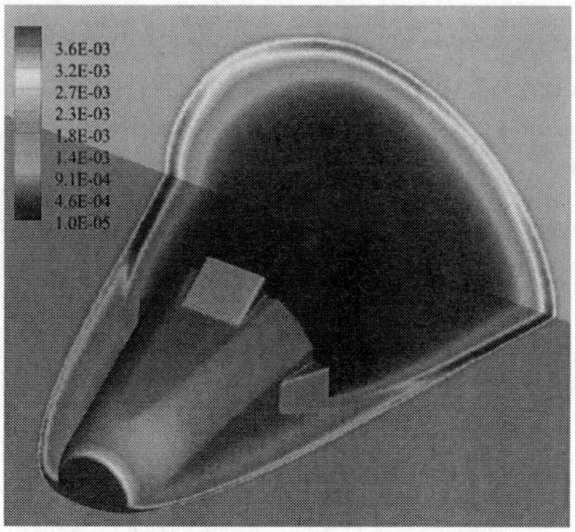

Figure 7. Density (in kg/m3) distribution on the surface and in the vicinity of the EXPERT reentry vehicle. Angle of attack of 0°.

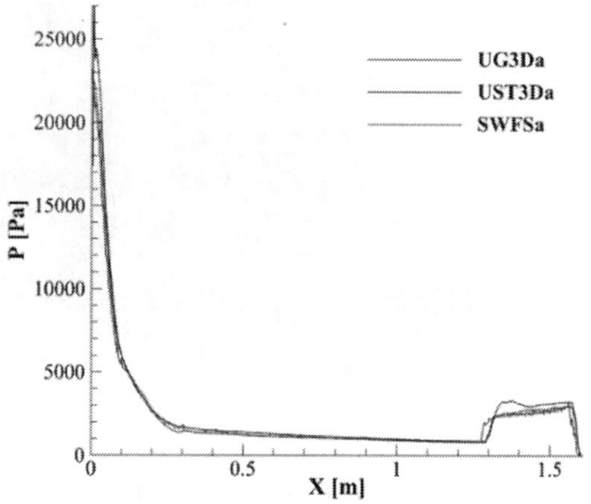

Figure 8. Pressure distribution on the surface of the EXPERT reentry vehicle. Angle of attack of 0°.

Figures 9 and 10 show density distribution on the surface and in the vicinity of the vehicle. The vehicle travels at an altitude of 50 km with the speed of V_∞ = 5 km/s at angles of attack of 10° and 15°.

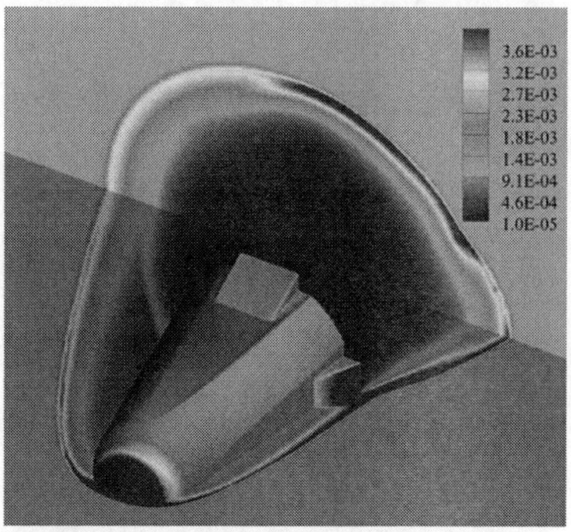

Figure 9. Density (in kg/m^3) distribution on the surface and in the vicinity of the EXPERT reentry vehicle. Angle of attack of 10°.

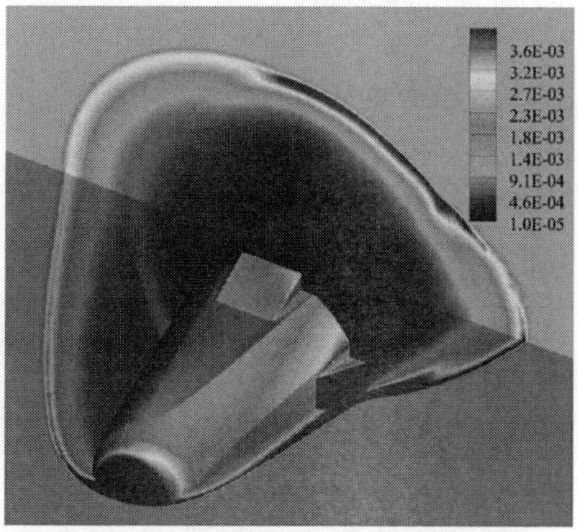

Figure 10. Density (in kg/m^3) distribution on the surface and in the vicinity of the EXPERT reentry vehicle. Angle of attack of 15°.

Figures 11–16 show pressure distribution on the surface of the vehicle. Figures 11–14 show diagrams with the comparison of results for UST3D and UG3D computer codes. The calculations were carried out using the same mesh.

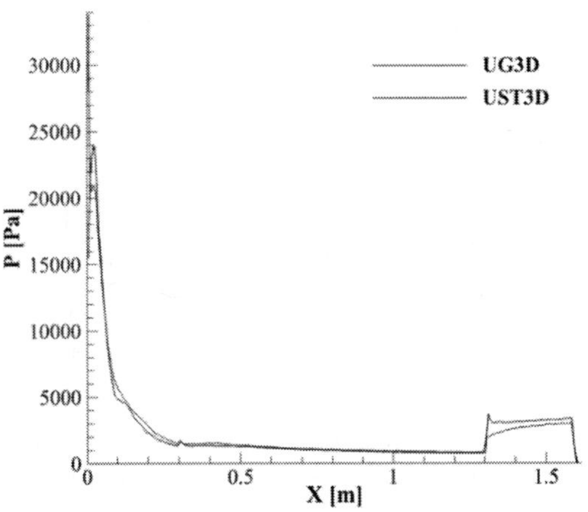

Figure 11. Pressure distribution on the surface of the EXPERT reentry vehicle. Angle of attack of 0°.

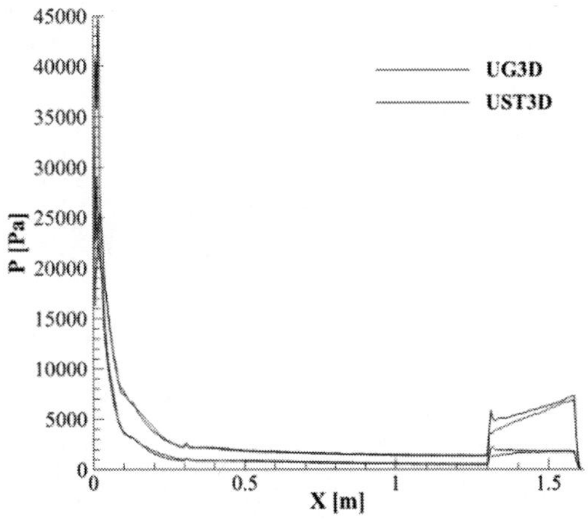

Figure 12. Pressure distribution on the surface of the EXPERT reentry vehicle. Angle of attack of 6°.

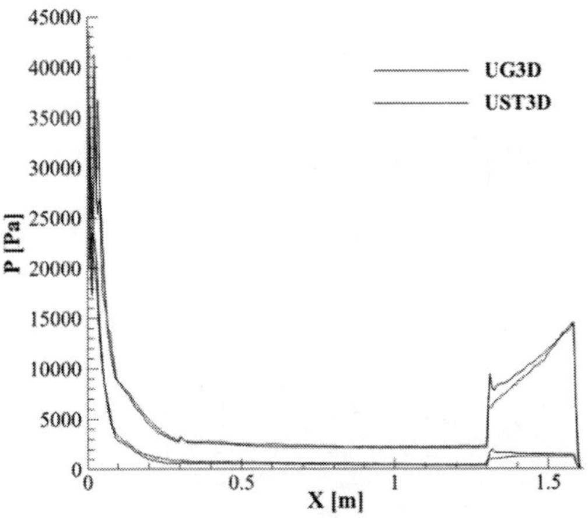

Figure 13. Pressure distribution on the surface of the EXPERT reentry vehicle. Angle of attack of 10°.

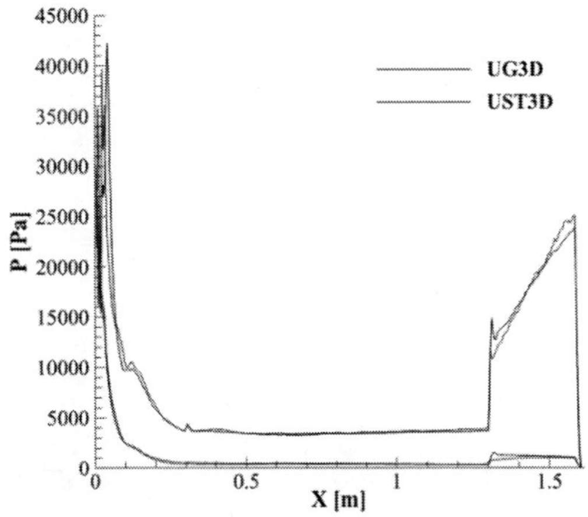

Figure 14. Pressure distribution on the surface of the EXPERT reentry vehicle. Angle of attack of 15°.

Figures 15 and 16 show the comparison of calculation results obtained with and without mesh adaptation using the UG3D computer code.

Strong fluctuations are observed on the head part of the vehicle surface when calculating without mesh adaptation.

The use of adaptive mesh increases the accuracy of the result obtained when calculating aerothermodynamics of flight vehicles (figure 17). But this leads to additional requirements for geometry construction and mesh generation [14].

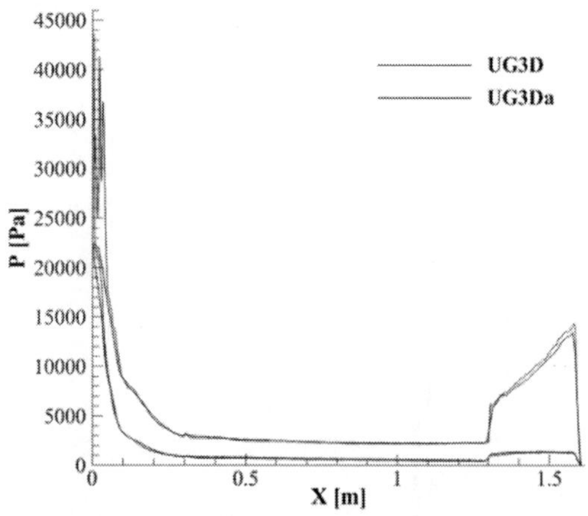

Figure 15. Pressure distribution on the surface of the EXPERT reentry vehicle. Angle of attack of 10°.

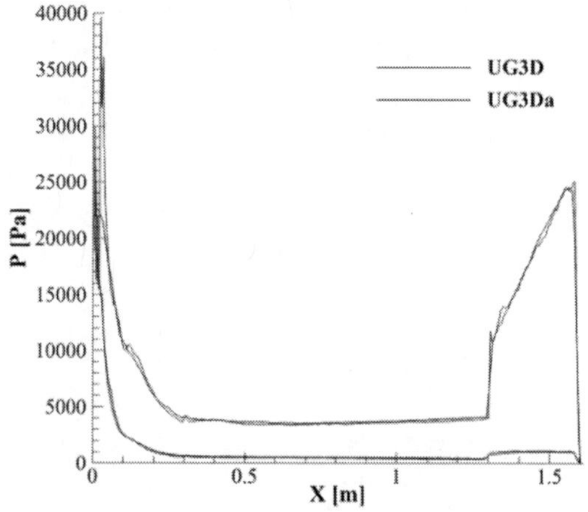

Figure 16. Pressure distribution on the surface of the EXPERT reentry vehicle. Angle of attack of 15°.

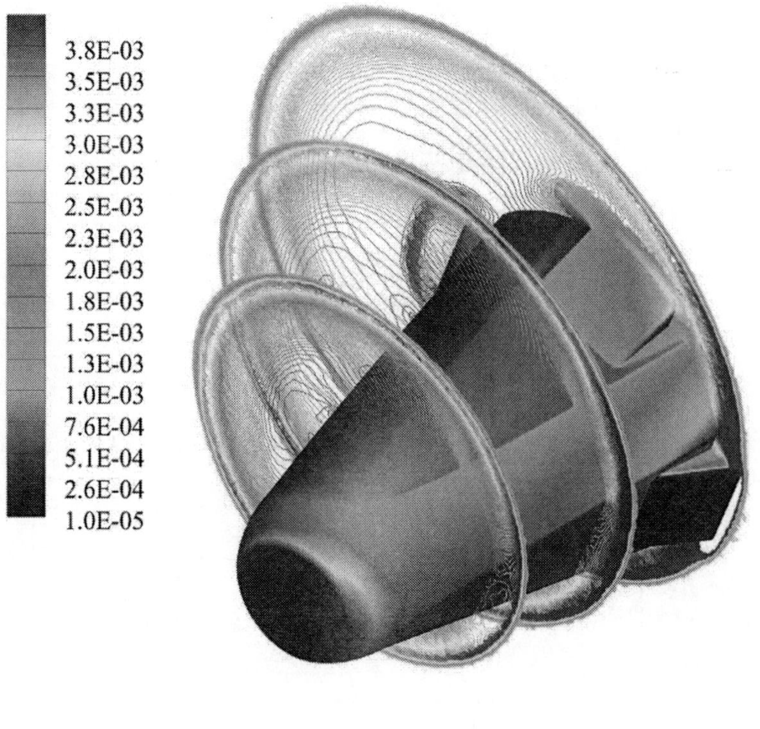

Figure 17. Density (in kg/m^3) distribution on the surface of the EXPERT reentry vehicle.

Figures 18 and 19 show behavior diagrams of aerodynamic coefficients of lift and drag forces as a function of the angle of attack. The results are calculated using four computational codes: UST3D, UG3D, INTEG and SolidWorks Flow Simulation.

INTEG code was developed at the Institute of Theoretical and Applied Mechanics RAS. The computer code is based on the local bridge method and is designed to calculate aerodynamic characteristics.

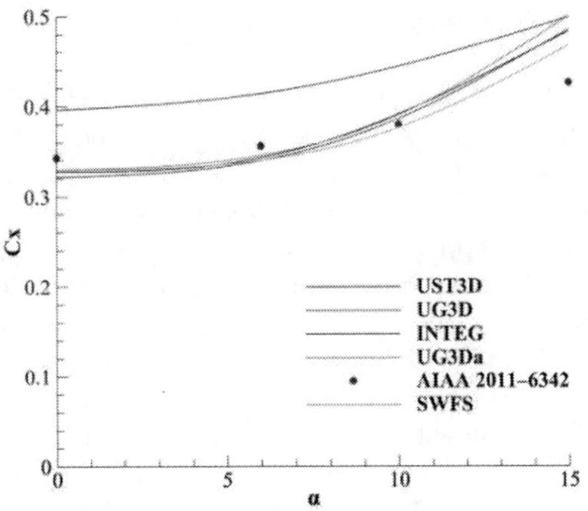

Figure 18. Behavior of aerodynamic drag force coefficient as a function of the angle of attack of the EXPERT reentry flight vehicle. Angles of attack of 0°,6°,10°,15°.

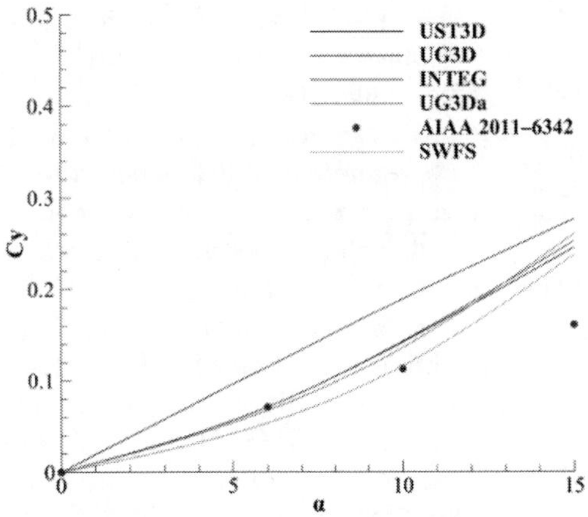

Figure 19. Behavior of aerodynamic lift force coefficient as a function of the angle of attack of the EXPERT reentry flight vehicle. Angles of attack of 0°,6°,10°,15°.

The difference between the results obtained and the data given in the paper is due to the fact that simulation of vehicle flights in dense atmosphere at high speeds requires the consideration of various physical and chemical processes in the air [6, 14, 15].

Conclusion

The paper demonstrates the necessity to refine the mesh on the head shock wave front and in the areas of high gradients. It leads to additional requirements for construction of geometry and computational mesh used in the calculation of aerothermodynamics of flight vehicles.

In the future, it is planned to develop an automated mesh adaptation.

The presented study was partially supported by Russian State Assignment under contract No. AAAA-A17-117021310372-6.

References

[1] http://www.esa.int/ESA
[2] Adamov N P, Kharitonov A M, Mazhul I I, Vasenyov L G, Zapryagaev V I, Zvegintsev V I and Muylaert J M 2008 Investigations of Aerogasdynamics of Re-Entry Ballistic Vehicle Expert *International Conference on Methods of Aerophysical Research*
[3] Kharitonov A M, Adamov N P, Chirkashenko V F, Mazhul I I, Shpak S I, Shiplyuk A N, Vasenyov L G, Zvegintsev V I and Muylaert J M 2012 Aerothermodynamics of Expert Ballistic Vehicle at Hypersonic Speeds *Progress in Flight Physics* **3** pp 277–294
[4] Brazier J P, Schramm J M, Paris S and Gawehn T 2015 An overview of HyFIE Technical Research Project: cross testing in main European hypersonic wind tunnels on EXPERT body 50th *3AF International Conference on Applied Aerodynamics*
[5] Muylaert J, Cipollini F, Walpot L, Ottens H 2005 Flight Experiments for Hypersonic Vehicle Development Expert *RTO AVT* RTO–EN–AVT–116
[6] Martinez Barrio A, Sudars M, Gavira J, Aulisio R, Ratti F, Massobrio F, Walpot L, Passarelli G, Thoemel J, and Thirkettle A 2011 EXPERT – The ESA EXPERIMENTAL Re–Entry Test–Bed Trajectory and mission design *AIAA Guidance Navigation and Control Conference*

AIAA 2011–6342

[7] Surzhikov S T 2017 Validation of computational code UST3D by the example of experimental aerodynamic data *J. of Physics: Conference Series* **815** 012023

[8] Kotov M A, Kryukov I A, Ruleva L B, Solodovnikov S I, and Surzhikov S T 2015 Supersonic Air Flows Around Some Geometrical Primitives 33rd *AIAA Applied Aerodynamics Conference* AIAA 2015–3012

[9] Zheleznyakova A L and Surzhikov S T 2013 Application of the method of splitting by physical processes for the computation of a hypersonic flow over an aircraft model of complex configuration *High Temperature* **51** pp 816–829

[10] Surzhikov S T 2011 Radiative–Convective Heat Transfer of a Spherically Shaped Space Vehicle in Carbon Dioxide *High Temperature* **49** pp 92–107

[11] Shang J S and Surzhikov S T 2010 Simulating Nonequilibrium Flow for Ablative Earth Reentry *J. of Spacecraft and Rockets* **47** pp 806–815

[12] Surzhikov S T, Seleznev R K, Tretyakov P K and Zabaykin V A 2014 Unsteady Thermo-Gasdynamic Processes in Scramjet Combustion Chamber with Periodical Input of Cold Air 50th *AIAA/ASME/SAE/ASEE Joint Propulsion Conference* AIAA 2014–3917 p 25

[13] Godunov S K 1959 A finite-difference method for the numerical computation and discontinuous solutions of the equations of fluid dynamics *Mat. Sb.* **47** pp 271–306

[14] Surzhikov S T 2013 Convective heating of small-radius spherical blunting for relatively low hypersonic velocities *High Temperature* **51** pp 231–245

[15] Surzhikov S T 2017 Spatial Multiphysics Models of the Radiation Gas Dynamics of Super Orbital Re–Entry Space Vehicles. *J. of Physics: Conference Series* **837** 012021

IOP Conf. Series: Journal of Physics: Conf. Series **1009** (2018) 012005 doi:10.1088/1742-6596/1009/1/012005

Numerical Study of the Supersonic Base Flow around Aircraft Model

Ya V Khankhasaeva, V E Borisov and A E Lutsky

Keldysh Institute of Applied Mathematics, Moscow, Russia

hanhyana@mail.ru, narelen@gmail.com, lutsky@kiam.ru

Abstract. This paper deals with a numerical study of the base part of the supersonic flow past an aircraft model. The simulation was carried out using the software developed by the authors for calculating 3D turbulent flow of viscous compressible gas in the framework of the unsteady Reynolds averaged Navier–Stokes equations (URANS) with the Spalart–Allmaras (SA) and Menter's SST turbulence models. Flight modes with $M = 2.5$ and angle of attack $\alpha = 0°, 3°$ were considered. A comparison is made of the effect of the choice of the turbulence model on the base flow.

1. Introduction

Numerical simulations of gas-dynamic processes around an aircraft is an integral stage of its design, allowing to obtain aerodynamic characteristics with small (relative to field experiments) costs. At the same time, due to the complex design of modern aircrafts, as well as high (super- and hypersonic) flight speeds, the structure of the emerging currents is quite complicated due to the presence of narrow zones of large gradients, boundary layers, shock waves, etc. Correct calculation of all these features makes high demands on the quality of the computational model.

One of the major problems in the supersonic flow around the aircraft is the simulation of currents in extended separation zones in the base part of the flow in order to determine the base pressure and wake parameters downstream. The base flow features have a significant effect on the aerodynamic characteristics of the aircraft. In particular, base drag makes a significant contribution to the total drag of the aircraft. Schematically, separated flow in the base region of the supersonic flow around a blunt cone is shown in figure 1. In front of the axisymmetric body (1), a detached shock wave (2) is formed. Boundary layer (3) adjacent to the body surface separates from the rear edge of the cone forming a viscous mixing layer (7). At the trailing edge of the cone in the fan of rarefaction waves (4) the flow turns toward the axis. As the free shear layer approaches the axis of symmetry, a recompression process (5) occurs which eventually realigns the flowfield with the axis. Current line (6) separates the recirculation zone (9) from the rest of the stream which forms a turbulent trace (8) after the point of reattachment. Because of the complexity of making reliable measurements mainly in the separated region, few experimental works were devoted to supersonic base flows. Works [1,2] contain experimental data of base pressure coefficients for bodies with different aft forms, with and without wings and fins, etc., for different Mach numbers and angles of attack. In [3] experiments were performed in order to study all features of axisymmetric base flow past a cylinder. The paper [4] is devoted to the same topic but with a non-zero angle of attack. In [5, 6] hypersonic flow around aircrafts of different configurations for a wide range of Mach numbers and angles of attack, including the accompanying base flow, was experimentally investigated. In [7] a generalization of a large number of experimental data obtained for a wide spectrum flight speeds is made. The main types of base ledges encountered in practice are considered, the flow pattern behind the flat, axially symmetric, three-dimensional ledges, including flow asymmetry, caused by the presence of an angle of attack, is investigated.

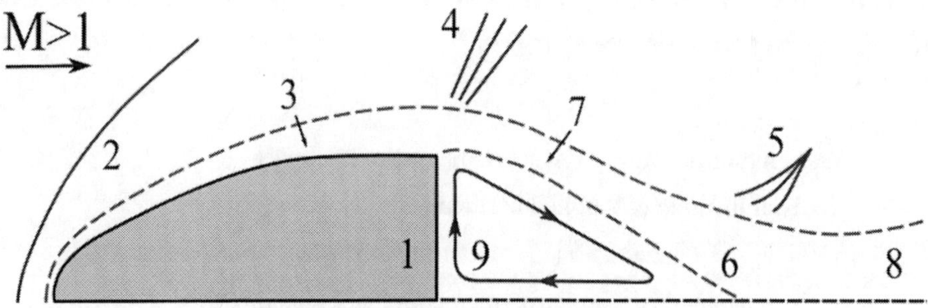

Figure 1. Scheme of the base flow.

Also in recent years a large number of numerical studies of base flow have been carried out [8-25]. Correct modeling of such flows is a difficult task, especially when calculating the flow around an aircraft in a complete layout. The following works are devoted to the numerical study of base flow in various formulations and with different numerical approaches, such as: (a) Reynolds averaged Navier–Stokes equations (RANS) with various turbulent models [8-16, 23], (b) Large Eddy Simulations (LES) [11,13,17, 21], Direct Numerical Simulation (DNS) [18], RANS/LES hybrid methods [11-13,19,20], Detached Eddy Simulations (DES) and its modifications DDES, ZDES [11,15,16,19, 21-23]. All methods have their own advantages and limitations.

The RANS methods had been developed to predict such important flow characteristics as the force, moment and velocity, etc., but were not intended to model complex unsteady coherent structures, pressure fluctuations or turbulent stresses inherent in base flow. However, such methods are widely used for analyzing the afterbody design options. LES method works well for resolving large scale unsteady typically time- and geometry-dependent flows but requires a significantly better grid resolution. Also, sub-grid models for the boundary layer and compressible flows are still not sufficiently well-developed to accurately predict supersonic base flows. Hybrid methods RANS/LES and DES use combinations of the methods described above depending on the flow characteristics. The turbulence model used in any of these methods should simulate near-wall turbulence with good accuracy.

This work presents the results of numerical study of the supersonic flow around the prototype aircraft TRK, particularly its base flow. Flight modes with M = 2.5 and angles of attack $\alpha = 0°$, $3°$ were considered. The simulation was carried out using the software developed by the authors for calculating 3D turbulent flow of viscous compressible gas in the framework of the unsteady Reynolds averaged Navier–Stokes equations (URANS) with the Spalart–Allmaras (SA) and Menter's SST turbulence models. A comparison is made of the effect of the choice of the turbulence model on the base flow.

2. Problem statement

The problem considered is the supersonic flow around aircraft TRK (figure 2). Aircraft consists of a pointed cylinder, two wings and four fins. The radius of the cylinder is taken as the unit of length. The radius of curvature of the nose is $r = 0.03$.

Calculations were performed for the Mach number M = 2.5 and angles of attack $\alpha = 0°$ and $3°$. The Reynolds number was $\mathrm{Re} = 1.0 \times 10^7$ [m^{-1}],

Calculation area was a truncated cone, in the center of which the device was located. A hexagonal block-structured grid with 1 596 672 elements was used. The lateral conical boundaries were located at a sufficient distance from the body, so that they did not intersect with the shock waves. On the wall, an adiabatic boundary condition was used.

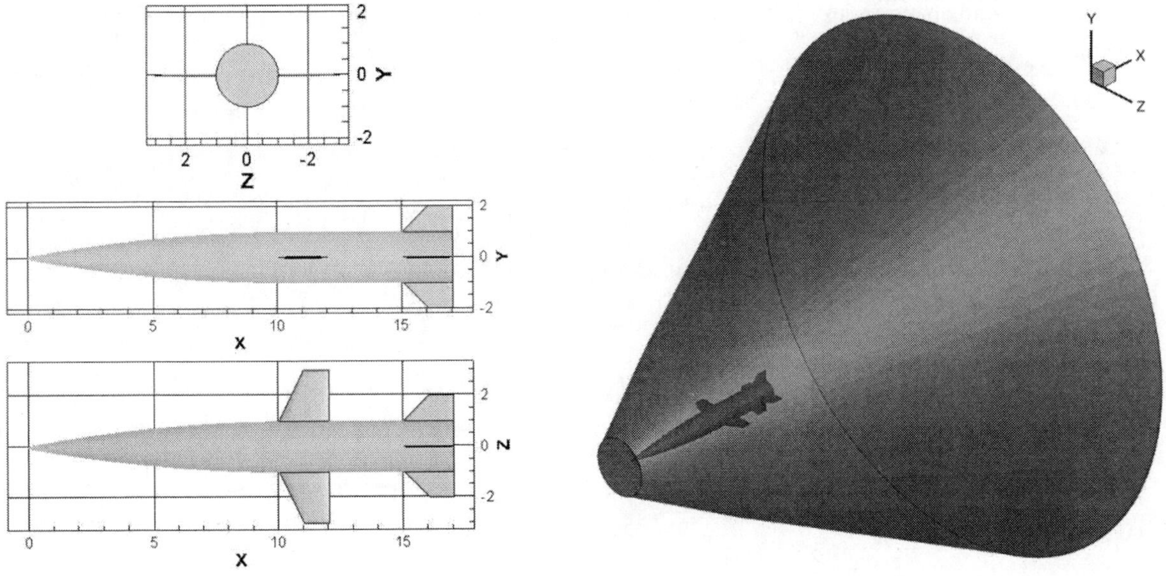

Figure 2. Aircraft model and calculation area.

2.1. Basic equations

For calculating flows of viscous compressible gas the unsteady Reynolds averaged Navier–Stokes equations (URANS) are used which in Cartesian coordinate system $(x_1 = x, x_2 = y, x_3 = z)$ have the following form:

$$\frac{\partial \mathbf{q}}{\partial t} + \frac{\partial \mathbf{f}_j}{\partial x_j} = \frac{\partial \mathbf{g}_j}{\partial x_j} \tag{1}$$

where the summation is over $j = 1,2,3$; $\mathbf{q}$ – vector of conservative variables; $\mathbf{f}_j$ и $\mathbf{g}_j$ – vectors of in viscid and viscous fluxes, respectively:

$$\mathbf{q} = \begin{pmatrix} \rho \\ \rho u_1 \\ \rho u_2 \\ \rho u_3 \\ \rho E^* \end{pmatrix}, \quad \mathbf{f}_j = \begin{pmatrix} \rho u_j \\ \rho u_j u_1 + \delta_{1,j} p \\ \rho u_j u_2 + \delta_{2,j} p \\ \rho u_j u_3 + \delta_{3,j} p \\ u_j \left(\rho E^* + p \right) \end{pmatrix}, \quad \mathbf{g}_j = \begin{pmatrix} 0 \\ \tau_{1j} \\ \tau_{2j} \\ \tau_{3j} \\ \tau_{ij} u_i + h_j \end{pmatrix}.$$

Here ρ – density; u_j – velocity vector u components; τ_{ij} – components of the viscous stress tensor; δ_{ij} – Kronecker tensor; E^* – total energy of turbulent flow:

$$E^* = E + k,$$

where E – total energy of the mean flow, k – kinetic energy of turbulent pulsations. The thermodynamic pressure p is calculated from the equation of state of the perfect gas:

$$p=(\gamma-1)\rho\left(E-\frac{1}{2}\sum_{j=1}^{3}u_j^2\right).$$

Here γ – adiabatic index, for air $\gamma=1.4$. Components of the viscous stress tensor and heat flux vector

$$\tau_{ij}=\mu_{eff}\left(\frac{\partial u_i}{\partial x_j}+\frac{\partial u_j}{\partial x_i}\right)-\frac{2}{3}\mu_{eff}\frac{\partial u_l}{\partial x_l}\delta_{i,j}-\frac{2}{3}\rho k\delta_{i,j},$$

$$h_j=\lambda_{eff}\frac{\partial T}{\partial x_j}.$$

In the above equations, T is the gas temperature and is determined from formula

$$T=\frac{p}{\rho R},$$

where R – gas constant, for air $R=287\ J/(\mathrm{kg}\cdot\mathrm{K})$.

The "effective" value of the coefficients of viscosity and thermal conductivity is defined as

$$\mu_{eff}=\mu+\mu_t,$$

$$\lambda_{eff}=C_p\left(\frac{\mu}{\mathrm{Pr}}+\frac{\mu_t}{\mathrm{Pr}_t}\right).$$

Here C_p – coefficient of specific heat of gas at constant pressure; Pr и Pr_t – laminar and turbulent Prandtl number, for air $\mathrm{Pr}=0.7$; $\mathrm{Pr}_t=0.9$. Molecular viscosity is determined by following power-law dependence:

$$\frac{\mu}{\mu_0}=\left(\frac{T}{T_0}\right)^s,$$

where $s=0.76$, μ_0 — molecular viscosity at a reference temperature T_0.

The value of μ_t denotes the additional turbulent viscosity, which is determined by the turbulence model.

The initial and boundary conditions for the system of equations (1) are established in the standard way [26].

2.2. *Spalart-Allmaras turbulence model*

The version of the one-parameter Spalart-Allmaras turbulence model (SA) for compressible flows [27] with Edwards modification [28] is used in this paper. Within the framework of this model, the averaged value of the kinetic energy of turbulent pulsations cannot be found directly, by virtue of which it is assumed $k=0$.

The turbulent viscosity is given by

$$\mu_t=\rho\tilde{\nu}f_{v1},\quad f_{v1}=\frac{\chi^3}{\chi^3+C_{v1}^3},\quad \chi\equiv\rho\frac{\tilde{\nu}}{\mu}$$

where $\tilde{\nu}$ – model variable, which is determined from the basic equation of the model

$$\frac{\partial\rho\tilde{\nu}}{\partial t}+\nabla\cdot(\rho\tilde{\nu}\mathbf{u})=\rho(P_{\tilde{\nu}}-D_{\tilde{\nu}}+T_{\tilde{\nu}})+\frac{1}{\sigma_{\tilde{\nu}}}\nabla\cdot\left[(\mu+\mu_t)\nabla\tilde{\nu}\right]+$$
$$+\frac{C_{b2}}{\sigma_{\tilde{\nu}}}\rho(\nabla\tilde{\nu})^2-\frac{1}{\sigma_{\tilde{\nu}}\rho}(\mu+\rho\tilde{\nu})\nabla\rho\cdot\nabla\tilde{\nu}. \qquad (2)$$

Values $P_{\tilde{\nu}}$ and $D_{\tilde{\nu}}$ responsible, respectively, for the production and dissipation of turbulence, and $T_{\tilde{\nu}}$ for the determination of the laminar-turbulent transition in the boundary layer, are written as

$$P_{\tilde{\nu}} = C_{b1}\left(1-f_{t2}\right)\tilde{S}_{\tilde{\nu}}, \quad D_{\tilde{\nu}} = \left(C_{w1}f_w - \frac{C_{b1}}{k_r^2}f_{t2}\right)\left[\frac{\tilde{\nu}}{s}\right]^2, \quad T_{\tilde{\nu}} = f_{t1}\left(\Delta U\right)^2,$$

$$f_w = g\left[\frac{1+C_{w3}^6}{g^6+C_{w3}^6}\right]^{1/6}, \quad g = r + C_{w2}\left(r^6 - r\right).$$

Here ΔU – module of the difference between the velocities in the flow and the nearest point of the laminar-turbulent transition; d is the distance from the solid wall.

In the Edwards modification of the Spalart-Allmaras model [28], the quantities $\tilde{S}$ and r have the form

$$\tilde{S} = \sqrt{S}\left[\frac{1}{\chi} + f_{v1}\right], \quad S = \sum_{i,j=1}^{3}\left(\frac{\partial u_i}{\partial x_j} + \frac{\partial u_j}{\partial x_i}\right)\frac{\partial u_i}{\partial x_j} - \frac{2}{3}\left(\frac{\partial u_k}{\partial x_k}\right)^2,$$

$$r = \tanh\left(\frac{\tilde{\nu}}{\tilde{S}k_r^2 d^2}\right)\Big/\tanh\left(1.0\right).$$

The remaining values are constants of the SA model and are presented in table 1.

Table 1. Constants of SA turbulence model.

$\sigma_{\tilde{\nu}}$	k_r	C_{b1}	C_{b2}	C_{v1}	C_{w1}	C_{w2}	C_{w3}
$\frac{2}{3}$	0.41	0.1335	0.622	7.1	$C_{w1} = \frac{C_{b1}}{k^2} + \frac{1+C_{b2}}{\sigma}$	0.3	2.0

When modeling a completely turbulent boundary layer, taking into account f_{t1} and f_{t2} does not make any significant changes to the solution, so they are usually neglected [27].

2.3. Menter's SST turbulence model
This work uses a modified version of the Menter's two-parameter model SST (Shear Stress Transport) [29], so called SST–2003.

The turbulent viscosity is given by

$$\mu_t = \frac{\rho k}{\max\left(\omega, SF_2/\alpha\right)}$$

where, as before, k is the kinetic energy of turbulent pulsations, and ω is specific dissipation rate. These quantities are determined from the equations of the model:

$$\frac{\partial \rho k}{\partial t} + \nabla\cdot\left(\rho k\mathbf{u}\right) = \nabla\cdot\left(\left(\mu+\sigma_k\mu_t\right)\nabla k\right) + P_k - \beta^*\rho\omega k, \tag{3}$$

$$\frac{\partial \rho\omega}{\partial t} + \nabla\cdot\left(\rho\omega\mathbf{u}\right) = \nabla\cdot\left(\left(\mu+\sigma_\omega\mu_t\right)\nabla\omega\right) + \gamma_s\frac{\rho}{\mu_t}P_k - \beta\rho\omega^2 + 2\left(1-F_1\right)\frac{\rho\sigma_{w_2}}{\omega}\nabla k\nabla\omega. \tag{4}$$

The generator term P_k is calculated as

$$P_k = \min\left(P, 10\beta^*\rho\omega k\right)$$

Here

$$P=\mu_t S^2,\quad S=\sqrt{2S_{ij}S_{ij}},\qquad S_{ij}=\frac{1}{2}\left(\frac{\partial u_i}{\partial x_j}+\frac{\partial u_j}{\partial x_i}\right).$$

The interface functions of the SST model are determined according to formulas:

$$F_1=\tanh\left(\arg_1^4\right),\qquad \arg_1=\min\left[\max\left(\frac{\sqrt{k}}{\beta^*\omega d},\frac{500\mu}{\rho\omega d^2}\right),\frac{4\rho\sigma_{w_2}k}{CD_{kw}d^2}\right],$$

$$CD_{kw}=\max\left(\frac{2\rho\sigma_{w_2}}{\omega}\nabla k\nabla\omega,1.0\times10^{-10}\right),$$

$$F_2=\tanh\left(\arg_2^2\right),\qquad \arg_2=\max\left(\frac{2\sqrt{k}}{\beta^*\omega d},\frac{500\mu}{\rho\omega d^2}\right),$$

where d is the distance from the solid wall.
The coefficients of the SST model are defined as:

$$\phi=F_1\phi_1+(1-F_1)\phi_2,\qquad \phi=\sigma_k,\sigma_w,\gamma_s,\beta$$

The remaining values are model constants and are presented in table. 2.

Table 2. Constants of SST turbulence model.

σ_{k_1}	σ_{k_2}	σ_{w_1}	σ_{w_2}	β_1	β_2	γ_{s_1}	γ_{s_2}	β^*	α	α_1	α_2	α_3
0.85	1.0	0.5	0.856	0.075	0.0828	5/9	0.44	0.09	0.31	1.0	0.4	0.2

3. Numerical simulation results

Numerical simulations were carried out using the software developed by the authors for calculating 3D turbulent flow of viscous compressible gas *on high-performance computing systems.* The basis of the computational core of the software complex was the mathematical model described above based on the URANS equations with the Spalart–Allmaras and Menter's SST turbulence models. Approximation of the equations over space was carried out by the finite volume method with the reconstruction schemes of the 2nd (TVD) or the 3rd (WENO3) order of accuracy, both the explicit and the implicit scheme based on the LU-SGS method were used for the time approximation. Simulations were carried at the Keldysh Institute of Applied Mathematics on supercomputers K–100 and K–60 using 24 to 84 cores.

3.1. Flow at angle of attack case α = 0°

As shown in figure 3, there is a bow shock and the shock waves from wings and fins. Base flow (figure 4) corresponds to the described in introduction. Between the shock wave and the boundary layer there is a region of weakly viscous flow. The boundary layer separated at the trailing edge of the body forming a viscous mixing layer which constrains the recirculation region. At the trailing edge the flow turns toward the axis through the fan of rarefaction waves. At some distance from the base reattachment of the free shear layer takes place. A reattachment shock is formed. After the attachment point, a turbulent wake develops.

In general, the results obtained for SA and SST are comparable. The main differences between them are observed in the wake of base flow (figure 5). The maximum turbulent viscosity for SA is

0.020947, for SST it is 0.018235. In region after the body SA has slightly higher turbulent viscosity values.

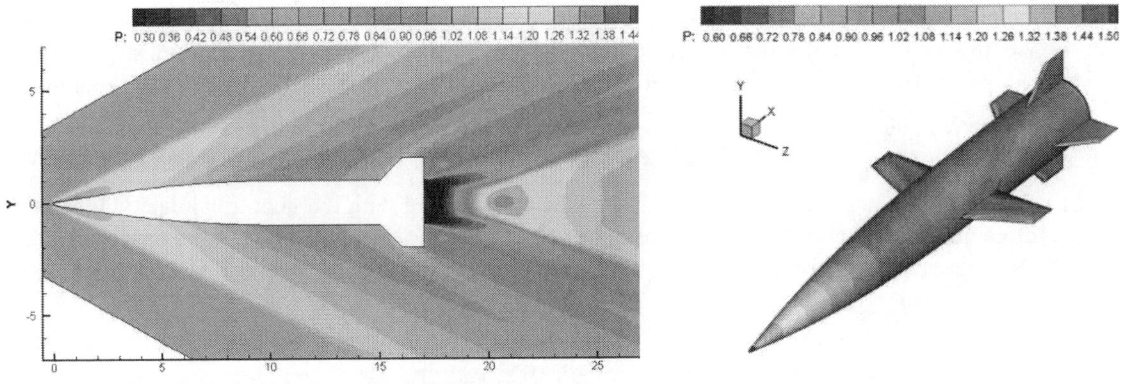

Figure 3. Pressure distribution at α = 0°, slice $z = 0$ and model surface, SA.

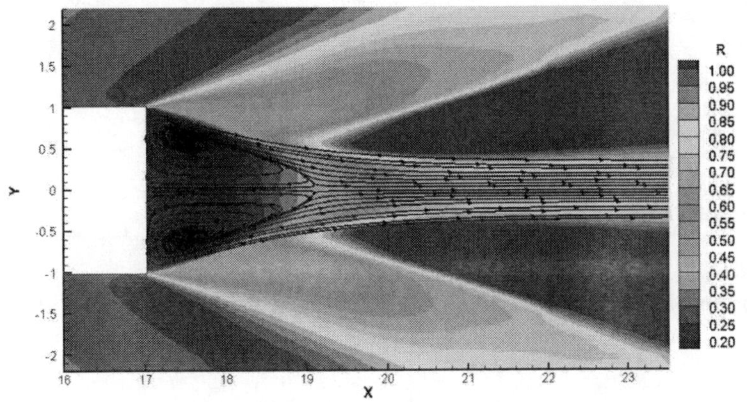

Figure 4. Density distribution and streamlines in the base region, α = 0°, SA.

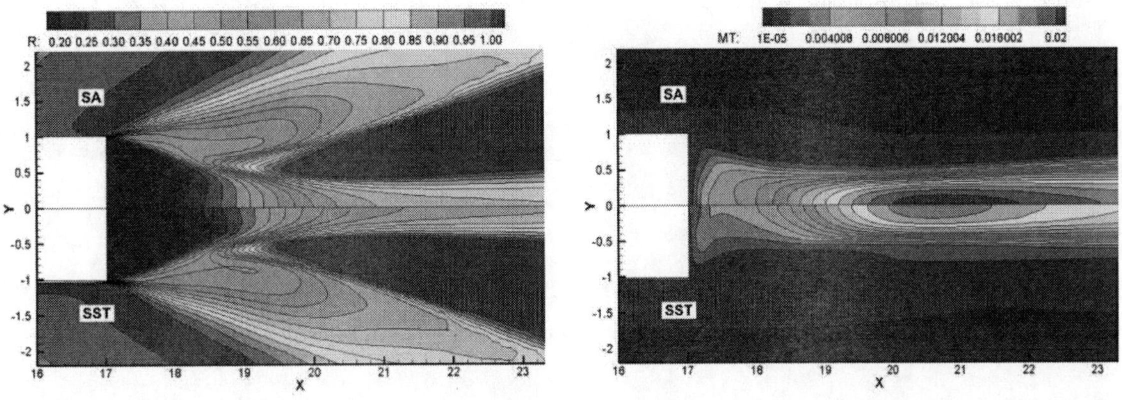

Figure 5. Density distribution (left) and turbulent viscosity (right) in the base region, α = 0°, SA (y>0) and SST (y<0).

Due to the lack of experimental data on the aircraft model considered in this paper, comparison of the obtained numerical information will be performed with experimental data [3] and data [11]

IOP Conf. Series: Journal of Physics: Conf. Series **1009** (2018) 012005 doi:10.1088/1742-6596/1009/1/012005

obtained with RANS and SA. Despite the fact that in this paper the base cut differs somewhat from that considered in [3] and [11] due to the presence of fins, such a comparison is meaningful. Since the simulation was carried out in the framework of URANS with SA and SST turbulence models, as in [11], the obtained base pressure, streamwise velocity component along the axis of symmetry from the base and the location of the reattachment point (figure 6) do not match the experimental data [3], where $C_{P_B} = -0.102$. However, this can also be the effect of fins, since, according to [4], the presence of fins at zero angle of attack reduces base pressure. Also for a body similar to that considered in this paper, with the exception of the absence of wings, $C_{P_B} = -0.116$ [4]. The difference between C_{P_B} from this work and [2] is not as great as it is for [3]. The coincidence with the numerical data [11] is good, both in the quantitative and qualitative sense. The observed change in the value of the base pressure as the flow approaches the base in the recirculation zone is due to the overestimated values of the longitudinal velocity (figures 6-8).

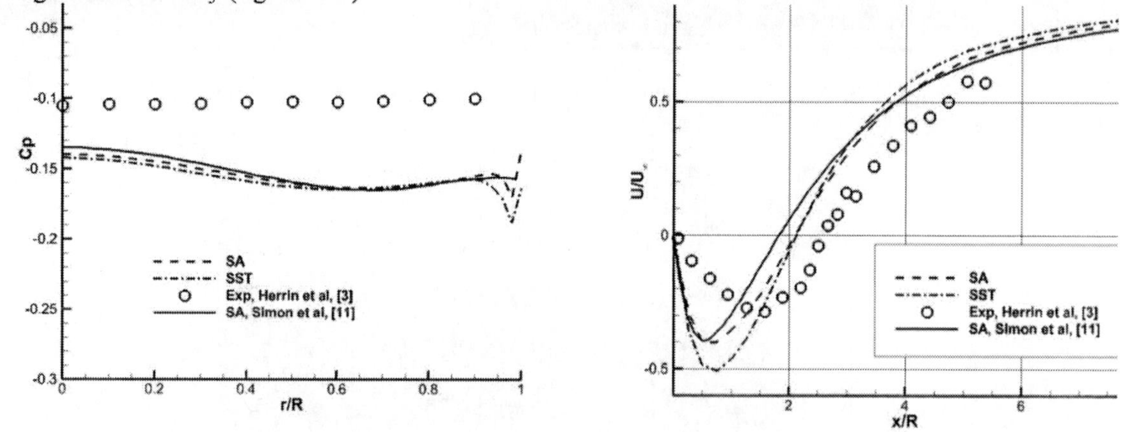

Figure 6. Base pressure coefficient (left) and streamwise velocity along the axis of symmetry from the base (right), comparison with experimental data [3] and numerical data [11] with SA.

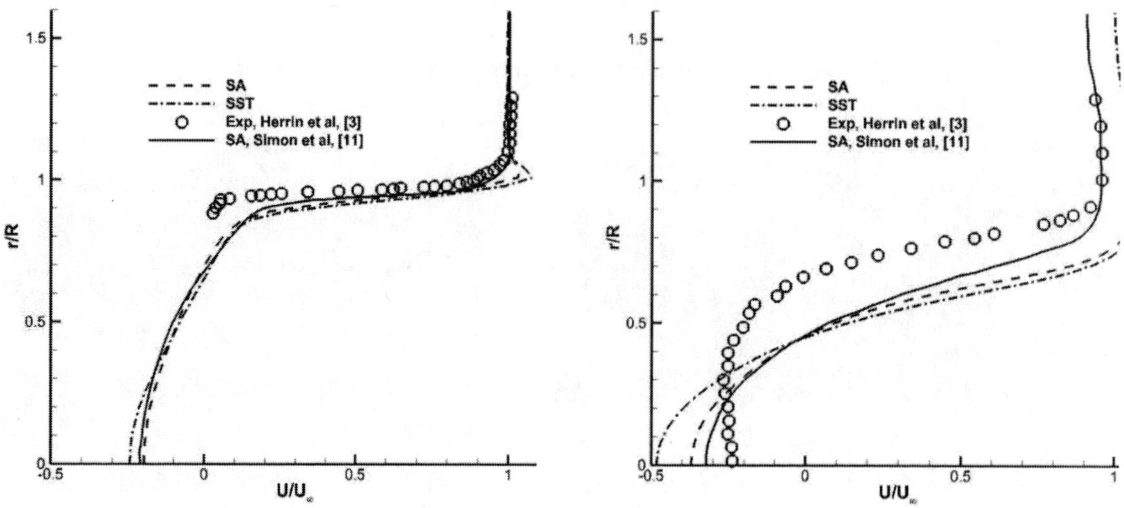

a b

Figure 7. Streamwise velocity, slices l/R = 0.1575 (a); 0.9449 (b), comparison with experimental data [3] and numerical data [11] with SA.

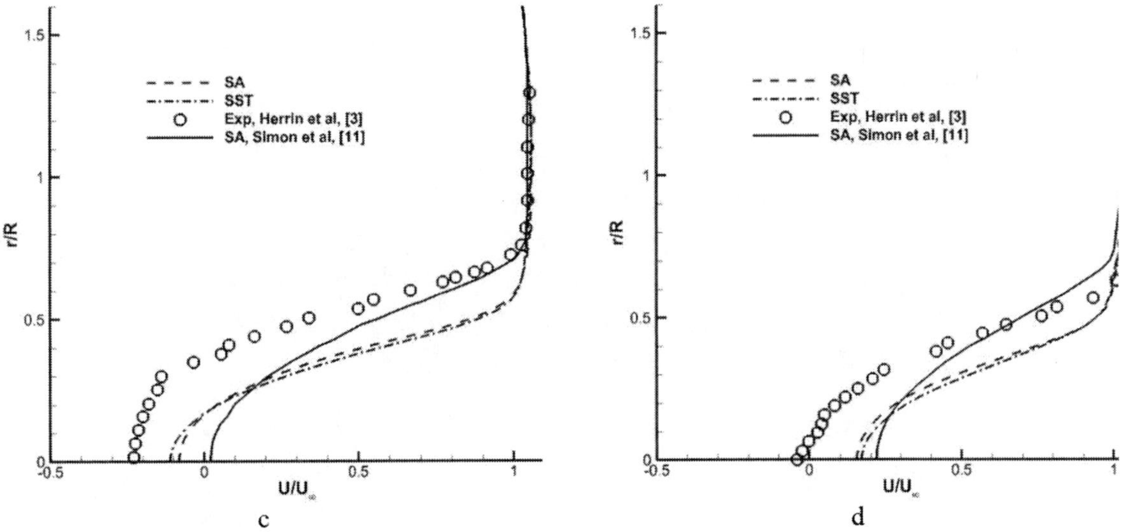

Figure 8. Streamwise velocity, slices l/R = 1.8898 (c); 2.5197 (d), comparison with experimental data [3] and numerical data [11] with SA.

It can also be seen that the SA data is in better agreement with the data from [11] than SST because [11] also uses SA model. The shift of the reattachment point from the data [11], perhaps, is the effect of fins.

Comparison of the streamwise velocity with [3] and [11] in the slice l/R = 0.1575 (l is the distance from the base) (figure 7a), where the boundary layer has just separated from the body, shows that RANS simulations have a somewhat overestimated streamwise velocity at the base cut as compared with experiment. In the mixing layer, the differences are not as big. An overstated streamwise velocity is also observed for the remaining slices l/R = 0.9449 (figure 7b), 1.8898 (approximate start of the recompression zone) (figure 8c) and 2.5197 (near reattachment point) (figure 8d).

Fins also affect the base flow. One of the obvious expressions of this influence is the change in the wake shape behind the body (figure 9), which now has a diamond-shaped cross section.

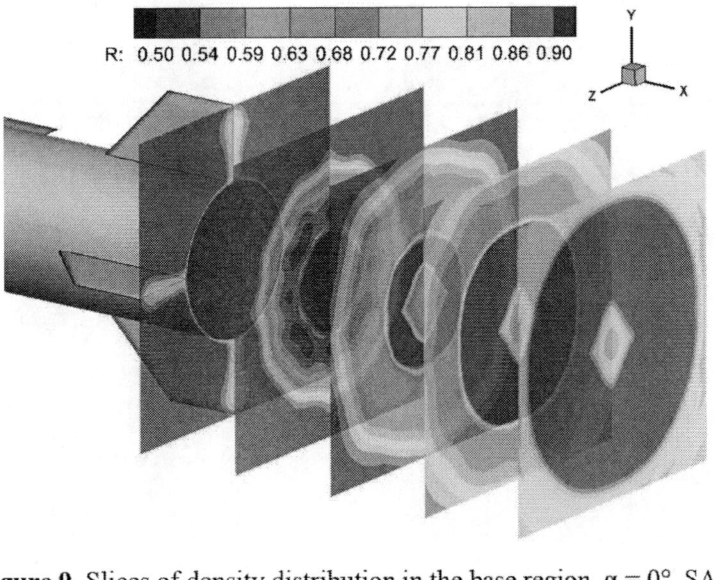

Figure 9. Slices of density distribution in the base region, $\alpha = 0°$, SA.

IOP Conf. Series: Journal of Physics: Conf. Series **1009** (2018) 012005 doi:10.1088/1742-6596/1009/1/012005

3.2. Flow at angle of attack case α = 3°

In general, the structure of the flow varies little with increasing angle of attack. The differences lie in the fact that rarefaction occurs above the upper surface and pressure increases on the lower surface (figure 10). The intensity of the head shock wave on the lower part of the nose increases, on the upper part decreases. Lift force appears.

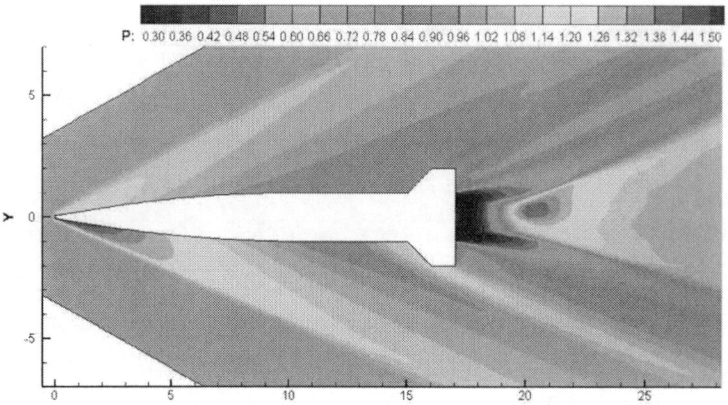

Figure 10. Pressure distribution, α = 3°, slice $z = 0$, SA.

Base flow becomes asymmetric in slice z = 0 (figures 11, 12). The observed base flow structures are qualitatively similar to those found in zero-angle-of-attack but are rotated to an angle roughly corresponding to the angle of attack α = 3°. Length of the recirculation region becomes slightly shorter, reattachment point shifts closer to the body.

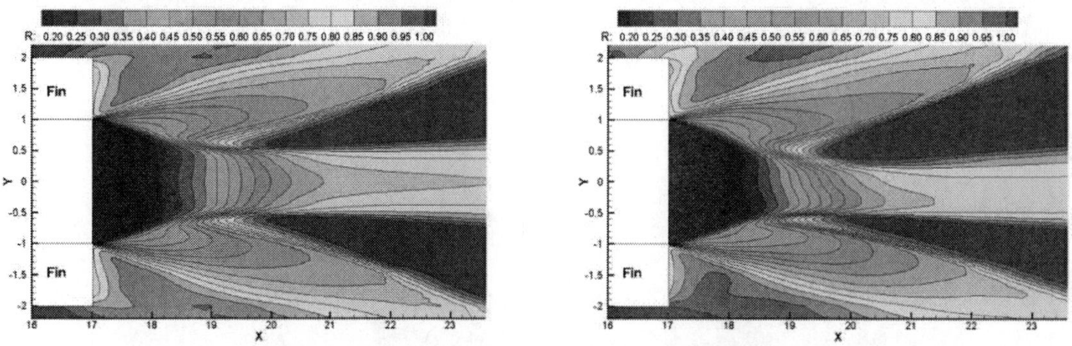

Figure 11. Density distribution, α = 0° (left) and α = 3° (right), slice $z = 0$, SA.

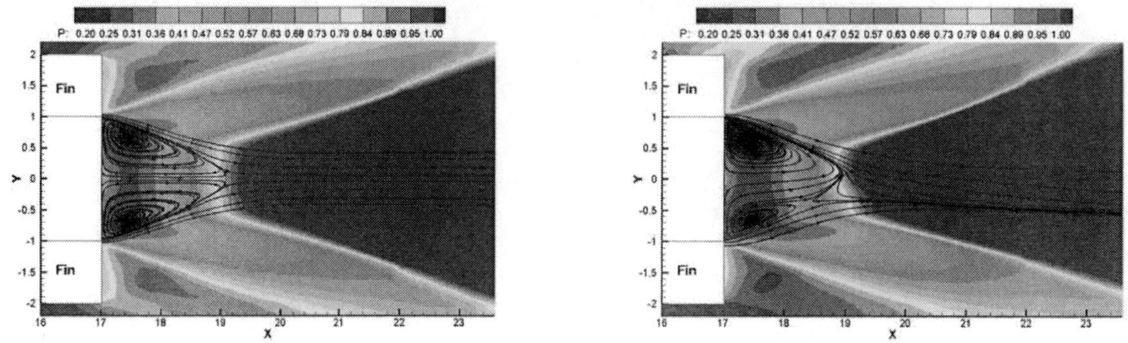

Figure 12. Pressure distribution and streamlines, α = 0° (left) and α = 3° (right), slice $z = 0$, SA.

As can be seen from figure 13, base pressure coefficient for α = 3° is lower than for α = 0°, which agrees with [2,4], where it is stated that with increasing angle of attack base pressure coefficient decreases. For SST turbulence model, the situation is exactly the same and differs a bit from results by SA, that is why it is not given here.

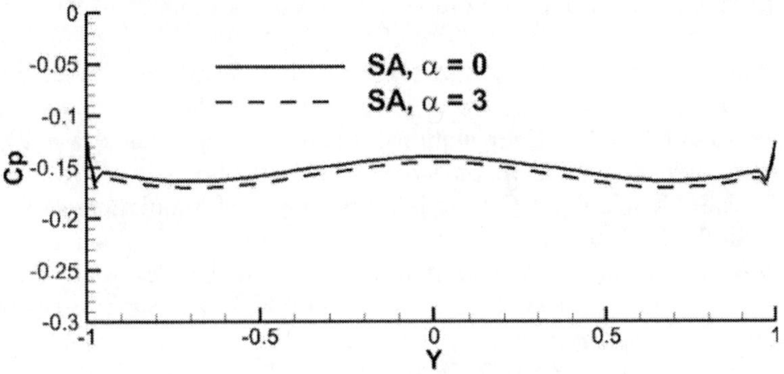

Figure 13. Base pressure coefficient, α = 0° и α = 3°, slice $z = 0$, SA.

4. Conclusions

In this work numerical study of supersonic flow around aircraft model TRK, particularly its base flow, in full layout for several flight modes was performed. The simulation was carried out in the framework of the unsteady Reynolds averaged Navier–Stokes equations (URANS) with the Spalart–Allmaras (SA) and Menter's SST turbulence models. The results of the calculations were compared with the available experimental data and the work of other authors. The qualitative and quantitative agreement of the results was demonstrated. The effect of the turbulence model used on the final result is studied separately; it is shown that for the given problem both models are comparable.

Acknowledgments

This work was supported by the Russian Foundation for Basic Research, projects No. 16-31-00399, 17-08-00909.

References

[1] Love E S 1957 Base Pressure at Supersonic Speeds on Two-Dimensional Airfoils and on Bodies of Revolution with and without Turbulent Boundary Layers, *NACA TN* 3819

[2] Moore F G, Wilcox F and Hymer T 1992 Improved Empirical Model for Base Drag Prediction on Missile Configurations Based on New Wind Tunnel Data *NSWCDD/TR– 92/509*

[3] Herrin J L and Dutton J C 1994 Supersonic Base Flow Experiments in the Near Wake of a Cylindrical Afterbody *AIAA Journal* **32 1** pp 77–83

[4] Boswell B and Dutton J 2001 Flow Visualizations and Measurements of a Three-Dimensional Supersonic Separated Flow *AIAA Journal* **39 1** pp 113–21

[5] Surzhikov S, Borovoi V and Skuratov A 2004 Study of Convective Heating of Segmental-Conical Martian Descent Vehicle in Shock Wind Tunnel *34th AIAA Fluid Dynamics Conference and Exhibit, Fluid Dynamics and Co-located Conferences AIAA* 2004-2634

[6] Borovoy V Y, Egorov I V, Skuratov A S and Struminskaya I V 2010 Peculiarities of flow and heat transfer in the base area of interplanetary probes *TsAGI Science Journal* **41 3** pp 227-57

[7] Lavrukhin G N and Popovich K F 2017 Aerodynamics of jet nozzles Vol. 2 Gas flow around base cutoffs (Moscow: Fizmatlit) (in Russian)

[8] Suzen Y B and Hoffmann K A 1998 Investigation of supersonic jet exhaust flow by one- and two-equation turbulence models *AIAA Paper* 98–0322

[9] Tucker P K and Shyy W 1993 A Numerical Analysis of Supersonic Flow over an Axisymmetric

Afterbody *AIAA Paper* 93-2347
[10] Papp J L and Ghia K N 2001 Application of the RNG Turbulence Model to the Simulation of Axisymmetric Supersonic Separated Base Flow *AIAA Paper* 2001-0727
[11] Simon F, Deck S, Guillen P and Sagaut P 2006 Reynolds-Averaged Navier-Stokes/Large-Eddy Simulations of Supersonic Base Flow *AIAA Journal* **44** **11** pp 2578-90
[12] Baurle R A, Tam C J, Edwards J R and Hassan H A 2003 Hybrid Simulation Approach for Cavity Flows: Blending, Algorithm and Boundary Treatment Issues *AIAA Journal* **41** **8** pp 1463–80
[13] Kawai S and Fujii K 2005 Computational Study of Supersonic Base Flow Using Hybrid Turbulence Methodology *AIAA Journal* **43** **6** pp 1265–75
[14] Simon F, Deck S, Guillen P and Cayzac R 2006 Numerical Simulations of Projectile Base Flow *AIAA Paper* 2006-1116
[15] Forsythe J R, Hoffmann K A, Cummings R M and Squires K D 2002 Detached-Eddy Simulation with Compressibility Corrections Applied to a Supersonic Axisymmetric Base Flow *Journal of Fluids Engineering* **124** pp 911–23
[16] Xiao Z and Fu S 2009 *Acta Mech Sin* **25**: 471
[17] Fureby C, Nilsson Y and Andersson K 1999 Large-Eddy Simulation of Supersonic Base Flow *AIAA Paper 99-0426*
[18] Sandberg R D and Fasel H F 2005 Direct Numerical Simulations of Transitional Supersonic Base Flows *AIAA Paper* 2005-98
[19] Strelets M 2001 Detached eddy simulation of massively separated flows *AIAA paper 2001–0879*
[20] Fureby C and Kupiainen M 2003 Large Eddy Simulation for Supersonic Axisymmetric BaseFlow *Turbulent Shear Flow Phenomena, TSFP3. Sendai, Japan*
[21] Deck S, Garnier E 2005 Detached and Large Eddy Simulation of Unsteady Side-Loads Over An Axisymetric Afterbody *Fifth European Symposium on Aerothermodynamics for Space Vehicles* **563**
[22] Simon F, Deck S, Guillen P, Sagaut P, Merlen A 2007 Numerical simulation of the compressible mixing layer past an axisymmetric trailing edge *Journal of Fluid Mechanics* **591** pp 215-53
[23] Garbaruk A, Shur M, Strelets M and Travin A 2009 Supersonic base flow *DESider — A European Effort on Hybrid RANS-LES Modelling. Notes on Numerical Fluid Mechanics and Multidisciplinary Design* **103** pp197-206
[24] Surzhikov S T 2016 Aerodynamics of the re-entry spacecraft Stardust within the hypersonic flight *Herald of the Bauman Moscow State Tech. Univ., Mech. Eng.* **3** pp 4-22
[25] Zheleznyakova A, Kuzenov V, Petrusev A and Surzhikov S 2010 The modelling of hydrodynamics for two different models of space vehicle *Physical-Chemical Kinetics in Gas Dynamics* **9**
http://chemphys.edu.ru/issues/2010-9/articles/144/
[26] Bykov L V, Molchanov A M, Scherbakov M A and Yanyshev D S 2015 Computational mechanics of continuous media in the problems of aviation and space technology (Moscow: Lenand) (in Russian)
[27] Allmaras S R, Johnson F T and Spalart P R 2012 Modifications and Clarifications for the Implementation of the Spalart-Allmaras Turbulence Model *Seventh International Conf. on CFD (ICCFD7), Big Island, Hawaii*
[28] Edwards J R and Chandra S 1996 Comparison of Eddy Viscosity-Transport Turbulence Models for Three-Dimensional, Shock-Separated Flowfields *AIAA Journal* **34** 4 pp 756–63
[29] Menter F R, Kuntz M and Langtry R 2003 Ten Years of Industrial Experience with the SST Turbulence Model *Turbulence, Heat and Mass Transfer* **4** pp 625–32

Numerical simulation of a flow over blunted cylinder-flare configuration

A S Kryuchkova[1,2]

[1] Ishlinsky Institute for Problems in Mechanics RAS,
Vernadsky prospekt 101(1), Moscow, 119526, Russia
[2] Moscow Institute of Thermal Technology,
Berezovaya avenue 10(14), Moscow, 127273, Russia

E-mail: kryuchkova.arina.s@gmail.com

Abstract. Numerical modeling of a flow over blunted cylinder-flare configuration using UST3D code was performed. Flow variables fields and aerodynamic drag coefficient distribution over a wide range of Mach numbers were obtained. Good agreement of the numerical results with experimental data was observed. The numerical simulation can serve as a supplementary validation of the UST3D computer code which proved to be a reliable instrument for aerodynamic calculations of widespread configurations.

1. Introduction

The paper presents results of numerical modeling of a flow over blunted cylinder-flare configuration. Thanks to the availability of an extensive experimental foundation of aerodynamics for simple bodies it has become possible to conduct corresponding calculations and hence to validate numerical approaches by using well-known cases. The experiment considered in this paper was carried out in the Small Ballistic Tunnel (SBT) of the N E Zhukovsky Central Aerohydrodynamic Institute (TsAGI) [1]. Model aerodynamic drag coefficients as well as flow structure (figure 1) were investigated.

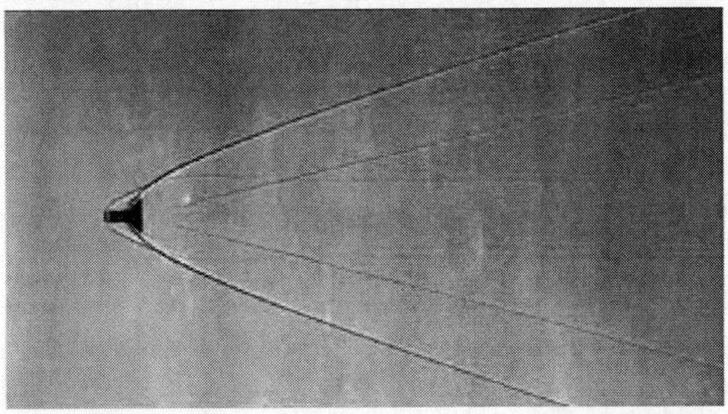

Figure 1. Shadowgraph of a typical flow structure experimentally obtained in SBT.

Models of blunted cylinder-flare shape were shot to the test section at speeds from 500 up to 2000 m/s. The Mach numbers of the test section had the values of 2.5, 3.0, and 3.5. In this way the resulting

Mach numbers of the incoming flow made up values from 6 to 14. The method has allowed investigation of flow characteristics in a wide range of Mach numbers. For experimental modeling of total Mach numbers from 1.5 to 6 the models were launched to a pressure chamber with undisturbed air. Reynolds numbers in the experiments were changing with the flow rate from 10^6 up to $20{\cdot}10^6$. The measurements were taken for zero angle of attack. The model geometry is shown in figure 2. Dimensions were determined with aid of the following relationships: d/2R = 1.0 (semi-spherical leading part); aspect ratio L/d = 1.5; cone half-angle θ = 15°; the model fullness was characterized by the ratio of cylinder to middle diameters d/D = 0.333.

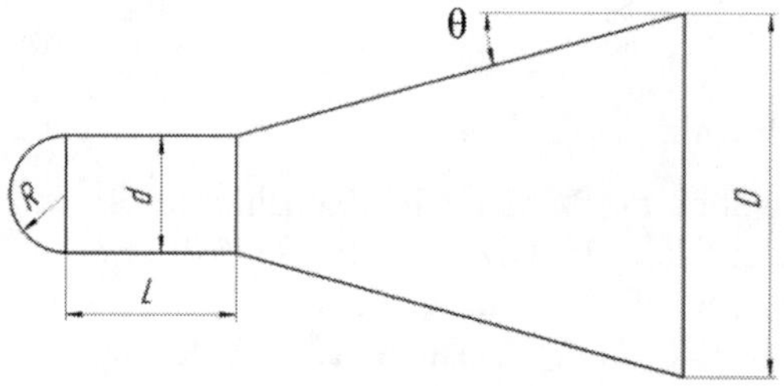

Figure 2. Model geometry.

2. Numerical method

UST3D (Un-Structured Tetrahedral 3-Dimentional) code was used for computational modeling of the problem [2,3]. The software solves spatial Navier-Stokes equations using modificated method of physical processes splitting [4,5,6]. In Cartesian coordinate system the equations have the form of:

$$\frac{\partial U}{\partial t}+\frac{\partial Ec}{\partial x}+\frac{\partial Fc}{\partial y}+\frac{\partial Gc}{\partial z}=\frac{\partial Ev}{\partial x}+\frac{\partial Fv}{\partial y}+\frac{\partial Gv}{\partial z}, \tag{1}$$

where $U=(\rho, \rho u, \rho v, \rho w, \rho E)^T$ is conservative variables vector; $Ec=(\rho u, \rho u^2+p, \rho uv, \rho uw, \rho uE+pu)^T$, $Fc=(\rho v, \rho uv, \rho v^2+p, \rho vw, \rho vE+pv)^T$, $Gc=(\rho w, \rho uw, \rho vw, \rho w^2+p, \rho wE+pw)^T$ are convective terms projections; $Ev=(0, \tau_{xx}, \tau_{yx}, \tau_{zx}, u\tau_{xx}+v\tau_{yx}+w\tau_{zx}-q_x)^T$, $Fv=(0, \tau_{xy}, \tau_{yy}, \tau_{zy}, u\tau_{xy}+v\tau_{yy}+w\tau_{zy}-q_y)^T$, $Gv=(0, \tau_{xz}, \tau_{yz}, \tau_{zz}, u\tau_{xz}+v\tau_{yz}+w\tau_{zz}-q_z)^T$ are viscous terms projections; ρ is density; u, v, w – velocity projections; E – specific energy, τ_{ij} – viscous tensor components; q_i – heat flux projections.

The flow is supposed to be laminar. For the equations closure the ideal gas law is used:

$$p=(\gamma-1)\,\rho\,[E-\tfrac{1}{2}(u^2+v^2+w^2)], \tag{2}$$

where $\gamma = 1.4$ – heat capacity ratio of air.

The UST3D solver is adapted for unstructured tetrahedral grids. The computational code consists of a preprocessor, a solver and a postprocessor. The preprocessor stocks all necessary data concerning computational cells topology. The solver actually realizes numerical computations and the postprocessor records the results output in a convenient format for graphical visualization. Time step is chosen automatically according to condition on final discrepancy limitation.

The calculi were performed on two tetrahedral unstructured meshes counting 417 384 and 668 675 cells. The mesh density was higher round the model walls, so that the layer between the body and the bow shock wave numbered at least 6 cells in the direction normal to the boundary. Computational domain is patterned in figure 3. Boundary conditions are also indicated in the figure, where: 1 – inflow boundary condition; 2 – flow at infinity; 3 – supersonic exit; 4 – viscous wall. Symmetry boundary condition which coincides with the pattern plane was used to reduce computational costs.

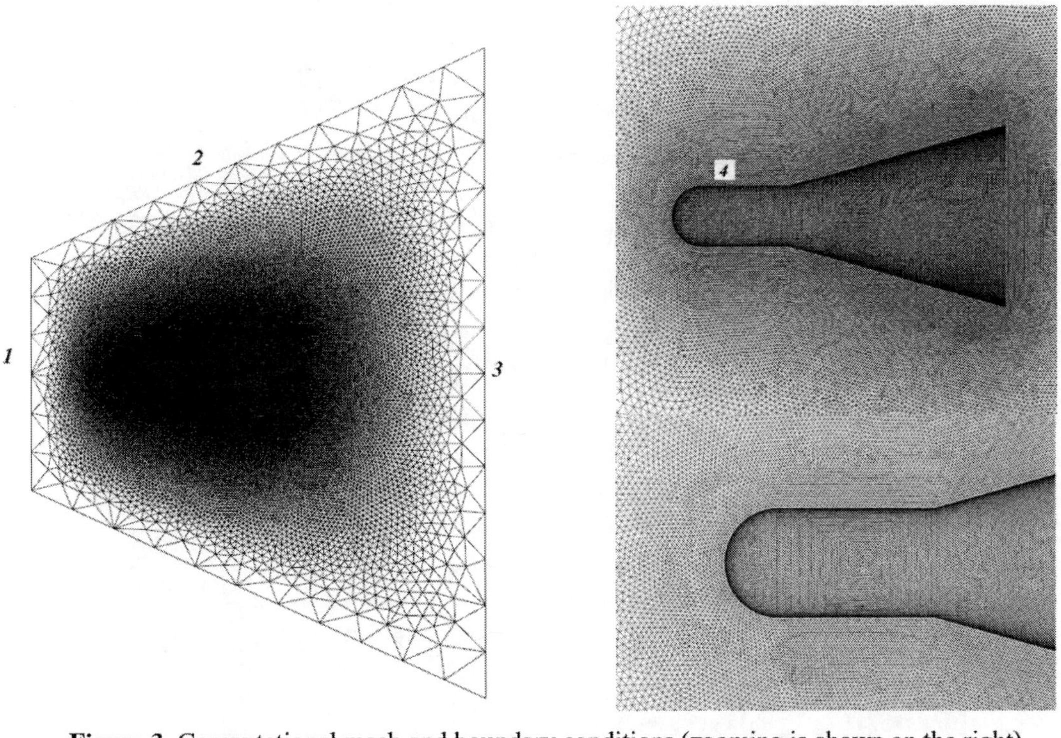

Figure 3. Computational mesh and boundary conditions (zooming is shown on the right).

3. Results and discussion

In the numerical modeling the middle diameter was assumed to be 200 cm (approaching to real flight vehicle dimensions). Nominal flow conditions were set to: $p_\infty = 40000 erg/cm^3$, T_∞=219K, T_{surf}=300K.

The resulting flow structure includes typical elements such as bow shock, oblique shock near the generating line break, and transient flow in the bottom area. It was found that for this configuration there is no separation zone in the whole range of Mach numbers, which agrees with experimental observations [1].

Calculated drag coefficient comparison with experimental data from [1] is shown in figure 4. Drag coefficient is related to the product of incident flow dynamic pressure and body midsection area (note that it coincides with axial force coefficient at zero angle of attack). According to the study drag coefficient decreases with Mach number growth and stabilizes when Mach number mounts to 7. Good agreement with the experimental curve is observed. The computational accuracy for both meshes is approximately the same and does not exceed the experimental data spread. Flow structure and its transient behavior are illustrated in figure 5.

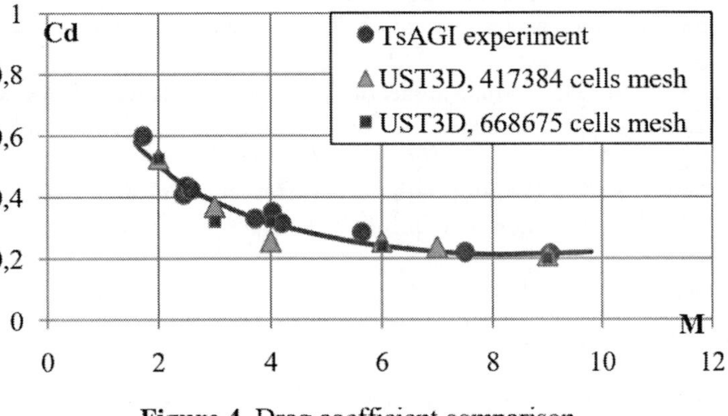

Figure 4. Drag coefficient comparison.

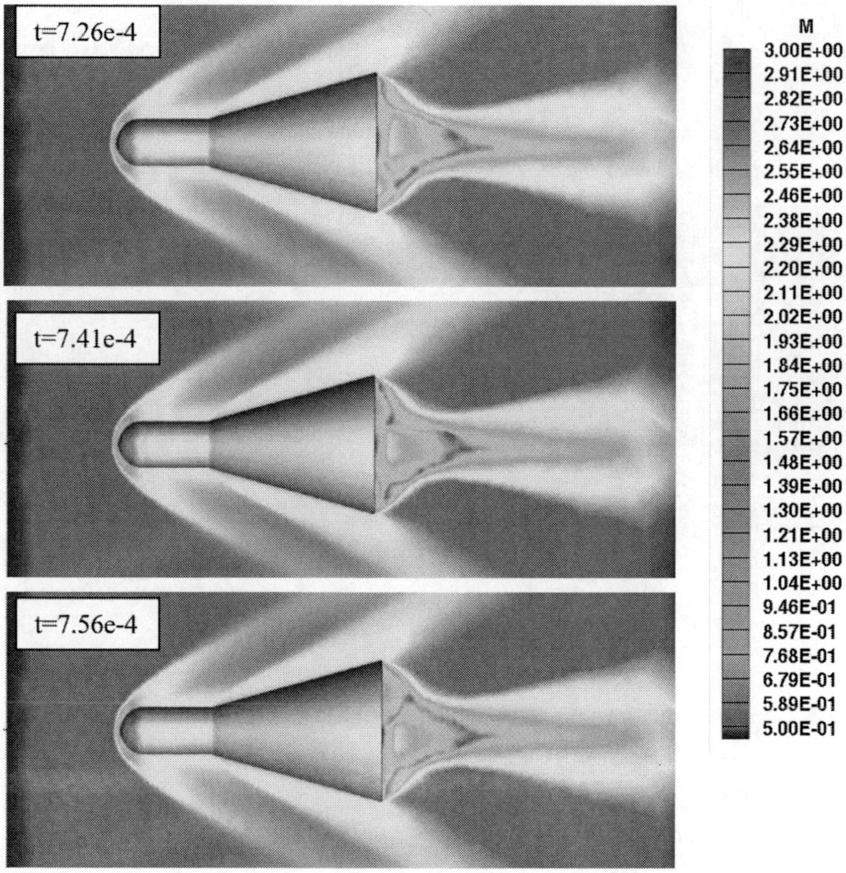

Figure 5. Mach number field for the case of $M_\infty = 3$ in various moments of time.

One more factor testifying the accuracy of the numerical modeling of a flow over a body of revolution is bow shock deviation. The shock shape and its distance from the body are mostly determined by the shape of the body leading part. Since the leading part of the considered configuration is semi-spherical, the shock deviation distance on the axis of symmetry ε can be assessed by shock deviation tables and curves obtained for a sphere. There is a set of widely known papers dealing with sphere bow shock deviation [7, 8].

Comparison of the computed deviation distance on the axis of symmetry with experimental and numerical results given in [7] is shown in figure 6. The numerical simulation is noted to give a satisfactory estimation of bow shock deviation.

After validating numerical method with help of the tests mentioned above a set of computations of a flow under various angles of attack (5°, 10°, 15°, and 20°) for $M_\infty = 3$ was performed. The purpose of the computations was to demonstrate angle of attack effect on drag force as well as to verify UST3D stability in more complex problems. Computed drag coefficient dependence on the angle of attack is presented in figure 7. The drag coefficient is shown to increase while angle of attack rises and it can be seen that there is still no stalling in the considered angle of attack range. Flow structure for $M_\infty = 3$ and angle of attack $\alpha = 20°$ is shown in figure 8.

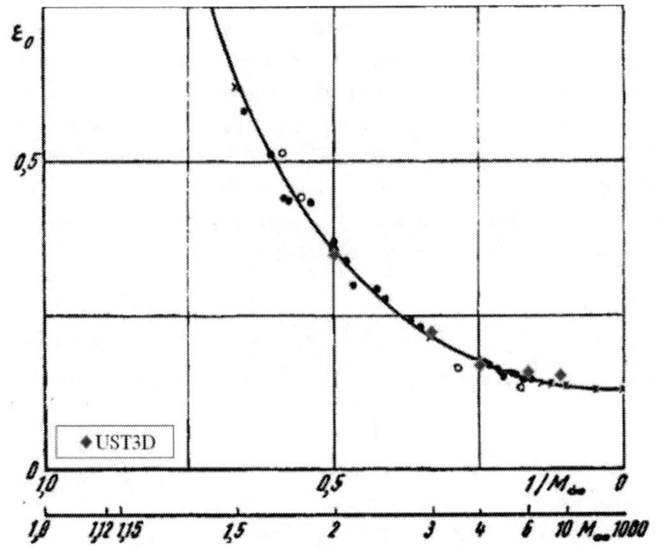

Figure 6. Normalized bow shock deviation distance $\varepsilon_0 = \varepsilon / R$ obtained in UST3D (red diamonds) compared to numerical data from [7] (black curve) and [9] (crosses) as well as to experimental data from [10] (white circles) and [11] (black points).

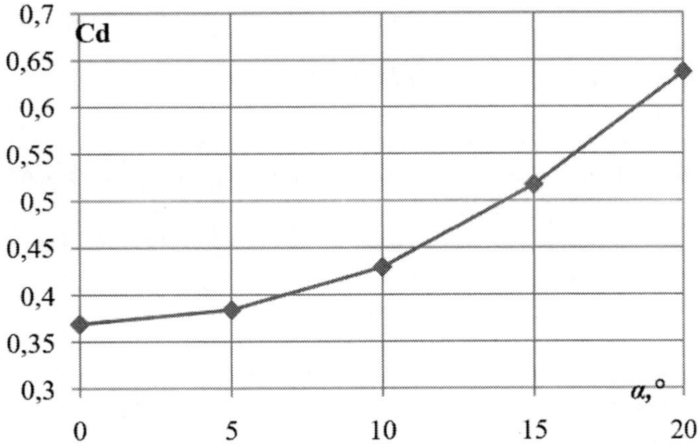

Figure 7. Angle of attack effect on drag coefficient for M_∞ =3.

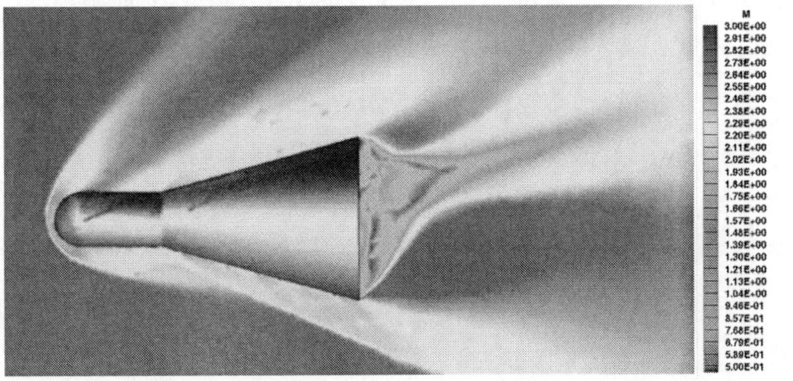

Figure 8. Mach number field for the case of M_∞ =3 and angle of attack α = 20°.

4. Conclusion

Computer modeling of a flow over blunted cylinder-flare configuration using UST3D code was carried out. Comparison of resulting computational and experimental drag coefficient distributions over a wide range of Mach numbers has shown good accuracy of numerical method predictions at zero angle of attack. Furthermore, it was demonstrated that UST3D program estimates the bow shock deviation distance reasonably well. An additional computation with non-zero incident flow angle of attack was performed to illustrate angle of attack effect on drag force and on the flow structure. The numerical simulation has provided an extra UST3D code validation and has testified the program to be a robust instrument for aerodynamic calculations of widespread configurations.

Acknowledgments

Author is grateful to professor S T Surzhikov and to the staff of Laboratory of Radiation Gas Dynamics (Ishlinsky Institute for Problems in Mechanics RAS) for useful discussions on the problems considered in the study.

References

[1] Krasil'shchikov A P and Guryashkin L P 2007 *Experimental investigation of bodies of revolution in hypersonic flows* (Moscow: PhysMathLit) p 208 (in Russian)

[2] Surzhikov S T 2017 Validation of computational code UST3D by the example of experimental aerodynamic data *J. Phys.: Conf. Ser.* **815** 012023

[3] Zheleznyakova A L and Surzhikov S T 2013 Application of the Method of Splitting by Physical Processes for the Computation of a Hypersonic Flow over an Aircraft Model of Complex Configuration *High Temperature* **51** (6) pp 816–829

[4] Ewans M V and Harlow F H 1957 The particle-in-cell method for hydrodynamic calculations *Los Alamos Scientific Lab. Rept.* N LA–2139 Los Alamos

[5] Belotserkovsky O M and Davydov Yu M 1982 *Method of the Large Particles for Gas Dynamics* (Moscow: Nauka) p 391(in Russian)

[6] Marchuk G I 1988 *Splitting methods* (Moscow: Nauka) p 263 (in Russian)

[7] Bazzhin A P, Blagosklonov V I, Minailos A N and Pirogova S V 1971 Supersonic flow of a perfect gas over a sphere *Uch. Zap. TsAGI* **2** (3).

[8] Krasil'shchikov A P and Podobin V P 1968 Experimental study of sphere aerodynamic characteristics in free flight up to M ~ 15 *Fluid Dynamics* **3** (4) pp 132–134

[9] Lyubimov A N and Rusanov V V 1970 *Gas Flows Over Blunt Bodies* (Moscow: Nauka) p 668 (in Russian)

[10] Lipman G V and Roshko A 1960 *Elements of Gas Dynamics* (Moscow: IL) p 352 (in Russian)

[11] Maslennikov V G 1967 Study of detached shock wave location for supersonic flow past ellipsoids of revolution in gases with different intramolecular structure *Aerophysical Studies of Supersonic Flows* (Moscow, Leningrad: Nauka) p 241 (in Russian)

IOP Conf. Series: Journal of Physics: Conf. Series **1009** (2018) 012007 doi:10.1088/1742-6596/1009/1/012007

Analysis of the high speed gas flow over a sphere in the range of Mach numbers 2–12

N A Kharchenko[1,2,3], **M A Kotov**[1]

[1] Ishlinsky Institute for Problems in Mechanics RAS, Vernadsky prospect 101(1), Moscow, 119526, Russia

[2] Tactical Missiles Corporation JSC, Ilicha street 7, Korolev, Moscow region, 141080, Russia

[3] Moscow Physical–Technical Institute, Institutskiy per. 9, Dolgoprudny, Moscow region, 141701, Russia

E–mail: kharchenko.na@phystech.edu

Abstract. In this paper, the results of analysis of the high speed gas flow over a sphere are presented. Numerical investigations were carried out using the UST3D code developed at the Ishlinsky Institute for Problems in Mechanics of RAS (IPMech RAS). The code implements numerical simulation of aerothermodynamics of aircrafts in the context of integration of the full system of Navier – Stokes equations on unstructured mesh using the perfect gas model.

1. Introduction

During many years the investigations of the drag of the sphere moving in the gas flow remains relevant. The papers [1–7] are devoted to investigation of this problems.

The main goal of this work is the validation of the UST3D code in a context of analysis of a high speed flow over a sphere [8]. The code-calculated data were validated against experimental data obtained during study of aerodynamic characteristics of a sphere in a free flight [9]. Tests were performed in the aeroballistic research facility of the Central Aerohydrodynamic Institute (TsAGI) [10] in the range of Mach numbers 1.5 – 15.2 and in the range of Reynolds numbers 10^6–10^7.

During the study, a large amount of experimental data concerning the drag coefficient Cx was obtained. Experiments were performed as follows: the sphere was shot from the launch tube with a caliber of 14.5 mm in the direction opposite to the supersonic flow in the wind tunnel [11]. The diameter of spherical model used was about 10 mm.

2. Computational model

The UST3D code is designed for numerical investigation of aerothermodynamics of high-speed aircrafts and can be applied in a wide range of Mach numbers, altitudes, and angles of attack [8, 12]. The code is based on solving the three-dimensional unsteady system of Navier – Stokes equations (1) – (4) [13, 14].

$$\frac{\partial \mathrm{w}}{\partial t}+\frac{\partial F^{x}(\mathrm{w})}{\partial x}+\frac{\partial F^{y}(\mathrm{w})}{\partial y}+\frac{\partial F^{z}(\mathrm{w})}{\partial z}=\frac{\partial G^{x}(\mathrm{w})}{\partial x}+\frac{\partial G^{y}(\mathrm{w})}{\partial y}+\frac{\partial G^{z}(\mathrm{w})}{\partial z}, \tag{1}$$

IOP Conf. Series: Journal of Physics: Conf. Series **1009** (2018) 012007 doi:10.1088/1742-6596/1009/1/012007

$$\mathrm{w} = \begin{pmatrix} \rho \\ \rho u \\ \rho v \\ \rho w \\ \rho E \end{pmatrix}, \quad (2)$$

$$F^x = \begin{pmatrix} \rho u \\ \rho u^2 + p \\ \rho uv \\ \rho uw \\ \rho uE + pu \end{pmatrix}, \; F^y = \begin{pmatrix} \rho v \\ \rho uv \\ \rho v^2 + p \\ \rho vw \\ \rho vE + pv \end{pmatrix}, \; F^z = \begin{pmatrix} \rho w \\ \rho uw \\ \rho vw \\ \rho w^2 + p \\ \rho wE + pw \end{pmatrix}, \quad (3)$$

$$G^x = \begin{pmatrix} 0 \\ \tau_{xx} \\ \tau_{yx} \\ \tau_{zx} \\ u\tau_{xx} + v\tau_{yx} + w\tau_{zx} - q_x \end{pmatrix}, \; G^y = \begin{pmatrix} 0 \\ \tau_{xy} \\ \tau_{yy} \\ \tau_{zy} \\ u\tau_{xy} + v\tau_{yy} + w\tau_{zy} - q_y \end{pmatrix},$$

$$G^z = \begin{pmatrix} 0 \\ \tau_{xz} \\ \tau_{yz} \\ \tau_{zz} \\ u\tau_{xz} + v\tau_{yz} + w\tau_{zz} - q_z \end{pmatrix}, \quad (4)$$

The complete set of Navier – Stokes equations is used in conjunction with the equation of state for ideal gas (5).

$$p = (\gamma - 1)\rho U = (\gamma - 1)\rho\left[E - \frac{1}{2}\left(u^2 + v^2 + w^2\right)\right], \quad (5)$$

The UST3D code performs numerical integration of this system by using the method of splitting on physical processes. The method is explicit and has the first order approximation with respect to time and the second order approximation with respect to space. Solution of the problem is obtained as the result of relaxation [8, 12].

As one of approaches to calculation of flows through interface boundaries of mesh cells used in the UST3D, the HLLE solver was implemented [15]. The HLLE solver was used in order to prevent emergence of the carbuncle-type effect during computations. Emergence of this effect is a characteristic for bodies with large-radius blunt noses.

In all Godunov-type methods, the flow through the cell interfaces is calculated from the Riemann problem solution [16]. Using the exact Riemann solver proposed by S. K. Godunov requires a costly iterative procedure. However, a large number of approximate Riemann solvers are available now, which makes the S. K. Godunov's approach quite efficient for solving the problems of unstructured meshes. In this work, the HLLE method is used as an approximate Riemann solver [15].

The procedure of approximate Riemann solver for flux computation (well-known as HLL) was proposed by Harten, Lax, and van Leer [17]. The method is based on consideration of a simplified two-wave configuration without taking into account the contact discontinuity. The key difference of the HLLE (Harten – Lax – van Leer – Einfeldt) method is the Roe-averaged density of the sound speed used to find the maximum and minimum wave speeds.

The flux of the HLLE method is determined from the following relation [18]:

$$F_{HLLE} = \begin{cases} F_L & \text{при} \quad 0 \le S_L \\ \dfrac{S_R F_L - S_L F_R + S_L S_R (w_R - w_L)}{S_R - S_L} & \text{при} \quad S_L \le 0 \le S_R \\ F_R & \text{при} \quad 0 \ge S_R \end{cases}, \tag{6}$$

Speeds of the left and right shock waves S_L and S_R are determined from the minimum and maximum disturbance speeds [18]:

$$S_L = V_L - c_L, \quad S_R = V_R + c_R, \tag{7}$$

$$S_L = \min\left\{\bar{V} - \bar{c}, V_L - c_L\right\}, \quad S_R = \max\left\{\bar{V} + \bar{c}, V_R + c_R\right\}, \tag{8}$$

$$\bar{V} = \frac{\sqrt{\rho_L} V_L + \sqrt{\rho_R} V_R}{\sqrt{\rho_L} + \sqrt{\rho_R}}, \quad \bar{c} = \left[(\gamma - 1)\left(H - \frac{1}{2}\bar{V}^2\right)\right]^{\frac{1}{2}}, \quad \bar{H} = \frac{\sqrt{\rho_L} H_L + \sqrt{\rho_R} H_R}{\sqrt{\rho_L} + \sqrt{\rho_R}}, \tag{9}$$

The pressure distribution over the interface can be calculated from the following relations:

$$\bar{p} = 0.5(p_L + p_R) - 0.5(V_R - V_L)\bar{\rho}\bar{c},$$

$$\bar{\rho} = 0.5(\rho_L + \rho_R), \quad \bar{c} = 0.5(c_L + c_R), \tag{10}$$

Hight viscosity of this method allows eliminating the carbuncle-type effects during the computation.

3. Problem statement for computing flows around the sphere

For numerical simulation of the flow, a three-dimensional surface of the sphere was created with the diameter of 10 mm.

In this work, simulation of three-dimensional flow field over a sphere was performed using unstructured tetrahedral meshes The computational mesh was refined towards the surface of the sphere. The number of triangles on the surface was 36 354 elements, and the overall number of tetrahedrons in the volume was 2 920 576 cells (figure 1).

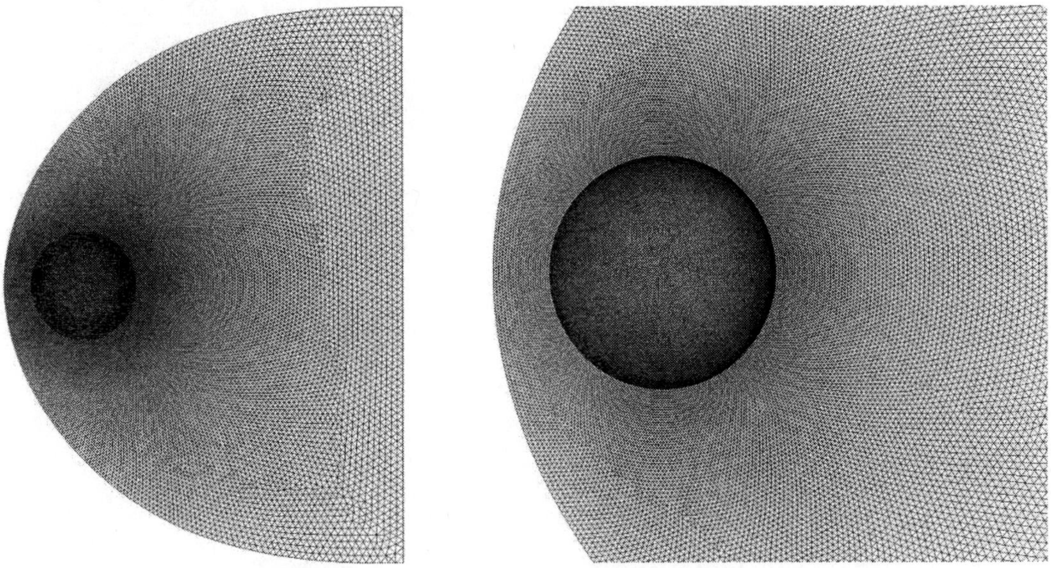

Figure 1. Volume computational mesh.

4. Results obtained

A series of computations of a flow over a sphere was carried out for Mach numbers from 2 to 12, in increments of one Mach. The pressure and density parameters in incident flow were $p_\infty = 0.405 \cdot 10^4$ Pa and $\rho_\infty = 0.646 \cdot 10^{-1}$ kg/m^3 respectively at $Re = 10^4$–10^5. The results of computations for $M = 6$ are shown in figures 2 – 4. Results obtained with the UST3D code where the HLLE solver is implemented are presented in the right-hand side of figures 2 – 4. Evidently, carbuncle-type effects are absent in the leading shock wave, when fluxes through computational cells are calculated using the HLLE method.

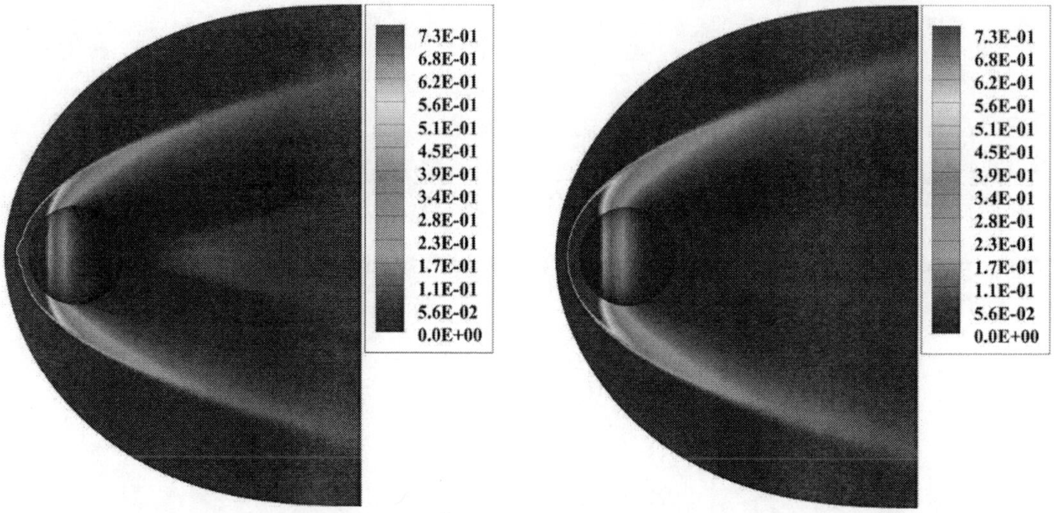

Figure 2. Pressure distribution ($p/10^5$) over the surface and in the vicinity of the sphere.

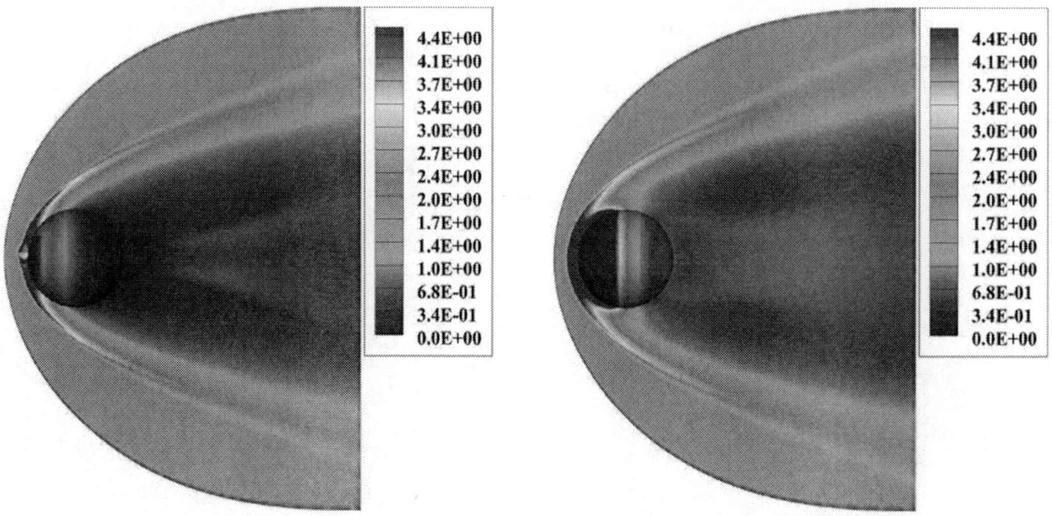

Figure 3. Density distribution (ρ/ρ_∞) over the surface and in the vicinity of the sphere.

IOP Conf. Series: Journal of Physics: Conf. Series **1009** (2018) 012007 doi:10.1088/1742-6596/1009/1/012007

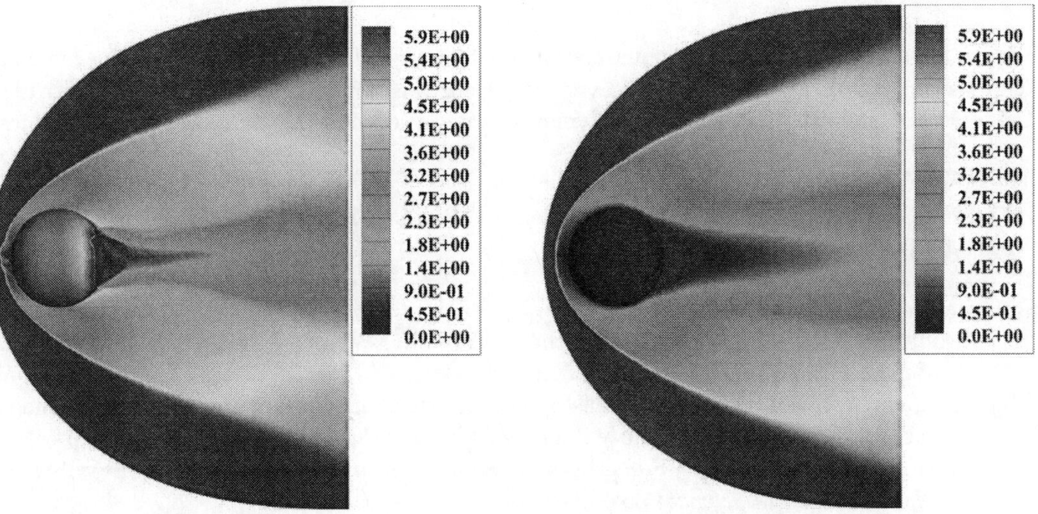

Figure 4. Mach number distribution over the surface and in the vicinity of the sphere.

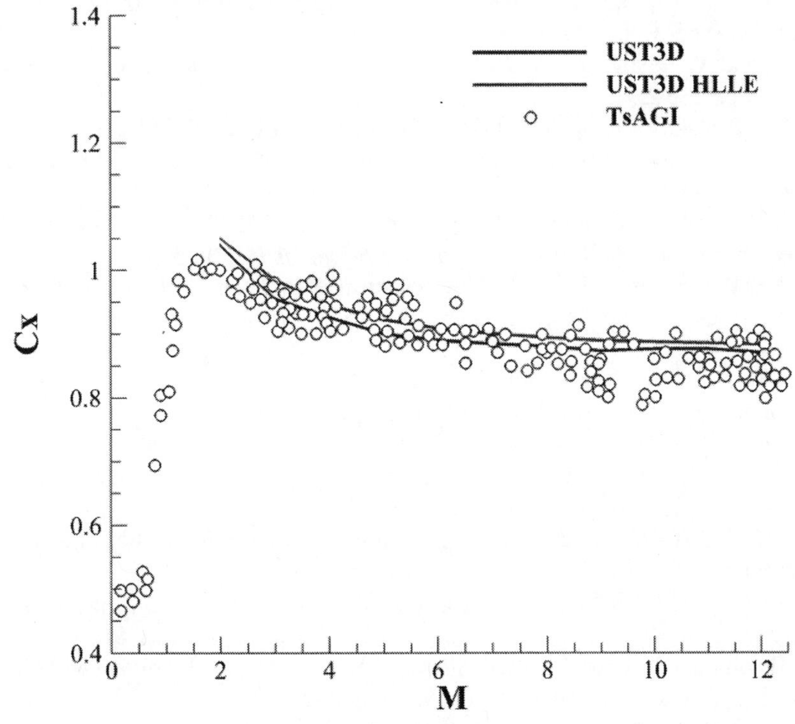

Figure 5. Change of the aerodynamic drag coefficient of the sphere as a function of the Mach number.

From figure 5 one can see that obtained calculated data are in a close agreement with the experimental data [9, 11]. Difference in obtained values of the drag coefficient *Cx* can occur due to the presence of carbuncle-type effect in computations.

Conclusion

Within the framework of the computer code validation a good agreement between the calculated and the experimental data for the air flow over the sphere was obtained. This is important for the future UST3D code applications to aerothermodynamics analysis of aircraft with large-radius blunt noses.

The presented study was partially supported by Russian State Assignment under contract No. AAAA-A17-117021310372-6 and partially supported by Russian Foundation for Basic Research grant No. 16-01-00379.

References

[1] Hodjes A J 1957 The Drag Coefficient of Very High Velocity Spheres *J. of the Aeronautical Sciences* pp 755–758
[2] Charters A C and Thomas R N 1945 The Aerodynamic Performance of Small Spheres from Subsonic to High Supersonic Velocities *Journal of the Aeronautical Sciences* pp 468–476
[3] May A and Witt W R 1953 Free-Flight Determinations of the Drag Coefficients of Spheres *J. of the Aeronautical Sciences* pp 635–638
[4] Bailey A B and Hiatt J 1972 Sphere Drag Coefficients for a Broad Range of Mach and Reynolds Numbers *AIAA J.* **10** pp 1436–1440
[5] Lyons W C, Brady J J and Levensteins J 1964 Hypersonic Drag, Stability and Wake Data for Cones and Spheres *AIAA J.* **2** pp 1448–1456
[6] Ross F W and Willmarth W W 1971 Some Experimental Results on Sphere and Disk Drag *AIAA J.* **9** pp 285–291
[7] Kotov M A, Kryukov I A, Ruleva L B, Solodovnikov S I, and Surzhikov S T 2015 Supersonic Air Flows Around Some Geometrical Primitives 33rd *AIAA Applied Aerodynamics Conf.* AIAA 2015–3012
[8] Surzhikov S T 2017 Validation of computational code UST3D by the example of experimental aerodynamic data *J. of Physics: Conference Series* **815** 012023
[9] Krasilshikov A P and Podobin V P 1968 Experimental research of free flight ball aerodynamic characteristics up to the Mach numbers M ~ 15. *Fluid Dynamics* **4** pp 190–193
[10] http://www.tsagi.ru
[11] Krasilshikov A P and Guryashkin L P 2007 Bodies of revolution experimental research in hypersonic flows (Moscow: Fizmatlit) p 208 (in Russian)
[12] Zheleznyakova A L and Surzhikov S T 2013 Application of the method of splitting by physical processes for the computation of a hypersonic flow over an aircraft model of complex configuration *High Temperature* **51** pp 816–829
[13] Surzhikov S T 2013 Convective heating of small-radius spherical blunting for relatively low hypersonic velocities *High Temperature* **51** pp 231–245
[14] Surzhikov S T 2017 Spatial Multiphysics Models of the Radiation Gas Dynamics of Super Orbital Re–Entry Space Vehicles. *J. of Physics: Conference Series* **837** 012021
[15] Einfeldt B, Munz C D, Roe P L and Sjögreen B 1991 On Godunov–Type Methods near Low Densities *J. of Computational Physics* **92** pp 273–295
[16] Godunov S K 1959 A finite-difference method for the numerical computation and discontinuous solutions of the equations of fluid dynamics *Mat. Sb.* **47** pp 271–306
[17] Harten A, Lax P D and van Leer B 1983 On Upstream Differencing and Godunov–Type Schemes for Hyperbolic Conservation Laws *SIAM Review* **25** pp 35–61
[18] Toro E F 2009 Riemann Solvers and Numerical Methods for Fluid Dynamics *Springer* p 724

IOP Conf. Series: Journal of Physics: Conf. Series **1009** (2018) 012008 doi:10.1088/1742-6596/1009/1/012008

Asymmetric wake from two closely located cylinders within the framework of one-dimensional model

G V Gembarzhevskii [1, 2], A K Lednev [1] and K Yu Osipenko [1, 3]

[1] Ishlinsky Institute for Problems in Mechanics RAS, Vernadsky prospect 101(1), Moscow, 119526, Russia
[2] Moscow Aviation Institute (National Research University), Moscow, Russia
[3] National Research University, Moscow State University of Civil Engineering (NRU, MGSU), Moscow, Russia

E-mail: gvgemb@ipmnet.ru

Abstract. Simple oscillatory model for wake hydrodynamics is constructed with the aim of physical modelling of complex object - turbulent plasma wake in glow discharge. A wake from not very closely located parallel cylinders assumed to be of the shape of the Von Karman streets pair, or a modified streets pair. Each street modelled by an oscillator of Van der Pole type. The specific features of our model are the follows: first - account for the nonlinear interaction of Von Karman streets in the area of their formation; second - explicit account for the oscillation frequency dependence on oscillation amplitude of the streets. Several global modes and intermittent regimes of the wake were revealed within the framework of the model. The main accent of this article is a description of two-frequency asymmetric mode of such two-cylinder wake.

1. Introduction

Low temperature plasma flows are of interest for fundamental science as well as in industrial applications such as fast-flow lasers, combustion ignition, electro-chemical reactors [1]. In this paper main interest concentrated around quasi-stationary and quasi-homogeneous flows with high level of heat and mass transfer [2, 3]. One opportunity to realize such flows is the use of a wake of cylinders row with relatively low pitch: $L/D \leq 3$; the latter condition is aimed to provide the flow quasi-homogeneity. However, it is well known that such turbulent wakes are very unstable flows, especially in the case of the cylinders close pitch $1.5 \leq L/D \leq 2$ where intermittent regime is observable as a rule [4], when quasi-stationary flow is desirable. Ipso facto some optimization of the wake hydrodynamics is required. And, in parallel with experimental optimization, a theoretical one is useful. For the latter use some wake model preferably simple is desirable.

The object of our main interest is plasma wakes in non-self-sustained glow discharge, which is typical for constant power industrial lasers [5]. In our case using of a certain wake model is not only desirable but is necessary one since flow visualization is unavailable, only velocity signal is acceptable [6, 7]. Accordingly, simple relevant model of the wake hydrodynamic is constructed: [8, 9, 6]. It was found that the model contains five one-frequency modes including asymmetric mode of biased wake with two von Karman streets of different intensities but equal oscillation frequencies. Some wake modes intermits in appropriate range of the model parameters. Furthermore, our model reproduces a two-frequency asymmetric wake mode consisted of two Von Karman streets of distinct intensities and different frequencies as well. This mode is well observable in visualization frames and in velocity

IOP Conf. Series: Journal of Physics: Conf. Series **1009** (2018) 012008 doi:10.1088/1742-6596/1009/1/012008

spectra both, for example [10]. This article mainly concerns characteristic features of this two-frequency mode within the framework of the model.

2. Formulation of one dimensional wake model

Plasma flows of elevated turbulence intensity are used in fast-flow constant-power lasers (see for example [2]). Usually such flows are formed by some network of obstacles that may be a row of cylinders. Here generic flow is a wake produced by solitary cylinder being the Von Karman vortex street for wide range of the flow Reynolds number. The second step is a consideration of a cylinders pair wake, and then adding more cylinders in raw arrangement. If the distance between cylinders axes is intermediate, the whole (laminar) wake expected to be a set of synchronized Von Karman streets that are deformed slightly due to their interaction in the domain of their formation. Within the framework of a reduced-order modeling it is common practice to represent a wake from solitary cylinder by a model of Landau-Stuart type [11] or by a similar model of Van der Pole oscillator [12, 13]. Van der Pole equation is a starting point for our wake model from cylinders group as well:

$$\left(1+\Delta\left(\rho^2-4\right)\right)\frac{d^2X}{dt^2}+X-\varepsilon\left[1-X^2-\lambda Y^2-lXY\right]\frac{dX}{dt}=h_X \tag{1}$$

$$\left(1+\Delta\left(r^2-4\right)\right)\frac{d^2Y}{dt^2}+Y-\varepsilon\left[1-Y^2-\lambda X^2-lXY\right]\frac{dY}{dt}=h_Y \tag{2}$$

Here the dependent variables X, Y represent governing parameters of two Von Karman streets behind the first and the second cylinder, respectively. The interplay between neighboring streets in the area of their formation is accounted for by the term in square brackets (with λ and l multipliers). With the aim of the model generality the form of interplay term selected is nonlinear one, whereas some linear term in model equations is usually used for the interplay modeling, see for example [11]. (Justification of the latter generally accepted option assumed to be a weakness of the streets interaction for not very closely located cylinders.) The second peculiarity of our model is an explicit account for dependence of oscillation frequency of the street on its amplitude (denoted as ρ and r, respectively), which is accomplished by the nonlinear term in brackets. This complication of the model was made to account for essential splitting of the oscillation frequencies for different wake modes. Usually this effect of the wake oscillation amplitude on its frequency is neglected within the framework of oscillatory wake modeling [12, 13]. Right-hand sides in equations (1) and (2) represent a source of stochasticity for turbulent wake (a certain analog of Langevin force). This model contains four parameters, namely: $\varepsilon, \Delta, \lambda, l$. The Van der Pole parameter is small $\varepsilon \ll 1$ in agreement with experimental data [14]. The value of quantity Δ is also small $|\Delta| \sim \varepsilon \ll 1$. The Krylov-Bogolyubov method [15] of slow varying amplitudes and phases of oscillations have to be used for the equations (1) and (2) solving because of parameter ε smallness. The substitution (3) into (1), (2) have to be used.

$$X=\rho\cos P_X;\quad Y=r\cos P_Y \tag{3}$$

Corresponding truncated (averaged) equations of the model are (4) – (7).

$$\Phi_\rho\frac{d\rho^2}{dt}-\varepsilon\rho^2\left[1-\frac{\rho^2+\lambda r^2\left(2-\cos(2P)\right)+l\rho r\cos(P)}{4}\right]=h_\rho \tag{4}$$

$$\Phi_r\frac{dr^2}{dt}-\varepsilon r^2\left[1-\frac{r^2+\lambda\rho^2\left(2-\cos(2P)\right)+l\rho r\cos(P)}{4}\right]=h_r \tag{5}$$

$$\frac{dP_X}{dt} - \frac{1}{\sqrt{1+\Delta(\rho^2-4)}} + \frac{\varepsilon}{8\Phi_\rho}\left[\lambda r^2 \sin(2P) + l\rho r \sin(P)\right] = h_\theta \quad (6)$$

$$\frac{dP_Y}{dt} - \frac{1}{\sqrt{1+\Delta(r^2-4)}} - \frac{\varepsilon}{8\Phi_r}\left[\lambda \rho^2 \sin(2P) + l\rho r \sin(P)\right] = h_\psi \quad (7)$$

Here notation P for the independent variable the total phase difference of oscillation in two Von Karman streets was introduced

$$P = P_X - P_Y \quad (8)$$

and notation Φ_β, as follows:

$$\Phi_\rho = 1 + \Delta\left(0.5\rho^2 - 4\right) \quad (9)$$

$$\Phi_r = 1 + \Delta\left(0.5r^2 - 4\right) \quad (10)$$

Using independent variable P the system of four model equations (4), (5), (6), and (7) was reduced to the system of three equations (4), (5), and (11).

$$\frac{dP}{dt} - \frac{1}{\sqrt{1+\Delta(\rho^2-4)}} + \frac{1}{\sqrt{1+\Delta(r^2-4)}} + \\ + \frac{\varepsilon}{8}\left[\lambda\left(\frac{r^2}{\Phi_\rho} + \frac{\rho^2}{\Phi_r}\right)\sin(2P) + l\rho r\left(\frac{1}{\Phi_\rho} + \frac{1}{\Phi_r}\right)\sin(P)\right] = h_P \quad (11)$$

Accordingly, the model phase space appears to be a three-dimensional one.

Moreover it turned out that modeling of the wake from solitary cylinder with the help of modified Van der Pole equation

$$\left(1+\Delta\left(\rho^2-4\right)\right)\frac{d^2X}{dt^2} + X - \varepsilon\left[1 - X^2\right]\frac{dX}{dt} = 0 \quad (12)$$

(after transformation to truncated equations, in assumption: $|\Delta| \sim \varepsilon << 1$) is equivalent to modeling by means of the Landau-Stuart equation. The latter option has firm ground in the Navier-Stokes system. Accordingly, our wake model has hydrodynamic substantiation as well.

3. Two-frequency wake mode

Our model contains one-frequency global laminar modes of the wake as stationary stable points of equation system (4), (5), (11) with zero right-hand side. There are three symmetric modes among them characterized by equal amplitudes of oscillation in the streets: $\rho = r$, and two asymmetric modes with $\rho \neq r$. Among asymmetric one-frequency modes, the first one is a mode with one street of the pair completely damped, while the second global mode with $\rho \neq r, \rho \neq 0, r \neq 0$ [6, 9]. From visualization of cylinders close pair wakes it is well known that this wake usually composed of two streets of different strength and different oscillation frequency (see for example [11]). Accordingly, the question arises whether our model contain such two-frequency mode of the form (13) as a solution of the system (4), (5) and (11) with zero right-hand sides.

$$\rho^2 \approx const, \quad r^2 \approx const, \quad P \approx (const)t \quad (13)$$

Putting $\varepsilon = 0$ as a first step in solving the system (4), (5), (11) one may deduce: such mode exists in line $\lambda = 0.5$ under the condition: $\rho^2 + r^2 = 4$. Furthermore, this mode is marginally stable along the direction: $\rho^2 + r^2 = 4$ and is stable in opposite direction. For the second step, taking into account first non vanishing terms to small parameters ε and Δ, a complicated system of nonlinear equations may be deduced for desired configuration of the mode. Its approximate solution is as follows

$$R^2 \approx 4 \tag{14}$$

$$N \approx \pm\frac{1}{2}\sqrt{\frac{l^2+\lambda^2}{l^2+128[(1/\lambda)-2](\Delta/\varepsilon)^2}} \tag{15}$$

$$\Omega \approx \mp 2\Delta\sqrt{\frac{l^2+\lambda^2}{l^2+128[(1/\lambda)-2](\Delta/\varepsilon)^2}} \tag{16}$$

Here collective variables [6, 7] of the wake were used, namely: $R^2 = \rho^2 + r^2$ for entire oscillation intensity, and $N = \frac{1}{2}(\rho^2 - r^2)/(\rho^2 + r^2)$ for asymmetry of the wake. Quantity Ω is a mean splitting of oscillation frequencies of the two Von Karman streets, which made up the total wake. According to the solution (13), (14), and (15) this frequency splitting Ω is proportional to product of the model parameter Δ and the variable N. High values of the frequency splitting correspond to high values of the wake asymmetry: $\Omega \approx 4\Delta N$, accordingly maximum splitting value is restricted as $\Omega < 2\Delta$. Maximum values of Ω are achievable in the process: $\lambda \to 0.5$; $\lambda < 0.5$.

An example of the wake stabilization process on two-frequency mode is presented in figure 1.

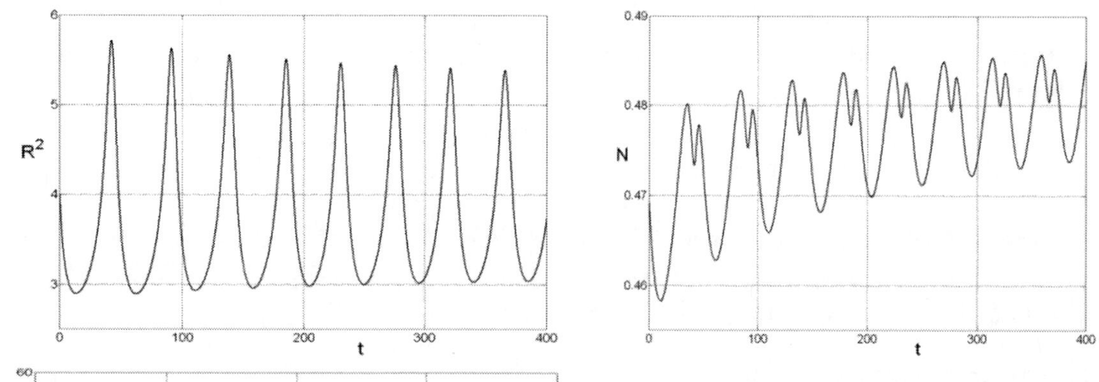

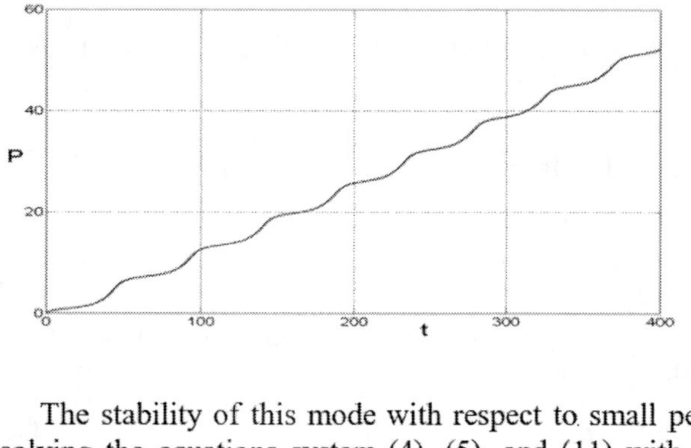

Figure 1. Time dependence of collective variables of the wake relaxing to two-frequency asymmetric mode in the case $\varepsilon = 0.2; \Delta = -0.12; \lambda = 0.49; l = 3$: the entire intensity R^2; the asymmetry of the wake N; the total phase shift for oscillations in two streets P.

The stability of this mode with respect to small perturbations was studied directly, by numerical solving the equations system (4), (5), and (11) with initial conditions according to (13), (14), and arbitrary P_0. It was revealed that two-frequency mode is stable to small perturbations in some aria of

IOP Conf. Series: Journal of Physics: Conf. Series **1009** (2018) 012008 doi:10.1088/1742-6596/1009/1/012008

the model parameters. It covers narrow interval $\lambda - 0.5 << 1$ and some segment of quantity $|l|$ while within the framework of the above rough analysis, this mode is marginally stable one.

It is easily seen from the wake parameters dynamics illustrated in figure 1 that all the characteristics of the wake are far from simple constants or linear functions of time as in rough approximation (13), but mean values of collective variables are in good agreement with approximate prediction (14), (15), and (16). Relaxation time, being proportional to ε^{-1}, is not long for this choice of quasi-equilibrium initial conditions.

4. Discussion and brief conclusion

Developed one-dimensional model contains some symmetric and asymmetric wake global modes (laminar modes). In studying turbulent wakes the main question is how to distinguish wake relaxing modes if one disposes only the velocity oscillograms, moreover turbulent – "noisy" oscillograms. For the case of two closely-located cylinders the main question is how to distinguish between one- and two-frequency asymmetric wake modes. Here the key feature of any mode is its phase trajectory. Below relevant examples of trajectory projection on P, N plane are shown in figure 2. Accordingly, qualitative difference of the P, N dependence for one- and two-frequency asymmetric wake modes is evident.

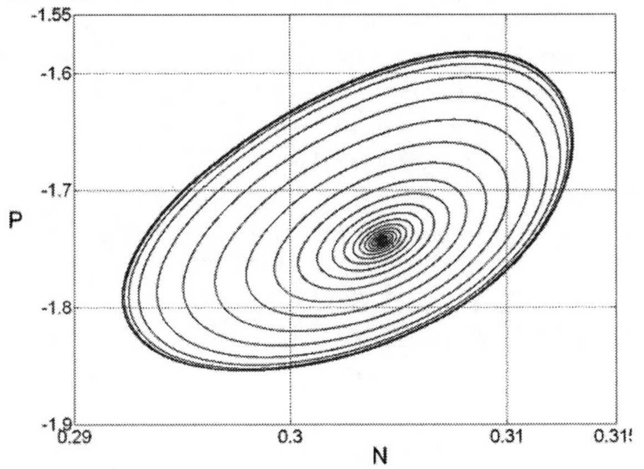

a

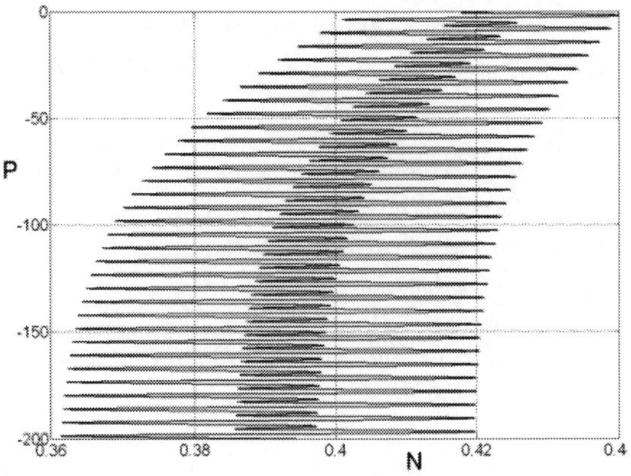

b

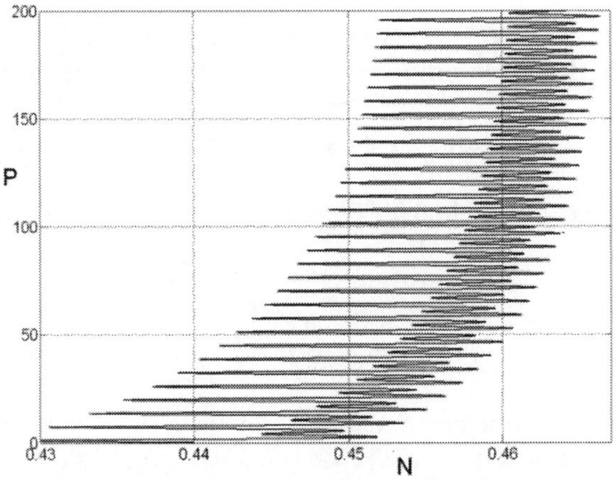

c

Figure 2. Comparison of the wake relaxation dynamics: towards one-frequency asymmetric mode in the case :

(a) – $\varepsilon = 0.12; \Delta = 0.12; \lambda = 0.34; l = 3$;

(b) – towards two-frequency asymmetric mode in the case $\varepsilon = 0.12; \Delta = 0.12; \lambda = 0.49; l = 3$,

(c) – towards two-frequency asymmetric mode in the case $\varepsilon = 0.12; \Delta = -0.12; \lambda = 0.49; l = 3$.

1. Summarizing, our one-dimensional model of cylinders pair wake is comparatively rich in different wake regimes. Respectively, one may hope of the model capability to reproduce the observed

features of gas and plasma wakes in our experiment. Moreover, the model is consistent with Landau-Stuart model of solitary cylinder laminar wake, which has firm ground in Navier-Stokes equations.

2. Two circumstances have to be taken into account while estimating the model parameters from experimental (or simulation) data. Magnitudes of the first pair of the model parameters (ε, Δ) are conveniently assessable from data for solitary cylinder. Information on phase difference of oscillation in the two Von Karman streets ($P_{(t)}$) composed cylinders pair wake would be useful for evaluating the parameters second pair (λ, l).

5. Acknowledgements

The authors are grateful to E.V. Teodorovich for fruitful discussions and to RFBR for financial support of the first part of this work by grant No 13-01-00742-a.

6. References

[1] Surzhikov S T 2013 *Computational Physics of Electric Discharge in Gas Flows* (Berlin/Boston: Walter de Gruyter Gmbh)

[2] Panchenko V Ya, Zavalov Yu N, Galushkin M G, Grishaev R V, Golubev V S and Dubrov V D 2006 The development of turbulence in the active medium of a fast-flow gas-discharge laser *Laser Physics* **16** 1 pp 40-51

[3] Huang H and Wang Y 2010 Kinetic modeling and optimum design of the discharge tube for the CO_2 laser with computational fluid dynamics method *Optical Engineering* **49** 11 114201

[4] Bearman P W, Wadcock A J 1973 The interaction between a pair of circular cylinders normal to a streem *J Fluid Mech* **61** **(**3) p 499

[5] Gembarzhevskii G V, Generalov N A, Kosunkin V D 1987 Anomalous change in velosity pulsations of a turbulent gas flow under the effect of a gas discharge *Technical Physics Letters* **13** pp 176-177

[6] Gembarzhevskii G V, Lednev A K and Osipenko K Yu 2017 Mode intermittence in a wake from two cylinders *Journal of Physics: Conf Series* **825** 012020

[7] Gembarzhevskii G V, Generalov N A, Kosunkin V D 1993 A sensor of turbulent gas flow rate pulsations in a glow discharge *Prib Tekh Eksp* **36**(3) pp 149-155 (in Russian)

[8] Gembarzhevskii G V, Lednev A K 2013 Reduced order model for effect of pulsation energy redistribution in cylinders wake induced by glow discharge *Proceedings of the 12-th International warkshop on Magneto-Plasma Aerodynamics* (Moscow: Joint Institute for High Temperatures RAS) pp 67-70

[9] Gembarzhevskii G V, Lednev A K and Osipenko K Yu 2015 Simulation of evolution of the two cylinders plasma wake under the electric discharge influence *Technical Physics Letters* **41** 12 pp 1132–1135

[10] Sumner D 2010 Two circular cylinders in cross-flow: A review *Journal of Fluids and Structures* **26** pp 849-899

[11] Peschard I and Le Gal P 1996 Coupled Wakes of Cylinders *Physical Review Letters* **77** 15 pp 3122-3125

[12] Simiu E, Scanlan R H 1978 Wind effect on structures: An introduction to wind engineering (New York: John Willey & Sons)

[13] Landa P S 1996 *Nonlinear oscillations and waves in dynamical systems* (Dordrecht-Boston-London: Kluver Academic Publ)

[14] Gembarzhevskii G V 2009 The effect of flow structure transformation induced by electric discharge and its simple model *Contributed papers of VI Int Conf on Plasma Physics and Plasma Technology (Minsk, Belarus)* vol 1 (Minsk: B I Stepanov Institute of Physics NAS) p 27

[15] Bogolyubov N, Mitropolskii Y A 1961 *Asymptotical methods in the theory of nonlinear oscillations* (Gordon & Breach) p 537

IOP Conf. Series: Journal of Physics: Conf. Series **1009** (2018) 012009 doi:10.1088/1742-6596/1009/1/012009

Aerodynamics and Flight Dynamics Simulation of Basic Finner Supersonic Flight in Aeroballistic Experiment

G S Makeich and I A Kryukov

Ishlinsky Institute for Problems in Mechanics RAS, Vernadsky prospect 101(1), Moscow, 119526, Russia

E-mail: mac_george@mail.ru

Abstract. This paper presents some results of aerodynamics and flight dynamics modeling of Basic Finner supersonic flight in an aeroballistic experiment [1]. Three dimensional surface based on the experiment description was formed. Unstructured tetrahedral volume mesh is formed after it. Aerodynamics computations were done by software package for CFD numerical computation of free form high-speed vehicle ug3D developed in the Institute for Problems in Mechanics. Obtained from CFD calculations aerodynamic coefficients became initial data for unguided Basic Finner flight dynamics calculations. Flight dynamics computations were done by developed solver MODIN for typical high speed vehicles flight dynamics calculations [2]. Six degrees of freedom motion without considering Earth rotation and form are adopted. Results of experimental and obtained numerical aerodynamic coefficients and flight parameters data comparison are also presented.

1. Introduction

The key objective of this paper is the demonstration of developed software packages abilities to perform a full technological cycle of supersonic vehicles computer simulation from 3D surface geometry creation to flight trajectories calculations.

The main goal of this paper is a description of fully coupled computational fluid dynamics (CFD)/Six degree-of-freedom motion (6-DoF) computations of finned projectiles at high supersonic velocities as well as a description of some results of developed computer codes of aerodynamics and flight dynamics calculations validation based on aeroballistic experiments data [1].

Obtained from CFD calculations aerodynamic coefficients were compared to experimental database as a part of developed CFD ug3D solver validation but also became initial data for flight dynamics calculations. 6-DoF motion computations were made by developed solver MODIN using calculated and experimental aerodynamic coefficients as initial data. This sequence of calculations allows evaluating the accuracy of both ug3D and MODIN in their application area and quality of «virtual vehicle» computer simulation with regard to aeroballistic experiment data.

2. Object description

Basic Finner configuration has been well studied as a calibration model for different types of aeroballistic ranges and wind tunnels and as a reference projectile for dynamic stability research is considered [1]. Thereby many experimental data of aerodynamic coefficients and flight dynamics parameters from wind tunnels and aeroballistic ranges are available.

IOP Conf. Series: Journal of Physics: Conf. Series **1009** (2018) 012009 doi:10.1088/1742-6596/1009/1/012009

The object of research is a symmetrical projectile consisting of a cone nose blunted by the small-radius sphere 10 degrees and a 0.03 m caliber cylindrical fuselage body [1]. Four rectangular fins with a simple blunted wedge profile and a 0.09 m wingspan located in the tail. Total length of the object is 0.3 meters.

There are experimental data for different fin cant angles of 0°, 2°, 4° for spinning motion creation. Within the framework of this paper in 3D computer surface creation and aerodynamics simulation only 0° fin cant is considered. Flight dynamics computations were done with both 0° and 4° cant angles. Model configuration is presented at figure 1 (dimensions in calibers).

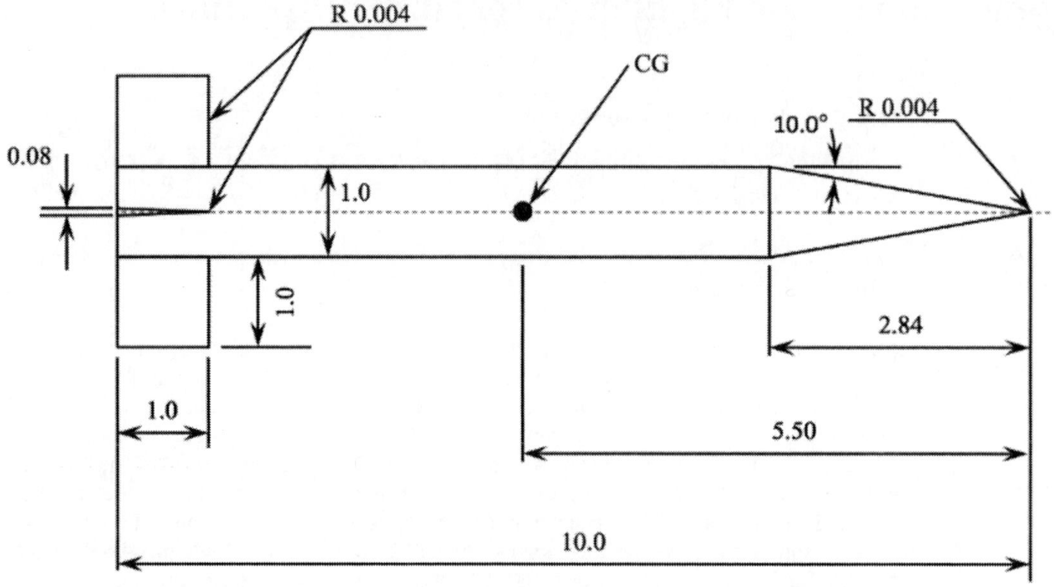

Figure 1. Model configuration (dimensions in calibers 1 cal = 0.03 meters).

Reference length and reference area for further aerodynamic coefficients calculations are $l_{ref} = 0.03$ m and $S_{ref} = 0.0007$ m^2 respectively. Finner barycentre for aerodynamic moments evaluation is located at the symmetry axis at the distance of $x_{c.g.} = 0.165$ m from the projectile's nose. Mass inertial characteristics for flight dynamics calculations are presented in the table 1. They do not change during the flight since in the aeroballistic experiment. We assume that Basic Finner is a body of constant mass moving by inertia without thrust.

Table.1. Basic Finner mass inertial characteristics.

Projectile's mass, m kg	1.58
Axial moment of inertia, I_x kg·m^2	0.000192
Transverse moments of inertia, I_y, I_z kg·m^2	0.00979

3. Surface geometry creating and volume mesh generation

For the Basic Finner computer simulation CAD surface was created considering all the features based on the object's geometry (figures 2–5).

Next step is a volume mesh generation. 3D CAD geometry was exported to a mesh generator [3-5]. Developed CFD solver ug3D can use both structured and unstructured grids, but in this case a 2.7 million cells unstructured tetrahedral volume grid was generated (figure 6). Grid has a cell refinement near the sphere-blunted cone nose (figure 7), cone-cylinder junction (figure 8) and fins blunted leading edges (figure 9).

IOP Conf. Series: Journal of Physics: Conf. Series **1009** (2018) 012009 doi:10.1088/1742-6596/1009/1/012009

In the ambient flow area maximum cell size is 28 mm. Tetrahedron's size near the sphere-blunted cone nose tip, cone-cylinder junction and fins blunted leading edges is 0.09 mm. Generated meshes quality was evaluated by the common criteria [6-8]. Separately the aspect ratio for any triangular cell of the surface mesh and any tetrahedral cell of the volume mesh does not exceed 4 and 6 respectively. Equiangle skew for triangle elements is less than 0.7 and not more than 0.8 for tetrahedral cells. Generated mesh models meet the requirements for effective CFD solver ug3D performance.

Figure 2. Total view of Basic Finner CAD surface.

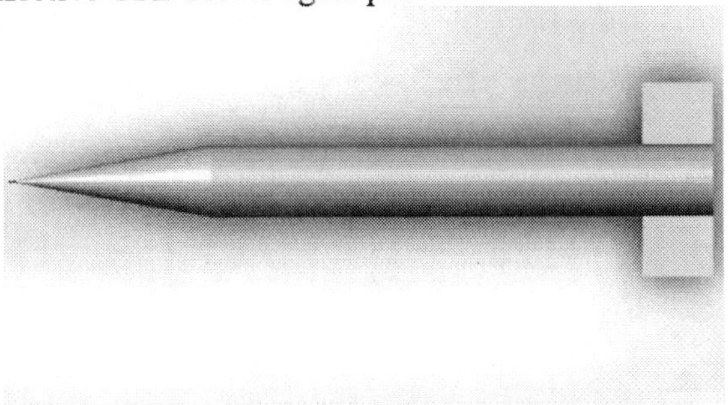

Figure 3. Side view of Basic Finner CAD surface.

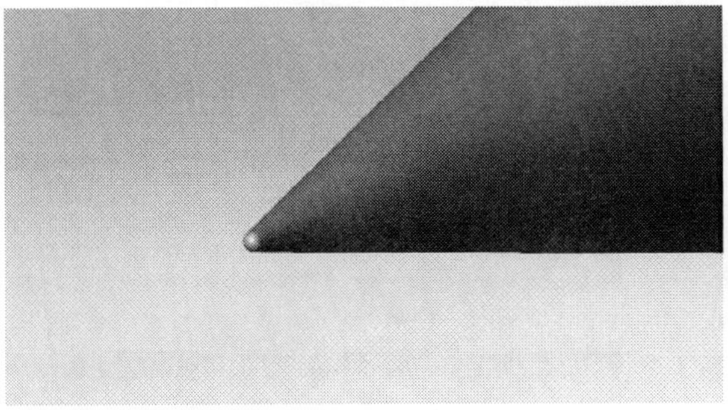

Figure 4. Basic Finner's blunt nose CAD surface.

Figure 5. Blunt fin CAD surface.

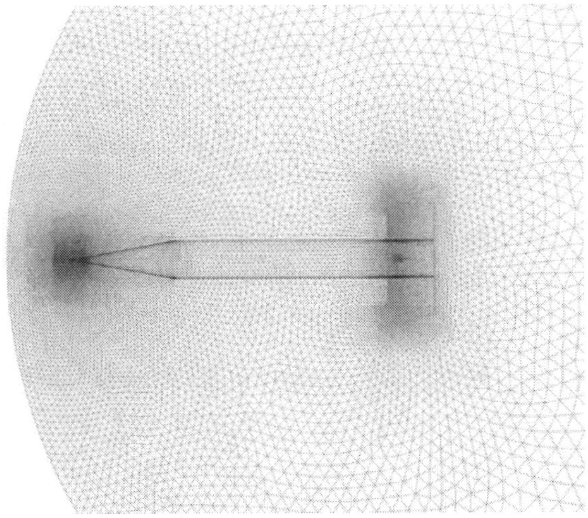

Figure 6. Generated mesh general view.

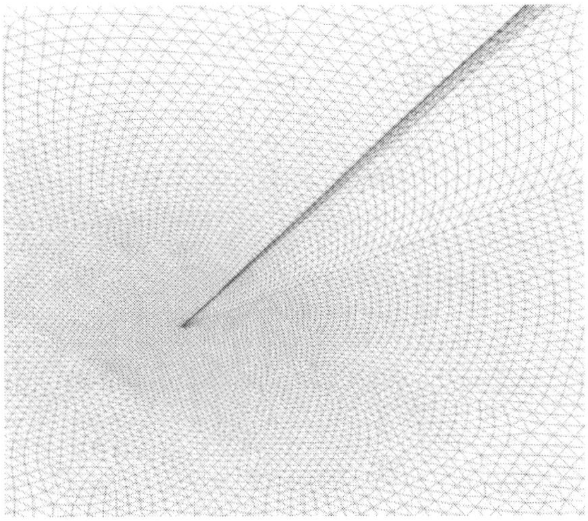

Figure 7. Cell refinement near the nose.

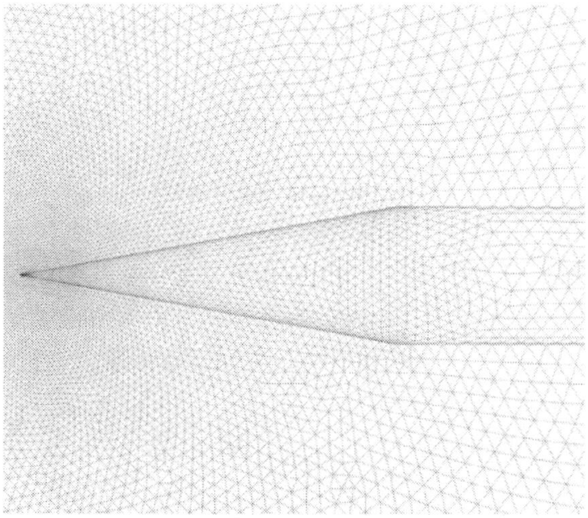

Figure 8. Cell refinement near the cone-cylinder junction.

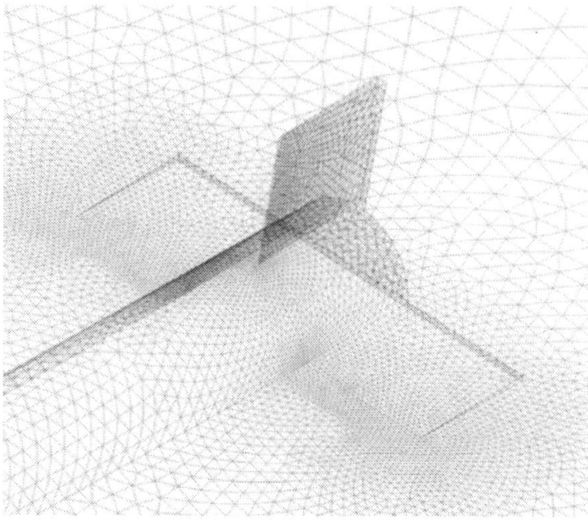

Figure 9. Cell refinement near the fins.

4. Aerodynamics simulation

Basic Finner aerodynamics calculations were made by a software package developed in the Institute for Problems in Mechanics Russian Academy of Science for numerical aerodynamics simulation of complex-form high-speed vehicles ug3D [9–10].

4.1. Numerical method

For numerical simulation of aerodynamics of Basic Finner we assume that the flow is governed by the compressible Navier-Stokes equations which describe the conservation of mass, momentum, and energy in a viscous fluid. The working fluid is assumed a calorically perfect gas, and the ratio of specific heats is assumed to be 1.4. For viscous flow problems, Sutherland's law is used to compute the molecular viscosity and a constant laminar Prandtl number of 0.72 is used to compute the thermal conductivity.

The flow solver ug3D is a density based cell center finite volume Navier-Stokes solver and can be considered as a Godunov-type solver. In order to achieve second order spatial accuracy, a piecewise linear reconstruction MUSCL approach is used [11].

The inviscid numerical fluxes may be computed using a variety of exact or approximate Riemann solvers. We implemented the most popular solvers in our flow solver. For the work shown here, most of the computations have been performed using the Advective Upstream Splitting Method (AUSM) of Liou and Steffen [12]. This flux function is rather computationally cheap and is applicable for viscous flow calculation.

The gradient needed for the linear reconstruction can be computed by either the cell based Green-Gauss approach or the least-square method [11]. Usually the weighted least square method is used in ug3D as the default reconstruction method. In order to prevent any overshoot for the reconstructed variables and the generation of oscillations in the regions of large gradients or shocks, the limiter of Barth and Jespersen [11], Venkatakrishnan's limiter [13], and Michalak and Ollivier-Gooch limiter [14] are incorporated into the ug3D solver.

The velocity and temperature gradients for viscous fluxes are calculated using either a Green-Gauss or least-squares method at all grid cell centroids as described in above and then averaged to obtain the gradients at the cell faces. However, it was demonstrated that this approach would lead to the decoupling of the solution on quadrilateral or hexahedral grids [15]. In order to reduce the decoupling error of the method the correction [16] is applied.

For temporal discretization, both the second and third-order accurate explicit TVD Runge-Kutta methods [17] are available. A spatially varying time step is used, with a blending of inviscid and viscous time-step constraints.

4.2. Initial data for aerodynamics modeling

Besides the generated tetrahedral unstructured grid, we need to determine free stream parameters to start aerodynamical simulation. Since one of the main goals of this work is to validate developed computer codes for aerodynamics and flight dynamics calculations, the main interest is provided by experimental data obtained in aeroballistic range experiments. Specialty of this type of experiments is in the free uncontrolled flight of the object after its launch from the gun in the aero-ballistic range with a certain initial velocity. Schematic view of the aeroballistic range is presented at figure 10.

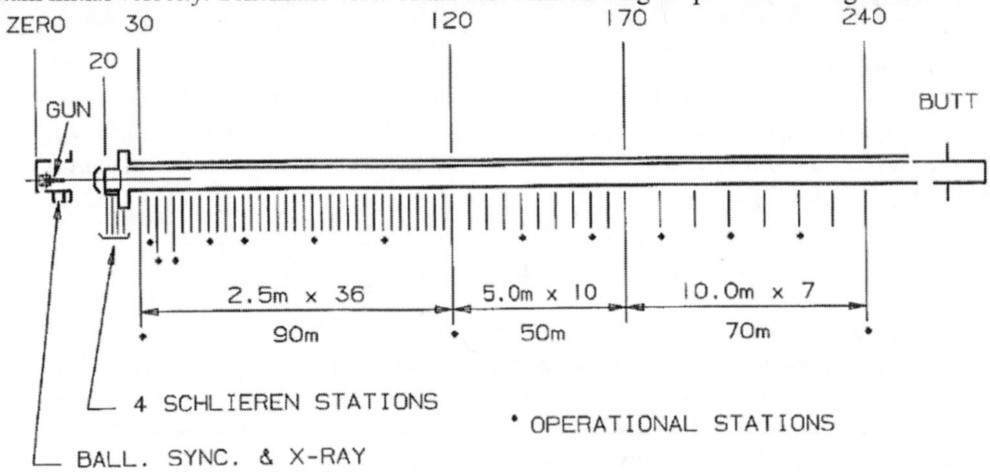

Figure 10. Schematic view of DREV Aeroballistic Range [18].

Used experimental aerodynamic characteristics are determined by 6-DoF reduction from the free-flight trajectories measured in the Aeroballistic Range DREV (Defense Research Establishment Valcartier) [18]. Initially moving in the gun barrel, the object is located in the sabot, which splits up into 2 pieces after exit the barrel due to its aerodynamic profiling. Duration of the free flight is limited by the length of Aeroballistic Range (approximately 220 m) and depends on the initial velocity (figures 11-12). Initial velocity is set by the solid fuel mass putting in the sabot and in the DREV subsonic as well as high supersonic launches were conducted with the range of initial velocities of 0 – 1500 m/s.

Figure 11. Photo of DREV Aeroballistic Range [18].

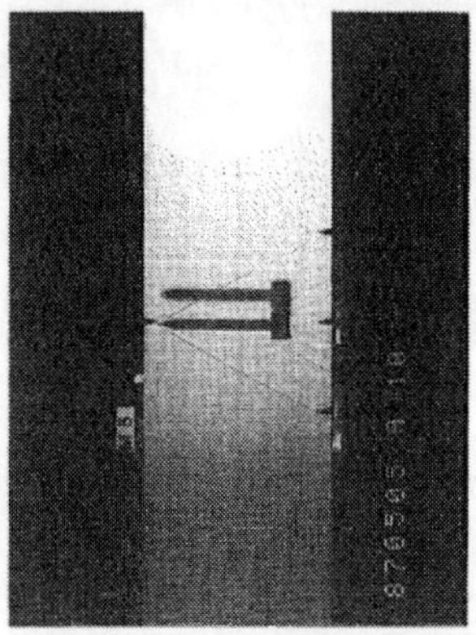

Figure 12. Typical shadowgraph of aeroballistic experiment [18].

Due to developed software packages specificity five shots with high supersonic initial velocities were chosen from the wide range of available free-access experimental data. Appropriate initial free stream conditions of each selected shots are presented in table 2.

Table.2. Initial conditions of selected shots.

Shot number	17	05	26	23	01
Mach number, M_∞	2.413	2.97	3.337	4.127	4.42
Sound velocity, a m/s	343.41	343.85	343.52	343.83	343.8
Density, ρ kg/m^3	1.17843	1.17484	1.20303	1.18066	1.18122

4.3. Aerodynamics calculation results

In accessible experimental data sources aerodynamic normal force and pitch moment coefficients are given as an angle of attack (AoA) derivative. For each selected shot two calculations were done for angles of attack values of α = 0° and α = 5° to validate ug3D code. The coefficients in this case are determined by formulas (1):

$$c_y^\alpha = \frac{c_y}{\sin\alpha},\ m_z^\alpha = \frac{m_z}{\sin\alpha} \tag{1}$$

Three-dimensional gas-dynamic distributions for different Mach numbers and angles of attack are presented as an illustration of computation results. For example, density and Mach number distributions for minimum considered value of initial velocity $M_\infty = 2.41$ with angle of attack α = 5° and for maximum considered value of initial velocity $M_\infty = 4.42$ with 0° angle of attack are presented at figures 13–14 and 15–16 respectively.

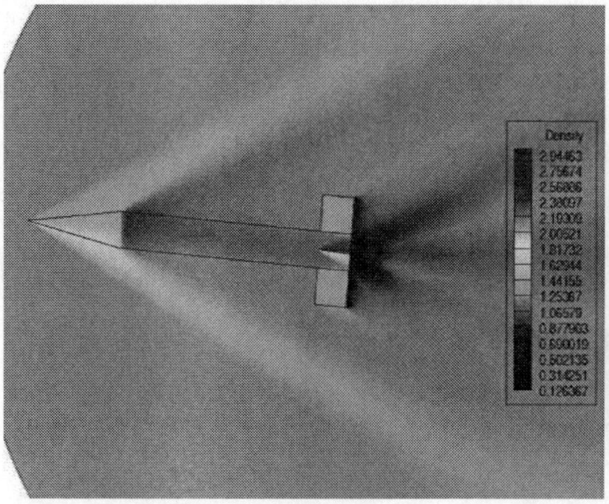

Figure 13. Density distribution $M_\infty = 2.41$, α = 5°.

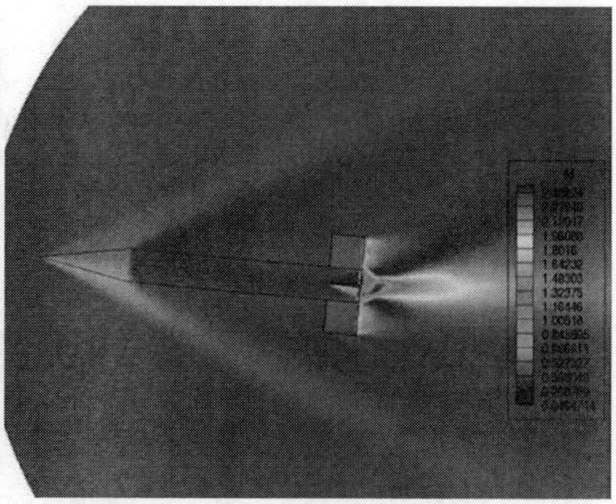

Figure 14. Mach number distribution, $M_\infty = 2.41$, α = 5°.

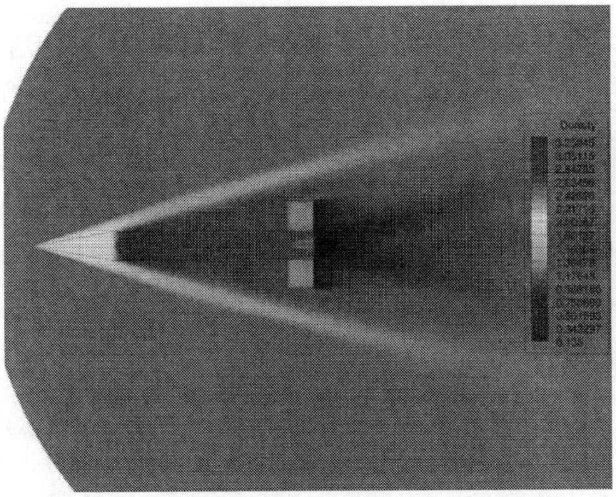

Figure 15. Density distribution $M_\infty = 4.47$, $\alpha = 0°$.

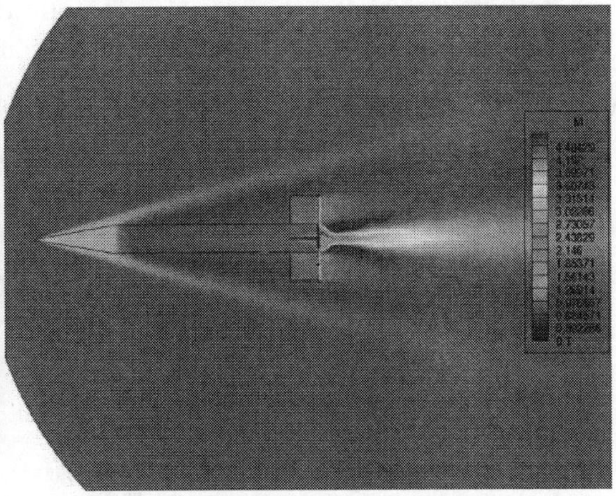

Figure 16. Mach number distribution, $M_\infty = 4.47$, $\alpha = 0°$.

The figures 17–18 show density gradient compared to analytical solution [19] of the attached conical shock wave deflection angle. This comparison is correct since the projectiles nose is a cone with a very small blunt radius. The attached conical shock wave deflection angle can be calculated through the equation

$$ctg(\theta) = tg(\beta)\left[\frac{(\gamma+1)}{2}\frac{M_\infty^2}{M_\infty^2 \sin^2\beta - 1} - 1\right], \tag{2}$$

where β – attached shock wave deflection angle; θ – half-cone angle; γ – heat capacity ratio.

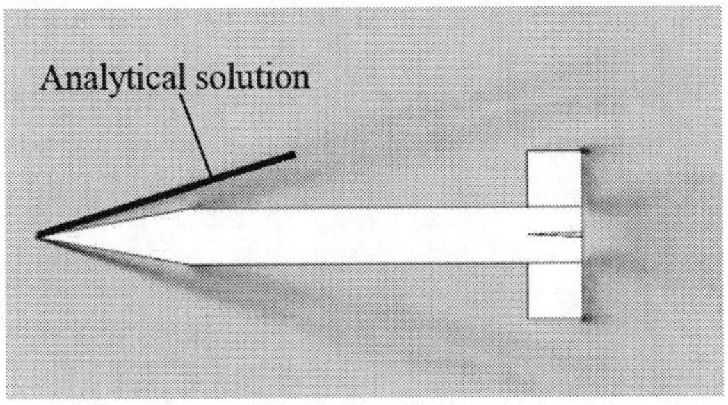

Figure 17. Comparison of density gradient with analytical solution of the shock wave deflection angle. $M_\infty = 4.47$, $\alpha = 0°$.

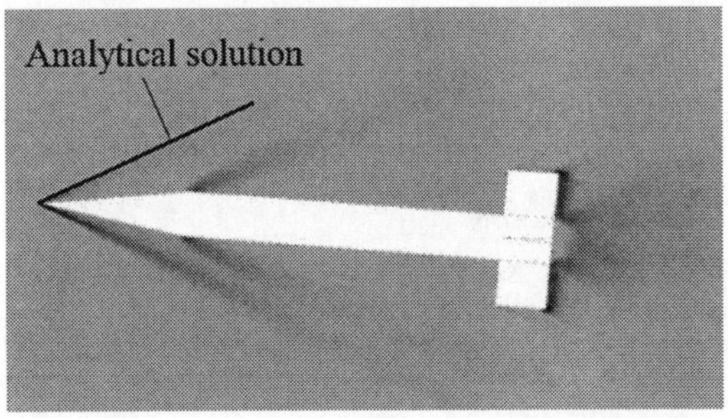

Figure 18. Comparison of density gradient with analytical solution of the shock wave deflection angle. $M_\infty = 2.41$, $\alpha = 5°$.

Analytical results are presented in table 3. There is an excellent agreement of numerical calculations results with the sharp cone analytical solution despite three-dimensional surface used in CFD calculations has a blunted nose.

Table.3. Results of analytical solution for the attached conical shock wave deflection angle.

Shot number	17	01
Mach number, M_∞	2.413	4.42
Conical shock wave deflection angle, β °	29.203	16.951

IOP Conf. Series: Journal of Physics: Conf. Series **1009** (2018) 012009 doi:10.1088/1742-6596/1009/1/012009

The figures 19–21 show results of ug3D solver validation comparing the obtained results of aerodynamics simulation with selected experimental data. Comparison of aerodynamic integral characteristics such as axial force, normal force and pitch moment coefficients is presented. Above-mentioned coefficients are determined at a body-referenced frame. Body-referenced frame aerodynamic forces coefficients are related with airspeed reference frame coefficients by the following equations:

$$c_{xa} = c_x \cos\alpha + c_y \sin\alpha$$
$$c_{ya} = c_y \cos\alpha - c_x \sin\alpha \tag{3}$$

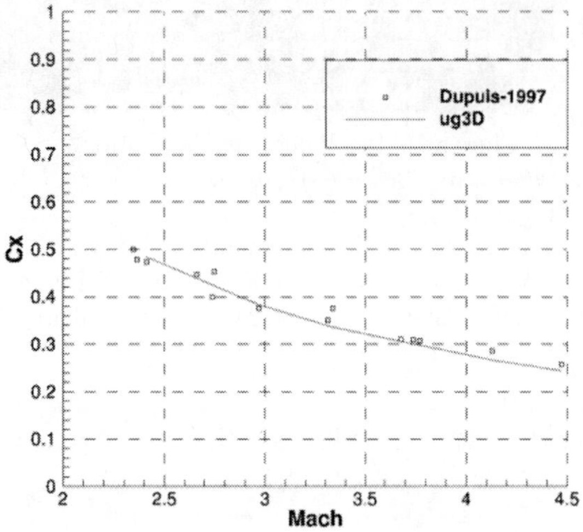

Figure 19. Comparison of calculated axial force coefficient with experimental data.

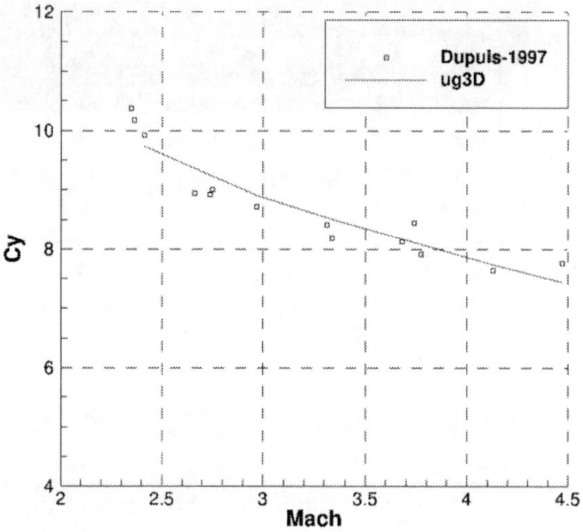

Figure 20. Comparison of calculated normal force coefficient with experimental data.

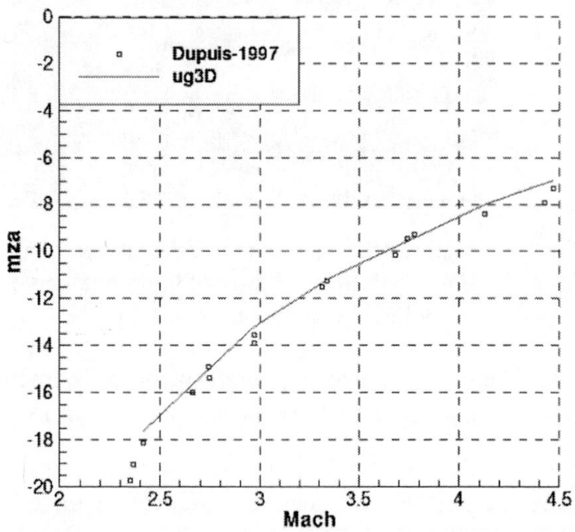

Figure 21. Comparison of calculated pitch moment coefficient with experimental data.

There is an excellent agreement between calculated axial force coefficient and experimental data. Computed normal force and pitch moment coefficients derivatives approximate experimental data satisfactorily. Computational error of axial force, normal force and pitch moment coefficients derivatives with regard to experimental data does not exceed 10%, 5% and 10% respectively.

5. Flight dynamics simulation

Flight dynamics computations were carried out with the solver MODIN for typical high-speed vehicles flight dynamics cases. During the sabot separation from the projectile in aeroballistic experiment, the object is inevitably exposed to external forces. This determines the initial angular velocity of detached projectile and oscillatory character of projectile's further motion.

5.1. Equations of motion

Since in this case the projectile under consideration is an uncontrolled vehicle performing an oscillatory and rotation motion in all planes, it is necessary to use the spatial model of object's motion with 6 degrees of freedom. Because of trajectory's limitation by the Aeroballistic Range 220 meters length (figure 10) and because the characteristic flight time does not exceed 0.2 seconds, it makes sense not to consider Earth's form and angular rotation influence and atmosphere parameters variations. Because of considering the object as a rigid body the effect of elastic deformations on aerodynamic coefficients is neglected.

Thus 6-DoF equations of motion of the unguided projectile without thrust take the following form (4) [20]:

$$\begin{cases} m\dot{V} = -X_a - mg\sin\theta; \\ mV\dot{\theta} = Y_a\cos\gamma_a - Z_a\sin\gamma_a - mg\cos\theta; \\ -mV\cos\theta\dot{\psi}_c = Y_a\sin\gamma_a + Z_a\cos\gamma_a; \\ I_x\dot{\omega}_x = M_{x\Sigma} - (I_z - I_y)\omega_y\omega_z; \\ I_y\dot{\omega}_y = M_{y\Sigma} - (I_x - I_z)\omega_z\omega_x; \\ I_z\dot{\omega}_z = M_{z\Sigma} - (I_y - I_x)\omega_x\omega_y; \\ \dot{\psi} = \left(\omega_y\cos\gamma - \omega_z\sin\gamma\right)/\cos\vartheta; \\ \dot{\vartheta} = \omega_y\sin\gamma + \omega_z\cos\gamma; \\ \dot{\gamma} = \omega_x - \dot{\psi}\sin\vartheta; \\ \dot{L} = V\cos\theta\cos\psi_c; \\ \dot{H} = V\sin\theta; \\ \dot{Z} = -V\cos\theta\sin\psi_c. \end{cases} \tag{4}$$

Where X_a, Y_a, Z_a – dimensional aerodynamic drag, lift and side forces in airspeed referenced frame; M_x, M_y, M_z – dimensional aerodynamic roll yaw and pitch moments; θ – flight path angle; ψ_c – course angle; γ, ψ, ϑ – Euler angles (roll, yaw and pitch); g – gravitational acceleration; L, H, Z – center of gravity coordinates; γ_a – kinematic roll angle, determined by equations (5) by the Euler angles, angles of attack α and slip β.

$$\begin{aligned} &\sin\beta = \cos\theta\,[\sin\vartheta\sin\gamma\cos(\psi_c - \psi) - \cos\gamma\sin(\psi_c - \psi)] - \sin\theta\cos\vartheta\sin\gamma; \\ &\sin\alpha = \{\cos\theta[\sin\vartheta\cos\gamma\cos(\psi_c - \psi) + \sin\gamma\sin(\psi_c - \psi)] - \sin\theta\cos\vartheta\cos\gamma\}/\cos\beta; \\ &\sin\gamma_a = [\cos\alpha\sin\beta\sin\vartheta - \cos\vartheta(\sin\alpha\sin\beta\cos\gamma - \cos\beta\sin\gamma)]/\cos\theta. \end{aligned} \tag{5}$$

5.2. Initial data for flight dynamics modeling

In order to calculate the spatial motion of symmetric unguided projectile without thrust (according to (4)) it is necessary to possess initial data on the aerodynamic coefficients of axial, normal forces along with roll, pitch moments and damping roll and pitch moments. Projectile's mass, axial and transversal inertia moments and center of gravity location are also required.

Mass inertia characteristics are shown in table 1. Calculated within the framework of this paper and experimentally obtained aerodynamic coefficients of axial and normal forces and pitch moment are

IOP Conf. Series: Journal of Physics: Conf. Series **1009** (2018) 012009 doi:10.1088/1742-6596/1009/1/012009

presented in figures 19–21. Roll, yaw and pitch damping moments as well as roll moment due to fin cant $\delta = 4°$ were taken from experimental aerodynamic database (figures 22–25), because in this paper aerodynamics calculations were made without damping and roll motion.

It should be noted that the experimental damping moments coefficients calculations use non-dimensional angular velocity rates. The dimensional damping moment is determined from the damping coefficient by the following equation

$$M^{\omega} = m^{\bar{\omega}} \cdot \bar{\omega} \cdot qS_{ref} l_{ref}, \tag{6}$$

where $\bar{\omega} = \omega \cdot l_{ref} / 2V$.

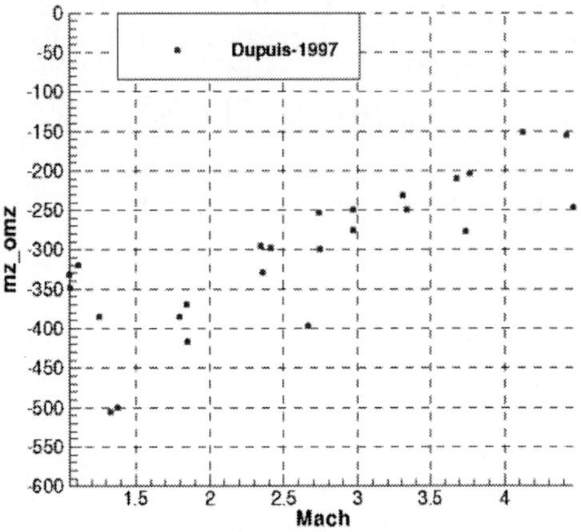

Figure 22. Pitch damping moment as a function of Mach number.

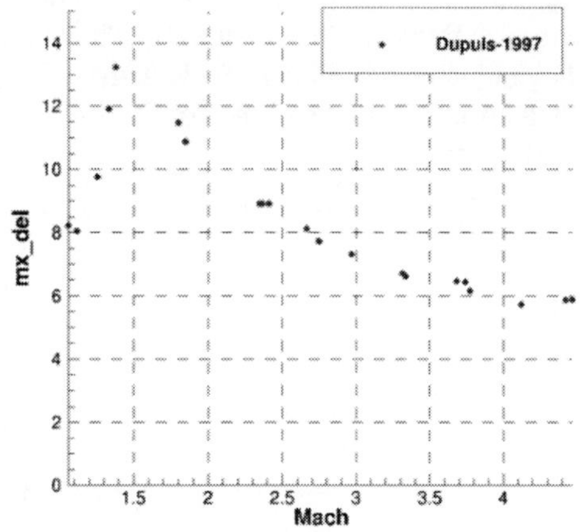

Figure 23. Roll moment due to fin cant as a function of Mach number.

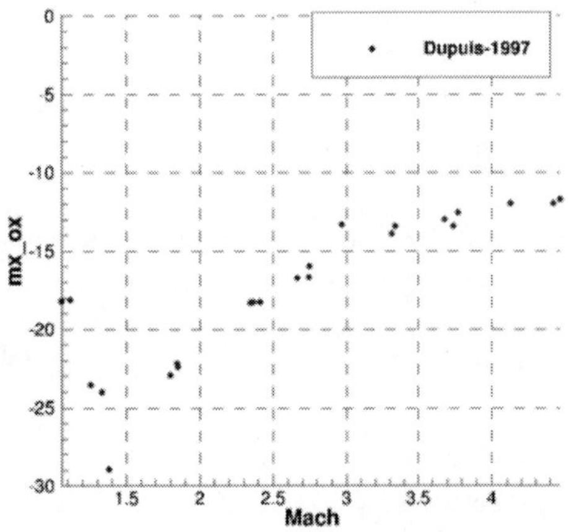

Figure 24. Roll damping moment coefficient as a function of Mach number.

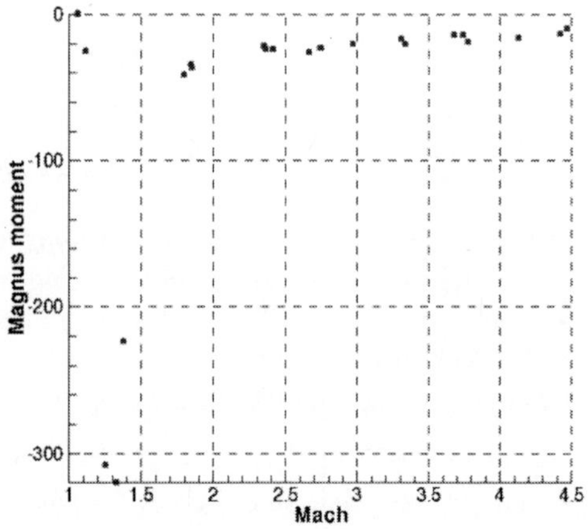

Figure 25. Magnus moment coefficient as a function of Mach number.

For the developed flight dynamics solver MODIN validation two experimental shots from the table 2 were chosen. One of them (shot number 26) with a roll motion produced by fin cant angle $\delta = 4°$, and the other – number 01 without fin cant.

Dependences of the total angle of attack, and pitch and yaw angles to range of flight are presented in [1]. However, to set the initial state vector in accordance with equations of motion (4) we need to know initial values of the flight path and course angles as well as roll, yaw and pitch angular velocities. Table 3 shows chosen values of initial state vector variables for flight dynamics simulation of two experimental shots.

Table. 3. Initial state vector.

Shot number	01	26
Mach number	4.42	3.33
Fin cant angle δ, °	0	4
Flight path angle θ, °	0.125	0.1
Course angle ψ_c, °	-3	0
Pitch angle ϑ, °	-0.2	0
Yaw angle ψ, °	-3.5	0
Roll angular velocity ω_x,°/c	0	6303
Yaw angular velocity ω_y,°/c	-25	-3
Pitch angular velocity ω_z,°/c	-13	0

5.3. Flight dynamics calculation results

As a result of uncontrolled finned projectile flight dynamics calculations, comparisons of calculated trajectory parameters such as pitch, yaw angles and total angle of attack with experimental data are presented. The influence of using calculated and experimental initial aerodynamic characteristics in yaw and pitch channels on computed and experimental trajectories agreement is evaluated.

Figures 26 – 28 present obtained numerical results comparison with experimental data for shot 01 without rotation motion. Here and further, the blue line illustrates results of 6-DoF motion simulated with experimental aerodynamic characteristics, the red line - with calculated ones. The experimental data are marked with circles. Equation for total angle of attack is presented below

$$\alpha_{tot} = \sqrt{\alpha^2 + \beta^2} \, . \tag{1}$$

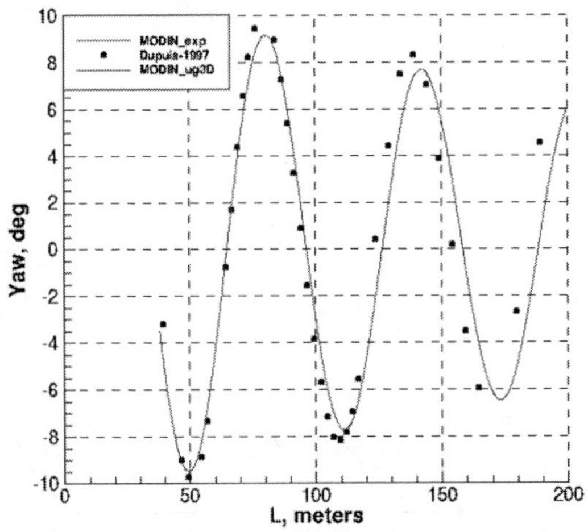

Figure 26. Simulated and measured yaw angle of a non-spinning case with respect to range of flight (Shot 01).

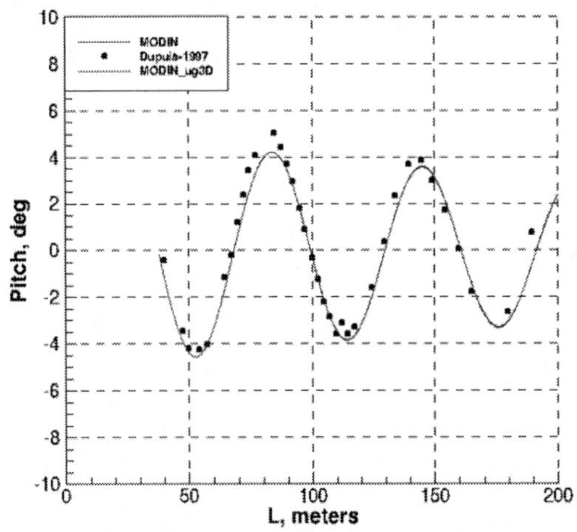

Figure 27. Simulated and measured pitch angle of a non-spinning case with respect to range of flight (Shot 01).

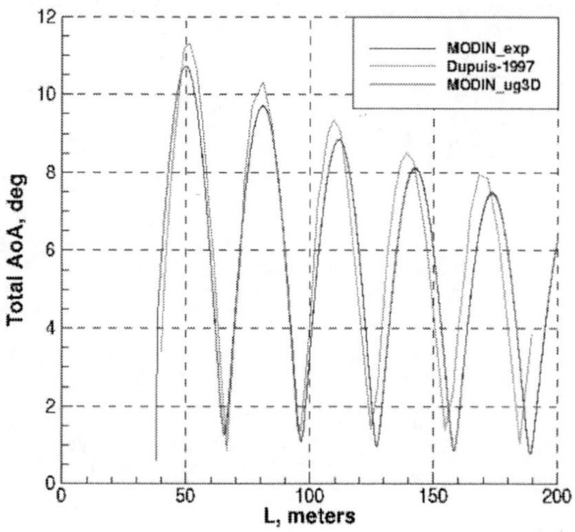

Figure 28. Simulated and measured total angle of attack of a non-spinning case with respect to range of flight (Shot 01).

High output values of Euler angles and total AoA are observed in non-rotation case but the motion is stationary with all angles oscillation damping. Excellent agreement between calculated and experimental data is also registered.

For the shot number 26 with rotation caused by fin cant angle $\delta = 4°$ we can see a slightly worse match between experimental and calculated trajectory parameters, but the character of parameters changing by time and the oscillation period are quite the same. Lower values of Euler angles and total AoA due to rotational motion are also observed. High quality of obtained results in all considered cases (at figures 26 – 31 blue and red lines almost completely merge) allows to draw a conclusion about the legitimacy of application of the developed software packages for coupled CFD/6–DoF computations for aeroballistic experiments modeling and prediction of objects aerodynamic characteristics and flight dynamics parameters at unattainable regimes for experimental facilities.

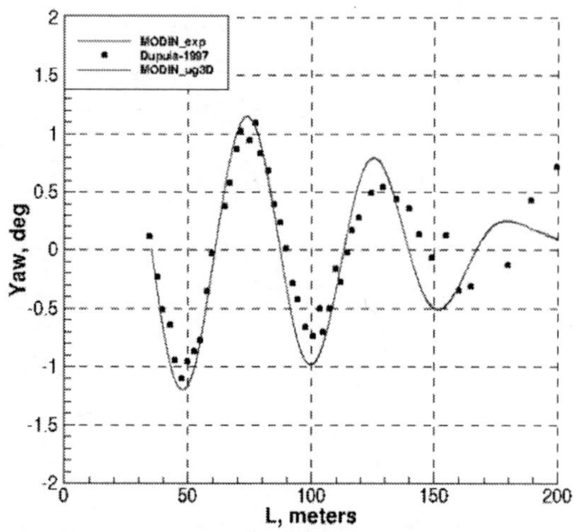

Figure 29. Simulated and measured yaw angle of a spinning case with respect to range of flight (Shot 26).

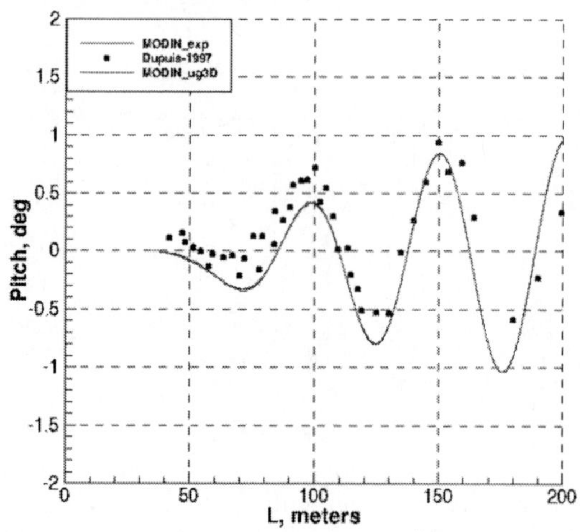

Figure 30. Simulated and measured pitch angle of a spinning case with respect to range of flight (Shot 26).

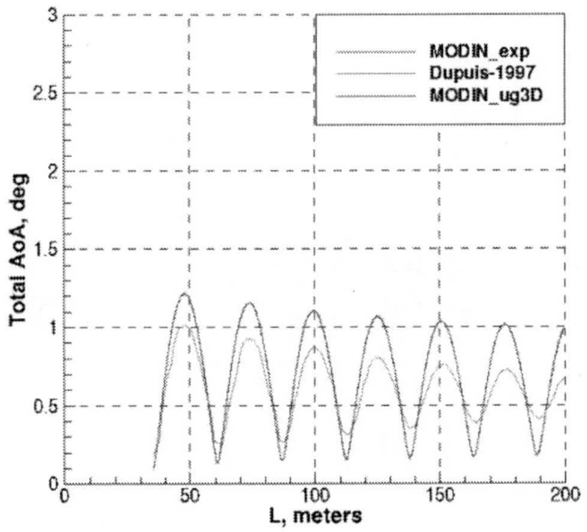

Figure 31. Simulated and measured total angle of attack of a spinning case with respect to range of flight (Shot 26).

5.4. Roll motion stability analysis

Within the framework of this paper special attention is focused on dynamic stability of rolling motion analysis. The value of the Magnus moment coefficient which is increasing due to angular velocity growth as the projectile flies down the range plays an important role in dynamic stability analysis. The steady spin rates of 4° fin cant is approximately 125 degrees per meter [1]. In spin stabilized projectiles ballistics oscillatory motion determination to fast nutational and slow precessional motion is accepted [21]. Firstly, the preliminary analysis was done using the established ballistics equations for yaw damping factors of the nutation and precession arms (8). The criterion of object's dynamic stability is that damping factors are negative.

$$\lambda_N = \frac{\rho S_{ref}}{4m}\left[-c_{ya}\left(1-\frac{1}{\sigma}\right)+\frac{k_2^{-2}}{2}\left(1+\frac{1}{\sigma}\right)m_z^{\bar{\omega}_z}+\frac{k_1^{-2}}{\sigma}m_{magnus}\right];$$
$$\lambda_P = \frac{\rho S_{ref}}{4m}\left[-c_{ya}\left(1+\frac{1}{\sigma}\right)+\frac{k_2^{-2}}{2}\left(1-\frac{1}{\sigma}\right)m_z^{\bar{\omega}_z}-\frac{k_1^{-2}}{\sigma}m_{magnus}\right]; \qquad (8)$$
$$k_1^{-2}=\frac{md^2}{I_x}, k_2^{-2}=\frac{md^2}{I_y}, \sigma=\sqrt{1-\frac{1}{s_g}}, s_g=\frac{2I_x^2\omega_x^2}{\pi I_y \rho m_z^\alpha V^2 d^3}.$$

For shot number 26 the nutation and precession damping factors were calculated. With the angular velocity growth the precession motion loses its stability earlier. Corresponding precession damping factor versus angular velocity ratio is presented at figure 32. In this case the projectile loses its stability approximately at 120 degrees per meter while nutation factor is still less than zero. In the experiment the aeroballistic range length is finite and the projectile doesn't have enough time to grow angular velocity for dynamic stability lost which is obtained at the range of 200 – 210 meters approximately. This fact explains that there are no divergent oscillations observed in shot number 26 in Aeroballistic Range experiments – the rolling angular velocity just did not have enough time to reach the instability area.

So now we can take a look at the trajectory parameters such as total AoA and yaw angle after the projectile losing its stability (figures 34-35). Character of the total angle of attack evolution (figure 34) before reaching 200 meters range is definitely stable with damped oscillations. After reaching the instability spin rates the total AoA median curve have an increasing form. Herewith the oscillations remain stable. This suggests that after the loss of stability of slow precession oscillatory motion the fast nutational motion is still stable at the same roll angular velocities. The yaw angle evolution (figure 35) before the 200 meters borderline is definitely stable as well. But it became divergent after the borderline. These numerical results confirm the preliminary analysis conclusion that 120 degrees per meter is the limit spin rate for dynamically stable motion.

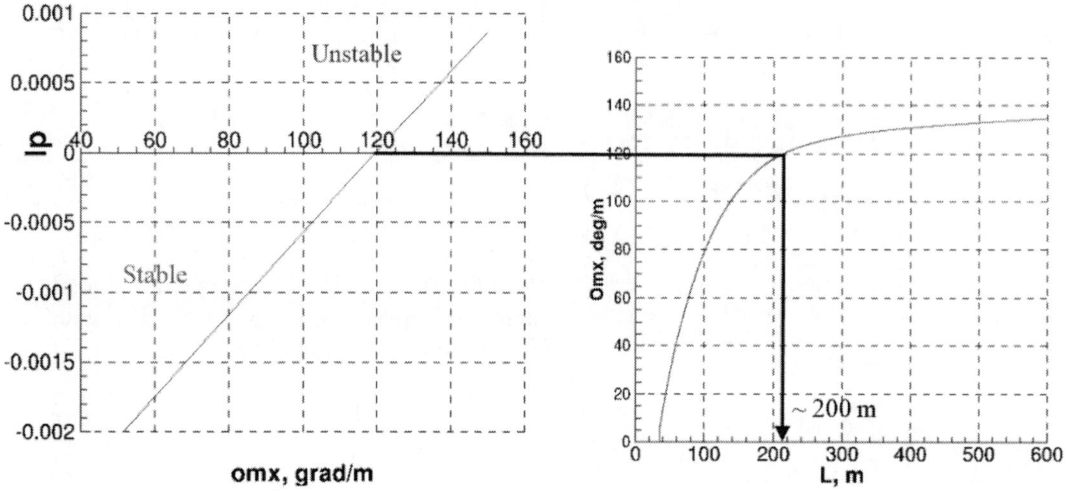

Figure 32. Precession damping factor versus roll angular velocity (Shot 26).

Figure 33. Roll angular velocity with respect to range of flight (Shot 26).

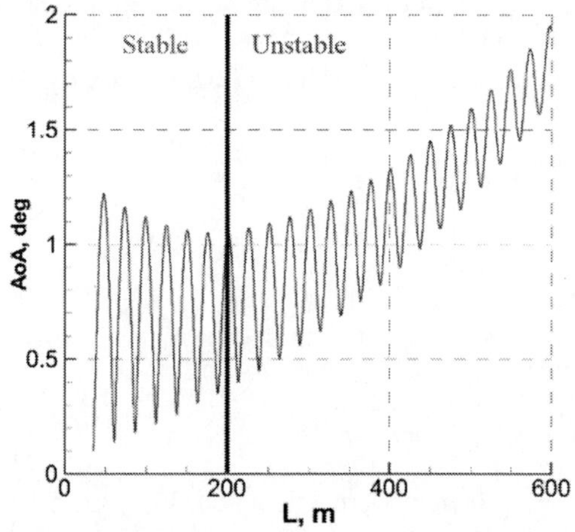

Figure 34. Total angle of attack with respect to range of flight (Shot 26).

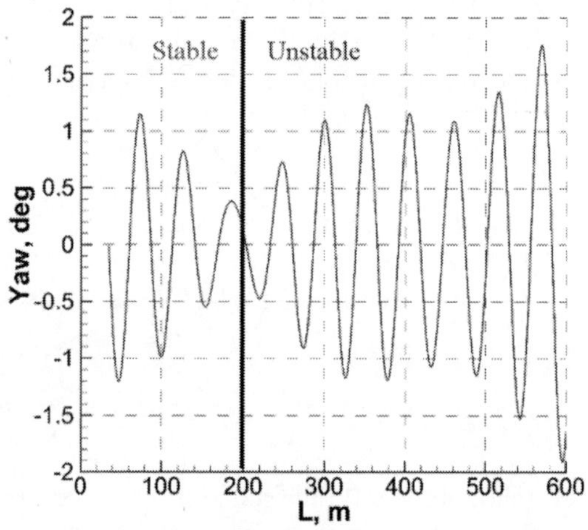

Figure 35. Yaw angle with respect to range of flight (Shot 26).

6. Conclusion

This paper demonstrates abilities of developed software packages to perform full technological cycle of high supersonic vehicles computer simulation from three-dimensional surface geometry creation to flight trajectory calculations. To provide CFD calculations 3D CAD surface was created and unstructured tetrahedral volume mesh was generated.

Developed CFD ug3D solver was successfully validated by comparison of Basic Finner aerodynamic coefficients of axial, normal force and pitch moment with experimental data obtained in DREV Aeroballistic Range [18].

Developed 6–DoF motion MODIN solver was successfully validated by comparison of trajectory parameters of yaw, pitch and total angle of attack evolution with Aeroballistic range experimental data using fully experimental aerodynamic database as well as obtained by ug3D solver Basic Finner aerodynamics calculations.

For Basic Finner configuration without fin cant angle there is an excellent agreement between calculated and experimental trajectory parameters. For a case with rotation caused by fin cant angle $\delta = 4°$ a slightly worse match between experimental and calculated trajectory parameters, but the character of parameters evolution in time and the oscillation period are quite the same. High quality of obtained results in spinning and non-spinning cases allows to draw a conclusion about the legitimacy of application of the developed software packages for coupled CFD/6-DoF computations for aeroballistic experiments modeling and prediction of objects aerodynamic characteristics and flight dynamics parameters at unattainable regimes for experimental facilities.

Dynamic stability of rolling motion analysis was carried out as well. Numerical results of spinning motion confirm the preliminary analysis conclusion that 120 degrees per meter is the boundary spin rate for dynamically stable motion.

Thus it may be concluded that developed ug3D and MODIN solvers allows to obtain a great accuracy results in their application area, and they can be successfully applied at coupled CFD/6-DoF calculations of high supersonic vehicles what is confirmed by comparison with established experimental data.

The presented study was partially supported by Russian State Assignment under contract No. AAAA-A17-117021310372-6.

References

[1] Dupuis A D and Hathaway W 1997 Aeroballistic range tests of the basic finner reference projectile at supersonic velocities *DREV-TM-9703*

[2] Makeich G S, Kryukov I A and Obnosov B V 2016 Preliminary version of flight dynamics solver for typical hypersonic aircraft *Physical-Chemical Kinetics in Gas Dynamics* **17** iss. 3 (http://chemphys.edu.ru/issues/2016-17-3/articles/668/)

[3] Borovikov S N, Ivanov I E and Kryukov I A 2005 Constrained Delaunay tetrahedralization for bodies with curved boundaries *Comput. Math. Math. Phys* **45** 8 1358-73

[4] Borovikov S N, Kryukov I A and Ivanov I E 2005 An approach for Delaunay tetrahedralization of bodies with curved boundaries *Proc. of the 14th Int. Meshing Roundtable* (San Diego) pp 221-237

[5] Borovikov S N, Kryukov I A and Ivanov I E 2006 Method of tetrahedral mesh generation for bodies with curvilinear boundaries *Numerical Modeling of Space Plasma Flows (Astronomical Society of the Pacific Conf. Ser.* vol. 359*)* ed N V Pogorelov and G P Zank (San Francisco) pp 295-301

[6] Zheleznyakova A L 2017 Numerical study of aerothermodynamic characteristics of the supersonic airliner prototype in cruising flight mode *Physical-Chemical Kinetics in Gas Dynamics* **18** iss. 1 (http://chemphys.edu.ru/issues/2017-18-1/articles/672/)

[7] Liseikin V D 1999 Grid generation methods (Berlin: Springer)

[8] Thompson J F, Soni B K and Weatherill N P 1998 Handbook of grid generation (Boca Raton: CRC Press)

[9] Borovikov S N, Ivanov I E and Kryukov I A 2006 Modeling of test ideal gas dynamics problems using tetrahedral meshes *Mat. Model.* **18** 8 37-48 (Russian. English summary)

[10] Kotov M A, Kryukov I A, Ruleva L B, Solodovnikov S I and Surzhikov S T 2014 Multiple flow regimes in a single hypersonic shock tube experiment *AIAA Paper* 2014-2657

[11] Barth T J and Jespersen D C 1989 The design and application of upwind schemes on unstructured meshes *AIAA Paper* 1989-0366

[12] Liou M S and Steffen C J Jr. 1993 A new flux splitting scheme *J. Comput. Phys.* **107** 23-39

[13] Venkatakrishnan V 1995 Convergence to steady state solutions of the Euler equations on unstructured grids with limiters *J. Comput. Phys.* **118** 120-130

IOP Conf. Series: Journal of Physics: Conf. Series **1009** (2018) 012009 doi:10.1088/1742-6596/1009/1/012009

[14] Michalak C and Ollivier-Gooch C 2012 Accuracy preserving limiter for the high-order accurate solution of the Euler equations *J. Comput. Phys.* **228** 8693-9711

[15] Haselbacher A and Blazek J 2000 Accurate and efficient discretization of Navier – Stokes equations on mixed grids *AIAA J.* **38** 11 2094–2102

[16] Weiss J M, Maruszewski J P and Smith W A 1999 Implicit solution of preconditioned Navier–Stokes equations using algebraic multigrid *AIAA J.* **37** 1 29–36

[17] Shu C-W and Osher S 1989 Efficient implementation of essentially non-oscillatory shock-capturing schemes II *J. Comput. Phys.* **83** 32–78

[18] Dupuis A D and Drouin G 1989 The DREV aeroballistic and data analysis system *15th Aerodynamic Testing Conf.* (San Diego)

[19] Fabrikant N Y 1964 Aerodinamika. Obshchiy kurs [Aerodynamics. General course] (Moscow: Nauka) p 815 (in Russian)

[20] Bochkarev A F 1985 Aeromechanika samoleta [Aircraft aeromechanics] (Moscow: Mashinostroenie) p 361 (in Russian)

[21] Projectile Design Analysis System (PRODAS),1992 *Arrow Tech Associates Inc.*, DREV contract No. W7701-0-1227, March 1992

IOP Conf. Series: Journal of Physics: Conf. Series **1009** (2018) 012010 doi:10.1088/1742-6596/1009/1/012010

Shock wave - boundary layer interaction in a long shock tube

I A Kryukov[1,2] **and I E Ivanov**[2,3]

[1] Ishlinsky Institute for Problems in Mechanics RAS, Vernadsky prospekt 101(1), Moscow, 119526, Russia
[2] Moscow Aviation Institute (National Research University), Moscow, 121993, Russia
[3] Lomonosov Moscow State University, Moscow, 119991, Russia

E-mail: kryukov@ipmnet.ru

Abstract. The numerical simulation of influence of viscous effects on flow inside shock tubes are described. We considered three shock tubes with different length-to-height ratio: 1 to 1 (Daru & Tenaud [6] case), 8 to 1 (extended Daru & Tenaud case) and 116 to 1 (HAST-1 [10-11] tube). Some details of different stages of the shock wave-boundary layer interaction are discussed.

1. Introduction

Shock tubes are a convenient tool for experimental study of fast-flowing high-speed flows. Interest in them arose long ago and proceeds to present day [1, 2]. This is due, first of all, to the complexity and high cost of realization of full-scale tests in high-speed flows. Recently, special attention is spared to carrying out high-precision experiments on the shock tubes for verification and validation of CFD computer codes intended for the numerical simulation of hypersonic flows [3].To improve the accuracy of the experimental results obtained, a clear understanding of the processes occurring in the shock tubes is needed. First of all, this is the influence of viscous effects on the propagation of gas-dynamic discontinuities inside the shock tubes. This influence was studied experimentally [3,4] and theoretically [5]. Theoretical studies were conducted on the basis of approximate models such as quasi one-dimensional gas-dynamics and boundary layer equations.

At the present time, the development of computer technologies makes it possible to numerically study the processes inside the shock tubes on the basis of the complete Navier-Stokes equations on very detailed grids and to obtain a detailed structure of the flow in the shock tubes. Most of these numerical studies of the flow in shock tubes concentrate on the investigation of the wave structure that occurs when a shock wave is reflected from a wall. In [6] one such variant was proposed as a benchmark solution and was studied in detail on fine grids for several Reynolds numbers (200, 1000, 5000, etc.) [6–9]. This variant is characterized by a very small length of the shock tube, which makes it impossible to separate the interaction of various gas-dynamical discontinuities with the boundary layer and with each other.

Real shock tubes are characterized by an appreciably larger length in comparison with a diameter. For example, in the case of HAST–1 [10–11] the total length of the high and low pressures chambers is 9.5 m, but the diameter is 0.08 m. This leads to the facts that:

1). The time of propagation of the shock wave from the diaphragm to the shock tube end is large enough for a noticeable influence of viscous effects on the flow inside the shock wave.

2). The time intervals between the arrival of various gas-dynamic discontinuities to the shock tube end increase, which allows us to separate the various processes occurring near the shock tube end and study it more accurately.

2. Mathematical model and numerical method

We shall study the flow in a shock tube using three mathematical models:

1). One-dimensional unsteady Euler equations for a perfect gas;

2). One-dimensional unsteady Navier-Stokes equations. In this case, viscosity and thermal conductivity along the shock tube are additionally taken into account.

3). Two-dimensional unsteady Navier-Stokes equations. As in the Daru and Tenaud paper [6], the Reynolds number, determined from the shock wave height, is 200, and the Prandtl number is 0.72. At this Reynolds number, the boundary layer on the shock tube wall is rather thick, so in this case, one can use computational grids of medium dimension for a reliable resolution of the majority of flow features [6–9].

In all 2D calculations presented, computational grids consisting of square cells as in [6-9] were used. This makes it possible to carry out calculations without the appearance of a transverse numerical instability ("carbuncle" effect), which often arises in the calculation of the propagation of gas-dynamic discontinuities in longitudinally stretched cells. The use of square cells makes it very difficult to perform numerical simulation of real-size shock tubes due to the very large ratio of the shock tubes length to its height. In most of the calculations, the grid steps were selected as $\Delta x = \Delta y = 1/1000$. Therefore, in this paper, the maximum length-to-height ratio used for calculations was 8 to 1, like on a grid of 8000×500 cells.

The systems of equations described above were solved by the numerical method [12, 13]. It is based on the Godunov high-resolution method and uses the MUSCL approach to improve the accuracy of the method with respect to spatial variables. Depending on the reconstruction procedure used, this is the TVD/ENO/WENO type method. Most of the calculations were carried out with the 5th order accuracy reconstruction procedure [14] or [15].

The viscous terms were discretized with a second order central-difference scheme. The Runge-Kutta method of the third order of accuracy was used for time approximation [16].

Adiabatic boundary conditions were set on the walls, and the boundary conditions for the symmetry of the flow were set on the symmetry line.

3. Numerical code verification

For our numerical code verification we present some results of numerical simulation of the 2-D viscous shock tube problem. This problem was studied by Daru and Tenaud [6] and Sjögreen and Yee [7]. The 3D version of the problem was proposed in [8]. In the problem a tiny, very short shock tube with square cross section is considered. In 2D the shock tube is a 2D box $0 \le x,\ y \le 1$. A membrane located at $x = 0.5$ separates two different states of the gas. At time zero the membrane is removed and wave interaction occurs. The initial state is given as follows:

$$(\rho, v_1, v_2, p)_L = (120, 0, 0, 1120/\gamma), \quad (\rho, v_1, v_2, p)_R = (1.2, 0, 0, 1.2/\gamma). \tag{1}$$

$\gamma = 1.4$ and the Prandtl number is 0.73. The two Reynolds numbers considered are 200 and 1000. The viscosity is assumed to be constant and independent of temperature, so Sutherland's law is not used.

Figure 1 shows the numerical solution of the problem obtained on a 2000×1000 mesh for Re = 200. Figure 1 contains numerical Shlirien image (the density gradient countours). A comparison with the benchmark solution [6–9] shows that in this case the key features of the flow dynamics generated by the interaction of the shock wave and the boundary layer are captured well.

The next, we consider the case Re = 1000. For such Reynolds number, vortical structures become more complex. Small vortices are generated just downstream of the boundary layer separation, and shocklets are present at many locations inside the large vortices. Figure 2 presents density gradient magnitude contours at non-dimensional time $t_{\text{end}} = 1$ predicted on mesh 3000×1500. Presented results agree reasonably well with those reported by Daru and Tenaud [6] using a 2000×1000 mesh and those of Sjögreen and Yee [7] using a 4000×2000 mesh.

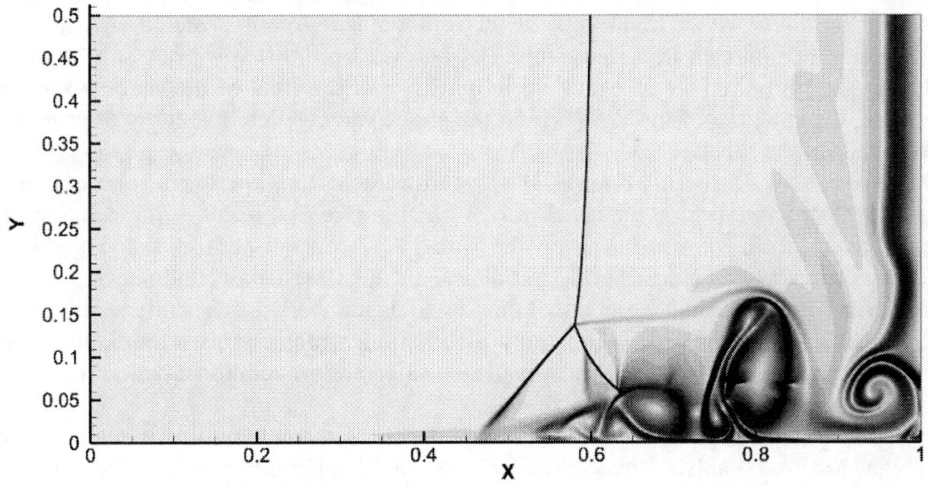

Figure 1. Density gradient contours for Reynolds number Re = 200 at t=1.

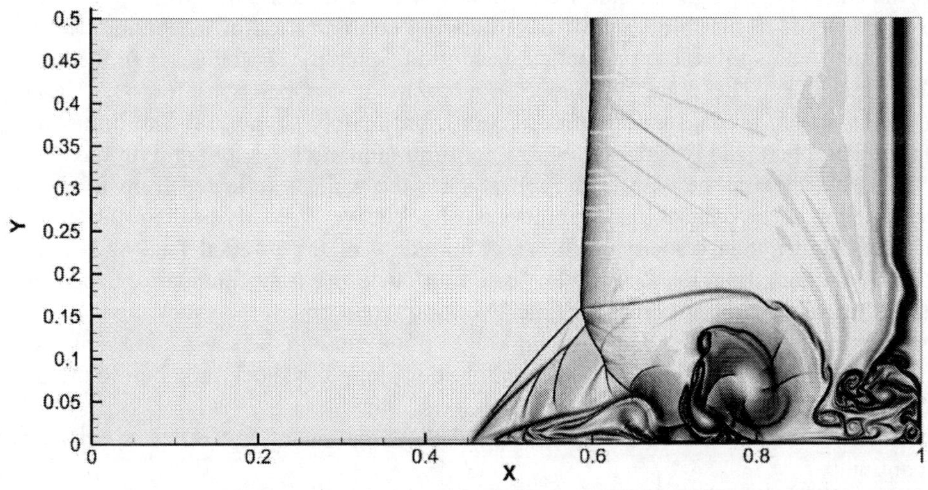

Figure 2. Density gradient contours for Reynolds number Re = 1000 at t=1.

Obtained solutions of these two problems show that the flow in shock tube has complex 2-D structure due to shock wave interaction with boundary layers. Usual 1-D numerical simulation of the flow can't reproduce all flow details and therefore has limited applicability.

Comparing these figures, it can be noted that an increase in the Reynolds number leads to a noticeable complication in the flow structure. A large number of small vortex structures appear, for the resolution of which it is necessary to refine of the computational grid in both longitudinal and transverse directions. It is an important reason why the Daru and Tenaud benchmark problem is usually solved on grids with square cells. Therefore the need for a strong refinement limits the possibility of numerical simulation of the flow in shock tubes of large length with high accuracy.

4. Shock wave propagation in a tube

Let us consider the flow inside the shock tube in more detail. Usually, the rupture of the diaphragm occurs non-uniformly and its further "opening" cannot be complete. This introduces additional perturbations in the flows within shock tube at the initial stage. A very detailed description of the process of opening a diaphragm in a shock tube can be found in [18]. However, at a large distance from the diaphragm, all perturbations decay fairly quickly and the front of the shock wave becomes plane. Therefore, the structure of the flow behind the shock wave is determined by the perturbations arising on the walls of the shock tube.

With sensitive tuning of the interference–shadow instruments in experiment, one can observe a characteristic "net" structure behind the shock wave, which arises as a result of the addition of many random perturbations from the roughness on the wall [17]. Contact surfaces that are part of this structure form vortices which can generate turbulence in the flow behind the shock wave. In our calculations, given below, we shall assume that the Reynolds number is sufficiently small and there is no transition to turbulence. The second important assumption, which confirms the laminar nature of the flow, is the smoothness of the shock tube walls, so there are no additional "micro-perturbations" that can initiate the transition.

A similar assumption is made about the initial rupture of the diaphragm: the diaphragm rupture is modeled by an instant complete "disappearance" of the diaphragm, which does not introduce additional perturbations into the flow.

If viscous effects are not taken into account, then in shock tube, after a diaphragm rupture, a flow close to the one-dimensional solution of the Riemann problem is realized [19]. As a rule, the ratio of pressures in the chambers of high pressure (to the left of the diaphragm) and low pressure (to the right of the diaphragm) leads to the formation of the following configuration of discontinuities: the shock wave moves to the right, followed by a contact discontinuity separating the gases in the chambers of high and low pressures. The main difference in pressure is compensated for by a large fan of rarefaction waves, which propagates in both directions, but mainly to the left. The initial state in the Daru and Tenaud problem [6] gives the same flow configuration after a diaphragm rupture.

Part of the rarefaction fan moving to the left reaches the wall, is reflected from it and begins to move in the direction of the contact discontinuity and shock wave. Velocity of the reflected part of the rarefaction fan is higher than velocity of the right boundary of the original fan. As result, under a certain initial conditions, the reflected fan can "catch up" with the right boundary of the original fan and begin to influence the flow structure near the "working" right end of the shock tube.

It will be convenient to consider the following stages of the flow in the shock tube separately:

1). Motion of the initial configuration of discontinuities along the shock tube before the interaction of the incidental shock with the wall.

2). Bifurcation of reflected shock wave before the interaction of the reflected shock with the contact discontinuity.

3). The interaction of the reflected-shock bifurcation with the contact discontinuity until the arrival of the right rarefaction wave from the fan.

For numerical simulation a shock tube with a length-to-height ratio of 8 to 1 was considered. The high-pressure chamber occupied the left part of the shock tube $x \in [0, 0.5)$, and the low-pressure chamber occupied the right one $x \in (0.5, 8]$. To construct a computational grid, we used the grid step $\Delta x = \Delta y = 1/1000$. So one-dimensional calculations were performed on a grid of 8000 cells, and two-dimensional calculations were performed on a grid of 8000 × 500. The initial state of the gas in the shock tube corresponds to the Daru and Tenaud problem (1).

If the flow in a shock tube is considered in the 1D formulation or the 2D inviscid formulation, then the flow parameters behind the shock front must be constant up to the contact surface, the speed of which is equal to the gas velocity behind the shock wave. This is clearly seen in figure 3, where the left part of the figure shows the density along the shock tube axis at some time, and the right part shows the pressure. It is seen that in one-dimensional calculations the density and pressure between the shock waves and the contact discontinuity are constant.

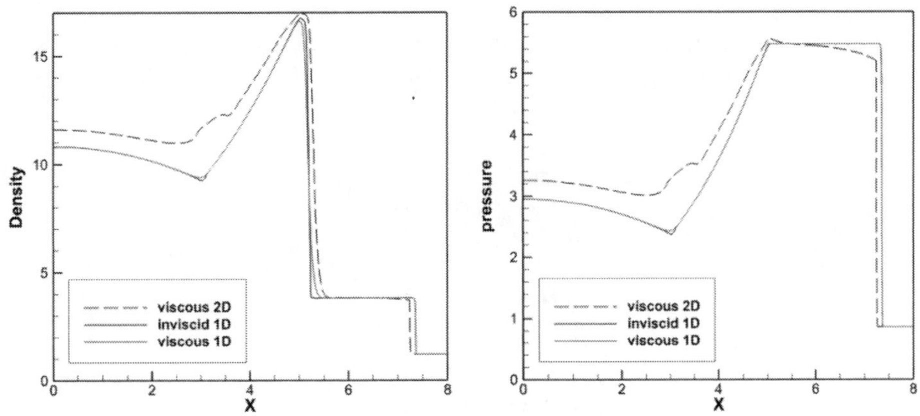

Figure 3. Comparison of viscous 2D calculation with 1D calculations.

In reality, the flow parameters behind the shock wave do not remain constant, as is clearly seen in the pressure plot in figure 3 in the case of a two-dimensional viscous calculation. This is due to the fact that the boundary layer on the walls of the shock tube grows over the entire interval from the shock wave to the contact discontinuity (see figure 4).

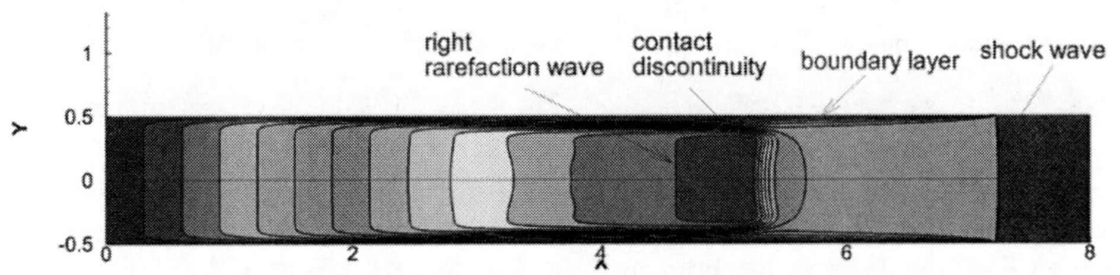

Figure 4. Configuration of discontinuities in a shock tube.

From figure 4 it can be easily seen that the growth of the boundary layer leads to a change in the "effective cross-section" of the shock tube and, consequently, to a change in the parameters in the inviscid core of the flow [20]. The flow velocity increases smoothly from the shock wave to the contact surface [21], the temperature and pressure increase. These changes can be estimated from the approximate theory [5].

From figure 4 it can be also seen that in the viscous 2D case the shock speed decreases, and the contact surface speed increases in comparison with the 1D case when these speeds are constant.

We shall note that in long shock tubes, the boundary layers on the side walls can merge. In this case, one-dimensional calculations are not very suitable for estimating parameters in shock tubes.

It should also be noted that taking into account viscous effects in the longitudinal direction in the 1D approximation in this case has very little effect on the flow in the shock tube, as is clearly seen in figure 3.

5. Bifurcation of reflected shock wave

When the shock wave is reflected from a rigid wall in a shock tube, it propagates along an inhomogeneous flow. Mainly this flow consists of two layers: the inviscid core and the boundary layer formed after the incident shock wave.

Griffiths predicted [22], and Mark confirmed experimentally [23] that flow regimes with the bifurcation of reflected shock can arise in the viscous flow, and moreover the characteristic distance from the wall to the splitting point of the reflected shock is greater than the thickness of the boundary layer. The bifurcation of reflected shock in a shock tube can be obtained numerically too. Some complicated variants of the bifurcation of reflected shock in the Daru and Tenaud shock tube are shown in figures 1 and 2. Schematically, a lambda-shape like bifurcated shock wave pattern is shown in figure 5.

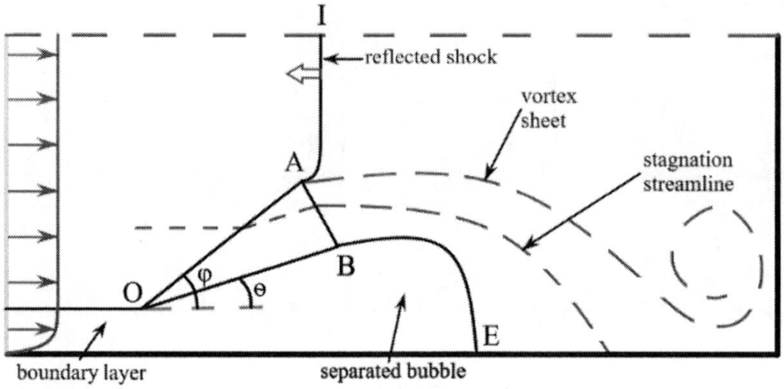

Figure 5. Typical bifurcated shock wave pattern.

After reflection the stagnation pressure of the boundary-layer becomes lower than the pressure behind the reflected shock wave. As result the fluid cannot pass under the reflected shock and a separated flow region appears. This separated region propagates upstream with the reflected shock (figure 5).

The problem of calculating the wave structure arising during bifurcation is not completely solved. The main problem is the obtaining of a convergent solution when the grid is refined [9] in the case of large Reynolds numbers. This was one of the reasons for choosing Re = 200 in this numerical simulation. The temperature field and the Mach number contour lines are shown in figure 6 after the shock reflection. The lambda-shape like bifurcated shock structure is clearly visible, which completely corresponds to figure 5. In addition, in figure 6 on the left, the contact surface is also visible, moving towards the reflected shock wave.

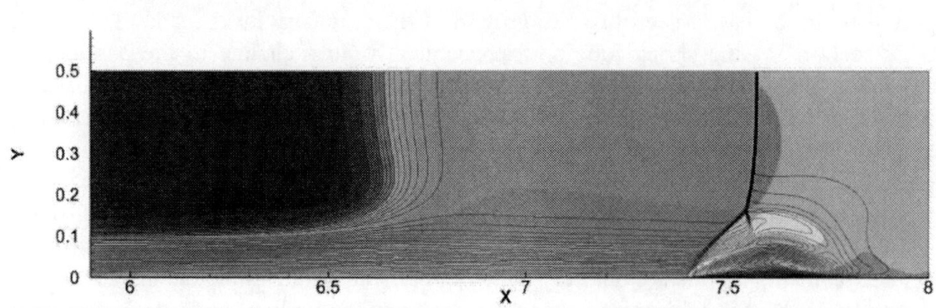

Figure 6. Temperature field and Mach number contour lines after the shock reflection.

A comparison of density (on the left) and pressure (on the right) distributions along the axis of the shock tube is shown in figure 7. It can be seen that the boundary layer affects most of all on the position of the contact surface and the flow behind it. The positions of the shock wave in one-dimensional and two-dimensional calculations are generally in good agreement, despite the bifurcation of reflected shock in the 2D case.

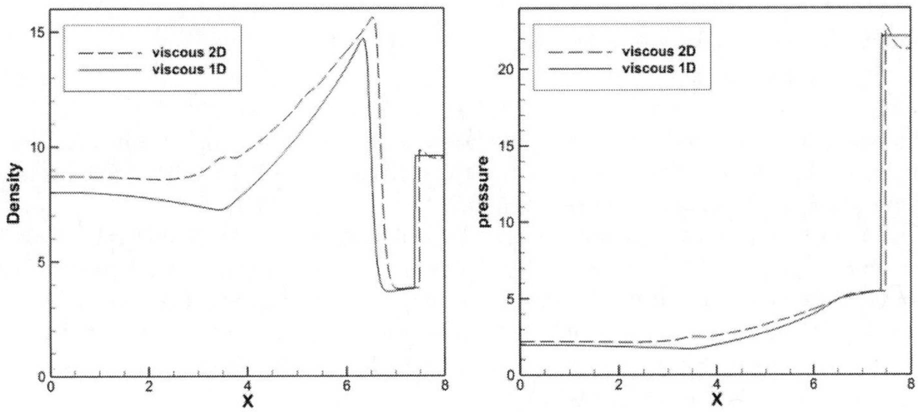

Figure 7. The flow in the shock tube after the shock reflection. Comparison of viscous 2D calculation with 1D calculation.

It should be noted that there is a slight decreasing of pressure near the end of the shock tube in the two-dimensional case in comparison with the one-dimensional calculation.

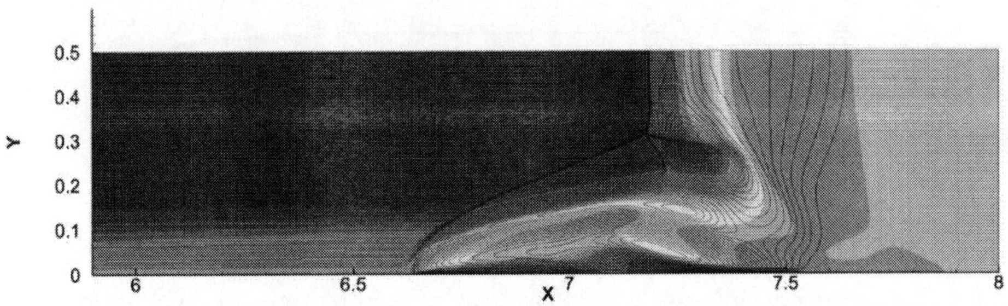

Figure 8. Flow field near the right end of shock tube after the shock and contact surface interaction.

Figure 8 shows the flow field near the right end of shock tube just after the shock and contact surface interaction. The temperature field and the Mach number contour lines are drawn in figure 8. It can be seen that the bifurcation of shock grow significantly and the triple point moves in the direction of the axis. The size of the bifurcation of shock is larger than in the case of original Daru & Tenaud problem.

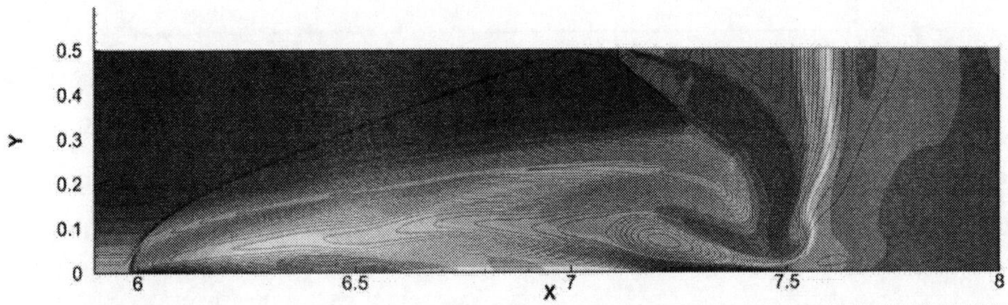

Figure 9. Flow field near the right end of shock tube after the bifurcation junction.

IOP Conf. Series: Journal of Physics: Conf. Series **1009** (2018) 012010 doi:10.1088/1742-6596/1009/1/012010

The speed of the triple point movement is quite high and very soon the triple point reaches the tube axis (see figure 9). It can be seen the bifurcation shock junction. The flow near the shock tube end becomes two-dimensional.

6. Numerical simulation of the flow in a real shock tube

In real shock tubes such as described in [10–11] the length is considerably large than the tube height. The total length of driver and driven parts of HAST–1 [10–11] equals to 9.32 m and the diameter of this shock tube is 0.08 m. It means that the length-to-diameter ratio is 116.5. With such large ratio the possibility exists that the boundary layers on the tube walls are closed. This may affect the structure of the shock wave and the dynamics of its propagation.

Such large length-to-height ratio introduces additional computational difficulties. It doesn't allow to generate a computational grid with the same resolution as for Daru and Tenaud problem above. We used a 5000×200 uniform mesh in our numerical simulation. The comparison this solution with the solution on a 4000×150 mesh shows that all base features of the flow are very similar on the both meshes. In the driver tube we set the pressure 20 atm and the pressure in the driven tube was 10^{-3} atm. The temperature in both tubes was 300°K.

Figure 10 shows the density gradient magnitude contours at different times. It should be noted that in this figure essentially different length scales along different axes are used. On the top part of the figure the density gradient contours at a time shortly before the shock wave interaction with the wall are shown. It can be seen considerable growth of the boundary layer which takes up a third of the tube radius. The remaining three pictures correspond to three different times after the shock wave reflection from the wall.

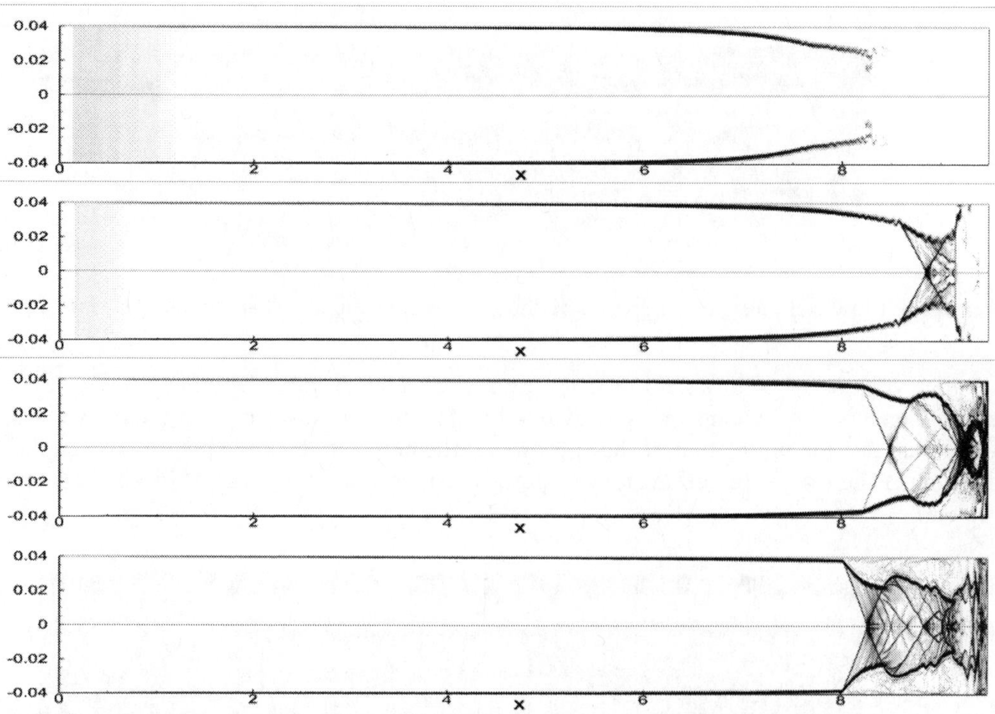

Figure 10. Density gradient magnitude contours for a 2-D viscous shock tube problem in HAST at different times.

The figure illustrates that the flow after the shock wave reflection from the wall becomes very complex. It's clearly seen that most small eddies which were resolved in Daru and Tenaud problem aren't resolved in the presented solution. Insufficiency of the grid resolution becomes even more pronounced when reflected from the left wall rarefaction fan reaches the right end of the tube and begins to interact with the reflected shock wave.

7. Conclusion

The numerical simulation of influence of viscous effects on flow inside shock tubes are described. We considered three shock tubes with different length-to-height ratio: Daru & Tenaud [6] case (1 to 1 ratio), extended Daru & Tenaud case (8 to 1 ratio) and HAST-1 tube [10–11] (116 to 1 ratio). The Daru & Tenaud case was used as a benchmark problem. Good agreement between the obtained results and well-known solutions was demonstrated. The extended Daru & Tenaud case allowed us to separate different stages of the unsteady flow in shock tube and compare results of 1D and 2D calculations.

The numerical simulation of the flow in HAST-1 shows the significant influence of viscous effects on the flow.

Acknowledgments

This work was partially supported by Government program of basic research of Russian academy of science (contract #AAAA-A17-117021310372-6) and partially supported by Russian Foundation for Basic Research grants # 16-01-00444 and 16-38-60185.

References

[1] Hertzberg A 1956 *Jet Propulsion* **6** 7 549–554
[2] Stalker R J 2006 *The Aeronautical Journal* **110** 1103 21-39
[3] Holden M S and Wadhams T P 2003 *AIAA Paper* No. 2003-3641
[4] Bazhenova T V, Naboko I M and Nemkov R G 1969 In: *Shock tubes* Toronto 60–67
[5] Mirels H 1966 *Phys. Fluids* **9** 10 1907–1912
[6] Daru V and Tenaud C 2001 *Comp. Fluids* **30** 89–113
[7] Sjögreen B and Yee H C 2003 *J. Comput. Phys.* **185** 1–26
[8] Yoon S-H, Kim C and Kim K H 2008 *J. Comput. Phys.* **227** 6001–6043
[9] Zhou G, Xu K and Liu F 2018 *Physics of Fluids* **30** 1 016102
[10] Kryukov I, Kotov M, Ruleva L, Solodovnikov S and Surzhikov S 2013 *AIAA Paper* 2013–2931
[11] Kotov M, Ruleva L, Solodovnikov S, Kryukov I and Surzhikov S 2014 *AIAA Paper* 2014–2657
[12] Glushko G S, Ivanov I E and Kryukov I A 2010 *Math. Models Comput. Simul.* **2** 407–422
[13] Ivanov I E and Kryukov I A 1996 *Matematicheskoe Modelirovanie* **8** 6 47–55
[14] Suresh A and Huynh H T 1997 *J. Comput. Phys.* **136** 1 83–99
[15] Kim K H and Kim C 2005 *J. Comput. Phys.* **208** 2 527–569
[16] Shu C-W and Osher S 1989 *J. Comput. Phys.* **83** 32–78
[17] Shugaev F V and Shtemenko L S 1998 *Propagation and Reflection of Shock Waves* (Singapore:World Scientific) p 244
[18] Igra O and Seiler F 2015 *Experimental Methods of Shock Wave Research* (Vienna:Springer) p 478
[19] Godunov S 1959 *Sbornik Mathematics* **47** 271–306
[20] Hartunian R A and Marrone P V 1961 *Phys. Fluids* **4** (5) 535–543
[21] Bazhenova T V 1973 *Shock Waves in Real Gases* (Vienna: Springer) p 78
[22] Griffith W C and Bleakney W 1954 *Amer. J. Phys.* **22** 597–612
[23] Mark H 1957 *J.Aeron. Sci.* **24** (4) 304–306

IOP Conf. Series: Journal of Physics: Conf. Series **1009** (2018) 012011 doi:10.1088/1742-6596/1009/1/012011

Effect of reaction products kinetic on stability of thermonuclear spark ignition

A Yu Mikulin[1], S I Glazyrin[1,2] and S E Kuratov[1]

[1] Dukhov Research Institute of Automatics (VNIIA), Sushchevskaya 22, 127055, Moscow, Russia

[2] Institute for Theoretical and Experimental Physics, NRC Kurchatov Insitute, Bolshaja Cheremushkinskaya 25, 117259, Moscow, Russia

mikandr90@mail.ru

Abstract. Thermonuclear spark instability driven by reaction products momentum transfer is considered in this paper. General theory of such instability was developed by Lykov [1] based on one-group particles transport equation solving. Main dimensionless parameter of this problem is l/λ, where λ is the perturbation wavelength and l is the reaction product mean free path. In two limiting cases of l/λ ratio we recalculated perturbations growth rates in a more precise manner, obtaining reaction products distribution function through Fokker-Planck equation. While our results are generally consistent with those of Lykov's work, instability growth rates predicted in this investigation are somewhat smaller than the previously reported values. By performing a sequence of hybrid kinetic-fluid simulations in code FRONT3D, we demonstrate their better agreement with our asymptotic predictions. However, it was shown that in a wide range of l/λ ratio the instability growth rates are well described by Lykov's theory. We apply Champan-Enskog method to describe plasma ions deviations away from Maxwellian distributions, which appears from interaction with higly non-maxwellian reaction products. We derive an equation, that connects concentration of non-equilibrium particles with such parameters as ion momentum relaxation time and reaction rate. As it was shown, for typical ICF parameters, non-equilibrium part does not exceed several percents even at the burn peak.

1. Introduction

Hydrodynamic instabilities of thermonuclear spark surface still remain one of the main problem in ICF [2]. According to general theory of spark ignition [3], hot spot threshold energy, after which ignition is achieved, defined through heat balance equation $W_{fus} - W_r - W_{ec} - W_{ad} \geq 0$. Here W_{fus}, W_r, W_{ec} – are hot spot volume weighted rates of thermonuclear heating, radiation and electron conductivity cooling respectively; W_{ad} - mean rate of purely hydrodynamic cooling due to adiabatic expansion. Since W_{ec} value is directly proportional to hot spot surface area, it could strongly increase due to hot spot spherical symmetry violation.

Main contribution to hot spot perturbations growth is caused by Rayleigh-Taylor instability, acting in deceleration phase of ICF targets implosion [4]. During this stage the hot spot ion temperature climbs to many kiloelectronvolts, so reaction products mean free path becomes close to hot spot own length scale [5]. Nowadays the lack of agreement between ICF experiments and numerical simulations [6, 7, 8] is particularly associated with kinetic regime of reaction products energy transfer, therefore this question has received considerable attention [9]. Numerical simulations with alpha particles transport treated with multi-group diffusion model shows that re-deposition of alpha particles energy into the cold fuel shell leads to polishing effect and stabilization of high-mode perturbations[10].

In this work we are focusing on perturbation evolution caused by reaction products momentum transfer. Previous research [1] has attempted to describe reaction product transfer by employing one-group transfer equation. Although one-group approximation allows to obtain instability increment for arbitrary l/λ ratios, it neglects the effect of stopping power varying during reaction product deceleration. Therefore we were strongly motivated to recalculate the part of previous results [1] in a more accurate and precise manner.

2. Growth rates

Let us consider two superposed incompressible, inviscid, irrotational, and immiscible liquids, in one of which reactions occurs (active medium). OXY plane is unperturbed mediums interface. Z axis is up, so active and passive mediums lie in half-space $z>0$ and in half-space $z<0$ respectively. The goal of this section is to describe evolution of a single mode interface perturbation.

2.1. Small l/λ ratio

In this case reaction products produce non-zero momentum flux (and volume force respectively) only in very thin "Knudsen" layers adjacent to the interface. Since mean free paths in both medias are small and perturbation wavelength is large, every element of interface could be then considered like an element of radial surface with its own curvature.

While asymptotic smallnes of Knudsen layers allows to neglect their width, potential flow of active and passive medias is described by a Laplace equation for the hydrodynamic potentials φ_i,

$$\Delta\varphi_i = 0, \tag{1}$$

where $\vec{\upsilon} = \nabla\varphi$ and the indices $i = 1,2$ indicate active and passive media.

On the boundary between the mediums, described by its z displacement, the normal component of the velocity must be continuous. Effect of volume force on flow can be written in terms of the pressure jump across the interface that gives the equation

$$\rho_{act}\left[\frac{\partial\varphi_{act}}{\partial t} + \frac{1}{2}\left(\nabla\varphi_{act}\right)^2\right]_{z=\eta} + \delta P_{z=\eta} = \rho_{pass}\left[\frac{\partial\varphi_{pass}}{\partial t} + \frac{1}{2}\left(\nabla\varphi_{pass}\right)^2\right]_{z=\eta} \tag{2}$$

To compute pressure jump δP it is necessary to find reaction products distribution function $f(r,\upsilon,\mu,\theta)$ in cylindrical shells shown in figure 1, where μ is the cosine of angle between velocity and radius-vector and θ is the cosine of angle between velocity and Z axis. δP is then

$$\delta P = M_{RP}\iint \mu f(r,\upsilon,\mu,\theta)\,d\vec{\upsilon}\,dr \tag{3}$$

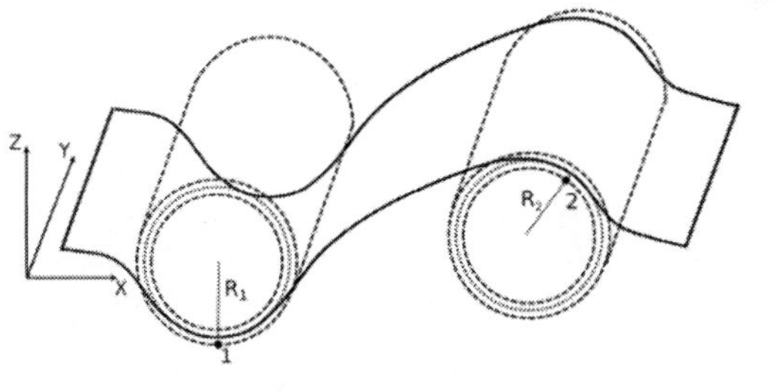

Figure 1. Interface perturbation curvature. Solid lines – perturbed interface, dotted lines – imaginary cylindrical continuations of interface, space between dashed lines – imaginary regions of nonzero acceleration. Figure shows only two cylinders for clarity, thought every point of the interface is associated with its own cylinder.

However, due to pressure additivity, δP could be obtained from more simple consideration of spherical shell system with the same radii, shown in figure 2.

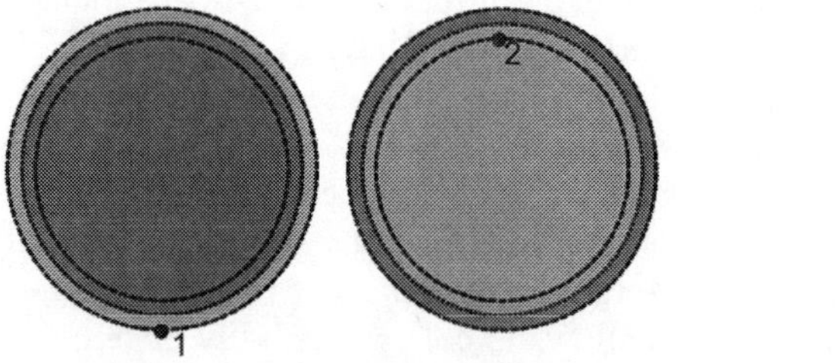

Figure 2. Considered spherical geometry. Shell radii and reference points are equal to radii of cylindrical shells and points, represented in figure 1.

In case of spherical symmetry the distribution function $f(r,\upsilon,\mu,\theta)$ of reaction products is governed by the stationary Fokker-Planck equation with isotropic source term.

$$\upsilon\left(\mu\frac{\partial f}{\partial r}+\frac{1-\mu^2}{r}\frac{\partial f}{\partial \mu}\right)=-\frac{1}{\upsilon^2}\frac{\partial}{\partial \upsilon}\left(\upsilon^2 a(\upsilon)f\right)+q\frac{\delta(\upsilon-\upsilon_0)}{4\pi\upsilon^2}, \quad (4)$$

where only the dynamical friction term is retained for the present discussion. In our formulation coefficients of function $a(\upsilon)$ could be varying from shell to shell and q is not equal to zero only for active medium shells. Inhomogeneous term in equation(4) could be[11] extracted in boundary condition and reduce to Cauchy problem for homogenous equation. For uniform sphere radius R of active medium it results in

$$\begin{cases}\upsilon\left(\mu\frac{\partial f}{\partial r}+\frac{1-\mu^2}{r}\frac{\partial f}{\partial \mu}\right)+\frac{1}{\upsilon^2}\frac{\partial}{\partial \upsilon}\left(\upsilon^2 a(\upsilon)f\right)=0\\ f(r,\upsilon_0,\mu)=-\frac{q}{4\pi\upsilon_0^2 a(\upsilon_0)}\theta(R-r)\end{cases} \quad (5)$$

Its solution could be obtained by the method of characteristics:

$$f(r,\upsilon,\mu)=-\frac{q}{4\pi\upsilon^2 a(\upsilon)}\theta\left(R-r\left[1-\mu^2+\left(\mu-\frac{1}{r}\int_{\upsilon_0}^{\upsilon}\frac{\upsilon d\upsilon}{a(\upsilon)}\right)^2\right]^{1/2}\right) \quad (6)$$

Since distribution functions are found in all other spherical shells, one can calculate pressure jump. It's known[11], that deceleration of reaction products due to interaction with plasma ions could be neglected comparing its deceleration by plasma electrons at initial stages of thermonuclear burning. Thus, reaction product deceleration could be written in following form:

$$\frac{d\upsilon}{dt}=a(\upsilon)=-\upsilon\frac{8\sqrt{\pi}Z_{RP}^2 e^4 n_e\sqrt{m_e}}{3(\kappa T_e)^{3/2}M_{RP}}\ln\Lambda_{e,RP}, \quad (7)$$

where indices RP indicates reaction products quantities. By providing rather cumbersome integration, we obtain pressure jump for spherical systems, that were shown in figure 2:

$$\begin{cases}\delta P_1^s=\frac{qM_{RP}\upsilon_0 l_{act}}{6}\left(1+\frac{l_{act}-l_{pass}}{8R}\right)\\ \delta P_2^s=\frac{qM_{RP}\upsilon_0 l_{act}}{6}\left(1-\frac{l_{act}-l_{pass}}{8R}\right)\end{cases} \quad (8)$$

Minimum curvature of ellipse, resulting from section of cylinder with radius R by plane, makes an angle of α with element of cylinder, is $R/\sin^2\alpha$. Providing angle integration, we obtain pressure jump in cylindrical systems.

$$\begin{cases} \delta P_1 = \dfrac{qM_{RP}\upsilon_0 l_{act}}{6}\left(1+\dfrac{l_{act}-l_{pass}}{16R}\right) \\ \delta P_2 = \dfrac{qM_{RP}\upsilon_0 l_{act}}{6}\left(1-\dfrac{l_{act}-l_{pass}}{16R}\right) \end{cases} \tag{9}$$

Since curvature radius is approximately inversely proportional to second spatial derivative, two equations of system (9) could be reduced to one, connecting interface perturbation and pressure jump on it:

$$\delta P(x,t) = \frac{qM_{RP}\upsilon_0 l_{act}}{6}\left(1-\eta''(x,t)\frac{l_{pass}-l_{act}}{16}\right) \tag{10}$$

This is similar to capillary wave problem[12], so corresponding perturbation growth rate is

$$\gamma = \left[\frac{1}{96}\frac{qM_{RP}\upsilon_0 l_{act}\left(l_{act}-l_{pass}\right)k^3}{\left(\rho_{pass}+\rho_{act}\right)}\right]^{1/2} \tag{11}$$

Before comparing with previous results[1] we have to notice, that in one-group approximation reaction product mean free path has the following definition:

$$l_{1G}^{-1} = -3M_{RP}\left[\frac{a(\varepsilon_0)}{\varepsilon_0} - \frac{1}{\varepsilon_0^3}\int_0^{\varepsilon_0}\varepsilon\frac{d}{d\varepsilon}\left(\varepsilon a(\varepsilon)\right)d\varepsilon\right] \tag{12}$$

Corresponding growth rate for the function $a(\upsilon)$ taken in the form(7) is

$$\gamma_{1G} = \left[\frac{25}{432\pi}\frac{qM_{RP}\upsilon_0 l_{act}\left(l_{act}-l_{pass}\right)k^3}{\left(\rho_{pass}+\rho_{act}\right)}\right]^{1/2} \tag{13}$$

2.2. *Large l/λ ratio*

When perturbation wavelength is much smaller than reaction product mean free paths, their distribution functions stay close to their unperturbated values. Besides, due to strong surface localization of gravitational waves (here waves on mediums surfaces in external gravitational field, not waves of gravitational field itself), one can use $f(0)$ instead of $f(z)$ for the acceleration field evaluation. Computing $f(0)$ with neglecting deceleration on ions and using classical theory of gravitational waves we obtain the following dispersion relation:

$$\gamma = \left[\frac{qM_{RP}\upsilon_0 k\left(l_{pass}-l_{act}\right)}{4l_{pass}\left(\rho_{pass}+\rho_{act}\right)}\right]^{1/2} \tag{14}$$

It should be compared with growth rates, obtained[1] in one-group approximation

$$\gamma_{1G} = \left[\frac{qM_{RP}\upsilon_0 k\left(l_{pass}-l_{act}\right)}{\pi l_{pass}\left(\rho_{pass}+\rho_{act}\right)}\right]^{1/2} \tag{15}$$

Parameter space, where both growth rates(11,14) are real is the parameter space of interface instability. Nontrivial result here consists in instability development while vectors of density gradient and acceleration are codirectional.

3. Numerical simulations

Validation of growth rates, obtained in previous section were provided by numerical simulation in Euler code FRONT3D [13]. Account of reaction products propagation was provided by computing of their fluxes from adjacent cells closer than free path maximum to given cell. In all our simulations ratio between reaction products mean free path in active and passive media is $l_{act} = 2l_{pass}$. To simulate both considered limit cases, we varied l_{pass} / λ ratio from 0.04 to 10.24.

Analysis of data from figure 3 and figure 4 shows the justice of predicted asymptotic for instability growth rates. Since the results obtained in [1] are also in good agreement with presented numerical simulations, it may be concluded that the theory based on one-group kinetic equation solving is really good applicable to this problem. Some differences could be easy eliminated by correction multiplicators including.

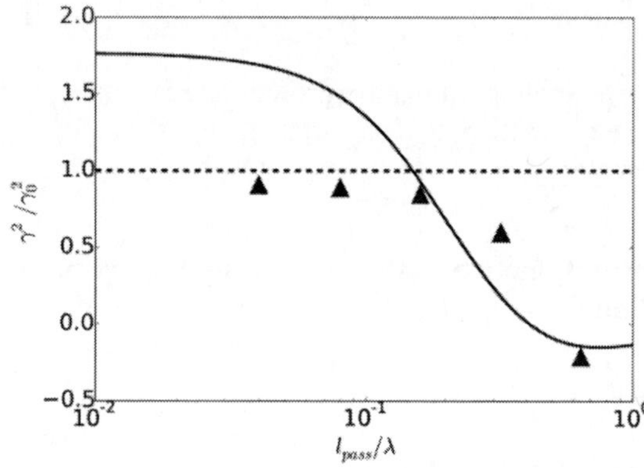

Figure 3. Comparing squares of numerical growth rates (triangles) with present (dotted line) and previous reported results[1] (solid) for small values of l/λ ratio. Values normalized by $\gamma_0^2 = \dfrac{1}{96}\dfrac{qM_{RP}\upsilon_0 l_{act}\left(l_{act} - l_{pass}\right)k^3}{\left(\rho_{pass} + \rho_{act}\right)}$.

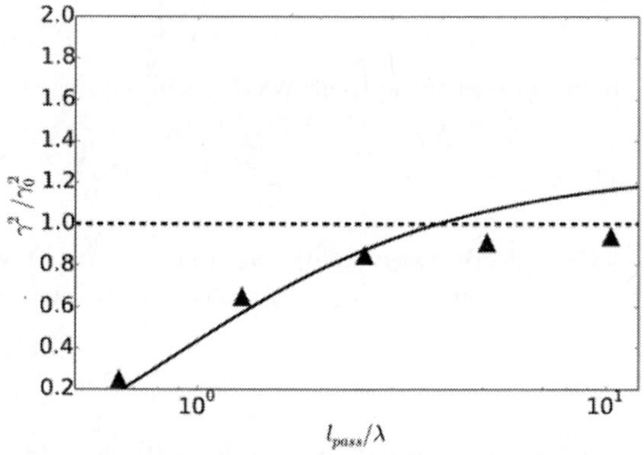

Figure 4. Comparing squares of numerical growth rates (triangles) with present (dotted line) and previous reported results[1] (solid) for large values of l/λ ratio. Values normalized by $\gamma_0^2 = \dfrac{1}{4}\dfrac{qM_{RP}\upsilon_0\left(l_{act} - l_{pass}\right)k}{l_{pass}\left(\rho_{pass} + \rho_{act}\right)}$.

4. Hydrodynamic approximation applicability

Due to interaction with energetic reaction products, plasma ions and electrons is falling out of equilibrium. Since ion relaxation time is much greater, than electron relaxation time [14], it is reasonable to consider non-equilibrium of plasma ions only.

Let us write the kinetic equation for plasma ion distribution function in case of spatial homogeneity and absence of external force:

$$\frac{\partial f_i}{\partial t} = C\left(f_i, f_e\right) + C\left(f_i, f_i\right) + C\left(f_i, f_{RP}\right) \tag{16}$$

In the RHS of equation(16) appear collision terms, model the effect of collisions between plasma ions with themselves and all other species. Ion-ion relaxation time relates to ion-electron relaxation as approximately $\sqrt{m_e T_i}\big/\sqrt{M_i T_e}$, so the first term could be neglected. Third term models the collisions between ions and reaction products and is given by the following Fokker-Planck form:

$$C\left(f_i, f_{RP}\right) = \frac{K}{\upsilon^2}\frac{\partial}{\partial \upsilon}\left(F\upsilon + D\frac{\partial f}{\partial \upsilon}\right), \tag{17}$$

where $F(\upsilon) = \dfrac{4\pi M_i}{M_{RP}}\int_0^{\upsilon} du u^2 f_{RP}(u)$ and $D(\upsilon) = \dfrac{4\pi}{3\upsilon}\int_0^{\upsilon} du u^4 f_{RP}(u) + \dfrac{4\pi\upsilon^2}{3}\int_{\upsilon}^{\infty} du u f_{RP}(u)$ [15].

Constant K is $\frac{4\pi Z_i^2 Z_{RP}^2 e^4}{M_i^2}\ln\Lambda_{RP,i}$, where $\ln\Lambda_{RP,i}$ – Coulomb logarithm for collisions between plasma ions and reaction products[16].

The solution of f_i can be carried out via the Chapman-Enskog expansion of equation (16). We will restrict ourselves only to a first order expansion. Plasma ions distribution function then brakes into sum of Maxwell distribution $f_{i,0}$ and non-equilibrium addition $f_{i,1}$. Since Maxwell distribution vanishes collision terms, equation (16) becomes:

$$\frac{\partial f_{i,0}}{\partial t}+\frac{\partial f_{i,1}}{\partial t}=C\left(f_{i,0},f_{i,1}\right)+C\left(f_{i,0},f_{RP}\right) \tag{18}$$

Being focused on intensive burn stages, we consider the case of high plasma temperatures (hundreds keV). In this situation reaction products deceleration by plasma electrons could be neglected and therefore their distribution function is equal to $\frac{q}{4\pi\beta}\theta\left(\upsilon_0-\upsilon\right)$, where $\beta=\frac{4\pi Z_i^2 Z_{RP}^2 e^4 n_i}{M_i M_{RP}}\ln\Lambda_{i,RP}$: Besides generation of non-equilibrium particles, collisions with reaction products leads to evolution of equilibrium distribution, that results in temperature growth $\frac{\partial f_{i,0}}{\partial t}=-\frac{3}{2}f_{i,0}\left[1-\frac{M_i\upsilon_i^2}{3\kappa T}\right]\frac{1}{T}\frac{dT}{dt}$.

Evaluating the collision term for ions and reaction products(17) results in:

$$\begin{gathered}\widehat{C}\left(f_{i,0},f_{RP}\right)=C\left(f_{i,0},f_{RP}\right)_{\upsilon_i<\upsilon_0}-\frac{\partial f_{i,0}}{\partial t}=\frac{Kqf_{i,0}}{\beta}\left\{\frac{4}{15}\left(\frac{M_i\upsilon_i^2}{2\kappa T}\right)^2+\right.\\ \left.+\left[\frac{2}{3}\frac{M_i\upsilon_i^2}{2\kappa T}-1\right]\left[\frac{M_i\upsilon_0^2}{\kappa T}-\frac{M_i\upsilon_i^2}{\kappa T}-\frac{M_i}{M_{RP}}-\frac{3}{2}\frac{1}{T}\frac{dT}{dt}\frac{\beta}{Kq}\right]\right\}\end{gathered} \tag{19}$$

Integration $C\left(f_{i,0},f_{RP}\right)$ with $\frac{1}{2}M_i\upsilon_i^2$ over ion velocity space gives the expected temperature growth rate:

$$\frac{3}{2}n_i\kappa\frac{dT}{dt}=\frac{qM_{RP}\upsilon_0^2}{2} \tag{20}$$

Therefore collision term, corresponding to non-equilibrium particles generation is:

$$\widehat{C}\left(f_{i,0},f_{RP}\right)=\frac{Kqf_{i,0}}{\beta}\left\{\frac{4}{15}\left(\frac{M_i\upsilon_i^2}{2\kappa T}\right)^2+\left[\frac{2}{3}\frac{M_i\upsilon_i^2}{2\kappa T}-1\right]\left[-\frac{M_i\upsilon_i^2}{\kappa T}-\frac{M_i}{M_{RP}}\right]\right\} \tag{21}$$

Being function of ion velocity, collision term(21) can be positive or negative. Computing threshold ion velocity $\upsilon_{i,TH}$, which makes collision term(21) zero and using the fact of particle number conversation, we can write the expression for non-equilibrium particles concentration growth rate:

$$\int_0^{\upsilon_{i,TH}}\widehat{C}4\pi\upsilon_i^2 d\upsilon_i=q\Omega\left(M_i,M_{RP}\right) \tag{22}$$

Calculation $\Omega\left(M_i,M_{RP}\right)$ for deuterium and tritium collisions with alpha particles gives ~0.99 and ~0.75 respectively. To evaluate collision term $C\left(f_{i,0},f_{i,1}\right)$, models collisions between equilibrium and non-equilibrium ions we use τ-approximation. Therefore we can finally obtain an equation, connects concentration of non-equilibrium particles with such parameters as reaction rate and ion momentum relaxation time [14].

IOP Conf. Series: Journal of Physics: Conf. Series **1009** (2018) 012011 doi:10.1088/1742-6596/1009/1/012011

$$\frac{dn_{NE}}{dt} = -\frac{n_{NE}}{\tau} + q\Omega \tag{23}$$

Comparison of non-equilibrium particles concentration, obtained by equation(23), solved with concentrations, obtained from numerical solution of Fokker-Planck equation in burning DT-plasma[17] is shown in figure 5.

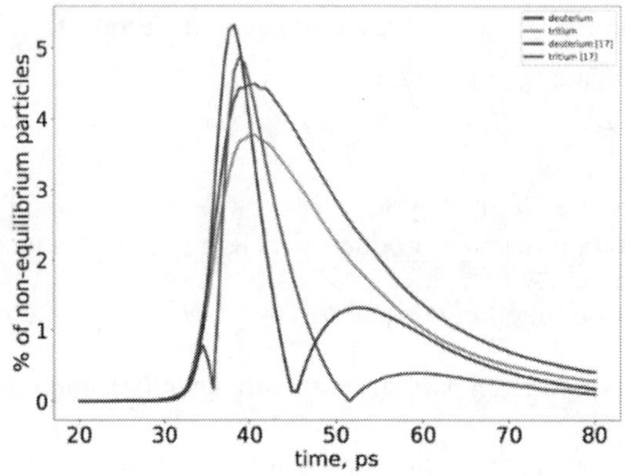

Figure 5. Results of present work and previously reported data[17]

One can conclude, that considered approach of non-equilibrium particles number estimation leads to result, that is similar to results from full numerical kinetic simulations.

5. Conclusion

The paper presents description of instability, driven by energetic reaction products momentum transfer. Analytical result was justified by FRONT3D code simulations. Besides it was shown, that previously reported growth rates, obtained from one-group kinetic treatment of reaction products, are also in rather good agreement. So one could confirm the previous work [1] conclusions about long-wavelength spark instability in ICF targets.

In the framework of purely kinetic analysis of this problem the equation that describes concentration of non-equilibrium particles dependence on such parameters as plasma ions momentum relaxation time and thermonuclear reaction rate was obtained. In case of infinite burning DT–plasma, its solution gives concentrations of non-equilibrium plasma ions that are close to results of direct numeric solution of Fokker-Planck kinetic equation.

References

[1] Lykov V 1994 *AIP Conference Proceedings* **318** pp 403–412
[2] Kishony R and Shvarts D 2001 *Physics of Plasmas* **8** 4925-4936
[3] Basko M 1990 *Nuclear Fusion* **30** 2443
[4] Sanz J and Betti R 2005 *Physics of Plasmas* **12** 042704
[5] Krokhin O N and Rozanov V B 1973 *Soviet Journal of Quantum Electronics* **2** 393
[6] Clark D, Hinkel D, Eder D, Jones O, Haan S, Hammel B, Marinak M, Milovich J, Robey H, Suter L *et al.* 2013 *Physics of Plasmas* **20** 056318
[7] Clark D, Marinak M, Weber C, Eder D, Haan S, Hammel B, Hinkel D, Jones O, Milovich J, Patel P *et al.* 2015 *Physics of Plasmas* **22** 022703
[8] Clark D, Weber C, Milovich J, Salmonson J, Kritcher A, Haan S, Hammel B, Hinkel D, Hurricane O, Jones O *et al.* 2016 *Physics of Plasmas* **23** 056302
[9] B E Peigney, O Larroche, and V Tikhonchuk 2014 *Physics of Plasmas* **21** 122709
[10] Nishiguchi A, Nakagawa S, Takahashi H, Takabe H and Mima K 1994 *AIP Conference Proceedings* **311** pp 256–268
[11] Gus'kov S Y, Krokhin O N and Rozanov V B 1975 *Soviet J. of Quantum Electronics* **4** 895

IOP Conf. Series: Journal of Physics: Conf. Series **1009** (2018) 012011 doi:10.1088/1742-6596/1009/1/012011

[12] Landau L D and Sykes J 1987 *Fluid mechanics* (New York: Pergamon)
[13] Glazyrin S 2014 *Astrophysics and Space Science* **350** pp 683-689
[14] Silin V 1971 *Introduction to kinetic gas theory* (Moscow: Nauka)
[15] Huang C K, Molvig K, Albright B J, Dodd E S, Vold E L, Kagan G and Hoffman N M 2017 *Physics of Plasmas* **24** 022704
[16] Huba J D 2004 *NRL Plasma formulary* (Washington, DC: Naval Research Laboratory)
[17] Michta D, Graziani F, Luu T and Pruet J 2010 *Physics of Plasmas* **17** 012707

IOP Conf. Series: Journal of Physics: Conf. Series **1009** (2018) 012012 doi:10.1088/1742-6596/1009/1/012012

Comparison of the stabilized jellium model and the model with fixed point ions

D S Shidlovski[1], **A E Mukhanov**[1]

[1]Dukhov Research Institute of Automatics (VNIIA), Sushchevskaya str. 22, Moscow, 127055, Russia

dmitry.shidlovski@yandex.ru

Abstract. A numerical analysis of the electron density and the electrostatic potential distribution in spherically symmetric bubbles of ionized hydrogen of nanometer size was made using density functional theory (DFT). Two models are considered: the stabilized jellium model, which allows to carry out one-dimensional computations for the spherically symmetric case, and the model with fixed point ions, where calculations are carried out in 3D geometry. For calculation in the point ions model Octopus and CP2K codes were used. One-dimensional computations in the jellium model require much less computational resources that allows to carry out calculations for systems with large number of electrons (about 10^5). A conclusion about applicability of the stabilized jellium model for calculations of such systems. These calculations would be useful for studying inhomogeneities of the electric field, which affect ion motion during rapid compression of gas bubbles.

1. Introduction

In this work a numerical analysis of the electron density and the electrostatic potential distribution in spherically symmetric bubbles of ionized hydrogen of nanometer size with concentration $10^{29}\div10^{31}$ m^3 was made using density functional theory (DFT). These calculations would be useful for studying inhomogeneities of the electric field, which affect ion motion during rapid compression of gas bubbles. Two models were used. The first one is the stabilized jellium model [1], where ions are presented as a smooth positive charge distribution, that allows to carry out one-dimensional computations for the spherically symmetric case. The second one is the model with fixed point ions, which are homogeneously distributed inside the sphere and every ion is represented as a pseudopotential. In this case calculations are carried out in 3D geometry. Rough but allowing to carry out fast computations with large number of particles jellium model is compared to detailed but resource-intensive the point ion model to make a conclusion about its applicability. All computations were carried out for homogeneous spherically symmetric bubbles of hydrogen.

2. Description of used models

2.1. Stabilized jellium model

For spherically symmetric computations in the stabilized jellium model one-dimensional Kohn-Sham [2] equations are solved with the exchange-correlation functional PZ (Perdew, Zunger) [3] in the local density approximation (LDA). Single electron wavefunctions are presented as a product of radial and spherical wavefunctions. Kohn-Sham equation for radial wavefunctions:

$$\left(-\frac{1}{2}\frac{d^2}{dr^2}+V_{KS}(r)+\frac{l(l+1)}{2r^2}\right)P_{nl}(r)=E_{nl}P_{nl}(r), \tag{1}$$

$$V_{KS}=V_{ion}+V_H+V_{xc}+\langle\delta v\rangle_{WS}, \tag{2}$$

with border conditions

$$P_{nl}(0)=0, \qquad P_{nl}(\infty)=0, \tag{3}$$

where V_{KS} is the effective Kohn-Sham potential, V_H and V_{ion} are electrostatic potentials arising from the electronic density and the jellium, respectively, V_{xc} is the exchange-correlation potential. In this section all electrostatic potentials are potentials for electrons, so they are multiplied by −1.

Equation (1) is solved self-consistently using simple iteration method, at this a new value for electron density is obtained by mixing with values from previous iterations. A number of computed radial wavefunctions P_{nl}, i.e a number of filled shells is much less than a number of electrons, for example, there are only ~1000 shells for 10^5 electrons.

In the stabilized jellium model there is a correction for the effective Kohn-Sham potential $\langle\delta v\rangle_{WS}$, which is the difference between potential of the homogeneously charged ball with ion charge V(r) and the model pseudopotential averaged over the volume per ion Z/n_{ion}

$$\langle\delta v\rangle_{WS}=\frac{3}{4\pi r_0^3}\int_0^{r_0} dr 4\pi r^2[\omega(r)-V(r)], \tag{4}$$

$$r_0=Z^{1/3}r_s=Z^{1/3}\left(\frac{3}{4\pi n_{ion}}\right)^{1/3}, \tag{5}$$

where Z is the ion charge, n_{ion} is the positive charge concentration, V(r) is the potential of the homogeneously charged ball, $\omega(r)$ is the model pseudopotential of interaction between an electron and an ion with charge Z:

$$\omega(r)=\begin{cases}-Z/r, & r>r_c\\ 0, & r<r_c\end{cases} \tag{6}$$

Further when we talk about the electrostatic potential in the jellium model it is the sum:

$$V_{ES}=V_{ion}+V_H+\langle\delta v\rangle_{WS}, \tag{7}$$

For computations an own code was used, which is specially created for spherically symmetric calculations using DFT. The program is checked by reproduction of results in papers [4] and [5]. Computations were carried out on grid with 5000 points and a border condition $P_{nl}(R_{max})=0$, where R_{max} is several times more than the bubble radius, instead of the condition at infinity (3). In this work only calculations with the homogeneous ion distribution are considered. In the stabilized jellium model it looks like this: the ionic jellium density, which is also the positive charge density, equals the average density $\bar{n}_e=N(3/4\pi R^3)$ inside the sphere and 0 outside.

2.2. *Discrete ion model*

For computations with point ions Octopus [6] code and CP2K [7] code were used with the Hartwigsen-Goedecker-Hutter (HGH) [8] pseudopotential for ions. The modeling was carried out on three-dimensional cubic grid with spacing 0.2-0.4 a.u. (atomic units or bohr radius). In every case for the point ion model series of computations were carried out with different positions of ions, which are randomly homogeneously distributed inside the sphere. Distribution of ions inside the sphere is presented in figure 1. At the same time, all ions are fixed and do not change their position, regardless of the forces acting on them.

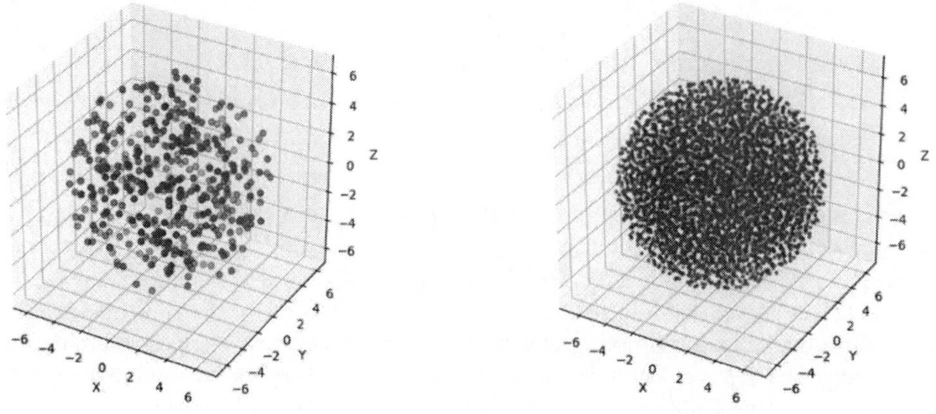

Figure 1. Distribution of 450 (a) and 4500 (b) ions inside the sphere of radius 6.6 a.u.

So far as output of Octopus code for some quantity is the set of values on three-dimensional grid and electron density such as potentials are very inhomogeneous, to get the radius dependence and to compare the results to the jellium model averaging over directions is needed. Averaging over directions for potential carried out with weights equal to density to the 1/3 power.

$$n_e(r_k) = \frac{\int_{|r|-r_k<a/2} n_e(r)\mathrm{d}^3r}{\int_{|r|-r_k<a/2} \mathrm{d}^3r}, \tag{8}$$

$$V_{\mathrm{ES}}(r_k) = \frac{\int_{|r|-r_k<a/2} V_{\mathrm{ES}}(r){n_e}^{1/3}(r)\mathrm{d}^3r}{\int_{|r|-r_k<a/2} {n_e}^{1/3}(r)\mathrm{d}^3r}, \quad r_k = ka, \tag{9}$$

where k is a nonnegative integer, step a selected to be equal to the step of three-dimensional grid, which was used for calculation. In this way we get average values in several points. However, as a result of such averaging values at points near the center have large deviations, which depend on ion positions. Density and potentials obtained are very inhomogeneous and their values near the center depend strongly on whether the ion has got there or not as a result of a random distribution. This is connected to the fact, that averaging for every r_k is made over the area with volume $\sim r_k^2$ and near the center this area is very small.

With homogeneous random distribution of ions inside the sphere some of them could be very close to each other, as a result there would be large inhomogeneities of the electrostatic potential. The ion distribution should corresponds to the correlation function, so ions can be placed keeping the minimal distance between them as a simple analogue to such function.

Since individual ions do not act on themselves, to get the potential acting on ion we should exclude the potential of this ion from the electrostatic potential. For this the potential of the closest ion is excluded from the electrostatic potential in every point of three-dimensional grid. In figures 2–3 shown distribution of the full electrostatic potential and of the potential acting on individual ions in hydrogen bubbles of radius 6.6 a.u. with 450 and 4500 ions, that corresponds to density $2{,}5\cdot10^{30}$ m^{-3} and $2{,}5\cdot10^{31}$ m^{-3}.

IOP Conf. Series: Journal of Physics: Conf. Series **1009** (2018) 012012 doi:10.1088/1742-6596/1009/1/012012

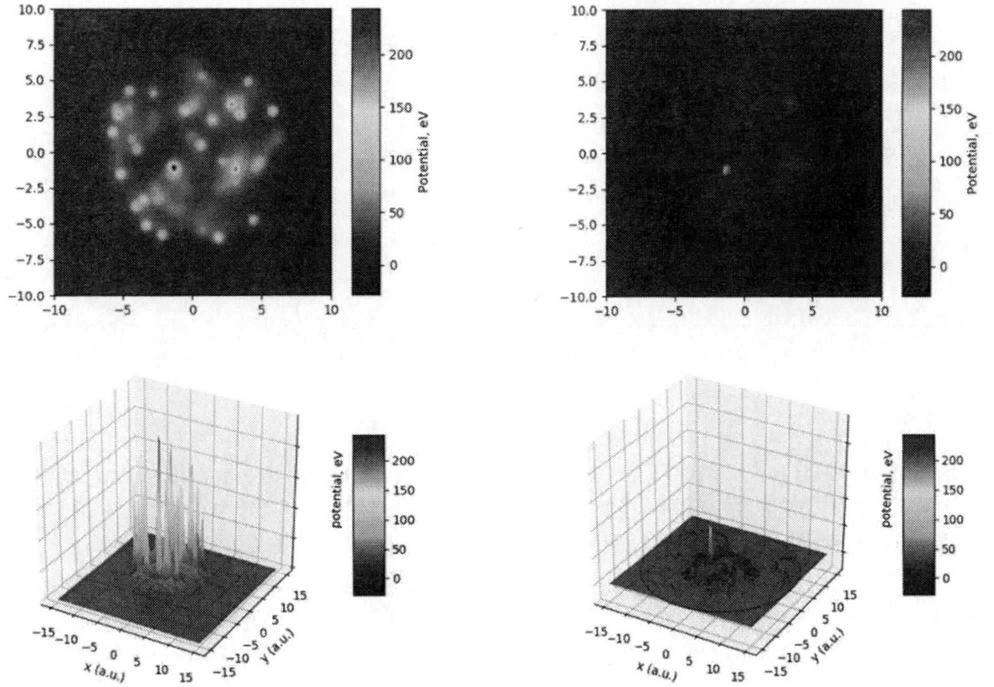

(a) (b)

Figure 2. Distribution of the full electrostatic potential and potential, which acts on point ions, in the plane XY intersecting centre of the bubble with the radius 6.6 and the number of particles 450 shown two different ways.

Calculation for 450 ions was made with completely random ion distribution, but for 4500 with minimal distanse between ions, which is close to the average distance between them. In the first case there are large inhomogeneities, in the second they are smaller. Also extracting a potential of the closest ion in every point a potential dispersion become less as well as value difference between the nearest points.

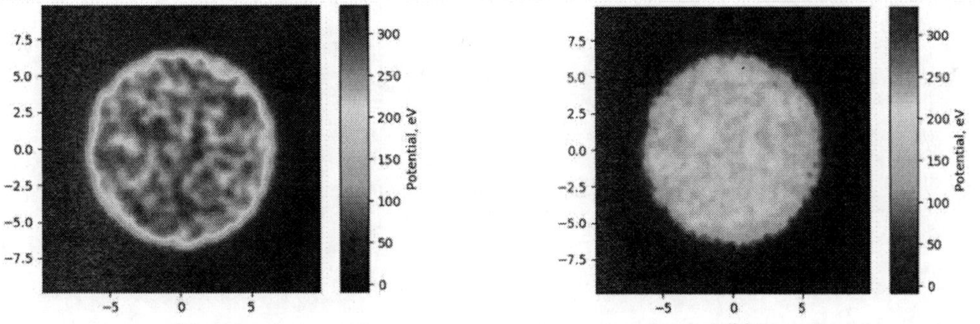

(a) (b)

Figure 3. The same as figure 2 but for 4500 particles.

Figure 4 shows full electrostatic potentials and potentials acting on point ions averaged by directions, which are obtained using Octopus code with point ions for hydrogen bubble of radius 6.6 bohr and with 450 particles. There are displayed results for 15 random configurations of ions inside the sphere averaged using formulas (8), (9) and also its average value. One of these configurations was shown in figure 2.

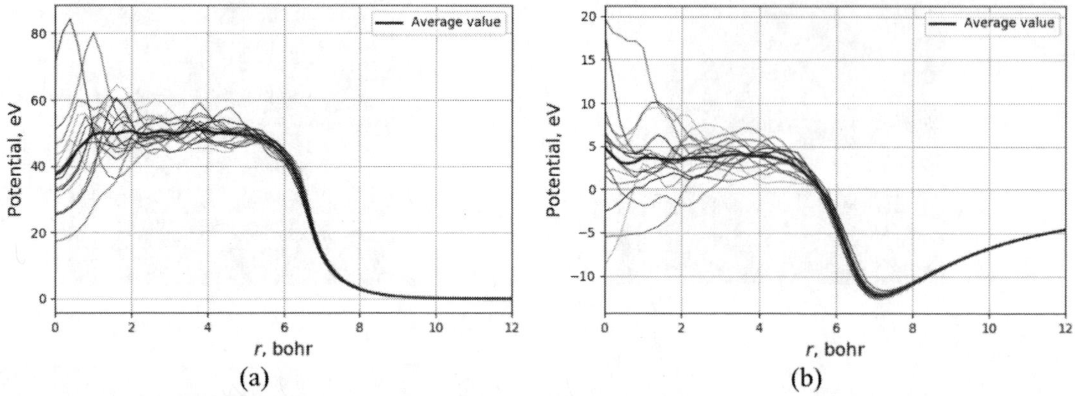

Figure 4. The full electrostatic potential (a) and the potential excluding contribution of the closest ion (b) in hydrogen bubble of radius 6.6 a.u. and with 450 particles obtained using code Octopus. Potentials for all ion configurations shown separately by thin lines, average value over 15 configurations shown by thick line.

The electrostatic potential for ions averaged over directions differs from the full potential by a constant value. So further the full electrostatic potentials will be used for a comparison with the jellium model.

3. Comparison to the jellium model

In figure 5 the results shown for the electron density and the electrostatic potential obtained from the stabilized jellium model are compared to the results obtained from the point ions model using Octopus and CP2K codes. Averaging for the point ions model was made over 5 and 20 configurations, respectively. The computations were made for the hydrogen bubble of radius 17 a.u. with 450 particles. There were used two different programs for the point ions model to check the correctness of the results.

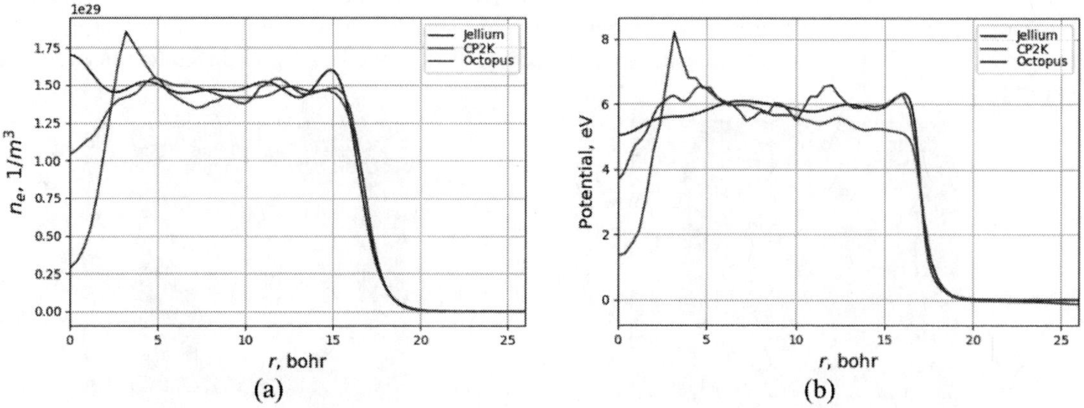

Figure 5. The electron density (a) and the electrostatic potential (b) in the jellium model compared to the results obtained for the point ions model using Octopus and CP2K codes.

In figures 6–8 shown the results for systems of radius 8.5 and 6.6 a.u. with 450 ions and also of radius 6.6 a.u. with 4500 ions, obtained for the jellium model and the point ions model using Octopus code. In case of the point ions model computations were carried out for 5, 15 and 6 ion configurations, respectively.

IOP Conf. Series: Journal of Physics: Conf. Series **1009** (2018) 012012 doi:10.1088/1742-6596/1009/1/012012

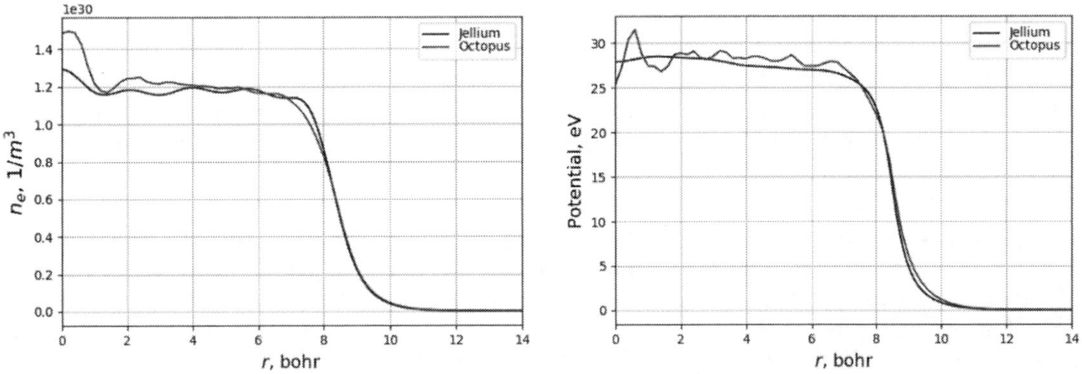

(a) (b)

Figure 6. The electron density (a) and the electrostatic potential (b) in the jellium model compared to the point ions model for the hydrogen bubble of radius 8.5 a.u. with 450 ions, that is corresponds to density $1{,}2{\cdot}10^{30}$ m^{-3} .

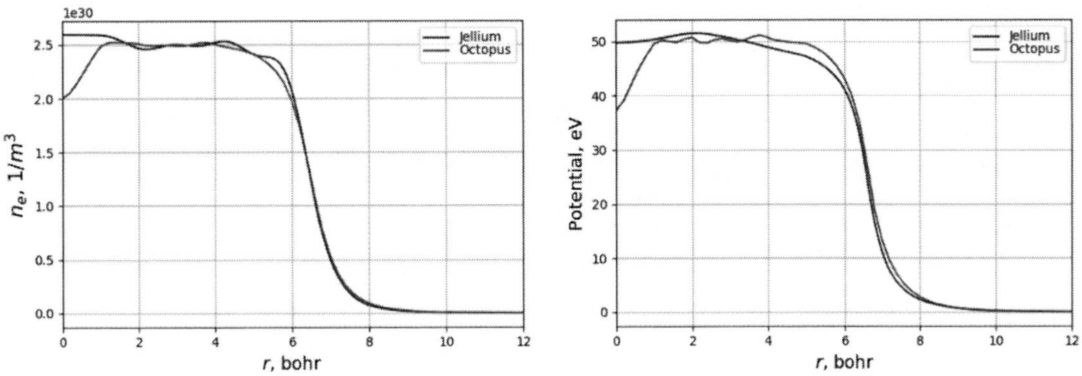

(a) (b)

Figure 7. The same as figure 6 but for radius 6.6 a.u.

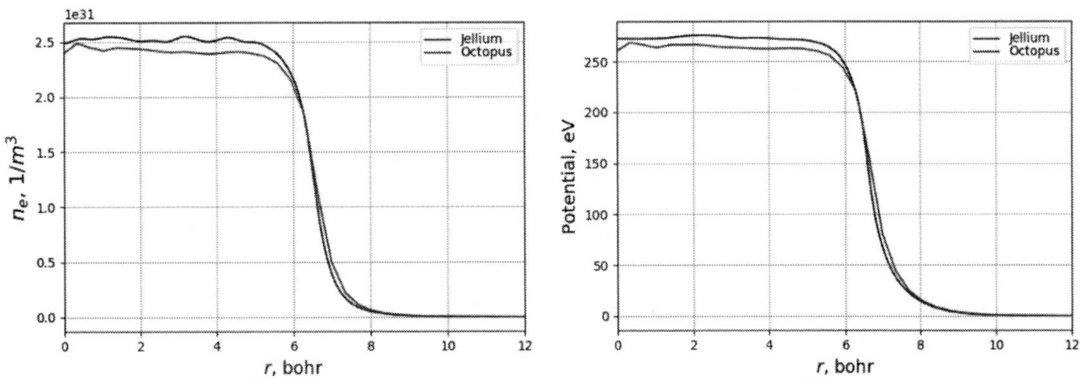

(a) (b)

Figure 8. The same as figure 6 but for radius 6.6 a.u. and with 4500 ions.

As we can see average values in the jellium model and in the point ions model are in a good agreement with each other. The deviations near the system center for the point ions model are connected with the fact that averaging in the center made over small volume. Distributions of the electron density and potentials are very inhomogeneous, so average value near the center strongly depends on specific ion configuration in this area. However, in the case of the point ion model oscillations of density and potential have random behavior connected with ion distribution. So determine oscillations, which are presented in the jellium model, are not visible in the results for the point ions model. Probably, averaging over bigger number of ion configurations is needed.

4. Conclusion

The numerical researches carried out show that a simple model of the stabilized jellium is quite useful for calculations of the electron density and the electric field in such spherically symmetric systems. One-dimensional computations in the jellium model require much less computational resources than three-dimensional computations with point ions. That fact allows to carry out calculations for systems with large number of electrons (about 10^5) in less time than three-dimensional require. In this way the jellium model could be used to calculate the electrostatic potential, which acts on ion motion during the gas bubbles of nanometer and micron size compression.

Acknowledgements
This work was supported by a grant of the President of the Russian Federation to support the leading scientific schools NSh-6579.2016.2.

References

[1] Perdew J P, Tran H Q and Smith E D 1990 *Phys. Rev. B* **42(18)** 11627–11636
[2] Kohn W and Sham L J 1965 *Phys. Rev.* **140(4A)** A1133–A1138
[3] Perdew J P and Zunger A 1981 *Phys. Rev. B* **23(10)** 5048–5079
[4] Ekardt W 1984 *Phys. Rev. B.* **29** 1558–1564
[5] Koch E and Gunnarsson O 1996 *Phys. Rev. B.* **54** 5168–5177
[6] Andrade X *et al.* 2015 *Phys. Chem. Chem. Phys.* **17(47)** 31371–31396
[7] VandeVondele J *et al.* 2005 *Comput. Phys. Commun.* **167** 103–128
[8] Hartwigsen C, Goedecker S and Hutter J 1998 *Phys. Rev. B* **58(7)** 3641–3662
[9] Shidlovski D S 2017 *Journal of Physics: Conference Series* **798** 012093
[10] Perdew J P 1995 *Progress in Surface Science* **48** 245–259
[11] Kiejna A 1999 *Progress in Surface Science* **61** 85–125

Inhomogeneous Spatial Distribution of Electrons in a Compressed Gas Bubble of Submicron Size

S E Kuratov[1], **D S Shidlovski**[1], **S I Blinnikov**[1]

[1] Dukhov Research Institute of Automatics (VNIIA), Sushchevskaya str. 22, Moscow, 127055, Russia

E-mail: ser.evg.kuratov@gmail.com, dmitry.shidlovski@yandex.ru

Abstract. The inhomogeneous spatial distribution of the macroscopic amount of electrons (N $<10^9$) in a spherical potential well is considered. It is shown that the electron distribution has two spatial scales. The first scale is the Fermi length. By effect, we mean the existence of the second scale, which is related with the radius of the potential well and much larger than the first scale. Two systems are considered: a system of free electrons in a spherical well and electrons in a compressed gas bubble. Based on the analysis of the distribution function of a system of free semiclassical electrons, we demonstrate the cause of the appearance of the spatial heterogeneity. This result is confirmed by a numerical summation of exact solutions of the wave functions of electrons in an infinite potential well. It is shown that there is an analogous effect for the spatial distribution of electrons in compressed hydrogen gas bubble of submicron size (<0.1 μm). This system can be realized in the thermonuclear experiments. Numerical simulation of the electron distribution was carried out by the DFT method. The limiting factors of the effect are considered. The aim of problem considered in the article deals with the analysis of possible physical processes occurring in the thermonuclear experiments.

Keywords: spatial distribution, hydrogen gas bubble, DFT method.

1. Introduction

We analyze the inhomogeneous spatial distribution of electrons in a strongly compressed gas bubble (H_2, D_2) of submicron size. The characteristic values of the thermodynamic quantities of the compressed gas are the following: $\rho_{gas} \sim (10\text{-}30)$ g / cm^3, Ne $\sim 10^{30\text{-}31}$ 1 / m^3, $E_F \sim (20\text{-}100)$ eV, $T_i \sim T_e \sim (0.1\text{-}1)$ eV. All electrons are ionized and degenerated, and the ions are the classical nonideal gas.

This article demonstrates that there is an inhomogeneous spatial distribution of electrons under certain conditions. The characteristic heterogeneity scale is of the same order of the well bubble radius magnitude. The origin of this heterogeneity is the influence of quantum shell effects affecting the electron gas.

It is known, that the density of the electron distribution has the oscillating behavior because of the presence of shell corrections in a spherically symmetric potential. For example, in an atom the electron density distribution has an oscillatory character. The deviation from the average value of the distribution density can reach up to 20% [1]. In the atom, the number of electrons is relatively small (N $<10^3$).

IOP Conf. Series: Journal of Physics: Conf. Series **1009** (2018) 012013 doi:10.1088/1742-6596/1009/1/012013

In the present article it will be shown that the heterogeneity of the distribution is also present for a macroscopic number of electrons ($N < 10^9$) in a spherical potential well.

Section 1 of the article is devoted to the theoretical analysis of the distribution of a system of free semiclassical electrons in a spherical potential well. This analysis shows the cause of the appearance of the heterogeneity in the system. It also demonstrates that the spatial scale of the heterogeneity is of the same order of the well radius and can be greater than the distance between the particles by several orders.

Section 2 presents the effective computational technique that allows to calculate the electron density by numerical summation of exact solutions for the wave functions of free electrons in a potential well. The calculations of various systems with the electron number of up to 1 billion particles are presented in this article. These results confirm the existence of the effect.

Section 3 presents the results of DFT calculations of the electron distribution in a compressed gas bubble. We used the jellium model for hydrogen clusters (the number of atoms from 4000 to 100,000). In the conclusion we consider the limiting factors of the effect.

2. A system of free semiclassical electrons in an infinite spherical well. The theoretical analysis

It is important to analyze the distribution of free semiclassical electrons. All electrons are degenerated in this system and their Fermi energy is greater than the thermal and Coulomb energies. Also, electrons have significant orbital angular momentum and therefore they have semiclassical behavior. So this task setting is close to the real situation. The solution of the problem gives us the possibility to obtain analytical dependences of the electron distribution on the system parameters. It is important for the interpretation of the numerical results.

The semiclassical approach was widely used for the analysis of metal clusters [2], for the calculation of the nuclei energy spectrum [3], for the calculation of the electron concentration oscillations in the atom.

To determine the electron concentration, we use the Green's function representation $G(r'',r',e)$ for electrons using the semiclassical approximation [3].

$$G(r'',r',e) = G_0 - \frac{1}{(2\pi h^5)^{1/2}} \sum_\alpha \left\{ p_\rho D^{1/2} \exp\left(\frac{i}{h} S_\alpha(r'',r',e) + iv - \frac{i\pi}{4}\right)\right\}_\alpha \quad (1)$$

S_α- classical action integral

$$S_\alpha = \int_{r'}^{r''} p_\alpha \, \mathrm{d}l_\alpha \quad (2)$$

$$G_0(r'',r',e) = -\frac{m}{2\pi h^2 |\boldsymbol{r}'' - \boldsymbol{r}'|} exp\left\{\frac{i}{h} |\boldsymbol{r}'' - \boldsymbol{r}'| p(r)\right\} \quad (3)$$

$$\boldsymbol{r} = \frac{\boldsymbol{r}' + \boldsymbol{r}''}{2}, \boldsymbol{p}' = \frac{\mathrm{d}S_\alpha(\boldsymbol{r}',\boldsymbol{r}'',e)}{\mathrm{d}\boldsymbol{r}'}, \boldsymbol{p}'' = \frac{\mathrm{d}S_\alpha(\boldsymbol{r}',\boldsymbol{r}'',e)}{\mathrm{d}\boldsymbol{r}''} \quad (4)$$

$$D = det\begin{bmatrix} \frac{\mathrm{d}p'_\rho}{\mathrm{d}\rho''} & \frac{\mathrm{d}p'_\rho}{\mathrm{d}z''} & \frac{\mathrm{d}p'_\rho}{\mathrm{d}e} \\ \frac{\mathrm{d}p'_z}{\mathrm{d}\rho''} & \frac{\mathrm{d}p'_z}{\mathrm{d}z''} & \frac{\mathrm{d}p'_z}{\mathrm{d}e} \\ \frac{\mathrm{d}t_\alpha}{\mathrm{d}\rho''} & \frac{\mathrm{d}t_\alpha}{\mathrm{d}z'} & 0 \end{bmatrix} \quad (5)$$

$$t_\alpha(r',r'') = \frac{dS_\alpha(r',r'',e)}{de} = \int_{r'}^{r''} \frac{1}{|\dot{\boldsymbol{r}}|} dl_\alpha \tag{6}$$

In (1) the summation is extended over all real classical trajectories α connecting r' and r'', t_α is the time of the motion along the trajectory α.

The electron concentration is determined from the following expression

$$n(r) = -\frac{2}{\pi} \int_{-\infty}^{\varepsilon_F} ImG(r,r,e)de \tag{7}$$

In our case, to calculate the electron concentration (integral in (7)), it is necessary to take into account a variety of non-periodic orbits having coincident initial and final coordinates. The examples of such trajectories are shown in figure 1.

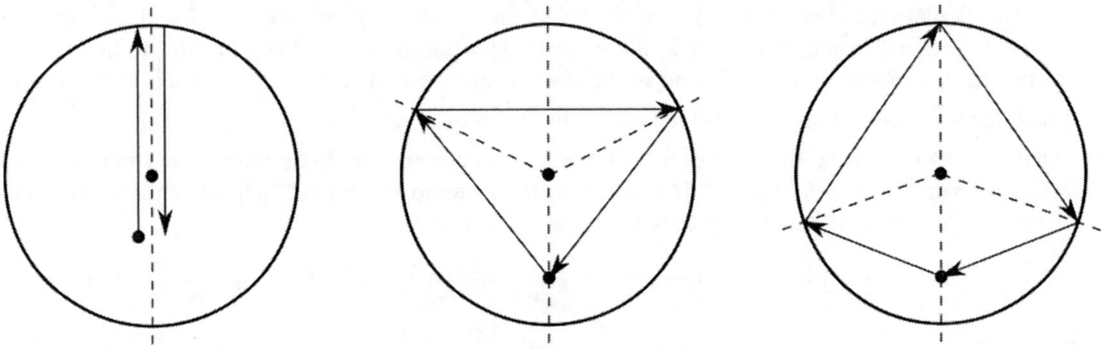

Figure 1. The non-periodic trajectories of the electron.

To determine the cause of the appearance of the spatial heterogeneity we estimate the expected effect by the order of the magnitude.
In our case

$$S_\alpha \sim pR_0L_\alpha\left(\frac{r'}{R_0},\frac{r''}{R_0}\right), \tag{8}$$

where $L\alpha$ is the length of the trajectory α. So

$$D \sim \frac{m^2}{R_0 p} F\left(\frac{r'}{R_0},\frac{r''}{R_0}\right). \tag{9}$$

The expression for Green function has the following form:

$$G(r'',r',e) \sim \frac{1}{h^{\frac{5}{2}}} p \left(\frac{m^2}{R_0 p}\right)^{1/2} \left(F\left(\frac{r'}{R_0},\frac{r''}{R_0}\right)\right)^{1/2} \exp\left(\frac{i}{h} pR_0L_\alpha\left(\frac{r'}{R_0},\frac{r''}{R_0}\right)\right) \tag{10}$$

$$\Delta n(r) \sim \int_{-\infty}^{\varepsilon_F} ImG(r,r,e)de \tag{11}$$

For the integral calculation, we use the large value ($\sim 10^3$) of the exponent in (10). We obtain the following:

$$\Delta n(r) \sim Im\left[-i\frac{\left(F\left(\frac{r'}{R_0},\frac{r''}{R_0}\right)\right)^{1/2}}{L_\alpha\left(\frac{r'}{R_0},\frac{r''}{R_0}\right)}\left(\frac{\mathrm{n}}{\mathrm{R}_0^3}\right)^{1/2}\exp\left(\frac{i}{h}pR_0L_\alpha\left(\frac{r'}{R_0},\frac{r''}{R_0}\right)\right)\right] \tag{12}$$

$$\frac{\Delta n(r)}{n} \sim \left(\frac{1}{\mathrm{R}_0^3\mathrm{n}}\right)^{1/2}\sin\left(\frac{1}{h}p_FR_0L_\alpha\left(\frac{r'}{R_0},\frac{r''}{R_0}\right)\right) \tag{13}$$

To analyze the heterogeneity scale of the spatial distribution of the electron we calculate the electric field potential

$$\varphi(r) = \frac{4\pi e\int_0^r \Delta n(r)r^2\mathrm{d}r}{4\pi\varepsilon_0 r} \tag{14}$$

To calculate the electric field potential we use the stationary phase method. There are two possible cases. In the first case the function L_α has no stationary points on the interval of integration, and so the potential has the obvious oscillating behavior. So the spatial distribution of the electron has one heterogeneity scale. This scale is the Fermi length $\frac{1}{h}p_F\mathrm{r}$.
In the second case, the function L_α has stationary points. So the distribution of the potential has the second spatial scale relating with trajectory having a minimal length. Such trajectory is presented in figure 2. The value of the second spatial scale is order of $R_0/2$

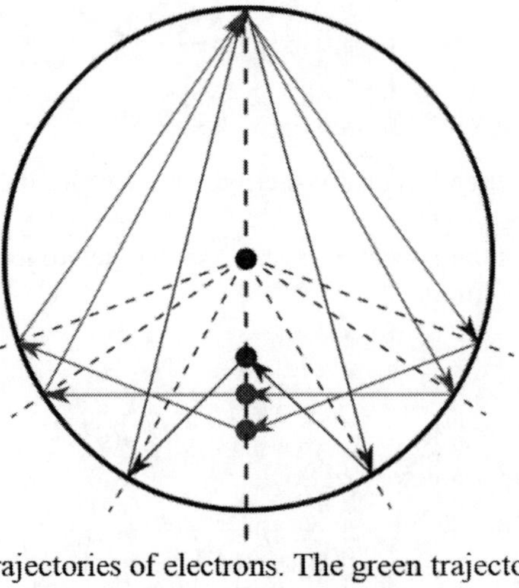

Figure 2. Some possible trajectories of electrons. The green trajectory has a minimal length.

For the electric field potential we have the following estimation:

$$\varphi(r) \sim \frac{e}{\varepsilon_0 R_0}\left(R_0 n_0^{1/3}\right) \tag{15}$$

3. Numerical calculation of spatial distribution of free electrons in the infinite spherical potential well

To simulate small plasma bubbles with the electron number up to $N=10^8$ we consider a very simple model. This model consists of the spherical potential well with infinite walls and free electrons. All electrons do not interact with each other. The distribution of electron density and potential is very similar to those in small metal clusters [4], where the calculations were carried out for the jellium model (SJBM). But calculations for metal clusters using jellium model were performed earlier only for the number of electrons up to 6000 [5]. The model with an infinite potential well is very crude and can misinterpret the behavior electrons, so these two models must be compared for a large number of electrons.

If one considers a large number of N electrons in a cubic box, then their distribution quickly goes to the semiclassical regime and turns out to be uniform. In contrast, in a spherically symmetric potential in many situations, the density of the electron distribution shows an oscillating behavior along the radius. This can be seen in the solutions by Hartree-Fock, Hartree-Fock-Dirac methods, and in Thomas-Fermi approximation [1,2] and in DFT [4]. Such oscillating behavior is observed, not only for electrons, but also for other particles, and for Coulomb as well as for other interaction potentials between particles (Yukawa potential, hard core, etc.) [6]. In such a situation it is interesting to explore how the quasi-classical regime will proceed with growth of N, how will oscillations behave in the radial distribution of electrons, and accordingly, what orders of electric field may be expected for a local violation of the electrical neutrality of the plasma.

In this section we investigate the distribution of N electrons in the ground state in a spherically symmetric potential well with impermeable (i.e. infinitely high walls). This problem allows a complete analytical solution (see, for example, books by Fluegge and Messiah [7, 8]). The transition to a large number of $N \gtrsim 10^8$ electrons in this problem allows one to trace in detail the oscillating behavior of the electron density in the ground state, to see the difference from the quasi-classical Thomas-Fermi solution and to estimate the magnitude of the electric field, which can arise during adiabatic compression of a cavity containing a degenerate plasma with a similarly large number of electrons. Adiabaticity will be understood here in the spirit of P. Paradoxov [9].

In the subsection 3.1, the starting point of the problem is the statement of the problem in a spherically symmetric potential well, and a long-known exact solution is written down for a well with infinitely high walls. Examples of the oscillating behavior of the electron density are given in the section 3.2. The section 3.3 illustrates the possible distribution of the electrical potential fields under different assumptions about the concentration of a homogeneous distribution ions, partially compensating the field of electrons.

3.1. Solution for the wave function of an electron in a spherical well with impermeable walls

Here we reproduce the notation from Fluegge's book [7]: Schrödinger equation in the problem of determining the energy levels of a particle in a spherically symmetric potential. A box with infinite walls and a radius R admits separation of variables, and the solution can be written in the form

$$u(r,\theta,\varphi) = \frac{1}{r}\chi_{n_r,l}(r)Y_{l,m}(\theta,\varphi), \tag{16}$$

where $Y_{l,m}(\theta,\varphi)$ is a spherical function. Radial wave function χ satisfies the equation

$$\chi''_{n_r,l} + \left[k^2 - \frac{l(l+1)}{r^2}\right]\chi_{n_r,l} = 0, \quad k^2 = \frac{2mE}{\hbar}, \tag{17}$$

The function $\chi(r)$ corresponds to different energy levels E in the dependence from the orbital angular momentum l and the radial quantum number n_r, since k will be fixed by the boundary conditions. This equation has a solution $\chi_{n_r,l}(r) = C_{n_r,l}\, r j_l(r x_{n_r,l}/R)$, where $j_l(x)$ is called a spherical Bessel function of the order l, $x_{n_r,l}$ are roots of the Bessel function $J_{l+1/2}(x)$, $C_{n_r,l}$ is a normalization constant.

The expression for the electron density for the filled shells:

$$\rho_e(r) = \frac{1}{4\pi R^3}\sum_{n_r,l} g_e(2l+1)\, j_l^2\left(\frac{x_{n_r,l}r}{R}\right)\left(j_l^2(x_{n_r,l}) - j_{l-1}(x_{n_r,l})j_{l+1}(x_{n_r,l})\right)^{-1}. \tag{18}$$

It immediately shows that if the physical radius of the cavity R is changed for any set of filled states the density varies self-similarly, simply as $f(x)/R^3$, where $f(x)$ is determined by the sum on orbitals in (18).

Potential is computed by the formula

$$V(r) = 4\pi\left[\frac{1}{r}\int_0^r \rho_c(x)x^2dx + \int_r^R \rho_c(x)xdx\right]. \tag{19}$$

Here, the charge density is $\rho_c = |e|(\rho_i - \rho_e)$, the elementary charge is e, the electron density $\rho_e(r)$ was taken from the calculation using the formula (18), and for ions the simplest approximation was applied $\rho_i(r) = \text{const}(r)$. The constant was chosen either from the electroneutrality condition within the radius R, or was chosen so as to minimize the potential difference inside a sphere with radius R.

3.2. Density distribution for various values of radius R

The results for ρ are self-similar when the radius of the cavity R is changed. For illustration, we fix the average value of electron density assuming that inside R we have a deuterium plasma at a density of $5\ \text{g/cm}^3$. The Fermi energy is then 47.9 eV.

Figure 3 shows the density calculated directly from the formula (18) for $R = 3.958$ nm. Figure 4 shows the density calculated from the formula (18) for $R = 39.85$ nm near the outer border on a large scale.

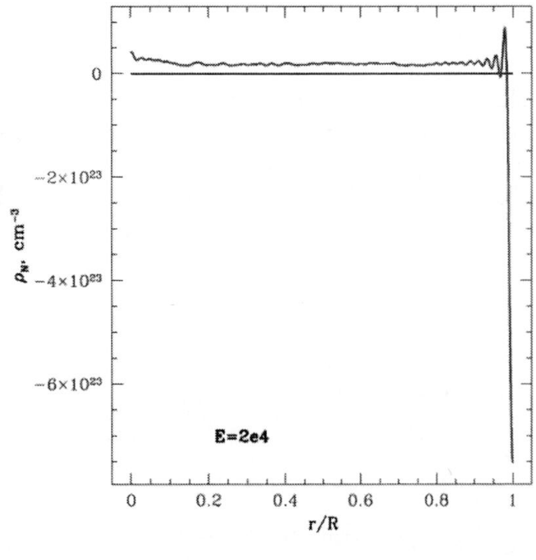

Figure 3. Distribution of the number density of states for $R = 3.958$ nm (dimensionless energy $E = 2 \cdot 10^4$).

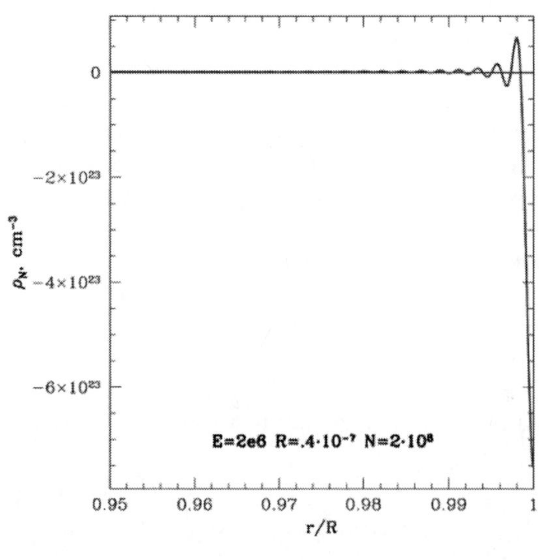

Figure 4. Distribution of the number density of states for $R = 39.85$ nm (dimensionless energy $E = 2 \cdot 10^6$) at the edge.

3.3. Distribution of the potential of electric field

In this section we show the potential distributions $\varphi(r)$.

Figure 5 shows the potential distribution calculated by the formula (19) for $R = 3.958$ nm, homogeneous distribution ρ_i with complete compensation of the electron charge. In this case, the potential difference is large. On such a chart, no oscillations of the potential are seen, which give density oscillations, since density is the second derivative of φ. But some oscillations become visible if we choose the compensating distribution of the ions ρ_i, where the potential difference is minimal - see figure 6. Here ρ_i is increased to compensate for the negative charge in the central zones, since there ρ_e is higher than the mean ρ_e over the entire volume. The value of the excess density, sufficient to minimize the potential difference, in this case, 2.48% of the original density.

IOP Conf. Series: Journal of Physics: Conf. Series **1009** (2018) 012013 doi:10.1088/1742-6596/1009/1/012013

From these graphs, we see in one case a potential difference of order ~15V, in another ~1.5kV. What kind of potential difference can be expected in reality? This is the most important issue of the problem. It is necessary to understand that we have here a strong idealization: in a box with high mirror walls is filled with N electrons that have no potential except the walls to feel. This means, that there are N of classical singly charged ions, which compensated all the charges, and there are no potential differences!

If the ions were distributed strictly uniformly, compensating only full charge of electrons, it would be great potential, as in the figure 5. There is the potential well for ions in the center: there is an excess of electrons, and they pull positive charges toward the center.

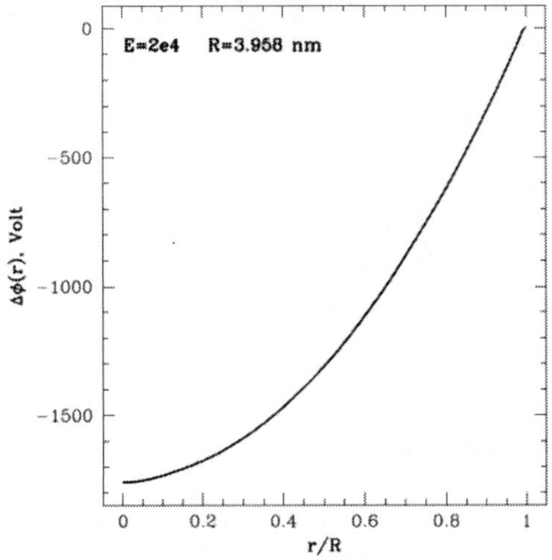

Figure 5. Distribution of potential for $R = 3.958$ nm (dimensionless energy $E = 2 \cdot 10^4$) in case of uniform distribution ρ_i with full compensation of electron charge.

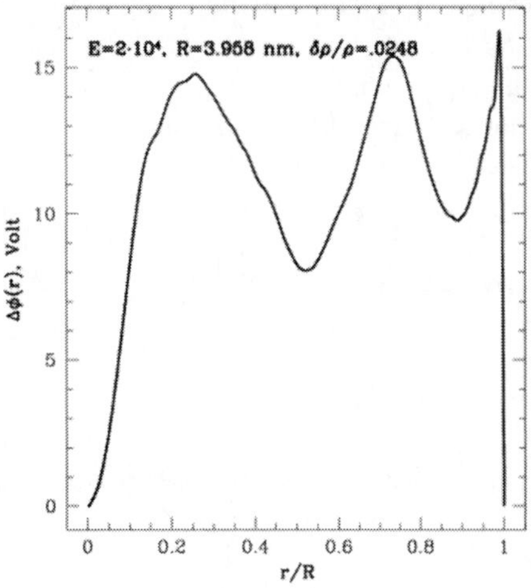

Figure 6. Distribution of potential for the same set as in figure 6, but with ρ_i fitted to minimize the difference of potentials.

4. Numerical simulation of the nonuniform electron distribution in a compressed gas bubble of nanometer size

4.1. Description of the jellium model

Numerical simulation of the electron and potential distribution is made using density functional theory (DFT) in the spherical jellium background model (SJBM). In the jellium model ions are represented as a continuous fixed distribution of a positive charge. In this form the spherically symmetric distribution of ions and their potential is possible, that allows to reduce the problem to one-dimensional. Here the jellium model modification is used, which is so called stabilized jellium model [10,11] with a correction, that takes into account an average difference between the jellium and point ions.

When the external potential has spherical symmetry, as in section 3.1, one-electron wavefuction can be decomposed as (16) to a radial $\chi_{n_r,l}(r)$ and a spherical $Y_{l,m}(\theta,\varphi)$ wavefunction, if we suppose that the electron distribution also has a spherical symmetry. For the radial wavefunction Kohn-Sham equation [12] has the following form (in atomic units):

$$\left(-\frac{1}{2}\frac{\mathrm{d}^2}{\mathrm{d}r^2} + V_{\mathrm{KS}} + \frac{l(l+1)}{2r^2}\right)\chi_{n_r,l}(r) = E_{n_r,l}\chi_{n_r,l}(r), \tag{20}$$

$$V_{\mathrm{KS}} = V_{\mathrm{ion}} + V_{\mathrm{H}} + V_{\mathrm{xc}} + \langle\delta v\rangle_{\mathrm{WS}} = V_{\mathrm{ES}} + V_{\mathrm{xc}}, \tag{21}$$

where V_{KS} is the effective Kohn-Sham potential, V_{H} and V_{ion} are electrostatic potentials arising from the electronic density and the ionic jellium, respectively, V_{xc} is the exchange-correlation potential,

$\langle \delta v \rangle_{\mathrm{WS}}$ is the stabilized jellium correction. In this section all electrostatic potentials are potentials for electrons, so they are multiplied by -1.

Equations for the single electron wavefunction are solved self-consistently using simple iteration method, at this electrons with equal values of quantum numbers n and l have the same radial wavefunctions, so electron density equals

$$n_{\mathrm{e}}(r) = \sum_{n_r,l} N_{n_r,l} \frac{P_{n_r,l}^2(r)}{4\pi r^2}. \tag{22}$$

Here $N_{n_r,l}$ is a number of electrons in every shell, for fully occupied shells $N_{n_r,l} = 2(2l+1)$, for partially occupied, if there are more than one, such values used, that their energy eigenvalues $E_{n_r,l}$ will be equal. As a result, the number of the single electron wavefunction, that should be found, is much less than the electron number.

For spherically symmetric computations in the stabilized jellium model an own code was used [13], which was solving one-dimensional Kohn-Sham equations with the exchange-correlation functional PZ (Perdew, Zunger) [14] in the local density approximation (LDA).

In this work only calculations with the homogeneous ion distribution are considered. In the stabilized jellium model it looks like this: the ionic jellium density, which is also the positive charge density, equals the average density $\bar{n}_{\mathrm{e}} = N(3/4\pi R^3)$ inside the sphere and 0 outside.

4.2. Calculation results

A series of calculations was carried out using the jellium model for ionized hydrogen bubbles with different density values of homogeneous ionic jellium and radius values. The results are obtained for density 10^{30}, 10^{31} and 10^{32} m^{-3}, three radius values used for each, parameters are listed in table 1.

Because the ionic jellium density is constant and approximately equal to the average electron density in a bubble, as the electron density oscillations we can use difference between the electron density and the ionic jellium density $\Delta n = n_{\mathrm{e}} - n_{\mathrm{ion}}$.

To get the electrostatic potential oscillations ΔV_{ES} we can, for example, extract from V_{ES} average value, which is approximately expressed in terms through E_{F}:

$$\Delta V_{\mathrm{ES}} = V_{\mathrm{ES}} + V_{\mathrm{xc}}(n_{\mathrm{ion}}) + \left(E_{\mathrm{F}}(n_{\mathrm{ion}}) - E_{n_r,l}^{\max}\right), \tag{23}$$

since

$$V_{\mathrm{ES}} + V_{\mathrm{xc}} = V_{\mathrm{KS}} \approx -E_{\mathrm{F}}. \tag{24}$$

Figures 7–9 show the difference between the electron density and the ionic jellium density Δn and also the electrostatic potential oscillations, obtained using formula (23), for one radius value at each density.

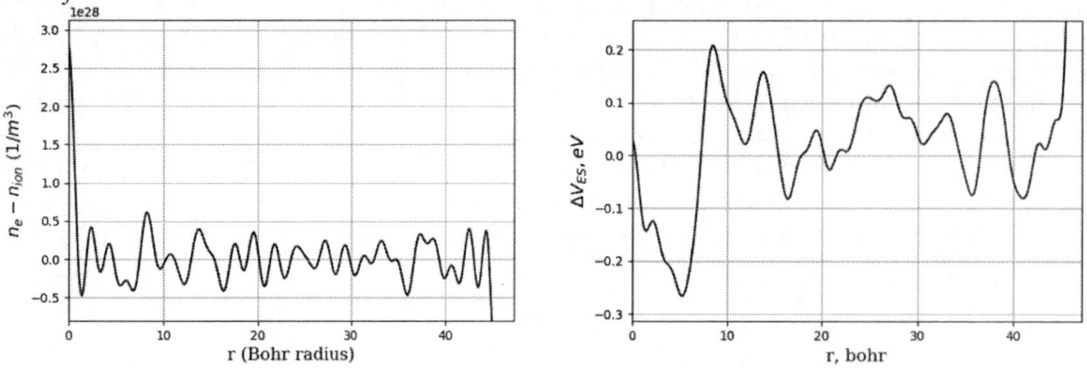

(a) (b)

Figure 7. The difference between the electron density n_{e} and the jellium density n_{ion} (a), the electrostatic potential oscillations ΔV_{ES} (b) for density n_{ion} equal to 10^{30} m^{-3} and radius 2.5 nm.

In graphs we can see, that, if we do not take into account fluctuations close to the center, double amplitudes of the electron density oscillations are in a good agreement with values, which can be obtained from (13):

$$\frac{2\Delta n}{n} \sim \frac{2}{(R_0^3 n)^{1/2}} = 2\left(\frac{4\pi}{3N}\right)^{1/2} \approx \frac{4}{N^{1/2}}. \tag{25}$$

A comparison of values $2\Delta n/n$ and $4/N^{1/2}$ for all cases is presented in table 1. For all values the difference is about 2 times, a value of $2\Delta n/n$ is taken approximately from graphs.

As can be clearly seen from the table 1, the amplitude of the electrostatic potential oscillations decrease with a number of electrons increase. Also in many cases the electrostatic potential oscillations can be observed, which have a scale of the order of the system size, unlike small oscillations with a scale of the order of the interatomic distance or in other words h/p_F. Especially this is noticeably for the density 10^{32} m^{-3}.

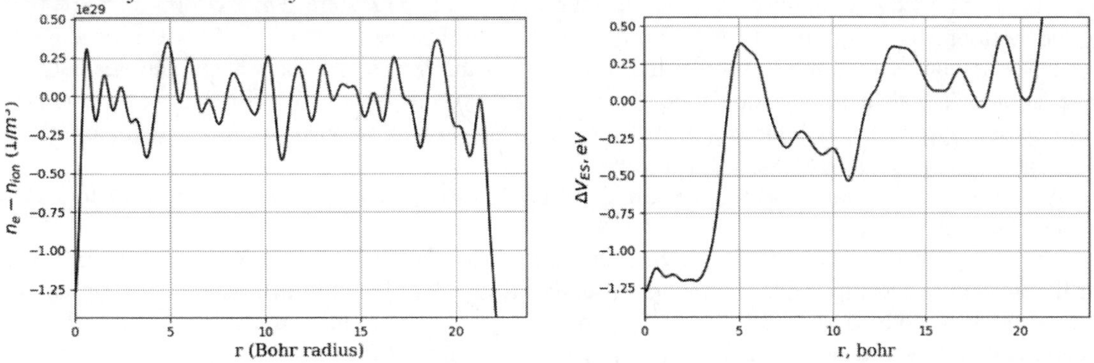

(a) (b)

Figure 8. The same as the previous figure for density 10^{31} m^{-3} and radius 1.25 nm.

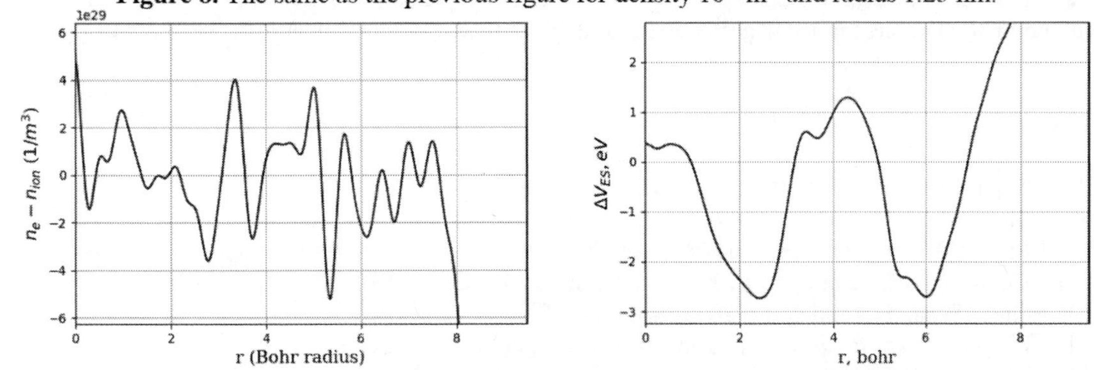

(a) (b)

Figure 9. The same as the previous figure for density 10^{32} m^{-3} and radius 0.5 nm.

Table 1. A comparison of values $4/N^{1/2}$ and double relative oscillation amplitudes $2\Delta n/n$ obtained in the numerical simulation(n is the density in m^{-3}, R is the radius in nm, N is the number of electrons).

n, m^{-3}		10^{30}			10^{31}			10^{32}	
R, nm	1.0	2.0	2.5	0,5	1.0	1.25	0.25	0.5	0.625
N	4188	33510	65450	5236	41888	81812	6544	52360	102266
$4/N^{1/2}$	0.062	0.022	0.016	0.055	0.020	0.014	0.050	0.017	0.013
$2\Delta n/n$	0.030	0.012	0.0085	0.022	0.010	0.0075	0.020	0.009	0.007

5. Conclusion

Analysis of the spatial distribution of electrons in mesoscopic system is presented in this paper. We consider two mesoscopic systems: a macroscopic number of free quasiclassical electrons in a spherical potential well and a compressed gas bubble.

From analysis of first system we show that the electron distribution has the spatial scale which is of the order of the well radius magnitude . The appearance of this effect is caused by influence of the shell corrections in a spherically symmetric potential. We obtained the analytical estimation for magnitude of the spatial heterogeneity of electron distribution

$$\frac{\Delta n(r)}{n} \sim \left(\frac{1}{\mathrm{R}_0^3 \mathrm{n}}\right)^{1/2}$$

The compressed gas bubble of submicron size is more realistic system. This system can be realized in the thermonuclear experiments. Numerical simulation of the electron distribution was carried out by the DFT method. Jellium model used in the study is relevant to describe the realistic physical systems The numerical calculation demonstrate the existence of the spatial distribution of electrons in compressed hydrogen gas bubble too. Account the spatial distribution of electrons is important for description of the thermonuclear experiments.

A number of factors limit and weaken the effect considered in the article. These factors have different origin.

The first of these is associated with the relaxation of the classical ionic system under the influence of the generated electric field. The motion of classical ions will tend to neutralize the arising charge. The estimates, following from an analysis of the gas-dynamic motion of ions, show that the neutralization time is order of 0.2 ns. Essentially this is the time of the effect's existence.

The second factor is associated with the interaction of almost free electrons with ions and with each other. After the collision the electron forgets about the spherical well. Therefore, the analyzed effect is limited by the bubble radius magnitude. The radius estimated in this way is of order of 0.1 μm.

Authors thank Y.E.Lozovik for very helpful comments.

References

[1] Kirzhnits D A, Lozovik Y E and Shpatakovskaya G V 1975 *Soviet Physics Uspekhi* **18** 649
[2] Shpatakovskaya G V 2012 *Physics-Uspekhi* **55** 429
[3] Strutinsky V M and Magner A G 1976 (in Russian) *Physica elementarnich chastic* **7**(2) 356-418
[4] Ekardt W 1984 *Phys. Rev. B.* **29** 1558–1564
[5] Koch E and Gunnarsson O 1996 *Phys. Rev. B.* **54** 5168–5177
[6] Gu F, Wang H J and Li J T 2012 *Phys. Rev. E.* 2012 **85** 056402.
[7] Flugge Z 1974 *Problems in quantum mechanics* (Moscow: Science)
[8] Messia A 1979 *Quantum mechanics* vol. 1 (Moscow: Science)
[9] Paradoxov P 1966 (in Russian) *Soviet Physics Uspekhi* **89** 707–709
[10] Perdew J P 1995 *Progress in Surface Science* **48** 245–259
[11] Perdew J P, Tran H Q and Smith E D 1990 *Phys. Rev. B* **42**(18) 11627–11636
[12] Kohn W and Sham L J 1965 *Phys. Rev.* **140**(4A) A1133–A1138
[13] Shidlovski D S 2017 *Journal of Physics: Conference Series* **798** 012093
[14] Perdew J P and Zunger A 1981 *Phys. Rev. B* **23**(10) 5048–5079

IOP Conf. Series: Journal of Physics: Conf. Series **1009** (2018) 012014 doi:10.1088/1742-6596/1009/1/012014

Modeling and Optimization of Layer-by-Layer Structures

S A Lychev[1,2], G V Kostin[1], K G Koifman[2] and T N Lycheva[1]

[1]Ishlinsky Institute for Problems in Mechanics RAS, Moscow, Russia
[2]Bauman Moscow State Technical University, Moscow, Russia

E-mail: lychevsa@mail.ru

Abstract. In the present paper a differential-geometric approach is developed to modeling of residual stresses in layered (LbL) structures obtained as a result of successive curing of thin layers of material. The objects of modeling are the structures obtained by sequential adsorption of a large number of thin layers. During this assembling, the internal (residual) stresses appear in the multilayered structure while local deformations turn out to be incompatible. This leads to accumulation of residual stresses and distortion of the final shape of the LbL structure. To reduce these factors, geometric compensation is used, the calculation of which is extremely laborious with a large number of layers. Geometric methods allow us to implement fast algorithms for obtaining the compensation. They are based on the "smoothing" of the multilayer LbL structure and its representation by introducing a smooth body-manifold with non-Euclidean connection that characterizes the incompatibility of deformations. Connection is determined from the solution of the evolutionary problem, which formalizes the course of the technological process. The classical fields of the mechanics of a continuous medium are put in correspondence with their non-Euclidean counterparts. As an example, model problems for cylindrical LbL elastic structures are considered. A model problem of structural optimization is solved to determine the optimal strategy of the technological process.

1. Introduction

Recently, the technique of obtaining thin films by layer-by-layer assembling (LbL) has been intensively developing. It was initially aimed at improving the quality of coatings and other nanostructural objects. Lately great advance in understanding the internal interaction between layers and particles have been made. This gave impetus to the development of LbL technologies that nowadays have reached substantial sophistication in structure and variety of components [1, 2]. At present one should expect significant research efforts to be directed at fine-tuning and optimization of complex multilayered structure as well as at design of specific stack architecture for improving their functionalities.

One can describe LbL assembly as the sequential adsorption of positively and negatively charged species or layers that are held in the compound by a wide range of intermolecular interactions, such as ion–ion, Van der Waals, hydrophobic interactions, *etc*. [2]. In this regard the LbL structures as a rule are self-stressed. This feature, from one hand, can be qualified as undesirable, and the efforts of developers should be directed towards reducing internal stresses. From the other hand, one is able use this property of LbL structures for the benefit and design them with a predetermined distribution of internal stresses for improving their specific features. All this leads to the formulation of problems of structural optimization [3, 4]. Note that the approach of small deformations seems to be too rough for the majority of LbL structures and comprehensive modeling can be provided in the framework of nonlinear mechanics only. Moreover one has to have an appropriate tool for modeling incompatible local deformations, similar to them in nonlinear dislocation mechanics. Such a tool is available, and this is the Geometrical Mechanics [5, 6]. The effectivity of differential-geometrical approach was demonstrated in classical works [7–12] as well as in a number of resent investigations [13–16]. The present work is aimed at the development of methods of Geometrical Mechanics in the application to the modeling of LbL structures.

Some words should be said about major ideas of Geometrical Mechanics. In the classical elasticity theory, deformations are defined relatively to a reference shape which is assumed to be free from stresses [17]. As it was already mentioned, LbL structures do not possess such shape, so classical approach cannot be applied for their analysis. It is possible however to find stress-free shape in more general sense. To do this one can define reference shape as an image of considered LbL structure with respect to an embedding into non-Euclidean space with specific, material connection [18–21]. For simple solids this connection becomes affine. This fact has simple geometrical explanation. Indeed it is easy to see formal correspondence between the mathematical description of incompatible deformations and the geometry of spaces endowed with affine connection. In simple materials the transformation of each elementary volume into a uniform state is carried out by a linear transformation. By assumption these transformations are determined by a set of continuous functions. The moving frame method proposed by E. Cartan for constructing of the space with affine connection also uses a smooth field of linear transformations that acts on coordinate frame [22–24]. Thus, these two approaches have much in common and the geometric point of view may be applied for modeling of solids whose local deformations do not satisfy the compatibility conditions, particularly, LbL solids. Connection that is determined in such a way is called material connection [20]. There are two basic ways to introduce a material connection. One can use directly the correspondence between local deformations that brings infinitesimally part of the body into stress-free state and linear transformations that acts on coordinate frame. This approach results in flat space with non-trivial torsion, called Weitzenböck space or teleparallel space [22]. Another way is based on the correspondence between a local metric defined by local deformation and the construction of a Riemannian space by given distribution for the metric. Below we use both approaches. For simple bodies these approaches give equivalent state description because the state of simple material is defined by only symmetrical part of local deformation (in virtue of frame indifference principle [19]).

Since the present work affects the theory of parallelizable manifolds we will briefly discuss here the purely geometric aspects [23, 24]. We have pointed out seven basic concepts and give below brief comments to them.

1°. *Global frame fields.* Let $\mathfrak{M}$ be a n-dimensional smooth manifold. Recall that a global frame field is a n-tuple $(e_i, i=1,\dots,n)$ of vector fields on $\mathfrak{M}$ that are linearly independent at each point $x\in\mathfrak{M}$. It follows from this assumption that they must be globally non-zero. A n-dimensional smooth manifold is said to be parallelizable if one can define a global frame field on it. One can regard global frame field as a global section of the principal fiber bundle $GL(\mathfrak{M})\to\mathfrak{M}$ and, consequently, treat it as a set of n smooth maps $e_i:\ \mathfrak{M}\to GL(\mathfrak{M})$ such that (e_i) defines a basis for the tangent vector space $T_x\mathfrak{M}$ for every x. Another way for characterizing the frame (e_i) is to say that it represents a linear isomorphism $(e_i):\ \mathbb{R}^n\to T_x\mathfrak{M},\ \ (v^1,\dots,v^n)\mapsto v^i e_i$. The inverse isomorphism

$$(\theta^i):\ T_x\mathfrak{M}\to\mathbb{R}^n,\quad v\mapsto(v^1,\dots,v^n)=\left(\theta^1(x)v,\dots,\theta^n(x)v\right)$$

defines the coframe field at x and, consequently, a global section of the principal fiber bundle $GL^*(\mathfrak{M})\to\mathfrak{M}$. The above mentioned linear isomorphisms are inverse each to other. Thus, every frame field has a unique reciprocal coframe field (θ^i), which can be defined by the reciprocal property $\theta^i\left(e_j\right)=\delta_i^j$. At this stage we can already establish a correspondence between geometrical and continuum mechanical approaches. If a 3-dimensional manifold $\mathfrak{M}$ represents reference shape in classical sense, i.e. stress free shape that can be embedded into physical Euclidean space, then the field of frames represents elementary fibers that are the object for measuring by any kind of deformation measures. In more general non-Euclidean approach such considerations remain valid only locally [25, 26]. These questions will be discussed in section 2.

2°. *Holonomic frames.* Since every smooth manifold can be covered by coordinate charts, which together constitute an atlas and, in turn, every coordinate chart (U,φ), $\varphi : x \mapsto (x^i)$ defines the natural local frame and coframe fields

$$e_i = \frac{\partial}{\partial x^i}, \quad \theta^i = dx^i,$$

then every smooth manifold admit a local frame field and a local coframe field. The natural local frame and coframe fields are holonomic, i.e.

$$\left[e_i, e_j\right] = 0, \quad d\theta^i = 0, \tag{1}$$

where $[\cdot,\cdot]$ are Lie brackets [23]. This property is equivalent to the integrability conditions for the local frame field. The vanishing of 1-forms $d\theta^i$ means that there are n functions x^i on U such that $\theta^i = dx^i$. Moreover, linearly independence of 1−forms θ^i leads to the fact that the map $x: \ U \to \mathbb{R}^n$ is a diffeomorphism. From the point of view adopted in continuum mechanics it means the existence of a displacement field that transforms the stress-free shape into some deformed (actual) shape.

3°. *Nonholonomic frames.* If $\left[e_i, e_j\right]$ and $d\theta^i$ are non-vanishing for some (i,j) then (e_i) represents a nonholonomic frame field. It cannot be directly integrable into the natural frame field of any coordinate chart. However the local frame fields are integrable in more general sense $e_i = \Omega_i^j \partial_j$, where Ω_i^j defines a smooth transition function: $\Omega : U \to GL(n)$. Dually, $\theta^j = \mho_i^j dx^i$. Transition function has clear mechanical interpretation: it represents (may be up to some deformation in classical sense) the implants that assemble the body from geometrically incompatible parts [27].

The coframe field defines a basis for the vector space $\Lambda_x^2(\mathfrak{M})$ of 2-forms through the tuple $\left(\theta^i \wedge \theta^j\right)_{1 \le i < j \le n}$ [23, 28]. We can express the exterior derivative of 1-forms θ^i in terms of that basis:

$$d\theta^i = -\frac{1}{2} c^i_{jk}(x)\,\theta^j \wedge \theta^k,$$

where the functions $c^i_{jk} : U \to \mathbb{R}$ are referred to as structural functions of the coframe field θ^i. Moreover it can be shown that

$$\left[e_i, e_j\right] = c^k_{ij} e_k, \quad d\theta^i\left(e_i, e_k\right) = -\theta^i\left(\left[e_i, e_k\right]\right) = -c^i_{jk}, \tag{2}$$

Relations (2) generalize the relations (1). They allow us to formalize in geometric terms the idea of the stress free reference shape as a set of geometrically incompatible elementary (infinitesimal) volumes. These questions will be discussed in section 3.

4°. *Parallelism.* The pure geometric idea of parallel vector fields plays an important role in the theory being developed, since it allows one to formalize the idea of a material parallel transfer. Recall that a vector field v on $\mathfrak{M}$ is said to be parallel iff its components v^i are constant functions. Locally, a vector $v \in T_x\mathfrak{M}$ is parallel to a vector $v' \in T_y\mathfrak{M}$, $x \neq y$ iff $v^i = v'^i$. Suchwise a vector field is parallel iff all of its values, i.e. vectors from tangent spaces, are pair-wise parallel to each other. In the framework of continuum mechanics one can describe deformation by observing the distortion of the material fiber during its transfer along the curve, provided that the preimage of this fiber in the reference form remains parallel to itself. All that is required to generalize classical representations is to assume that the reference form is embedded in a space with a non-Euclidean rule of parallel transport.

5°. *Connection.* To introduce a connection that facilitates such parallel translation we assume that the frame field itself is parallel. Let v be a vector field:

$$v = v^i \partial_i = \bar{v}^i e_i, \quad \bar{v}^i = \mho^i_j v^j.$$

Some calculations give

$$d\bar{v}^i = d\mho^i_j v^j + \mho^i_j dv^j = \mho^i_j \nabla v^j, \nabla v^j = dv^j + \Gamma^i_j v^j, \Gamma^i_j = \Omega^i_k d\mho^k_j = \Gamma^i_{jk} dx^k, \Gamma^i_{jk} = \Omega^i_k \frac{\partial \mho^k_j}{\partial x^k}.$$

Note that since $d\left(\mho^i_k \Omega^k_j\right) = d\delta^i_j = 0$ then $\Gamma^i_j = -d\Omega^i_k \mho^k_j$ also. Suppose that v is parallel. According to aforementioned definition it means that v^i are constant functions and hence $d\bar{v}^i = 0$ and $\nabla v^j = 0$. In the case of general vector field v

$$\nabla v = \Omega^i_j \nabla v^j \otimes \partial_i = \nabla \bar{v}^i \otimes e_i.$$

Thus, $\Gamma^i_j = 0$ relative to the nonholonomic frame field e_i and one can say that geometry as observed in that frame field is "nonholonomic Euclidean" [22]. The Γ^i_j is called 1-form teleparallelism connection. In the framework of continuum mechanics this kind of connection formalizes local transformations of representative (infinitesimal) part to the uniform state, so-called crystal state [27]. This question will be discussed in section 5.

6°. *Torsion and curvature.* Special properties of so defined material connection can be characterized by torsion and curvature of it. Recall that torsion can be defined by 2-form [28]

$$\mathfrak{T}^i = d\theta^i + \Gamma^i_k \wedge \theta^k,$$

that in general does not vanish. Note that if one has two parallel vector fields v and w then

$$\mathfrak{T}^i(v,w) e_i = -[v,w].$$

It can be shown that curvature 2-form [28]

$$\mathfrak{R}^i_j = d\Gamma^i_j + \Gamma^i_k \wedge \Gamma^k_j$$

that corresponds to teleparallel connection vanishes. Indeed, direct calculations shows that [12]

$$\mathfrak{R} = d\mho \wedge d\Omega + (d\mho\Omega) \wedge (d\mho\Omega) = d\mho \wedge d\Omega + (d\mho) \wedge (\Omega d\mho\Omega) = 0.$$

7°. *Metric.* Finally let us pay some attention to the second, metrical way for introducing the material connection. Let $\mathfrak{M}$ be a parallelizable manifold. We can define on $\mathfrak{M}$ the volume element V and the metric g according to

$$V = \theta^1 \wedge \theta^2 \wedge \theta^3, \quad g = g_{ij} \theta^i \otimes \theta^j.$$

This allows us to determine Levi-Civita connection [28]

$$\tilde{\Gamma}^i_{jk} = \frac{1}{2} g^{im} \left(\partial_j g_{mk} + \partial_k g_{mj} - \partial_m g_{jk}\right).$$

However the Levi-Civita connection does not have vanishing curvature. If the metric g is identified with local deformation measures then so defined connection can be treated as material. This question will be discussed in section 4.

Some words about notation. In the paper we use mathematical terminology and definitions that are consistent with those in [24, 30]. In particular, if $\mathfrak{M}$ is a smooth manifold, then symbols $T_p\mathfrak{M}$ and $T_p^*\mathfrak{M}$ for each point denote respectively tangent and cotangent spaces to $\mathfrak{M}$ at a point $p \in \mathfrak{M}$. The tangent and cotangent bundles are denoted by $T\mathfrak{M}$ and $T^*\mathfrak{M}$ respectively. The symbol $\otimes$ stays for the abstract tensor product. We denote by $\langle v, u\rangle$ the value of the following bilinear operation

$$\langle \cdot, \cdot \rangle : T_p^*\mathfrak{M} \times T_p\mathfrak{M} \to \mathbb{R}, \quad \langle v, u\rangle := v(u).$$

Some words should be mentioned about vector bundles and sections. If $(\mathfrak{C}, \mathfrak{M}, \pi)$ is a vector bundle [24] with base $\mathfrak{M}$, total space $\mathfrak{C}$, and projection $\pi : \mathfrak{C} \to \mathfrak{M}$, then by a *section* [24] of such the bundle we mean a mapping $s : \mathfrak{M} \to \mathfrak{C}$ that satisfies the following condition: $\pi \circ s = \mathrm{I}$, in which $\mathrm{I} : \mathfrak{M} \to \mathfrak{M}$ is the identity map. We use the term "field" as a synonym for "section". Note that the construction of vector bundle allows one to define the notion of smooth section similarly to the notion of smooth mapping between manifolds. It is convenient instrument that is used in the modern physics [28].

The following notations are used for sets of smooth sections:

a) $F(\mathfrak{M})$, the set of all smooth scalar functions $f : \mathfrak{M} \to \mathbb{R}$;

b) $\mathrm{Vec}(\mathfrak{M})$, the set of all smooth vector fields $u : \mathfrak{M} \to T\mathfrak{M}$;

c) $\mathrm{CVec}(\mathfrak{M})$, the set of all smooth covector fields $v : \mathfrak{M} \to T^*\mathfrak{M}$.

The mapping $\langle \cdot, \cdot \rangle$ induces bilinear operations $\llcorner$ and $\lrcorner$, which are acting on dyads as follows:

$$\llcorner : (T \otimes \vartheta, v) \mapsto T \otimes \vartheta \llcorner v = \langle \vartheta, v\rangle T, \quad \lrcorner : (v, \vartheta \otimes T) \mapsto v \lrcorner \vartheta \otimes T = \langle \vartheta, v\rangle T.$$

Here T is a smooth section of a vector bundle, and $v \in \mathrm{Vec}(\mathfrak{M})$, $\vartheta \in \mathrm{CVec}(\mathfrak{M})$ are vector and covector fields.

If $\mathfrak{M}$ and $\mathfrak{N}$ are smooth manifolds, $\varkappa : \mathfrak{M} \to \mathfrak{N}$ is a smooth mapping, and T, Q are respectively smooth sections of vector bundles over $\mathfrak{M}$ and $\mathfrak{N}$, then the symbols $\varkappa_* T$ and $\varkappa^* Q$ denote pushforward and pullback of these fields [30]. For a given Riemannian metric musical isomorphisms are denoted by $(\cdot)^\flat : T\mathfrak{M} \to T^*\mathfrak{M}$ and $(\cdot)^\sharp : T^*\mathfrak{M} \to T\mathfrak{M}$ [24].

Symbol $[a_{ij}]$ denotes a matrix with elements a_{ij}.

2. Body and Physical Space

A *body* $\mathfrak{B}$ is a set with continuum cardinality. Its elements $\mathfrak{X}$, $\mathfrak{Y}$, $\mathfrak{Z}$, ..., are referred to as *material points*. This set is endowed with smooth n-dimensional manifold structure [24]. That is, $\mathfrak{B}$ is a Hausdorff second-countable topological space with a fixed maximal atlas $A_{\max}$. Each chart from this atlas is represented by an ordered pair (U, φ), in which $U \subset \mathfrak{B}$ is an open set, and $\varphi : U \to O$ is a homeomorphism between U and some open set $O \subset \mathbb{R}^n$. We refer to φ as *coordinate mapping*. For each two charts (U, φ) and (V, ψ) from $A_{\max}$ one has that either **i)** $U \cap V = \varnothing$, or **ii)** $U \cap V \neq \varnothing$ and the mapping $\psi \circ \varphi^{-1}|_{\varphi(U \cap V)} : \varphi(U \cap V) \to \psi(U \cap V)$ is a C^∞-diffeomorphism.

A physical space is a formalization of the ideal "laboratory". In the present paper we consider the physical space as a 3-dimensional Euclidean point space $\mathcal{E}$ with the associated translation vector space $\mathcal{V}$. That is, the following Weyl axioms hold [31]: there exists a mapping $w:\mathcal{E}\times\mathcal{E}\to\mathcal{V}$, $(x,y)\mapsto\overrightarrow{xy}$, such that

a) for each $x,y,z\in\mathcal{E}$ *Chasles' identity* holds: $\overrightarrow{xy}+\overrightarrow{yz}+\overrightarrow{zx}=\mathbf{0}$;

b) for each $x\in\mathcal{E}$ the mapping $w(x,\cdot):\mathcal{E}\to\mathcal{V}$, $y\mapsto\overrightarrow{xy}$ is bijective.

The 3–dimensional vector space $\mathcal{V}$ is endowed with an inner product $\cdot:\mathcal{V}\times\mathcal{V}\to\mathbb{R}$.

A Cartesian frame $(o,(\mathbf{i}_k)_{k=1}^3)$ is chosen in $\mathcal{E}$. Here $o\in\mathcal{E}$ is the origin and $(\mathbf{i}_k)_{k=1}^3$ is an orthonormal basis of $\mathcal{V}$, i.e., $\mathbf{i}_p\cdot\mathbf{i}_k=\delta_{pk}$, for each $p,k=1,2,3$. The basis which is dual to $(\mathbf{i}_k)_{k=1}^3$, is denoted by $(i^k)_{k=1}^3$, i.e., i^k are elements of the dual vector space $\mathcal{V}^*$ and $\langle i^k,\mathbf{i}_p\rangle=\delta_{kp}$, for each $p,k=1,2,3$.

The space $\mathcal{E}$ has a trivial smooth manifold structure. Its minimal atlas $A_{\min}$ consists of one chart $(\mathcal{E},c)$, where $c:\mathcal{E}\to\mathbb{R}^3$ is Cartesian arithmetization:

$$\forall x\in\mathcal{E}:\quad c(x):=((x-o)\cdot\mathbf{i}_1,(x-o)\cdot\mathbf{i}_2,(x-o)\cdot\mathbf{i}_3).$$

In what follows, the tangent space $T_x\mathcal{E}$ to $\mathcal{E}$ at any point x is identified with the translation space $\mathcal{V}$.

The physical space $\mathcal{E}$ contains all shapes of every body[1] $\mathfrak{B}$. These are images $\varkappa(\mathfrak{B})$ of smooth embeddings $\varkappa:\mathfrak{B}\to\mathcal{E}$. Such a mapping is referred to as configuration. The set of all configurations is denoted by $\mathfrak{C}(\mathfrak{B})$.

Remark 1. A body $\mathfrak{B}$ is observable in the physical space $\mathcal{E}$ only through its shapes. This fact was allegorically described by M. Epstein in [6]: *"The body itself dwells in the Platonic world of differentiable manifolds, of which we can only see the manifestations in the world of phenomena, amenable to perception, in this case, in the guise of configurations."* ♣

Since each configuration $\varkappa\in\mathfrak{C}(\mathfrak{B})$ is a homeomorphism onto its image, $\varkappa(\mathfrak{B})$, it is convenient to introduce a new mapping $\hat{\varkappa}$, which is obtained by the restriction of the codomain of $\varkappa$ into $\varkappa(\mathfrak{B})$. That is[2],

$$\hat{\varkappa}:\mathfrak{B}\to\varkappa(\mathfrak{B}),\quad \hat{\varkappa}(\mathfrak{X})=\varkappa(\mathfrak{X}).$$

This mapping is a homeomorphism. Relations between the body, the physical space and configurations are illustrated in figure 1.

Remark 2. Formalization of the body as a three-dimensional smooth manifold was provided by W. Noll and C.-C. Wang in [20, 21]. In their considerations the smooth structure was generated by configurations that are mappings from the body into the physical space. M. Gurtin and A. Murdoch in [32] developed theory of elastic material surfaces (that are two-dimensional smooth manifolds) by using Noll's approach. Since that time the mathematical theory of smooth manifolds was deeply developed, but key ideas in mechanical interpretation are the same. ♣

[1] It is assumed that $\dim\mathfrak{B}\leqslant\dim\mathcal{E}$.

[2] In the present paper we use the following notation. Let $f:X\to Y$ be a mapping. Then by $\hat{f}$ we denote the mapping $\hat{f}:X\to f(X)$, that assigns to each $x\in X$ a value $\hat{f}(x):=f(x)$.

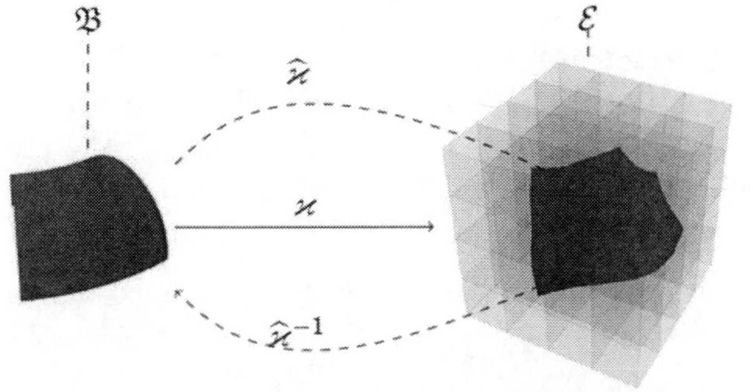

Figure 1. A body $\mathfrak{B}$, the physical space $\mathcal{E}$, a configuration $\varkappa$.

Since the body $\mathfrak{B}$ is not observable by itself, neither are material fibers. If some configuration $\varkappa \in \mathfrak{C}(\mathfrak{B})$ is chosen, then for each material fiber $\mathfrak{c}:\mathbb{I} \to \mathfrak{B}$ and its infinitesimal element $u \in T\mathfrak{B}$, the images $\varkappa \circ \mathfrak{c}(\mathbb{I})$ and $T\varkappa(u)$ of the composition $\varkappa \circ \mathfrak{c}:\mathbb{I} \to \mathcal{E}$ and the tangent map $T\varkappa : T\mathfrak{B} \to T\mathcal{E}$ are observable. Thus, the tangent map $T\varkappa$ plays the same role as the conventional deformation gradient. Thereby, we refer to $T\varkappa$ as configuration gradient.

3. Material Uniformity and Inhomogeneity

Let $\mathfrak{X}$ be a material point of a body $\mathfrak{B}$. Local configuration at $\mathfrak{X}$ formalizes the idea of infinitesimal localization: if the "observer" is placed into infinitesimal neighborhood of the point $\mathfrak{X}$, then it can distinguish configurations only within this neighborhood. Assume that this localization has the first order: the "observer" cannot distinguish configurations that have the same values at $\mathfrak{X}$ and the same linear approaches. Mathematical formalization of this idea may be provided as follows. Let $\sim_{\mathfrak{X}}$ be the following equivalence relation on the set $\mathfrak{C}(\mathfrak{B})$:

$$\forall \varkappa_1, \varkappa_2 \in \mathfrak{C}(\mathfrak{B}): \quad (\varkappa_1 \sim_{\mathfrak{X}} \varkappa_2) \Leftrightarrow ((\varkappa_1(\mathfrak{X}) = \varkappa_2(\mathfrak{X})) \wedge (T_{\mathfrak{X}}\varkappa_1 = T_{\mathfrak{X}}\varkappa_2)).$$

Following W. Noll [20] we refer to each equivalence class $\mathrm{cl}_{\mathfrak{X}}$ as local configuration at $\mathfrak{X} \in \mathfrak{B}$. A local configuration $\mathrm{cl}_{\mathfrak{X}}$ contains such configurations that are not distinguishable from each other by the imaginary "observer".

Denote the set of all local configurations at $\mathfrak{X}$ by $\mathcal{L}_{\mathfrak{X}}$. Each local configuration $\mathrm{cl}_{\mathfrak{X}}$ generates linear mapping $\mathbf{K}_{\mathfrak{X}} : T_{\mathfrak{X}}\mathfrak{B} \to \mathcal{V}$ as follows. Choose some configuration $\varkappa \in \mathrm{cl}_{\mathfrak{X}}$ and define $\mathbf{K}_{\mathfrak{X}}$ as $\mathbf{K}_{\mathfrak{X}} := T_{\mathfrak{X}}\varkappa$. According to the definition of the relation $\sim_{\mathfrak{X}}$, the mapping $\mathbf{K}_{\mathfrak{X}}$ does not depend on the choice of a representative $\varkappa \in \mathrm{cl}_{\mathfrak{X}}$ and is fully determined by the class $\mathrm{cl}_{\mathfrak{X}}$. It maps infinitesimal neighborhood of the point $\mathfrak{X}$, i.e., tangent space $T_{\mathfrak{X}}\mathfrak{B}$, into infinitesimal neighborhood of the point $x = \varkappa(\mathfrak{X})$, $\varkappa \in \mathrm{cl}_{\mathfrak{X}}$. By definition, there is a one-to-one correspondence between such mappings and elements from $\mathcal{L}_{\mathfrak{X}}$. In the paper we identify a local configuration $\mathrm{cl}_{\mathfrak{X}}$ with the corresponding linear mapping $\mathbf{K}_{\mathfrak{X}}$.

Let $T\mathcal{E} \otimes T^*\mathfrak{B} = \bigcup_{(\mathfrak{X},x)\in\mathfrak{B}\times\mathcal{E}} T_x\mathcal{E} \otimes T^*_{\mathfrak{X}}\mathfrak{B}$. Introduce a section $\mathbf{K} : \mathfrak{B} \to T\mathcal{E} \otimes T^*\mathfrak{B}$ as follows:

$$\forall \mathfrak{X} \in \mathfrak{B}: \quad (\mathbf{K}_{\mathfrak{X}} \in \mathcal{L}_{\mathfrak{X}}).$$

We refer to the section $\mathbf{K}$ as *reference* [20]. By its definition, the reference $\mathbf{K}$ is a two-point tensor field with dyadic decomposition $\mathbf{K} = K^k_{\alpha}\mathbf{i}_k \otimes E^{\alpha}$, in which (E^{α}) is a coframe field on the body $\mathfrak{B}$.

The material at any point $\mathfrak{X} \in \mathfrak{B}$ is characterized by its response. Let R be a set of response descriptors, i.e., a set of objects that are determined by[3] "measurers". The body $\mathfrak{B}$ is *simple* [20], if there is a mapping $\mathcal{R}_{\mathfrak{X}} : \mathcal{L}_{\mathfrak{X}} \to R$ for any material point $\mathfrak{X}$. In the paper we consider only simple bodies. Each of them is *materially uniform*, i.e., the material at any two of its points is the same. The mathematical formalization of such an idea is based on the notion of *material isomorphism* [20]. That is, let $\mathfrak{X}, \mathfrak{Y} \in \mathfrak{B}$ be material points. An invertible linear mapping $\Phi_{\mathfrak{X}\mathfrak{Y}} : T_{\mathfrak{Y}}\mathfrak{B} \to T_{\mathfrak{X}}\mathfrak{B}$ is called as *material isomorphism* from $T_{\mathfrak{Y}}\mathfrak{B}$ onto $T_{\mathfrak{X}}\mathfrak{B}$, if the equality

$$\mathcal{R}_{\mathfrak{X}}(\mathbf{K}_{\mathfrak{X}}) = \mathcal{R}_{\mathfrak{Y}}(\mathbf{K}_{\mathfrak{X}} \Phi_{\mathfrak{X}\mathfrak{Y}}),$$

holds for each $\mathbf{K}_{\mathfrak{X}}$ from $\mathcal{L}_{\mathfrak{X}}$. Let the set of all material isomorphisms from $T_{\mathfrak{Y}}\mathfrak{B}$ onto $T_{\mathfrak{X}}\mathfrak{B}$ denote by $g_{\mathfrak{X}\mathfrak{Y}}$. Thus, the simple body is materially uniform if $g_{\mathfrak{X}\mathfrak{Y}} \neq \varnothing$ for all material points $\mathfrak{X}, \mathfrak{Y} \in \mathfrak{B}$.

It follows from the material uniformity that there exists a *uniform reference* $\mathbf{K}^R$ [20], i.e., such a reference that the linear mapping

$$\Phi_{\mathfrak{X}\mathfrak{Y}} := [\overset{R}{\mathbf{K}}_{\mathfrak{X}}]^{-1} \overset{R}{\mathbf{K}}_{\mathfrak{Y}},$$

is a material isomorphism from $T_{\mathfrak{Y}}\mathfrak{B}$ onto $T_{\mathfrak{X}}\mathfrak{B}$ for all material points $\mathfrak{X}, \mathfrak{Y} \in \mathfrak{B}$. According to the definition, the following equality holds for a uniform reference $\mathbf{K}^R$: $\mathcal{R}_{\mathfrak{X}}(\mathbf{K}^R_{\mathfrak{X}}) = \mathcal{R}_{\mathfrak{Y}}(\mathbf{K}^R_{\mathfrak{Y}})$. This means that the response of the body is the same at each material point. Another consequence is that images of the linear mappings $\mathbf{K}^R_{\mathfrak{X}}$ coincide and are represented by a vector subspace of $\mathcal{V}$ of the dimension $\dim \mathfrak{B}$. Thus, each infinitesimal neighborhood $T_{\mathfrak{X}}\mathfrak{B}$ may be obtained from such a space by means of $[\overset{R}{\mathbf{K}}_{\mathfrak{X}}]^{-1}$.

A uniform reference $\mathbf{K}^R$ generates a family $\mathrm{Ref}_{\mathfrak{B}} = \{\mathbf{K}^R_{\mathfrak{X}}\}_{\mathfrak{X} \in \mathfrak{B}}$ of local configurations, that are values of $\mathbf{K}^R$. If a representative $\varkappa^R_{\mathfrak{X}}$ is chosen from each $\mathbf{K}^R_{\mathfrak{X}}$, then one arrives at the family $\overline{\mathrm{Ref}}_{\mathfrak{B}} = \{\varkappa^R_{\mathfrak{X}}\}_{\mathfrak{X} \in \mathfrak{B}}$ of *global* configurations

$$\varkappa^R_{\mathfrak{X}} : \mathfrak{B} \to \mathcal{E}, \quad \varkappa^R_{\mathfrak{X}} : \mathfrak{Y} \mapsto \varkappa^R_{\mathfrak{X}}(\mathfrak{Y}),$$

such that each of them transforms an infinitesimal neighborhood of a point $\mathfrak{X}$ into a *uniform shape* [25].

Physically, a materially uniform simple body consists of infinitesimal representative volumes with the following property. Each of them, if being separated from others and placed in "testing machine", gives the same response on the same deformation. It leads to the idea that each representative volume of such a body may be obtained from one fixed subject to the corresponding local deformation. This fixed infinitesimal object is referred to as *archetype* and the corresponding mapping as *implant*. Let $\mathbf{K}^R$ be a uniform reference. Fix a material point $\mathfrak{X}_0 \in \mathfrak{B}$. Thus, the material isomorphism $\Phi_{\mathfrak{X}\mathfrak{X}_0} = [\overset{R}{\mathbf{K}}_{\mathfrak{X}}]^{-1} \overset{R}{\mathbf{K}}_{\mathfrak{X}_0}$ from $T_{\mathfrak{X}_0}\mathfrak{B}$ onto $T_{\mathfrak{X}}\mathfrak{B}$ allows us to obtain $T_{\mathfrak{X}}\mathfrak{B}$ from $T_{\mathfrak{X}_0}\mathfrak{B}$ for each $\mathfrak{X} \in \mathfrak{B}$. The tangent space $T_{\mathfrak{X}_0}\mathfrak{B}$ and the linear mapping $\mathbf{P}_{\mathfrak{X}} = \Phi_{\mathfrak{X}\mathfrak{X}_0}$ are viewed respectively as an archetype and an implant. It is M. Epsteins' standpoint [5], in which the archetype is considered as a material object. By W. Nolls' standpoint [20] the archetype is a spatial object. Thus, it is represented by $\mathbf{K}^R_{\mathfrak{X}_0}[T_{\mathfrak{X}_0}\mathfrak{B}]$. In such consideration the implant is represented by the composition $\Phi_{\mathfrak{X}\mathfrak{X}_0}[\overset{R}{\mathbf{K}}_{\mathfrak{X}_0}]^{-1}$, i.e., $\mathbf{P}_{\mathfrak{X}} = [\overset{R}{\mathbf{K}}_{\mathfrak{X}}]^{-1}$.

---

[3] In particular, R may consist of second order tensors that represent values of the Cauchy stress tensor.

If $\mathbf{K}^R$ is a uniform reference and there exists a configuration $\varkappa \in \mathfrak{C}(\mathfrak{B})$, such that $\mathbf{K}^R_{\mathfrak{X}} = T_{\mathfrak{X}}\varkappa$ for each material point $\mathfrak{X} \in \mathfrak{B}$, then the body $\mathfrak{B}$ is *homogeneous*. In this case one can apply $\varkappa^R_{\mathfrak{X}} := \varkappa$ and consider the global uniform shape $\varkappa(\mathfrak{B})$ as a reference shape. In the contrary, the body $\mathfrak{B}$ is *inhomogeneous* if such the configuration does not exist. It means that there does not exist a global uniform shape for the body $\mathfrak{B}$ in the Euclidean physical space $\mathcal{E}$. Figure 2 illustrates the case of an inhomogeneous body. A red cube represents an infinitesimal neighborhood in an uniform state.

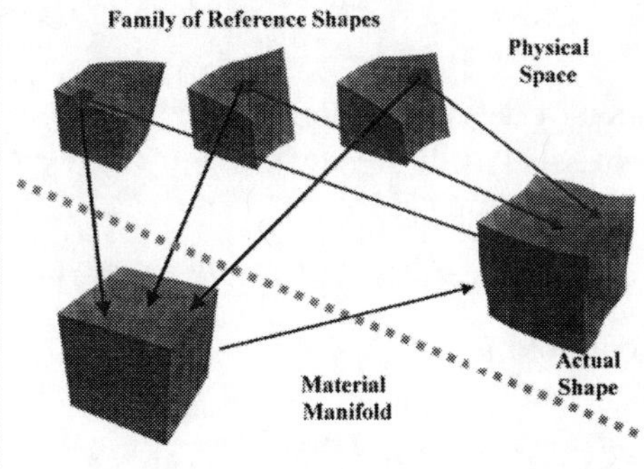

Figure 2. Materially uniform inhomogeneous body.

Remark 3. Instantaneous material composition of LbL structure may be represented by a materially uniform body $\mathfrak{B}$. In this case, one can divide the body $\mathfrak{B}$ into a collection of "joined" layers. Each layer is homogeneous, but as a whole, the body is inhomogeneous. ♣

The key idea of the geometric approach is *to give the body $\mathfrak{B}$ a role of the fictional reference shape and to derive equations in material description relatively to it*. Since these equations use the notion of parallel translation, one needs to define the affine connection on the body $\mathfrak{B}$, which should also reflect material properties of the body. We refer to such a connection as *material connection*.

4. Synthesizing of the Material Metric

Using elements of the family $\overline{\mathrm{Ref}}_{\mathfrak{B}}$ one can obtain the collection $\{(\mathfrak{B}, \mathbf{G}^{(\mathfrak{X})})\}_{\mathfrak{X}\in\mathfrak{B}}$ of Riemannian spaces with metrics $\mathbf{G}^{(\mathfrak{X})}$ that are defined relatively to the embeddings $\varkappa^R_{\mathfrak{X}}$. Thus,

$$\forall \mathfrak{Y} \in \mathfrak{B} \quad \forall u, v \in T_{\mathfrak{Y}}\mathfrak{B}: \quad \mathbf{G}^{(\mathfrak{X})}|_{\mathfrak{Y}}(u,v) := [T_{\mathfrak{Y}}\varkappa^R_{\mathfrak{X}}(u)] \cdot [T_{\mathfrak{Y}}\varkappa^R_{\mathfrak{X}}(v)].$$

The corresponding Levi-Civita connection is Euclidean. Let us synthesize Riemannian metric, non-Euclidean in general case. Assume that the section $\mathbf{G} : \mathfrak{B} \to T^*\mathfrak{B} \otimes T^*\mathfrak{B}$, defined by a formula $\mathbf{G}_{\mathfrak{X}} := \mathbf{G}^{(\mathfrak{X})}|_{\mathfrak{X}}$, is smooth. Each of values $\mathbf{G}^{(\mathfrak{X})}|_{\mathfrak{X}}$ represents a symmetric positive definite bilinear form. Hence, $\mathbf{G}$ is a Riemannian metric. Since it is defined on the body $\mathfrak{B}$, we refer to $\mathbf{G}$ as *material metric*. By the definition, the material metric is combined from the family of Riemannian metrics $\{\mathbf{G}^{(\mathfrak{X})}\}_{\mathfrak{X}\in\mathfrak{B}}$, i.e.,

$$\forall \mathfrak{X} \in \mathfrak{B} \quad \forall u, v \in T_{\mathfrak{X}}\mathfrak{B}: \quad \mathbf{G}_{\mathfrak{X}}(u,v) = [T_{\mathfrak{X}}\varkappa^R_{\mathfrak{X}}(u)] \cdot [T_{\mathfrak{X}}\varkappa^R_{\mathfrak{X}}(v)].$$

In terms of the uniform reference $\mathbf{K}^R$ this relation takes the form [20]

$$\forall \mathfrak{X} \in \mathfrak{B} \quad \forall u, v \in T_{\mathfrak{X}}\mathfrak{B}: \quad \mathbf{G}_{\mathfrak{X}}(u,v) = [\mathbf{K}^R_{\mathfrak{X}}(u)] \cdot [\mathbf{K}^R_{\mathfrak{X}}(v)]. \tag{3}$$

The material metric $\mathbf{G}$ defines Levi-Civita connection. The coefficients $\Gamma^{\gamma}_{\alpha\beta}$ of the latter in a coordinate frame (∂_α) may be calculated by using the following expressions

$$\Gamma^{\gamma}_{\alpha\beta} = \frac{G^{\gamma\delta}}{2}\left(\partial_\alpha G_{\beta\delta} + \partial_\beta G_{\alpha\delta} - \partial_\delta G_{\alpha\beta}\right),$$

where $G_{\alpha\beta} = g_{sp} K^s_\alpha K^p_\beta$, $[G^{\alpha\beta}] = [G_{\alpha\beta}]^{-1}$ and $g_{sp} = \mathbf{e}_s \cdot \mathbf{e}_p$. Here $(\mathbf{e}_k)^3_{k=1}$ is a local frame field for some curvilinear coordinate system in $\mathcal{E}$.

5. Synthesizing of Material Connection

5.1. Affine connection

Now we switch on to the question of defining material connection. Let us review some aspects from the affine connection theory. In what follows the symbol $\mathfrak{M}$ denotes a smooth n-dimensional manifold. Affine connection ∇ is a mapping [33]

$$\nabla : \mathrm{Vec}(\mathfrak{M}) \times \mathrm{Vec}(\mathfrak{M}) \to \mathrm{Vec}(\mathfrak{M}), \quad (u, v) \mapsto \nabla_u v,$$

for which the following conditions hold:

a) $\forall u, v, w \in \mathrm{Vec}(\mathfrak{M})$: $\nabla_{u+v} w = \nabla_u w + \nabla_v w$;

b) $\forall u, v \in \mathrm{Vec}(\mathfrak{M})$ $\forall f \in F(\mathfrak{M})$: $\nabla_{fu} v = f \nabla_u v$;

c) $\forall u, v, w \in \mathrm{Vec}(\mathfrak{M})$: $\nabla_u (v + w) = \nabla_u v + \nabla_u w$;

d) $\forall u, v \in \mathrm{Vec}(\mathfrak{M})$ $\forall f \in F(\mathfrak{M})$: $\nabla_u (fv) = f \nabla_u v + (uf) v$ (*Leibniz rule*).

In d) uf denotes the action of a vector field u on a scalar function f. Here u is identified with *derivation* [24]. The vector field $\nabla_u v$ represents *covariant derivative* of v among u.

In a fixed (nonholonomic) frame field $(e_i)^n_{i=1}$, $e_i \in \mathrm{Vec}(\mathfrak{M})$, and the corresponding coframe field $(\vartheta^i)^n_{i=1}$, $\vartheta^i \in \mathrm{CVec}(\mathfrak{M})$, the mapping ∇ is represented by n^3 scalar mappings $\Gamma^i_{jk} = \langle \vartheta^i, \nabla_{e_j} e_k \rangle \in F(\mathfrak{M})$.

Remark 4. Since $\nabla_{\partial_i} \partial_j = \Gamma^k_{ij} \partial_k$, for vector fields $u = u^i \partial_i$ and $v = v^j \partial_j$, one has the following coordinate representation of covariant derivative $\nabla_u v$:

$$\nabla_u v = u^i \left(\partial_i v^k + v^j \Gamma^k_{ij}\right) \partial_k = u^i \nabla_i v^k \partial_k,$$

where $\nabla_i v^k = \langle dx^k, \nabla_{\partial_i} v \rangle$. If the field $\nabla v = \nabla_i v^k \partial_k \otimes dx^i$ is introduced then $\nabla_u v = \nabla v \llcorner u$. ♣

Let us obtain the transformation law of the mappings Γ^i_{jk}. Let the mappings $\Omega^i_j \in F(\mathfrak{M})$ define a smooth field of non-degenerated $n \times n$ matrices $[\Omega^i_j]$. There is a frame field related to this field according to

$$\mathfrak{M} \ni p \mapsto (e'_i|_p)^n_{i=1}, \quad e'_i = \Omega^j_i e_j.$$

The corresponding coframe field $\mathfrak{M} \ni p \mapsto (\vartheta'^i|_p)^n_{i=1}$ is expressed through "old" coframe field via the relation $\vartheta'^i = \mho^i_j \vartheta^j$, where $[\mho^i_j] = [\Omega^i_j]^{-1}$. Finally, let $e_i = A^j_i \partial_{x^j}$, where (x^i) are local coordinates on $\mathfrak{M}$. Then

$$\nabla_{e'_j} e'_k = \nabla_{(\Omega^s_j e_s)} (\Omega^q_k e_q) = \Omega^s_j \nabla_{e_s} (\Omega^q_k e_q) = \Omega^s_j \left\{\Omega^q_k \nabla_{e_s} e_q + e_s(\Omega^q_k) e_q\right\},$$

from which, since $\Gamma'^i_{jk} = \langle \vartheta'^i, \nabla_{e'_j} e'_k \rangle = \mho^i_m \langle \vartheta^m, \nabla_{e'_j} e'_k \rangle$ and $e_s(\Omega^q_j) = A^l_s \partial_{x^l} \Omega^q_j$, it follows that

$$\Gamma'^i_{jk} = \mho^i_m \Omega^s_j \left\{\Omega^q_k \langle \vartheta^m, \nabla_{e_s} e_q \rangle + \langle \vartheta^m, e_q \rangle A^l_s \partial_{x^l} \Omega^q_k\right\}.$$

Taking into account the relations $\Gamma^m_{sq} = \langle \vartheta^m, \nabla_{e_s} e_q \rangle$ and $\langle \vartheta^m, e_q \rangle = \delta^m_q$, we arrive at the final expression:

$$\Gamma'^i_{jk} = \Gamma^m_{sq} \mho^i_m \Omega^s_j \Omega^q_k + A^l_s \mho^i_m \Omega^s_j \partial_{x^l} \Omega^m_k. \tag{4}$$

The expression (4) defines the affine connection transformation rule in the general case: when one changes an "old" nonholonomic frame field into a "new" one. If the "old" frame field is a coordinate frame, then $A^l_s = \delta^l_s$ and the formula (4) transforms into

$$\Gamma'^i_{jk} = \Gamma^m_{sq} \mho^i_m \Omega^s_j \Omega^q_k + \mho^i_m \Omega^s_j \partial_{x^s} \Omega^m_k. \tag{5}$$

If both frame fields are coordinate frames, i.e. the affine connection transformation is generated by coordinate transformation, then $\Omega^i_j = \dfrac{\partial x^i}{\partial x'^j}$, where (x^i) are "old" coordinates and (x'^j) are "new" ones. In this case the relation (5) takes the form:

$$\Gamma'^i_{jk} = \Gamma^m_{sq} \frac{\partial x'^i}{\partial x^m} \frac{\partial x^s}{\partial x'^j} \frac{\partial x^q}{\partial x'^k} + \frac{\partial x'^i}{\partial x^m} \frac{\partial^2 x^m}{\partial x'^j \partial x'^k}.$$

Connection coefficients may have nonzero values in an Euclidean space when one considers curvilinear coordinates, and in more general spaces, such as Riemann and Cartan spaces. In order to distinguish the space of non-Euclidean geometry, it is convenient to use tensor fields [33] of the *torsion*

$$\mathfrak{T} : \mathrm{Vec}(\mathfrak{M}) \times \mathrm{Vec}(\mathfrak{M}) \to \mathrm{Vec}(\mathfrak{M}),$$

curvature

$$\mathfrak{R} : \mathrm{Vec}(\mathfrak{M}) \times \mathrm{Vec}(\mathfrak{M}) \times \mathrm{Vec}(\mathfrak{M}) \to \mathrm{Vec}(\mathfrak{M}),$$

and *nonmetricity*

$$\mathfrak{Q} : \mathrm{Vec}(\mathfrak{M}) \times \mathrm{Vec}(\mathfrak{M}) \times \mathrm{Vec}(\mathfrak{M}) \to F(\mathfrak{M}).$$

For a given Riemannian metric g these fields are defined by the relations

$$\mathfrak{T}(u, v) = \nabla_u v - \nabla_v u - [u, v],$$

$$\mathfrak{R}(u, v, w) = \nabla_u \nabla_v w - \nabla_v \nabla_u w - \nabla_{[u,v]} w,$$

$$\mathfrak{Q}(u, v, w) = g(\nabla_u v, w) + g(v, \nabla_u w) - u[g(v, w)],$$

for $u, v, w \in \mathrm{Vec}(\mathfrak{M})$. Here $[\cdot,\cdot] : \mathrm{Vec}(\mathfrak{M}) \times \mathrm{Vec}(\mathfrak{M}) \to \mathrm{Vec}(\mathfrak{M})$ denote Lie brackets.

Remark 5. Nonmetricity is a measure of affine connection compatibility with the Riemannian metric. If $\mathfrak{Q} = 0$ then affine connection is compatible with the metric. ♣

Remark 6. Torsion is a measure of affine connection non-symmetry. Let $(\partial_i)^n_{i=1}$ be a coordinate frame field on $\mathfrak{M}$. Then

$$\mathfrak{T}(u, v) = \nabla_u v - \nabla_v u - [u, v] = u^i v^j \left(\Gamma^k_{ij} - \Gamma^k_{ji}\right) \partial_k,$$

and one has that $\mathfrak{T} = \left(\Gamma^k_{ij} - \Gamma^k_{ji}\right) \partial_k \otimes dx^i \otimes dx^j$. It means that $\mathfrak{T} = 0$ if and only if $\Gamma^k_{ij} = \Gamma^k_{ji}$, for all i, j, k. ♣

Given fields of torsion, curvature and nonmetricity, they fully determine affine connection via the formula

IOP Conf. Series: Journal of Physics: Conf. Series **1009** (2018) 012014 doi:10.1088/1742-6596/1009/1/012014

$$\Gamma_{ijk} = 2\Gamma^{m}_{jk} g_{mk} = \partial_j g_{ik} + \partial_k g_{ij} - \partial_i g_{jk} + T_{ijk} - T_{jki} + T_{kij} + Q_{ijk} + Q_{jki} - Q_{kij}.$$

5.2. Weitzenböck connection

Now let us use Cartans' moving frame method to define a special connection. Assume that a smooth field of non-degenerated $n \times n$ matrices $[\Omega^i_j]$ is given. This field is generated by some scalar mappings $\Omega^i_j \in F(\mathfrak{M})$. Using the field $[\Omega^i_j]$ one can define the new frame field

$$\mathfrak{M} \ni p \mapsto (z_i|_p)^n_{i=1}, \quad z_i = \Omega^j_i \partial_j.$$

In particular, one has $\mathbf{z}_i = \Omega^j_i \mathbf{i}_j$ for the space $\mathcal{E}$. This case corresponds to the classical moving frame method: an orthonormal frame is deformed by means of $[\Omega^i_j]$ and becomes nonholonomic.

Let us obtain such an affine connection ∇ that the following relations hold:

$$\nabla_{z_i} z_j = 0, \quad i, j = 1, \ldots, n.$$

It is *Weitzenböck connection*. The corresponding space is the space with *absolute parallelism*. The connection coefficients relatively to the frame (z_i) are equal to zero: $\Gamma'^{i}_{jk} = 0$. According to the formula (5) we have the following system of equations relatively to the connection coefficients Γ^m_{sq}:

$$\Gamma^m_{sq} \mho^i_m \Omega^s_j \Omega^q_k + \mho^i_m \Omega^s_j \partial_s \Omega^m_k = 0.$$

Multiply both sides of the last equation on $\Omega^o_i \mho^j_l \mho^k_r$ and sum over i, j, k. Then one arrives at the expressions

$$\Gamma^m_{sq} \delta^o_m \delta^s_l \delta^q_r + \delta^o_m \delta^s_l \mho^k_r \partial_s \Omega^m_k = 0,$$

from which it is possible to establish the formula

$$\Gamma^m_{sq} = -\mho^k_q \partial_s \Omega^m_k.$$

The obtained relation can be transformed as follows

$$\Gamma^m_{sq} = -\mho^k_q \partial_s \Omega^m_k = -\mho^k_q \partial_s \Omega^r_k \delta^m_r = -\mho^k_q \partial_s \Omega^r_k \mho^i_r \Omega^m_i,$$

and then, using the relation $\partial_i \mho^k_j = -\mho^m_j (\partial_i \Omega^q_m) \mho^k_q$, we take the final result:

$$\Gamma^k_{ij} = \Omega^k_m \partial_i \mho^m_j. \tag{6}$$

5.3. Material connection

Now consider a uniform reference $\mathbf{K}^R$ of a three-dimensional materially uniform inhomogeneous body[4] $\mathfrak{B}$. It represents the field $\mathfrak{X} \mapsto \mathbf{K}^R_{\mathfrak{X}}$ of non-degenerate linear transformations. These transformations, in turn, generate the implant field $\mathfrak{X} \mapsto \mathbf{P}_{\mathfrak{X}} = [\mathbf{K}^R_{\mathfrak{X}}]^{-1}$. Choose some configuration $\tilde{\varkappa} \in \mathfrak{C}(\mathfrak{B})$. The shape $\tilde{\varkappa}(\mathfrak{B})$ represents a "model" of the body. In addition, one embeds all infinitesimal neighborhoods into $\tilde{\varkappa}(\mathfrak{B})$ by means of $T\tilde{\varkappa}$. Because of inhomogeneity, the shape $\tilde{\varkappa}(\mathfrak{B})$ is non-uniform (in particular, with internal stresses). The implant field $\mathbf{P}$, up to embedding,

---

[4] Here we restrict ourselves with the case $\dim \mathfrak{B} = \dim \mathcal{E}$. Then $\mathbf{K}^R_{\mathfrak{X}}$ is a vector space isomorphism. In the general case the key idea is the same.

IOP Conf. Series: Journal of Physics: Conf. Series **1009** (2018) 012014 doi:10.1088/1742-6596/1009/1/012014

generates the frame field $\mathbf{z}_i = P_i^j \mathbf{i}_j$ defined on the shape $\tilde{\varkappa}(\mathfrak{B})$. This field is nonholonomic (see figure 3).

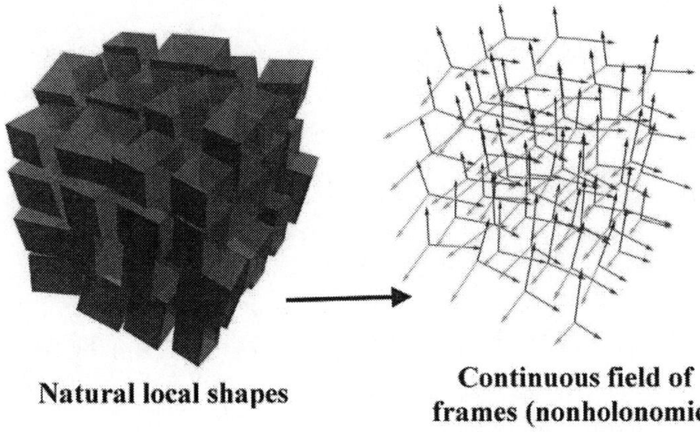

Figure 3. Correspondence between local shapes and space with absolute parallelism.

Parallelize the frame field $(\mathbf{z}_k)$ by means of Weitzenböck connection. According to (6) one gets

$$\Gamma_{ij}^k = P_m^k \partial_i [\mathbf{P}^{-1}]_j^m,$$

or, in terms of local configuration,

$$\Gamma_{ij}^k = [(\mathbf{K}^R)^{-1}]_m^k \partial_i [\mathbf{K}^R]_j^m.$$

The Weitzenböck space $(\tilde{\varkappa}(\mathfrak{B}), \nabla)$ represents the desired global uniform shape.

6. Strains and Stresses

6.1. Cauchy-Green tensors

Since material metric $\mathbf{G}: \mathfrak{B} \to T^*\mathfrak{B} \otimes T^*\mathfrak{B}$ has been synthesized in section 4, we can define the notion of the transposed configuration gradient. Let $\mathfrak{X} \in \mathfrak{B}$. We denote by a $T_{\mathfrak{X}}\varkappa^{\mathrm{T}}$ a unique linear map defined on $\mathcal{V}$ with values in $T_{\mathfrak{X}}\mathfrak{B}$, transpose to the tangent map $T_{\mathfrak{X}}\varkappa : T_{\mathfrak{X}}\mathfrak{B} \to \mathcal{V}$, i.e. the following relation is held for $T_{\mathfrak{X}}\varkappa^{\mathrm{T}}$:

$$\forall u \in T_{\mathfrak{X}}\mathfrak{B} \quad \forall v \in \mathcal{V}: \quad [T_{\mathfrak{X}}\varkappa(u)]\cdot v = \mathbf{G}(u, T_{\mathfrak{X}}\varkappa^{\mathrm{T}}(v)).$$

By using $T_{\mathfrak{X}}\varkappa$ and $T_{\mathfrak{X}}\varkappa^{\mathrm{T}}$ the following tensors can be constructed:

$$C_{\mathfrak{X}} := T_{\mathfrak{X}}\varkappa^{\mathrm{T}} \circ T_{\mathfrak{X}}\varkappa \in T_{\mathfrak{X}}\mathfrak{B} \otimes T_{\mathfrak{X}}^*\mathfrak{B}, \quad \mathfrak{X} \in \mathfrak{B},$$

$$\mathbf{B}_x := T_{\tilde{\varkappa}^{-1}(x)}\varkappa \circ T_{\tilde{\varkappa}^{-1}(x)}\varkappa^{\mathrm{T}} \in \mathcal{V} \otimes \mathcal{V}^*, \quad x \in \varkappa(\mathfrak{B}).$$

We refer to the tensors $C_{\mathfrak{X}}$ and $\mathbf{B}_X$ as right and left Cauchy-Green tensors [17]. In components,

$$C_{\mathfrak{X}} = G^{\alpha\beta}|_{\mathfrak{X}}\, g_{ij}|_{\varkappa(\mathfrak{X})} \left.\frac{\partial x^i}{\partial \mathfrak{X}^\beta}\right|_{\varphi(\mathfrak{X})} \left.\frac{\partial x^j}{\partial \mathfrak{X}^\gamma}\right|_{\varphi(\mathfrak{X})} \partial_{\mathfrak{X}^\alpha}|_{\mathfrak{X}} \otimes d\mathfrak{X}^\gamma|_{\mathfrak{X}}, \tag{7}$$

IOP Conf. Series: Journal of Physics: Conf. Series **1009** (2018) 012014 doi:10.1088/1742-6596/1009/1/012014

$$\mathbf{B}_x = G^{\alpha\beta}\,|_{\varkappa^{-1}(x)}\, g_{lj}\,|_x \left.\frac{\partial x^l}{\partial \mathfrak{X}^\beta}\right|_{\varphi(\varkappa^{-1}(x))} \left.\frac{\partial x^k}{\partial \mathfrak{X}^\alpha}\right|_{\varphi(\varkappa^{-1}(x))} \mathbf{e}_k\,|_x \otimes e^j\,|_x . \tag{8}$$

Here $\varphi:\mathfrak{X}\mapsto(\mathfrak{X}^1,\ldots,\mathfrak{X}^n)$ is a coordinate mapping on $\mathfrak{B}$, (x^i) are curvilinear coordinates on $\mathcal{E}$ with the local frame field $(\mathbf{e}_i)$, $(\mathfrak{X}^1,\ldots,\mathfrak{X}^n)\mapsto(x^1,x^2,x^3)$ is a coordinate representation of $\varkappa$, $[G^{\alpha\beta}]=[G_{\alpha\beta}]^{-1}$, $G_{\alpha\beta}=\mathbf{G}(\partial_\alpha,\partial_\beta)$, and $g_{ij}=\mathbf{e}_i\cdot\mathbf{e}_j$. Greek indices varies from 1 to $n=\dim\mathfrak{B}$, and Latin ones varies from 1 to 3. Thus, one has the sections

$$C:\mathfrak{B}\to T\mathfrak{B}\otimes T^*\mathfrak{B}, \quad \text{and} \quad \mathbf{B}:\varkappa(\mathfrak{B})\to T\mathcal{E}\otimes T^*\mathcal{E}.$$

6.2. Covector-valued exterior stress-forms

By a part $\mathfrak{P}$ of the body $\mathfrak{B}$ we mean a n-dimensional oriented smooth submanifold with the boundary in $\mathfrak{B}$ [24]. Let smooth sections $\mathbf{T}:\varkappa(\mathfrak{B})\to T^*\mathcal{E}\otimes T^*\mathcal{E}$ and $\mathbf{P}:\mathfrak{B}\to T^*\mathcal{E}\otimes T^*\mathfrak{B}$ correspond to the Cauchy and first Piola-Kirchhoff stress tensor fields.

Remark 7. The following interpretation for the section $\mathbf{T}$ takes place. Let $\mathfrak{P}$ be a part of the body $\mathfrak{B}$ and $\mathbf{n}:\partial\varkappa(\mathfrak{P})\to V$ be a unit normal vector field that is applied to the boundary of the shape $\varkappa(\mathfrak{P})$. Then the covector field $x\mapsto t_{\mathbf{n}_x}(x):=\mathbf{T}_x[\mathbf{n}_x]$ represents a classical stress vector field $t^\sharp_{\mathbf{n}}$. If $\mathbf{v}$ is a spatial velocity field then the scalar field $\langle t_{\mathbf{n}},\mathbf{v}\rangle = t^\sharp_{\mathbf{n}}\cdot\mathbf{v}$ represents a surface density of the power generated by contact forces on the boundary $\partial\varkappa(\mathfrak{P})$. The interpretation of the section $\mathbf{P}$ is quite cumbersome since the notion of oriented area element of the $\partial\mathfrak{B}$ requires clarification. This problem was considered in a variety of works [34, 35] and we follow the ideas that are given in them. ♣

The reasoning starting point is as follows. The exterior form is an object which can be integrated over a smooth manifold. Let the classical situation hold. The body has a stress free reference shape, that is the image of a configuration $\varkappa_R\in\mathfrak{C}(\mathfrak{B})$, and there is a motion, which may be represented by a family $\{\varkappa_t\}_{t\in\mathbb{T}}$ of configurations $\varkappa_t\in\mathfrak{C}(\mathfrak{B})$. Here $\mathbb{T}$ is a real axis interval. The power on the actual shape is represented by the integral

$$p=\int_{\partial\varkappa_t(\mathfrak{P})} \mathbf{v}\lrcorner\mathbf{T}\llcorner\mathbf{n}\,ds,$$

and the power on the reference shape can be written in the form

$$p=\int_{\partial\varkappa_R(\mathfrak{P})} \mathbf{V}\lrcorner\mathbf{P}\llcorner\mathbf{N}\,dS,$$

where $\mathbf{v}$, $\mathbf{V}$ are the spatial and material velocities and $\mathbf{n}$, $\mathbf{N}$ are normal vector fields. In the framework of these reasoning we consider the first Piola-Kirchhoff stress tensor as $\mathbf{P}:\varkappa_R(\mathfrak{B})\to T^*\mathcal{E}\otimes T^*\varkappa_R(\mathfrak{B})$. Note that $\mathbf{n}ds$ and $\mathbf{N}dS$ are oriented area elements, and, thus, can be considered as exterior forms. The idea is to view on $\mathbf{T}\llcorner\mathbf{n}ds$ and $\mathbf{P}\llcorner\mathbf{N}dS$ as to one-type objects that are covector-valued exterior forms. By returning to the general case, we introduce the sections

$$T=*_2\mathbf{T}, \quad P=*_2\mathbf{P}, \tag{9}$$

where $*_2$ is the Hodge star operation applied to the "second leg". Thus, one gets coordinatewise

$$T=T_{ab}e^a\otimes(*e^b), \quad P=P_{a\beta}e^a\otimes(*E^\beta),$$

where $\mathbf{T}=T_{ab}e^a\otimes e^b$, $\mathbf{P}=P_{a\beta}e^a\otimes E^\beta$, and e^i, E^α are coframe fields on $\mathcal{E}$ and $\mathfrak{B}$. Since [30]

$$*E^{\beta}=\sqrt{G}\sum_{\gamma=1}^{n}(-1)^{\gamma-1}G^{\beta\gamma}E^{1}\wedge\ldots\wedge E^{\gamma}\wedge\ldots\wedge E^{n},$$

$$*e^{b}=\sqrt{g}\sum_{j=1}^{3}(-1)^{j-1}g^{bj}e^{1}\wedge e^{j}\wedge e^{3},$$

we arrive at the following representations for (9):

$$T=\sum_{j=1}^{3}(-1)^{j-1}g^{bj}T_{ab}\sqrt{g}\,e^{a}\otimes\left(e^{1}\wedge e^{j}\wedge e^{3}\right), \tag{10}$$

$$P=\sum_{\gamma=1}^{n}(-1)^{\gamma-1}G^{\beta\gamma}P_{a\beta}\sqrt{G}\,e^{a}\otimes\left(E^{1}\wedge\ldots\wedge E^{\gamma}\wedge\ldots\wedge E^{n}\right). \tag{11}$$

Here $[g^{ij}]=[g_{ij}]^{-1}$, $g=\det[g_{ij}]$, $g_{ij}=\mathbf{e}_i\cdot\mathbf{e}_j$. Similarly, $[G^{\alpha\beta}]=[G_{\alpha\beta}]^{-1}$, $G=\det[G_{\alpha\beta}]$ and $G_{\alpha\beta}=\mathbf{G}(E_\alpha,E_\beta)$. Note that the power which is generated by stress forms is represented by the following expressions:

$$p=\int_{\partial\varkappa_t(\mathfrak{P})}\mathbf{v}\lrcorner T,\quad p=\int_{\partial\mathfrak{P}}\mathbf{V}\lrcorner P.$$

The analog of the classical Piola transformation is represented in terms of pullback [35]:

$$P=\varkappa^{*_2}T. \tag{12}$$

Here $\varkappa^{*_2}$ is the pullback provided by the configuration $\varkappa$ on the “second leg” of T, i.e.

$$P=T_{ab}e^{a}\otimes\varkappa^{*}(*e^{b}).$$

The exterior form $\varkappa^{*}(*e^{b})$ is defined on $\mathfrak{B}$.

6.3. *The particular case*

Consider the particular case: $n=3$. Obtain the representation of the Piola-Kirchhoff stress form through the Cauchy one by using (12). We begin with the relation (10) which has in our considerations, the form

$$T=T_{ab}\sqrt{g}\,dx^{a}\otimes\left(g^{b1}dx^{2}\wedge dx^{3}-g^{b2}dx^{1}\wedge dx^{3}+g^{b3}dx^{1}\wedge dx^{2}\right). \tag{13}$$

To use (12) we need a formula for $\varkappa^{*}(dx^{i}\wedge dx^{j})$. It follows from the definition of pullback that

$$\varkappa^{*}(dx^{i}\wedge dx^{j})=\frac{\partial\varkappa^{i}}{\partial\mathfrak{X}^{\alpha}}\frac{\partial\varkappa^{j}}{\partial\mathfrak{X}^{\beta}}d\mathfrak{X}^{\alpha}\wedge d\mathfrak{X}^{\beta}.$$

After substituting formula (13) into (12), taking into account the linearity of $\varkappa^{*}$ and the above formula for $\varkappa^{*}(dx^{i}\wedge dx^{j})$, we obtain the relation

$$P=JT_{ab}\sqrt{g}\,g^{bk}dX^{a}\otimes\left((\mathbf{F}^{-1})^{1}_{k}d\mathfrak{X}^{2}\wedge d\mathfrak{X}^{3}-(\mathbf{F}^{-1})^{2}_{k}d\mathfrak{X}^{1}\wedge d\mathfrak{X}^{3}+(\mathbf{F}^{-1})^{3}_{k}d\mathfrak{X}^{1}\wedge d\mathfrak{X}^{2}\right), \tag{14}$$

where $J=\det\left(\frac{\partial\varkappa^{i}}{\partial\mathfrak{X}^{k}}\right)$ and $\mathbf{F}^{-1}=(\mathbf{F}^{-1})^{l}_{k}\partial_{\mathfrak{X}^{l}}\otimes dx^{k}$ is the inverse to $\mathbf{F}=T\varkappa$. Since components for $\mathbf{F}^{-1}$ and its transpose, $\mathbf{F}^{-T}$ are related by the expression

$$g^{ki}(\mathbf{F}^{-1})^l_i = G^{li}(\mathbf{F}^{-T})^k_i,$$

the relation (14) may be rewritten in terms of $(\mathbf{F}^{-T})^k_i$. Now let us note that the expression (11) in our particular case takes the form

$$P = P_{ab}\sqrt{G}e^a \otimes \left(G^{b1}d\mathfrak{X}^2 \wedge \mathfrak{X}^3 - G^{b2}\mathfrak{X}^1 \wedge \mathfrak{X}^3 + G^{b3}\mathfrak{X}^1 \wedge \mathfrak{X}^2\right).$$

Then we get the final expression

$$P_{ak} = \sqrt{\frac{g}{G}}JT_{ab}(\mathbf{F}^{-T})^b_k, \tag{15}$$

in which $g = \det[g_{ij}]$, and $G = \det[G_{ij}]$.

Remark 8. In the case of incompressible material the factor $\sqrt{\frac{g}{G}}J$ in the expression (15) may be not equal to 1, in contrast to the classical Piola transformation. The reason is of the use of different coordinate maps on the reference and actual shapes. ♣

6.4. The Eshelby energy-momentum tensor

The Eshelby energy-momentum tensor can be generalized for the body $\mathfrak{B}$ as follows. The balance equation on the body has the form

$$\mathrm{Div}\mathbf{P}^{\sharp} + \rho_0 b_0^{\sharp} = 0. \tag{16}$$

Here $\mathbf{P}: \mathfrak{B} \to T^*\mathcal{E} \otimes T^*\mathfrak{B}$ is the first Piola-Kirchhoff stress tensor, $\mathbf{P}^{\sharp} = P^{a\kappa}\partial_{x^a} \otimes \partial_{\mathfrak{X}^\kappa}$ and[5]

$$\mathrm{Div}\mathbf{P}^{\sharp} = [\partial_\kappa P^{a\kappa} + P^{a\kappa}\Gamma^\beta_{\kappa\beta} + P^{b\kappa}F^c_\kappa \gamma^a_{bc}]\partial_{x^a},$$

where Γ^q_{kl} and γ^q_{kl} are Levi-Civita connection coefficients on $\mathfrak{B}$ and $\mathcal{E}$, respectively[6]; ρ_0 is the mass density and b_0 is the body force. Applying the contraction with $\mathbf{F}^{\mathrm{T}} = (\mathbf{F}^{\mathrm{T}})^\alpha_j \partial_{\mathfrak{X}^\alpha} \otimes dx^j$ to (16) gives us

$$\mathbf{F}^{\mathrm{T}} \llcorner \mathrm{Div}\mathbf{P}^{\sharp} + \rho_0 \mathbf{F}^{\mathrm{T}} \llcorner b_0^{\sharp} = 0. \tag{17}$$

Further we will use the bilinear operation $\Diamond$, that satisfies the following property:

$$\partial_{\mathfrak{X}^\alpha} \otimes dx^a \otimes d\mathfrak{X}^\beta \Diamond \partial_{x^b} \otimes \partial_{\mathfrak{X}^\kappa} = \langle dx^a, \partial_{x^b}\rangle\langle d\mathfrak{X}^\beta, \partial_{\mathfrak{X}^\kappa}\rangle \partial_{\mathfrak{X}^\alpha}.$$

The following formula can be verified directly:

$$\mathrm{Div}(\mathbf{F}^{\mathrm{T}} \llcorner \mathbf{P}^{\sharp}) = \mathbf{F}^{\mathrm{T}} \llcorner \mathrm{Div}\mathbf{P}^{\sharp} + (\nabla \mathbf{F}^{\mathrm{T}}) \Diamond \mathbf{P}^{\sharp}. \tag{18}$$

By using (18) in the equation (17) we have

$$\mathrm{Div}(\mathbf{F}^{\mathrm{T}} \llcorner \mathbf{P}^{\sharp}) - (\nabla \mathbf{F}^{\mathrm{T}}) \Diamond \mathbf{P}^{\sharp} + \rho_0 \mathbf{F}^{\mathrm{T}} \llcorner b_0^{\sharp} = 0.$$

---

[5] According to [17], the covariant derivative of a two point tensor field $Q = Q^{a\kappa}\partial_{x^a} \otimes \partial_{\mathfrak{X}^\kappa}$ has the representation $\nabla Q = Q^{a\kappa}_{|\beta}\partial_{x^a} \otimes \partial_{\mathfrak{X}^\kappa} \otimes d\mathfrak{X}^\beta$, in which $Q^{a\kappa}_{|\beta} = \partial_\beta Q^{a\kappa} + Q^{a\lambda}\Gamma^\kappa_{\lambda\beta} + Q^{b\kappa}F^c_\beta \gamma^a_{bc}$. It follows that $\mathrm{Div}Q = Q^{a\kappa}_{|\kappa}$.

[6] Let us remind the reader that, for convenience, in coordinate representations we use Greek indices for the body and Latin ones for the physical space.

Let $W:\mathfrak{B}\to\mathbb{R}$ be the elastic energy density defined on $\mathfrak{B}$. Subtracting the field[7] $(\nabla_{\mathfrak{B}}W)^{\sharp}$ from the left and right sides of the obtained equation leads to

$$\mathrm{Div}(\mathbf{F}^{\mathrm{T}}\lrcorner\mathbf{P}^{\sharp})-(\nabla_{\mathfrak{B}}W)^{\sharp}-(\nabla\mathbf{F}^{\mathrm{T}})\lozenge\mathbf{P}^{\sharp}+\rho_0\mathbf{F}^{\mathrm{T}}\llcorner b_0^{\sharp}=-(\nabla_{\mathfrak{B}}W)^{\sharp}. \tag{19}$$

Denote $I=\partial_{\mathfrak{X}^{\kappa}}\otimes d\mathfrak{X}^{\kappa}$ (the identity tensor on $\mathfrak{B}$). Since

$$\mathrm{Div}(WI^{\sharp})=\{G^{\alpha\beta}\partial_{\mathfrak{X}^{\beta}}W\}\partial_{\mathfrak{X}^{\alpha}}=(\nabla_{\mathfrak{B}}W)^{\sharp},$$

the expression (19) takes the final form

$$\mathrm{Div}(e)+(\nabla\mathbf{F}^{\mathrm{T}})\lozenge\mathbf{P}^{\sharp}-\rho_0\mathbf{F}^{\mathrm{T}}\llcorner b_0^{\sharp}=(\nabla_{\mathfrak{B}}W)^{\sharp}, \tag{20}$$

where $e=WI^{\sharp}-\mathbf{F}^{\mathrm{T}}\llcorner\mathbf{P}^{\sharp}$ is the *Eshelby energy-momentum tensor*.

Let $\tilde{W}$ be the elastic energy density represented as a function of points and fields defined on $\mathfrak{B}$ according to

$$W_{\mathfrak{X}}=\tilde{W}(\mathfrak{X};\mathbf{F}_{\mathfrak{X}},\ldots).$$

By following [27] let us define $\nabla_{\mathfrak{B}}W|_{\mathrm{impl}}$ as[8]:

$$\nabla_{\mathfrak{B}}W|_{\mathrm{impl}}=\nabla_{\mathfrak{B}}W-\nabla_{\mathfrak{B}}W|_{\mathrm{expl}},\quad \text{where } \nabla_{\mathfrak{B}}W|_{\mathrm{expl}}=\nabla_{\mathfrak{B}}\tilde{W}(\mathfrak{X};\mathbf{F}_{\mathfrak{X}_0},\ldots)\Big|_{\mathfrak{X}_0=\mathfrak{X}}.$$

In this case the relation (20) can be rewritten in the form [27]:

$$-\mathrm{Div}(e)=f_{\mathrm{int}}^{\sharp}+f_{\mathrm{ext}}^{\sharp}+f_{\mathrm{inh}}^{\sharp},$$

where $f_{\mathrm{int}}^{\sharp}=(\nabla\mathbf{F}^{\mathrm{T}})\lozenge\mathbf{P}^{\sharp}-\nabla_{\mathfrak{B}}W|_{\mathrm{impl}}^{\sharp}$ is the internal force, $f_{\mathrm{ext}}^{\sharp}=-\rho_0\mathbf{F}^{\mathrm{T}}\llcorner b_0^{\sharp}$ is the body force, and $f_{\mathrm{inh}}^{\sharp}=-\nabla_{\mathfrak{B}}W|_{\mathrm{expl}}^{\sharp}$ is the *Eshelby force*.

7. Models of LbL structures

The object of the present paper is a solid with variable composition, i.e. deformable body composed of a set of material points, variable in time. Layer-by-Layer structures may serve as example. Such solids can be represented by the family of bodies $\mathfrak{G}=\{\mathfrak{B}_{\alpha}\}_{\alpha\in\mathbb{I}}$, where $\mathbb{I}$ is a linearly ordered set of indexes. Without any loss of generality one may assume that $\mathbb{I}\subset\mathbb{R}$. Each element of the family $\mathfrak{G}$, namely a set $\mathfrak{B}_{\alpha}$, characterizes an instantaneous composition of the solid. The cardinality of the set $\mathbb{I}$ determines the type of the process. When $\mathrm{Card}\mathbb{I}=N<\infty$, the process may be qualified as discrete, while $\mathrm{Card}\mathbb{I}=\aleph_1$ corresponds to continuous process [36]. Without loss of generality it is assumed that $\mathbb{I}=\{1,\ldots,N\}$ in the case of discrete process, and $\mathbb{I}=[a_0,a_1[\subset\mathbb{R}$ in the case of continuous one.

We accept the following simplifying assumptions:

a) For each $\alpha\in\mathbb{I}$ the set $\mathfrak{B}_{\alpha}$ is a materially uniform simple body with inhomogeneities.

b) Layers are sequentially join to the solid and become the inner part of it. During the process material cannot be removed. Thus, the family $\mathfrak{G}$ is ordered by the subset relation, i.e.

---

[7] Note that $\nabla_{\mathfrak{B}}W$ is a covector field, and $(\nabla_{\mathfrak{B}}W)^{\sharp}$ is vector field.

[8] The expression $\nabla_{\mathfrak{B}}\tilde{W}(\mathfrak{X};\mathbf{F}_{\mathfrak{X}_0},\ldots)\Big|_{\mathfrak{X}_0=\mathfrak{X}}$ has the following sense: covariant derivative is taken from the field $\mathfrak{X}\mapsto\tilde{W}(\mathfrak{X};\mathbf{F}_{\mathfrak{X}_0},\ldots)$, where $\mathbf{F}_{\mathfrak{X}_0},\ldots$ are field values at a fixed point $\mathfrak{X}_0$. After calculation of covariant derivative all entries of $\mathfrak{X}_0$ are replaced by $\mathfrak{X}$.

$$\forall \alpha, \beta \in \mathbb{I}: \ (\alpha < \beta) \Rightarrow (\mathfrak{B}_\alpha \subsetneqq \mathfrak{B}_\beta).$$

In addition, for any $\alpha < \beta$ the body $\mathfrak{B}_\alpha$ is a "part" of the $\mathfrak{B}_\beta$ and, hence, $\mathfrak{B}_\alpha$ should have topology induced by[9] $\mathfrak{B}_\beta$.

Thus, it is convenient to consider bodies from $\mathfrak{G}$ as smooth submanifolds [24] of an ambient n-dimensional smooth manifold $\mathcal{M}$ that represents the total variable body composition. It means that $\mathcal{M}$ is sufficiently large to hold each element of the $\mathfrak{G}$. We refer to $\mathcal{M}$ as *material manifold*. In the case of continuous process we enable to represent each element of the family $\mathfrak{G}$, which corresponds to $\alpha \in \mathbb{I}$, as $\mathfrak{B}_\alpha := \mathfrak{B} \vee \mathfrak{B}_\alpha$. Here the symbol $\mathfrak{B}_\alpha$ represents an extra body, formed by extra material that inflowed to the substrate $\mathfrak{B}$. In the case of discrete process we obtain the following expression: $\mathfrak{B}_{s+1} := \mathfrak{B}_s \vee \mathfrak{L}_s$. Here this symbol $\mathfrak{L}_s$ corresponds to the layer that is attached to the previous assembly $\mathfrak{B}_s$. In both cases the symbol $\vee$ denotes the operation of *joining*, that assigns to each pair of bodies a new body, that contains both of them.

Although the partition of the body into layers and its assembling seems to be natural operations on the set of material points, constituting the body, their formal implementations have technical issues. In particular, the set difference of the body and its part results on a set, which is not open and therefore it cannot be a body. Moreover, after the union operation applied to two bodies with a partially coincide boundary, the intersection of boundaries would not be a part of the result (recall that we treat a body as an open set). Below we will represent our view on how to solve these problems.

7.1. Layers and their joining

Intuitive notions of extra body (or layer, in discrete case) and joining operation may be formalized as follows. Denote the topology of $\mathcal{M}$ by $\mathfrak{T}$. Since each three-dimensional body containing in $\mathcal{M}$ is an open set, and vice versa, it is sufficient to define the operation of extra body (or layer) detachment over open sets. We consider the case when one body (or "substrate") contains in another. Let $\mathfrak{M}$ and $\mathfrak{N}$ be n-dimensional smooth submanifolds of $\mathcal{M}$ (that is, bodies), and $\mathfrak{N} \subset \mathfrak{M}$. Then, the set $\mathfrak{N}$ is open in the topological space $\mathfrak{M}$. Therefore, the difference $\mathfrak{M} \setminus \mathfrak{N}$ represents a closed set relatively to the *induced topology* $\mathfrak{T}_\mathfrak{M}$ of $\mathfrak{M}$. Hence, one has that $\overline{\mathfrak{M} \setminus \mathfrak{N}} = \mathfrak{M} \setminus \mathfrak{N}$. We arrive at the following representation [37]:

$$\mathfrak{M} \setminus \mathfrak{N} = \mathrm{Int}_\mathfrak{M}(\mathfrak{M} \setminus \mathfrak{N}) \cup \partial_\mathfrak{M}(\mathfrak{M} \setminus \mathfrak{N}),$$

where the topological operations of closure, interior ($\mathrm{Int}_\mathfrak{M}$), and boundary ($\partial_\mathfrak{M}$) are considered in the topology $\mathfrak{T}_\mathfrak{M}$. The set $\mathrm{Int}_\mathfrak{M}(\mathfrak{M} \setminus \mathfrak{N})$ is open in $\mathcal{M}$. Since interior operation is defined as the union of all open sets containing in the given set, we have that

$$\mathrm{Int}_\mathfrak{M}(\mathfrak{M} \setminus \mathfrak{N}) = \mathrm{Int}(\mathfrak{M} \setminus \mathfrak{N}),$$

where Int is the *interior operation relatively to the topology of* $\mathcal{M}$. Denote

$$\mathfrak{I}(\mathfrak{M}, \mathfrak{N}) := \mathrm{Int}(\mathfrak{M} \setminus \mathfrak{N}).$$

It is the desired extra body (or, respectively, layer). Thus, if one introduce the set

$$A = \{(\mathfrak{M}, \mathfrak{N}) \in 2^\mathcal{M} \times 2^\mathcal{M} \,|\, \mathfrak{M} \in \mathfrak{T}, \quad \mathfrak{N} \in \mathfrak{T}_\mathfrak{M}\},$$

then the following operation is defined:

---

[9] In general case smooth manifolds $\mathfrak{B}_\alpha$, $\mathfrak{B}_\beta$ ($\alpha, \beta \in \mathbb{I}$, $\alpha \neq \beta$) have independent topological structures, that may differ one from other [25].

$$\mathfrak{I}: A \to \mathfrak{T}, \quad \mathfrak{I}: (\mathfrak{M}, \mathfrak{N}) \mapsto \mathrm{Int}(\mathfrak{M} \setminus \mathfrak{N}).$$

Note that if $\mathfrak{N} \subset \mathfrak{M}_1 \subset \mathfrak{M}_2$, then, by definition, $\mathfrak{I}(\mathfrak{M}_1, \mathfrak{N}) \subset \mathfrak{I}(\mathfrak{M}_2, \mathfrak{N})$.

The operation $\mathfrak{I}$ allows one to partition any n-dimensional body $\mathfrak{M}$ into a body $\mathfrak{N} \in \mathfrak{T}_{\mathfrak{M}}$ and the body $\mathfrak{I}(\mathfrak{M}, \mathfrak{N})$. The symbol $\vee$ indicates this partition and is referred to as *joining operation*:

$$\mathfrak{M} = \mathfrak{N} \vee \mathfrak{I}(\mathfrak{M}, \mathfrak{N}) := \mathfrak{N} \cup \mathrm{Int}(\mathfrak{M} \setminus \mathfrak{N}) \cup \partial_{\mathfrak{M}}(\mathfrak{M} \setminus \mathfrak{N}).$$

For a continuous process one has $\mathfrak{B}_{\alpha} := \mathfrak{B} \vee \mathfrak{B}_{\alpha}$, where $\mathfrak{B}_{\alpha} = \mathfrak{I}(\mathfrak{B}_{\alpha}, \mathfrak{B})$, and for a discrete one, $\mathfrak{B}_{s+1} := \mathfrak{B}_s \vee \mathfrak{L}_s$, in which $\mathfrak{L}_s = \mathfrak{I}(\mathfrak{B}_{s+1}, \mathfrak{B}_s)$. In addition, $\mathfrak{L}_0 := \mathfrak{B}_1$, where $\mathfrak{B}_1$ represents a substrate. Thus, $\mathfrak{B}_N = (...(\mathfrak{B}_1 \vee \mathfrak{L}_1) \vee ... \vee \mathfrak{L}_{N-2}) \vee \mathfrak{L}_{N-1}$.

Remark 9. The operation $\vee$ is not equal to the conventional set union and has the following physical sense, which we will explain in the case of discrete process. When we obtain sequentially new assembly $\mathfrak{B}_{s+1}$ from the previous assembly $\mathfrak{B}_s$ and the layer $\mathfrak{L}_s$, we have to include all interior points of $\mathfrak{B}_s$ and $\mathfrak{L}_s$ and, besides them, such points of common boundary of $\overline{\mathfrak{B}_s}$ and $\overline{\mathfrak{L}_s}$, which represent thin film between "gluing" solids. ♣

7.2. Discrete process

Let discrete LbL structure is represented by a family $\mathfrak{G} = \{\mathfrak{B}_s\}_{s=1}^{N}$ Using the idea of layering developed in subsection 7.1 allows one to formulate the following assumptions:

a) For each $s \in \{1, ..., N\}$ the set $\mathfrak{B}_s$ is a *three-dimensional inhomogeneous materially uniform simple body*. A stress-free state is meant by the uniform state.
b) For each $s \in \{1, ..., N\}$ one has decomposition $\mathfrak{B}_s = (...(\mathfrak{B}_1 \vee \mathfrak{L}_1) \vee ... \vee \mathfrak{L}_{s-2}) \vee \mathfrak{L}_{s-1}$. *Each layer $\mathfrak{L}_k$ is a three-dimensional materially uniform simple body that has stress-free shape in $\mathcal{E}$.*
c) *Stress-strain state of each $\mathfrak{B}_s$ depends on the previous one of $\mathfrak{B}_{s-1}$.*

The deformation of each layer $\mathfrak{L}_k$ can be described in the framework of the classical continuum mechanics. If $\varkappa_k^R, \varkappa_k \in \mathfrak{C}(\mathfrak{L}_k)$ are respectively the stress-free reference configuration and the actual configuration of a layer $\mathfrak{L}_k$, then the corresponding deformation is defined as follows

$$\gamma_k = \varkappa_k \circ [\varkappa_k^R]^{-1} : \varkappa_k^R(\mathfrak{L}_k) \to \varkappa_k(\mathfrak{L}_k).$$

There is a finite family $\{\gamma_k\}_{k=0}^{N-1}$ of deformations and a finite family of $\{\varkappa_k^R(\mathfrak{L}_k)\}_{k=0}^{N-1}$ of reference shapes. The body $\mathfrak{B}_s$ is observed in some actual configuration $\varkappa_{\mathfrak{B}_s} \in \mathfrak{C}(\mathfrak{B}_s)$ and the deformations of the layers can be agreed with this configuration by the requirement $\varkappa_k := \varkappa_{\mathfrak{B}_s}|_{\mathfrak{L}_k}$, for $k \in \{0, ..., s-1\}$. Thus, $\mathfrak{B}_s$ has one actual shape and a family of reference shapes. Each of this reference shapes has internal stresses, except the open subset, that corresponds to the reference shape of a layer.

The assumption c) asserts that each element of the family $\mathfrak{G}$ represents a some kind of "memory". That is, it stores the information about stress-strain state of the LbL structure at the corresponding instant. Thus, layer deformations cannot be independent: they obey some recurrent relations. In section 8 we will consider an example of such relation.

7.3. Continuous process

A some class of solids with continuous inhomogeneity can be singled out in the following way. We intend to model a situation, when material continuously inflows to a substrate. Let $\mathfrak{B}$ be a three-dimensional materially uniform homogeneous simple body, that represents a substrate, and let

$\mathfrak{G} = \{\mathfrak{B}_\alpha\}_{\alpha\in\mathbb{I}}$ be a family of three-dimensional materially uniform simple bodies with inhomogeneities that represents a solid with variable material composition. Here by a uniform state we mean a stress-free state. Each element of the family $\mathfrak{G}$ contains $\mathfrak{B}$. Assume that $\mathbb{I} = [a_0, a_1[$ and that $\mathfrak{B}_{a_0} = \mathfrak{B}$. According to subsection 7.1 for each $\alpha \in \tilde{\mathbb{I}} =]a_0, a_1[$ take $\mathfrak{B}_\alpha := \mathrm{Int}(\mathfrak{B}_\alpha \setminus \mathfrak{B})$. Thus, we have arrived at the relation $\mathfrak{B}_\alpha := \mathfrak{B} \vee \mathfrak{B}_\alpha$ and constructed the new family $\mathfrak{G} = \{\mathfrak{B}_\alpha\}_{\alpha\in\tilde{\mathbb{I}}}$ of three-dimensional materially uniform simple bodies with inhomogeneities. Now let us formulate the following assumption respectively to the family $\mathfrak{G}$:

$\mathfrak{a}$) *For each $\alpha \in \tilde{\mathbb{I}}$ there exists a smooth fiber bundle $(\mathfrak{B}_\alpha, \mathbb{J}_\alpha, \pi_\alpha, \mathfrak{F})$ with one-dimensional base $\mathbb{J}_\alpha$, total space $\mathfrak{B}_\alpha$ and projection $\pi_\alpha : \mathfrak{B}_\alpha \to \mathbb{J}_\alpha$. Each fiber $\mathfrak{S}_{\alpha;i} := \pi_\alpha^{-1}(\{i\})$, $i \in \mathbb{J}_\alpha$, is diffeomorphic to a typical fiber $\mathfrak{F}$, which is a smooth two-dimensional manifold. The smooth fiber bundle $(\mathfrak{B}_\alpha, \mathbb{J}_\alpha, \pi_\alpha, \mathfrak{F})$ satisfies the following property. There exists a family $\overline{\mathrm{Ref}}_{\mathfrak{B}_\alpha} = \{\varkappa^R_{\alpha;\mathfrak{X}}\}_{\mathfrak{X}\in\mathfrak{B}_\alpha}$ of configurations $\varkappa^R_{\alpha;\mathfrak{X}} \in \mathfrak{C}(\mathfrak{B}_\alpha)$ with the index set $\mathfrak{B}_\alpha$ and the following attribute. If $\mathfrak{X} \in \mathfrak{B}_\alpha$ then each point from a two-dimensional submanifold $\varkappa^R_{\alpha;\mathfrak{X}}(\mathfrak{S}_{\alpha;\pi_\alpha(\mathfrak{X})})$ in $\mathcal{E}$ has some uniform spatial neighborhood. Moreover, if $\mathfrak{X}, \mathfrak{Y} \in \mathfrak{S}_{\alpha;i}$ for some $i \in \mathbb{J}_\alpha$, then $\varkappa^R_{\alpha;\mathfrak{X}} = \varkappa^R_{\alpha;\mathfrak{Y}}$.*

The assumption $\mathfrak{a}$) asserts that $\mathfrak{B}_\alpha$ can be represented as a union of disjoint two-dimensional manifolds $\mathfrak{S}_i$. That is,

$$\forall \alpha \in \tilde{\mathbb{I}}: \quad \mathfrak{B}_\alpha = \bigcup_{i\in\mathbb{J}_\alpha} \mathfrak{S}_{\alpha;i}.$$

The mapping π_α, that generates each $\mathfrak{S}_{\alpha;i}$, is smooth and surjective. It follows from such a representation of $\mathfrak{B}_\alpha$: **i)** *there exists a family $\overline{\mathrm{Ref}}_{\mathfrak{B}_\alpha} = \{\varkappa^R_{\alpha;\mathfrak{X}}\}_{\mathfrak{X}\in\mathfrak{B}_\alpha}$, as in section 3*; **ii)** *not only $\varkappa^R_{\alpha;\mathfrak{X}}(\mathfrak{X})$ has a uniform neighborhood, but all points of the whole set $\varkappa^R_{\alpha;\mathfrak{X}}(\mathfrak{S}_{\alpha;\pi_\alpha(\mathfrak{X})})$ are satisfying this property*; **iii)** *configurations $\varkappa^R_{\alpha;\mathfrak{X}}$ that are correspond to material points, belonging to the same $\mathfrak{S}_{\alpha;i}$, coincide*. The idea of fibration is illustrated in figure 4.

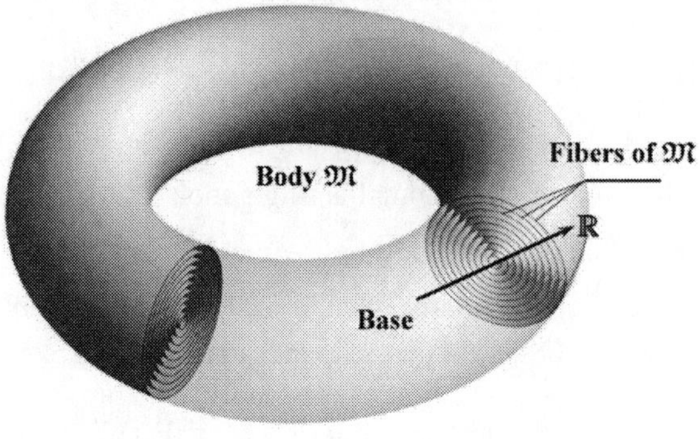

Figure 4. Fibration of a body $\mathfrak{M}$.

Recall that the body $\mathfrak{B}$, as a whole, has a uniform shape. Assume that

$\mathfrak{b}$) *For each $\alpha \in \tilde{\mathbb{I}}$ there exists a configuration $\mathring{\varkappa}_\alpha \in \mathfrak{C}(\mathfrak{B}_\alpha)$, such that $\mathring{\varkappa}_\alpha(\mathfrak{B})$ is a uniform shape and that the following property is satisfied. For any $\mathfrak{X} \in \mathfrak{B}_\alpha$ the deformation*

$$\tilde{\gamma}_{\alpha;\mathfrak{X}} = \hat{\varkappa}^R_{\alpha;\mathfrak{X}} \circ \{\hat{\mathring{\varkappa}}_\alpha|_{\mathfrak{B}_\alpha}\}^{-1} : \mathring{\varkappa}_\alpha(\mathfrak{B}_\alpha) \to \varkappa^R_{\alpha;\mathfrak{X}}(\mathfrak{B}_\alpha),$$

from the shape $\overset{\circ}{\varkappa}_\alpha(\mathfrak{B}_\alpha)$ *into the shape* $\overset{R}{\varkappa}_{\alpha;\mathfrak{X}}(\mathfrak{B}_\alpha)$ *is prescribed and depends on* $k \geqslant 1$ *free scalar parameters* $\rho^1_{\alpha;\mathfrak{X}},\dots,\rho^k_{\alpha;\mathfrak{X}} \in \mathbb{R}$.

$\mathfrak{c}$) *The mappings* $\tilde{b}^j_\alpha : \mathfrak{B}_\alpha \to \mathbb{R}$, *that are defined by the rule* $\tilde{b}^j_\alpha : \mathfrak{X} \mapsto \rho^j_{\alpha;\mathfrak{X}}$, *are smooth for all* $j = 1,\dots,k$ *and* $\alpha \in \tilde{\mathbb{I}}$.

The configuration $\overset{\circ}{\varkappa}_\alpha$, which has been introduced in assumption $\mathfrak{b}$), is referred to as intermediate configuration. Deformation parameters $\rho^1_{\alpha;\mathfrak{X}},\dots,\rho^k_{\alpha;\mathfrak{X}}$ determine its degrees of freedom. It is convenient to introduce the mappings $b^j_\alpha : \mathfrak{B}_\alpha \to \mathbb{R}$,

$$b^j_\alpha(\mathfrak{X}) = \begin{cases} 0, & \mathfrak{X} \in \mathfrak{B} \cup \partial_{\mathfrak{B}_\alpha}(\mathfrak{B}_\alpha \setminus \mathfrak{B}), \\ \tilde{b}^j_\alpha(\mathfrak{X}), & \mathfrak{X} \in \mathfrak{B}_\alpha, \end{cases}$$

that are defined on the whole $\mathfrak{B}_\alpha$. The set $\partial_{\mathfrak{B}_\alpha}(\mathfrak{B}_\alpha \setminus \mathfrak{B})$, which has been introduced in subsection 7.1, represents the set of jump points, and, in the physical sense, the thin layer between the substrate and the extra body. It follows from the homogeneity of the body $\mathfrak{B}$ that its material Riemannian metric is Euclidean. Thus, in general case, material metric in the whole body $\mathfrak{B}_\alpha$ would have points of discontinuity in $\partial_{\mathfrak{B}_\alpha}(\mathfrak{B}_\alpha \setminus \mathfrak{B})$. Hence, it is sufficient to provide our reasonings only for bodies $\mathfrak{B}_\alpha$.

Since the deformation $\tilde{\gamma}_{\alpha;\mathfrak{X}}$ is prescribed with accuracy up to parameters $\rho^j_{\alpha;\mathfrak{X}}$, the family $\{\overset{R}{\mathbf{K}}_\alpha\}_{\alpha\in\mathfrak{B}_\alpha}$ of uniform references can be obtained. To this end let us introduce local coordinates $(\xi^i_\alpha)^3_{i=1}$ on the intermediate shape $\overset{\circ}{\varkappa}_\alpha(\mathfrak{B}_\alpha)$. Using the map $\overset{\circ}{\varkappa}_\alpha$, it is possible to transfer these coordinates to the body $\mathfrak{B}_\alpha$. The frame and coframe on the body are denoted by $(\partial_{\xi^1_\alpha}, \partial_{\xi^2_\alpha}, \partial_{\xi^3_\alpha})$, and $(d\xi^1_\alpha, d\xi^2_\alpha, d\xi^3_\alpha)$ consequently. Despite the fact that local coordinates on the body and its intermediate shape have the same notations, the corresponding frames and coframes are different. This is indicated in the table 1.

Table 1. Correspondence between frames and coframes on the body and the intermediate shape.

	$\mathfrak{B}_\alpha$	$\overset{\circ}{\varkappa}_\alpha(\mathfrak{B}_\alpha)$
Coordinate map	$\overset{\circ}{H} \circ \hat{\overset{\circ}{\varkappa}}_\alpha : \mathfrak{X} \mapsto (\xi^1_\alpha, \xi^2_\alpha, \xi^3_\alpha)$	$\overset{\circ}{H} : x \mapsto (\xi^1_\alpha, \xi^2_\alpha, \xi^3_\alpha)$
Coordinates	$(\xi^1_\alpha, \xi^2_\alpha, \xi^3_\alpha)$	$(\xi^1_\alpha, \xi^2_\alpha, \xi^3_\alpha)$
Frame	$(\partial_{\xi^1_\alpha}, \partial_{\xi^2_\alpha}, \partial_{\xi^3_\alpha})$	$(\mathbf{e}_{\xi^1_\alpha}, \mathbf{e}_{\xi^2_\alpha}, \mathbf{e}_{\xi^3_\alpha})$
Coframe	$(d\xi^1_\alpha, d\xi^2_\alpha, d\xi^3_\alpha)$	$(e^{\xi^1_\alpha}, e^{\xi^2_\alpha}, e^{\xi^3_\alpha})$

Remark 10. Transferring local coordinates from intermediate shape to the body by means of the mapping $\hat{\overset{\circ}{\varkappa}}_\alpha$ is equivalent to the consideration of $\overset{\circ}{\varkappa}_\alpha(\mathfrak{B}_\alpha)$ as a "model" of $\mathfrak{B}_\alpha$. The body "dresses up" coordinate net taken from the set $\overset{\circ}{\varkappa}_\alpha(\mathfrak{B}_\alpha)$. ♣

For any point $\mathfrak{X} \in \mathfrak{B}_\alpha$, the tangent map (configuration gradient) $T_{\mathfrak{X}}\overset{\circ}{\varkappa}_\alpha : T_{\mathfrak{X}}\mathfrak{B}_\alpha \to \mathcal{V}$ has the following dyadic decomposition:

$$T_{\mathfrak{X}}\overset{\circ}{\varkappa}_\alpha = \mathbf{e}_{\xi^1_\alpha}|_x \otimes d\xi^1_\alpha|_{\mathfrak{X}} + \mathbf{e}_{\xi^2_\alpha}|_x \otimes d\xi^2_\alpha|_{\mathfrak{X}} + \mathbf{e}_{\xi^3_\alpha}|_x \otimes d\xi^3_\alpha|_{\mathfrak{X}},$$

where $x = \overset{\circ}{\varkappa}_\alpha(\mathfrak{X})$.

Introduce local coordinates $(\Xi^i_{\alpha;\mathfrak{X}})^3_{i=1}$ on the each shape $\varkappa^R_{\alpha;\mathfrak{X}}(\mathfrak{B}_\alpha)$. Let the coordinate representation of $\tilde{\gamma}_{\alpha;\mathfrak{X}}$ in the pair of coordinates $(\xi^i_\alpha)^3_{i=1}$ and $(\Xi^i_{\alpha;\mathfrak{X}})^3_{i=1}$ has the form

$$\tilde{\gamma}_{\alpha;\mathfrak{X}} \leftrightarrow (\gamma^1_{\alpha;\mathfrak{X}}, \gamma^2_{\alpha;\mathfrak{X}}, \gamma^3_{\alpha;\mathfrak{X}}).$$

Then the deformation gradient at point $y \in \overset{\circ}{\varkappa}_\alpha(\mathfrak{B}_\alpha)$ is represented by

$$T_y\tilde{\gamma}_{\alpha;\mathfrak{X}} = \left.\frac{\partial\gamma^i_{\alpha;\mathfrak{X}}(\xi^1_\alpha, \xi^2_\alpha, \xi^3_\alpha; b^1_\alpha(\mathfrak{X}), ..., b^k_\alpha(\mathfrak{X}))}{\partial\xi^j_\alpha}\right|_y \mathbf{e}_{\Xi^i_{\alpha;\mathfrak{X}}}|_{\tilde{\gamma}_{\alpha;\mathfrak{X}}(y)} \otimes e^{\xi^j_\alpha}|_y .$$

Thus, the tangent map $T_{\mathfrak{Y}}\varkappa^R_{\alpha;\mathfrak{X}} : T_{\mathfrak{Y}}\mathfrak{B}_\alpha \to \mathcal{V}$ to $\varkappa^R_{\alpha;\mathfrak{X}}$ at point $\mathfrak{Y} \in \mathfrak{B}_\alpha$ may be calculated as a composition $T_{\mathfrak{Y}}\varkappa^R_{\alpha;\mathfrak{X}} = T_Y\tilde{\gamma}_{\alpha;\mathfrak{X}} \circ T_{\mathfrak{Y}}\overset{\circ}{\varkappa}_\alpha|_{\mathfrak{B}_\alpha}$:

$$T_{\mathfrak{Y}}\varkappa^R_{\alpha;\mathfrak{X}} = \left.\frac{\partial\gamma^i_{\alpha;\mathfrak{X}}(\xi^1_\alpha, \xi^2_\alpha, \xi^3_\alpha; b^1_\alpha(\mathfrak{X}), ..., b^k_\alpha(\mathfrak{X}))}{\partial\xi^j_\alpha}\right|_y \mathbf{e}_{\Xi^i_{\alpha;\mathfrak{X}}}|_y \otimes d\xi^j_\alpha|_{\mathfrak{Y}},$$

where $y = \varkappa^R_{\alpha;\mathfrak{X}}(\mathfrak{Y})$. Finally, the uniform reference $\mathbf{K}^R_\alpha : \mathfrak{X} \mapsto T_{\mathfrak{X}}\varkappa^R_{\alpha;\mathfrak{X}}$ is defined at point $\mathfrak{X} \in \mathfrak{B}_\alpha$ by the following expression:

$$\mathbf{K}^R_\alpha|_{\mathfrak{X}} = \left.\frac{\partial\gamma^i_{\alpha;\mathfrak{X}}(\xi^1_\alpha, \xi^2_\alpha, \xi^3_\alpha; b^1_\alpha(\mathfrak{X}), ..., b^k_\alpha(\mathfrak{X}))}{\partial\xi^j_\alpha}\right|_x \mathbf{e}_{\Xi^i_{\alpha;\mathfrak{X}}}|_x \otimes d\xi^j_\alpha|_{\mathfrak{X}}, \tag{21}$$

where $x = \varkappa^R_{\alpha;\mathfrak{X}}(\mathfrak{X})$. Thus, assumptions $\mathfrak{a})-\mathfrak{c})$ are sufficient to determine uniform reference up to mappings b^j_α.

By using obtained formula (21), the material metric $\mathbf{G}_\alpha$ defined by (3) is derived. The section

$$\mathbf{G}_\alpha(\mathfrak{X}) = \begin{cases} \left(\overset{\circ}{\varkappa}_\alpha\right)^* \mathbf{g}, & \mathfrak{X} \in \mathfrak{B} \cup \partial_{\mathfrak{B}_\alpha}(\mathfrak{B}_\alpha \setminus \mathfrak{B}), \\ \mathbf{G}_\alpha(\mathfrak{X}), & \mathfrak{X} \in \mathfrak{B}_\alpha, \end{cases} \tag{22}$$

represents the total material metric $\mathbf{G}_\alpha$ in $\mathfrak{B}_\alpha$. Here $\mathbf{g}$ is the Riemannian metric on $\mathcal{E}$ which is induced by the scalar product $(\cdot)$.

The functions b^j_α, which appear in (21), represent distortion parameters and are defined by physical mechanisms accompanying the continuous process. Among such mechanisms are shrinkage of adhered layers and inheritance of inhomogeneities. The latter implies that material properties of adhered layers do not change after joining. Complete assumptions $\mathfrak{a})-\mathfrak{c})$ by one more:

$\mathfrak{d})$ *For each* $j \in \{1, ..., k\}$, *if* $\alpha, \beta \in \mathbb{I}$, $\alpha < \beta$, *then*

$$\tilde{b}^j_\beta|_{\mathfrak{B}_\alpha} = \tilde{b}^j_\alpha.$$

This assumption expresses the idea of inheritance.

8. Example: Cylinder

8.1. Discrete process

Consider an example. In order to avoid heavy numerical computations, assume that this solid is an inhomogeneous hollow cylinder, which inhomogeneity appears during the process of its assembling from a number of thin unstressed hollow cylindrical parts. Its incompatibility can be clarified from technological point of view as follows. Hollow cylindrical layers join consequently and each layer undergo shrinkage at once after joining. Apply the following assumptions:

a) Layers $\mathfrak{L}_k$ have stress free reference configurations in three-dimensional Euclidean point space $\mathcal{E}$:

$$\varkappa_k^R : \mathfrak{L}_k \to \mathcal{E}, \quad k = 0, ..., N-1.$$

b) Images of the layers $\mathfrak{L}_k$ under configurations $\varkappa_k^R$, $\varkappa_k^R(\mathfrak{L}_k) \subset \mathcal{E}$, are represented by hollow cylinders with the inner radii R_i^k and the exterior radii R_e^k and common to all height h.

c) All bodies $\mathfrak{B}_s$, $s = 2, ..., N$, result by a sequential joining of layers $\mathfrak{L}_k$ to $\mathfrak{B}_1$. Image of any body $\mathfrak{B}_s$ under the actual configuration $\varkappa_s$ is a hollow cylinder. Cylinders $\varkappa_s(\mathfrak{L}_k)$ are coaxial. Denote by $r_{i,s}^k$ actual inner radius of k-th layer, and by $r_{e,s}^k$ actual exterior radius of k-th layer. The index s after comma designates the number of assembly.

d) Each layer before joining to assembly undergoes axisymmetric plane deformation.

e) Inner radius of the $\mathfrak{B}_1$ in reference configuration, R_i^0, is equal to ρ, and reference thicknesses $\Delta^k = R_e^k - R_i^k$, $k = 0, ..., N-1$, of the layers are prescribed. The other reference radii are unknown, but they can be determined from the following recurrence relation: the inner reference radius of the layer $\mathfrak{L}_{s+1}$ is equal to $R_i^{s+1} = S^{s+1} r_{e,s}^{s-1}$, where $0 < S^{s+1} < 1$ is the shrinkage coefficient, and $r_{e,s}^{s-1}$ is the exterior actual radius of the layer $\mathfrak{L}_{s-1}$.

These assumptions are schematically illustrated in figure 5.

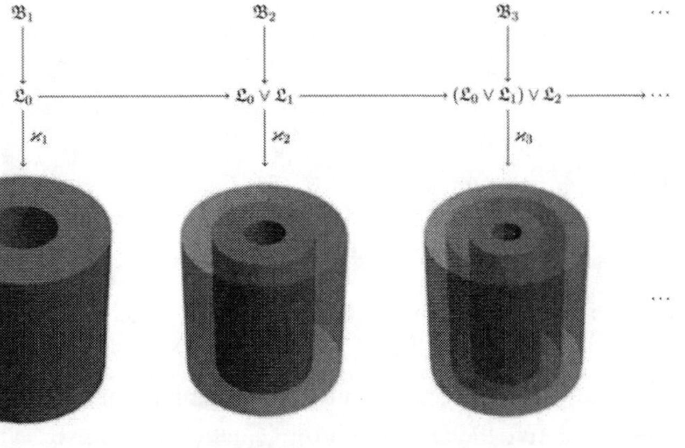

Figure 5. Cylindrical LbL structure.

Consider the assembly with the number s ($s \in \{1, ..., N\}$). Since $\varkappa_s(\mathfrak{B}_s)$ is a hollow cylinder, it is convenient to use the cylindrical coordinates (r, θ, z) that are related with the Cartesian coordinates via formulae:

$$x^1 = r\cos\theta, \quad x^2 = r\sin\theta, \quad x^3 = z.$$

The elements of the local frame $(\mathbf{e}_r, \mathbf{e}_\theta, \mathbf{e}_z)$, that corresponds to cylindrical coordinates, can be derived as $\mathbf{e}_c = (\partial x^k)/(\partial c)\mathbf{i}_k$. The index c designates one of the symbols r, θ, or z, and $x = x^k(r, \theta, z)\mathbf{i}_k$ is the position vector field, represented by the Cartesian frame. By (e^r, e^θ, e^z) we denote the coframe field $e^c : \varkappa_s(\mathfrak{B}_s) \to \mathcal{V}^*$, $e^c(\mathbf{e}_j) = \delta_j^c$, $c, j \in \{r, \theta, z\}$.

Let k be the layer index in the assembly with the number of layers s; $k \in \{0,\ldots,s-1\}$. According to the assumptions, stated above, the deformation of reference shape $\varkappa_R^k(\mathfrak{L}_k)$ into actual, $\varkappa_s(\mathfrak{L}_k)$, is simply "bloating". To this end we introduce cylindrical coordinates (R^k,Θ^k,Z^k) on the set $\varkappa_R^k(\mathfrak{L}_k)$. The deformation $d_k:\varkappa_R^k(\mathfrak{L}_k)\to\varkappa_s(\mathfrak{L}_k)$ belongs to the class of universal deformations [38] and can be represented in coordinate free notation as follows:

$$\mathbf{x}(\mathbf{X}_k)=\mathbf{e}_{R^k}\sqrt{(e^{R^k}\llcorner\mathbf{X}_k)^2+a_s^k}+\mathbf{e}_{Z^k}\otimes e^{Z^k}\llcorner\mathbf{X}_k.$$

In coordinates (R^k,Θ^k,Z^k) and (r,θ,z) it can be written as:

$$d_k:(R^k,\Theta^k,Z^k)\mapsto(r,\theta,z)=\left(\sqrt{(R^k)^2+a_s^k},\Theta^k,Z^k\right),$$

where a_s^k is the deformation parameter.

The deformation gradient $\mathbf{F}_k$, its transpose, $\mathbf{F}_k^{\mathrm{T}}$, and the left Cauchy-Green tensor $\mathbf{B}_k=\mathbf{F}_k\mathbf{F}_k^{\mathrm{T}}$ in coordinates (R^k,Θ^k,Z^k) and (r,θ,z) are written in the forms

$$\mathbf{F}_k=\frac{R^k}{\sqrt{(R^k)^2+a_s^k}}\mathbf{e}_r\otimes e^{R^k}+\mathbf{e}_\theta\otimes e^{\Theta^k}+\mathbf{e}_z\otimes e^{Z^k}, \tag{23}$$

$$\mathbf{F}_k^{\mathrm{T}}=\frac{R^k}{\sqrt{(R^k)^2+a_s^k}}\mathbf{e}_{R_k}\otimes e^r+\frac{(R^k)^2+a_s^k}{(R^k)^2}\mathbf{e}_{\Theta_k}\otimes e^\theta+\mathbf{e}_{Z_k}\otimes e^z,$$

$$\mathbf{B}_k=\frac{r^2-a_s^k}{r^2}\mathbf{e}_r\otimes e^r+\frac{r^2}{r^2-a_s^k}\mathbf{e}_\theta\otimes e^\theta+\mathbf{e}_z\otimes e^z. \tag{24}$$

In derivation of (23)–(24) we take into account that the metric tensors $\mathbf{G}_k$ and $\mathbf{g}$ on $\varkappa_R^k(\mathfrak{L}_k)$ and $\varkappa_s(\mathfrak{L}_k)$ have the following representations:

$$\mathbf{G}_k=e^{R^k}\otimes e^{R^k}+(R^k)^2e^{\Theta^k}\otimes e^{\Theta^k}+e^{Z^k}\otimes e^{Z^k},\quad \mathbf{g}=e^r\otimes e^r+r^2e^\theta\otimes e^\theta+e^z\otimes e^z.$$

Assume that layers are made of hyperelastic material. The constitutive relation has the form: $\mathbf{T}=-p\mathbf{I}+J_1\mathbf{B}+J_{-1}\mathbf{B}^{-1}$, where $\mathbf{I}$ is the identity tensor, p is the hydrostatic component, $J_1=(1+\beta)\mu/2$, $J_{-1}=(\beta-1)\mu/2$, and β and μ are the material constants. Taken into account the diagonality of $\mathbf{B}$, one can derive the expression for Cauchy stress tensor[10] in the layer with the number k:

$$\mathbf{T}_k^{\#}=T_k^{rr}\mathbf{e}_r\otimes\mathbf{e}_r+T_k^{\theta\theta}\mathbf{e}_\theta\otimes\mathbf{e}_\theta+T_k^{zz}\mathbf{e}_z\otimes\mathbf{e}_z,$$

$$T_k^{rr}=-p+J_1\frac{r^2-a_s^k}{r^2}+J_{-1}\frac{r^2}{r^2-a_s^k},\quad T_k^{\theta\theta}=-\frac{p}{r^2}+J_1\frac{1}{r^2-a_s^k}+J_{-1}\frac{r^2-a_s^k}{r^4},$$

$$T_k^{zz}=-p+J_1+J_{-1}.$$

It follows from the balance of linear momentum that

---

[10] Here we consider Cauchy stress tensor as a symmetric field $\mathbf{T}:\varkappa(\mathfrak{B})\to\mathcal{V}\otimes\mathcal{V}^*$. For a normal vector field $\mathbf{n}:\partial\varkappa(\mathfrak{B})\to\mathcal{V}$ we have the following representation for the stress vector: $\mathbf{t}_\mathbf{n}=\mathbf{T}\llcorner\mathbf{n}$.

$$\frac{\partial T_k^{rr}}{\partial r}+\frac{T_k^{rr}}{r}-rT_k^{\theta\theta}=0,\quad \frac{\partial T_k^{\theta\theta}}{\partial\theta}=0,\quad \frac{\partial T_k^{zz}}{\partial z}=0. \tag{25}$$

One can obtain components of $\mathbf{T}_k^{\#}$ by using expressions (25):

$$T_k^{rr}=\frac{\mu}{2}\left(\ln\frac{r^2-a_s^k}{r^2}-\frac{a_s^k}{r^2}\right)+p_{0,s}^k,\quad T_k^{\theta\theta}=\frac{T_k^{rr}}{r^2}+\frac{\mu}{r^2}\left(\frac{r^2}{r^2-a_s^k}-\frac{r^2-a_s^k}{r^2}\right),$$

$$T_k^{zz}=T_k^{rr}+\mu a_s^k\frac{r^2-(1+\beta)a_s^k/2}{r^2(r^2-a_s^k)}.$$

Here $p_{0,s}^k$ are the constants of integration.

The Cauchy stress tensor $\mathbf{T}_k$ is defined on each layer up to parameters a_s^k и $p_{0,s}^k$. Some boundary conditions can be satisfied by choosing the corresponding values of such the parameters. We assume that on inner, $r=r_{i,s}^0$, and on exterior, $r=r_{e,s}^{s-1}$, boundaries of assembly, the uniform hydrostatic loading $p_{i,s}$ and $p_{e,s}$ are given:

$$\mathbf{T}_1^{\#}\lrcorner e^r\Big|_{r=r_{i,s}^0}=p_{i,s}\mathbf{e}_r,\quad \mathbf{T}_s^{\#}\lrcorner e^r\Big|_{r=r_{e,s}^{s-1}}=p_{e,s}\mathbf{e}_r, \tag{26}$$

and also, that layers in actual shape are in ideal contact:

$$\mathbf{T}_k^{\#}\lrcorner e^r\Big|_{r=r_{e,s}^k}=\mathbf{T}_{k+1}^{\#}\lrcorner e^r\Big|_{r=r_{i,s}^{k+1}},\quad r_{e,s}^k=r_{i,s}^{k+1},\quad k=0,\dots,s-2. \tag{27}$$

Conditions $(26)-(27)$ can be satisfied if a_s^k and $p_{0,s}^k$, $k=0,\dots,s-1$, are solutions of the following system of non-linear equations:

$$\begin{gathered}\frac{\mu}{2}\left[\ln\frac{(R_i^0)^2}{(R_i^0)^2+a_s^0}-\frac{a_s^0}{(R_i^0)^2+a_s^0}\right]+p_{0,s}^0=p_{i,s},\\ \frac{\mu}{2}\left[\ln\frac{(R_e^{s-1})^2}{(R_e^{s-1})^2+a_s^{s-1}}-\frac{a_s^{s-1}}{(R_e^{s-1})^2+a_s^{s-1}}\right]+p_{0,s}^{s-1}=p_{e,s},\\ \frac{\mu}{2}\left[\ln\frac{(R_e^k)^2}{(R_e^k)^2+a_s^k}-\frac{a_s^k}{(R_e^k)^2+a_s^k}\right]+p_{0,s}^k=\frac{\mu}{2}\left[\ln\frac{(R_i^{k+1})^2}{(R_i^{k+1})^2+a_s^{k+1}}-\frac{a_s^{k+1}}{(R_i^{k+1})^2+a_s^{k+1}}\right]+p_{0,s}^{k+1},\\ (R_e^k)^2+a_s^k=(R_i^{k+1})^2+a_s^{k+1},\quad k=0,\dots,s-2.\end{gathered} \tag{28}$$

Solving the system (28) one can consistently exclude unknowns $p_{0,s}^k$ and arrive at the system of s equations relatively to the parameters a_s^k:

$$\begin{gathered}\prod_{k=0}^{s-1}\gamma_k\frac{\alpha_k+\nu_k x_s}{\beta_k+\nu_k x_s}=W_s\exp\left[\sum_{k=0}^{s-1}(1-\gamma_k)\frac{\alpha_k+\nu_k x_s-1}{(\alpha_k+\nu_k x_s)(\beta_k+\nu_k x_s)}\right],\\ a_s^k=a_s^0+A_k,\quad k=1,\dots,s-1,\end{gathered} \tag{29}$$

where $A_0=0$, $A_k=\sum_{p=0}^{k-1}((R_e^p)^2-(R_i^{p+1})^2)$, $k=1,\dots,s-1$, $\gamma_k=(R_i^k/R_e^k)^2$, $\alpha_k=1+A_k/(R_e^k)^2$, $\beta_k=\gamma_k+A_k/(R_e^k)^2$, $\nu_k=(R_e^0/R_e^k)^2$, $x_s=a_s^0/(R_e^0)^2$, $W_s=\exp\left[2\frac{p_{i,s}-p_{e,s}}{\mu}\right]$.

Conditions of *e)* with equations (28) allows one to determine the parameters a_s^k and $p_{0,s}^k$ recursively. Parameters a_s^k are determined from the system (29). Hereafter the parameters $p_{0,s}^k$ can be determined from the equations (28).

Since the Cauchy stress tensor $\mathbf{T}_k$ in each layer $\mathfrak{L}_k$ is determined, the first Piola-Kirchhoff stress tensor $\mathbf{P}_k$ is derived from the formula $\mathbf{P}_k = J_k \mathbf{T}_k \mathbf{F}_k^{-\mathrm{T}}$, where $J_k = \det \mathbf{F}_k$. For incompressible material $J_k = 1$. Thus $\mathbf{P}_k$ has the following dyadic decomposition (indexes corresponded to layer numbers are suppressed for the sake of brevity):

$$\mathbf{P}_k^{\#} = P_k^{rR} \mathbf{e}_r \otimes \mathbf{e}_R + P^{\theta\Theta} \mathbf{e}_\theta \otimes \mathbf{e}_\Theta + P^{zZ} \mathbf{e}_z \otimes \mathbf{e}_Z,$$

$$P^{rR} = \frac{\sqrt{R^2+a}}{R} T^{rr} = p_0 \frac{\sqrt{R^2+a}}{R} + \frac{\mu}{2} \frac{\sqrt{R^2+a}}{R} \left(\ln \frac{R^2}{R^2+a} - \frac{a}{R^2+a} \right),$$

$$P^{\theta\Theta} = T^{\theta\theta} = \frac{p_0}{R^2+a} + \frac{\mu}{2(R^2+a)} \left(\ln \frac{R^2}{R^2+a} - \frac{a}{R^2+a} \right) + \frac{\mu}{R^2+a} \left(\frac{R^2+a}{R^2} - \frac{R^2}{R^2+a} \right),$$

$$P^{zZ} = T^{zz} = p_0 + \frac{\mu}{2} \left(\ln \frac{R^2}{R^2+a} - \frac{a}{R^2+a} \right) + \mu a \frac{R^2 + (1-\beta) a/2}{R^2 (R^2+a)}.$$

We now proceed to construct a representation for the Eshelby energy-momentum tensor. To this end let us recall that the density of elastic energy stored in each layer $\mathfrak{L}_k$ is defined by the expression

$$W_k = \frac{J_1}{2} (I_1(\mathbf{B}_k) - 3) - \frac{J_{-1}}{2} (I_2(\mathbf{B}_k) - 3),$$

where $I_1(\mathbf{B}_k)$, $I_2(\mathbf{B}_k)$ are invariants of $\mathbf{B}_k$:

$$I_1(\mathbf{B}_k) = I_2(\mathbf{B}_k) = \frac{r^2 - a_{s-1}^k}{r^2} + \frac{r^2}{r^2 - a_{s-1}^k} + 1.$$

Therefore,

$$W_k = \frac{\mu (a_{s-1}^k)^2}{2 r^2 (r^2 - a_{s-1}^k)} = \frac{\mu (a_{s-1}^k)^2}{2 (R^k)^2 ((R^k)^2 + a_{s-1}^k)}.$$

Taking into account the expressions for the first Piola-Kirchhoff stress tensor and elastic energy, we obtain Eshelby energy-momentum tensor by formula

$$\mathbf{E} = W\mathbf{I} - \mathbf{F}^{\mathrm{T}} \mathbf{P}.$$

Thus, we arrive at the following decomposition:

$$\mathbf{E}^{\#} = E^{RR} \mathbf{e}_R \otimes \mathbf{e}_R + E^{\Theta\Theta} \mathbf{e}_\Theta \otimes \mathbf{e}_\Theta + E^{ZZ} \mathbf{e}_Z \otimes \mathbf{e}_Z,$$

$$E^{RR} = \frac{\mu a^2}{2R^2 (R^2+a)} - \frac{R}{\sqrt{R^2+a}} P^{rR},$$

$$E^{\Theta\Theta} = \frac{\mu a^2}{2R^4 (R^2+a)} - \frac{R^2+a}{R^2} P^{\theta\Theta}, \quad E^{ZZ} = \frac{\mu a^2}{2R^2 (R^2+a)} - P^{zZ}.$$

Let us make intermediate conclusions for the case of discrete LbL-assembling. Relations for strain and stresses were obtained in the framework of classical Euclidean continuum mechanics. Indeed, measures of strain and stresses are defined for each layer, and, in totality, form piecewise continuous mappings defined on actual shape. To satisfy boundary conditions one needs to solve the system (28). Nevertheless for a large amount of layers one may need a huge amount of time to solve the system. This must represent considerable computational problem. Thereby it may be preferable to model the appropriate continuous process. Methods for modeling such continuous processes are discussed below.

8.2. Continuous process

Consider a continuous assembling process. We do that in the framework of general reasonings given in subsection 7.3. Each set $\varkappa^{R}_{\alpha;\mathfrak{x}}(\mathfrak{S}_{\alpha;\pi_\alpha(\mathfrak{x})})$ represents a cylindrical surface, the whole body $\mathfrak{B}_\alpha$ is observed as a finite hollow cylinder in $\mathcal{E}$, and so do $\varkappa^{R}_{\alpha;\mathfrak{x}}(\mathfrak{B}_\alpha)$. Figure 6 illustrates the situation.

Let $\overset{\circ}{\varkappa}_\alpha \in \mathfrak{C}(\mathfrak{B}_\alpha)$ satisfy the assumption b) of subsection 7.3 (i.e., it is intermediate configuration), and, in addition, assume that $\overset{\circ}{\varkappa}_\alpha(\mathfrak{B}_\alpha)$ is a hollow cylinder. Thus, we endow the set $\overset{\circ}{\varkappa}_\alpha(\mathfrak{B}_\alpha)$ with cylindrical coordinates $(\overset{\circ}{R}_\alpha, \overset{\circ}{\Theta}_\alpha, \overset{\circ}{Z}_\alpha)$. The corresponding coordinate map is denoted by $\overset{\circ}{H}: X \mapsto (\overset{\circ}{R}_\alpha, \overset{\circ}{\Theta}_\alpha, \overset{\circ}{Z}_\alpha)$. The cylindrical coordinates are related with the Cartesian ones as follows:

$$X^1 = \overset{\circ}{R}_\alpha \cos\overset{\circ}{\Theta}_\alpha, \quad X^2 = \overset{\circ}{R}_\alpha \sin\overset{\circ}{\Theta}_\alpha, \quad X^3 = \overset{\circ}{Z}_\alpha.$$

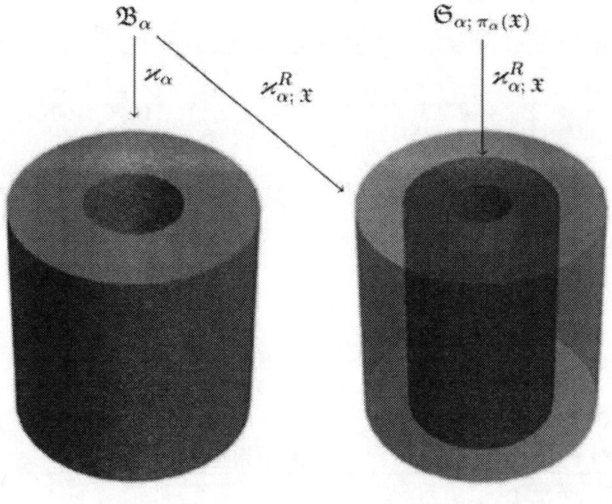

Figure 6. Actual shape $\varkappa_\alpha(\mathfrak{B}_\alpha)$ of the body $\mathfrak{B}_\alpha$ and reference shape $\varkappa^{R}_{\alpha;\mathfrak{x}}(\mathfrak{B}_\alpha)$.

Transfer the cylindrical coordinates $(\overset{\circ}{R}_\alpha, \overset{\circ}{\Theta}_\alpha, \overset{\circ}{Z}_\alpha)$ to the body $\mathfrak{B}_\alpha$ by means of $\overset{\circ}{\varkappa}_\alpha$. The frame and coframe on the body are denoted by $(\partial_{\overset{\circ}{R}_\alpha}, \partial_{\overset{\circ}{\Theta}_\alpha}, \partial_{\overset{\circ}{Z}_\alpha})$, and $(d\overset{\circ}{R}_\alpha, d\overset{\circ}{\Theta}_\alpha, d\overset{\circ}{Z}_\alpha)$, respectively.

Each of the cylinders $\varkappa^{R}_{\alpha;\mathfrak{x}}(\mathfrak{B}_\alpha)$ can be endowed with own cylindrical coordinates $(R_{\alpha;\mathfrak{x}}, \Theta_{\alpha;\mathfrak{x}}, Z_{\alpha;\mathfrak{x}})$. Assume that the deformation $\tilde{\gamma}_{\alpha;\mathfrak{x}}$ from subsection 7.3 has the following representation in the pair of coordinates $(\overset{\circ}{R}_\alpha, \overset{\circ}{\Theta}_\alpha, \overset{\circ}{Z}_\alpha)$ and $(R_{\alpha;\mathfrak{x}}, \Theta_{\alpha;\mathfrak{x}}, Z_{\alpha;\mathfrak{x}})$:

$$R_{\alpha;\mathfrak{x}} = \sqrt{(\overset{\circ}{R}_\alpha)^2 - \rho_{\alpha;\mathfrak{x}}}, \quad \Theta_{\alpha;\mathfrak{x}} = \overset{\circ}{\Theta}_\alpha, \quad Z_{\alpha;\mathfrak{x}} = \overset{\circ}{Z}_\alpha, \tag{30}$$

where $\rho_{\alpha;\mathfrak{X}} \in \mathbb{R}$. The uniform reference $\overset{R}{\mathbf{K}_\alpha}$ is defined via expression (21):

$$\overset{R}{\mathbf{K}_\alpha}|_{\mathfrak{X}} = \left.\frac{\overset{\circ}{R}_\alpha}{\sqrt{(\overset{\circ}{R}_\alpha)^2 - b_\alpha(\mathfrak{X})}}\right|_{\mathfrak{X}} \mathbf{e}_{R_{\alpha;\mathfrak{X}}}|_x \otimes d\overset{\circ}{R}_\alpha|_{\mathfrak{X}} + \mathbf{e}_{\Theta_{\alpha;\mathfrak{X}}}|_x \otimes d\overset{\circ}{\Theta}_\alpha|_{\mathfrak{X}} + \mathbf{e}_{Z_{\alpha;\mathfrak{X}}}|_x \otimes d\overset{\circ}{Z}_\alpha|_{\mathfrak{X}}, \tag{31}$$

where $x = \overset{R}{\varkappa}_{\alpha;\mathfrak{X}}(\mathfrak{X})$ and $(\mathbf{e}_{R_{\alpha;\mathfrak{X}}}, \mathbf{e}_{\Theta_{\alpha;\mathfrak{X}}}, \mathbf{e}_{Z_{\alpha;\mathfrak{X}}})$ is the coordinate frame corresponding to the coordinates $(R_{\alpha;\mathfrak{X}}, \Theta_{\alpha;\mathfrak{X}}, Z_{\alpha;\mathfrak{X}})$. The mapping $b_\alpha := b^1_\alpha$ represents a distortion parameter. From the property $\mathfrak{a}$) (see subsection 7.3) it follows that the coordinate representation of the mapping b_α depends on $\overset{\circ}{R}_\alpha$ only. Thereafter, we will use the symbol $\tilde{b}_\alpha$ for coordinate representation: $\tilde{b}_\alpha : \overset{\circ}{R}_\alpha \mapsto \tilde{b}_\alpha(\overset{\circ}{R}_\alpha)$.

We obtained the family $\{\tilde{b}_\alpha\}_{\alpha\in\mathbb{I}}$ of mappings $\tilde{b}_\alpha : \overset{\circ}{R}_\alpha \mapsto \tilde{b}_\alpha(\overset{\circ}{R}_\alpha)$ that are defined on finite interval $\mathbb{I}_\alpha \subset \mathbb{R}$. Let us choose intermediate configurations with the following property:

$$\forall \alpha_1, \alpha_2 \in \mathbb{I}: \ (\alpha_1 < \alpha_2) \Rightarrow (\tilde{b}_{\alpha_2}|_{\mathbb{I}_{\alpha_1}} = \tilde{b}_{\alpha_1}), \tag{32}$$

that corresponds to the $\mathfrak{d}$) from 7.3.

Remark 11. If $\alpha_1 < \alpha_2$ then $\overset{\circ}{\varkappa}_{\alpha_1}(\mathfrak{B}_{\alpha_1}) \subset \overset{\circ}{\varkappa}_{\alpha_2}(\mathfrak{B}_{\alpha_2})$ and stress-strain state at points in $\overset{\circ}{\varkappa}_{\alpha_2}(\mathfrak{B}_{\alpha_2})$, which are in $\overset{\circ}{\varkappa}_{\alpha_1}(\mathfrak{B}_{\alpha_1})$, is fixed. ♣

Now one can obtain the family $\{\mathbf{G}_\alpha\}_{\alpha\in\mathbb{I}}$ of material metrics:

$$\mathbf{G}_\alpha : \mathfrak{B}_\alpha \to T^*\mathfrak{B}_\alpha \otimes T^*\mathfrak{B}_\alpha, \quad \mathbf{G}_\alpha(u,v) = [\overset{R}{\mathbf{K}_\alpha} \llcorner u] \cdot [\overset{R}{\mathbf{K}_\alpha} \llcorner v].$$

In the explicit form:

$$\mathbf{G}_\alpha = \frac{(\overset{\circ}{R}_\alpha)^2}{(\overset{\circ}{R}_\alpha)^2 - \tilde{b}_\alpha(\overset{\circ}{R}_\alpha)} d\overset{\circ}{R}_\alpha \otimes d\overset{\circ}{R}_\alpha + \left((\overset{\circ}{R}_\alpha)^2 - \tilde{b}_\alpha(\overset{\circ}{R}_\alpha)\right) d\overset{\circ}{\Theta}_\alpha \otimes d\overset{\circ}{\Theta}_\alpha + d\overset{\circ}{Z}_\alpha \otimes d\overset{\circ}{Z}_\alpha. \tag{33}$$

Thus, the family $\{(\mathfrak{B}_\alpha, \mathbf{G}_\alpha)\}_{\alpha\in\mathbb{I}}$ of Riemannian spaces is obtained. Coefficients of the non-zero Levi-Civita connection ∇_α are represented by expressions (for the sake of brevity the indices α and $\circ$ are omitted):

$$\Gamma^1_{11} = \frac{-2\tilde{b}(r) + r\tilde{b}'(r)}{2(r^3 - r\tilde{b}(r))}, \quad \Gamma^1_{22} = \frac{(r^2 - \tilde{b}(r))(-2r + \tilde{b}'(r))}{2r^2}, \quad \Gamma^2_{12} = \Gamma^2_{21} = \frac{2r - \tilde{b}'(r)}{2r^2 - 2\tilde{b}(r)}.$$

The curvature of so-obtained connection is non-zero in general. Its Ricci invariant has the form[11]:

$$\mathrm{Ric}_\alpha = \frac{\overset{\circ}{R}_\alpha \tilde{b}''_\alpha(\overset{\circ}{R}_\alpha) - \tilde{b}'_\alpha(\overset{\circ}{R}_\alpha)}{(\overset{\circ}{R}_\alpha)^3}, \tag{34}$$

and reaches zero only when $\tilde{b}_\alpha(\overset{\circ}{R}_\alpha) = C_1(\overset{\circ}{R}_\alpha)^2 + C_2$. This mathematical fact reflects the specific features of such bodies: the incompatibility of deformations, the presence of intrinsic stresses.

The intermediate configuration $\overset{\circ}{\varkappa}_\alpha$ is used for the material metric synthesizing. Now apply

[11] Here prime $(\,')$ denote $d/d\overset{\circ}{R}_\alpha$.

$$b_\alpha(\overset{\circ}{R}_\alpha)=\begin{cases} 0, & \overset{\circ}{R}_{\alpha;i}<\overset{\circ}{R}_\alpha<\overset{\circ}{R}_{\alpha;e},\\ \tilde{b}_\alpha(\overset{\circ}{R}_\alpha), & \overset{\circ}{R}_\alpha>\overset{\circ}{R}_{\alpha;e}.\end{cases} \tag{35}$$

In this formula $\overset{\circ}{R}_{\alpha;i}$, $\overset{\circ}{R}_{\alpha;e}$ are prescribed the inner and outer radii of $\overset{\circ}{\varkappa}_\alpha(\mathfrak{B})$, respectively. The zero value corresponds to the substrate $\mathfrak{B}$, which is homogeneous simple body. Total material metric $\mathbf{G}_\alpha$ in $\mathfrak{B}_\alpha$ is defined via formula (22). Endow the set $\varkappa_\alpha(\mathfrak{B}_\alpha)$ with cylindrical coordinates $(r_\alpha,\theta_\alpha,z_\alpha)$. In the coordinates the deformation d_α from $\overset{\circ}{\varkappa}_\alpha(\mathfrak{B})$ to $\varkappa_\alpha(\mathfrak{B}_\alpha)$ is assumed to be in the form

$$d_\alpha:(\overset{\circ}{R}_\alpha,\overset{\circ}{\Theta}_\alpha,\overset{\circ}{Z}_\alpha)\mapsto(r_\alpha,\theta_\alpha,z_\alpha),\quad r_\alpha=\sqrt{(\overset{\circ}{R}_\alpha)^2+A_\alpha},\quad \theta_\alpha=\overset{\circ}{\Theta}_\alpha,\quad z_\alpha=\overset{\circ}{Z}_\alpha,$$

where $A_\alpha\in\mathbb{R}$. Hence, the deformation gradient is defined as

$$\mathbf{F}_\alpha=\frac{\overset{\circ}{R}_\alpha}{\sqrt{(\overset{\circ}{R}_\alpha)^2+A_\alpha}}\mathbf{e}_{r_\alpha}\otimes e^{\overset{\circ}{R}_\alpha}+\mathbf{e}_{\theta_\alpha}\otimes e^{\overset{\circ}{\Theta}_\alpha}+\mathbf{e}_{z_\alpha}\otimes e^{\overset{\circ}{Z}_\alpha}. \tag{36}$$

The configuration gradient $T_{\mathfrak{X}}\varkappa_\alpha$ can be represented as multiplicative decomposition $T_{\mathfrak{X}}\varkappa_\alpha=\mathbf{F}_\alpha\circ T_{\mathfrak{X}}\overset{\circ}{\varkappa}_\alpha$. According to (36) it has the form:

$$T_{\mathfrak{X}}\varkappa_\alpha=\frac{\overset{\circ}{R}_\alpha}{\sqrt{(\overset{\circ}{R}_\alpha)^2+A_\alpha}}\mathbf{e}_{r_\alpha}|_x\otimes d\overset{\circ}{R}_\alpha|_{\mathfrak{X}}+\mathbf{e}_{\theta_\alpha}|_x\otimes d\overset{\circ}{\Theta}_\alpha|_{\mathfrak{X}}+\mathbf{e}_{z_\alpha}|_x\otimes d\overset{\circ}{Z}_\alpha|_{\mathfrak{X}}, \tag{37}$$

where $x=\varkappa_\alpha(\mathfrak{X})$.

According to relations (7) and (8) one has formulae

$$C_\alpha=\frac{(\overset{\circ}{R}_\alpha)^2-b_\alpha(\overset{\circ}{R}_\alpha)}{(\overset{\circ}{R}_\alpha)^2+A_\alpha}\partial_{\overset{\circ}{R}_\alpha}\otimes d\overset{\circ}{R}_\alpha+\frac{(\overset{\circ}{R}_\alpha)^2+A_\alpha}{(\overset{\circ}{R}_\alpha)^2-b_\alpha(\overset{\circ}{R}_\alpha)}\partial_{\overset{\circ}{\Theta}_\alpha}\otimes d\overset{\circ}{\Theta}_\alpha+\partial_{\overset{\circ}{Z}_\alpha}\otimes d\overset{\circ}{Z}_\alpha,$$

$$\mathbf{B}_\alpha=\frac{r_\alpha^2-A_\alpha-b_\alpha(\sqrt{r_\alpha^2-A_\alpha})}{r_\alpha^2}\mathbf{e}_{r_\alpha}\otimes e^{r_\alpha}+\frac{r_\alpha^2}{r_\alpha^2-A_\alpha-b_\alpha(\sqrt{r_\alpha^2-A_\alpha})}\mathbf{e}_{\theta_\alpha}\otimes e^{\theta_\alpha}+\mathbf{e}_{z_\alpha}\otimes e^{z_\alpha}. \tag{38}$$

All is ready to calculate the stresses in the actual configuration $\varkappa_\alpha$. Using relation $\mathbf{T}=-p\mathbf{I}+J_1\mathbf{B}+J_{-1}\mathbf{B}^{-1}$, as in subsection 8.1, and taking into account (38) we obtain that

$$(\mathbf{T}_\alpha)^\#=T_\alpha^{r_\alpha r_\alpha}\mathbf{e}_{r_\alpha}\otimes\mathbf{e}_{r_\alpha}+T_\alpha^{\theta_\alpha\theta_\alpha}\mathbf{e}_{\theta_\alpha}\otimes\mathbf{e}_{\theta_\alpha}+T_\alpha^{z_\alpha z_\alpha}\mathbf{e}_{z_\alpha}\otimes\mathbf{e}_{z_\alpha}, \tag{39}$$

$$T_\alpha^{r_\alpha r_\alpha}(r)=B_\alpha+\mu\int_{r_{\alpha;i}}^{r}\left[\frac{\tau}{\tau^2-A_\alpha-b_\alpha(\sqrt{\tau^2-A_\alpha})}-\frac{\tau^2-A_\alpha-b_\alpha(\sqrt{\tau^2-A_\alpha})}{\tau^3}\right]d\tau, \tag{40}$$

$$T_\alpha^{\theta_\alpha\theta_\alpha}=\frac{T_\alpha^{r_\alpha r_\alpha}}{r_\alpha^2}+\frac{\mu}{r_\alpha^2}\left(\frac{r_\alpha^2}{r_\alpha^2-A_\alpha-b_\alpha(\sqrt{r_\alpha^2-A_\alpha})}-\frac{r_\alpha^2-A_\alpha-b_\alpha(\sqrt{r_\alpha^2-A_\alpha})}{r_\alpha^2}\right),$$

$$T_\alpha^{z_\alpha z_\alpha}=T_\alpha^{r_\alpha r_\alpha}+\mu\{A_\alpha+b_\alpha(\sqrt{r_\alpha^2-A_\alpha})\}\frac{r_\alpha^2-(1+\beta)\{A_\alpha+b_\alpha(\sqrt{r_\alpha^2-A_\alpha})\}/2}{r_\alpha^2(r_\alpha^2-A_\alpha-b_\alpha(\sqrt{r_\alpha^2-A_\alpha}))}.$$

Here B_α is the constant of integration that has appeared after integrating the equilibrium equation, and $r_{\alpha;i}$ is the actual inner radius. The corresponding Cauchy exterior 2-form is represented as follows

$$T_\alpha = r_\alpha T^{r_\alpha r_\alpha} e^{r_\alpha} \otimes \left(e^{\theta_\alpha} \wedge e^{z_\alpha}\right) - r_\alpha^3 T^{\theta_\alpha\theta_\alpha} e^{\theta_\alpha} \otimes \left(e^{r_\alpha} \wedge e^{z_\alpha}\right) + r_\alpha T^{z_\alpha z_\alpha} e^{z_\alpha} \otimes \left(e^{r_\alpha} \wedge e^{\theta_\alpha}\right).$$

Since the Cauchy stress tensor (39) is known, we can obtain the first Piola-Kirchhoff tensor by the formula (15). Here $\mathbf{F} = T\varkappa_\alpha$, $\mathbf{g} = e^{r_\alpha} \otimes e^{r_\alpha} + r_\alpha^2 e^{\theta_\alpha} \otimes e^{\theta_\alpha} + e^{z_\alpha} \otimes e^{z_\alpha}$. Hence,

$$J_\alpha = \frac{R_\alpha^\circ}{\sqrt{(R_\alpha^\circ)^2 + A_\alpha}}, \quad g_\alpha = r_\alpha^2 = (R_\alpha^\circ)^2 + A_\alpha, \quad G_\alpha = (R_\alpha^\circ)^2.$$

Next, we have that $\mathbf{F}^{-1} = \frac{\sqrt{(R_\alpha^\circ)^2 + A_\alpha}}{R_\alpha^\circ} \partial_{R_\alpha^\circ} \otimes e^{r_\alpha} + \partial_{\Theta_\alpha^\circ} \otimes e^{\theta_\alpha} + \partial_{Z_\alpha^\circ} \otimes e^{z_\alpha}$. Then, applying transpose, we arrive at the formula

$$\mathbf{F}_\alpha^{-\mathrm{T}} = \frac{R_\alpha^\circ \sqrt{(R_\alpha^\circ)^2 + A_\alpha}}{(R_\alpha^\circ)^2 - b_\alpha(R_\alpha^\circ)} \mathbf{e}_{r_\alpha} \otimes dR_\alpha^\circ + \frac{(R_\alpha^\circ)^2 - b_\alpha(R_\alpha^\circ)}{(R_\alpha^\circ)^2 + A_\alpha} \mathbf{e}_{\theta_\alpha} \otimes d\Theta_\alpha^\circ + \mathbf{e}_{z_\alpha} \otimes dZ_\alpha^\circ.$$

Thus, the first Piola-Kirchhoff tensor has the form:

$$\mathbf{P} = \frac{R_\alpha^\circ \sqrt{(R_\alpha^\circ)^2 + A_\alpha}}{(R_\alpha^\circ)^2 - b_\alpha(R_\alpha^\circ)} T^{r_\alpha r_\alpha} e^{r_\alpha} \otimes dR_\alpha^\circ +$$

$$+((R_\alpha^\circ)^2 + A_\alpha)((R_\alpha^\circ)^2 - b_\alpha(R_\alpha^\circ)) T^{\theta_\alpha\theta_\alpha} e^{\theta_\alpha} \otimes d\Theta_\alpha^\circ + T^{z_\alpha z_\alpha} e^{z_\alpha} \otimes dZ_\alpha^\circ.$$

Let us obtain the Eshelby energy-momentum tensor. Since

$$\mathbf{P}^\# = \frac{\sqrt{(R_\alpha^\circ)^2 + A_\alpha}}{R_\alpha^\circ} T^{r_\alpha r_\alpha} \mathbf{e}_{r_\alpha} \otimes dR_\alpha^\circ + T^{\theta_\alpha\theta_\alpha} \mathbf{e}_{\theta_\alpha} \otimes d\Theta_\alpha^\circ + T^{z_\alpha z_\alpha} \mathbf{e}_{z_\alpha} \otimes dZ_\alpha^\circ,$$

$$\mathbf{F}^{\mathrm{T}} = \frac{(R_\alpha^\circ)^2 - b_\alpha(R_\alpha^\circ)}{R_\alpha^\circ \sqrt{(R_\alpha^\circ)^2 + A_\alpha}} \partial_{R_\alpha^\circ} \otimes e^{r_\alpha} + \frac{(R_\alpha^\circ)^2 + A_\alpha}{(R_\alpha^\circ)^2 - b_\alpha(R_\alpha^\circ)} \partial_{\Theta_\alpha^\circ} \otimes e^{\theta_\alpha} + \partial_{Z_\alpha^\circ} \otimes e^{z_\alpha},$$

we arrive at

$$e = \frac{(R_\alpha^\circ)^2 - b_\alpha(R_\alpha^\circ)}{(R_\alpha^\circ)^2} \left\{W - T^{r_\alpha r_\alpha}\right\} \partial_{R_\alpha^\circ} \otimes \partial_{R_\alpha^\circ} +$$

$$+ \frac{1}{(R_\alpha^\circ)^2 - b_\alpha(R_\alpha^\circ)} \left\{W - ((R_\alpha^\circ)^2 + A_\alpha) T^{\theta_\alpha\theta_\alpha}\right\} \partial_{\Theta_\alpha^\circ} \otimes \partial_{\Theta_\alpha^\circ} + \left\{W - T^{z_\alpha z_\alpha}\right\} \partial_{Z_\alpha^\circ} \otimes \partial_{Z_\alpha^\circ}.$$

All the calculations carried out with the assumption that the family of reference shapes $\overline{\mathrm{Ref}_{\mathfrak{B}_\alpha}}$ is given. For example, with a hollow cylinder, this means that the family of functions $\{b_\alpha\}_{\alpha\in\mathbb{I}}$ is known. However, as a rule, they need to be found from the problem of the assembly process. Let us analyze this in detail.

Suppose that the following smooth functions are given: shrinkage $S:\mathbb{I} \to]0,1[$, internal $R_{\mathrm{int}}^\circ:\mathbb{I} \to \mathbb{R}_+$ and external $R_{\mathrm{ext}}^\circ:\mathbb{I} \to \mathbb{R}_+$ radii of the intermediate shape, internal $p_i:\mathbb{I} \to \mathbb{R}$ and external $p_e:\mathbb{I} \to \mathbb{R}$ hydrostatic loadings in actual configuration. The families $\{b_\alpha\}_{\alpha\in\mathbb{I}}$ and $\{A_\alpha\}_{\alpha\in\mathbb{I}}$ of deformation parameters are unknown. On the interior and exterior boundaries of $\varkappa_\alpha(\mathfrak{B}_\alpha)$ consider the following conditions:

$$T_\alpha^{r_\alpha r_\alpha}(r_{\alpha;i}) = p_i(\alpha), \quad T_\alpha^{r_\alpha r_\alpha}(r_{\alpha;e}) = p_e(\alpha),$$

where $T_\alpha^{r_\alpha r_\alpha}(r)$ is defined by (40). These equalities lead to the relation

$$\int_{\overset{\circ}{R}_{\text{int}}(\alpha)}^{\overset{\circ}{R}_{\text{ext}}(\alpha)} \left[\frac{1}{\xi^2 - b_\alpha(\xi)} - \frac{\xi^2 - b_\alpha(\xi)}{(\xi^2 + A_\alpha)^2} \right] \xi d\xi = \frac{p_e(\alpha) - p_i(\alpha)}{\mu}. \tag{41}$$

Another condition relates b_α with shrinkage S:

$$b_\alpha(\overset{\circ}{R}_{\text{ext}}(\alpha)) + A_\alpha = S(\alpha). \tag{42}$$

Both (41) and (42) allow us to determine A_α and b_α. According to (35) equation (41) takes the form

$$\int_{\overset{\circ}{R}_{\text{int}}(\alpha)}^{\overset{\circ}{R}_{\alpha;e}} \left[\frac{1}{\xi^2} - \frac{\xi^2}{(\xi^2 + A_\alpha)^2} \right] \xi d\xi + \int_{\overset{\circ}{R}_{\alpha;e}}^{\overset{\circ}{R}_{\text{ext}}(\alpha)} \left[\frac{1}{\xi^2 - \tilde{b}_\alpha(\xi)} - \frac{\xi^2 - \tilde{b}_\alpha(\xi)}{(\xi^2 + A_\alpha)^2} \right] \xi d\xi = \frac{p_e(\alpha) - p_i(\alpha)}{\mu},$$

or, since the first integral can be calculated directly,

$$\ln \frac{\overset{\circ}{R}_{\alpha;e}\sqrt{(\overset{\circ}{R}_{\text{int}}(\alpha))^2 + A_\alpha}}{\overset{\circ}{R}_{\text{int}}(\alpha)\sqrt{(\overset{\circ}{R}_{\alpha;e})^2 + A_\alpha}} - \frac{A_\alpha}{2}\left(\frac{1}{(\overset{\circ}{R}_{\alpha;e})^2 + A_\alpha} - \frac{1}{(\overset{\circ}{R}_{\text{int}}(\alpha))^2 + A_\alpha} \right) +$$

$$+ \int_{\overset{\circ}{R}_{\alpha;e}}^{\overset{\circ}{R}_{\text{ext}}(\alpha)} \left[\frac{1}{\xi^2 - \tilde{b}_\alpha(\xi)} - \frac{\xi^2 - \tilde{b}_\alpha(\xi)}{(\xi^2 + A_\alpha)^2} \right] \xi d\xi = \frac{p_e(\alpha) - p_i(\alpha)}{\mu}. \tag{43}$$

Equation (42) should be rewritten in terms of $\tilde{b}_\alpha$:

$$\tilde{b}_\alpha(\overset{\circ}{R}_{\text{ext}}(\alpha)) + A_\alpha = S(\alpha). \tag{44}$$

Thus, for this case one should solve the evolutionary problem (43), (44) to determine $\tilde{b}_\alpha$ and A_α.

8.3. Numerical results

To compare strain and stress distributions in discrete structurally inhomogeneous solids with similar distributions in solids with continuously distributed inhomogeneity consider following examples.

a) Hollow cylinder is assembled from a number of thin hollow cylindrical parts. We suppose that each part possesses a natural configuration. A sequence of assemblies with common to all total volume V and increasing number of parts (10, 30, 200 and 1000) is studied.
b) Hollow cylinder with volume V, the same as in a), and piecewise continuous distribution for deformation parameter b_α that is taken to be the continuous approximation of the sequence of deformation parameters a_s obtained for assembly from a) with maximal numbers of parts.
c) Hollow cylinder with volume V, the same as in a) and b), and piecewise continuous distribution for b_α that is obtained as numerical solution of evolutionary problem (43), (44).

From general considerations discussed above it is expected that stress and strain distributions in solids with discrete inhomogeneity tend to the corresponded stress and strain distributions in a solid with some piecewise continuous b_α, that can be obtained as a solution of evolutionary problem (43), (44) or can be represented as approximation of deformation parameters a_s of sufficiently dense assembly. The following calculations numerically illustrate this fact.

For calculations we took the following values of parameters [2]. For each of the assemblies inner radius ρ of the first cylindrical layer in reference configuration and its thickness Δ^0 are taken equal to $\rho = \Delta^0 = 2$ mkm. The reference thicknesses Δ^k, $k \geqslant 1$, for each of assemblies are taken equal to $\Delta^k = 22/(N-1)$ mkm, where N is the number of layers. The shrinkage coefficients are taken equal to $S^k = 0.7$. Suppose that inner and outer hydrostatic loadings, $p_{i,s}$ and $p_{e,s}$, vanish. In computations we put $\beta = 0.24$.

Strain and stress distributions are calculated by equations obtained in subsection 8.1. In figures 7 and 8 the numerical results are shown, where dotted line corresponds to 10 layers, dot-dashed line corresponds to 30 layers, dashed line corresponds to 200 layers. The case of 1000 layers is not shown on graphs because its difference from continuous distributions discussed below is graphically indistinguishable. Angle brackets in $Y_{<ij>}$ denote that the corresponding component of $\mathbf{Y}$ is considered in physical (with unit norm vectors) frame. The symbols $Y_{<rr>}$, $Y_{<\theta\theta>}$, $Y_{<zz>}$ correspondingly denote radial, circumferential and axial physical components of a tensor $\mathbf{Y}$. So do $Y_{<rR>}$, $Y_{<\theta\Theta>}$, $Y_{<zZ>}$, but relatively to the pair of cylindrical coordinate systems.

Results for continuously structural inhomogeneous solid that corresponds to considered above discrete cases can be obtained in the following way. Let the mapping $r \mapsto b_\alpha(r)$ given by

$$b_\alpha(r) = \begin{cases} -3.1651, & r < 3.5826, \\ -4.5681 + 1.3347r + 3.5255 \cdot 10^{-1} r^2 + 8.2975 \cdot 10^{-3} r^3 - 1.6201 \cdot 10^{-4} r^4, & r > 3.5826. \end{cases}$$

be the numerical approximation of the values for deformation parameters in the most dense fragmentation (1000 layers). Material metric in this case is Riemannian with nontrivial curvature. This fact is illustrated by the distribution of Ricci invariant (34) shown in figure 8f. Note that Ricci invariant is equal to zero on the part of material manifold corresponded to the initial body. Adhered material forms the part of the body with non-Euclidean material metrics.

The corresponding Cauchy stress tensor $\mathbf{T}^{\text{approx}}$ is obtained in subsection 8.2, and similarly, the corresponding first Piola-Kirchhoff stress tensor and Eshelby energy-momentum tensor. We considered the intermediate shape as actual and put A_α to be equal to zero. The corresponding distributions are shown in figures 7, 8a, 8c, 8e by solid lines.

The function

$$x \mapsto \delta(x) = \frac{T_{rr}^{\text{approx}}(x) - T_{rr}^{\text{discr}}(x)}{\max T_{rr}^{\text{discr}}},$$

estimates the difference between stress states in solids with discrete and continuous inhomogeneities. Plot of this function is shown in figure 9; label 1 designates 10 layers, label 2 designates 30 layers, label 3 designates 200 layers, and label 4 designates 1000 layers. In the right upper corner of the figure 9 the logarithms of absolute values for δ, i.e. $\ln|\delta|$, are shown.

9. Optimization Problem

Since analytical solution for the cylindrical problem stated in Section 8 is known, one can derive the intensity τ of normal stresses, which is defined by the formula

$$\tau := \sqrt{-3I_2(\operatorname{Dev}\mathbf{T})}. \tag{45}$$

Here,

$$I_2(\mathbf{Q}) = \frac{1}{2}\left(\operatorname{tr}^2\mathbf{Q} - \operatorname{tr}\mathbf{Q}\right)$$

IOP Conf. Series: Journal of Physics: Conf. Series **1009** (2018) 012014 doi:10.1088/1742-6596/1009/1/012014

is the second invariant of the 2-nd rank tensor $\mathbf{Q}$, and $\operatorname{Dev}\mathbf{T}$ is the deviator of stress tensor $\mathbf{T}$. In stress tensor components,

$$2\tau^2 = \left(T_{<rr>} - T_{<\theta\theta>}\right)^2 + \left(T_{<rr>} - T_{<zz>}\right)^2 + \left(T_{<zz>} - T_{<\theta\theta>}\right)^2,$$

where the symbols $T_{<rr>}$, $T_{<\theta\theta>}$, $T_{<zz>}$ denote respectively radial, circumferential, and axial physical components of the tensor $\mathbf{T}$, which is defined by the relations from subsection 8.1.

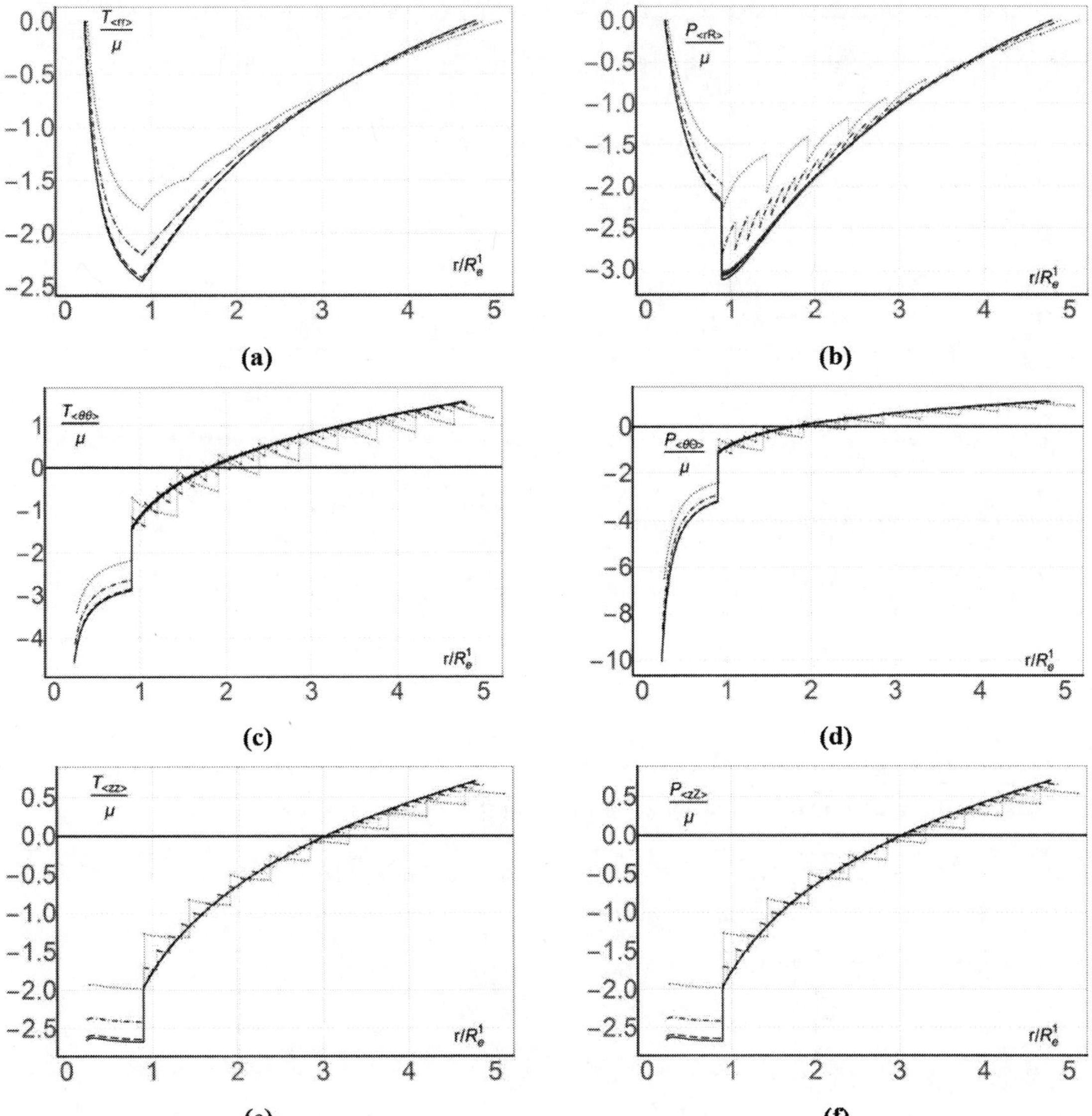

Figure 7. Relative physical components $T_{<ij>}/\mu$ and $P_{<ij>}/\mu$ of the Cauchy and first Piola-Kirchhoff stress tensors.

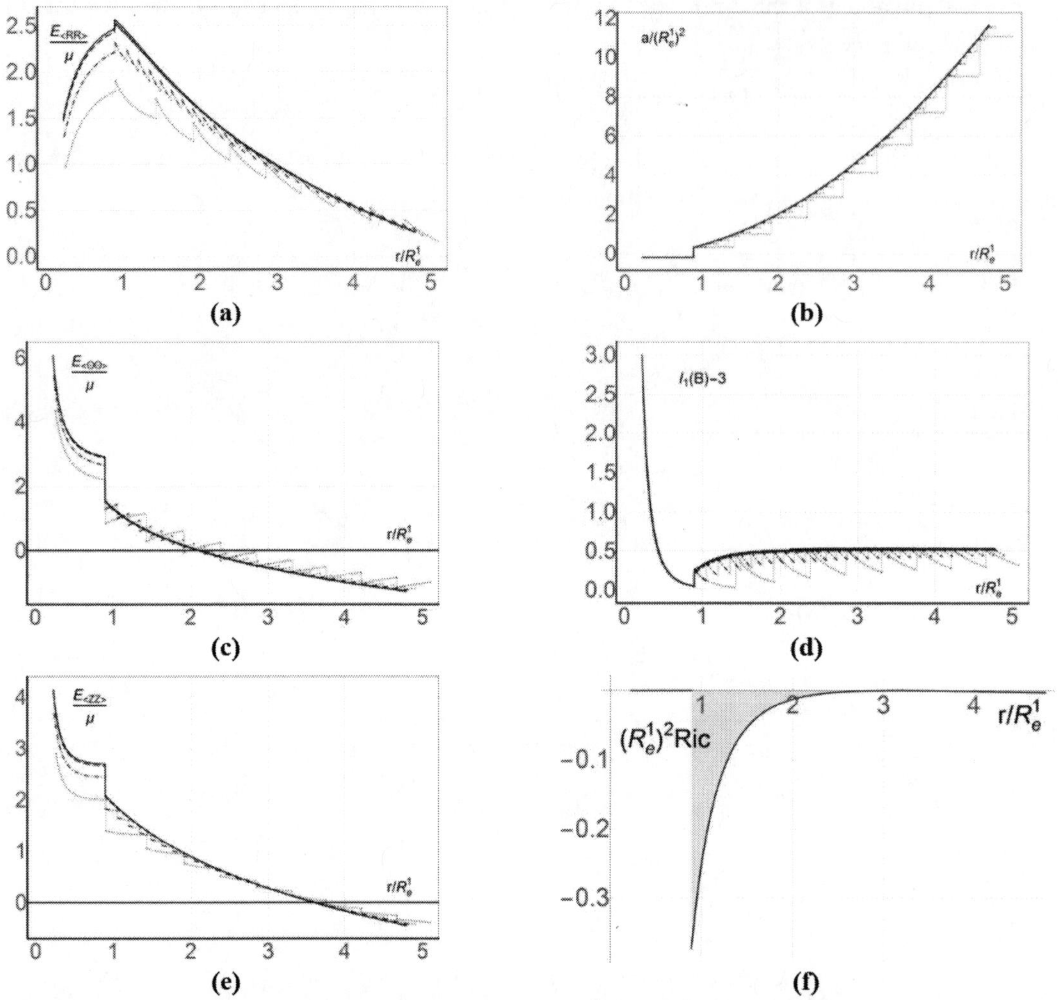

Figure 8. Relative physical components $E_{<ij>} / \mu$ of Eshelby energy-momentum tensor; distributions of deformation parameter, the first and Ricci invariants.

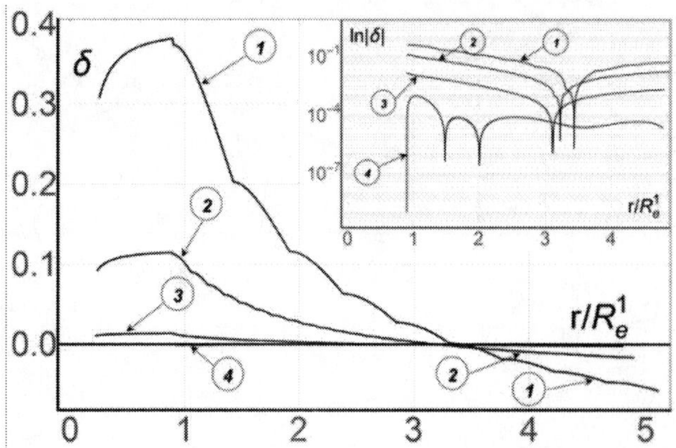

Figure 9. The estimation of the difference between stress states in solids with discrete and continuous inhomogeneities.

Let us consider internal hydrostatic loadings as optimization parameters. Thus, we introduce a set $\mathbb{R}^N$ of all N-tuples. Each tuple $p=(p_1,\dots,p_N)\in\mathbb{R}^N$ represents hydrostatic loadings for internal surfaces of the N layers constituting an assembly. Let external hydrostatic loadings, reference thicknesses, reference radii of the first layer, shrinkage and material parameters are given. Stress intensity τ defined by the formula (45) is an objective function depending on p.

The optimization problem is to find such a vector p that minimizes the maximal value of stress intensity $\tau_{\max}$ of the terminal assembly under given working inner pressure, i.e.

$$p^* = \arg\min_{p\in\mathbb{R}^N} \max_{X\in B} \tau(X,p). \tag{46}$$

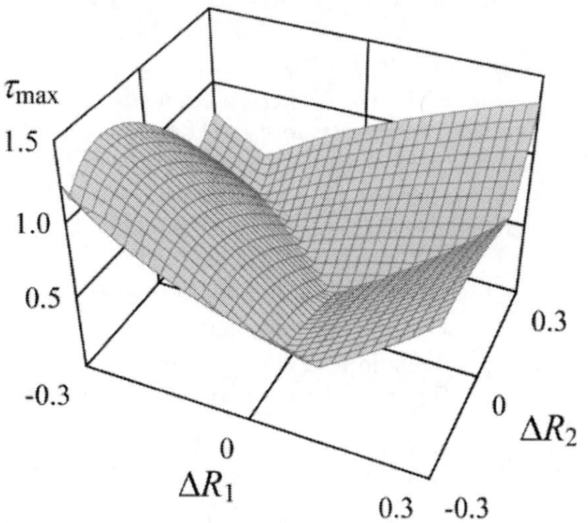

Figure 10. Stress intensity τ vs. incompatibility parameters ΔR_1 and ΔR_2.

The approaches applied for solving the optimization problem (46) require special attention, as the resulting functional is neither convex nor smooth. These properties can be demonstrated even on the example of the simplest assembling of two layers. In figure 10 the graph of the intensity function with respect to the radii incompatibility ΔR_1 and ΔR_2 is presented. The parameters ΔR_j for $j=1,2$ mean the differences between the outer radius R^e_{j-1} of the previous layer $j-1$ in the reference configuration and the inner radius R^i_j of the layer j. These gaps ($\Delta R_j<0$) or overlaps ($\Delta R_j>0$) define uniquely the resulting stress-strain stare of the structure and can be recalculate as functions of control tuple p. As seen from the figure the surface $\tau_{\max}(\Delta R_1,\Delta R_2)$ processes some folds, convex regions, and even visible local minima near the global one.

With great confidence, we can assume that the similar or more complicated patterns are inherent in the multidimentional case ($N>2$), Thus, it is unavoidable to search for an reasonable initial control vector in the attraction basing of the global minimum point in order to approach to the lowest level of stress intensity.

10. Conclusion Remarks

To summarize the work we note the following

- All basic concepts that form the basis of classical continuum mechanics can be generalized on non-Euclidean analysis.
- This approach allow to obtain continual models for multilayered LbL structures with large number of layers.

- For computations such continual models are much more effective then discrete models, based on classical (Euclidean) approach.
- Structural optimization of nonlinearly deformable LbL structures makes it possible to improve the quality of the design, but due to the substantial nonlinearity of the problem it is ambiguous and requires a delicate comprehension.

Acknowledgments
Authors acknowledge financial support from RSF 14-29-00198.

References
[1] Gibson I, Rosen D W and Stucker B 2010 *Additive Manufacturing Technologies* (Springer)
[2] Decher G and Schlenoff J B 2002 *Multilayer Thin Films. Sequential Assembly of Nanocomposite Materials* (Wiley –VCH)
[3] Dennis Jr J E and Schnabel R B 1996 *Numerical Methods for Unconstrained Optimization and Nonlinear Equations. Series: Classics in Applied Mathematics* (Philadelphia: SIAM)
[4] Sigmund O and Maute K 2013 Topology optimization approaches: A comparative review *Structural and Multidisciplinary Optimization* **48** pp 1031–55
[5] Epstein M 2010 *The Geometrical Language of Continuum Mechanics* (Cambridge University Press) p 312
[6] Epstein M and Elzanowski M 2007 *Material Inhomogeneities and Their Evolution: A geometric approach* (Springer Science & Business Media)
[7] Nye J 1953 Some geometrical relations in dislocated crystals *Acta metallurgica* **1** (2) pp 153–62
[8] De Wit R 1977 *Continual Theory of Disclinations* (Moscow: Mir Press Mekhanika Novoe v Zarubezh. Nauke)
[9] Kadic A and Edelen D 1983 *A Gauge Theory of Dislocations and Disclinations* (Berlin: Springer)
[10] Bilby B, Bullough R and Smith E 1955 Continuous distributions of dislocations: a new application of the methods of non-Riemannian geometry *Proc. of the Royal Society of London A: Mathematical, Physical and Engineering Sciences / The Royal Society* **231** pp 263–73
[11] Kröner E 1996 Dislocation theory as a physical feld theory *Meccanica* **31** pp 577–87
[12] Myasnikov V and Guzev M 2000 Non-Euclidean model of materials deformed at different structural levels *Physical Mesomechanics* **3** (1) pp 5–16
[13] Yavari A 2010 A geometric theory of growth mechanics *Journal of Nonlinear Science* **20** (6) pp 781–830
[14] Ozakin A and Yavari A 2010 A geometric theory of thermal stresses *Journal of Mathematical Physics* **51** (3) p 032902
[15] Yavari A and Goriely A 2012 Riemann-Cartan geometry of nonlinear disclination mechanics *Mathematics and Mechanics of Solids*
[16] Yavari A, Marsden J E and Ortiz M 2006 On spatial and material covariant balance laws in elasticity *Journal of Mathematical Physics* **47** (4)
[17] Marsden J E and Hughes T J 1994 *Mathematical Foundations of Elasticity* (Courier Corporation)
[18] Eckart C 1948 The thermodynamics of irreversible processes. IV. The theory of elasticity and anelasticity *Physical Review* **73** (4) pp 373–82
[19] Kondo K 1955 Non-Riemannian geometry of imperfect crystals from a macroscopic viewpoint *Memoirs of the unifying study of the basic problems in engineering science by means of geometry* **1** pp 6–17
[20] Noll W 1967 Materially uniform simple bodies with inhomogeneities *Archive for Rational Mechanics and Analysis* **27** (1) pp 1–32
[21] Wang C C 1967 On the geometric structures of simple bodies, a mathematical foundation for

the theory of continuous distributions of dislocations *Archive for Rational Mechanics and Analysis* **27** (1) pp 33–94

[22] Delphenich D H 2013 Selected papers on teleparallelism *www.neo-classical-physics.info*

[23] Eisenhart L P 2012 *Non-Riemannian geometry* (Courier Corporation)

[24] Lee J M 2012 *Introduction to Smooth Manifolds* (New York : Springer New York)

[25] Lychev S A and Koifman K G 2016 Geometric aspects of the theory of incompatible deformations Part I. Uniform configurations *Nanomechanics Science and Technology: An International Journal* **7** (3) pp 177–233

[26] Lychev S A and Koifman K G 2017 Geometric aspects of the theory of incompatible deformations for simple structurally inhomogeneous solids with variable composition *Far Eastern Mathematical Journal* **17** (2) pp 221–45

[27] Maugin G A 1993 *Material Inhomogeneities in Elasticity* (CRC Press) **3**

[28] Frankel T 2003 *The Geometry of Physics: an Introduction* (Cambridge University Press) 694

[29] Eshelby J D 1975 The elastic energy-momentum tensor *Journal of Elasticity* **5** (3–4) pp 321–35

[30] Abraham R, Marsden J E and Ratiu T 1988 *Manifolds, tensor analysis, and applications* (Springer Science & Business Media)

[31] Weyl H 2011 *Raum, Zeit, Materie* (Springer)

[32] Gurtin M E and Murdoch A I 1975 A continuum theory of elastic material surfaces *Arch. Ration. Mech. Anal.* **57** (4) pp 291–323

[33] Sternberg S 1964 *Lectures on Differential Geometry* Englewood (Cliffs, NJ: Prentice-Hall)

[34] Segev R and Rodnay G 1999 Cauchy's theorem on manifolds *Journal of Elasticity* **56** (2) pp 129–44

[35] Kanso E, Arroyo M, Tong Y, Yavari A, Marsden J G and Desbrun M 2007 On the geometric character of stress in continuum mechanics *ZAMP* **58** (5) pp 843–56

[36] Lychev S A and Manzhirov A V 2013 The mathematical theory of growing bodies. Finite deformations. (*J. Appl. Math. Mech.)* **77** (4) pp 421–32

[37] Lee J M 2011 *Introduction to Topological Manifolds* (New York: Springer)

[38] Truesdell C and Noll W 2004 *The non-Linear Field Theories of Mechanics* ed by Stuart S. Antman (Berlin Heidelberg: Springer)

IOP Conf. Series: Journal of Physics: Conf. Series **1009** (2018) 012015 doi:10.1088/1742-6596/1009/1/012015

Spectrograms of one-dimensional isotropic turbulence

V A Frost

Ishlinsky Institute for Problems in Mechanics, RAS, Pr. Vernadskogo, 101-1, Moscow, 119526, Russia

frost@ipmnet.ru

Abstract. An equation for describing the evolution of the spectrogram velocity amplitudes of individual wave components placed on a finite field, the interaction of the components in which satisfies the triad rule: $k = k' \pm k''$, is proposed. The details of the energy transport along the spectrum, the decay and power supply modes are considered. In the absence of dissipation and pumping the total intensity of the system that is analogy of energy is unchanged. The results of a numerical calculation of the transition to white noise (in the absence of dissipation), the decay regimes at various Reynolds numbers, and the transition to the equilibrium state for different pumping by the large-scale components, are obtained. The results obtained correspond to known experimental data obtained for real turbulent flows. In the one-dimensional case, the correlation function and the third moment that determines the structure change in the field structure upon decay is restored.

1. Introduction

The article is devoted to development of the method for receiving realization of the isotropic turbulent velocity field which can be used for testing various hypotheses of closuring of the statistical moment equations. Usage of concrete realization which represent the main characteristic of whole statistical ensemble allows to avoid the difficulties of closuring of the moment equations used for the description of turbulent fields.

The isotropic turbulent velocity field is the simplest example suitable for testing of various methods of creation of the closure methods of description of the turbulent phenomena. For the turbulent field there is a large number of experimental data on decay of turbulence in the uniform mean flow behind turbulence-producing grids that can be used for the analysis of methods of closuring. Assuming that a realization of the isotropic turbulent field is observed, it is possible to use these experimental data for testing of the hypotheses used for closuring. Analysis of three various hypotheses: K.Hasselmann [1], Yu.M. Lytkin and G.G. Chernykh [2] and N.I. Akatnov and E. N. Bystrova [3,4] , has shown [5] that it isn't possible to make the accurate choice of the most exact approach, most likely, because of incomplete compliance of experimental data for conditions of isotropic turbulence. Differences from the regularities corresponding to an isotropy at experiments with turbulence-produced grids of irregular shape [6-8] are clearly expressed.

Usage for our purposes a direct numerical integration of the Navier-Stokes equations is extremely expensive, and the necessary restriction of flow field significantly reduces area in which isotropy conditions are satisfied.

It would seem that to get the field corresponding to the isotropic turbulence, it is possible by setting a certain spectrum and having played the amplitudes and phases of large, but finite, number of combinations of three-dimensional harmonics that satisfy the continuity condition [9]. But the distribution of third moments that are an object of hypotheses of closing received in this case because of an extremity of a set of harmonicas and because of the possible requirements connected to limitation of area of a velocity field turn out very different from experimentally observable. When using the Navier-Stokes equation for calculation of time evolution of a field it is necessary to consider boundary conditions, which leads that the flow conditions is different from the needed conditions of isotropy.

The equations for various average values of parameters of the field such as average velocity, energy of turbulence, average dissipation, etc. (see for example [10]) and more difficult LES and RANS methods [11] that are applied only for flows in the conditions of far from an isotropy can't solve our problem.

The next way of receive and study of properties of the isotropic turbulent field of is offered.

The difficulties associated with the flow boundary, can be overcome by a transition to the spectral representation (to a wave space), and associated with no closure of the equations for averages, using the specific implementations of fields and corresponding selection of the initial conditions. The use of a concrete implementation leads to the fact that the role of the spectrum of the process now takes the spectrogram.

For the one-dimensional field the system of the equations describing evolution on time of a spectrogram is constructed in which interaction of separate harmonicas happens similar to the interactions described by Navier-Stokes's equation. The calculations illustrating development of the processes having the analogs in real turbulent streams are carried out.

The following designations and terminology are used:

i) common: ν – the dissipative factor (inverse Reynolds number);

ii) in the physical (configuration) space: $u=u(x,t)$ – turbulent velocity fluctuations; $U_m(x)$ – an average velocity field; $A_n(x)$ – an external source of fluctuations; $R(r)=\langle u(x)u(x+r)\rangle$ – of a u field correlation function; $T(r)=\langle u(x)u(x)u(x+r)\rangle$ – the moment of the third order of a field u;

iii) in the wave space: i – number of a wave; k_i – the wave number that represents the wavelength corresponding to i; A_i – coefficient in representation of a field $u=\sum A_i sin(k_i x)$; B_i – the coefficient in the representation of the average field; $a(k)$ – density of the amplitudes distribution; $S=\sum A(k)$ – total intensity of a field; k_d – shift of the initial distribution of amplitudes on k axis; σ – the parameter determining width of the initial distribution of amplitudes; S_0 – initial total intensity of a field; N_a – normalizing multiplier.

iv) terminology: instead of the term "spectrogram" is often used the term "spectrum"; for the coefficients A_i and B_i the term "amplitude" is used: for total intensity, sometimes for the purpose of abbreviation uses the term "energy".

As an example for creation of the equation for the spectrogram, the Burgers equation [12, 13] for some one-dimensional function $u(x,t)$ is considered with added the analog of some average field $U(x,t)$ what interaction with $u(x,t)$ can give to increase energy of field $u(x,t)$.

$$\frac{\partial}{\partial t}u(x,t)=u(x,t)\frac{\partial}{\partial x}u(x,t)+\frac{\partial}{\partial x}\big(u(x,t)U(x)\big)+\nu\frac{\partial^2}{\partial x^2}u(x,t) \quad (1)$$

The different forms of the equation, that similar to the equation (1), were considering with different conditions of the energy supplying to the flow [14–16]. From all results given in [12–16] it is necessary to mark following:

i) the existence of the exact solution of the equation (1) on finite and infinite intervals;

ii) the existence on a finite interval of x the critical Reynolds number $Re_m = 2\pi m / B_{2m}$ depending on the wave number m and the corresponding amplitude B_{2m} of a external field. If the Reynolds number exceeds their critical value the increase of amplitude A_m take place;

ii) the numerical calculation of development of fluctuation showed [13] that, having arisen, fluctuations grow during the initial stage of process then begin to decrease and eventually is disappearing;

iii) the requirements of conservation of energy can be fulfilled only using the corresponding boundary conditions which formulation is difficult for an infinite interval x, and for finite one it is is connected with restriction of the domain where the flow is isotropic.

2. Statement of the problem

On the interval of the positive integers $i = 1...,N$ the set of the wave numbers $k_i = ik_m$, where k_m is the scale determining the range of wave numbers, is define

The field $u(x,t)$ is considered in the form of decomposition on a set of orthogonal and normalized functions

$$u(x,t) = \sum_{j=1}^{N} A_j \phi_j(x) \tag{2}$$

For a limited set of wave numbers $\phi_k(x) = sin(kx)$ can be chosen.

The nonlinear term in the equation (1) describes the reorganization of the field after substitution in the equation of expression (2) becomes a sum of terms of type $sin(k_i x)cos(k_j x)$. Using a trigonometric ratio:

$$2sin(k')cos(k'') = sin(k' + k'') + sin(k' - k''), .$$

allows to write down an equation analog (1) as the system of the equations for $i = 1.,N$:

$$\frac{dA_i}{dt} = \frac{1}{2}\sum_{j=1}^{N} k_j(A_{i+j} + B_{i+j})A_j + \frac{1}{2}\sum_{j=1}^{N} k_j(A_{i-j} + B_{i-j})A_j - \sum_{j=1}^{N} k_j A_k A_j + \frac{dA_{n,i}}{dt} - \nu k_i^2 A_i \tag{3}$$

Here k_i – a wave number; $A_k = A_k(t)$ – the appropriate amplitude; B_k – coefficients in representation of a field of U; ν – inverse to the Reynolds number constructed on initial intensity of a field, a characteristic wave number, and the real coefficient ν defining dissipation; $dA_{n,i}/dt$ – intensity of an external source.

The third term of the right part is entered into the equation (3) for the saving value $\sum_{j=1}^{N} A_j$ in the case of $\nu = 0$ and the absence of the fields A_n and U

The first two items in the right part describe the arising or an increase in the intensity of the respective amplitudes, the third item describes the reduction of the existed nonzero amplitude A_i due to influence on the other amplitudes, and, finally, the last element of the right side describes the decrease in amplitude due to dissipation.

The result is the system (3) describes the change of the set of harmonic amplitudes of the implementation of the field $u(x,t)$. It is assumed that the properties of the solutions of this equation represent the properties of the whole set of possible implementations.

Considering the system (3) in the case when only one of $A_l \neq 0$, assuming that all other equal to zero, we get:

$$\frac{dA_l}{dt} = k_l A_l \left(U_{2l} - \nu k_l \right)$$

From where follows that the condition of stability will be: $Re_l = U_{2l} / \nu k_l = 1$.

Such form of critical number means that large–scale mode are the most sensitive, and the wave number of the arising fluctuation is two times less the wave number of the harmonics which are the cause of their occurrence.

Consideration of the system on a bounded set of wave numbers demands that when carrying out calculations only such transitions upon which all wave numbers k and $k+k'$ and $k-k'$ also belong to the (k_1, k_N) interval are considered, only in this case there is no energy removal from the considered region.

Use of a bounded part of the positive integers provides the energy conservation. In the absence of $U(x,t)=0$ and dissipation $\nu = 0$ the set (3) will be:

$$\frac{dA_i}{dt} = \frac{1}{2}\sum_{j=1}^{N} k_j A_{i+j} A_j + \frac{1}{2}\sum_{j=1}^{N} k_j A_{i-j} A_j - \sum_{j=1}^{N} k_j A_k A_j \tag{4}$$

Very similar equation can be received for a one–dimensional spectrum of turbulence when using the hypothesis for third moment of K. Hasselmann [1] for the Corrsin equation [17]. Produced in this work the transition from continuous functions to the functions described on a finite set of elements that is applicable also to equations similar to equation [17]. This transition, from continuous functions to functions on the finite sets of elements previously was used by us to calculate of the probability density function [18, 19].

The calculation is performed as a solution of the Cauchy problem for initial conditions for amplitudes and determination of the distributions of $B(k_i)$ and/or A_n and carried out by a method equivalent to the method of Runge–Kutta second order by time.

For calculation of a field in configuration space it is necessary to identify value $A(k)$ as $a(k)dk$ where $a(k)$ is density in the interval of $k \in (k_i, k_i + dk)$. Use of a ratio (2) due to a multiplicity of wave numbers leads to appearance of sharp bursts which can be avoided, selecting in a random way value k from values of an interval $k \in (k_i, k_i + dk)$, and selecting also in a random way the phase shift $d\phi \in (0, 2\pi)$ corresponding to this wave number.

3. White noise

In the absence of dissipation and energy supply the total intensity doesn't change. In this case there must be a stationary solution $A_i = invar(i)$ of the set (4) corresponding to the process called by "white noise", that is the process the correlation of which $\langle u(x)u(x+r)\rangle$ is delta–function $\delta(r)$. Substitution $A_i = invar(i)$ in the equation (4) shows that $A_i = invar(i)$ is the solution. Since this is limited diapason of wave numbers is considered in which there are no higher wave numbers, the correlation $\langle u(x)u'(x+r)\rangle \neq \delta(r)$. On a finite interval of k_i this solution leads to a correlation differing from delta–function and defined by the length of the interval used.

For obtaining a time evolution of amplitudes A_i the system (4) with initial conditions of $A_i = N_a \, exp\left(-(k_i - k_d)^2 / (2\sigma)\right)$ where k_d defines shift of distribution towards lagers k, and σ is width of the initial distribution, and value N_a is the initial total intensity of the field, is used. The parameter values are: $N_a = 1, \quad \sigma = 10, \quad k_d = 20$.

In figure 1 development of process of transition to finite distribution of amplitude A_i is pointed. Thick curves correspond to initial and finite distributions. The formation of the harmonicas with large wave numbers and the peaks at small k are noticeable, these peaks illustrate the emergence of the beats at low–wave numbers. Over time, these peaks arose, first rising and then smoothed.

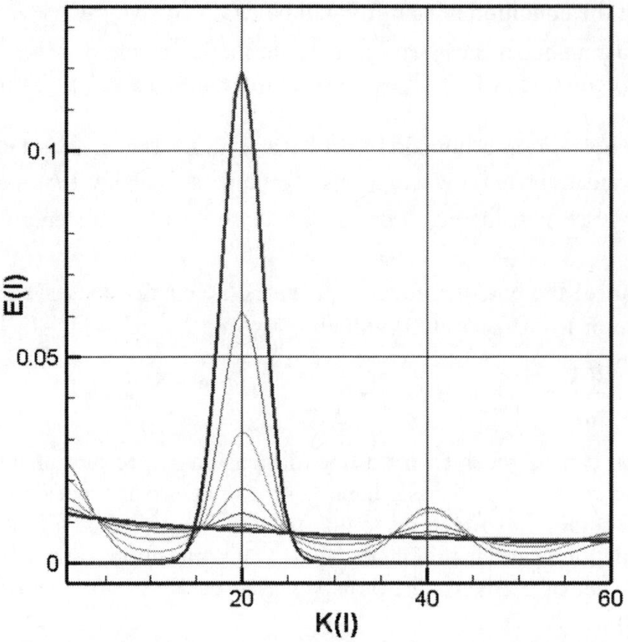

Figure 1. Time evolution of a spectrum in the course of tending to uniform distribution. The thick curves correspond to initial distribution and the distribution at the end of calculation.

4. The decay

The considered process corresponds to the conditions observed at a turbulent flow in the channel downstream behind the turbulence–produced grids, which formed the initial state of the turbulent field. In this case the initial state of the flow is modeled by an initial form of the spectrogram. Calculation of decay process is performed according to the equation (3) with A_n and B_i equal to zero, with the initial condition $A_i = N_a exp\left(-k_i^2/(2\sigma)\right)$, where σ determines a width of initial distribution, and N_a is determined by condition $\sum A_i = S_0$ where S_0 defines initial intensity of the field. The parameter values are: $S_0 = 1$, $\sigma = 10$.

The microscale λ is the parameter which being changing in the decay process and its changing characterizes the evolution of the field structure. In the configuration space its value is defined with correlation function $R(r) = \langle u(x)u(x+r)\rangle$:

$$\lambda^2 = -R(0) \Big/ \left.\frac{\partial^2 R}{\partial r^2}\right|_{r=0}$$

Analog of this relation in wave space (in our case of a finite set of wave numbers) will be:

$$\lambda^2 = \sum_{i=1}^{N} A_i \Big/ \sum_{i=1}^{N} k_i^2 A_i$$

The time evolution of the squared microscale λ^2, the inverse intensity $1/S$ and Reynolds number $Re = \lambda S/\nu$ are shown in figure 2. The dashed lines distinguish the intervals at which changes occur under the linear law. The temporal evolution of the total intensity S is shown in figure 3.

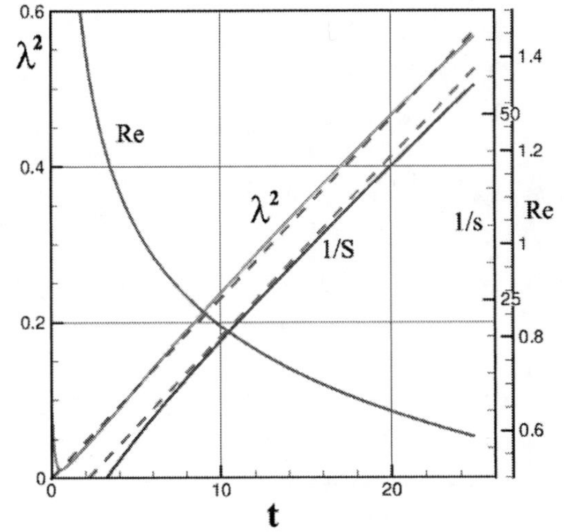

Figure 2. Decay process.

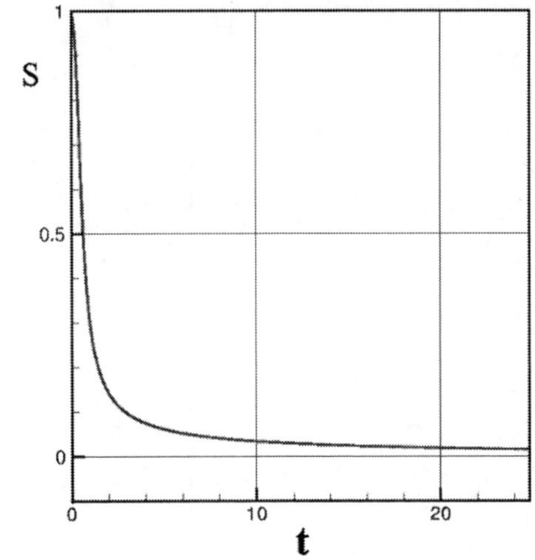

Figure 3. Time evolution of the total intensity S(t).

The experimental results of [20] (see figure 2 from [20]), are in good agreement with the results obtained in the present work. The coincidence of the curves λ^2 from [20] and our data is just amazing. In both cases there is a long–linearity of change of λ^2, and a small drop in the curves at the initial stage of the decay. Deviations from linearity at the final stage in figure 2 are observed when the value of λ^2 exceeds the observed experimental values.

Coincidence of results for the curves $1/S$ and $U^2/\langle u(x)^2 \rangle$ [20] is slightly worse. The linearity of $1/S$ begins a bit later than for the analogical dependence $U^2/\langle u(x)^2 \rangle$ and in addition at large times there is deviation from linearity. Whereas in the experimental dependences nonlinearity of the curve $U^2/\langle u(x)^2 \rangle$ at large distances from the grid are absent.

It should be noted that the regularities connected with the Reynolds number differ for real case and the considering case. The assumption of constancy of the Reynolds number under experimental conditions leads to the following equation for microscale λ^2:

$$\frac{d\lambda^2}{dt} = 10\nu,$$

what is confirmed by experimental data [20]. At the same time the similar result for settlement conditions forms out on condition of constancy of the combination $\lambda^2 S$ rather than the Reynolds number constancy. This leads to:

$$\frac{d\lambda^2}{dt} = \nu.$$

that corresponds to the monotonic decreasing the Reynolds number during the decay.

The time evolution of the Reynolds numbers Re_λ calculated from the results of several studies [20–23] are given in figure 4. The most part of curves can't be treated as a constancy of Reynolds number in decay process. Some results are, especially, received in [22] is very close to that presented in figure 2. The reason of wide range (5÷75) limiting values of Reynolds number isn't clear. For the unknown reason for large times the various values of Reynolds numbers are formed.

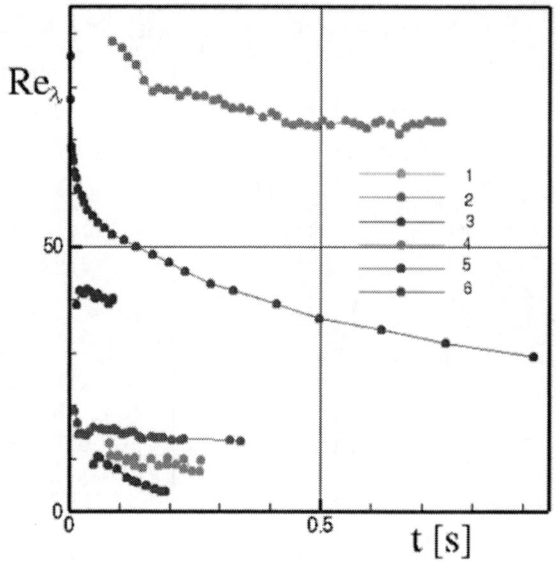

Figure 4. The time evolution of the Reynolds numbers: 1 – behind grids, made of parallel spaced cylindrical rods [21], 2 – standard grid [21], 3 – [21], 4 – [6], 5 – [22], 6 – the results of the calculation of the Navier–Stokes equations [23].

In the following section the influence of dissipation on decay process is studying.

4.1. Influence of the dissipation intensity ν on the microscale λ

Results of calculation of time evolution of microscale λ for various values of the dissipation intensity ν are given in figure 5. The rate of λ^2 increasing is proportional to the value ν. At small values ν the distinction from linear growth is noticeable. Results for the dependence λ^2 on νt are given in figure 6. In these coordinates the slope of the curves are identical, and the shift along ordinate axis is explained by differences in initial part of curves, that is illustrated figure 7 in where the respective initial stages of dependences from time are given.

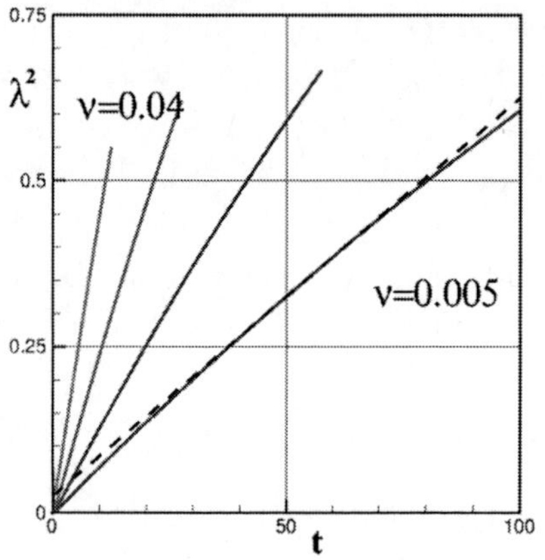

Figure 5. Time evolution of λ^2.

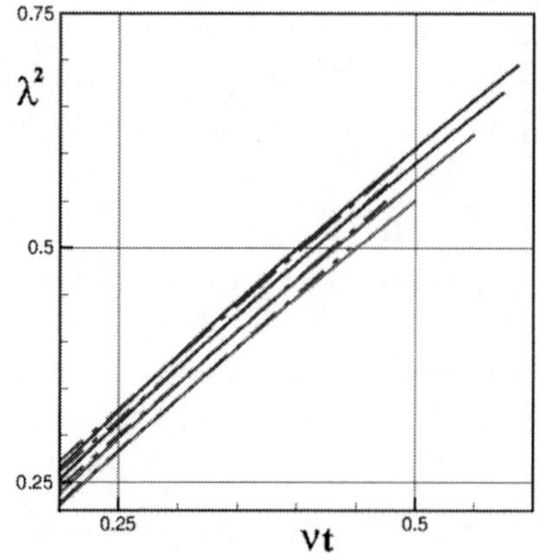

Figure 6. As figure 6 vs νt.

4.2. Spectrograms

Results of calculation of the change of values $A_k(t)$ in the decay process are given in the following figures. In figure 8 for one value ν the normalized spectra A_k/A_1 are given. On an initial stage the extension towards large wave numbers is forming, then amplitudes of large wave numbers are decreasing and the normalized spectrum A_k/A_1 ceases to change.

In figure 9 for a number of values ν changing in the range 0.005 till 0.04 the limited forms normalized spectra A_k/A_1 are given.

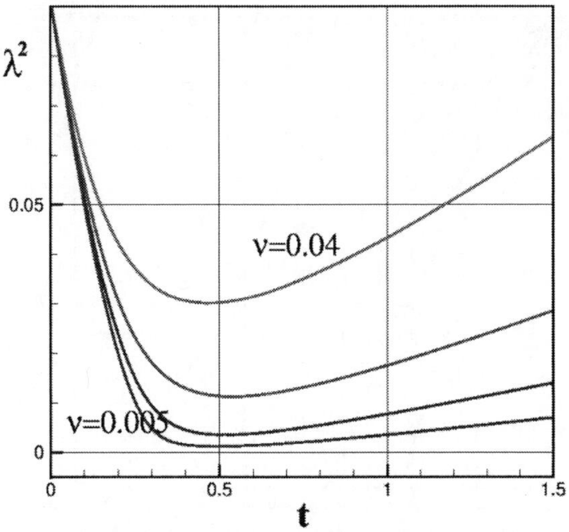

Figure 7. Initial stage of decay.

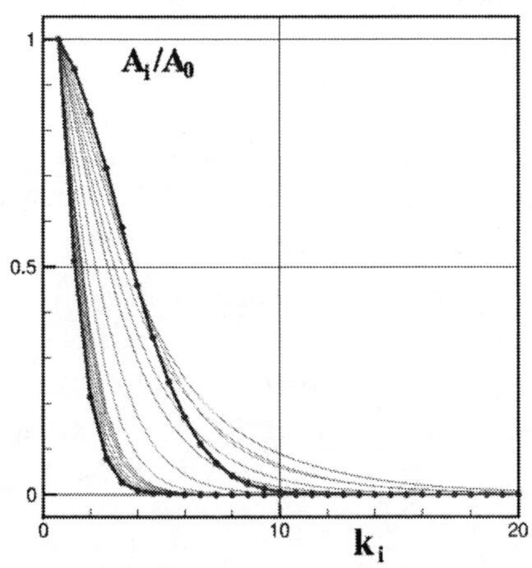

Figure 8. Change of the spectrum in the decay process, thick lines correspond to initial distribution and distribution at time of the end of calculation.

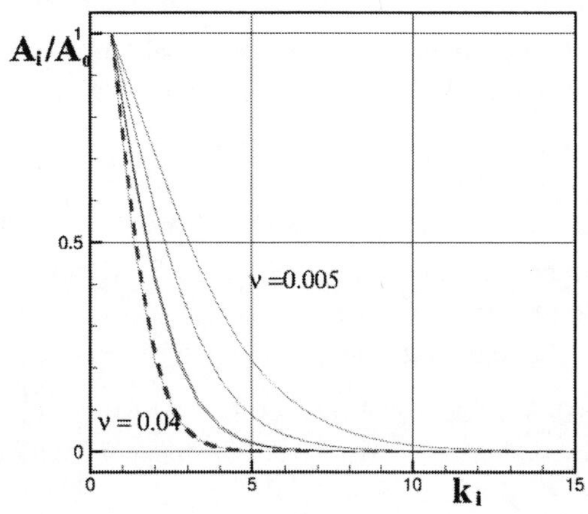

Figure 9. The normalized spectra at time of the end of calculation for various values ν.

The transformation

$$k_i'' = k_i^0 + \left(k_i - k_i^0\right)\left(\frac{\nu_i}{\nu_0}\right)^{-0.55}$$

allows to reduce all curves to a single one, in our case (figure 10) to dashed curve corresponding to the greatest ν.

In figure 10 the dependences of normalized spectra at time of the end of calculation are pointed as $log(A_k/A_1)$. Straight dashed lines select intervals of large wave numbers on which the viscosity plays a defining role. At large wave numbers all curves are well approximated by straight lines and the expression (5) leads to the following type of the spectrum at large wave numbers:

$$A_i \sim exp(-\nu^{0.55}k)$$

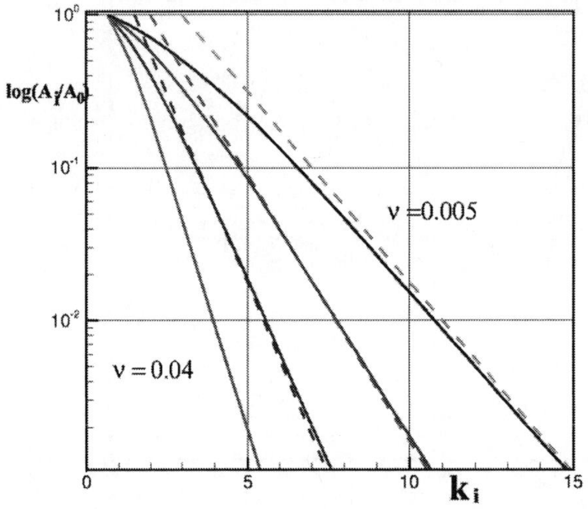

Figure 10. Too that figure 8 in semi logarithmic coordinates.

5. The influence of the external field and external forces

The stationary regimes can be received only at special impact to amplitude A_i which can be executed by several ways. The first method is as follow: at each step of calculation, produced an increase of certain amplitudes, which compensates for the change in the spectrum under the influence of dissipation. The quantity B_i characterizes distribution over wave numbers the intensity of such influence (a rate of pumping). This is model of the fan operation which continuously adds the energy into the flow.

The second method: the growth of quantity A_i proceeds due to the action of a certain stationary external field which spectral representation B_i is known .This way models the velocity perturbation growth due to an existence of the mean velocity gradient as it has in real flows.

5.1. Fan mode

The calculation was performed for zero initial distribution of values $A_i = 0$, which corresponds to the activation of the fan at the initial absence of flow. The distribution was set uniformly distributed on the 15 lower wave numbers and was normalized to set the intensity of the impact.

In figure 11 the regime of going to a steady state is displayed. Dependences of the quantities λ^2 and total intensity S are given.

The results can be explained by the existence of an initial time interval when the jet of the fan raised in the initially quiet medium consists at first from of the large enough fractions which size is specified by fan construction, and hence at first the turbulent viscosity is small and energy of system sharply increases. Amount of input in flow of energy far exceeds the energy which disappears under the action of turbulent viscosity. However, the disintegration of the jet leads to the emergence of small–scale formations, the growth of the turbulent viscosity occurs and the excess initially stored energy dissipated. The stationary condition in which all inputted energy equal to dissipated one is forming.

In figure 12 the initial and stationary spectra for the regime of modeling the fan are presented.

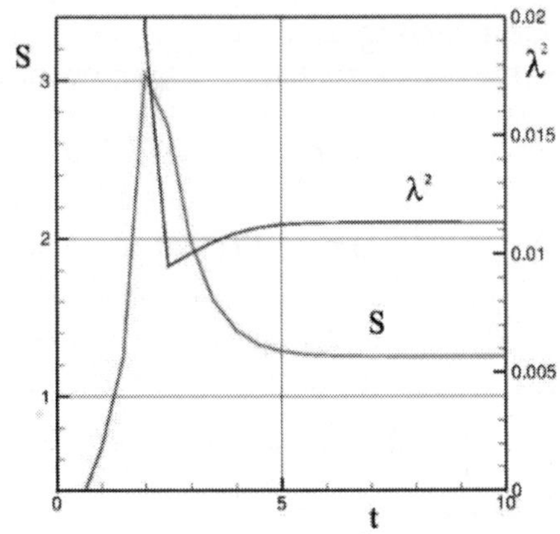

Figure 11. Time evolution of values of S and λ^2 for the fan mode.

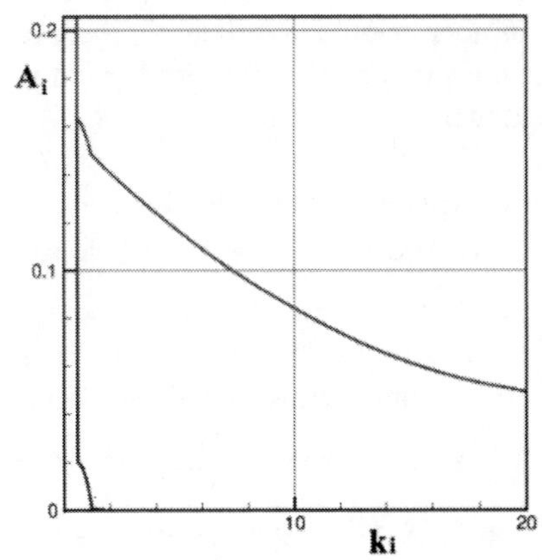

Figure 12. The spectra to initial time and the time of the end of calculation for the fan mode.

5.2. Influence of the external field

In this case the initial distribution $A_i = 0.01$ of was given. Distribution of the amplitudes of the external field B_i was specified, also as in the previous calculation, uniformly distributed on 15 lowest wave numbers, and the intensity of external field action was normalized.

In figure 13 the regime of transition to stationary state under the external field action is presented. In this case the display scale of the parameter λ^2 was changed with the aim to illustrate a small non–monotonicity of curves.

In figure 14 the initial and formed under the influence of the external field spectra are given.

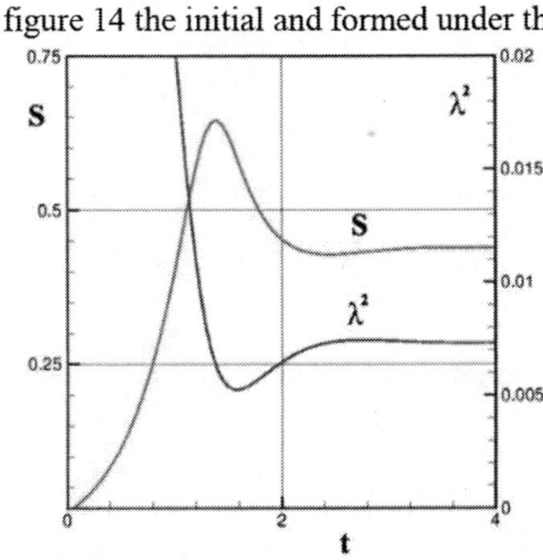

Figure 13. Time evolution of values of λ^2 and S (the case of the external field influence).

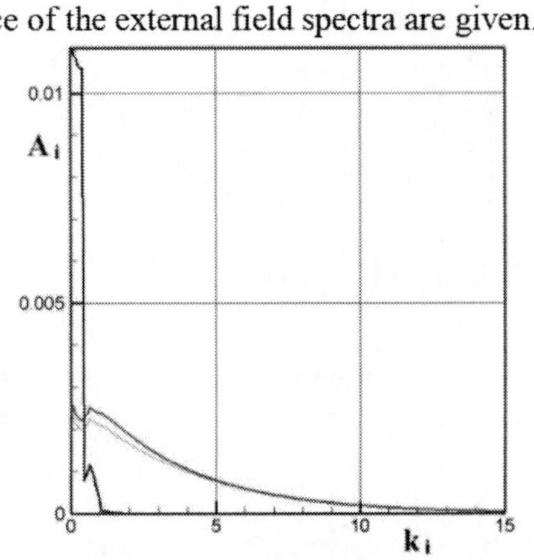

Figure 14. The spectra to initial time and the time of the end of calculation (the case of the external field influence)

In both analyzed cases (the fan mode and external field influence) the results practically coincide, except for a small bend down the curve u at the case when the influence of external field is considered.

In this case such a behavior of the curve u as above is explained due to an existence of initial period on which amplitudes of the lowest wave numbers firstly grow. It is also confirmed by the features of microscale change. The studied examples are typical illustrations of the initial conditions influence on the character of solution. The initial fields being specified in more wide range of the wave numbers would significantly increase an initial turbulent viscosity.

The considered examples are typical illustrations of the influence of initial conditions on the nature of phenomenon.

6. Configuration space. Third–order moments

The use for restoring a field in physical space of the relation $u(x,t)=\sum A_i(t)\sin(k_i x)$ due to the wave numbers multiplicity that is caused the accepted form of wave numbers interval leads to emergence of sharp velocity changes at separate points x. It is possible to avoid this if to recall that we deal with approximation of the continuous distribution of amplitudes $A(k,t)$ and, therefore, the values of A_i represent a contribution of the interval designated by k_i into the integral $\int a(k)dk$. This means $A_i = a(k_i)dk$. To avoid these difficulties, when calculating the field $u(x)$, one have to execute a random choice of the wave number k_i from the interval $(\mathrm{k}_i, k_i + dk_i)$. It is also possible to choose random phase shift from the range $(0,2\pi)$.

In figure 15 the example of realization of the field $u(x,t)$ is presented. In the next figure 16 the correlation function $R(r)=\langle u(x)u(x+r)\rangle$ and the statistical moment of the field $u(x,t)$ third–order $T(r)=\langle u(x)u(x)u(x+r)\rangle$, which in form nearly coincides with the derivative of correlation function are presented.

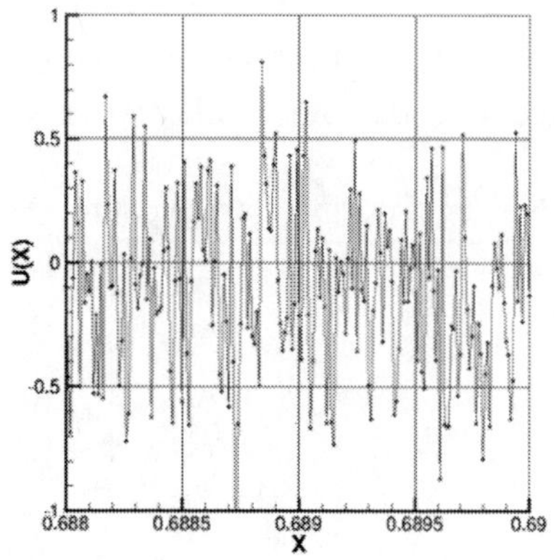

Figure 15. Example of realization of the field $u(x,t)$.

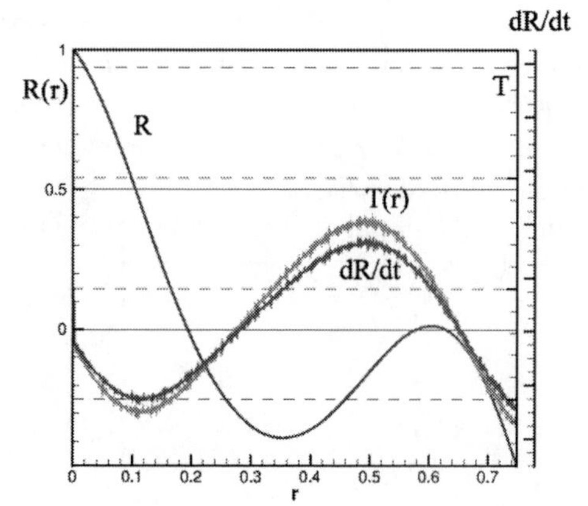

Figure 16. Correlation function $R(r)=\langle u(x)u(x+r)\rangle$, third–order moment $\mathrm{T}(r)=\langle u(x)u(x)u(x+r)\rangle$ and the derivative $dR(r)/dr$.

7. Conclusion

This research of the properties of isotropic turbulent phenomena is based on four main assumptions.

1. Using the equations for a certain implementation of random phenomenon to which, undoubtedly, liquid whirls belong allows to avoid the difficulties arising by transition from

the dynamic equations (for example Navier–Stokes equations) to the equations for averages when nonlinearity of the dynamic equations results in their no closure. It is supposed that the received description of separate implementation rather precisely corresponds to properties of a whole ensemble.

2. The transition to wave space, which is natural in case of studying the isotropic fields, enables one to avoid the difficulties related to existence of the boundary conditions being necessary in configuration space that violates, in principle, requirements of isotropy.
3. The select of the finite set of wave numbers located in a finite segment of the positive integers allows one to receive the discrete description of the process of the calculating in which the interaction of different components of the wave field does not lead to extension of the set of wave numbers.
4. And, perhaps, the main thing. The interaction of the wave-field amplitudes related to non linearity of dynamic equations is using the triad rule $k = k' + k''$.

Numerical solving the Cauchy problem allows to receive results for the regime of the dissipative decay of the field, the regimes of energy input directly or as a result of interaction with external field that, in many details, are in agreement with known results for real turbulent flows.

Especially it should be noted that the considered approach allows one to account for the influence of initial field structure which can have significant effect on the results received.

Acknowledgments
The author wishes to thank E.V. Teodorovich for fruitful discussions. The work was partially supported by Russian State Assignment under contract No. AAAA-A17-117021310385-6.

References

[1] Hasselmann K 1958 *Deutsche Hydrographisehe Zeitschrift*. B **11** H **5** 207-217
[2] Lytkin Y M and Chernykh G G 1976 *Dynamics of Continuous Media* [in Russian] 124-130
[3] Akatnov N I and Bystrova E N 1999 *High Temperatur* V **37** N **6** 895 903
[4] Akatnov N I and Bystrova E N 2000 *High Temperatur* V **38** N **4** 600-608
[5] Frost V 2016 Analysis of some methods of Karman-Howarth equation closure [in Russian] *Prepint IPMech RAS* №1120 24
[6] Krogstad P and Davidson P 2012 Near-field investigation of turbulence produced by multi-scale grids *J. Physics of fluids*. V **24(3)** 1-6
[7] Hurst D and Vassilicos J C Scalings and decay of fractal-generated turbulence 2007 *J. Physics of Fluids* V **19(3)** 1-31
[8] Valente P C and Vassilicos J C 2014 The non-equilibrium region of grid-generated decaying turbulence *J. Fluid Mechanics* V **744** N **10** 5-37
[9] Kraichnan R H 1970 Diffusion by a Random Velocity Field *J. Physics of Fluids* V **13** N **1** 22–31
[10] Launder B E and Spalding D B 1974 The numerical computation of turbulent flows *J. Computer Methods in Applied Mechanics and Engineering* V **3**(2) 269–289
[11] Secundov A N, Cheprasov S A and Yakubovsky K Y 2015 *High Temperature* V **53** N 5 709–712
[12] Burgers J M 1939 Mathematical examples illustrating relations occurring in the theory of turbulent fluid motion *Verh. KNAW, A/d Natuurkunde* **XVII** (2) 1-53
[13] Krasitskii V P. Nonlinear model equation for finite fluctuations 1967 in Proceedings of All–Union Symposium on the problems of turbulent flows, including geophysical applications Kiev 58-61
[14] Weinan E and Vanden E E Another note on forced burgers turbulence 2000 *J. Physics of Fluids* V **12** (1) 149-154
[15] Bec J, Frisch U and Khnanin K 2000 Kicked Burgers turbulence *J. Fluid Mech.* V **416** 239–267

[16] Frachebourg L and Martin H A 2000 Exact statistical properties of the Burgers equation *J. Fluid Mech.* V **417** 323–349
[17] Frost V 2008 in Proceedings of School-seminar "Aerophysics and physical mechanics of classical and quantum systems" IPMech RAS 171–175
[18] Frost V, Kaminsky V and Rabinovich A 2004 Modeling modeling of micromixing and the equation for correlation function in: Micromixing in Turbulent Reactive Flows. Moscow 51–56
[19] Frost V 2015 *Mathematics Models and Computer Simulation* V **7** (5) 430-438
[20] Stewart R W and Townsend A 1951 Similarity and self-preservation in isotropic turbulence *Phil. Trans. of the Royal Soc. A* V **243** (867) 359-386.
[21] Kostomakha V A 1985 in: *Dynamics of Continuous Media (in Russian)* IGIL SO AN SSSR, Novosibirsk **70** 92-104
[22] Huang M and Leonard A 1994 Power decay of homogeneous turbulence at low Reynolds number *J. Phys. Fluids* **6** (11) 3765-3775.
[23] Bennett J C and Corrsin S Small 1978 Reynolds number nearly isotropic turbulence in a straight duct and a contraction *J. Phys. Fluids* **21** (12) 2129-2140.

IOP Conf. Series: Journal of Physics: Conf. Series **1009** (2018) 012016 doi:10.1088/1742-6596/1009/1/012016

Analysis of the turbulent boundary layer on a flat plate at M=6÷8.8 with the use of NERAT-2D code and algebraic turbulence models

S T Surzhikov

Ishlinsky Institute for Problems in Mechanics RAS,Vernadsky prospekt 101(1), Moscow, 119526, Russia
Moscow Institute of Physics and Technology (State University), 9 Institutskiy per., Dolgoprudny, Moscow Region, 141701, Russia
Dukhov All-Russian Scientific Research Institute of Automatics, Moscow, Russia

surg@ipmnet.ru

Abstract. Experimental data on heat transfer of supersonic air flows to flat plate, which were analysed by Dilley A.D [NASA/CR-2001-210837, 2001], are compared for Mach numbers $M = 6 \div 8.8$ with the calculated data obtained using the Navier – Stokes flow model in conjunction with the analytical models of turbulent mixing of the Prandtl mixing length and the Baldwin-Lomax model. In the paper, the author's computer code NERAT-2D was used for numerical study.

1. Introduction

One of the key tasks of hypersonic aerothermodynamics is the heating of the surface of a high-speed aircraft.The primary problems of external aerothermodynamics are:

- heating blunted edges of hypersonic aircrafts, air intakes as well as steering controls;
- local heating of the surface in the interaction of incident shock waves with boundary layers;
- local heating in the laminar-turbulent transition (LTP) and turbulent heating of the developed surface.

A detailed calculation of the thermal regimes of the region of turbulent motion and the LTP zone in hypersonic flow past a flat plate was carried out in [1]. The studied speed range was $M = 6 \div 8.8$. The work was carried out as part of the development work of hypersonic aircraft Hyper-X. The advantage of this work is a detailed analysis of three series of experimental data [2–4] using numerical simulation. For this, the computer code CFL3DE [5] was used, in which three models were realized: the parabalized Navier – Stokes equations, the Navier – Stokes model of a viscous compressed layer, the Euler equations. Algorithmically implemented is an upwind, finite volume flow solver.

An analytical Baldwin – Lomax model [6] was used in [1]. The influence of grids used was studied in detail. The final recommendations of [1] for an adequate description of the experimental data consist in the need to provide a grid detail along the normal to the surface $y^+ < 2$, where $y^+ = \dfrac{y}{\nu_w}\sqrt{\dfrac{\tau_w}{\rho_w}}$, where: y is the coordinate normal to the surface; ν_w is the kinematic viscosity; τ_w is the frictional stress on a surface; ρ_w is the gas density near to surface.

In [1], solutions were obtained for very small sizes of the minimum step along the normal coordinate $\Delta y_1 \sim 0.5\times10^{-4}$ cm (0.5 μm), which corresponded to Reynolds numbers corresponding to the minimal step

$$\mathrm{Re}_{\Delta y_1} = \frac{\rho_\infty V_\infty \Delta y_1}{\mu_\infty} \sim 10$$

In this paper, the author's computer code NERAT-2D [7] is used for numerical analysis of experiments [2–4] (validation) and comparison with calculation results [1] (verification). This code has been repeatedly tested successfully with the example of physico-chemical aerothermodynamics [8]. Taking into account the urgency of studying the features of the structure of boundary layers at hypersonic velocities of flow above surfaces, experimental [2–4] and calculated data [1] were chosen to test the work of the NERAT-2D code together with the simplest algebraic model of turbulence.

The problem of convective heating of a flat plate with a sharp edge is classical for aerothermodynamics. This is because the flow model for a plane surface is very versatile with respect to applicability for other types of flows and other geometries. Nevertheless, in spite of the large number of papers devoted to this problem, the problem of convective heat transfer to a plate (or, in a more general formulation, on a surface) for M>6 has not yet been studied sufficiently, in particular regarding the reliable use of turbulent mixing models in aerothermodynamic computer codes .

This problem has been studied most extensively in experimental and computational studies, the results of which have been generalized in [9–11]. Approximate analytical solutions are also obtained for this problem. On the basis of these solutions and experimental data, approximate semi empirical relationships are proposed, which have become widely used in practical aerothermodynamics. These correlations have been repeatedly tested in a wide range of applications, and they underlie the sectoral techniques used with a high degree of reliability in the profile aerospace organizations. Multiple confirmation of the reliability of the calculated data resulted in the fact that modern computational models are primarily tested by comparing precisely with these correlations.

Despite the convincing results of the development of the theory of the boundary layer for a flat plate, there are a limited number of publications on the structure of the boundary layer at hypersonic flow velocities, in particular, flow regimes with a laminar-turbulent transition.

2. Approximate correlations based on criterial relations of boundary layer theory

The most widely used model of heat transfer is based on the introduction of an effective heat transfer coefficient [9].

For a *laminar* boundary layer

$$q_w = \left(\frac{\alpha_l}{c_p}\right)(H_e - h_w), \tag{1}$$

$$H_e = h_\delta + \frac{1}{2}\kappa V_\delta^2, \tag{2}$$

$$\left(\frac{\alpha_l}{c_p}\right) = 0.332\mathrm{Pr}_w^{-2/3}\rho_* V_\delta \sqrt{\frac{(2\nu+1)\mu_*}{\rho_* V_\delta x}}, \tag{3}$$

where q_w is the convective heat flux density; (α_l/c_p) is the effective heat transfer coefficient for a laminar boundary layer; H_e, h_δ are the adiabatic wall enthalpy and thermodynamic enthalpy, $h = e + (p/s)$; e is the specific internal energy.

To calculate the density of a convective heat flux for a laminar flow past a plate ($\nu = 0$) or an acute cone ($\nu = 1$), the following algorithm can be used:

IOP Conf. Series: Journal of Physics: Conf. Series **1009** (2018) 012016 doi:10.1088/1742-6596/1009/1/012016

1) First, the initial flow data is specified. For example, in the case under consideration, based on [1] one can use

$$c_p = 10^7 \text{ erg/(g×K)},\ M_\Sigma = 29 \text{ g/mole},\ T_\infty = T_\delta = 65.04 \text{ K},\ V_\infty = V_\delta = 9.696\times10^4 \text{ cm/s},$$

$$\gamma = 1.4,\ R_0 = 8.3145\times10^7 \text{ erg/(mole×K), Mach number } \mathrm{M}_\delta = 6, T_w = 320\text{ K}, \mathrm{Pr}_w = \mu_w c_p / \lambda_w, h_w = c_p T_w,$$

where indices δ and w indicate the flow parameters at the outer boundary of the boundary layer and at the surface.

2) Using these data, the enthalpy, temperature and density are calculated:

$$h_\delta = c_p T_\delta, h_w = c_p T_w,\ h_* = \frac{1}{2}(h_\delta + h_w) + 0.11\sqrt{\mathrm{Pr}_w} V_\delta^2,\ T_* = \frac{h_*}{c_p},\ \rho_* = \frac{p_\delta M_\Sigma}{R_0 T_*}$$

The viscosity is determined by Sutherland's formula

$$\mu_* = C_1 \frac{T_*^{3/2}}{T_* + 110.4},\quad C_1 = 1.458\times10^{-5} \text{ g/(cm×s×K}^{1/2})$$

3) Then calculations of q_w are made at each section in x according to formulas (1)–(3).

The friction stress was also calculated

$$\tau_w = 0.332\rho_* V_\delta^2 \sqrt{\frac{(2\nu+1)\mu_*}{\rho_* V_\delta x}} \tag{4}$$

The density of the convective heat flux and the frictional stress in the *turbulent* boundary layer is determined in an analogous sequence by the formulas

$$q_w = \left(\frac{\alpha_t}{c_p}\right)(H_e - h_w) = 0.0296\,\mathrm{Pr}^{-0.6} K_* \rho_\delta \mathrm{Re}_\delta^{-0.2}(H_e - h_w), \tag{5}$$

$$\tau_w = 0.0296 K_* \rho_\delta V_\delta^2 \mathrm{Re}_\delta^{-0.2}, \tag{6}$$

$$K_* = \left(\frac{\rho_*}{\rho_\delta}\right)^{0.8}\left(\frac{\mu_*}{\mu_\delta}\right)^{0.2},$$

$$\mathrm{Re}_\delta = \frac{\rho_\delta V_\delta x}{\mu_\delta},\ H_e = h_\delta + \frac{1}{2}\kappa V_\delta^2,\ \kappa = \mathrm{Pr}^{1/3}$$

The characteristic parameters ρ_* and μ_* are calculated using T_*, which, in its turn, is calculated with *reference* enthalpy

$$h_* = \frac{1}{2}(h_\delta + h_w) + 0.22(H_e - h_\delta) = \frac{1}{2}(H_\delta + H_w) - \left(\frac{1}{2} - 0.22\kappa\right)\frac{V_\delta^2}{2} = \frac{1}{2}(h_\delta + h_w) + \frac{0.22}{2}\kappa V_\delta^2,$$

$$H_\delta = h_\delta + \frac{V_\delta^2}{2}$$

After calculating the surface friction stress and the density of the convective heat flux, the dimensionless friction and heat transfer coefficients (Stanton numbers) are determined

$$C_f = \frac{\tau_w}{0.5\rho_\delta V_\infty^2}, \tag{7}$$

$$\mathrm{St} = \frac{q_w}{\rho_\delta V_\delta (H_e - h_w)} = \frac{q_w}{0.5\rho_\delta V_\delta^3} \tag{8}$$

3. Calculating model of surface flow based on the Navier – Stokes equations

The dynamics of a viscous, heat-conducting gas is calculated using the Navier – Stokes equations in a two-dimensional formulation. The calculated region includes the flow field in the undisturbed flow, behind the front of the shock wave formed behind the sharp edge of the streamlined plate and above its surface.

The initial system of equations is formulated as follows:

$$\frac{\partial \rho}{\partial t} + \mathrm{div}(\rho \mathbf{V}) = 0, \tag{9}$$

$$\frac{\partial \rho u}{\partial t} + \mathrm{div}(\rho u \mathbf{V}) = -\frac{\partial p}{\partial x} - \frac{2}{3}\frac{\partial}{\partial x}(\mu_{eff}\,\mathrm{div}\mathbf{V}) + \frac{\partial}{\partial y}\left[\mu_{eff}\left(\frac{\partial u}{\partial y} + \frac{\partial v}{\partial x}\right)\right] + 2\frac{\partial}{\partial x}\left(\mu_{eff}\frac{\partial u}{\partial x}\right), \tag{10}$$

$$\frac{\partial \rho v}{\partial t} + \mathrm{div}(\rho v \mathbf{V}) = -\frac{\partial p}{\partial y} - \frac{2}{3}\frac{\partial}{\partial y}(\mu_{eff}\,\mathrm{div}\mathbf{V}) + \frac{\partial}{\partial x}\left[\mu_{eff}\left(\frac{\partial u}{\partial y} + \frac{\partial v}{\partial x}\right)\right] + 2\frac{\partial}{\partial y}\left(\mu_{eff}\frac{\partial v}{\partial y}\right), \tag{11}$$

$$\rho c_p \frac{\partial T}{\partial t} + \rho c_p \mathbf{V}\mathrm{grad}T = \chi\frac{\partial p}{\partial t} + \chi \mathbf{V}\mathrm{grad}p + \mathrm{div}(\lambda_{eff}\,\mathrm{grad}T) + Q_V, \tag{12}$$

where x, y are the Cartesian coordinates; $\mathbf{V} = (u, v)$ is the flow velocity and its projections on the x and y axes; ρ, p are the density and pressure; $\mu_{eff} = \mu + \mu_t$ is the effective coefficient of viscosity, determined by the Boussinesq hypothesis; μ, μ_t are the dynamic coefficient of viscosity and coefficient of turbulent viscosity; c_p is the specific heat at constant pressure; T is the temperature; λ_{eff}, λ, λ_t are the effective coefficient of thermal conductivity, the coefficients of molecular and turbulent thermal conductivity, which was calculated using the turbulent Prandtl number $\mathrm{Pr}_t = \frac{\mu_t c_p}{\tau_t}$; Q_V is the power of heat release due to dissipative processes and heating from external energy sources.

The energy conservation equation is written here in a non-conservative form with respect to temperature (in the form of the Fourier – Kirchhoff equation). The system of equations (9) – (12) is used in conjunction with the equation of state of an ideal gas

$$p = \rho \frac{R_0}{M_\Sigma} T, \tag{13}$$

where R_0 is the universal gas constant; M_Σ is the molecular weight.

The boundary conditions give the unperturbed incident flow

$$\begin{aligned} x = 0: &\quad u = u_\infty,\ v = 0,\ T = T_\infty,\ p = p_\infty,\ \rho = \rho_\infty, \\ y = H: &\quad u = u_\infty,\ v = 0,\ T = T_\infty,\ p = p_\infty,\ \rho = \rho_\infty, \end{aligned} \tag{14}$$

and non-reflecting boundary conditions at the exit from the computational domain

$$x = L: \quad \frac{\partial u}{\partial x} = \frac{\partial v}{\partial x} = \frac{\partial T}{\partial x} = \frac{\partial \rho}{\partial x} = 0 \tag{15}$$

On the surface of a streamlined body, the sticking conditions are formulated as following:

$$x = 0: \qquad u = v = 0 \tag{16}$$

The design scheme of the problem being solved is shown in figure 1.

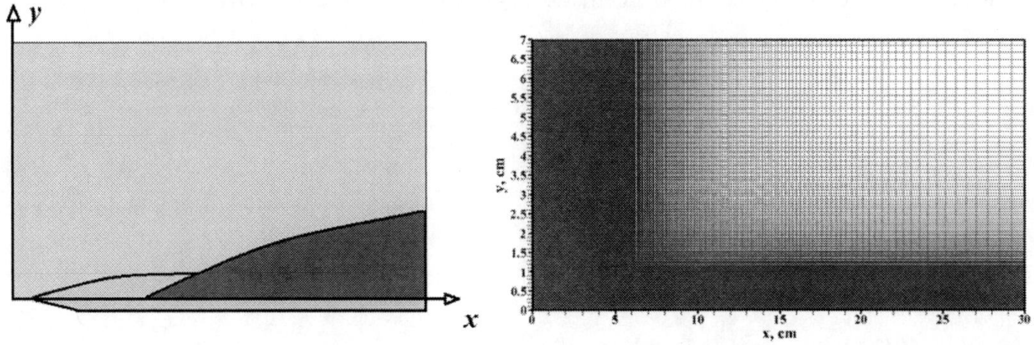

Figure 1. The scheme of the problem (on the left) and the calculated grid 701×401 (along and across the plate).

4. Algebraic models of turbulence

Two algebraic models of turbulent heating and friction of a plane surface were used. The laminar-turbulent transition was determined by specifying the critical Reynolds number

$$\mathrm{Re}_{x,LT} = \frac{\rho_\infty V_\infty x_{LT}}{\mu_\infty},$$

where x_{LT} is the longitudinal coordinate at which the laminar-turbulent transition began.

4.1. The Prandtl turbulent mixing model

This is one of the first and most researched models. In different variations, this model is presented, for example, in [12– 14].

Turbulent viscosity is determined by the well-known phenomenological relationship [15]

$$\mu_t = \rho L_m^2 |\Omega|, \tag{17}$$

where L_m is the mixing length of Prandtl; Ω is the vorticity function.

In the case of the simplest flow near the plate without a pressure gradient in the external flow, it is approximately assumed

$$|\Omega| = \left|\frac{\partial u}{\partial y}\right| \tag{18}$$

A two-layer model is considered, according to which the length of the Prandtl mixing path is determined by the formula:

$$L_m = \begin{cases} L_m^{in} = \kappa y \left[1 - \exp\left(-\dfrac{y^+}{A^+}\right)\right] & \text{at } \dfrac{y}{\delta} < 0.2 \\ L_m^{out} = 0.085\delta & \text{at } \dfrac{y}{\delta} > 0.2, \end{cases} \tag{19}$$

where δ is the thickness of dynamic boundary layer; κ=0.43 is the empirical constant; $A^+ = 26$;

$$y^{+}=\frac{y}{\nu_{w}}u_{\tau},\quad u_{\tau}=\sqrt{\frac{\tau_{w}}{\rho_{w}}},\quad \tau_{w}=\mu_{w}\left(\frac{\partial u}{\partial y}\right)_{w}, \tag{20}$$

where ν_w is the kinematic viscosity near the surface.

If we use the relationship between kinematic and dynamic viscosity $\mu=\rho\nu$, then

$$y^{+}=\frac{y}{\mu_{w}}\rho_{w}u_{\tau}=\frac{y}{\mu_{w}}\sqrt{\rho_{w}\tau_{w}}=y\sqrt{\frac{\rho_{w}}{\mu_{w}}\left(\frac{\partial u}{\partial y}\right)_{w}} \tag{21}$$

4.2. The Baldwin – Lomax model [6]

A two-layer model is considered. Each layer has its own turbulent viscosity. In the inner layer (closer to the surface)

$$\mu_{t,in}=\kappa\rho yD|\Omega|, \tag{22}$$

where $\kappa=0.4$; D is the van Driest damping function

$$D=1-\exp\left(-\frac{y^{+}}{A^{+}}\right),\quad A^{+}=26;$$

the vorticity function in the two-dimensional case (the projection of the vector $\mathrm{rot}\,\mathbf{V}$ on the z axis)

$$\Omega=\frac{\partial v}{\partial x}-\frac{\partial u}{\partial y}$$

In the outer layer

$$\mu_{t,out}=KC_{cp}\rho F_{wake}F_{Kleb}(y), \tag{23}$$

where $K=0.018$, $C_{cp}=1.6$,

$$F_{Kleb}(y)=\left[1+5.5\left(y\frac{C_{Kleb}}{y_{\max}}\right)^{6}\right]^{-1},$$

$$F_{wake}=y_{\max}F_{\max},\quad F(y)=y|\Omega|D,\quad C_{Kleb}=0.3, \tag{24}$$

where $y_{\max}$ is determined by the coordinate y, where $F(y)$ reaches its maximum, and $F_{\max}=F(y_{\max})$.

The turbulent viscosity, substituted into equations (10)–(11), is found as follows:

$$\mu_{t}=\begin{cases}\mu_{t,in}, & y\le y_{cross},\\ \mu_{t,out}, & y>y_{cross},\end{cases}$$

where y_{cross} is the coordinate y, at which $\mu_{t,in}=\mu_{t,out}$ the first time as y increases.

Other criteria are also used for the transition from the inner to the outer layer in the hypersonic flow (see, for example, [17, 18]).

We will especially emphasize that the use of algebraic turbulence models in conjunction with the complete Navier – Stokes model faces a number of problems that are not present in the solution of the problem in the classical boundary layer formulation. For example, the choice of boundary layer thickness in a supersonic flow with shock waves is discussed in [18].

IOP Conf. Series: Journal of Physics: Conf. Series **1009** (2018) 012016 doi:10.1088/1742-6596/1009/1/012016

5. Results of numerical simulation

The calculated data will be analyzed in the following order. First we make a comparison of the calculated and experimental data. The calculations are performed by correlation relationships and using the NERAT-2D code with two models of turbulent mixing. Then, specific features of the distributions of gas-dynamic functions in the laminar and turbulent boundary layers will be discussed.

In accordance with [1], the following conditions were set in the oncoming stream:

Table 1. Initial data.

Variant # / experiment reference	$\frac{T_w}{T_t}$	M_∞	T_∞ (K)	T_w (K)	ρ_∞ (g/cm^3)	p_∞ (erg/cm^3)	V_∞ (cm/s)
1/[2]	0.2	6	65.04	106.67	1.185×10^{-4}	2.209×10^{4}	9.696×10^{4}
2/[2]	0.6	6	65.04	320.0	1.185×10^{-4}	2.209×10^{4}	9.696×10^{4}
3/[3]	0.65	6	63.01	336.7	1.160×10^{-4}	2.100×10^{4}	9.540×10^{4}
4/[4]	0.11	7.12	240	299	1.120×10^{-4}	7.730×10^{4}	2.210×10^{5}
5/[4]	0.19	8.8	93.99	296.3	2.290×10^{-4}	6.170×10^{4}	1.710×10^{5}
6/[4]	0.26	7.0	105.0	296.0	4.290×10^{-4}	1.290×10^{5}	1.440×10^{5}

The initial data for the calculations were determined from the data of table 1 in [1] (by values M_∞, T_∞, Re_∞, T_w) by the successive application of the relations

$$\mu_\infty = 1.458\times10^{-5}\frac{T_\infty^{3/2}}{110.4+T_\infty}, \text{ g/(cm×s)},$$

$$a_\infty = \sqrt{\gamma\frac{T_\infty R_0}{M_\Sigma}} \text{ cm/s}, R_0 = 8.3145\times10^7 \text{ erg/(mole×K)}, \gamma = 1.4, M_\Sigma = 29 \text{ g/mole},$$

$$V_\infty = M_\infty a_\infty \text{ cm/s}, \ \rho_\infty = \frac{\mathrm{Re}_\infty \mu_\infty}{V_\infty}, \text{ g/cm}^3$$

The braking temperature was estimated from the relationships

$$T_0 = T_\infty + \frac{V_\infty^2}{2c_{p,\infty}}, \ c_{p,\infty} = \gamma c_{V,\infty}, \ c_{V,\infty} = \frac{R_0}{M_\Sigma(\gamma-1)}, \ \mathrm{Re}_\infty = \frac{\rho_\infty V_\infty}{\mu_\infty}\frac{1}{L_{scale}}, \ L_{scale} = 100 \text{ cm}$$

For reference, the conditions in the oncoming stream are given in table 2.

Table 2.

Variant	μ_∞ (g/(cm × s))	λ_∞ (erg/(cm × K × s))	Re_∞ (1/cm)
1	0.436×10^{-4}	0.608×10^{3}	2.64×10^{7}
2	0.436×10^{-4}	0.608×10^{3}	2.64×10^{7}
3	0.421×10^{-4}	0.486×10^{3}	2.64×10^{7}
4	0.155×10^{-3}	0.214×10^{4}	1.60×10^{7}
5	0.650×10^{-4}	0.789×10^{3}	6.02×10^{7}
6	0.728×10^{-4}	0.898×10^{3}	8.46×10^{7}

Figure 2 shows the distribution of the Stanton numbers St (8) calculated from the correlations (1), (5), and also using the author's code NERAT-2D with three different numbers $Pr_t = 0.6$; 0.7 and 1.0. Here we used the Prandtl mixing model. In the case under consideration, $Re_x = 4\times10^6$ was assumed and the transition from the laminar flow regime to the turbulent flow was assumed to be instantaneous.

We note a good agreement between the calculated densities of convective heat fluxes in both laminar and turbulent flows with experimental data [2]. The effect of the given Pr_t numbers on the density of convective heat fluxes for this variant is insignificant.

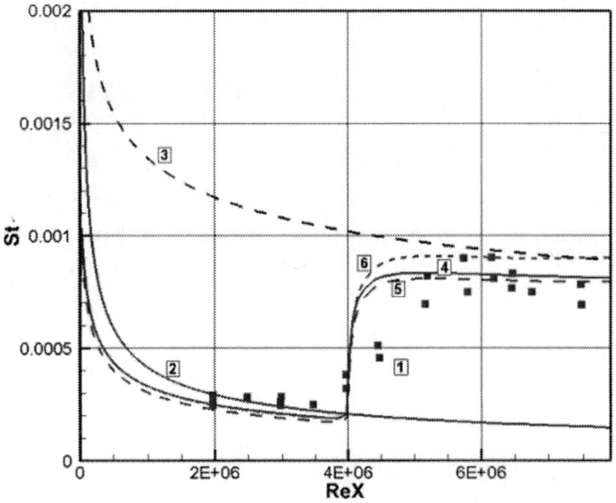

Figure 2. Distribution of the numbers St along the surface of the streamlined plate in the first variant of the calculation: 1 – experiment [2]; 2, 3 – calculation using the correlation relations (1) and (5); 4, 5, 6 – calculations of this work for $Pr_t = 0.6, 0.7$ and 1.0, respectively (NERAT–2D code with the Prandtl mixing model).

Figure 3 shows similar data for the second variant. The main difference between the 1st and 2nd variants is that in the second case the surface of the plate is noticeably heated, which corresponds more to the actual flight situation of the supersonic apparatus at $M \sim 6$. In the second case, the Pr_t number strongly affects on the heating intensity of plate. As before, the laminar-turbulent transition was not modeled.

Figures 4 and 5 show the results of modeling heat exchange using the Baldwin-Lomax model. Note that for a heated surface this model yields underestimated values of the numbers St at $Pr_t = 0.72$.

The third variant of calculation is practically similar to the first two (see figures 6 an 7). However, it corresponds to a different series of experiments and a somewhat higher surface temperature. The calculated data in figures 6 show somewhat worse agreement with the experimental data [3] when using both models of turbulent mixing. Note that the calculated data [1] also deviate from these experimental data.

The results of a comparison with the experimental data [4] on heat transfer in the range of Mach numbers $M = 7 \div 8.8$ are shown in figures 8– 10, respectively, for the calculated cases 4– 6.

In addition to comparing the data on the heating of the streamlined plate (the number St), the available data on the coefficient of surface friction were compared. Figures 11,12 give results of calculating C_f for the first variant by the correlation ratios and by the NERAT-2D code using the Prandtl mixing model and the Baldwin-Lomax model. In figures 13,14 compares analogous data with experiment [3] for the second calculated case using the Prandtl and Baldwin– Lomax models. In general, we note a satisfactory agreement of the calculated data with the correlation data and experiment.

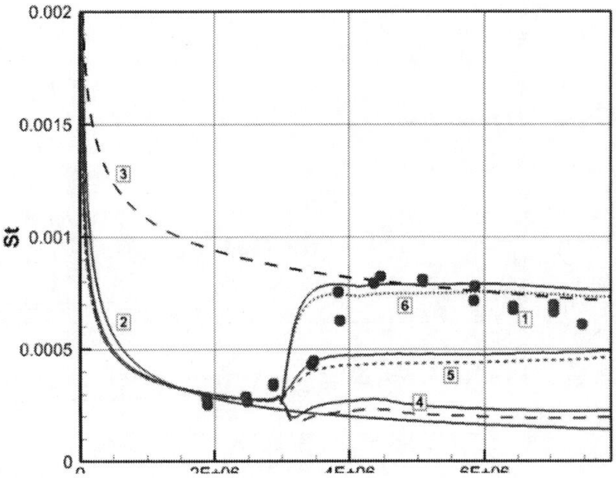

Figure 3. Distribution of the numbers St along the surface of the streamlined plate in the second variant of the calculation: 1 – experiment [2]; 2, 3 – calculation using the correlation relations (1) and (5); 4, 5, 6 – calculations of this work with $\Pr_t = 0.6, 0.7$ and 1.0, respectively (NERAT-2D code with Prandtl mixing model, dushed curves – convective heating, solid curves – with allowance for friction heating).

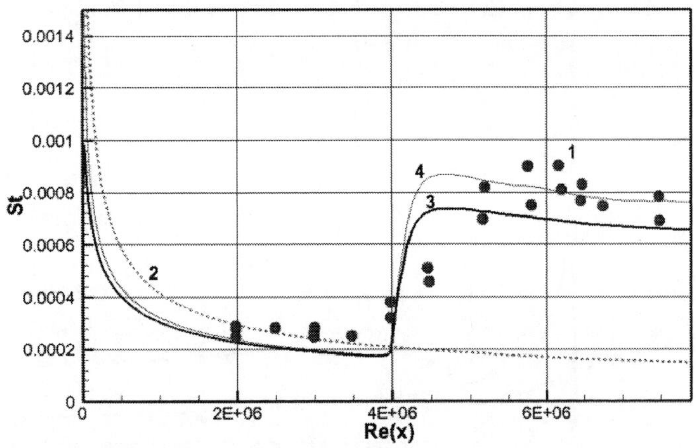

Figure 4. Distribution of the numbers St along the surface of the streamlined plate in the first version of the calculation: 1 – experiment [2]; 2 – calculation using the correlation relations (1) for the laminar boundary layer; 3,4 – calculations of this work using the NERAT–2D code with the Baldwin – Lomax model; 4 – total density of heat flow due to thermal conductivity and friction. $\Pr_t = 0.72$.

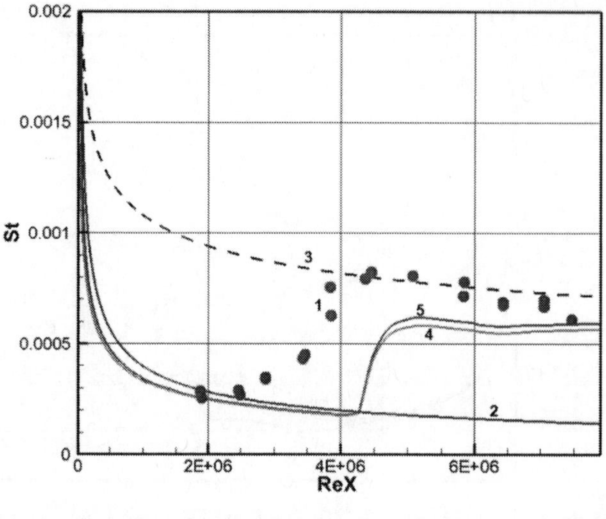

Figure 5. Distribution of the numbers St along the surface of the streamlined plate in the second version of the calculation: 1 – experiment [2]; 2, 3 – calculation using the correlation relations (1) and (5); 4, 5 – calculations of this work using the NERAT–2D code with the Baldwin– Lomax model; 5 is the total density of the heat flux due to thermal conductivity and friction.

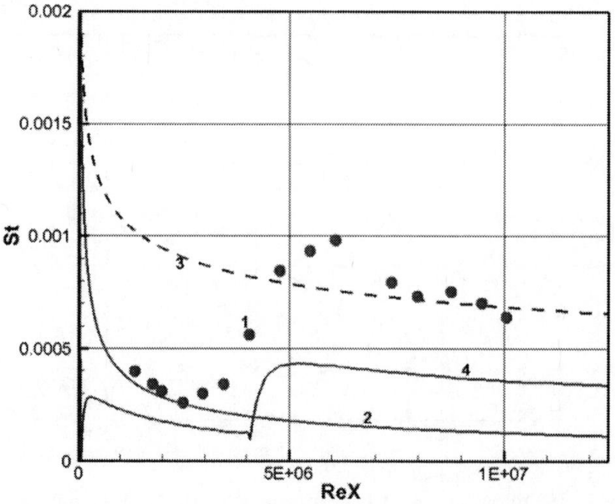

Figure 6. Distribution of the numbers St along the surface of the streamlined plate in the third version of the calculation: 1– experiment [3]; 2, 3 – calculation using correlation ratios for laminar (1) and turbulent (5) boundary layer; 4 – calculations of this work using the NERAT–2D code with the Baldwin– Lomax model. $\Pr_t = 0.72$.

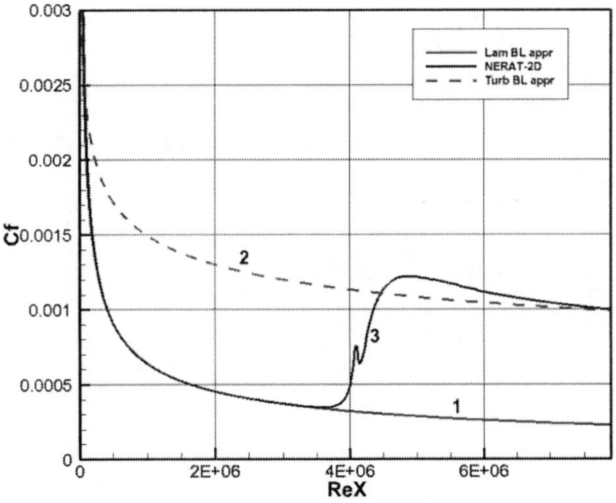

Figure 7. Distribution of the coefficient of surface friction C_f along the surface of the streamlined plate in the third variant of calculation: 1, 2 – calculation using correlation relations for laminar (1) and turbulent (5) boundary layer; 3 – calculations of this work using the NERAT-2D code with the Baldwin – Lomax model. $\Pr_t = 0.72$.

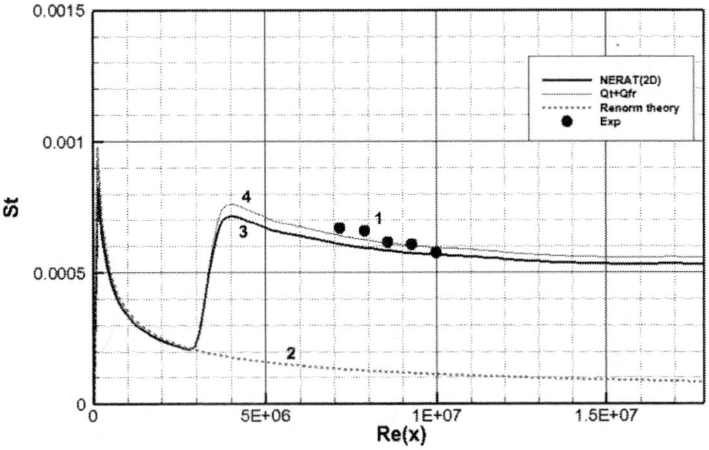

Figure 8. Distribution of the numbers St along the surface of the streamlined plate in the 4th variant of the calculation: 1 – experiment [4]; 2 – calculation using the correlation relations (1) for the laminar boundary layer; 3,4 – calculations of this work using the NERAT-2D convective component code and taking into account the friction heating (4) with the Baldwin – Lomax model. $\Pr_t = 0.7$.

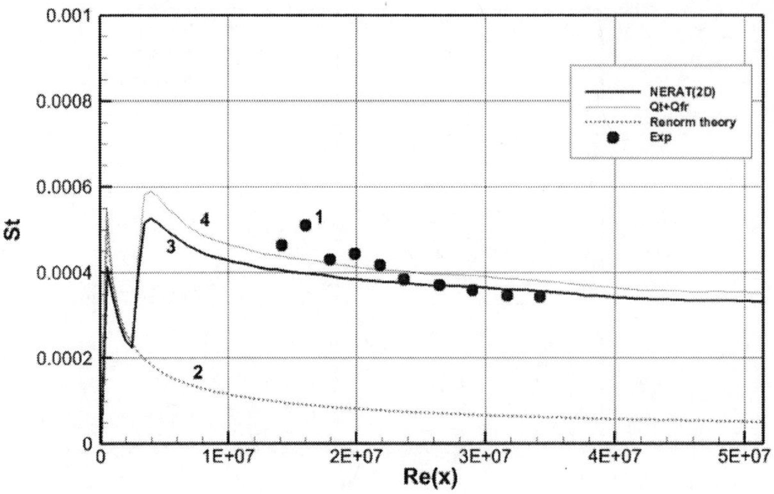

Figure 9. Distribution of the numbers St along the surface of the streamlined plate in the 5th version of the calculation: 1 – experiment [4]; 2 – calculation using the correlation relations (1) for the laminar boundary layer; 3,4 – calculations of this work using the NERAT-2D convective component code and taking into account the heating by the friction with the Baldwin – Lomax model.

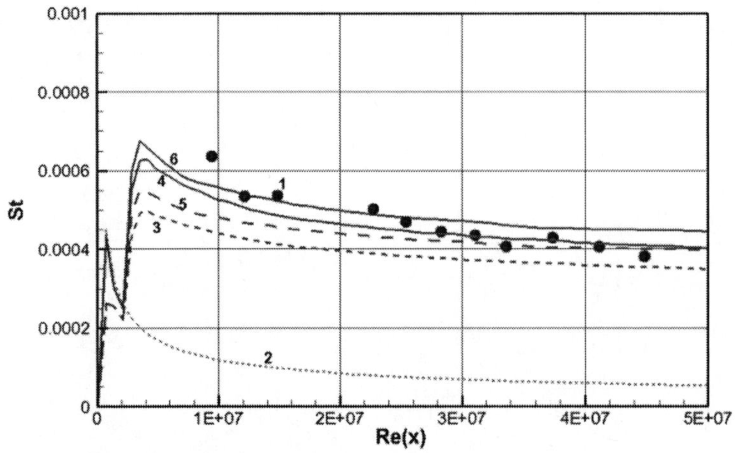

Figure 10. Distribution of the numbers St along the surface of the streamlined plate in the 6th version of the calculation: 1 – experiment [4]; 2 – calculation using the correlation relations (1) for the laminar boundary layer; 3– 6 – calculations of this work using the NERAT-2D convective component (3.5) and taking into account friction heating with the Baldwin – Lomax model (4.6). $\Pr_t = 0.7$ (3.4) and $\Pr_t = 1.0$ (5.6).

IOP Conf. Series: Journal of Physics: Conf. Series **1009** (2018) 012016 doi:10.1088/1742-6596/1009/1/012016

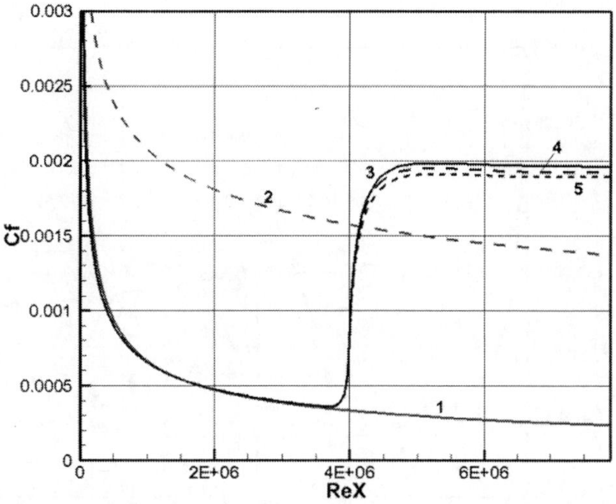

Figure 11. Distribution of the coefficient of surface friction C_f along the surface of the streamlined plate in the first variant of the calculation: 1, 2 – calculation using correlation ratios for laminar (1) and turbulent (5) boundary layer; 3, 4, 5 – calculations of this work using the NERAT-2D code with the Prandtl mixing path model. $\mathrm{Pr}_t = 0.6, 0.72, 1.0$.

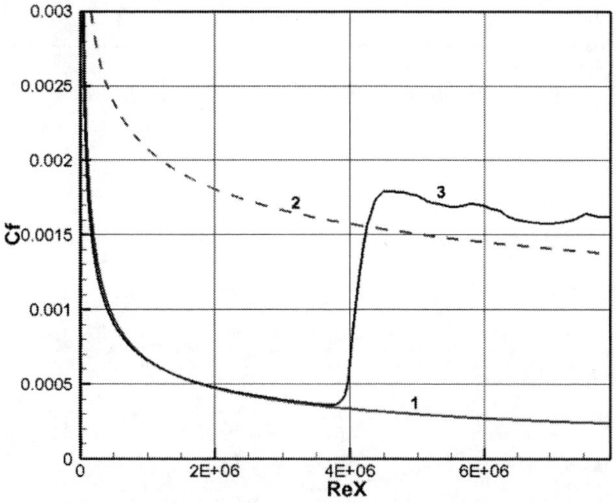

Figure 12. Distribution of the coefficient of surface friction C_f along the surface of the streamlined plate in the first variant of calculation: 1, 2 – calculation using correlation relations for laminar (1) and turbulent (5) boundary layer; 3 – calculations of this work using the NERAT-2D code with the Baldwin– Lomax model. $\mathrm{Pr}_t = 0.72$.

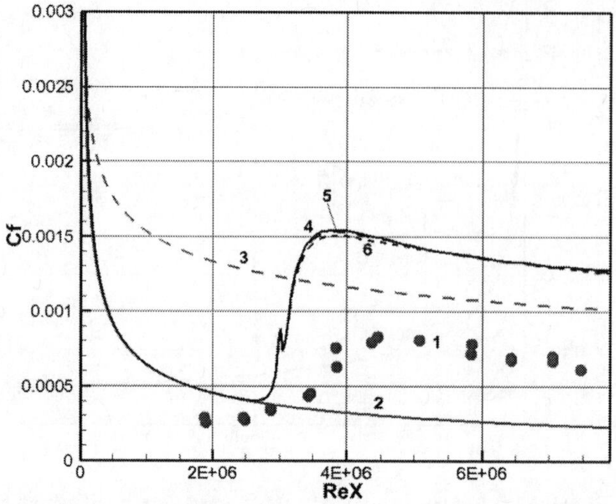

Figure 13. Distribution of the coefficient of surface friction C_f along the surface of the streamlined plate in the second variant of calculation: 1, 2 – calculation using correlation relations for laminar (1) and turbulent (5) boundary layer; 3, 4, 5 – calculations of this work using the NERAT-2D code with the Prandtl mixing path model. $\Pr_t = 0.6, 0.72, 1.0$.

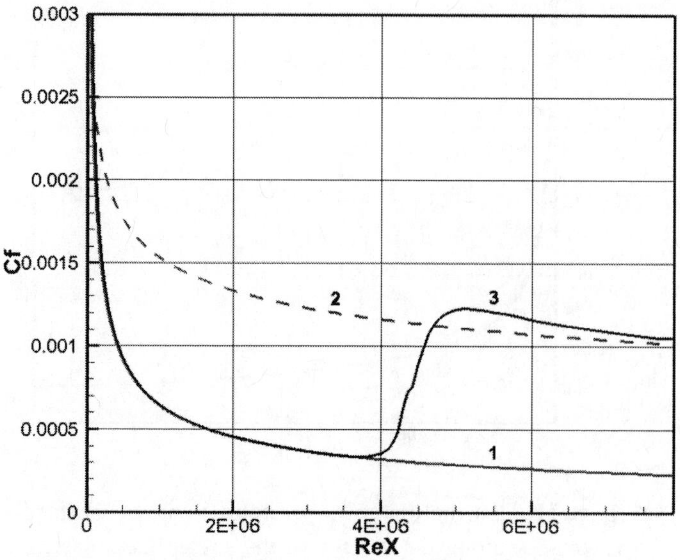

Figure 14. Distribution of the coefficient of surface friction C_f along the surface of the streamlined plate in the second variant of calculation: 1, 2 – calculation using correlation relations for laminar (1) and turbulent (5) boundary layer; 3 – calculations of this work using the NERAT-2D code with the Baldwin – Lomax model. $\Pr_t = 0.72$.

IOP Conf. Series: Journal of Physics: Conf. Series **1009** (2018) 012016 doi:10.1088/1742-6596/1009/1/012016

Now let us consider some features of the distribution of gas-dynamic functions. Figure 15 shows a two-dimensional pressure field above the surface of the streamlined plate. Its sharp edge has the coordinate $x_0 = 0.01\,\text{cm}$. The distributions of gas-dynamic functions along the normal to the surface were constructed in sections with coordinates

$$x_1 = 10\,\text{cm},\, x_2 = 18\,\text{cm},\, x_3 = 21\,\text{cm},\, x_4 = 24\,\text{cm},\, x_5 = 27\,\text{cm}.$$

Figures 16, *a*, *b* show temperature distributions in 5 sections along the surface of the plate. The first temperature profile corresponds to the laminar flow regime, and the subsequent 5 distributions correspond to the turbulent boundary layer. Figure 16, *b* shows the same distributions, but closer to the surface. It is clearly seen how the temperature gradients increase in the turbulent boundary layer, and as the displacement along the surface increases, the thickness of the temperature boundary layer increases. In this case, the temperature profiles are similar to the longitudinal velocity distributions (figure 17) with regions of the viscous sub layer, buffer zone, the zone of the developed turbulent boundary layer with a logarithmic velocity distribution and the outer part of the boundary layer well studied in the literature. We note that in this case the highest temperature in the boundary layer slightly exceeds the surface temperature.

Figure 18 shows the profiles of molecular and turbulent viscosity in six sections along the x axis. At $y < 2 < 10^{-2}$ cm, the molecular viscosity prevails in the viscous sub layer, and at large distances from the wall, the turbulent viscosity. In the viscous sub layer the turbulent viscosity $\mu_{t,in}$ is determined by the formula (22), and above the surface by the model $\mu_{t,out}$ (23). Corresponding distributions are shown in figure 19. From the indicated relations it is clear that vortex functions $|\Omega|$ are the determining functions, the profiles of which are shown in figure 20. We note the regions of nonmonotonicity of these functions due to the specific nature of the flow near the shock wave. The profiles of the function $F(y)$ (24), which determines the turbulent viscosity $\mu_{t,out}$ in the Baldwin– Lomax model are shown in figure 21.

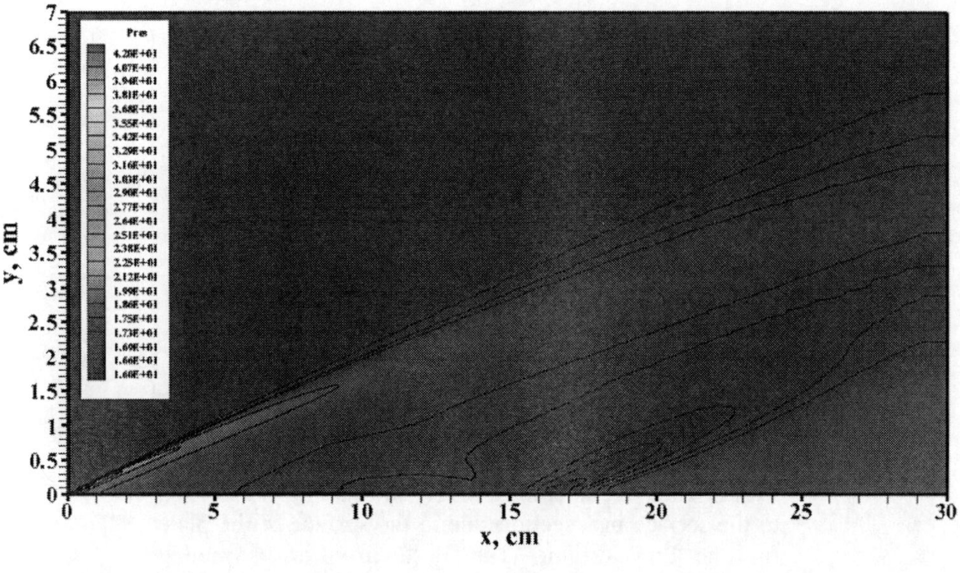

Figure 15. Pressure field above the plate surface for the 2nd variant.

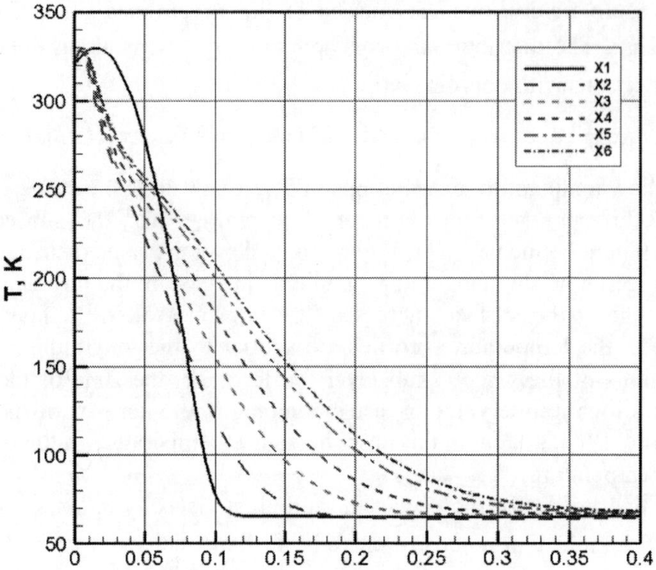

Figure 16, *a*. Temperature distribution along the normal to the surface in 6 sections along the surface of the plate for the 2nd variant. The Baldwin – Lomax model. $\Pr_t = 0.72$.

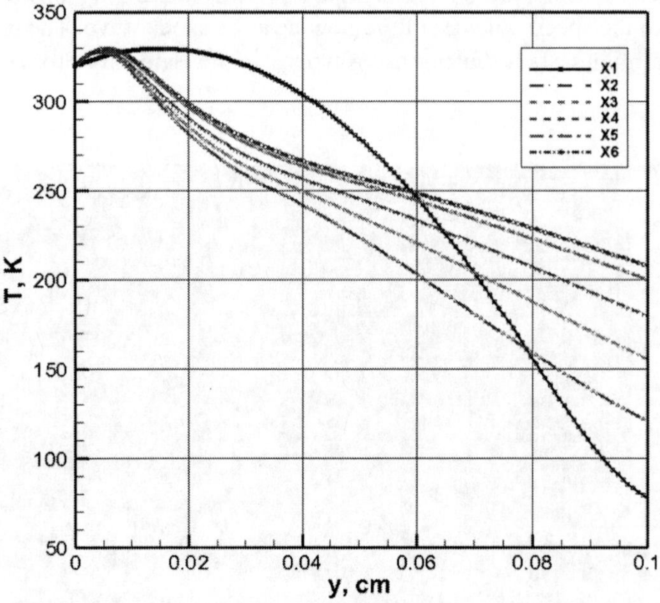

Figure 16, *b*. Temperature distribution along the normal to the surface in 6 sections along the surface of the plate for the 2nd variant. The Baldwin – Lomax model. $\Pr_t = 0.72$.

IOP Conf. Series: Journal of Physics: Conf. Series **1009** (2018) 012016 doi:10.1088/1742-6596/1009/1/012016

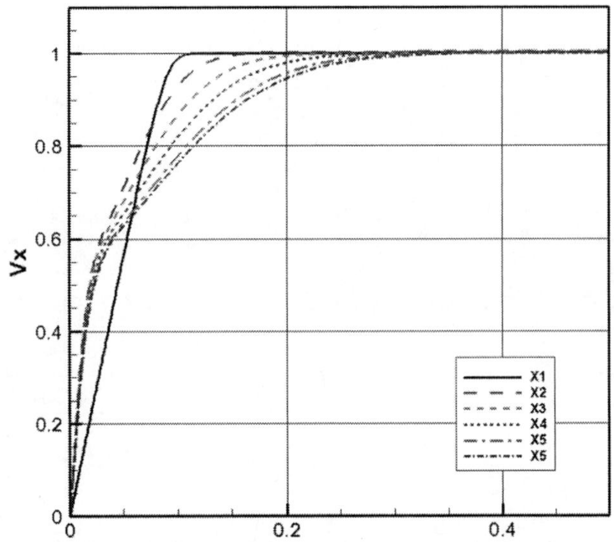

Figure 17. Distribution of longitudinal velocity in a direction normal to the surface in 6 sections along the x axis. The second variant. The Baldwin – Lomax model. $\Pr_t = 0.72$.

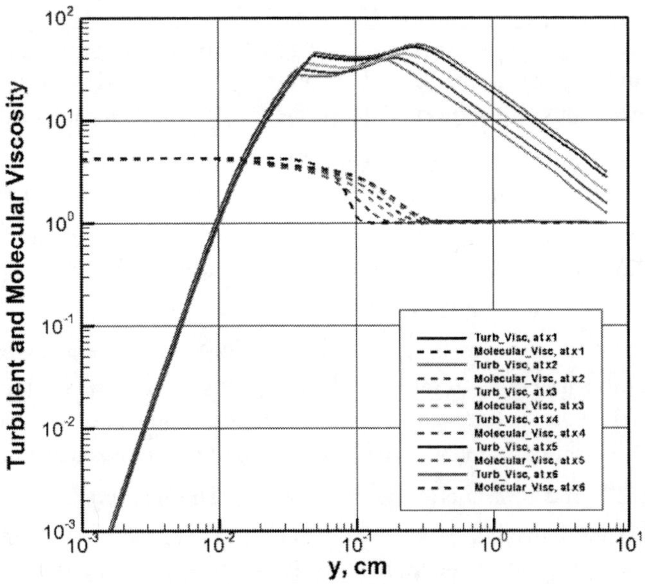

Figure 18. Distributions of the molecular and turbulent viscosity along the normal to the plate in 6 sections with respect to x. The second calculation variant. The Baldwin – Lomax model. $\Pr_t = 0.72$.

IOP Conf. Series: Journal of Physics: Conf. Series **1009** (2018) 012016 doi:10.1088/1742-6596/1009/1/012016

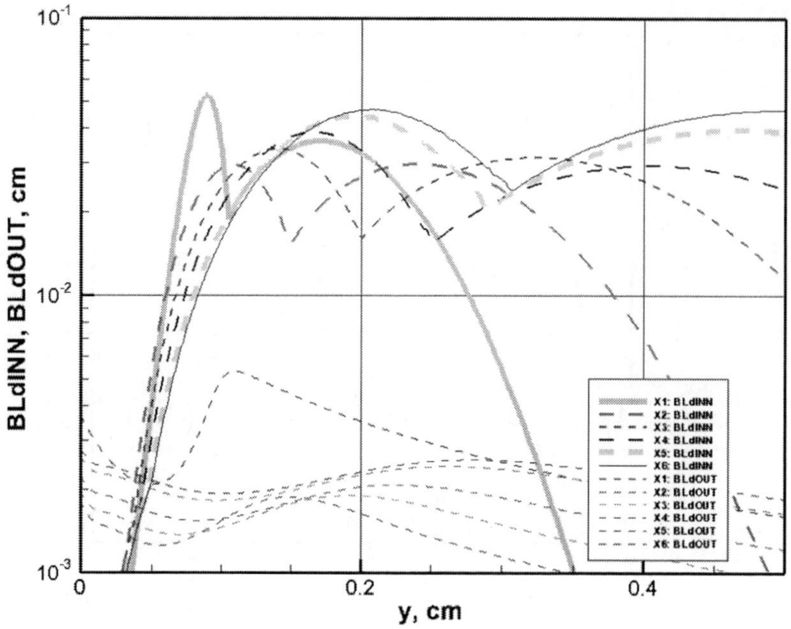

Figure 19. Distributions of two components of turbulent viscosity along the normal to the plate in 6 sections with respect to x. The second calculation variant. The Baldwin – Lomax model. $\Pr_t = 0.72$.

To the question of the difference between the results of modeling of turbulent mixing in the framework of the boundary layer model from the solution of the complete Navier – Stokes equations, we point out the distributions of the normal to surface velocity $V_y = v/V_\infty$ in different sections with respect to x and the pressure gradient. Such distributions are shown in figures 22 and 23. Despite the fact that the absolute values of V_y are about 100 times smaller than the characteristic value $V_x = u/V_\infty \approx 1$, it should be taken into account that the derivatives in the outer region can nevertheless be commensurable, which will affect the distribution of the vortex function Ω.

The distributions of the function y^+ (21) in different sections along the surface of the plate are shown in figure 24. The smallest value of y^+ near the surface is equal to $y^+_{\min} = 0.3$ cm.

Distributions of similar functions in the boundary layer using the Prandtl mixing length model are shown in figures 25– 31. Note that the data presented in these figures correspond to the case of a cold surface ($T_w/T_t = 0.2$). Particularly noteworthy is the oscillatory character of the behavior of turbulent viscosity (figures 27, 28) and the function $|\Omega|$ (figure 29). The temperature distributions (figure 25), longitudinal (figure 26) and transverse (figure 30) velocities are similar to the previous case.

These oscillations of the vortex function are related to a complex flow over the plate than is modeled in the boundary-layer models. A possible way to deal with these oscillations, which can lead to a general instability of the computational process, can be a limitation of the vortex function from below.

Another feature of the use of the Prandtl mixing length model is the condition (19), which is shown graphically in figure 28. It follows that for $y/\delta > 0.2$ the length of the mixing path remains constant in the outer part of the boundary layer.

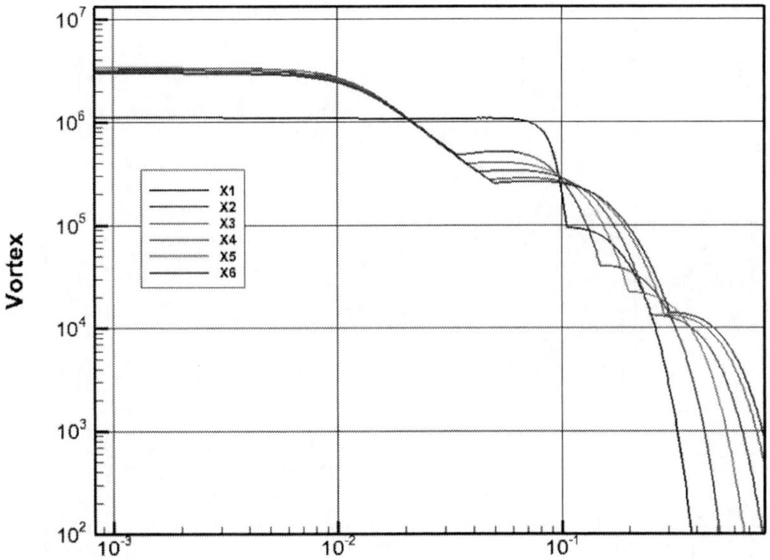

Figure 20. Distributions of the vortex functions along the normal to the plate in 6 sections with respect to x. The second calculation variant. The Baldwin – Lomax model. $\Pr_t = 0.72$.

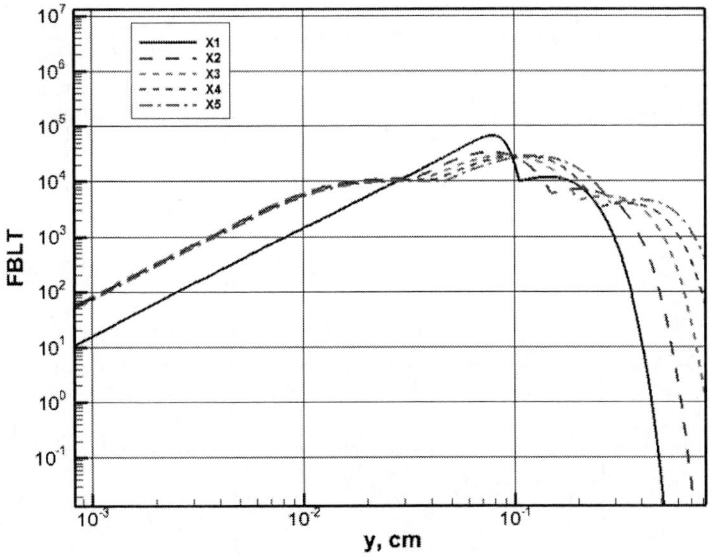

Figure 21. Distributions of functions $F(y) = y|\Omega|D$ along the normal to the plate in 6 sections with respect to x. The second calculation variant. The Baldwin – Lomax model. $\Pr_t = 0.72$.

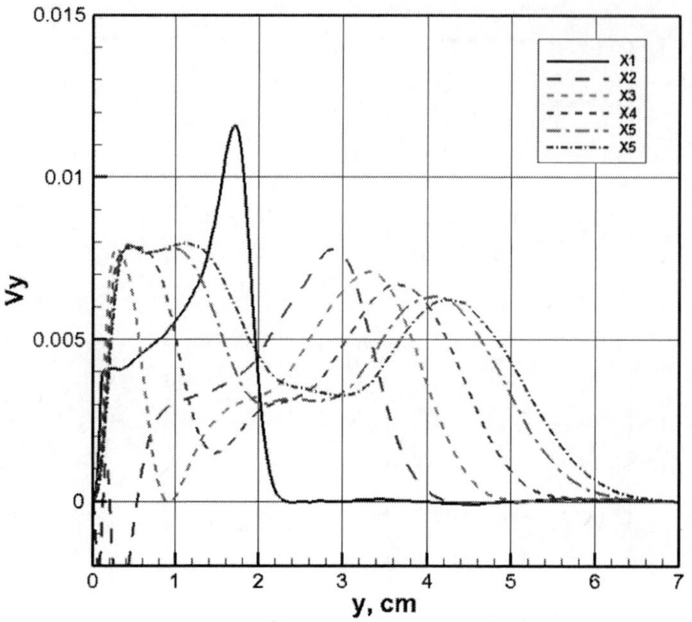

Figure 22. Transverse to the flow velocity $V_y = v/V_\infty$ in different sections along the x-axis. The second calculation variant. The Baldwin – Lomax model. $\mathrm{Pr}_t = 0.72$.

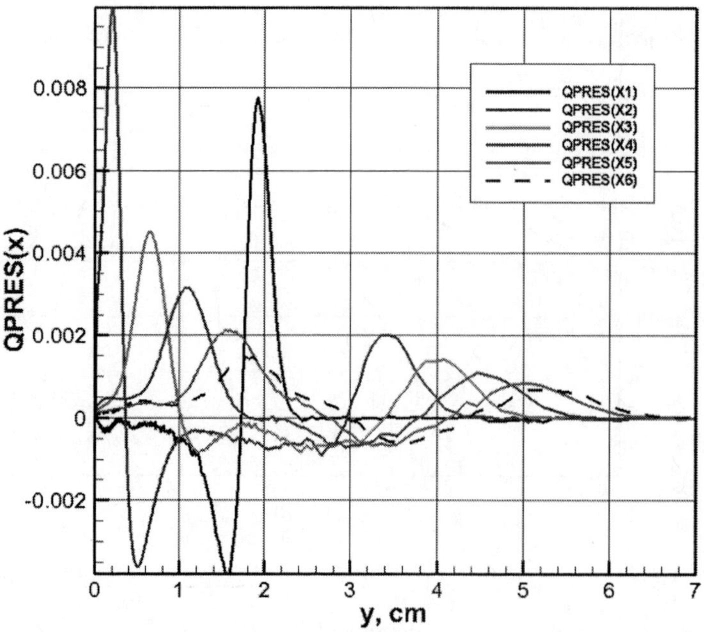

Figure 23. Distribution of functions $\mathrm{grad}(p)$ in different cross-sections along the x-axis The second variant. The Baldwin – Lomax model. $\mathrm{Pr}_t = 0.72$.

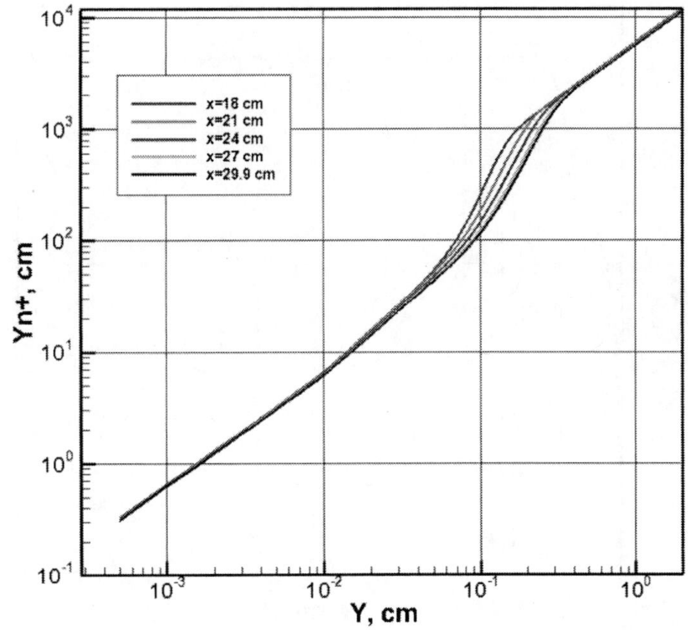

Figure 24. Distribution of functions y^+ in different cross-sections along the x-axis. The second variant. The Baldwin – Lomax model. $\mathrm{Pr}_t = 0.72$.

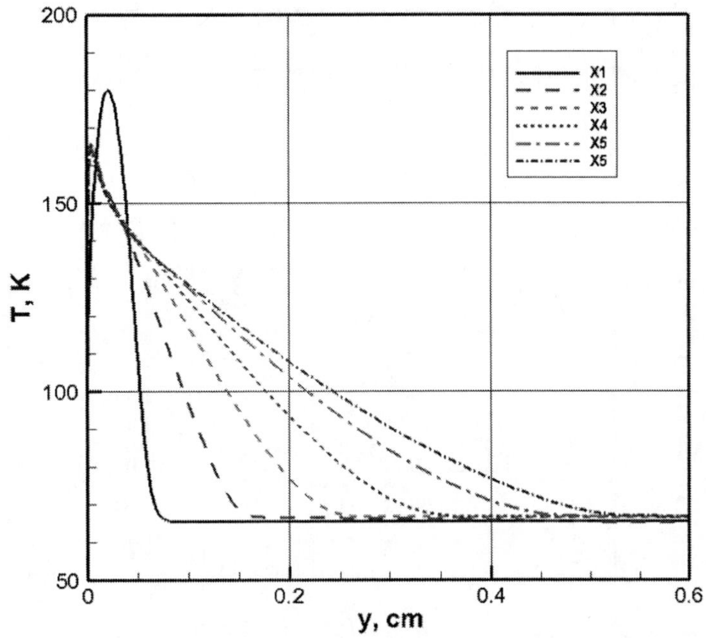

Figure 25. Temperature distribution along the normal to the surface in 6 sections along the surface of the plate for the 1st variant. Model of the Prandtl mixing path. $\mathrm{Pr}_t = 0.6$.

IOP Conf. Series: Journal of Physics: Conf. Series **1009** (2018) 012016 doi:10.1088/1742-6596/1009/1/012016

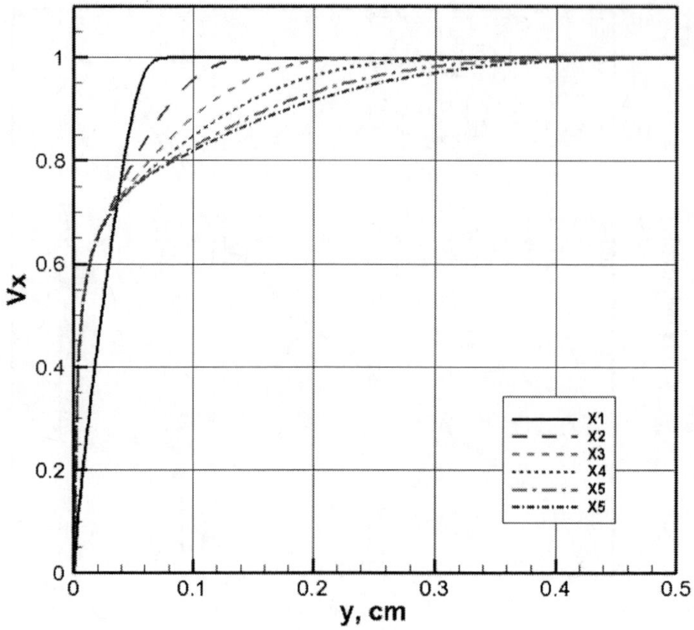

Figure 26. Distribution of longitudinal velocity along the normal to the surface in 6 sections along the surface of the plate for the 1st variant. Model of the Prandtl mixing path. $\Pr_t = 0.6$.

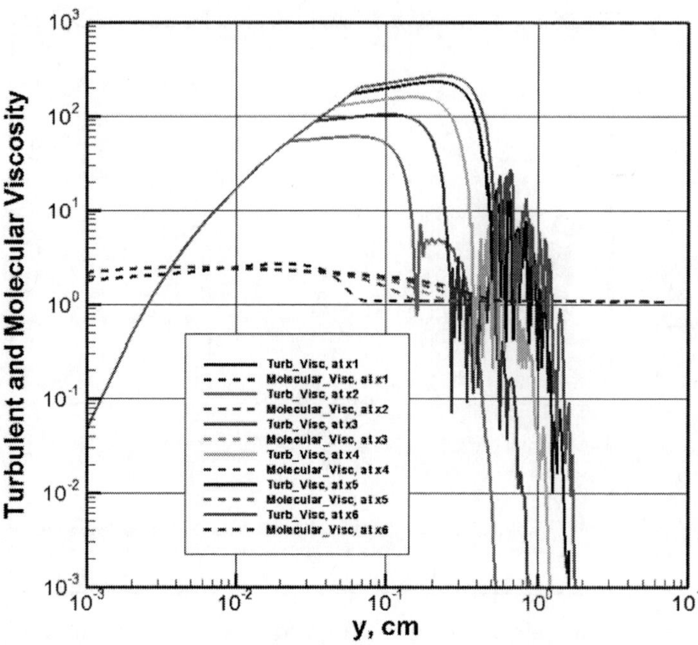

Figure 27. Distribution of turbulent and molecular viscosity along the normal to the surface in 6 sections along the surface of the plate for the first variant. Model of the Prandtl mixing path. $\Pr_t = 0.6$.

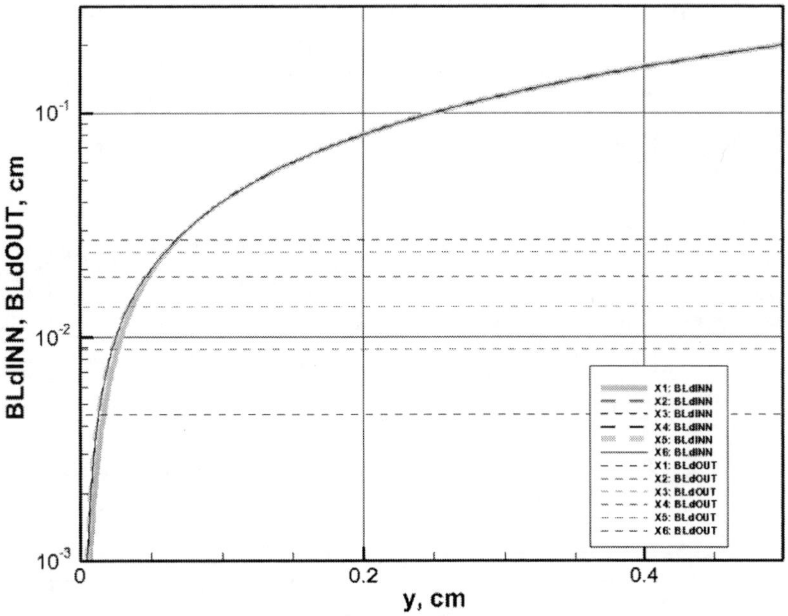

Figure 28. Distribution of functions along the normal to the surface in 6 sections along the surface of the plate for the 1st variant. Model of the Prandtl mixing path. $\mathrm{Pr}_t = 0.6$.

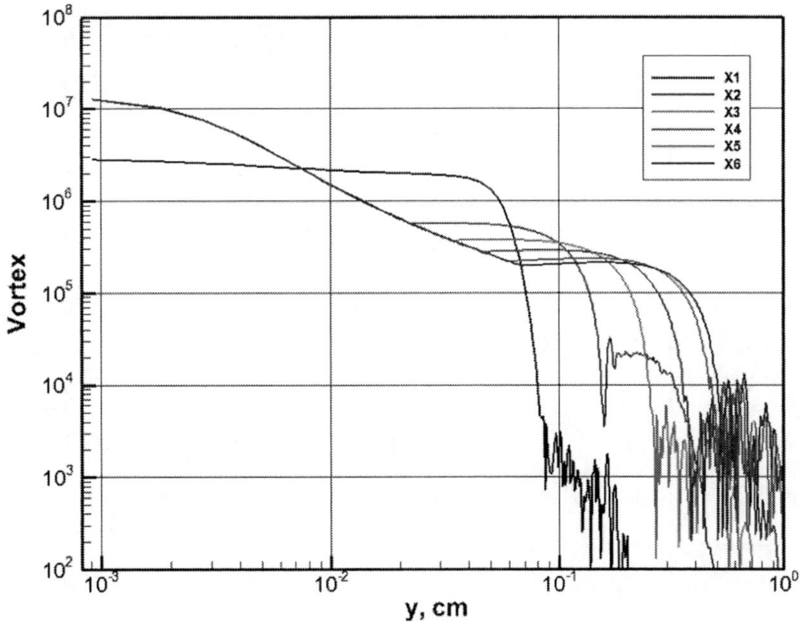

Figure 29. Distribution of the vortex of velocity along the normal to the surface in 6 sections along the surface of the plate for the 1st variant. Model of the Prandtl mixing path. $\mathrm{Pr}_t = 0.6$.

IOP Conf. Series: Journal of Physics: Conf. Series **1009** (2018) 012016 doi:10.1088/1742-6596/1009/1/012016

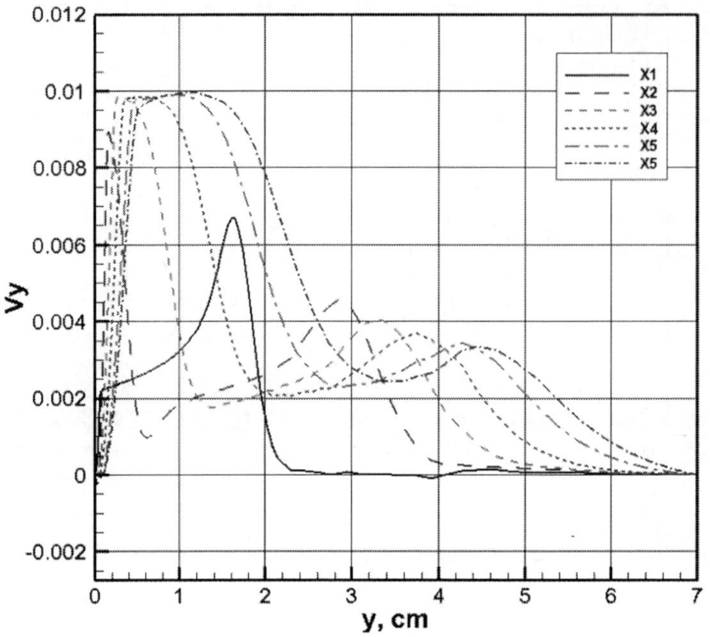

Figure 30. Distribution of the transverse velocity component along the normal to the surface in 6 sections along the surface of the plate for the 1st variant. Model of the Prandtl mixing path. $\Pr_t = 0.6$.

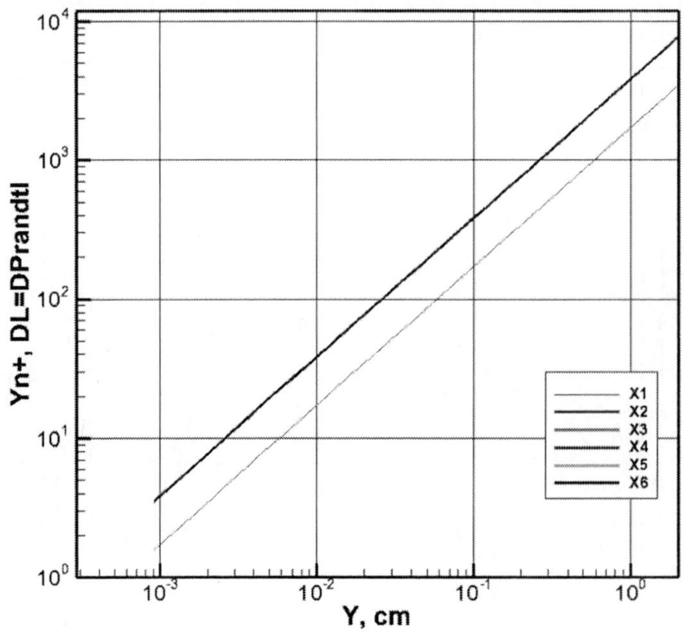

Figure 31. Distribution of the function y^+ along the normal to the surface in 6 sections along the surface of the plate for the 1st variant. Model of the Prandtl mixing path. $\Pr_t = 0.6$.

6. Conclusion

The experimental data on plate heat transfer in supersonic air flows are compared for Mach numbers $M = 6 \div 8.8$ with the calculated data obtained using the Navier – Stokes flow model in conjunction with the analytical models of turbulent mixing of the Prandtl mixing length and the Baldwin-Lomax model. The author's computer code NERAT-2D was used in the calculations. A satisfactory agreement with the experimental data is obtained, which gives grounds for recommending the use of these algebraic turbulent mixing models in conjunction with the Navier – Stokes model for more complex configurations of hypersonic flow, which was also recommended in other papers (see, for example, References 17,18).

Along with this, attention is drawn to the need for accurate application of boundary layer turbulent mixing models to the analysis of complete flow models.

Acknowledgments

The work was performed within the framework of the Government program of basic research of the Russian academy of sciences (contract # AAAA–A17–117021310372–6) and partially under support of the RFBR grant # 016 – 01– 00379.

References

[1] Dilley A D 2001 Evaluation of CFD Turbulent Prediction Techniques and Comparison with Hypersonic Experimental Data *NASA/CR*–2001–210837 26 p

[2] Bertram M H, Cary Jr A M and Whitehead Jr A H 1968 Experiments with Hypersonic Turbulent Boundary Layers on Flat Plate and Delta Wings *AGARD Specialists' Meeting on Hypersonic Boundary Layers and Flow Fields* (London, England)

[3] Bertram M H and Neal Jr L 1965 Recent Experiments in Hypersonic Turbulent Boundary Layer. *AGARD Specialists' Meeting on Recent Developments in Boundary Layer Research* (Italy)

[4] Wallace J E 1967 Hypersonic Turbulent Boundary Layer Studies at Cold Wall Conditions *Proc/ of the 1967 Heat Transfer and Fluid Mechanics Institute* (Stanford University Press) pp 427–451

[5] Walters R W, Reu T, McGrory W D and Hicks J W A 1988 longitudinally Patched Approach with Application to High Speed Flows *AIAA Paper* 88–0715 13 p

[6] Baldwin B S and Lomax H 1978 Thin Layer Approximation and Algebraic Model for Separated Turbulent Flows *AIAA Paper 78–0257* 8p

[7] Surzhikov S T 2013 Convective heating of small-radius spherical blunting for relatively low hypersonic velocities *High Temperature* **51**(2) pp 231–245

[8] Surzhikov S T 2016 Radiative Gas Dynamics of the Fire-II Superorbital Space Vehicle *Technical Physics* **61** (3) pp 349–359

[9] Zemljanskii B A, Lunev V V, Vlasov V I, Gorshkov A B, Zalogin G N, Kovalev R V, Marinin V P and Murzinov I N 2014 *Convective heating of flight crafts*M. (Fizmatlit) 330 p (in Russian)

[10] Tannehill J C, Anderson D A and Pletcher R H 1997*Computational Fluid Mechanics and Heat Transfer* (Taylor & Francis) 792p

[11] Date A W 2005 *Introduction to Computational Fluid Dynamics* (Cambridge University Press) 377p

[12] Cebeci T and Bradshaw P 2012 *Physical and Computational Aspects of Connective Heat Transfer* (Springer–Verlag) 486p

[13] Loitsjansky L G 2003 *Mechanics of Fluids and Gases* (M: "Dropha") 840 p (in Russian)

[14] Horstman C C 1987 Prediction of Hypersonic Shock-Wave/Turbulent-Boundary-Layer Interaction Flows *AIAA 87-1367* 10 p

[15] Shirazi S A and Truman C R 1987 Comparison of Algebraic Turbulence Model for PNS Predictions of Supersonic Flow Past a Sphere-Cone *AIAA 87-0544* 12 p

IOP Conf. Series: Journal of Physics: Conf. Series **1009** (2018) 012016 doi:10.1088/1742-6596/1009/1/012016

[16] Shang J S and Scherr S J 1986 Navier– Stockes Solution for a Complete Re–Entry Configuration *J. Aircraft* **23** (12) pp 881–888
[17] Cheatwood F M and Thompson R A 1993 The addition of algebraic turbulence modelling to program LAURA *NASA TM 107 758* 30p
[18] Visbal M and Knight D 1984 The Baldwin– Lomax Turbulence Model for Two-Dimensional Shock-Wave/Boundary-Layer Interaction *AIAA J* **22** (7) pp 921– 928

IOP Conf. Series: Journal of Physics: Conf. Series **1009** (2018) 012017 doi:10.1088/1742-6596/1009/1/012017

Results of investigations of non-isothermal turbulent flows based on stochastic equations of the continuum and equivalence of measures

A V Dmitrenko[1,2]

[1]Department of Thermal Physics, National Research Nuclear University «MEPhI», Kashirskoye shosse 31, Moscow, 115409, Russian Federation
[2]Department of Thermal Engineering , Russian University of Transport «MIIT», Obraztsova Street 9, Moscow, 127994, Russian Federation

AVDmitrenko@mephi.ru, ammsv@yandex.ru

Abstract. The review of main results of new stochastic theory of turbulence for non-isothermal flows is presented. Based on sets of the stochastic equations for the laws of conservation and for the equivalence of measures between deterministic process and random process, the review of results of an analytical expressions is presented for main parameters of non-isothermal turbulent flow in the tube depending on the initial fluctuation in the non-turbulent medium. It is shown that the derived formulas for the critical Reynolds numbers and critical point, for the velocity and temperature profiles, for the second-order correlation and the turbulent Prandtl number, and heat transfer coefficients are in satisfactory agreement with the classical experimental data for these flow characteristics.

1. Introduction

Searching for the strong solution of the Navier-Stokes equations or the creation of new theories and systems of equations, which allow to predict the phenomenon onset of turbulence out of deterministic movement are in the constant attention[1–11]. As known, the impact of initial fluctuations on the result of equations solution, which are used to describe turbulent processes, is significant. This is especially important for numerical modeling and for developing computational codes. However till today, numerous numerical solutions do not allow to define the essence of the phenomenon of turbulence even together with experimental results. Also these results do not allow to provide the answer, in which proposes to determine the physical regular pattern about the reason of onset of turbulence and present the mathematical method to describe the phenomenon on the basis of this physical regular pattern. In articles [12–25] the physical regularity of the equivalence of measures and the systems of the stochastic equations for the determination onset of turbulence in isothermal and non-isothermal and flows were presented. Moreover in the article [12-25], based on the definition of the equivalence of measures between the deterministic process and random process, the systems of stochastic equations of energy, mass and momentum were defined for next space-time areas :

1) the beginning of the generation of turbulence;
2) the generation of turbulence;
3) the diffusion;

4) the dissipation of turbulence.

2. Sets of Equations

According to [1], the physical process is represented as a non-equilibrium thermodynamic system with i-subsets, which is characterized by the values of the energy $\overset{i=n}{\underset{i=1}{U}}(E_i)_{g_{st}}$, the momentum $\overset{i=n}{\underset{i=1}{U}}(M_iU_i)_{g_{st}}$, and the mass $\overset{i=n}{\underset{i=1}{U}}(M_i)_{g_{st}}$. Here, U_i is the speed, $\overset{i=n}{\underset{i=1}{U}}(E_i)_{g_{st}}$ is the stochastic-field energy (index g_{st}); $\overset{i=n}{\underset{i=1}{U}}(E_i)_{col_{st}}$ is the fraction of the field energy, its deterministic component (index col $_{st}$) having the zero stochastic component of measure; and $\overset{i=n}{\underset{i=1}{U}}(E_i)_{st}$ is the fraction of the field energy, which is actually the stochastic field component (index st). Similarly, components of the momentum and the mass (ρ -the density) are determined. Stochastic equations of conservation, which were derived in [1-5] for non-isothermal conditions, take the form:
the equation of mass (continuity)

$$\frac{d(\rho)_{colst}}{d\tau} = -\frac{(\rho)_{st}}{\tau_{cor}} - \frac{d(\rho)_{st}}{d\tau}, \tag{1}$$

the momentum equation

$$\frac{d(\rho u_i)_{colst}}{d\tau} = div(\tau_{i,j})_{col_{st}} + div(\tau_{i,j})_{st} - \frac{(\rho U)_{st}}{\tau_{cor}} - \frac{d(\rho U)_{st}}{d\tau} + F_{col_{st}\ 1} + F_{st\ 1}, \tag{2}$$

and the energy equation

$$\frac{dE_{col\,st}}{d\tau} = div(\lambda\frac{\partial T}{\partial x_j} + u_i\tau_{i,j})_{col_{st}} + div(\lambda\frac{\partial T}{\partial x_j} + u_i\tau_{i,j})_{st} - \left(\frac{E_{st}}{\tau_{cor}}\right) - \left(\frac{dE_{st}}{d\tau}\right) + (u_iF)_{col_{st}\ 1} + (u_iF)_{st\ 1}. \tag{3}$$

Here, $\rho, \vec{U}, u_i, u_j, u_l, \mu, \tau, \tau_{i,j}$ are the density; the velocity vector; the velocity components in directions x_i, x_j, x_l (i, j, l = 1, 2, 3); the dynamic viscosity; the time; and stress tensor $\tau_{i,j} = P + \sigma_{i,j}$, $\delta_{ij}=1$ if i=j, $\delta_{ij}=0$ for i≠j. P is the pressure of liquid or gas; λ is the thermal conductivity; c_p and c_v are the specific heat at constant pressure and volume, respectively; F is the external force, and $\sigma_{ij} = \mu\left(\frac{\partial u_i}{\partial x_j} + \frac{\partial u_j}{\partial x_i}\right) - \delta_{ij}\left(\xi - \frac{2}{3}\mu\right)\frac{\partial u_l}{\partial x_l}$. Then for the non-isothermal motion of the medium, using the definition of equivalency measures between deterministic and random process in the critical point, the sets of stochastic equations of energy, momentum, and mass are defined for the next space-time areas: 1) the beginning of the generation (index 1,0 or 1); 2) generation (index 1,1); 3) diffusion (1,1,1) and 4) the dissipation of the turbulent fields. These results provide an opportunity to introduce the concept of the correlator, which is defined for the potential physical quantities and combinations (N, M). This correlator, in its structure, will determine the possible range of motion in space depending on the different combinations of (M, N) and the corresponding values that determine the correlation interval of space-time. According to [12,13], the correlator in space-time is

$$\lim_{m_i\to m_c; r_i\to r_c; \Delta\tau_i\to\tau_c}\left(D_{N,M}(m_i \to m_c; r_i \to r_c; \Delta\tau_i \to \tau_c)\right) = 0\ , \tag{4}$$

$$D_{N,M}(m_c;r_c;\tau_c)=\sum_i \lim_{m_i\to m_c}\lim_{r_i\to r_c}\lim_{\Delta\tau_i\to\tau_c}\left\{m\left(T^M Z^*\cap T^N Y^*\right)-R_{1_{T^M Z^* T^N Y^*}}\, m\left(T^M Z^*\right)\right\}. \quad (5)$$

Subscript j denotes the parameters m_{cj} ($j = 3$ means mass, momentum, and energy). For the case of the binary intersections, it was written that $X= Y + Z + W$. Here subscripts «cr» or «c» refer to critical point $r(x_{cr}, \tau_{cr})$ or r_c: the space-time point of the beginning of the interaction between the deterministic field and random field that leads to turbulence. In addition, subsets Y, Z, W are called extended in X, if the measures $m(Y)$, $m(Z)$, $m(W)$ have the properties:

$m(Y)=m(Y^*)=m(T^nY)+\bigcup_{k=0}^{k=n-1} m(T^k(G_1^{n-k}))$ and wandering subsets $\bigcup_{k=0}^{k=n-1} m(T^k(G_1^{n-k}))\subset Y$;

$m(Z)=m(Z^*)=m(T^nZ^*)+\bigcup_{k=0}^{k=n-1} m\left(T^k\left(G_2^{n-k}\right)\right)$ and wandering subsets $\bigcup_{k=0}^{k=n-1} m(T^k(G_2^{n-k}))\subset Z$;

$m(W)=m(W^*)=m(T^nW)+\bigcup_{k=0}^{k=n-1} m\left(T^k\left(G_3^{n-k}\right)\right)$ and wandering subset $\bigcup_{k=0}^{k=n-1} m(T^k(G_3^{n-k}))\subset W$.

The correlation function produces a set of equations that determine the equivalence of measures:

$$\left|m\left(T^M Z\right)\right|=\left(R_{T^M ZT^N Y}\right)_{n,m}\left|m\left(T^N Y\right)\right|,\ 0\le\left(R_{T^M ZT^N Y}\right)\le 1.$$

Here, $R_{T^M ZT^N Y}$ is the fractal correlation function assumed to be equal to unity to obtain analytical solutions. Thus, $|m(Z)|=\left(R_{ZTY}\right)_{n,m}|m(TY)|$ for the pair $(N, M)=(1,0)$, and $|m(TZ)|=\left(R_{TZTY}\right)_{n,m}|m(TY)|$ for $(N, M) = (1,1)$. Here, T^n is the conservative transformation of X for all n, such that there is $n > n_d$ when T^n is the dissipative transformation for $Y^*\subset X$ and $Z^*\subset X$.

In [12,23-25], for the transfer of substantial values F (the mass (the density-ρ), the momentum (ρU), the energy (E)) of deterministic (laminar) motion into random (turbulent) motion (for the area 1), the beginning of the generation of turbulence, the pair (N, M) = (1,0)), the equivalence of measures was written as

$$\left(d\Phi_{col,st}\right)_{1,0}=-R_{1,0}\left(\Phi_{st}\right),\ \left(\frac{d(\Phi)_{col,st}}{d\tau}\right)_{1,0}=-R_{1,0}\left(\frac{\Phi_{st}}{\tau_{cor}}\right). \quad (6)$$

In addition, for «correlator» $D_{N,M}\left(r_c;m_{ci};\tau_c\right)=D_{1,1}\left(r_c;m_{ci};\tau_c\right)$, the pair (N,M)=(1,1) and for the space-time area 2) the generation of turbulence $r_{c1}\left(x_c+\Delta x_0+\Delta x_1;\tau_c+\Delta\tau_0+\Delta\tau_1\right)-r_{c0}$,the author of [12, 25] wrote that

$$\left(d\Phi_{col,st}\right)_{1,1}=-R_{1,1}\left(d\Phi_{st}\right),\ \left(\frac{d(\Phi)_{col,st}}{d\tau}\right)_{1,1}=-R_{1,1}\left(\frac{d\Phi_{st}}{d\tau}\right). \quad (7)$$

$R_{1,0}$,$R_{1,1}$ are fractal coefficients. For example, to obtain the new analytical dependences, these coefficients are taken equal unit. For the integration of the set, we can write an expression for time interval τ^1_{cor} . Expressions for non-isothermal flows are

$$\left(\tau^0_{cor}\right)_{1U,P}=\frac{L}{\left(\left(E_{st}\right)_{U,P}/\rho\right)^{1\backslash 2}},\quad \left(\tau^0_{cor}\right)_{2U,P}=\frac{L^2}{(\nu)},\quad \left(\tau^0_{cor}\right)_{3U,P}=\frac{\nu}{\left(\left(E_{st}\right)_{U,P}/\rho\right)}, \quad (8)$$

IOP Conf. Series: Journal of Physics: Conf. Series **1009** (2018) 012017 doi:10.1088/1742-6596/1009/1/012017

$$\left(\tau_{cor}^{0}\right)_{1T}=\frac{L_T}{\left(\left(E_{st}\right)_{u_j}/\rho\right)^{1/2}},\qquad \left(\tau_{cor}^{0}\right)_{2T}=\frac{L_T^2}{\left(\lambda/\rho\cdot c_p\right)},\qquad \left(\tau_{cor}^{0}\right)_{3T}=\frac{\lambda/\rho\, c_p}{\left(\left(E_{st}\right)_{u_j}\right)}$$

$$\left(\tau_{cor}^{0}\right)_{1T}=\left(\frac{L_T}{\left(\left(u_j^2\right)_{st}\right)^{1/2}}\right)=\left(\frac{L}{\left(\left(u_i^2\right)_{st}/2\right)^{1/2}}\right)\left(\frac{\left(\left(u_i^2\right)_{st}/2\right)^{1/2}}{\Pr\cdot\left(\left(u_j^2\right)_{st}\right)^{1/2}}\right)=\left(\tau_{cor}^{0}\right)_{1U,P}\left(\frac{\left(\left(u_i^2\right)_{st}/2\right)^{1/2}}{\Pr\cdot\left(\left(u_j^2\right)_{st}\right)^{1/2}}\right). \tag{9}$$

For the correlation time we have the following representations for the case of non-isothermal flow: Further, $L = L_{U,P} = L_U$ is the scale of turbulence. Indexes ($_{U,P}$) and ($_U$) refer to the velocity field and index ($_T$) refers to the temperature field. L_y on $x_2 = y$, or L_x, $x_1 = x$. Here, x_1 and x_2 are coordinates along and normal to the wall. Here $L_T = \dfrac{L}{\Pr}$, $\Pr = \dfrac{\rho\cdot\nu\cdot C_p}{\lambda}$ is the Prandtl number.

3. The critical Reynolds number

The space-time area 1) is the area of the beginning of the generation of turbulence. On the basis of equivalence of measures, let us define the expressions for the critical Reynolds number, the velocity profiles, temperature, and the second-order correlations for the non-isothermal processes. For the area 1) $r_{c0}(x_c+\Delta x_0;\tau_c+\Delta\tau_0)-r_c$, referring the pair (N, M) =(1,0), we have a set of equations of mass, momentum, and energy:

$$\left(\frac{d(\rho)_{col,st}}{d\tau}\right)_{1,1}=-\frac{\rho_{st}}{\tau_{cor}}$$

$$\begin{cases}\left(\dfrac{d(\rho\vec{U})_{col,st}}{d\tau}\right)_{1,1}=-\left(\dfrac{(\rho\vec{U})_{st}}{\tau_{cor}}\right); \\ div(\tau_{i,j})_{col,st2}=\dfrac{(\rho\vec{U})_{st}}{\tau_{cor}}\end{cases}. \tag{10}$$

$$\begin{cases}\left(\dfrac{d(E)_{col,st}}{d\tau}\right)_{1,1}=-\left(\dfrac{(E)_{st}}{\tau_{cor}}\right)_{1,1}; \\ div(\lambda\dfrac{\partial T}{\partial x_j}+u_i\tau_{i,j})_{col,st2}=\left(\dfrac{(E)_{st}}{\tau_{cor}}\right)_{1,1}\end{cases}$$

According to [14,15], the motion is determined by a quadratic equation of the velocity profile and the temperature profile as a function of the fourth power of the vertical coordinate assuming constant physical properties of the medium. Then let us define that $Ec=\dfrac{U_0^2}{c_p(T_0-T_w)}$ is the Eckert number, $T_T=|T_{st}|/(T_0-T_w)$, $T_u=\sqrt{\dfrac{\Sigma\,(u_i^2)_{st}}{U_0^2}}$. Also T_0 and T_w are the temperatures along the pipe axis and on the wall; R, U_0, and u_1 are the pipe radius and velocities along the axis and along x_1. Thus, the critical point is

$$\left(\frac{x_2}{R}\right)=\left[\frac{1}{4}\frac{(Est)_{UP}}{U_0^2}\left(\frac{R}{L_U}\right)\frac{Ec\cdot\Pr}{1+Ec\cdot\Pr}\left(1+\frac{(Est)_T}{(Est)_{UP}}\right)\right]^{\frac{1}{3}}=\left[\frac{1}{4}\frac{(Est)_{UP}}{U_0^2}\left(\frac{R}{L_U}\right)\frac{Ec\cdot\Pr}{1+Ec\cdot\Pr}\left(1+\frac{2\cdot T_T}{Tu^2\cdot Ec}\right)\right]^{\frac{1}{3}}. \tag{11}$$

The first critical Reynolds number using (8) – (11) is defined as:

$$\mathrm{Re}_{d(T,U)} = \left(6\left(\frac{1}{2}\right)^{1/3}\left(\frac{U_0}{\sqrt{E_{st}/\rho}}\right)^{5/3}\left(\frac{L_U}{R}\right)^{1/3}\right)\cdot\left(\frac{\left[1+\frac{2T_T}{Ec\cdot Tu^2}\right]^{2/3}}{\left[1+\left(\frac{\Pr\cdot\left((u_j^2)_{st}\right)^{1/2}}{\left((u_i^2)_{st}\right)^{1/2}}\right)\frac{2T_T}{Ec\cdot Tu^2}\right]}\frac{\left(1+2\frac{1}{\Pr\cdot Ec}\right)}{\left(1+\frac{1}{\Pr\cdot Ec}\right)^{2/3}}\right). \quad (12)$$

Index (T,U) refers to the non-isothermal flow. The first bracket in (13) is an expression of the critical Reynolds number for an isothermal process. The second bracket determines the effects of the temperature field (Pr, Ec), the turbulence intensities (Tu, T_T) and $(u_j)_{st}/(u_i)_{st}$ as well on the critical Reynolds number. It is seen that a decrease in T_{st} and an increase in wall cooling lead to an increase in the critical Reynolds number Notably, if Eckert number $\mathrm{Ec}^{-1} = c_p\left(T_w - T_0\right)/U_0^2 = -0.01$, and $T_T = T_U = 0.01 \div 0.03$, Pr = 0.72, then the critical Reynolds number is increased by a factor of $\sim 1.5 \div 1.9$. This fact is in satisfactory agreement with the experimental data presented in [26–30].

4. The velocity and temperature profiles for turbulent flow in the pipe

The space-time area 2) is the area of the generation of turbulence. For the turbulence generation region $r_{c1}(x_c + \Delta x_0 + \Delta x_1; \tau_c + \Delta\tau_0 + \Delta\tau_1) - r_{c0}$,((N, M) = (1,1), «correlator» $D_{N,M}\left(r_c; m_{ci}; \tau_c\right) = D_{1,1}\left(r_c; m_{ci}; \tau_c\right)$, the set of stochastic equations was defined as [15, 16]

$$\left(\frac{d(\rho)_{col,st}}{d\tau}\right)_{1,1} = -\left(\frac{d\rho_{st}}{d\tau_{cor}}\right)$$

$$\begin{cases}\left(\frac{d(\rho\vec{U})_{col,st}}{d\tau}\right)_{1,1} = -\left(\frac{d(\rho\vec{U})_{st}}{d\tau}\right); \\ div(\tau_{i,j})_{col,st2} = \frac{d(\rho\vec{U})_{st}}{d\tau}\end{cases}. \quad (13)$$

$$\begin{cases}\left(\frac{d(E)_{col,st}}{d\tau}\right)_{1,1} = -\left(\frac{d(E)_{st}}{d\tau}\right)_{1,1}; \\ div(\lambda\frac{\partial T}{\partial x_j} + u_i\tau_{i,j})_{col,st2} = \left(\frac{d(E)_{st}}{d\tau}\right)_{1,1}\end{cases}$$

It is known that the experimental study of the averaged characteristics of fully developed turbulence showed that the velocity profiles have affine similarity. Therefore, in the case of the flow in the pipe [16-18], the final equation for the expression for the index of the velocity profile «n» has the form

$$\left|2\frac{1-n}{n^2}\sqrt{\left(\frac{R}{x_2}\right)^4\left(\frac{x_2}{R}\right)^{(2/n)}}\right| = \left|\left(\mathrm{Re}_{st} - \frac{1}{\mathrm{Re}_{st}}\right)\right|^{0.5} \quad (14)$$

The final equation for the expression for the index of the temperature profile «n_T» has the form

$$\left|\frac{\left(\left(\frac{2-n}{n^2}\right)\left(\frac{x_2}{R}\right)^{(2/n)-4}\right)_U+\left(\frac{1-n_T}{2n_T^2}\right)_T\frac{1}{Pr}\left(\frac{(T_0-T_w)}{U^2/2c_p}\right)\left(\frac{x_2}{R}\right)^{(1/n_T)-4}}{6+12\frac{1}{Pr}\left(\frac{(T_0-T_w)}{U^2/2c_p}\right)}\right|=\left(\left(Re_{st}\right)_U\right)\frac{\left\{1+2\frac{F\cdot T_T}{Ec\cdot Tu^2}\right\}}{1+2\frac{T_T}{Ec\cdot Tu^2}} \tag{15}$$

According to [16], there is agreement between the values obtained using equations (14), (15) $n \approx 7$ and $n_T \approx 8$ with the experimental data [26-30] for the indexes of profiles n and n_T.

5. Profiles of turbulent characteristics

The space-time area 3) is area of the diffusion of turbulence. For the critical point

$$q=\left(\left[\sqrt{E_{st}/\rho}\right]_0\right)^2\left(Re_{st,0}-\frac{1}{Re_{st,0}}\right)=(E_{st}/\rho)_0\left(Re_{st,0}-\frac{1}{Re_{st,0}}\right), \tag{16}$$

where q is the kinetic energy of the turbulence at the critical point. Then quantity $q/(u^*)^2$ is equaled to

$$\frac{q}{(u^*)^2}=0{,}144*\left(\frac{\left[\sqrt{E_{st}/\rho}\right]_0}{U_m}\right)^2 Re_\delta\left(Re_{st,0}-\frac{1}{Re_{st,0}}\right)=0{,}144*10*0{,}0001*8000*10\approx 11{,}52 \tag{17}$$

For the first critical number $Re_x=\frac{x_1U_0}{\nu}\approx 3*10^5$ and for the initial turbulent Reynolds number $Re_{st,0}$, we determine the value of $Re_\delta=\frac{\delta U_0}{\nu}$. Index 0 corresponds to the initial turbulence. Re_δ is the Reynolds number calculated on the thickness of the boundary layer δ. Then turbulent stresses q_{xy} are

$$\frac{q_{xy}}{(u^*)^2}=(0{,}14\div 0{,}16)\frac{q}{(u^*)^2}=0{,}155*0{,}144*\left(\frac{\left[\sqrt{E_{st}/\rho}\right]_0}{U_m}\right)^2 Re_\delta\left(Re_{st,0}-\frac{1}{Re_{st,0}}\right)=0{,}155*11{,}36\approx 1{,}8 \tag{18}$$

The value in (18) is in satisfactory agreement with the data [25-30]. The range of possible values $(q_{xy}/q)=0{,}08\div 0{,}2$ are in an agreement with Klebanoff's experimental values [26-30]. Similarly, for each point of the space, performances of profiles $\frac{q}{(u^*)^2}$ and $\frac{q_{xy}}{(u^*)^2}$ are defined.

6. Spectral and correlation functions depending on initial turbulence. Fractal equation by Landau

The space-time area 3) is area of the diffusion of turbulence. For the space-time area of 3) diffusion, according to [12, 17-19] we have two fractal equations. The first equation is written as

$$\frac{d(E_{st})_j}{d\tau}=-(R_{zTz})_{1;(1,0)}\frac{(E_{st})_j}{\delta\tau} \tag{19}$$

Then the solution may be written as

$$\left(E_{st}\right)_j = C\exp\left[-\frac{(R_{zTz})_{1;(1,0)}}{2}\left(\frac{|\tau|}{\tau_{cor1}}\right)^2\right] \tag{20}$$

and the spectral function is equaled to

$$E(w)_j = \frac{C}{\sqrt{\pi\alpha}}\exp\left[-\frac{w^2}{4\alpha}\right] , \tag{21}$$

$$C = \left(E_{st}\right)_{\tau^1_{cor}} = \left(E_{st}\right)_{\tau^0_{cor}}\left|\mathrm{Re}_{st} - \frac{1}{\mathrm{Re}_{st}}\right|_{\tau^0_{cor}} .$$

The second fractal equation is written as

$$\frac{d\left(E_{st}\right)_j}{d\tau} = (R_{zTz})_{1;(1,0)}\frac{\left(E_{st}\right)_j}{\delta\tau} . \tag{22}$$

This expression is the same as the equation for the square of the amplitude of oscillations $|A|^2$ in the theory of the turbulence by Landau:

$$\frac{d|A|^2}{dt} = 2\cdot\gamma\cdot|A|^2 - \delta\cdot|A|^4 . \tag{23}$$

Solutions of equation (22) were written as [18, 19]

$$\left(E_{st}\right)_j = C_L \exp\left[(R_{zTz})_{1;(1,0)}\left(\frac{|\tau|}{\tau^1{}_{cor}}\right)\right], \tag{24}$$

$$\left(E_{st}\right)_j = C_L \exp\left[\frac{(R_{zTz})_{1;(1,0)}}{2}\left(\frac{|\tau|}{\tau^1{}_{cor}}\right)^2\right]. \tag{25}$$

7. The turbulent Prandtl number as a function of initial turbulence

The space-time area 3) is area of the diffusion of turbulence. It is known that the turbulent Prandtl number $\mathrm{Pr_T}$ is determined by an expression [25–34], which in the particular case of a forced flow has the form

$$\mathrm{Pr}_T = \left(\frac{u_i'u_j'}{u_j'T'}\right)\left(\frac{dT/dx_j}{du_i/dx_j}\right) = \left(\frac{u_1'u_2'}{u_2'T'}\right)\left(\frac{dT/dx_2}{du_1/dx_2}\right) \tag{26}$$

Here $q_{xy} = u_i'u_j'$ and $q_T = u_j'T'$ are the turbulent stress and the turbulent heat flux, u_i', u_j', T' are fluctuations of values u_i , u_j , T . For the turbulent flow in the pipe $\mathrm{Pr_T}$ is [18]

$$\mathrm{Pr}_T = \frac{1}{3.18}\left(\frac{T_u}{T_T}\right)\left(\frac{n}{n_T}\right)\left(\frac{x_2}{R}\right)^{(n-n_T)/(n\cdot n_T)} \tag{27}$$

and for the turbulent flow on the flat plate, it is

$$\mathrm{Pr}_T = \frac{1}{3.18}\mathrm{Pr}^{1/n_T}\left(\frac{T_u}{T_T}\right)\left(\frac{n}{n_T}\right)\left(\frac{x_2}{R}\right)^{(n-n_T)/(n\cdot n_T)} \quad (28)$$

Notably, for values n = 7, n_T = 8, T_T =0.001— 0.01, Tu = 0.005—0.04, $x_2/R = x_2/\ \delta$= 0.05—0.1, Pr = 0.05—7.0 we have Pr_T = 0.4—1.2.

8. Heat transfer coefficient

According to [20, 21], using above derived solutions, the Nu number has the form for the water flow under the assumption that $\mathrm{Pr}_L = 3$, $\mathrm{Pr}_W = 4$ (the heat transfer from fluid to the wall). Then the equation for Nu was written as

$$Nu_d = 0.0077\,\mathrm{Pr}^{7/12}\,\mathrm{Re}_d^{\ 7/8} = 0.0145\,\mathrm{Re}_d^{\ 7/8} \quad (29)$$

Indexes L and W refer to the parameters of liquid and at the wall, correspondingly. For the air flow with Pr = 0.7, the Nu number was written as

$$Nu_d = 0.00854\,\mathrm{Pr}^{7/12}\,\mathrm{Re}_d^{\ 7/8} = 0.0069\,\mathrm{Re}_d^{\ 7/8} \quad (30)$$

Checking the formulas (29) and (30) shows satisfactory agreement with the experimental data [25–30].

9. Conclusions

The review of main results of new stochastic theory of turbulence for non-isothermal flows is presented. The sets of stochastic equations for the laws of conservation and for the equivalence of measures between the deterministic process and a random process were derived. The review of results of analytical expressions is presented for the main parameters of a non-isothermal turbulent flow in the tube depending on initial fluctuations in the medium. It is shown that the derived formulas for the critical Reynolds numbers and the critical point, for the velocity and temperature profiles, for the second-order correlation, for the turbulent Prandtl number and for the heat transfer number are in satisfactory agreement with the classical experimental data for these flow characteristics. The derived equations determine not only the nature of turbulent processes but also allow us to explain the scatter of the experimental data due to different initial conditions. In addition, new equations provide a more accurate analysis of numerical studies and experiments.

Acknowledgements

This work was supported by the program of increasing the competitive ability of National Research Nuclear University MEPhI (agreement with the Ministry of Education and Science of the Russian Federation of August 27, 2013, Project No.02.a03.21.0005).

Dedicated to the memory of Academician Yu. A. Ryzhov.

References

[1] Kolmogorov A N 1940 *Dokl. Akad. Nauk* **26** (1) 6
[2] Kolmogorov A N 1958 *Dokl. Akad. Nauk* **119** No. 5 861
[3] Kolmogorov A N 1959 *Dokl. Akad. Nauk* **124** No. 4 754
[4] Kolmogorov A N 2004 *Usp. Mat. Nauk* **59** (1) (355) 5
[5] Landau L D 1944 *Dokl. Akad. Nauk* **44** (8) 339
[6] Lorenz E N 1963 *J. Atmos. Sci.* **20** 130
[7] Ruelle D and Takens F 1971 *Communs. Math. Phys.* **20** 167p
[8] Klimontovich Yu L 1989 *Usp. Fiz. Nauk* **158** B1 59p
[9] Arnol'd V I 1990 *Theory of Catastrophes* (Nauka: Moscow)
[10] Haller G 1999 *Chaos Near Resonance* (Springer: Berlin)

[11] Orzag S A and Kells L C 1980 *J. Fluid Mech.* **96** (1) 159p
[12] Dmitrenko AV 2013 *Dokl. Phys.* **58** (6) 228 doi.org/10.1134/s1028335813060098
[13] Dmitrenko A V 2007 *Dokl. Phys.* **52** (7) 384 doi.org/10.1134/s1028335807120166
[14] Dmitrenko A V 2014 Adv. Stud. in Theoret. Phys **8** (25) 1101 doi.org/10.12988/astp.2014.49131
[15] Dmitrenko A V 2015 J.of Eng. Phys.and Thermophys.**88**(6)1569 doi:10.1007/s10891-015-1344-x
[16] Dmitrenko A V 2016 *Heat TransferResearch* **47** (1) 338 doi:10.1615/HeatTransRes.014191
[17] Dmitrenko A V 2016 *Int. J. Fluid Mech. Res.* **43** (3) 82 doi:10615/InterJFluidMechRes.v43.i3.
[18] Dmitrenko A V 2016 *Int. J. Fluid Mech. Res.* **43** (2) 182 doi:10.1615/InterJFluidMechRes.v43.i2
[19] Dmitrenko A V 2017 *Contin. Mechan. and Thermod.* **29** (1) 1 doi:10.1007/s00161-016-0514-1
[20] Dmitrenko A V 2017 *Contin.Mechan. and Thermod.* **29** (6) 1197 doi.org/10.1007/s00161-017-0566x
[21] Dmitrenko A V 2017 *Heat Trans. Res.* **49**(15) 1195; doi:10.1615/HeatTransRes.2017018750.
[22] Dmitrenko AV2017 *J.of Eng. Phys.and Thermophys.***90** (6) 1288 doi.org/10.1007/s10891-017-1685-8
[23] Dmitrenko A V 2013 *Regular Coupling between Deterministic (Laminar) and Random (Turbulent) Motions-Equivalence of Measures Scientifi c Discovery Diploma* No. 458 registration No. 583 of December 2
[24] Dmitrenko A V 2013 *Theory of Equivalent Measures and Sets with Repeating Denumerable Fractal Elements. Stochastic Thermodynamics and Turbulence. Determinacy–Randomness Correlator* [in Russian] (Galleya-Print: Moscow) 226p https://search.rsl.ru/ru/record/01006633402
[25] Dmitrenko A V 2008 *Fundamentals of heat and mass transfer and hydrodynamics of single-phase and two-phase media.Criterial integral statistical methods and direct numerical simulation.* (Galleya print: Moscow) 398p http://search.rsl.ru/ru/catalog/record/6633402
[26] Davidson P A 2004 Turbulence (Oxford Univ. Press) 678p
[27] Millionshchikov M D 1969 Turbulent flow in boundary layers and in pipes (Nauka: Moscow)
[28] Schlichting H 1968 *Boundary-Layer Theory* (6th Edition McGraw-Hill)
[29] Monin A S and Yaglom A M 1971 *Statistical Fluid Mechanics* (M.I.T. Press)
[30] Hinze J O 1975 *Turbulence* (McGraw-Hill)
[31] Dmitrenko A V 1997 *33th AIAA/ASME/SAE/ASEE AIAA Paper97-2911*; doi: 10.2514/6.1997-2911; http://arc.aiaa.org/doi/abs/10.2514/6.1997-2911.
[32] Dmitrenko A V 1998 *34thAIAA/ASME/SAE/ASEE AIAA Paper98-3444*; doi:10.2514/6.1998-3444; http://arc.aiaa.org/doi/abs/10.2514/6.1998-3444
[33] Dmitrenko A V 1993 *Aviats. Tekh.* No.1 39p
[34] Dmitrenko A V 1986 *Proc.11th Conf. Young Scientists* Moscow Physicotechnical Institute Part 2 Moscow (1986) pp 48–52. Deposited at VINITI 08.08.86 No. 5698-B86

IOP Conf. Series: Journal of Physics: Conf. Series **1009** (2018) 012018 doi:10.1088/1742-6596/1009/1/012018

Normal glow discharge between curvilinear electrodes

S T Surzhikov

Ishlinsky Institute for Problems in Mechanics RAS, Vernadsky prospekt 101(1), Moscow, 119526, Russia
Moscow Institute of Physics and Technology (State University), 9 Institutskiy per., Dolgoprudny, Moscow Region, 141701, Russia
Dukhov All-Russian Scientific Research Institute of Automatics, Moscow, Russia

surg@ipmnet.ru

Abstract. A diffusion-drift calculation model of a normal glow discharge between curvilinear electrodes is presented. The verification of the created computational model has been performed using the example of a normal glow discharge between flat electrodes.
Results of numerical simulation are presented of normal glow discharge between electrodes in the form of spherical segments at pressure $p = 5$Torr and emf of power supply in the region of $1000 \div 1750$ V.

1. Introduction

A normal steady-state glow discharge (NGD) is investigated in the laboratory between flat (disk) electrodes spaced a few centimeters apart. In the normal glow discharge a column of the gas-discharge plasma should be located far from the boundaries of the electrodes. Such a discharge burns in the self-organizing mode of transverse dimensions, and the current density at the cathode is predicted with good accuracy by the one-dimensional theory of Engel and Steenbeck [1-3]

$$V_n = \frac{3B}{A}\ln\left(1+\frac{1}{\gamma}\right), \text{V},$$

$$\frac{j}{p^2} = 5.92\times10^{-14}\frac{AB^2\mu_i p(1+\gamma)}{\ln(1+\gamma^{-1})}, \text{A/(cm}^2\times\text{Torr)},$$

$$d_n p = \frac{3.78}{A}\ln(1+\gamma^{-1}), \text{cm}\times\text{Torr},$$

where: V_n, j_n, d_n are the voltage drop on the gas-discharge gap, current density, thickness of the cathode layer of a normal glow discharge; p is the pressure in Torr; μ_i is the mobility of ions; γ is the is the coefficient of secondary electron emission. It is assumed that the coefficient of impact ionization of molecules by electrons α is approximated as follows

$$\alpha = \alpha\left(\frac{E}{p}\right) = Ap\exp\left(-\frac{B}{E/p}\right),\ \text{cm}^{-1},$$

where A, B are the approximating coefficients; E is the electric field strength.

The increase in the total current through the gas-discharge gap (typical values from units to tens of mA) in such a glow discharge leads to an increase in the transverse dimensions of the current column, but the current density in the near-axis region remains practically unchanged. This phenomenon is called the law of normal current density [1, 2] and has been studied in classical physics of gas discharge for more than 100 years [3].

Since the 80s of the last century, the symmetrical normal glow discharge is the object of study by methods of computational physics [4–6]. Numerical simulation of this type of discharge proved to be very useful for the development of computational methods and algorithms for mathematical modeling. Subsequently, the problems were investigated of calculating the heating of a neutral gas, the dynamics of a glow discharge in a transverse and longitudinal magnetic field with induction up to ~ 1.0 T, taking into account the physical kinetics of excitation of N_2 vibrational states in collision with electrons [2, 7].

Localization of the discharge column of a normal glow discharge between flat electrodes is achieved in a computational experiment by setting the initial conditions. In a real physical experiment, it is not easy to achieve a stationary state far from the electrode boundaries [8].

The subject of this paper is a computational study of the structure of a normal glow discharge between convex electrodes. Schematic of the task is shown in figure 1, as well as an external electric circuit consisting of a constant-current power supply with an emf ε and ohmic resistance R_0. The discharge gap is placed in a vacuum chamber. The investigated range of molecular nitrogen pressures is $p = 1 \div 10$ Torr.

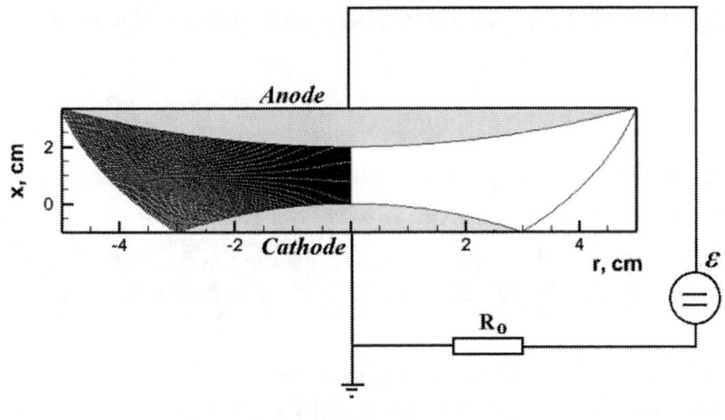

Figure 1. Schematic of the task.

2. Equations for the diffusion-drift model in curvilinear geometry

The equations of the diffusion-drift model used to simulate a normal glow discharge are obtained from the system of equations for a multi-fluid and a multi-temperature partially ionized gas mixture (a detailed derivation of these equations is given in [2])

$$\text{div}\mathbf{E} = -\varepsilon^*\left(n_+ - n_e\right) \tag{1}$$

$$\frac{\partial n_e}{\partial t} + \text{div}\mathbf{\Gamma}_e = \alpha\left|\mathbf{\Gamma}\right| - \beta n_+ n_e \tag{2}$$

IOP Conf. Series: Journal of Physics: Conf. Series **1009** (2018) 012018 doi:10.1088/1742-6596/1009/1/012018

$$\frac{\partial n_+}{\partial t} + \mathrm{div}\boldsymbol{\Gamma}_+ = \alpha|\boldsymbol{\Gamma}| - \beta n_+ n_e, \tag{3}$$

where n_+, n_e are the concentrations of ions and electrons; $\mathbf{E}$ is the vector of electric field strength with electric potential φ

$$\mathbf{E} = -\mathrm{grad}\varphi, \tag{4}$$

$\boldsymbol{\Gamma}_e$, $\boldsymbol{\Gamma}_+$ are the vectors of electron and ion flux densities,

$$\boldsymbol{\Gamma}_e = -n_e\mu_e\mathbf{E} - D_e\mathrm{grad}n_e,\ \ \boldsymbol{\Gamma}_+ = -n_+\mu_+\mathbf{E} + D_+\mathrm{grad}n_+, \tag{5}$$

μ_e, μ_+ are the mobilities of electrons and ions; D_e, D_+ are the diffusion coefficients of electrons and ions; α, β are the coefficient of impact ionization of a neutral gas by electrons (the 1st Townsend coefficient) and the recombination coefficient; $\varepsilon^* = 4\pi e = 1.81\times10^{-6}$ V×cm.

Boundary conditions for the solution of the boundary value problem (1)–(3) are

$$\begin{aligned}
x=0:&\quad \Gamma_{e,x} = \gamma\Gamma_{+,x},\quad \frac{\partial n_+}{\partial x} = 0,\quad \varphi = 0,\\
x=H:&\quad \frac{\partial n_e}{\partial x} = 0,\quad n_+ = 0,\quad \varphi = V,\\
r=0:&\quad \frac{\partial n_e}{\partial r} = \frac{\partial n_+}{\partial r} = \frac{\partial \varphi}{\partial r} = 0,\\
r=R_c:&\quad \frac{\partial n_e}{\partial r} = \frac{\partial n_+}{\partial r} = \frac{\partial \varphi}{\partial r} = 0,
\end{aligned} \tag{6}$$

where V is the voltage drop at the electrodes, determined from the electric circuit equation

$$\frac{\varepsilon - V}{eR_0} = 2\pi\int_0^{R_a} \Gamma_{e,n_{s,a}} r\mathrm{d}r \tag{7}$$

For curvilinear electrodes, the integral in (7) should be calculated along the surface of the electrode with projection of the electron flux density ($\Gamma_{e,n_{s,a}}$) on the local normal to the surface $n_{s,a}$.

Condition (7) determines the electric current through the anode. For a stationary glow discharge, the total electric current through the gas-discharge gap is a constant value and can be determined from the following equation:

$$\frac{\varepsilon - V}{eR_0} = 2\pi\int_0^{R_a} \Gamma_{+,n_{s,c}} r\mathrm{d}r,$$

where $n_{s,c}$ is the local normal to surface of cathode.

We also note that when analyzing the transient processes of establishing a stationary state, one should take into account the effect of accumulation of charges on electrodes (see [4] for more details).

The closing relations for the problem being solved are given for molecular nitrogen [2]

$$\mu_e p = 4.4\times10^5,\ \mu_i p = 1.45\times10^3,\ \text{Torr×cm}^2\text{/V×s},\ \beta = 2\times10^{-7},\ \text{cm}^3\text{/s},$$

$$\frac{\alpha}{p} = A\exp\left(-\frac{B}{E/p}\right),\ \text{1/(cm×Torr)},\ A = 12,\ \text{(cm×Torr)}^{-1},\ \ B = 342,\ \text{V/(cm×Torr)},$$

$$D_e = \mu_e T_e,\ D_+ = \mu_i T_+,\ T_e = 1\,\text{eV},\ T_+ = 0.026\,\text{eV}$$

The construction of a computational model of a normal glow discharge in the computational domain with curvilinear boundaries requires the introduction of a curvilinear coordinate system (ξ,η) connected by single-valued transformations with the original cylindrical coordinate system (x,r). The method of transforming the solved system of equations that is adapted to curvilinear surfaces is given in [10] (Ch.14). Below these transformations are given for the system of the Poisson equation and equations of the diffusion-drift model.

3. Transformation of the Poisson equation for the electric potential and the continuity equations for electrons and ions

The initial Poisson equation in an orthogonal cylindrical coordinate system is formulated as following:

$$\frac{\partial}{\partial x}\left(\frac{\partial\varphi}{\partial x}\right)+\frac{\partial}{\partial r}\left(\frac{\partial\varphi}{\partial r}\right)+\frac{\chi}{r}\frac{\partial\varphi}{\partial r}=\Phi(r,x), \tag{8}$$

where $\Phi(r,x)=4\pi e(n_+-n_e)$, $\chi=0$ for plane discharge, $\chi=1$ for axisymmetric discharge.

Input coordinate system (ξ,η) must be uniquely linked to the original coordinate system (x,r), therefore (8) can be rewritten in the form (hereinafter $y=r$)

$$\xi_x\frac{\partial}{\partial\xi}\left(\xi_x\frac{\partial\varphi}{\partial\xi}+\eta_x\frac{\partial\varphi}{\partial\eta}\right)+\eta_x\frac{\partial}{\partial\eta}\left(\xi_x\frac{\partial\varphi}{\partial\xi}+\eta_x\frac{\partial\varphi}{\partial\eta}\right)+\xi_y\frac{\partial}{\partial\xi}\left(\xi_y\frac{\partial\varphi}{\partial\xi}+\eta_y\frac{\partial\varphi}{\partial\eta}\right)+\eta_y\frac{\partial}{\partial\eta}\left(\xi_y\frac{\partial\varphi}{\partial\xi}+\eta_y\frac{\partial\varphi}{\partial\eta}\right)+$$

$$+\frac{1}{y}\left(\xi_y\frac{\partial\varphi}{\partial\xi}+\eta_y\frac{\partial\varphi}{\partial\eta}\right)=\Phi\left[\xi(x,y),\eta(x,y)\right]=\Phi \tag{9}$$

Division on Jacobian of the transformation J and differentiation by parts allows us to obtain the equation

$$\frac{\partial}{\partial\xi}\left(\frac{\xi_x^2}{J}\frac{\partial\varphi}{\partial\xi}\right)-\xi_x\frac{\partial\varphi}{\partial\xi}\frac{\partial}{\partial\xi}\left(\frac{\xi_x}{J}\right)+\frac{\partial}{\partial\xi}\left(\frac{\xi_x\eta_x}{J}\frac{\partial\varphi}{\partial\eta}\right)-\eta_x\frac{\partial\varphi}{\partial\eta}\frac{\partial}{\partial\xi}\left(\frac{\xi_x}{J}\right)+$$

$$+\frac{\partial}{\partial\eta}\left(\frac{\xi_x\eta_x}{J}\frac{\partial\varphi}{\partial\xi}\right)-\xi_x\frac{\partial\varphi}{\partial\xi}\frac{\partial}{\partial\eta}\left(\frac{\eta_x}{J}\right)+\frac{\partial}{\partial\eta}\left(\frac{\eta_x^2}{J}\frac{\partial\varphi}{\partial\eta}\right)-\eta_x\frac{\partial\varphi}{\partial\eta}\frac{\partial}{\partial\eta}\left(\frac{\eta_x}{J}\right)+$$

$$+\frac{\partial}{\partial\xi}\left(\frac{\xi_y^2}{J}\frac{\partial\varphi}{\partial\xi}\right)-\xi_y\frac{\partial\varphi}{\partial\xi}\frac{\partial}{\partial\xi}\left(\frac{\xi_y}{J}\right)+\frac{\partial}{\partial\xi}\left(\frac{\xi_y\eta_y}{J}\frac{\partial\varphi}{\partial\eta}\right)-\eta_y\frac{\partial\varphi}{\partial\eta}\frac{\partial}{\partial\xi}\left(\frac{\xi_y}{J}\right)+$$

$$+\frac{\partial}{\partial\eta}\left(\frac{\xi_y\eta_y}{J}\frac{\partial\varphi}{\partial\xi}\right)-\xi_y\frac{\partial\varphi}{\partial\xi}\frac{\partial}{\partial\eta}\left(\frac{\eta_y}{J}\right)+\frac{\partial}{\partial\eta}\left(\frac{\eta_y^2}{J}\frac{\partial\varphi}{\partial\eta}\right)-\eta_y\frac{\partial\varphi}{\partial\eta}\frac{\partial}{\partial\eta}\left(\frac{\eta_y}{J}\right)+$$

$$+\frac{\chi}{yJ}\left(\xi_y\frac{\partial\varphi}{\partial\xi}+\eta_y\frac{\partial\varphi}{\partial\eta}\right)=\frac{\Phi}{J} \tag{10}$$

Taking into account the following relations between components of the Jacobian of the transformation

$$y_\eta=\frac{\xi_x}{J},\quad -y_\xi=\frac{\eta_x}{J},\quad -x_\eta=\frac{\xi_y}{J},\quad x_\xi=\frac{\eta_y}{J}, \tag{11}$$

notice, that

$$-\xi_x \frac{\partial\varphi}{\partial\xi} y_{\eta\xi} - \eta_x \frac{\partial\varphi}{\partial\eta} y_{\eta\xi} + \xi_x \frac{\partial\varphi}{\partial\xi} y_{\eta\xi} + \eta_x \frac{\partial\varphi}{\partial\eta} y_{\xi\eta} +$$

$$+\xi_y \frac{\partial\varphi}{\partial\xi} x_{\eta\xi} + \eta_y \frac{\partial\varphi}{\partial\eta} x_{\eta\xi} - \xi_y \frac{\partial\varphi}{\partial\xi} x_{\eta\xi} - \eta_y \frac{\partial\varphi}{\partial\eta} x_{\xi\eta} \equiv 0, \qquad (12)$$

therefore

$$\frac{\partial}{\partial\xi}\left[\frac{\left(\xi_x^2+\xi_y^2\right)}{J}\frac{\partial\varphi}{\partial\xi}\right] + \frac{\partial}{\partial\xi}\left[\frac{\left(\xi_x\eta_x+\xi_y\eta_y\right)}{J}\frac{\partial\varphi}{\partial\eta}\right] + \frac{\partial}{\partial\eta}\left[\frac{\left(\xi_x\eta_x+\xi_y\eta_y\right)}{J}\frac{\partial\varphi}{\partial\xi}\right] +$$

$$+\frac{\partial}{\partial\eta}\left[\frac{\left(\eta_x^2+\eta_y^2\right)}{J}\frac{\partial\varphi}{\partial\eta}\right] + \frac{\chi}{yJ}\left(\xi_y\frac{\partial\varphi}{\partial\xi}+\eta_y\frac{\partial\varphi}{\partial\eta}\right) = \frac{\Phi}{J} \qquad (13)$$

The continuity equations for the electrons and ions (2) and (3) are similar, so consider the transformation of only the equation for determining the electron concentrations. In equation (2), we use the relation for the electron flux density (5), as well as the relation between the electric field strength and the electric potential (4). We rewrite (2) in the variables (ξ,η), then divide it by Jacobian J and using partial differentiation, receive

$$\frac{\partial}{\partial t}\left(\frac{n_e}{J}\right) + \frac{\partial}{\partial\xi}\left(n_e\mu_e\frac{\xi_x^2}{J}\frac{\partial\varphi}{\partial\xi}\right) - n_e\mu_e\xi_x\frac{\partial\varphi}{\partial\xi}\frac{\partial}{\partial\xi}\left(\frac{\xi_x}{J}\right) +$$

$$+\frac{\partial}{\partial\xi}\left(n_e\mu_e\frac{\xi_x\eta_x}{J}\frac{\partial\varphi}{\partial\eta}\right) - n_e\mu_e\eta_x\frac{\partial\varphi}{\partial\eta}\frac{\partial}{\partial\xi}\left(\frac{\xi_x}{J}\right) + \frac{\partial}{\partial\eta}\left(n_e\mu_e\frac{\xi_x\eta_x}{J}\frac{\partial\varphi}{\partial\xi}\right) - n_e\mu_e\xi_x\frac{\partial\varphi}{\partial\xi}\frac{\partial}{\partial\eta}\left(\frac{\eta_x}{J}\right) +$$

$$+\frac{\partial}{\partial\eta}\left(n_e\mu_e\frac{\eta_x^2}{J}\frac{\partial\varphi}{\partial\eta}\right) - n_e\mu_e\eta_x\frac{\partial\varphi}{\partial\eta}\frac{\partial}{\partial\eta}\left(\frac{\eta_x}{J}\right) + \frac{\partial}{\partial\xi}\left(n_e\mu_e\frac{\xi_y^2}{J}\frac{\partial\varphi}{\partial\xi}\right) - n_e\mu_e\xi_y\frac{\partial\varphi}{\partial\xi}\frac{\partial}{\partial\xi}\left(\frac{\xi_y}{J}\right) +$$

$$+\frac{\partial}{\partial\xi}\left(n_e\mu_e\frac{\xi_y\eta_y}{J}\frac{\partial\varphi}{\partial\eta}\right) - n_e\mu_e\eta_y\frac{\partial\varphi}{\partial\eta}\frac{\partial}{\partial\xi}\left(\frac{\xi_y}{J}\right) + \frac{\partial}{\partial\eta}\left(n_e\mu_e\frac{\xi_y\eta_y}{J}\frac{\partial\varphi}{\partial\xi}\right) - n_e\mu_e\xi_y\frac{\partial\varphi}{\partial\xi}\frac{\partial}{\partial\eta}\left(\frac{\eta_y}{J}\right) +$$

$$+\frac{\partial}{\partial\eta}\left(n_e\mu_e\frac{\eta_y^2}{J}\frac{\partial\varphi}{\partial\eta}\right) - n_e\mu_e\eta_y\frac{\partial\varphi}{\partial\eta}\frac{\partial}{\partial\eta}\left(\frac{\eta_y}{J}\right) + \frac{n_e\mu_e}{yJ}\left(\xi_y\frac{\partial\varphi}{\partial\xi}+\eta_y\frac{\partial\varphi}{\partial\eta}\right) =$$

$$= \frac{\partial}{\partial\xi}\left(D_e\frac{\xi_x^2}{J}\frac{\partial n_e}{\partial\xi} + D_e\frac{\xi_x\eta_x}{J}\frac{\partial n_e}{\partial\eta}\right) - D_e\left(\xi_x\frac{\partial n_e}{\partial\xi}+\eta_x\frac{\partial n_e}{\partial\eta}\right)\frac{\partial}{\partial\xi}\left(\frac{\xi_x}{J}\right) +$$

$$+\frac{\partial}{\partial\eta}\left(D_e\frac{\xi_x\eta_x}{J}\frac{\partial n_e}{\partial\xi} + D_e\frac{\eta_x^2}{J}\frac{\partial n_e}{\partial\eta}\right) - D_e\left(\xi_x\frac{\partial n_e}{\partial\xi}+\eta_x\frac{\partial n_e}{\partial\eta}\right)\frac{\partial}{\partial\eta}\left(\frac{\eta_x}{J}\right) +$$

$$+\frac{\partial}{\partial\xi}\left(D_e\frac{\xi_y^2}{J}\frac{\partial n_e}{\partial\xi} + D_e\frac{\xi_y\eta_y}{J}\frac{\partial n_e}{\partial\eta}\right) - D_e\left(\xi_y\frac{\partial n_e}{\partial\xi}+\eta_y\frac{\partial n_e}{\partial\eta}\right)\frac{\partial}{\partial\xi}\left(\frac{\xi_y}{J}\right) +$$

$$+\frac{\partial}{\partial\eta}\left(D_e\frac{\xi_y\eta_y}{J}\frac{\partial n_e}{\partial\xi}+D_e\frac{\eta_y^2}{J}\frac{\partial n_e}{\partial\eta}\right)-D_e\left(\xi_y\frac{\partial n_e}{\partial\xi}+\eta_y\frac{\partial n_e}{\partial\eta}\right)\frac{\partial}{\partial\eta}\left(\frac{\eta_y}{J}\right)+$$

$$+\frac{\chi D_e}{Jy}\left(\xi_x\frac{\partial n_e}{\partial\xi}+\eta_y\frac{\partial n_e}{\partial\eta}\right)+\frac{\alpha}{J}|\mathbf{\Gamma}_e|-\frac{\beta}{J}n_en_+ \tag{14}$$

By virtue of relations (11), this equation is simplified as follows

$$\frac{\partial n_e}{\partial t}\left(\frac{n_e}{J}\right)+\frac{\partial}{\partial\xi}\left(n_e\mu_e\frac{\xi_x^2+\xi_y^2}{J}\frac{\partial\varphi}{\partial\xi}\right)+\frac{\partial}{\partial\eta}\left(n_e\mu_e\frac{\eta_x^2+\eta_y^2}{J}\frac{\partial\varphi}{\partial\eta}\right)+$$

$$+\frac{\partial}{\partial\xi}\left(n_e\mu_e\frac{\xi_x\eta_x+\xi_y\eta_y}{J}\frac{\partial\varphi}{\partial\eta}\right)+\frac{\partial}{\partial\eta}\left(n_e\mu_e\frac{\xi_x\eta_x+\xi_y\eta_y}{J}\frac{\partial\varphi}{\partial\xi}\right)+\frac{n_e\mu_e}{yJ}\left(\xi_y\frac{\partial\varphi}{\partial\xi}+\eta_y\frac{\partial\varphi}{\partial\eta}\right)=$$

$$=\frac{\partial}{\partial\xi}\left(D_e\frac{\xi_x^2+\xi_y^2}{J}\frac{\partial n_e}{\partial\xi}\right)+\frac{\partial}{\partial\eta}\left(D_e\frac{\eta_x^2+\eta_y^2}{J}\frac{\partial n_e}{\partial\eta}\right)+$$

$$+\frac{\partial}{\partial\xi}\left(D_e\frac{\xi_x\eta_x+\xi_y\eta_y}{J}\frac{\partial n_e}{\partial\eta}\right)+\frac{\partial}{\partial\eta}\left(D_e\frac{\xi_x\eta_x+\xi_y\eta_y}{J}\frac{\partial n}{\partial\xi}\right)+$$

$$+\frac{\chi D_e}{yJ}\left(\xi_y\frac{\partial n_e}{\partial\xi}+\eta_y\frac{\partial n_e}{\partial\eta}\right)+\frac{\alpha}{J}|\mathbf{\Gamma}_e|-\frac{\beta}{J}n_en_+ \tag{15}$$

To simplify this equation, we introduce the notation

$$a=\frac{\eta_x^2+\eta_y^2}{J},\quad b=\frac{\xi_x^2+\xi_y^2}{J},\quad g=\frac{\xi_x\eta_x+\xi_y\eta_y}{J} \tag{16}$$

After this, the equation of continuity for electrons in the new coordinate system takes the form

$$\frac{\partial}{\partial t}\left(\frac{n_e}{J}\right)+\frac{\partial}{\partial\xi}\left(n_e\mu_e b\frac{\partial\varphi}{\partial\xi}\right)+\frac{\partial}{\partial\eta}\left(n_e\mu_e a\frac{\partial\varphi}{\partial\eta}\right)+\frac{\partial}{\partial\xi}\left(n_e\mu_e g\frac{\partial\varphi}{\partial\eta}\right)+$$

$$+\frac{\partial}{\partial\eta}\left(n_e\mu_e g\frac{\partial\varphi}{\partial\xi}\right)+\frac{n_e\mu_e}{yJ}\left(\xi_y\frac{\partial\varphi}{\partial\xi}+\eta_y\frac{\partial\varphi}{\partial\eta}\right)=$$

$$=\frac{\partial}{\partial\xi}\left(D_e b\frac{\partial n_e}{\partial\xi}\right)+\frac{\partial}{\partial\eta}\left(D_e a\frac{\partial n_e}{\partial\eta}\right)+\frac{\partial}{\partial\xi}\left(D_e g\frac{\partial n_e}{\partial\eta}\right)+\frac{\partial}{\partial\eta}\left(D_e g\frac{\partial n_e}{\partial\xi}\right)+$$

$$+\frac{\chi D_e}{yJ}\left(\xi_y\frac{\partial n_e}{\partial\xi}+\eta_y\frac{\partial n_e}{\partial\eta}\right)+\frac{\alpha}{J}|\mathbf{\Gamma}_e|-\frac{\beta}{J}n_en_+ \tag{17}$$

We note that, in spite of the considerable complication of the equation in introduced coordinate system (ξ,η), the construction of the finite-difference scheme for (17) proves to be a simpler problem than for the initial equation (2).

For the purpose of constructing a finite-difference scheme, we use the grid template shown in figure 2. To obtain the finite-difference relations by the finite-volume method the integrating of equation (9) is used on the volume selected in figure 2

$$Vol_{i,j} = \int_{\eta_{i-1/2}}^{\eta_{i+1/2}} \mathrm{d}\eta \int_{\xi_{j-1/2}}^{\xi_{j+1/2}} \{\ \} \mathrm{d}\xi = p_i q_j, \tag{18}$$

where
$$p_i = \frac{1}{2}(\eta_{i+1} - \eta_{i-1}), \quad q_j = \frac{1}{2}(\xi_{j+1} - \xi_{j-1}) \tag{19}$$

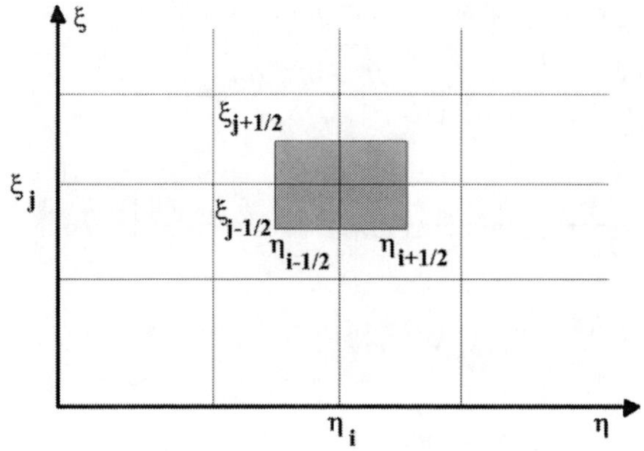

Figure 2. Difference grid template.

This leads to the five-point finite-difference equation of the following form

$$A_{i,j}U_{i-1,j} + B_{i,j}U_{i+1,j} - C_{i,j}U_{i,j} + \overline{A}_{i,j}U_{i,j-1} + \overline{B}_{i,j}U_{i,j+1} + F_{i,j} = 0 \tag{20}$$

where
$$U_{i,j} = \varphi_{i,j},$$

$$A_{i,j} = \frac{a_{i-1/2,j} + g_{i-1/2,j}}{p_i p_i^-}, \quad B_{i,j} = \frac{a_{i+1/2,j} + g_{i+1/2,j}}{p_i p_i^+} + \chi\left(\frac{\eta_y}{yJ}\right)_{i,j}\frac{1}{p_i},$$

$$\overline{A}_{i,j} = \frac{b_{i,j-1/2} + g_{i,j-1/2}}{q_j q_j^{-1}}, \quad \overline{B}_{i,j} = \frac{b_{i,j+1/2} + g_{i,j+1/2}}{q_j q_j^+} + \chi\left(\frac{\xi_y}{yJ}\right)_{i,j}\frac{1}{q_j},$$

$$F_{i,j} = \frac{\varepsilon^*}{J_{i,j}}(N_+ - N_-)\frac{N_0 L^2}{\varepsilon} \tag{21}$$

For the convenience of numerical implementation, the functions and variables are dimensionless as follows

$$N_+ = \frac{n_+}{N_0},\ N_e = \frac{n_e}{N_0},\ U = \frac{\varphi}{\varepsilon},\ X = \frac{x}{L},\ Y = \frac{y}{L},$$

where N_0 is the characteristic concentration of electrons in the positive column of the glow discharge ($N_0 = 10^9\ \mathrm{cm}^{-3}$); L is the distance between the electrodes.

In (20), (21) we used the variables

$$a_{i\pm1/2,j} = \frac{1}{2}(a_{i,j} + a_{i\pm1,j}), \quad b_{i,j\pm1/2} = \frac{1}{2}(b_{i,j} + b_{i,j\pm1}),$$

$$g_{i\pm1/2,j}=\frac{1}{2}\left(g_{i,j}+g_{i\pm1,j}\right),\quad g_{i,j\pm1/2}=\frac{1}{2}\left(g_{i,j}+g_{i,j\pm1}\right),$$

$$p_i^+=\eta_{i+1}-\eta_i,\quad p_i^-=\eta_i-\eta_{i-1},\quad q_j^+=\xi_{j+1}-\xi_j,\quad q_j^-=\xi_j-\xi_{j-1}$$

When obtaining finite-difference equations for electron concentrations, the continuity equation (17) is integrated according to the volume shaded in figure 2 and over a time interval $\tau=t^{p+1}-t^p$

$$Vol_{i,j}^{p,p+1}=\int_{t_p}^{t_{p+1}}\mathrm{d}t\int_{\eta_{i-1/2}}^{\eta_{i+1/2}}\mathrm{d}\eta\int_{\xi_{j-1/2}}^{\xi_{j+1/2}}\{\ \}\mathrm{d}\xi=\tau p_i q_j.$$

This leads to a five-point equation of the form (20), where

$$U_{i,j}=n_{e,i,j},$$

$$A_{i,j}=\frac{R_L^+}{p_i}+\frac{\left[D_e(a+g)\right]_{i-1/2,j}}{p_i p_i^-},\quad B_{i,j}=-\frac{R_R^-}{p_i}+\frac{\left[D_e(a+g)\right]_{i+1/2,j}}{p_i p_i^+}+\frac{\chi}{p_i}\left(\frac{\eta_y D_e}{yJ}\right)_{i,j},$$

$$C_{i,j}=\frac{1}{\tau J_{i,j}}+A_{i,j}+B_{i,j}+\overline{A}_{i,j}+\overline{B}_{i,j}+\left(\frac{\mu_e\eta_y}{yJ}\right)_{i,j}\frac{\varphi_{i+1,j}-\varphi_{i-1,j}}{2p_i}+\left(\frac{\mu_e\xi_y}{yJ}\right)_{i,j}\frac{\varphi_{i,j+1}-\varphi_{i,j-1}}{2q_j},$$

$$\overline{A}_{i,j}=\frac{S_L^+}{q_j}+\frac{\left[D_e(b+g)\right]_{i,j-1/2}}{q_j q_j^-},\quad \overline{B}_{i,j}=-\frac{S_R^-}{q_j}+\frac{\left[D_e(b+g)\right]_{i,j+1/2}}{q_j q_j^+}+\frac{\chi}{q_j}\left(\frac{\xi_y D_e}{yJ}\right)_{i,j},$$

$$F_{i,j}=\frac{U_{e,i,j}^p}{\tau J_{i,j}}+\dot{\omega}_e,\quad \dot{\omega}_e=\left(\frac{\alpha}{J}\left|\boldsymbol{\Gamma}_e\right|-\frac{\beta}{J}n_e n_+\right)\frac{L^2}{\varepsilon\mu_e N_0}\tag{22}$$

Here we use the following notation for the projection of the electron velocity in the coordinate system (ξ,η)

$$R_e=V_e+Z_e,\quad S_e=V_e+W_e,\quad V_e=\mu_e a\frac{\partial\varphi}{\partial\eta},\quad U_e=\mu_e b\frac{\partial\varphi}{\partial\xi},\quad Z_e=\mu_e q\frac{\partial\varphi}{\partial\xi},\quad W_e=\mu_e g\frac{\partial\varphi}{\partial\eta}$$

Note that the velocity components V_e and U_e correspond to the drift velocities along the coordinate lines, and the velocity components Z_e and W_e correspond to the contribution of the corresponding cross components. The finite-difference equation for ions is determined by the same relationships with the difference that

$$V_+=-\mu_e a\frac{\partial\varphi}{\partial\eta},\quad U_+=-\mu_e b\frac{\partial\varphi}{\partial\xi},\quad Z_+=-\mu_e q\frac{\partial\varphi}{\partial\xi},\quad W_+=-\mu_e g\frac{\partial\varphi}{\partial\eta}$$

in accordance with the definition of charged particle flux densities (5).

Stability in the numerical integration of finite-difference equations of continuity of electrons and ions is provided by the following relationships:

$$R_{R,L}^{\pm}=\frac{1}{2}\left(R_{R,L}\pm\left|R_{R,L}\right|\right),\quad S_{R,L}^{\pm}=\frac{1}{2}\left(S_{R,L}\pm\left|S_{R,L}\right|\right)$$

and also the one-sided derivatives of the drift and diffusion terms in (21) and (22), proportional to $1/(yJ)$.

4. Results of numerical modelling of a normal glow discharge

Based on the calculation model described in the previous section, a computer code was created to calculate the two-dimensional structure of a normal glow discharge. Trial test calculations were performed using the example of a well-studied normal glow discharge in molecular nitrogen between plane electrodes at $p = 5$ Torr, $\varepsilon = 1000$ V, $R_0 = 300$ kΩ, $\gamma = 0.33$, $L = 2$ cm [2]. Two-dimensional fields of concentrations of ions and electrons for axisymmetric ($\chi = 1$) and flat ($\chi = 0$) glow discharges are shown in figure 3. The data presented are in good agreement with the results of calculations [2]. The calculations were performed on a 71×71 grid. In these figures, the cathode and anodic layers are well identified, as well as the quasi-neutral positive column.

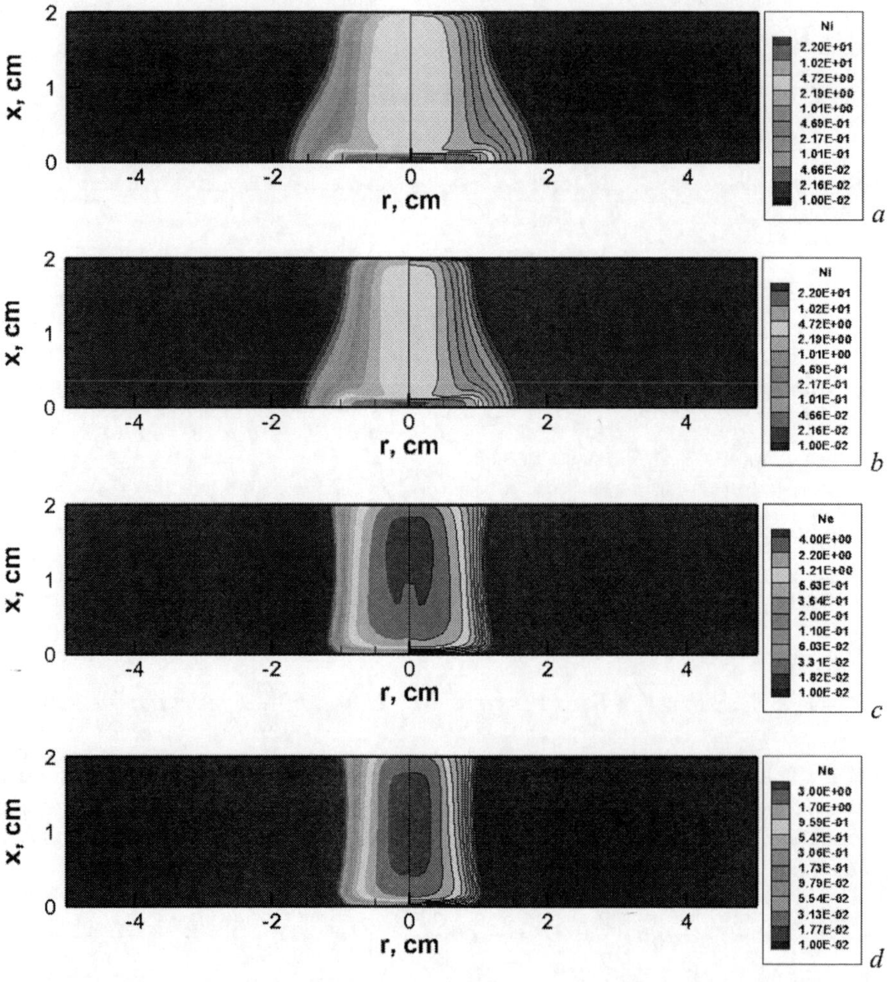

Figure 3. Distribution of the concentrations of ions (*a, b*) and electrons (*c, d*) in an axisymmetric (a, c) and flat (*b, d*) discharges between plane electrodes at $p = 5$ Torr, $\varepsilon = 1000$ V, $R_0 = 300$ kΩ, $\gamma = 0.33$, $L = 2$ cm.

Figures 4 and 5 show the axial distributions of electron and ion concentrations along the axis of cylindrical and planar discharges, as well as ionization and recombination rates. The latter are normalized by $N_0\mu_e\varepsilon/L^2$, having the dimension 1/(cm^3×s). The concentrations of ions in the cathode layer for axisymmetric and flat discharges are close, and the concentration of charged particles in the positive column of the axisymmetric discharge is somewhat higher. It is seen from figure 5 that the cathode and anode layers are the main regions of generation of charged particles.

Calculations of an axisymmetric normal glow discharge line between curvilinear electrodes were performed for similar conditions discussed above, but in the range $\varepsilon = 1000 \div 1750$ V. The radius of curvature of the cathode and the anode were $R_c = 5$ cm and $R_a = 10$ cm, respectively. Configuration of the computational domain with the number of nodes 141 × 141 is shown in figure 1.

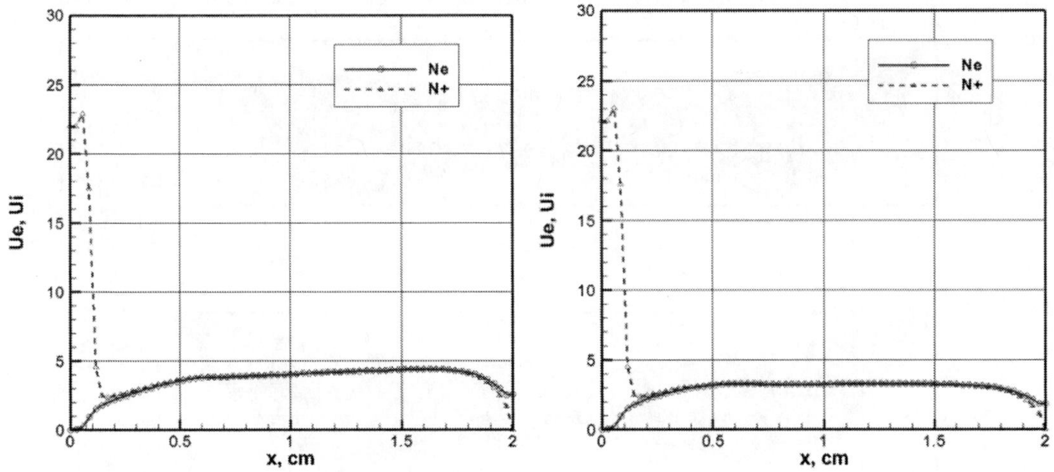

a *b*

Figure 4. The distribution of electron and ion concentrations along the symmetry axis (*a*) and the symmetry plane (*b*) for the conditions in the discharge in figure 3.

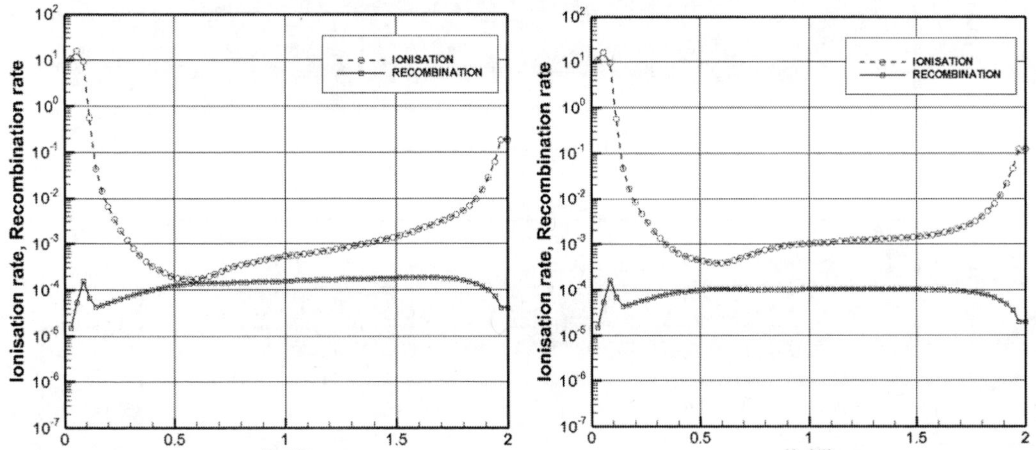

a *b*

Figure 5. Distribution of ionization and recombination rates along the symmetry axis (*a*) and the symmetry plane (*b*) for the conditions in the discharge in figure 3.

Figures 6 and 7 show the fields of ion and electron concentrations, electric potential and reduced field E/p in discharges at $\varepsilon = 1000$ V and $\varepsilon = 1750$ V, respectively. With increasing ε, the transverse dimensions of the glow discharge column increase.

Also, as in the case of a discharge between plane electrodes, the cathode layer (region of increased ion concentration) is characterized by a potential well of the electric potential (figures 6, *c* and 7, *c*) and increased values of the reduced field (figures 6, *d* and 7, *d*).

The axial distributions of the concentrations of charged particles in the two glow discharges, the distribution of the electric potential φ of the axial component of the electric field and the reduced field are shown in figures 8-9. Figures 10, *a*, *b* show the rates of ionization and recombination. We note the main differences from the normal glow discharge between flat electrodes. In the axial distribution of electron and ion concentrations, in a positive column, a local maximum appears.

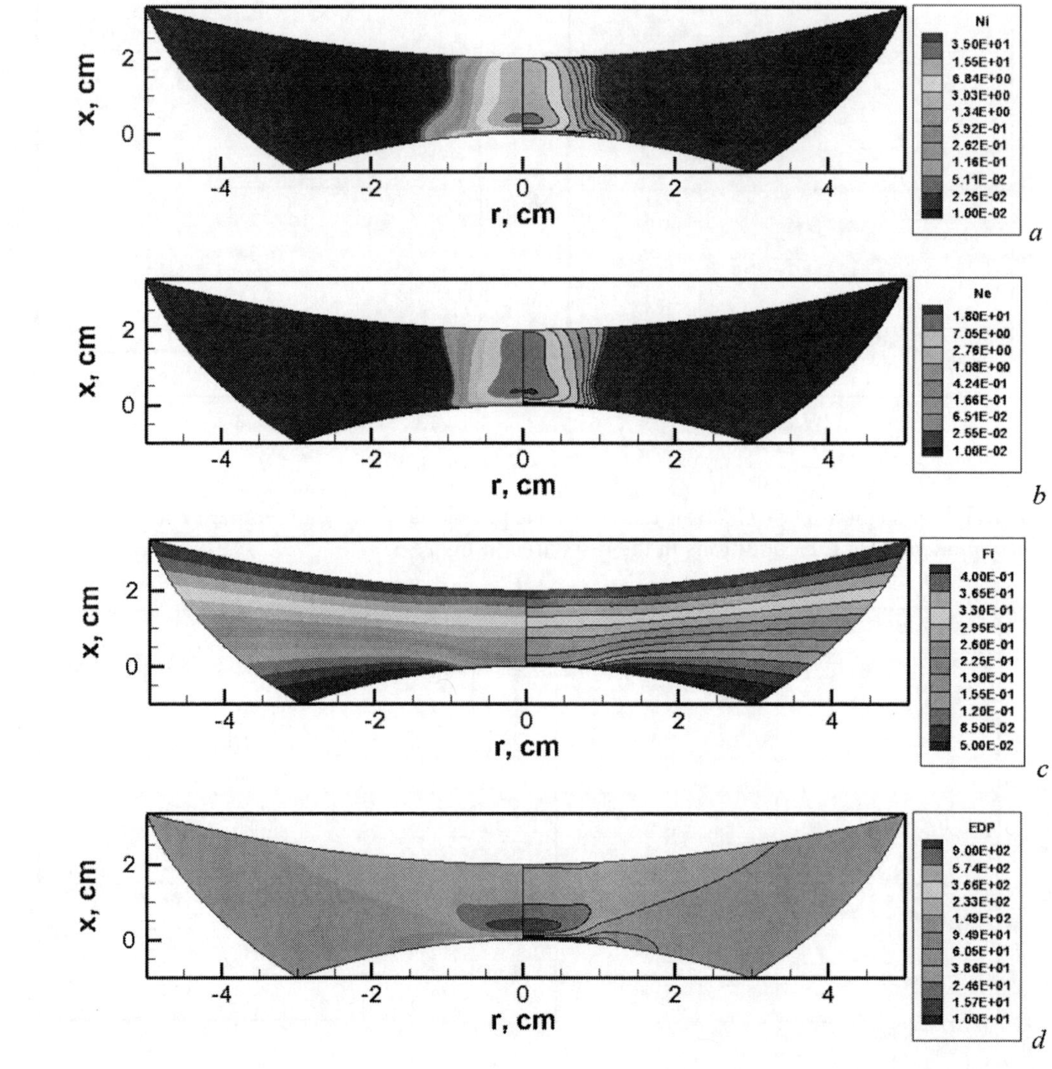

Figure 6. Distribution of concentrations of ions $N_i = n_+/N_0$ (*a*) and electrons $N_e = n_e/N_0$ (*b*), electric potential (*c*) and reduced field $EDP = E/p$ in V /cm×Torr (*d*) between segmental electrodes at $p = 5$ Torr, $\varepsilon = 1000$ V, $R_0 = 300$ kΩ, $\gamma = 0.33$, $L = 2$ cm.

An analysis of the distributions of the rates of ionization and recombination (figure 10, *a*, *b*) shows that recombination prevails in the vicinity of this local maximum. The rate of ionization in the cathode layer is several times greater than the corresponding value for the discharge between plane electrodes (compare figures 10, *a* and 5, *a*). With increasing ε, the concentration of ions and electrons in the positive column increases (figure 8, *a*, *c*).

An important result of the preliminary study of the structure of a normal glow discharge between curved electrodes was the demonstration of a satisfactory fulfillment of the law of the normal current density in figure 11. It can be seen that variation of ε in the range of $1000 \div 1750$ V resulted in an insignificant change in the current density at the cathode and anode.

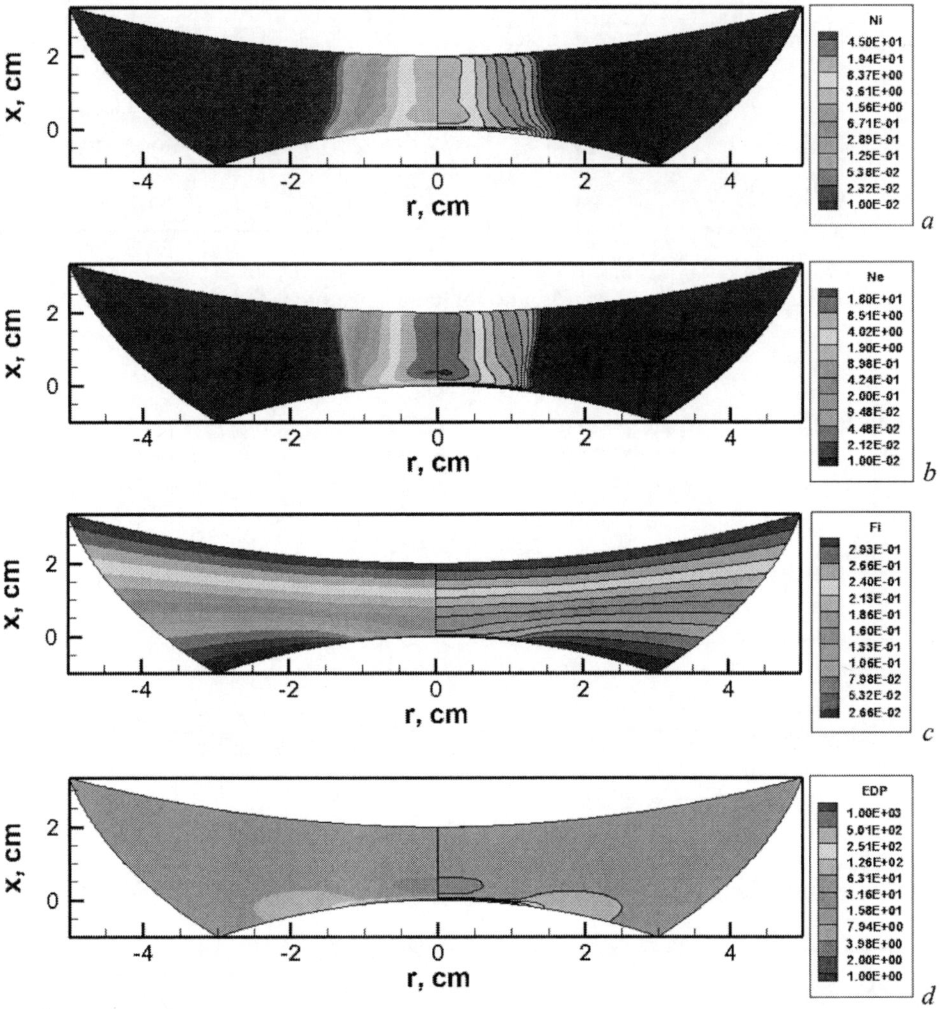

Figure 7. Distribution of concentrations of ions $N_i = n_+/N_0$ (*a*) and electrons $N_e = n_e/N_0$ (*b*), electric potential (*c*) and reduced field $EDP = E/p$ in V /cm×Torr (*d*) between segmental electrodes at $p = 5$ Torr, $\varepsilon = 1750$ V, $R_0 = 300$ kΩ, $\gamma = 0.33$, $L = 2$ cm.

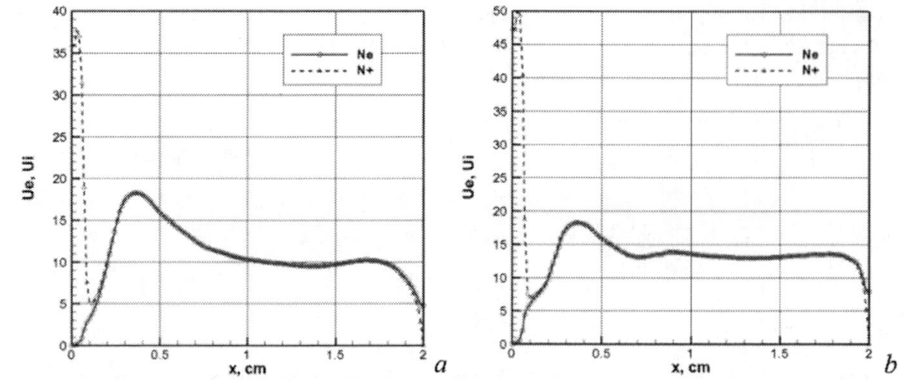

Figure 8. The distribution of electron and ion concentrations along the symmetry axis (*a*) and the symmetry plane (*b*) for $\varepsilon = 1000$ V (*a*) и $\varepsilon = 1750$ V (*b*).

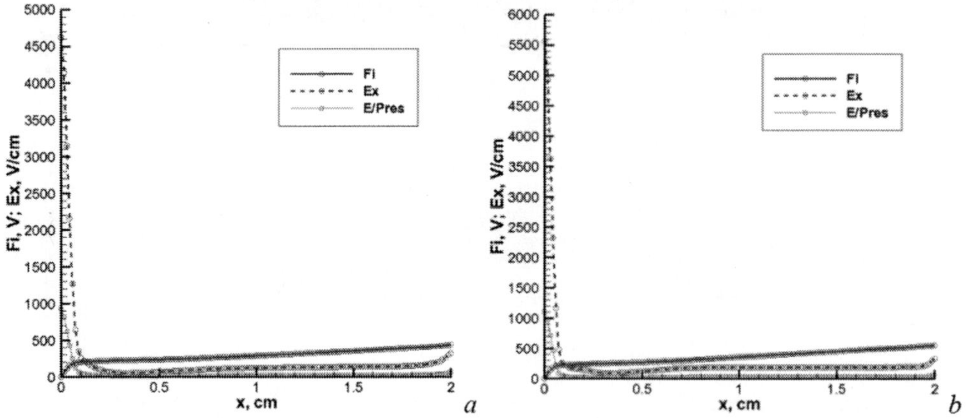

Figure 9. The axial distributions of the electric potential $Fi = \varphi$ (in V), the axial component of the electric field strength E_x, V/cm, and the reduced field E/p (in V/cm×Torr) along the symmetry axis at $\varepsilon = 1000$ V (*a*) and $\varepsilon = 1750$ V (*b*).

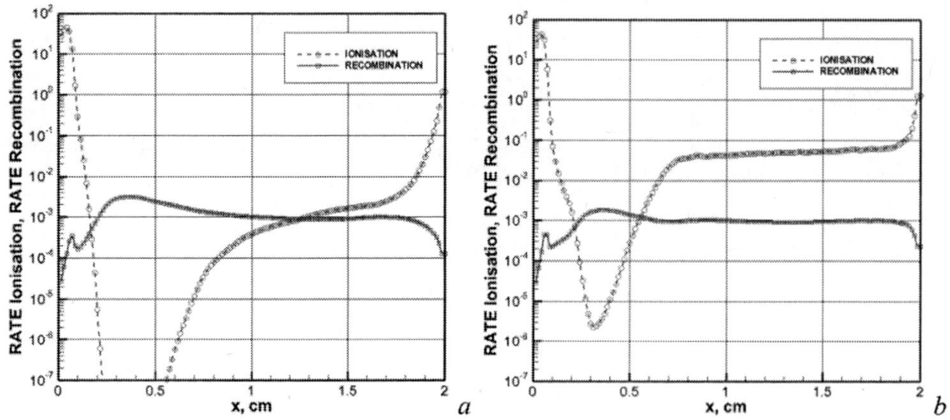

Figure 10. Axial distributions of the rate of ionization and recombination at $\varepsilon = 1000$ V (*a*) and $\varepsilon = 1750$ V (*b*), referred to the value $N_0 \mu_e \varepsilon / L^2$.

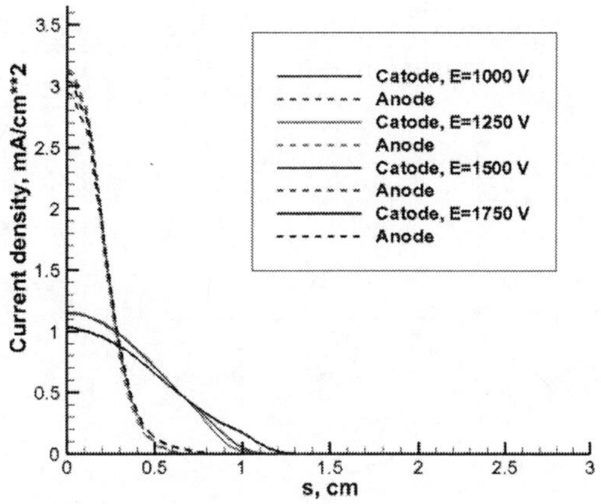

Figure 11. Current density at the cathode and anode as ε varies in the range $\varepsilon = 1000 \div 1750\,\text{V}$.

Conclusion

A diffusion-drift calculation model of a normal glow discharge between curvilinear electrodes is presented.

The verification of the created computational model is performed using the example of a normal glow discharge between flat electrodes.

Results of numerical simulation are presented of normal glow discharge between electrodes in the form of spherical segments with radii $R_c = 5\,\text{cm}$ and $R_a = 10$ cm at $p = 5$ Torr, $\varepsilon = 1000 \div 1750\,\text{V}$, $R_0 = 300\,\text{k}\Omega$, $\gamma = 0.33$, distance between electrodes $L = 2$ cm.

A satisfactory fulfillment of the law of the normal current density for the investigated discharge is shown.

Acknowledgments

The work was supported by the Russian Science Foundation grant # 16-11-10275.

References

[1] Raizer Yu P 1991 *Gas Discharge Physics* (Springer-Verlag) 449 p

[2] Surzhikov S T 2013 *Computational Physics of Electric Discharges in Gas Flows* Walter de Gruyter GmbH (Berlin/Boston) 428 p

[3] Engel A Von, Steenbeck M 1932 *Elektrische Gasentladungen* (Berlin:Springer)

[4] Gladush G G and Samokhin A A 1981 A numerical investigation of current lacing on the electrodes in a glow discharge *J. of Applied Mechanics and Technical Physics* **22** Issue 5 pp 608–615

[5] Petrusev A S, Surzhikov S T and Shang J S 2006 Chemical processes in air glow discharge for aerospace applications *AIAA Paper 06-1460* 11 p

[6] Petrusev A S, Surzhikov S T and Shang J S 2006 A two-dimensional model of glow discharge in view of vibrational excitation of molecular nitrogen *High Temperature* **44** (6) pp 814–822

[7] Surzhikov S T and Shang J S 2004 Two-component plasma model for two-dimensional glow discharge in magnetic field *J. of Computational Physics* **199** pp 437–464

IOP Conf. Series: Journal of Physics: Conf. Series **1009** (2018) 012018 doi:10.1088/1742-6596/1009/1/012018

[8] Kotov M A, Kozlov P V, Ruleva L B, Solodovnikov S I, Surzhikov S T and Tovstonog V A 2017 The spectral characteristic investigations of normal glow discharge *J. of Physics: Conf. Series* **815** 012026

IOP Conf. Series: Journal of Physics: Conf. Series **1009** (2018) 012019 doi:10.1088/1742-6596/1009/1/012019

Determination of the direction of the interstellar medium motion relative to the Sun by measurements of interstellar pickup ion helium

S V Chalov
Ishlinsky Institute for Problems in Mechanics RAS, Moscow,119526, Pr. Vernadskogo 101-1

E-mail: chalov@ipmnet.ru

Abstract. One of the main methods to measure the direction of the interstellar medium flow relative to the Sun is to measure parameters of interstellar pickup helium atoms at the Earth orbit or at its nearest neighbourhood. Pickup ions are formed in the interplanetary medium through photoionization of interstellar helium atoms. Charge exchange in the region of interaction of the solar wind with interstellar medium no effect on distribution of the atoms. Thus, the helium ions in the inner regions of the heliosphere carry information about the interstellar medium, and measurements of charged particles on spacecraft are much simpler than neutral atoms directly. Here, however, when one try to determine the direction of the interstellar flow exactly, some problems of "transport" nature can occur. These problems can sometimes result in unexpected "discoveries". All these aspects are discussed in the paper.

Introduction

The local interstellar medium (LISM) surrounding the Sun is partially ionized gas, which according to current knowledge moves with the relative speed of about 26.3 km s^{-1}. The main LISM components are hydrogen atoms, while the number density of protons is at least twice less. The collision of the supersonic solar wind with the interstellar flow gives rise to an interaction region, which includes the internal (termination) and external shocks at which the solar wind and the interstellar medium, respectively, decelerate, and the heliopause that separates the charged components of the both medium. It charged components responsible for the formation of the shock waves since mean free paths of the interstellar atoms are comparable to the size of the heliosphere and it is determined mainly by charge-exchange between hydrogen atoms and protons. The charge-exchange has a great influence on the structure of the interaction region and on physical processes in the outer heliosphere. For instance, as a result of the charge-exchange, the supersonic solar wind decelerates and heats up, the termination shock and the heliopause approach the Sun [1]. In addition, a specific very "hot" population of interstellar pickup ions is formed in the heliosphere, which in many respects changes thermodynamic properties of the solar wind [2].

The region inside the heliopause is a kind of a filter for the interstellar atoms with a large charge-exchange cross section (hydrogen, oxygen). But the interstellar helium atoms penetrate almost freely into the inner heliosphere since the charge-exchange process can be ignored. Thus, measurements of atomic helium parameters on spacecraft near the Earth's orbit can give important information about "pure" LISM, without invoking complex multidimensional numerical models of the heliosphere.

The only force that affects the motion of helium atoms is the gravitational attraction of the sun, since the radiation pressure is negligible. The clear manifestation of the solar attraction is the gravitational focusing of helium atom. The focusing results if formation of a region of enhanced atomic density of helium. This region is located on the downwind side of the Sun and is called the gravitational focusing cone (figure 1).

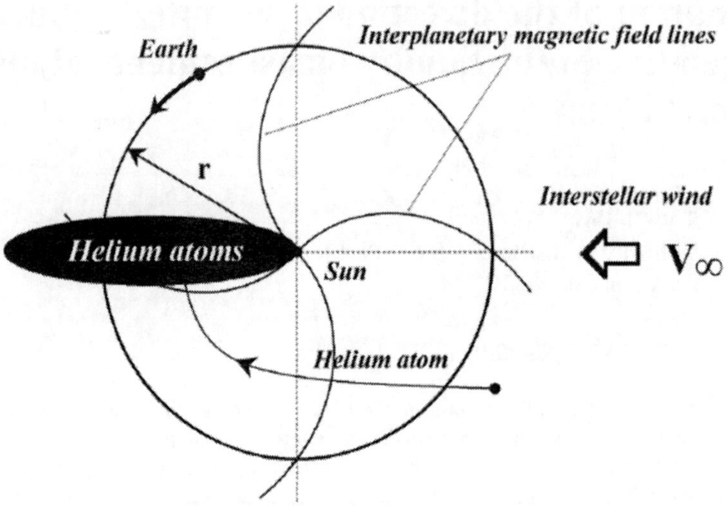

Figure 1. Gravitational helium cone is shown schematically.in the ecliptic plane.

In the inner heliosphere helium atoms undergo photoionization. As a result, helium ions (pickup ions) are formed. Usually it is once ionized atoms He+. Since the local production rate of pickup ions is proportional to the atomic density, in addition to the gravitational cone of helium atoms, a similar structure consisting of pickup helium ions must exist. One can expect that the positions of the density maxima and the geometric sizes for the two structures are identical. However, as was pointed out in [3], the deviation of the pickup-ion velocity distribution from an isotropic one can cause a relative angular displacement between the charged and neutral helium cones. Below we discuss this effect in more detail.

Pickup ion transport

1.1. Spatial distribution of helium atoms

In general, numerical models are needed to describe precisely the parameters of the helium atoms through the heliosphere. For some purpose, however, it is sufficient to use simple analytical models. The simplest one is the so-called cold model in the frame of which interstellar atoms are considered as a flow of neutral particles with zero temperature. Up to the moment of ionization atoms follow planar hyperbolic orbits under the action of gravitational attraction. The spatial distribution of the atoms in the interstellar flow $\mathbf{V}_\infty$. If Φ is the off-axis angle then the impact parameters identifying a direct and indirect orbit which pass through the point r, Φ:

$$p_{1,2} = \left| \left[\left(\frac{1}{2} r \sin\varphi \right)^2 + \frac{r}{C}(1+\cos\varphi) \right]^{1/2} \pm \frac{1}{2} r \sin\varphi \right|. \quad (1)$$

In the frame of the cold model, the number density of interstellar helium atoms in the heliocentric system of coordinates is given by [4]

$$\frac{n(r,\varphi)}{n_\infty}=\frac{\left[(r\sin\varphi)^2+4r(1+\cos\varphi)/C\right]^{-1/2}}{r\sin\varphi}\left\{p_1^2\exp\left[-\frac{r_E^2P_{loss}(r_E)(\pi-\varphi)}{V_\infty p_1}\right]+\right.$$
$$\left.+p_2^2\exp\left[-\frac{r_E^2P_{loss}(r_E)(\pi+\varphi)}{V_\infty p_2}\right]\right\}, \qquad (2)$$

where n_∞ is the helium number density in the LISM, $P_{loss}(r_E)$ is the ionization rate at 1 AU, $C=V_\infty^2/GM$, G is the gravitational constant, M is the mass of the Sun.

Equation (2) gives the spatial distribution of helium atoms rather precisely, however, for positions far enough away from the downwind focusing cone only. It is not correct near the cone since it gives an infinitely large density of particles at its axis. This effect is connected with the fact that the thermal velocities of atoms were not taken into account. In [5] a modified cold model for helium atoms in the heliosphere was proposed. The model takes into account thermal spreading of trajectories near the axis of the cone. In the frame of the modified model the number density is given as before by equation (2) at $\phi>\phi_c$, where ϕ_c is the half angle of the focusing cone:

$$\varphi_c=\frac{2}{3}\left(\frac{2k_BT_\infty}{\pi mV_\infty^2}\right)^{1/2}. \qquad (3)$$

Here T_∞ is the temperature helium in the LISM, m is the mass of the atom, and k_B is the Boltzmann constant. For angles $\varphi\le\varphi_c$, assuming Maxwellian velocity distribution atoms in the LISM, authors of [5] obtained

$$\frac{n(r,\varphi)}{n_\infty}=\left(\frac{\pi GMm}{rk_BT_\infty}\right)^{1/2}\exp\left[-\frac{\pi r_E^2P_{loss}(r_E)}{(2GMr)^{1/2}}\right](1-D\varphi^2), \qquad (4)$$

where

$$D=\frac{2\pi mV_\infty^2}{27k_BT_\infty}. \qquad (5)$$

1.2. Properties of helium pickup ions

Interaction of the pickup ions with solar wind electromagnetic fluctuations results in pitch-angle scattering. At the initial moment in the solar wind rest frame the motion of pickup ions originating in the region of the neutral helium cone is the of gyration and streaming along a magnetic field line towards the Sun. However, in the case of weak scattering this velocity distribution is anisotropic and the bulk of the ions move towards the Sun. This effect leads to a systematic shift (in the ecliptic plane) of the pickup ion cone relative to the LISM velocity vector in the direction of motion of the Earth around the Sun (figure 2). The angle of the shift depends on the mean free path of ions parallel to the magnetic field line $\lambda_{\|}$. At sufficiently low value of $\lambda_{\|}$ when the velocity distribution is almost isotropic, the shift is absent.

All these effects can be described by the general transport equation for velocity distribution function of pickup ions. We will consider helium cone in the ecliptic plane. If V_{SW} is the solar wind speed, $\Omega=2.7\times10^{-6}\,s^{-1}$, components of the solar magnetic field are

$$B_r = B_{r_E}\left(\frac{r_E}{r}\right)^2, \qquad B_\theta = 0, \quad B_\varphi = -B_{r_E}\frac{r_E^2\Omega}{rV_{SW}}, \tag{6}$$

Here we use heliocentric spherical coordinate system (r,θ,φ) with $\theta = \pi/2$ (θ – colatitude).

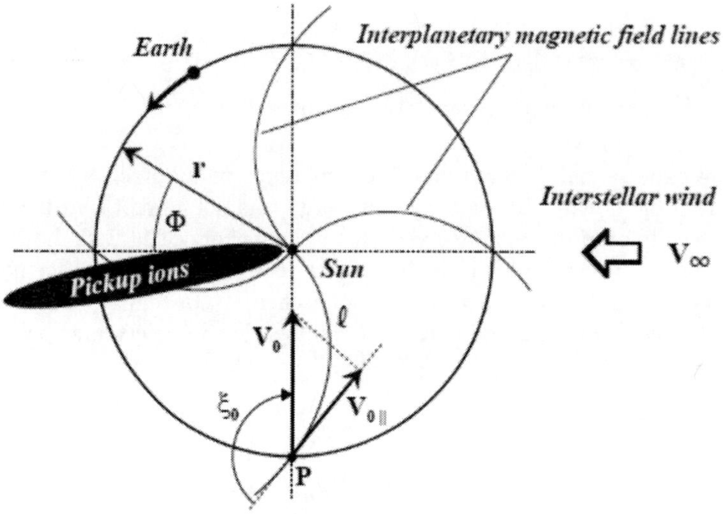

Figure 2. Schematic representation of the He^+ pickup cone in the ecliptic plane. Kinematics of newly created pickup ions is also shown. The initial velocity of a pickup ion in the solar wind rest frame directed towards the Sun. The motion of the pickup ion is a combination of cyclotron rotation around a magnetic field line l and motion along this line with velocity $\mathbf{V}_{0\parallel}$. As a result, pickup ions, on average, move in the same direction as the Earth moves around the Sun.

If we introduce the variable $\chi = \cos\psi$, where $\psi(r)$ is the angle between the vectors $\mathbf{V}_{SW}$ and $\mathbf{B}$, we obtain:

$$b_r = \chi, \qquad b_\varphi = -\mathrm{sign}(\chi)\left(1-\chi^2\right)^{1/2}. \tag{7}$$

Here $\mathbf{b} = \mathbf{B}/B$. Then in the case of the outward pointing magnetic field

$$\chi(r) = \left[1 + (\Omega r / V_{SW})^2\right]^{-1/2} \tag{8}$$

The differential equation for the velocity distribution function of pickup ions in the ecliptic plane can then be written in the following form:

$$\frac{\partial f}{\partial t} + (V_{SW} + v\mu\chi)\frac{\partial f}{\partial r} - \frac{v\mu\left(1-\chi^2\right)^{1/2}}{r}\frac{\partial f}{\partial \phi} + \left[\frac{1-3\mu^2}{2}\frac{1-\chi^2}{r} - \frac{1-\mu^2}{r}\right]V_{SW}v\frac{\partial f}{\partial v} + \\ + \frac{1-\mu^2}{2}\left[\frac{v}{r^2}\frac{\mathrm{d}}{\mathrm{d}r}\left(r^2\chi\right) + \frac{2V_{SW}}{r}\mu - 3\mu V_{SW}\frac{1-\chi^2}{r}\right]\frac{\partial f}{\partial \mu} = \hat{S}f + Q(t,r,\phi,v,\mu). \tag{9}$$

In equation (7) v and $\mu = \cos\xi$ are the speed and the cosine of the ion pitch angle ξ in the solar wind rest frame. Note that equation (9) written for nonstationary case to make it easier to imagine the general form of this equation, while we consider everywhere here only stationary solutions. The right hand side of the equation contains the scattering operator $\hat{S}$ applied to function f and the source term, which describes the influx of freshly ionized particles. The scattering operator describes pitch-angle scattering of pickup ions by solar wind fluctuations and velocity diffusion (see details in [6]).

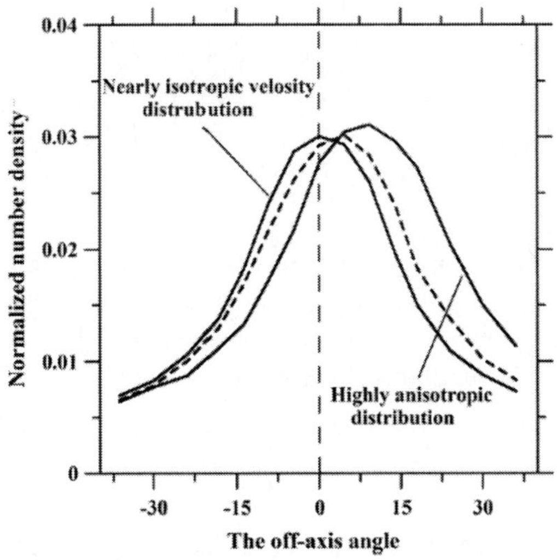

Figure 3. Spatial distributions of pickup ions in the helium focusing cone near the axis of symmetry in the ecliptic plane at 1AU. Different curves correspond to different values of the parallel mean free path: 0.1 (nearly isotropic velocity distribution), 0.3 and 1 AU (highly anisotropic distribution).

Numerical results

The authors of [7] have analyzed data from various spacecraft at different times on the interstellar wind flowing through the heliosphere. They concluded that temporal changes in the ecliptic longitude of the flow direction with time statistically indicated by the data available in the refereed literature at the time of paper writing. The data based on the measurements of pickup helium focusing cone in 2000 with the ACE and in 2010 with the STEREO spacecraft. For these events, the shift cone in longitude is a few degrees in the direction of increasing. The authors explain this shift by the turbulent nature of the interstellar cloud that surrounds the heliosphere. There are, however, another more natural way of explanation of the shift. In 2000, solar activity was near maximum, while in 2010 the spacecraft was in extended solar minimum conditions, beginning in 2007. The transport of charged particles, as mentioned earlier, essentially depends on the solar activity. Near maximum, the level of electromagnetic fluctuations in the solar wind is high, pitch-angle scattering of ions is very effective and, as a result, the mean free path of these ions is low. Under minimum solar wind conditions, the picture is opposite.

Figure 3 shows the spatial distributions of pickup ions in the helium focusing cone near the axis of symmetry in the ecliptic plane at 1 AU. The distributions were obtained by solving equation (9) for different values of $\lambda_{\|}$. The mean free path depends on the level of fluctuations in the solar wind, which determine the value of the scattering operator $\hat{S}$ (see details in [6]). For small value of $\lambda_{\|}$, the velocity distribution function of pickup ions is close to isotropy and positions of the neutral helium and pickup cone axis are coincides (no shift). During solar minimum (quiet solar wind) $\lambda_{\|}$ is large and the velocity distribution function is highly anisotropic. In this case one can expect a large angular shift of the pickup helium cone axis relative to neutral helium cone axis in the direction of the Earth's movement around the Sun.

Figure 4 shows the same as figure 3 but at different distances from the Sun; 0.3, 0.5 and 1 AU. In addition, the figures differ in the scale on the x-axis. Nearly isotropic distributions with $\lambda_{\|}$ = 0.1 AU (dashed curves) and highly anisotropic ones with $\lambda_{\|} = 1$ AU are shown. Number density of pickup ions increases with approach to the sun due to the increase of the neutral helium number and increasing the frequency of photoionization.

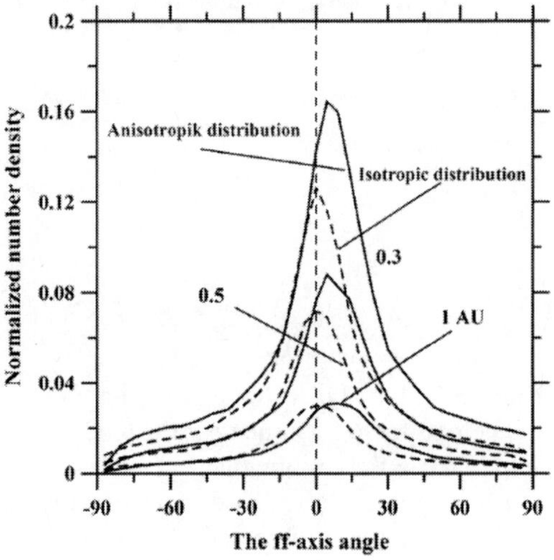

Figure 4. Spatial distributions of pickup ions in the helium focusing cone near the axis of symmetry in the ecliptic plane at 0.3, 0.5 and 1 AU. Different values of $\lambda_{\|}$ are used: 0.1 (dashed curves) and (solid curves).

Conclusions

We demonstrate in the present paper that attempts to use focusing cone consisting of the helium pickup ions should be performed with great caution. The fact that the axis of symmetry of this cone is different from the axis of symmetry of the atomic cone, which determines the direction of the interstellar wind. Due to transport effects pickup ion cone shift relative to atomic cone in the direction of the Earth's movement around the Sun. The value of the shift is several degrees and depends on the solar activity.

Acknowledgments

The work was supported in part by RFBR grant # 16–01–00305.

References

[1] Baranov V B and Malama Y G 1993 *J. Geophys. Res.* **98** 15157–15165
[2] Isenberg P A 1987 *J.Geophys. Res.* **92** 1067–1073
[3] Möbius E, Rucinski D, Hovestsdt D and Klecker B 1995 *Astron. Astrophys.* **301** 505–519
[4] Blum P W and Fahr H J 1970 *Astron. Astrophys.* **4** 280
[5] Feldman W C, Lange J J and Scherb F 1972 *Interstellar helium in interplanetary space, in the solar wind* Ed C P Sonnett, P J Coleman Jr, J M Wilcox (NASA SP–308) p 684
[6] Chalov S and Fahr H J 1999 *Solar Phys.* **187** 123–144
[7] Frisch P C, Bzowski M, Drews C, Leonard T, Livadiotis G, McComas D J, Möbius E, Schwadron N and Sokoll G M 2015 *Astrophys. J.* **801** 61 doi:10.1088/0004-637X/801/1/

IOP Conf. Series: Journal of Physics: Conf. Series **1009** (2018) 012020 doi:10.1088/1742-6596/1009/1/012020

Determination of the Sun's charge by the parameters of heavy ions in the solar wind

P I Vysikaylo and M A Korotkova

Moscow Region State University, Radio street 10A, Moscow, 105005, Russia

E-mail: filvys@yandex.ru

Abstract. As an asymptotic paradox, the presence of a constant uncompensated positive charge of the Sun is investigated. According to the type of ions experimentally detected in the solar wind, the total charge of the Sun is calculated. The formulated model makes it possible to explain the anomalous heating of the solar corona by a constant electric field and the presence in the solar wind of several times ionized heavy ions. Within the framework of this model, the energies of the solar wind particles in the Earth's region are calculated. The calculated parameters of the Sun and the solar wind are compared with the values known from experiments and the proposed mathematical model is thereby verified.

1. Introduction

At the moment, a large amount of data on the ion composition of the solar wind (SW) has been accumulated, and this information is constantly being updated. The following parameters are known: 1) the elements present in the SW; 2) the degree of ionization of the atoms – Z; 3) the mass of the ions - m that form the SW [1] and 4) the SW velocity near the Earth. Usually in mathematical models, the Sun and the SW are considered to be neutral plasma structures [2, 3]. However, the Sun, like any plasma structure that has free electrons in it, is charged with a positive charge [4]. This is due to the greater mobility of electrons with respect to positive ions, which because of their mass have less mobility. The presence of the same charge at the Sun and the ionic components of the SW leads to the appearance of Coulomb repulsive forces, which, with a sufficient charge of the Sun, can compensate gravitational attraction forces and form streams of positively charged ions from the Sun. The charge of the Sun arises and is maintained as a result of the processes of maxwellization of the electron distribution function, therefore in the Sun there will always be electrons possessing a sufficiently high energy in order to leave even the positively charged Sun forever. In this case, the shielding of the positive charge of the Sun at the Debye length is impossible due to the absence of the opposite charge. In the proposed spherically symmetric model, the solar positive charge is not compensated in the entire heliosphere and negative charge in the form of escaped electrons can be concentrated in planets and cosmic dust, thereby conditioning the Coulomb forces of mass cumulation to the Sun. It is known that the Earth is negatively charged and the field from its charge is of the order of 100 V/m on the Earth's surface. The electric field strength outside the Sun does not depend on the nature of the charge distribution over its volume. Since the electromagnetic interaction is stronger than gravitation, even the absence of a small number of electrons is sufficient for the formation of the SW made of protons, alpha particles and even 6-8 times ionized iron atoms. In the proposed model, all processes on the

Sun's surface are considered averaged over time and do not affect the SW at distances more than 3 solar radii.

We believe that the phenomenon of the SW is analogous to the well-known phenomenon of the flow of a charge from a charged needle in air. If a sufficiently high (breakdown) voltage is applied to the charged needle, then an electric or ionic wind arises, caused in the case of a negative charge by the flow of electrons from the needle, their adhesion to air molecules, and the subsequent movement of negative ions in the field of negative charge of the needle. In the case of a positive charge of the needle, the breakdown voltage will ionize the air molecules surrounding the needle and the resulting positive ions will move away from the needle due to the field of positive charge. A similar electric and ionic wind has a sufficient flow strength to deflect the candle flame, which is demonstrated in the known experiment. The same phenomena occur also in the Franklin's wheel, where an electric or ionic wind arising on spaced needles causes the wheel to rotate after a sufficient voltage is applied to the needles. It can be said that there is a kind of jet movement in which the air ions surrounding the needle are a discarded mass.

These considerations make it possible to modify the electric wind model to describe the nature of the SW and to investigate it for an explanation of all the accumulated paradoxes in the motion of the SW against the forces of gravity of the Sun.

According to the foregoing, the anomalous heating of the solar corona to temperatures of the order of a million degrees can be explained within the framework of the proposed model by simple heating of the electrons by the field of the charged Sun, similarly to how electrons are heated in a conventional gas discharge plasma. Moreover, as will be shown below, the minimum ratio of Z/m of iron ions makes it possible to estimate the minimal magnitude of the solar charge. In this case, ions with smaller Z/m fall on the Sun, as the gravitational forces exceed the Coulomb reflection forces, or such ions have not yet been detected experimentally in the SW.

2. Comparison of Coulomb and gravitational forces between different objects

For two identical particles with masses m and charges e, the ratio of the electric potential Ke^2/r to the gravitational potential Gm^2/r is Ke^2/Gm^2. In the case of an electron, this value is equal to $4.17{\cdot}10^{42}$, also known as the Eddington number [5]. If we compare the Coulomb and gravitational forces between a proton and an electron, we get $Ke^2/Gm_pm_e \approx 2.27{\cdot}10^{39}$ (Dirac number) [5]. Here m_p, m_e are the masses of the proton and electron, respectively. However, if we consider different objects of the cosmos, such as planets, stars, etc., rather than elementary particles, one must take into account that the ordinary nuclear matter consists of nucleons (protons and neutrons), which are 1836 times heavier than electrons. Therefore, when estimating the mass of macroworld structures, we can neglect the mass of electrons that leave it.

When comparing the forces of Coulomb repulsion and gravitational attraction between different parts of a charged gravitating structure, according to the laws of Coulomb and Newton, the forces $K{\cdot}(\alpha_{i1}{\cdot}N{\cdot}e){\cdot}(n{\cdot}\alpha_{i2}{\cdot}e)/R^2$ and $G{\cdot}(N{\cdot}m_p){\cdot}(n{\cdot}m_p)/R^2$ should be compared. Here, $K = 1/4\pi\varepsilon_0$, where ε_0 is the electric constant, α_{i1} is the degree of uncompensated bulk charge of protons in the whole structure, N is the average nucleon density in the sphere of the structure of radius R, α_{i2} is the degree of uncompensated bulk charge of protons in the investigated part of the structure, n is the mean density of nucleons in this part of the structure. The degree of uncompensated bulk charge of a structure can be represented as $\alpha_{i1}= n_i/N$, where n_i is the average density of positively charged ions (that are not compensated by electrons) which form a bulk charge of the whole positively charged structure, and N is the average density of the total number of nucleons determining the mass of the structure.

With this approach [4], the Eddington and Dirac numbers are modified to the number:

$$Ke^2/Gm_e^2 = 4.17{\cdot}10^{42} \rightarrow Ke^2/Gm_em_p = 2.27{\cdot}10^{39} \rightarrow K{\cdot}\alpha_{i1}{\cdot}\alpha_{i2}{\cdot}e^2/Gm_p^2 = 1. \quad (1)$$

The condition of equality of the forces of gravity and Coulomb forces leads to instability of the star, galaxy and other structures in a manner analogous to the rotational 2D instability. On the charged gravitating structure and its separated part will act, on the one hand, the forces of gravitational attraction, and on the other hand, the Coulomb repulsive forces (figure 1).

Condition (1) allows us to determine:

$$((\alpha_{i1}\cdot\alpha_{i2})^{0,5})^* = \alpha_i^* = 0.9\cdot10^{-18}. \quad (2)$$

α_i^* is the critical degree of a violation of neutrality (VN) of the structure when it ceases to contract under the action of gravitational forces and, as in the case of rotational instability, parts of a charged quasineutral plasma in the form of a plasma wind begin to be ejected from its surface.

The parameter α_i^* is determined by two parameters: α_{i1} - the VN parameter of the structure as a whole, and α_{i2} - the VN parameter of the element which investigated on the Coulomb instability. An element can have the value α_{i2} = 1, 1/2, 1/3 or less. In the case of the proton, $\alpha_{i2} = 1$ (since the ratio Z/m = 1), the instability will begin at the VN parameter for the entire star $\alpha_{i1} \geq 0.81\cdot10^{-36}$. At these values of α_{i1} and α_{i2}, the value of α_i will reach the critical degree of VN, see relation (2). The VN parameters do not depend on the size and other physical parameters of the structure. Their magnitudes are small, but the electromagnetic interaction is much stronger than the gravitational one. When the star reaches $\alpha_{i1} \geq 0.81\cdot10^{-36}$, for the proton, Coulomb repulsive forces will dominate gravitational attraction for a positively charged star of any type [4].

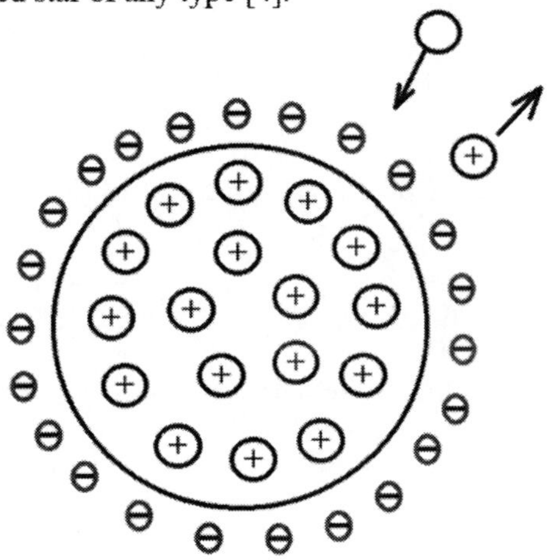

Figure 1. Model of the Sun as a positively charged structure with a weak violation of neutrality.

3. Calculations of the parameters of the Sun and solar wind

As an example of calculating the parameters of stars with allowance for VN, we calculate the positive charge of the Sun, which is necessary for the reflection of protons. It will be further shown that the magnitude of the solar charge in our model directly depends on the ratio Z/m, which takes into account the real ion composition of the solar wind. Knowing the mass of the Sun, and therefore the number of nucleons in the Sun – $1.2\cdot10^{57}$, the radius of the Sun – $R_S = 6.96\cdot10^8$ m and the VN parameter for the proton reflection – $\alpha_{i1} \approx 0.81\cdot10^{-36}$, we find the solar charge and the electric field magnitude on its surface. These quantities will be equal to $q_S = e\cdot\alpha_{i1}\cdot N_S = 154$ C and $E_S = Ke\cdot\alpha_{i1}\cdot N_S/R_S^2 = 2.86\cdot10^{-6}$ V/m, respectively. The number $\alpha_{i1} \approx 0.81\cdot10^{-36}$ is negligible, but a violation of neutrality at this level leads to the formation of a solar wind even without taking into account the generation of Alfven waves.

We shall characterize the energy of charged plasma particles in the chromosphere and solar corona by the parameter *E*/*N*. *E* is the electric field magnitude in the region of the charged Sun, and *N* is the particle density in this region. This parameter introduced A. Stoletov and later used J.S. Townsend in his work on the study of gas-discharge plasma. The reduced electric field, measured in Td (Townsend), *E*/*N* (where *N* is the density of hydrogen atoms or number of nucleons), significantly depends on the density of heavy plasma particles and rapidly grows from the photosphere ($h = 0$) to the chromosphere ($0 < h < 10000$ km). We take the relation of *N* with the height from [6], taking into account that the electron concentration is approximately equal to the concentration of nucleons. On the basis of these data, we calculate the relation of *E/N* with the height (table 1).

Table 1. The values of N_H, cm^{-3} and *E/N*, Td as a function of height above the surface of the Sun.

h [km] / $r[R_S]$	N_H, cm^{-3}	*E/N*, Td	E/N^*, Td	E/N^{**}, Td
0	$3.98{\cdot}10^{15}$	$7.19{\cdot}10^{-7}$	*E/N* x2 or x4	*E/N* x9
1 000	$3.16{\cdot}10^{13}$	$9.05{\cdot}10^{-5}$		
2 000	$6.31{\cdot}10^{12}$	$4.53{\cdot}10^{-4}$		
3 000	$1.99{\cdot}10^{12}$	$1.44{\cdot}10^{-3}$		
4 000	$7.94{\cdot}10^{11}$	$3.6{\cdot}10^{-3}$		
6 000	$2.51{\cdot}10^{9}$	1,14		
8 000	$1{\cdot}10^{9}$	2.86		
10 000	$6.31{\cdot}10^{8}$	4.53		
15 000	$1.99{\cdot}10^{8}$	13.78		
70 000	$7.94{\cdot}10^{7}$	29.75		
280 000	$1.26{\cdot}10^{7}$	115.79		
420 000	$5.01{\cdot}10^{6}$	223.31		
700 000/2	$1.58{\cdot}10^{6}$	452.28		
1 400 000/3	$3.98{\cdot}10^{5}$	798		
2 800 000/5	$6.31{\cdot}10^{4}$	1812		
6 200 000/10	$1{\cdot}10^{4}$	2942		
13 000 000/20	$2.51{\cdot}10^{3}$	2973		26757
44 000 000/65	$1{\cdot}10^{3}$	701		
150 000 000/215	$6.31{\cdot}10^{2}$	97.7		

It can be seen from the table 1 that in the photosphere region, where the concentration of plasma particles is large, the parameter *E/N* is very small, but in the chromosphere this parameter grows rapidly with the height and reaches breakdown values (several tens of Td) in the corona at the height of 15 000 km, which apparently leads to a substantial heating of the electrons already in this region of the positively charged Sun. According to the above model, even in the lower corona, a non-equilibrium plasma is formed in the external electric field of the positively charged Sun with *E/N* values much larger than the breakdown fields in the plasma of completely ionized hydrogen. This generates the SW into outer space.

When analyzing the ion composition of the solar wind, we are interested in the minimum ratio of the ion charge to its mass, Z/m. The smaller this ratio, the greater the positive charge of the Sun for the equality of Coulomb repulsive forces and the gravitational forces of attraction for a given ion. Since helium ions are observed in the SW, this means that all the parameters of the Sun are 2 (if helium is ionized twice, Z/m = 1/2) or 4 times (if helium is ionized 1 time, Z/m = 1/4) greater (see table 1 – *E/N**). Data from the ACE spacecraft [1] for 2011 indicate the presence of the following ions in the composition of the SW: C4+, O5+, Ne8+, Mg6+, Si6+, Fe6+. The last iron ion has a minimum value of Z/m = 0.107 among those observed in the SW. This means, in comparison with the calculations of the effective charge of the Sun for the SW consisting of protons, it is necessary to increase the charge of the Sun by 9 times, so that the heavy iron ions Fe6+ are repelled by Coulomb forces. Such an increase in charge leads to a corresponding change in the electric field magnitude and the parameter *E/N* (see table 1 - *E/N***).

4. Calculation of energy of solar wind particles in the Earth's region

The Coulomb potential energy of a proton born at a distance r from the Sun will be:

$$W = K{\cdot}q_S{\cdot}e/r \quad (3)$$

This potential energy can be completely changed to the kinetic energy of a charged particle far from the Sun if the particle does not experience inelastic collisions that substantially change its kinetic energy. For protons born in the Earth's orbit, $W = 1.5{\cdot}10^{-18}$ J (9.2 eV), where r is the astronomical unit (the distance from the positively charged Sun to the Earth) if the solar charge is 154 K.

From the experimental studies of the SW in the region of the Earth, it is known that the velocity of protons and ions in it reaches from $V = 400$ km/s to 1000 km/s [2]. Consequently, these protons came from a region close to the Sun. Knowing from the experiments the average speed of the solar wind near the Earth - V, we can estimate the kinetic energy of the protons in the solar wind:

$$W = m{\cdot}V^2/2 = 1.34 \cdot 10^{-16} \div 8.36 \cdot 10^{-16}\ \text{J}\ (835 \div 5\ 200\ \text{eV}) \tag{4}$$

Such energy can be obtained by protons born in the corona of the positively charged Sun and accelerated by its charge to such velocities in the region of the Earth. In this case, the mean free path, in which protons flying to the Earth, do not experience collisions with other nucleons, can be estimated from the formula:

$$L = 1/(\sigma N) \tag{5}$$

σ is the effective collision cross section of protons, which for energies of the order of 1 keV is equal to 2×10^{-17} cm^2 [7]. In this case, protons born at distances of the order of 2 800 000 km to 6 000 000 km (5 ÷ 10 solar radii) will arrive to the Earth with a small number of collisions. At these distances, their potential energy at birth is about 3 keV, if the solar charge is of the order of 1000 K, i.e. protons, alpha particles and even six times ionized iron ions are reflected from the Sun (figure 1).

If we consider that the Sun reflects alpha particles, iron ions, then we get an increase in the energy of positively charged particles in the region of the Earth also by two, four and nine times, respectively. Thus, experimental observations in the interplanetary space of the mass and charge compositions of the SW ions already provide valuable information not only about the solar atmosphere, but also about the average parameters of the Sun, such as its total charge, the electric field magnitude in the corona, and make a significant contribution to the verification of mathematical models of processes on the Sun.

Conclusion

For the first time, a specific mechanism is proposed for heating the solar wind plasma in its corona and heliosphere in the constant Coulomb field of the Sun, as a positively charged structure. In this model, we combine such phenomena as electric wind, plasma wind and solar wind into a single phenomenon caused by a violation of neutrality and substantial heating of plasma particles in the electric field of a charged structure (a charged needle, the Sun, etc.). As proved by analytical calculations, having determined in the experiments the ion composition of the solar wind and finding the minimum value of the ratio Z/m, according to the method proposed in this paper, it is possible to calculate the minimum electrical parameters of the Sun itself, its corona and analytically calculate the energy parameters of the solar wind in the region of the Earth. The smaller the Z/m ratio, the greater the positive charge of the Sun for the equality of Coulomb repulsive forces and gravitational attraction forces for a given ion. The ion velocity in the solar wind is also essentially dependent on this parameter and on the place where the ion was born. This determines the variance of the solar wind velocity from the ratio Z/m.

References

[1] Database on the ion composition of the solar wind http://www.srl.caltech.edu/ACE/ASC/level2/lvl2DATA_SWICS-SWIMS.html
[2] Parker E N 1964 The Solar Wind *Scient. American* **210 (4)** 66
[3] Podgornyi I M and Sagdeev R Z 1970 *Sov. Phys. Usp.* **12** 445–462
[4] Vysikaylo P I 2012 *Surf. Engin. Appl. Electrochem.* **48** 212
[5] Born M 1969 Atomic physics (London: Blackie & Son Ltd.) p 269

IOP Conf. Series: Journal of Physics: Conf. Series **1009** (2018) 012020 doi:10.1088/1742-6596/1009/1/012020

[6] Martynov D Ya 1971 Kurs obshchey astrofiziki (Moscow:Science gl.red. Phys. mat. lit) p 74
[7] Janev R K and Presnyakov L P 1981 Collision processes of multiply charged ions with atoms *Phys. Rep.* **70** 1–107

IOP Conf. Series: Journal of Physics: Conf. Series **1009** (2018) 012021 doi:10.1088/1742-6596/1009/1/012021

Normal glow discharge investigation in inert gases and nitrogen between flat electrodes with 20 mm of discharge gap

P V Kozlov, M A Kotov, L B Ruleva, S I Solodovnikov and S T Surzhikov

Ishlinsky Institute for Problems in Mechanics RAS, Vernadsky prospekt 101(1), Moscow, 119526, Russia

E-mail: kalevala@mail.ru, ruleva@ipmnet.ru

Abstract. Experimental studies of normal glow discharge (NGD) phenomenon have been carried out in the NGD facility at the Institute for Problems in Mechanics RAS. For helium, argon and nitrogen at the pressure values of 1-10 Torr, the dependences of the characteristic dimensions of the discharge current glow region are obtained. The results of the measurements were compared with calculations performed according to the formulas of Engel and Steinbeck's theory [1, 2]. Spectral measurements of radiation from different regions of normal glow in nitrogen for a pressure of 10 Torr are carried out.

1. Introduction

Glow discharge is the most used type of discharge in various applications, such as gas-discharge light sources, lasers, microelectronic technologies (etching, spraying thin films). It also belongs to the most studied gas discharges [1, 2]. However, the computational and theoretical description of the glow discharge is still insufficiently developed. This is due to the complex structure of the glow discharge and in practice the absence of unified calculation models that allow describing all elements of the glow discharge structure [3].

The Institute for Problems in Mechanics RAS (IPMech RAS) has significant experience in theoretical and numerical study of different types of discharges [3, 4]. This work continues the cycle of experimental studies of the normal glow discharge for the so-called “infinite” plane electrodes described in [5, 6]. The term "infinite" electrode is used to emphasize that the boundary effects on the electrodes do not affect the discharge structure. This allows to provide more accurate comparison of experimental data with theoretical models and numerical two-dimensional calculations of stationary discharge in a wide range of currents (spot sizes), taking into account diffusion [5].

Registration of electrical parameters, discharge form and spectral properties of radiation from in various regions is the main purpose of this investigation. The results of the study of the discharge shape and current-voltage characteristics for nitrogen, argon and helium were present for pressures $P = 1$, 3, 5, 6 and 10 Torr. Radiation spectra’s from different regions of glow discharge were obtained in nitrogen for a pressure of 10 Torr.

2. Experimental setup

Detailed description of the experimental setup and methods for measuring current and glow discharge voltage, pressure parameters in the discharge chamber, types of emission spectra are described in [5, 6].

In contrast to [5, 6], changes were made in the scheme of the installation, increasing the accuracy of the experiment. Namely, the ballast resistance equal to 300kOhm was selected; dividers (up to 5 kV) were manufactured and calibrated to measure the voltage of the power supply and the voltage on the electrodes; baratron in the pressure range P = 0.1÷11 Torr was shielded; the leakage of the discharge chamber was brought to the onset no worse than 10^{-3} torr/min. A window with a diameter of 50 mm and KU–1 glass was installed for spectral measurements in the range up to λ = 200 nm in the central part of one of the acrylic windows.

3. The results of the experiments

The methodology of the experiments is similar to that set out in the [5, 6]. For a fixed pressure (P = 1, 3, 5, 6 and 10 Torr) in the investigated gases (Ar, He or N_2) the discharge current varied from 1 to 10 mA. Voltage source (E), the discharge current and the voltage across the electrodes were recorded with a digital storage oscilloscopes. At the same time video recording of the glow discharge category was carried out.

After synchronization of the video and oscillograms of the current, a number of video frames were processed. With the help of the Universal Desktop Ruler program [7] the characteristic dimensions of the luminous region at the cathode were determined. These measurements allow to construct the dependence of the cathode spot diameter on the magnitude of the discharge current and to compare them with the calculations performed by the formulas of the theory of Engel and Steenbeck [1–3]. Empirical constants and mobility for different gases in [2] are shown in table 1.

Table 1. Empirical constants and mobility for different gases in [2]

s	A	B	μ_i (Gas)/ μ_i (N2)
N2	8.8	275	1 P>1 Torr
N2	12	342	1; P=1 Torr
Ar	12	180	1.1
He	3	34	5

Experimentally measured (symbols) and calculated dependences (solid lines) for the characteristic size of the luminous pre-cathode region for gases Ar, He and N_2 for different pressures and distance 20 mm between the flat electrodes are shown in figures 1, 2.

The dependence of the power of the discharge and the volume of the luminous discharge region on the time for nitrogen at a pressure of 3 Torr present on figure 3. Voltage - current characteristics for the series of experiments presented in Fig. 1 are show in figures 4, 5.

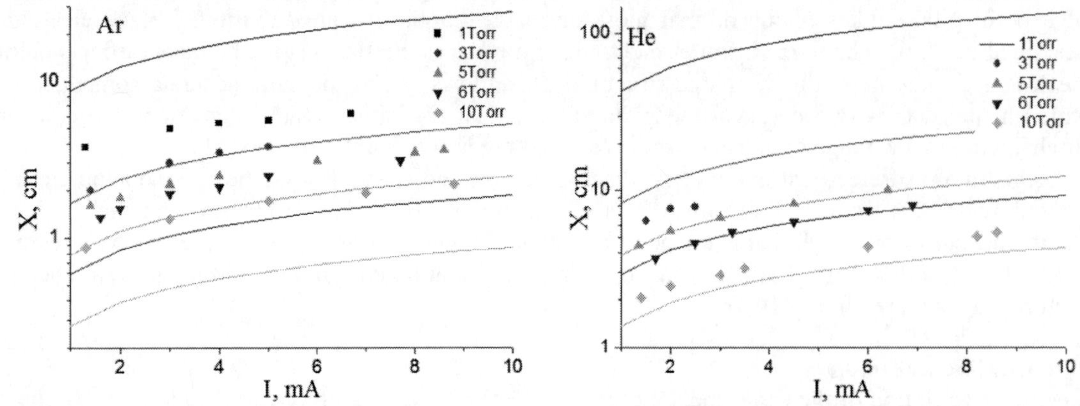

Figure 1. Discharge characteristic size dependence of the luminous pre-cathode region in Ar (left), He (right).

IOP Conf. Series: Journal of Physics: Conf. Series **1009** (2018) 012021 doi:10.1088/1742-6596/1009/1/012021

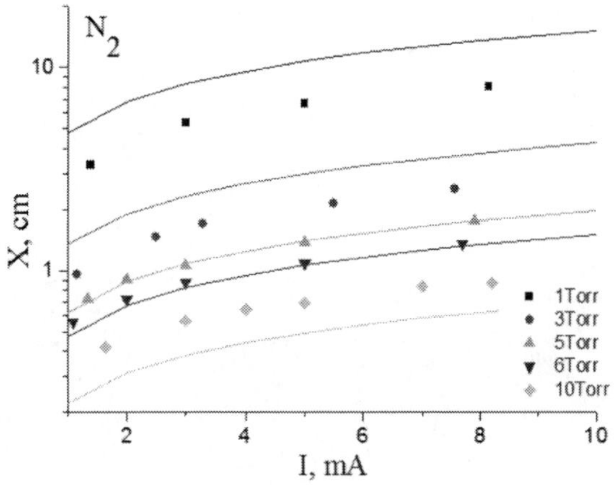

Figure 2. Discharge characteristic size dependence of the luminous pre-cathode region in N_2.

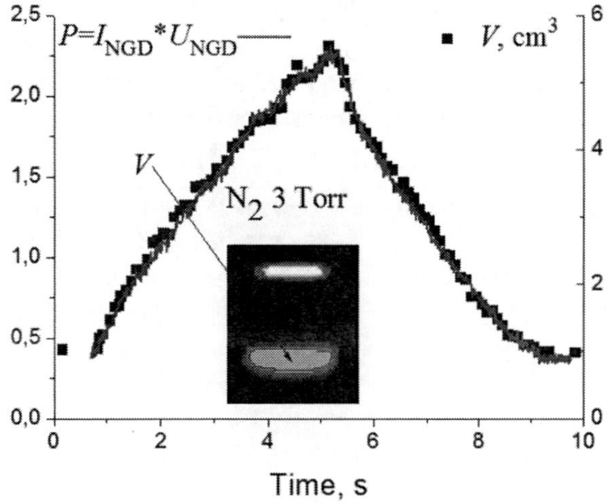

Figure 3. The dependence of the power of the discharge and the volume of the luminous discharge region on the time for nitrogen at a pressure of 3 Torr.

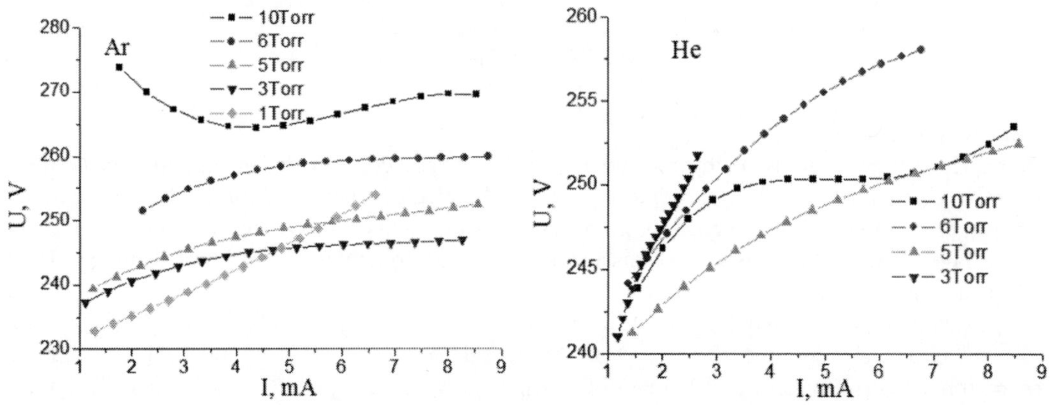

Figure 4. Current-voltage characteristics in Ar and He for different pressures.

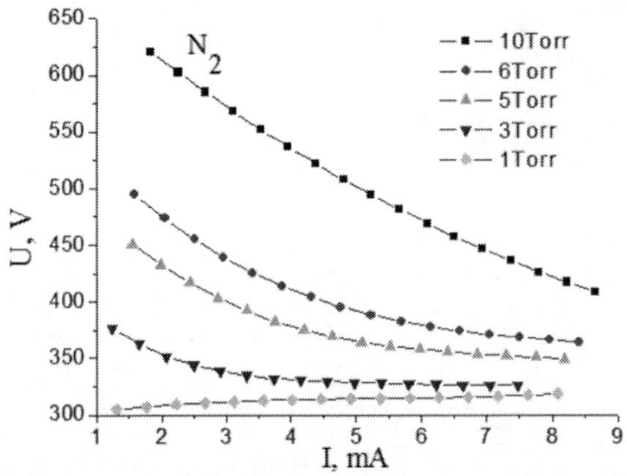

Figure 5. Current-voltage characteristic in N_2 for different pressures.

The dependence of the EMF of power supply and the discharge current for the ballast resistance of 300 kOhm are shown in table 2.

Table 2. The dependence of EMF and discharge current for ballast resistance of 300 kOhm.

Gas	*P*, Torr	$\mathcal{E} = V+R*I$
N_2	10	$\mathcal{E}$=752+255000*I
N_2	6	$\mathcal{E}$=577+269000*I
N_2	5	$\mathcal{E}$=515+275500*I
N_2	3	$\mathcal{E}$=423+283000*I
N_2	1	$\mathcal{E}$=351+293000*I
Ar	10	$\mathcal{E}$=311+292000*I
Ar	6	$\mathcal{E}$=296+291500*I
Ar	5	$\mathcal{E}$=284+292000*I
Ar	3	$\mathcal{E}$=279+292500*I
Ar	1	$\mathcal{E}$=267+296600*I
He:	10	$\mathcal{E}$=318+291000*I
He	6	$\mathcal{E}$=311+292000*I
He	5	$\mathcal{E}$=316+292000*I
He	3	$\mathcal{E}$=303+305000*I

4. Spectral measurements

Spectra of different regions were obtained for the discharge column in nitrogen at a pressure of 10 Torr and a current of 5.3 mA. The distance between the electrodes was 20 mm. While the diameter of the positive column did not exceed 10 mm, the measurements allowed us to obtain spectra of the cathode layer and positive column. Emission spectra were obtained at different distances from the cathode are shown in figure 6.

Spectrum analysis showed that the cathode has radiation of N2 molecular bands (C B). The spectra also allow one to estimate the vibrational and rotational temperatures of the gas in the discharge. The blue regions of the experimental studies of normal glow discharge at the IPMech RAS have been carried out.

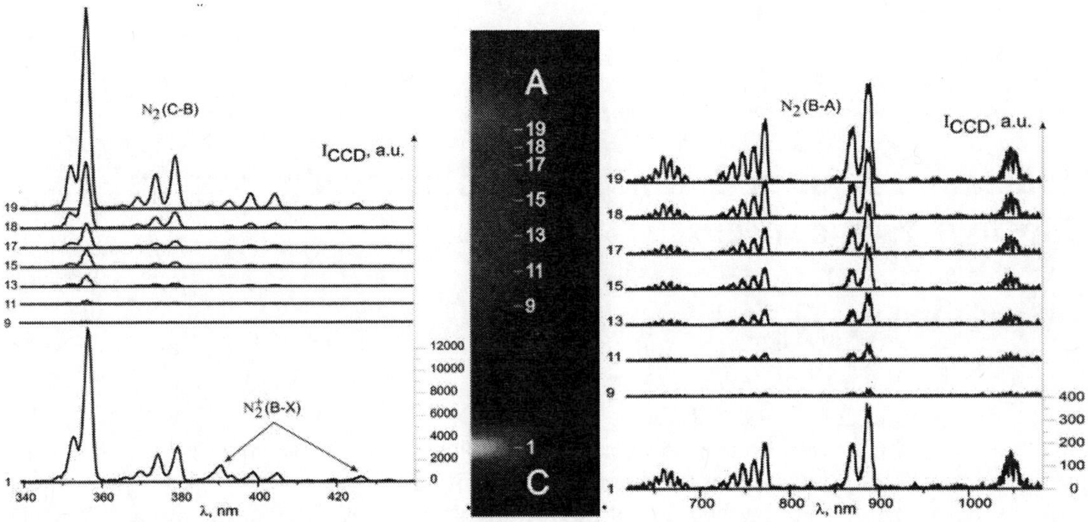

Figure 6. Emission spectra received from different regions of the discharge. The numbers to the left of the spectra and the photo indicate the distance in mm from the cathode. Relative intensities are indicated on the right side.

Conclusion

For helium, argon and nitrogen at the pressures of 1÷10 Torr, the dependences of the characteristic dimensions of the glow region on the discharge current were obtained. Spectral measurements of radiation from different regions of normal glow discharge in nitrogen for a pressure of 10 Torr were carried out. Further experiments will make it possible to validate the computational models describing this phenomenon.

Acknowledgments

The presented study was supported by the Russian Science Foundation project #16-11-10275.

References

[1] Engel A V 1997 Ionized gases *Oxford University Press*

[2] Raizer Y P 1991 *Gas discharge physics* (Springer-Verlag:Berlin Heidelberg)

[3] Surzhikov S T 2013 Computational Physics of Electric Discharges in Gas Flows *De Gruyter Studies in Mathematical Physics* **7**

[4] Raizer Yu P, Surzhikov S T 1988 Two-dimensional structure in a normal glow-discharge and diffusion effects in cathode and anode spot formation *High Temperature* **26** pp 304–311

[5] Kotov M A, Kozlov P V, Ruleva L B, Solodovnikov S I, Surzhikov S T and Tovstonog V A 2017 The spectral characteristic investigations of normal glow discharge *J. of Physics: Conference Series* **815** 012006

[6] Kotov M A, Kozlov P V, Ruleva L B, Solodovnikov S I and Surzhikov S T 2017 Experimental study of the electrical and spectroscopic characteristics of glow discharge 48th *AIAA Plasmadynamics and Lasers Conference AIAA Paper* 2017–3157

[7] http://avpsoft.ru/

Ionization of air in flow around a blunt wedge at relatively low hypersonic speeds

S T Surzhikov

Ishlinsky Institute for Problems in Mechanics RAS, Vernadsky prospekt 101(1), Moscow, 119526, Russia
Moscow Institute of Physics and Technology (State University), 9 Institutskiy per., Dolgoprudny, Moscow Region, 141701, Russia
Dukhov All-Russian Scientific Research Institute of Automatics, Moscow, Russia

surg@ipmnet.ru

Abstract. Calculations are performed for investigation of air ionization near to surface of blunted wedges with half-angles $\varphi = 3^0$ and 9^0 and radii of blunting of $R_n = 1.5$ and 2.0 cm, moving at speeds of $3 \div 6$ km/s in an air atmosphere with unperturbed gas parameters corresponding to altitudes of $20 \div 60$ km.

1. Introduction

The movement of hypersonic aircraft leads not only to a strong heating of their constructive elements, but also to ionization of the gas in the compressed layer near the streamlined surface, which is a very important aspect of the hypersonic flight of an aircraft, since it leads to a radio communication disruption. At superorbital velocities of spacecraft returning to Earth ($V_\infty > 9$ km/s) the kinetic energy of the colliding particles proves to be sufficient for direct shock ionization of gas particles. At relatively low hypersonic velocities ($V_\infty < 7$ km/s) the question of the primary cause of the appearance in the gas flow of a significant number of electrons. In a number of computational and theoretical studies [1– 6] it was established that the most probable cause of primary ionization is the process of associative ionization , which is at least the two-stage process

$$\mathrm{N} + \mathrm{O} \rightarrow \mathrm{NO}^* \rightarrow \mathrm{NO}^+ + \mathrm{e}^-,$$

and at this the atoms N and O appear as a result of the dissociation of air components, i.e. molecular nitrogen and molecular oxygen in a mutual collision. In this case, the dissociation of molecules is often not due just to their collision, but is accompanied by the excitation of internal degrees of freedom, in the first place – the vibrations of nuclei in molecules. Therefore, there is a complex multi-stage nature of this process.

Calculations and experiments [7] showed a high probability of realizing this scenario of ionization development at the surface of the body flying at moderate hypersonic speed. The experiments were performed on a shock tube for the velocities of shock waves in the range $V_\infty = 4.7 \div 6.7$ km/s. These experiments showed a very high intensity of the radiation of a heated air plug behind the front of the shock wave in the region of the vacuum ultraviolet spectrum (wavelength $\lambda \sim 190 \div 220$ nm) was detected, which confirms the appearance of NO molecules in the flow of excited electronic states $A^2\Sigma^+$, $B'^2\Delta$, $B^2\Pi$, $C^2\Pi$, $D^2\Sigma^+$ [8].

IOP Conf. Series: Journal of Physics: Conf. Series **1009** (2018) 012022 doi:10.1088/1742-6596/1009/1/012022

Numerical simulation results of the processes of physicochemical kinetics behind the front of the shock wave and the spectral emissivity of the air showed good agreement with the radiation intensity observed in the experiment. The used kinetic model of associative ionization [9], applied in [4,5], showed a high probability of associative ionization.

This is the rationale for choosing a model of physicochemical kinetics for the study of ionization processes at relatively low hypersonic velocities. However, it should be borne in mind that the use of this model at two speeds of experimental hypersonic blunted cones ~7.4 km/s [4] and 5.6 km/s [4, 10], has shown the necessity of modifying the rate constants of associative ionization with decreasing velocity.

Thus, the kinetic model underlying the study of ionization processes at the blunt wedge surface is based on the process of associative ionization, which includes the dissociation of N_2 and O_2 molecules, the formation of NO molecules as a result of chemical reactions (primarily due to collisional association of N and O atoms), the excitation of electronic states of NO* with the subsequent decomposition of NO* into NO^+ and e^-.

This kinetic model is used in this work to analyze the gas ionization in the vicinity of a blunt wedge in a wide range of velocities $V_\infty = 3 \div 6$ km with the parameters of the incident flow corresponding to altitudes $H = 20 \div 60$ km. Calculations are performed for wedges with half-angles $\varphi = 3^0$ and 9^0 for two radii of blunting $R_n = 1.5$ and 2.0 cm.

2. Results of numerical simulation of ionization processes in flow past a blunt wedge

The scheme of the problem being solved is shown in figure 1. The matrix of the calculated variants is given in table 1, where in addition to the initial data, the braking temperature is presented

$$T_0 = T_\infty + \frac{V_\infty^2}{c_p} = T_\infty\left(1 + \frac{\gamma - 1}{2} M_\infty^2\right)$$

Table 1. Initial conditions for the calculations.

p_∞, erg/cm^3	ρ_∞, g/cm^3	T_∞, K	V_∞, cm/s	M_∞	T_0, K
		H=20 km			
55300	0.889×10^{-4}	217	3×10^5	11.2	4565
			4×10^5	15	7957
		H=30 km			
12000	0.184×10^{-4}	227	3×10^5	11	4564
			4×10^5	14.6	7949
		H=40 km			
2870	0.400×10^{-5}	250	3×10^5	10.2	4591
			4×10^5	13.9	7979
			5×10^5	17.4	12334
		H=50 km			
798	0.103×10^{-5}	271	3×10^5	10.1	4628
			4×10^5	13.4	8029
			5×10^5	16.8	12401
		H=60 km			
220	0.310×10^{-6}	247	3×10^5	10.2	4584
			4×10^5	14.2	7967
			5×10^5	17.5	12318

The author's computer code NERAT–2D was used in the calculations. The method of integrating the system of Navier – Stokes equations for the motion of a viscous, heat-conducting, chemically reacting gas is described in detail in [4, 10].

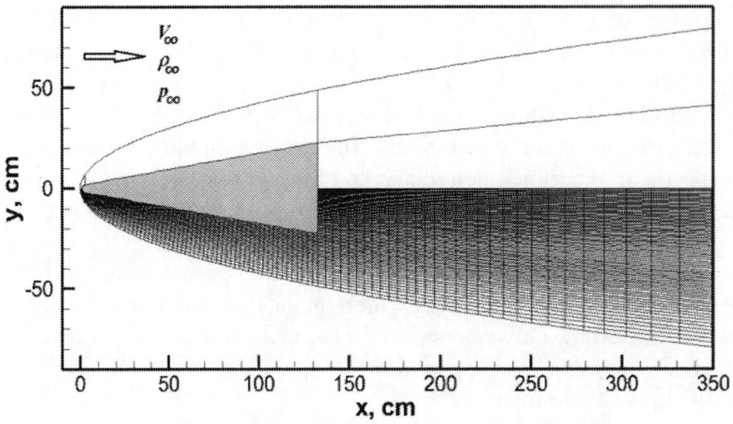

Figure 1. Schematic of the task and calculation grid.

Figures 2–4 show the results of calculations of flow field above a blunted wedge with a half angle $\varphi = 9^0$ at a velocity of $V_\infty = 4\,\text{km/s}$ at a height of $H = 20\,\text{km}$. We note the characteristic features of gas flow under the indicated conditions.

Figure 2, *a* shows the fields of translational temperatures and vibrational temperatures of N_2. The greatest values of translational temperature are achieved in the vicinity of the frontal blunting of the wedge. Here the temperature is close to the braking temperature $T \sim 4500\,\text{K}$. It is significant that the vibrational temperature N_2 differs markedly from the translational temperature in the near wake, where a flow of rarefaction is realized (see figure2, *b*). In figure 2, *c* electron concentration fields are shown for the wedge velocity of $V_\infty = 4\,\text{km/s}$ at an altitude $H = 20\,\text{km}$. Recall that in the flight experiment RAMC-II [11- 13] for the diagnosis of the degree of ionization of air in the compressed layer microwave sensors with a frequency $f_L \sim 1\,\text{GHz}$ and $f_X \sim 10\,\text{GHz}$ were used. With the use of these sensors, electron concentrations of $n_e \sim 10^{10}\,\text{cm}^{-3}$ were detected. Note that the critical electron concentration required to reflect the radio signal can be predicted from formula

$$n_{e,cr} = 1.24 \times 10^4 f_{MHz}^2 = 1.24 \times 10^{10} f_{GHz}^2,\ \text{cm}^{-3}$$

In the case considered in figure 2, *c*, the electron concentration in the compressed layer above the surface of the wedge reaches $n_e \sim 10^{11}\,\text{cm}^{-3}$. In this case, the distribution of electron and NO^+ ion concentrations along the thickness of the compressed layer is shown in figure 4, *a*, from which it is clearly seen that the thickness of the ionized region reaches almost 10 cm, and the degree of ionization falls insignificantly along the wedge generatrix.

The distributions of the translational and vibrational temperatures N_2 over the height of the compressed layer in these same sections (figure 4, *b*) show a good thermalization of the internal degrees of freedom. The distribution of the concentrations of atomic nitrogen (the result of N_2 dissociation) and NO molecules (formed as a result of the chain of chemical transformations) in figure 3, *a*, as well as the concentrations of O atoms and O_2 molecules in figure 3, *b*, give an idea of the degree of dissociation of air molecules and the configuration of the field of NO molecules, which serve as the primary source of electrons.

Figures 5- 7 show the results of calculations of aerophysics over a blunted wedge moving at a height of $H = 20\,\text{km/s}$ at a speed of $V_\infty = 3\,\text{km}$, that is, at a reduced speed of 1 km/s compared to the previous variant. The expected drop in translational and vibrational temperatures in the compressed layer is observed (figure5, *a*). We note a noticeable decrease in the concentration of electrons in the compressed layer (figure 5, *c*) and NO^+ ions. From the distributions of ion concentrations over the thickness of the compressed layer in different sections along x (figure 7, *a*), it is evident that the largest electron concentration does not exceed $n_e \sim 10^{11}\,\text{cm}^{-3}$. This means that with a further decrease in the velocity of the blunted wedge, the electron concentrations will not reach their critical value.

IOP Conf. Series: Journal of Physics: Conf. Series **1009** (2018) 012022 doi:10.1088/1742-6596/1009/1/012022

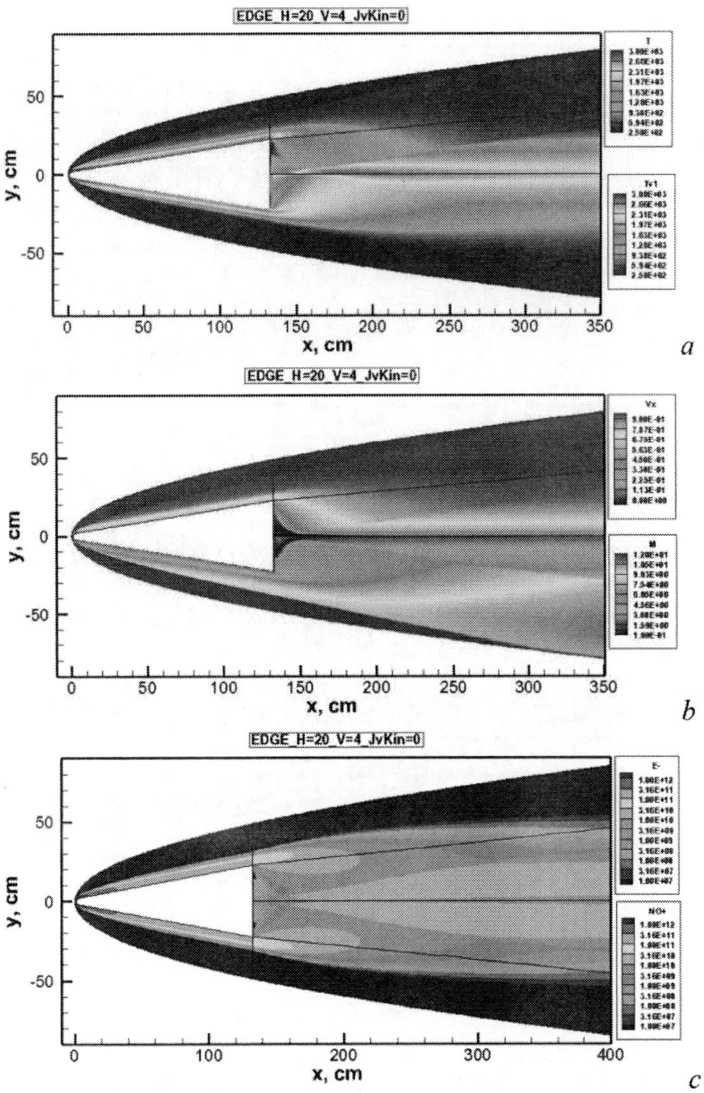

Figure 2. The fields of the translational temperature T(*a*, above) and the vibrational temperature T_V of N_2 (*a*, from below), the longitudinal velocity $Vx = u/V_\infty$ (*b*, from above) and the Mach number (*b*, from below), the electron concentration (*c*, from above) and NO^+ ions (*c*, from below). The temperature is in K, the concentration is in cm^{-3}; $H = 20\,\text{km}$, $V_\infty = 4$ km/s.

When the altitude of the flight is increased up to $H = 30\,\text{km}$, the nonequilibrium effects are more clearly manifested. Figures 8–10 and 11–13 show the results of calculations for aerophysics of a blunted wedge with a smaller half angle, $\varphi = 3^0$. At $V_\infty = 4\,\text{km/s}$ a faster decrease in the translational and vibrational temperatures along the wedge generates, but the concentration of electrons and NO^+ ions near the blunting is higher, $n_e \sim 10^{12}\,\text{cm}^{-3}$. With a decrease in velocity to $V_\infty = 3$ km/s (figures 11–13), the temperature and concentration fall faster than at $H = 20\,\text{km}$. The maximum electron concentration in the compressed layer decreases up to $n_e \sim 10^{10}\,\text{cm}^{-3}$ at half length of the wedge. At this altitude, there is a noticeable difference between translational and vibrational temperatures along the entire lateral surface (figures 10 and 13).

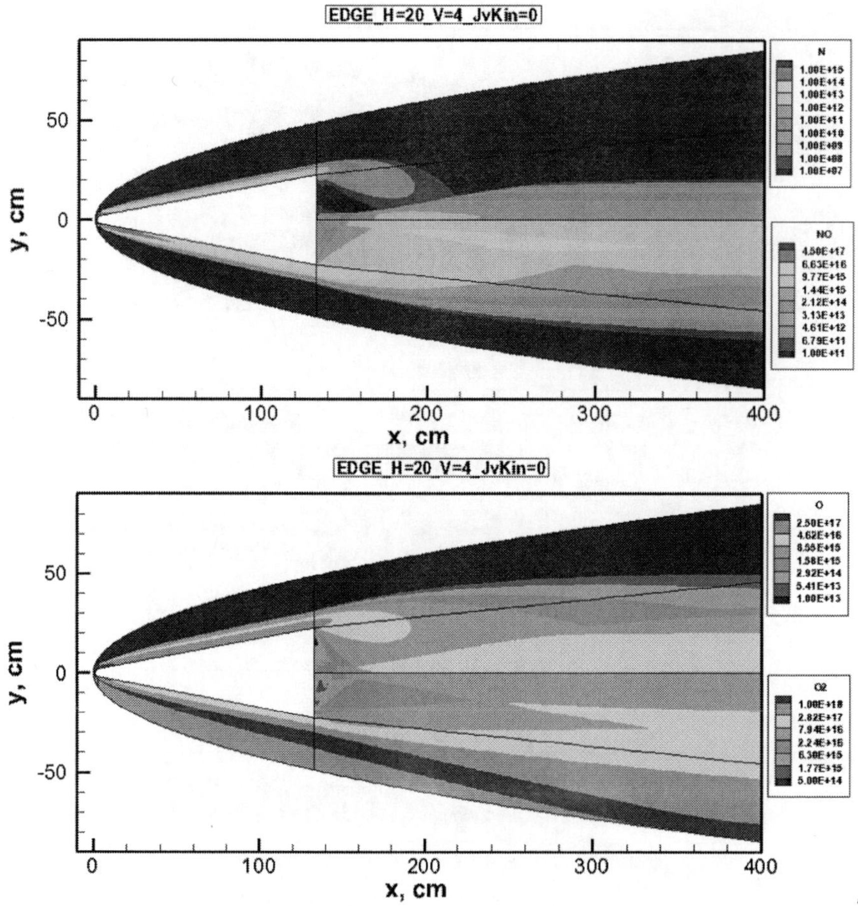

Figure 3. Fields of concentrations of N (*a*, above) and N_2 (*a*, below) molecules,atoms O (*b*, above) and molecules O_2 (*b*, below). Concentrations in cm^{-3}; $H = 20\,km$, $V_\infty = 4\,km/s$.

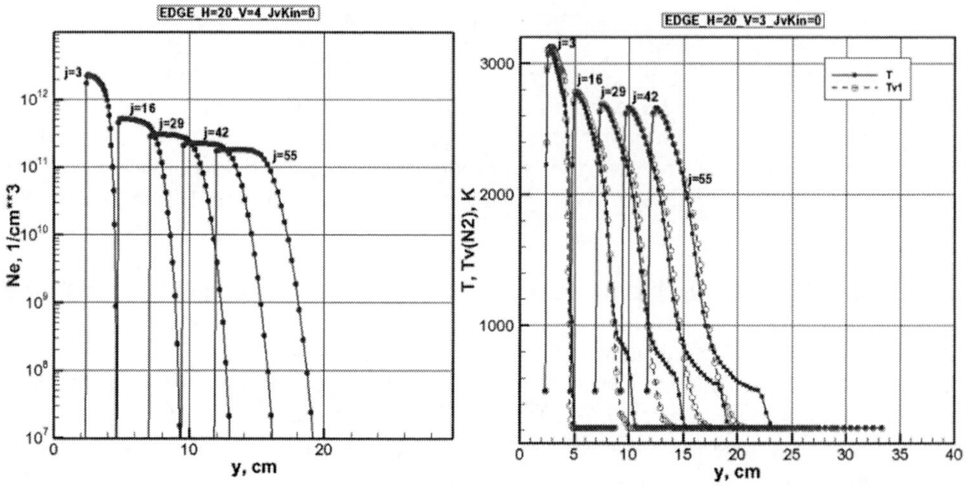

Figure 4. Axial distributions of the electron concentration (left), as well as translational and vibrational temperatures N_2 in 5 sections along the *x* axis ($x_3 = 4.92$ cm, $x_{16} = 19.5$ cm, $x_{29} = 34.2$ cm, $x_{42} = 49$ cm, $x_{55} = 64$ cm). $H = 20\,km$, $V_\infty = 4\,km/s$.

IOP Conf. Series: Journal of Physics: Conf. Series **1009** (2018) 012022 doi:10.1088/1742-6596/1009/1/012022

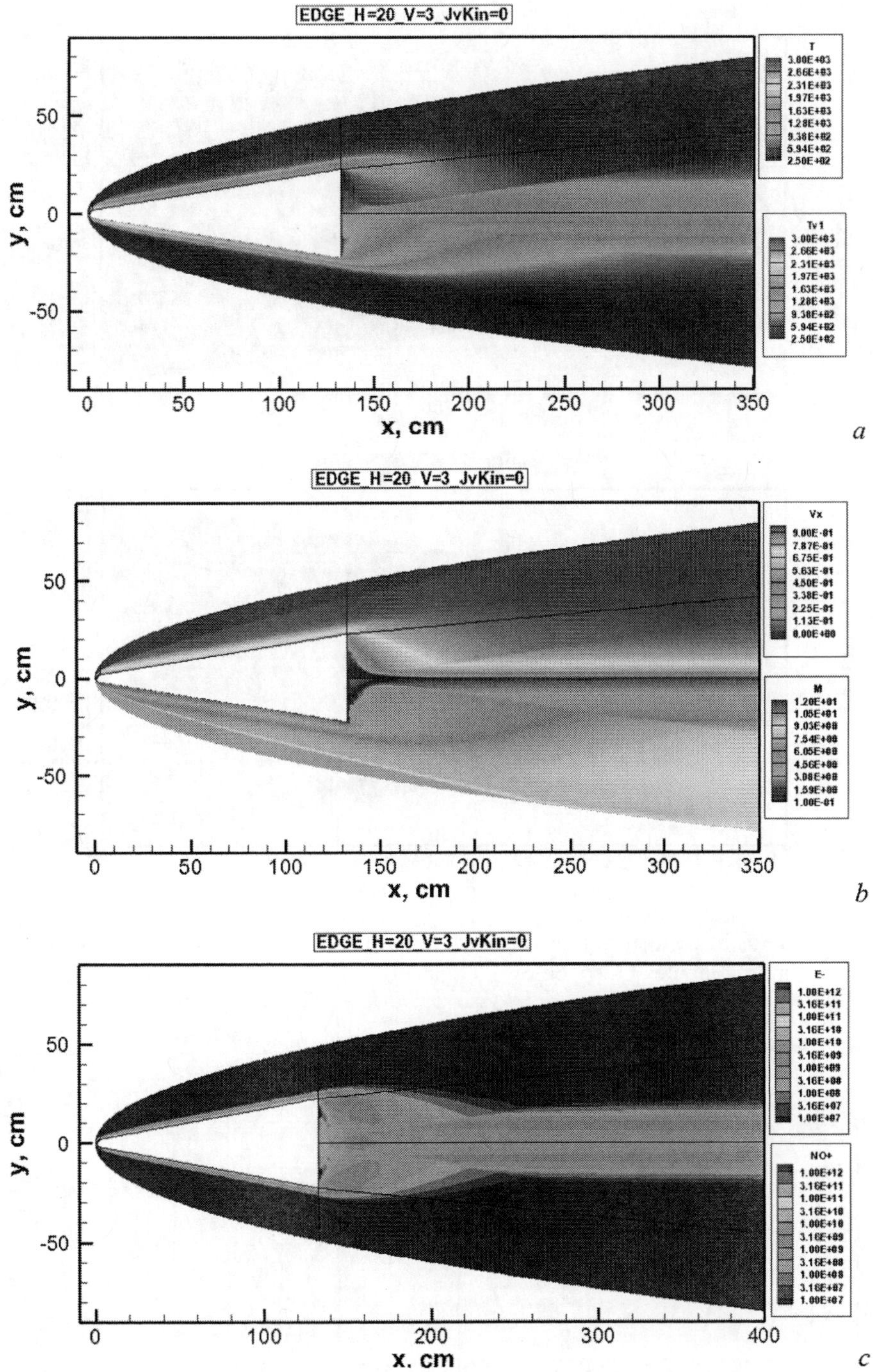

Figure 5. The fields of the translational temperature T (*a*, above) and the vibrational temperature T_V of N_2 (*a*, from below), the longitudinal velocity $Vx = u/V_\infty$ (*b*, from above) and the Mach number (*b*, from below), the electron concentration (*c*, from above) and NO^+ ions (*c*, from below). The temperature is in K, the concentration is in cm^{-3}; $H = 20$ km, $V_\infty = 3$ km/s.

IOP Conf. Series: Journal of Physics: Conf. Series **1009** (2018) 012022 doi:10.1088/1742-6596/1009/1/012022

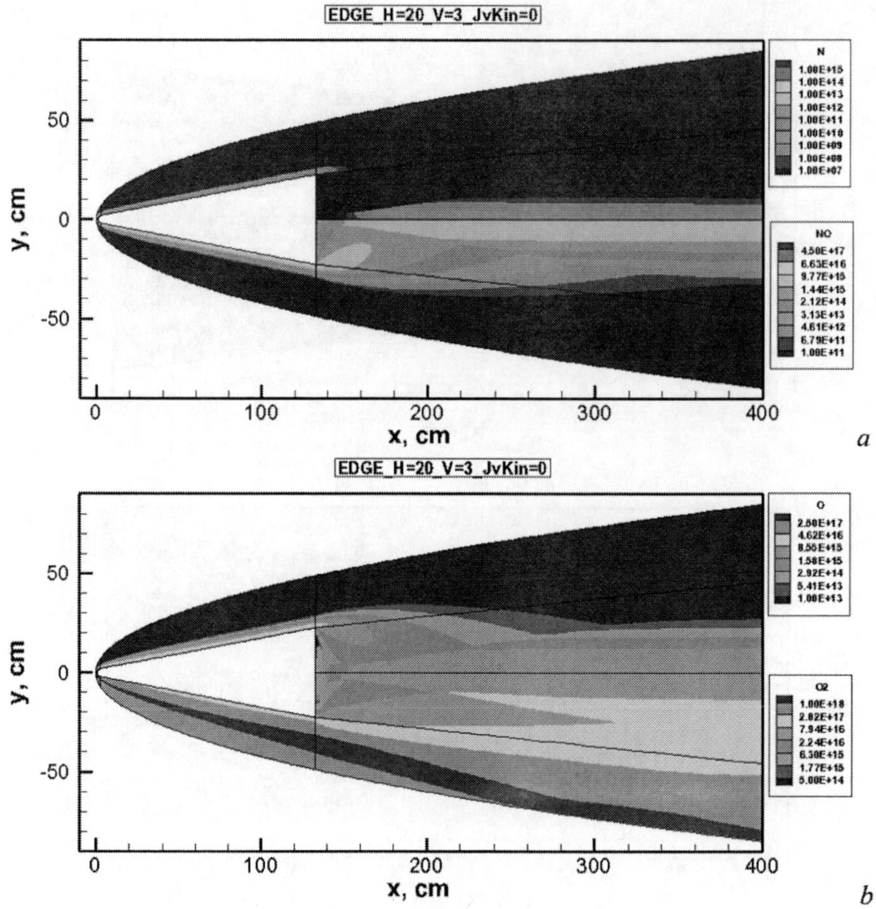

Figure 6. Fields of concentrations of N (*a*, above) and N_2 (*a*, below) molecules, atoms O (*b*, above) and molecules O_2 (*b*, below). Concentrations in cm^{-3}; $H = 20$ km, $V_\infty = 3$ km/s.

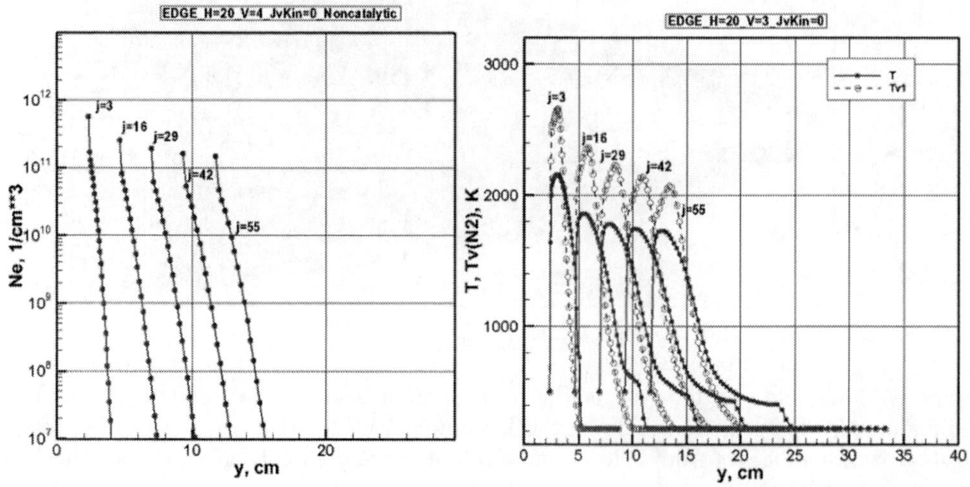

Figure 7. Axial distributions of the electron concentration (left), as well as translational and vibrational temperatures N_2 in 5 sections along the *x* axis. $H = 20$ km, $V_\infty = 3$ km/s.

IOP Conf. Series: Journal of Physics: Conf. Series **1009** (2018) 012022 doi:10.1088/1742-6596/1009/1/012022

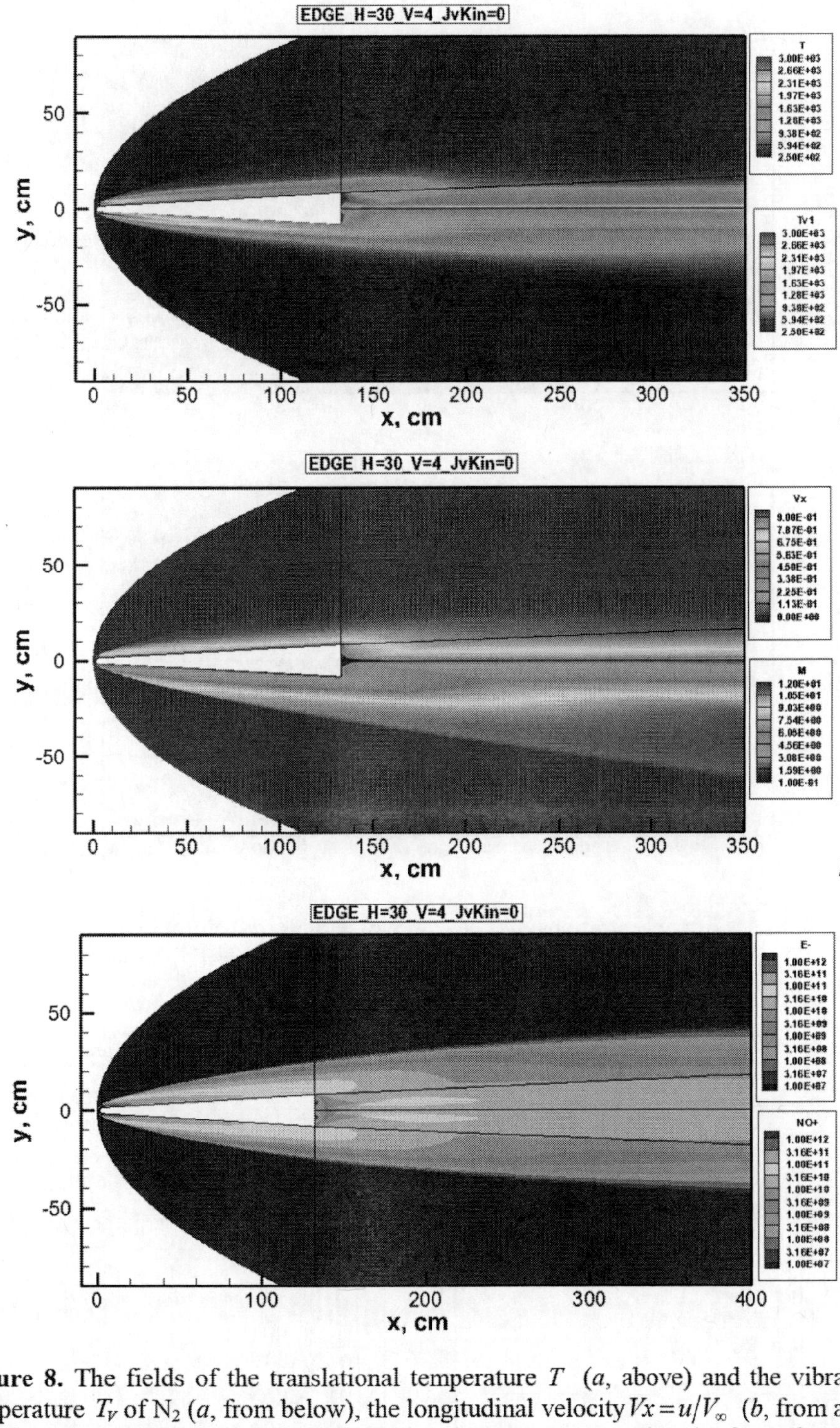

a

b

c

Figure 8. The fields of the translational temperature T (*a*, above) and the vibrational temperature T_V of N_2 (*a*, from below), the longitudinal velocity $Vx = u/V_\infty$ (*b*, from above) and the Mach number (*b*, from below), the electron concentration (*c*, from above) and NO^+ ions (*c*, from below). The temperature is in K, the concentration is in cm^{-3}; $H = 30\text{km}$, $V_\infty = 4\,\text{km/s}$.

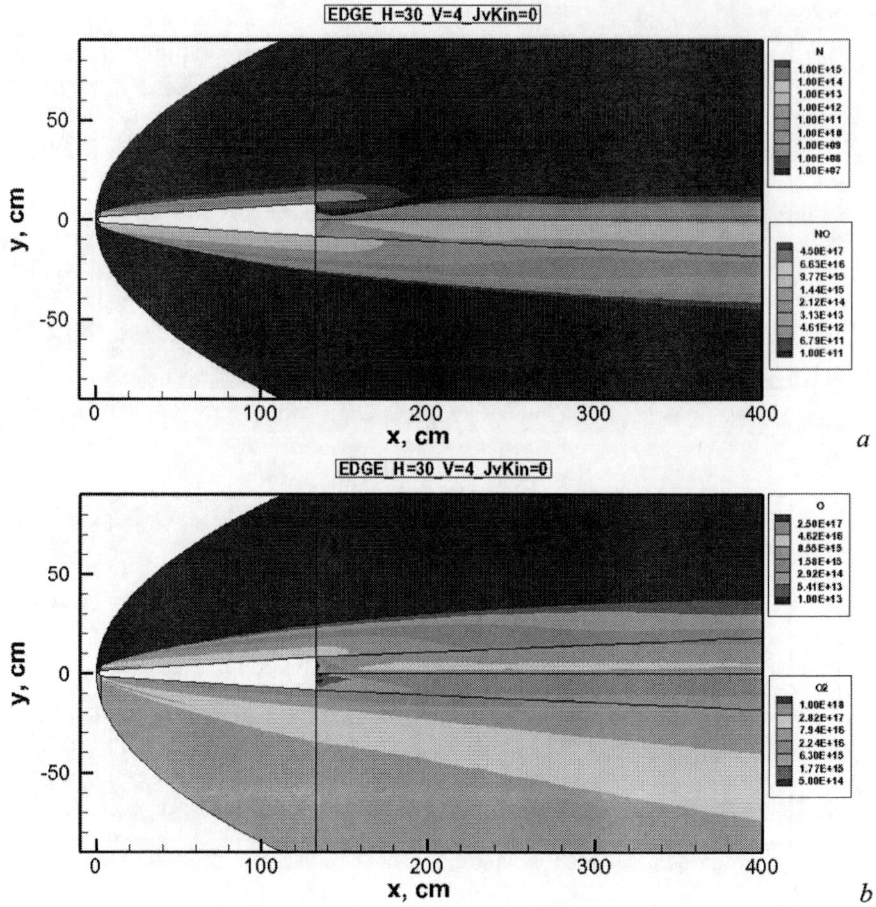

Figure 9. Fields of concentrations of N (a, above) and N_2 (a, below) molecules, atoms O (b, above) and molecules O_2 (b, below). Concentrations in cm^{-3}; $H = 30$ km, $V_{\infty} = 4$ km/s.

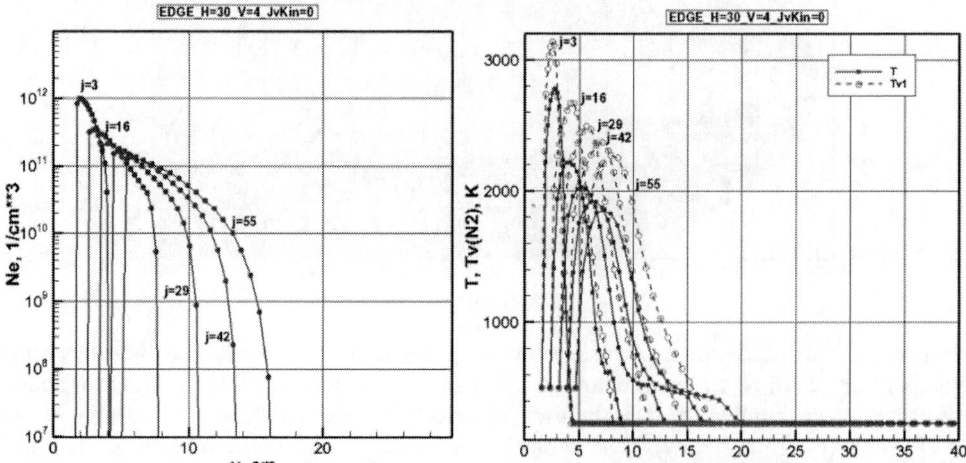

Figure 10. Axial distributions of the electron concentration (left), as well as translational and vibrational temperatures N_2 in 5 sections along the x axis. $H = 30$ km, $V_{\infty} = 4$ km/s.

IOP Conf. Series: Journal of Physics: Conf. Series **1009** (2018) 012022 doi:10.1088/1742-6596/1009/1/012022

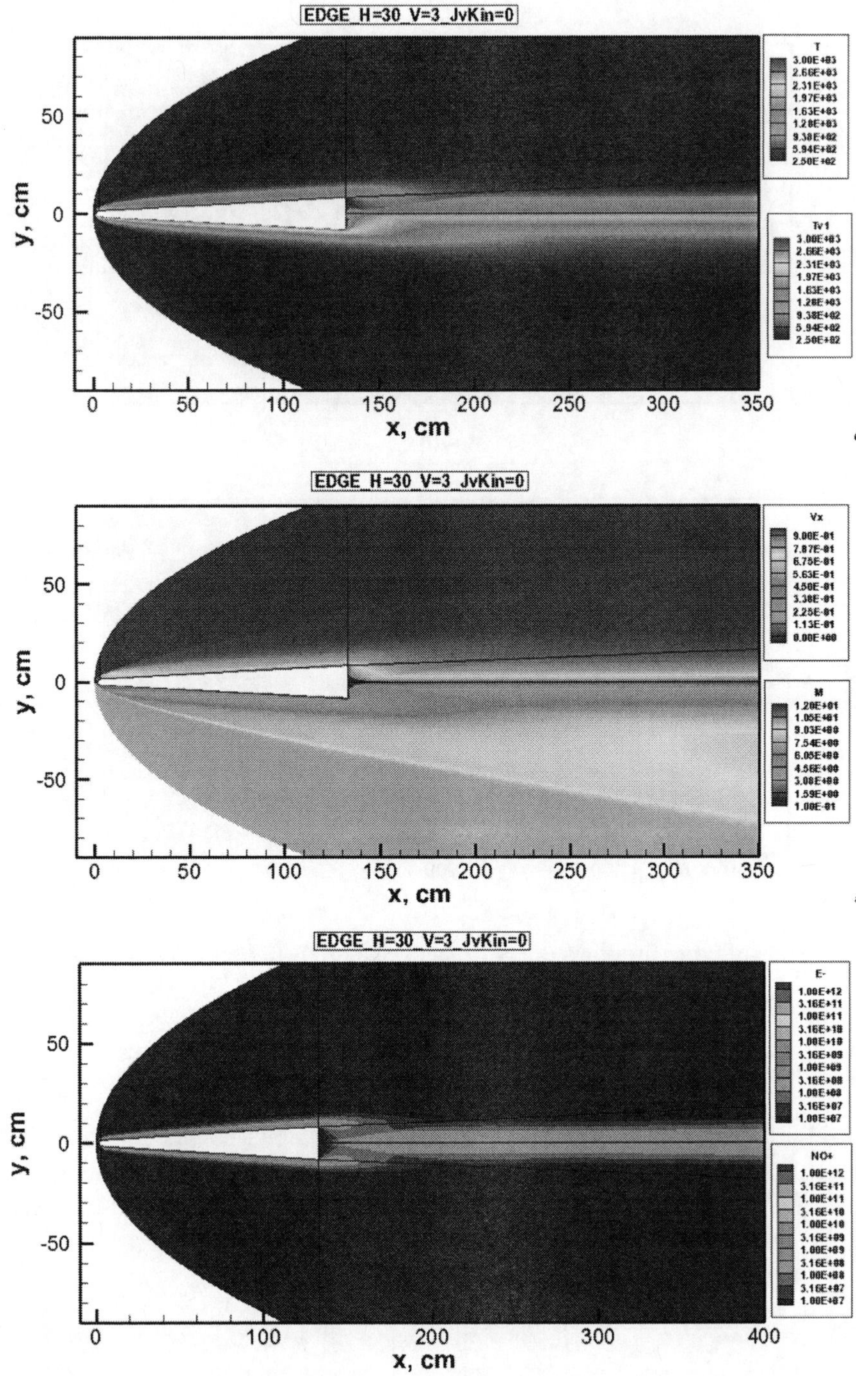

a

b

c

Figure 11. The fields of the translational temperature T (*a*, above) and the vibrational temperature T_V of N_2 (*a*, from below), the longitudinal velocity $Vx = u/V_\infty$ (*b*, from above) and the Mach number (*b*, from below), the electron concentration (*c*, from above) and NO^+ ions (*c*, from below). The temperature is in K, the concentration is in cm^{-3}; $H = 30$ km, $V_\infty = 3$ km/s.

IOP Conf. Series: Journal of Physics: Conf. Series **1009** (2018) 012022 doi:10.1088/1742-6596/1009/1/012022

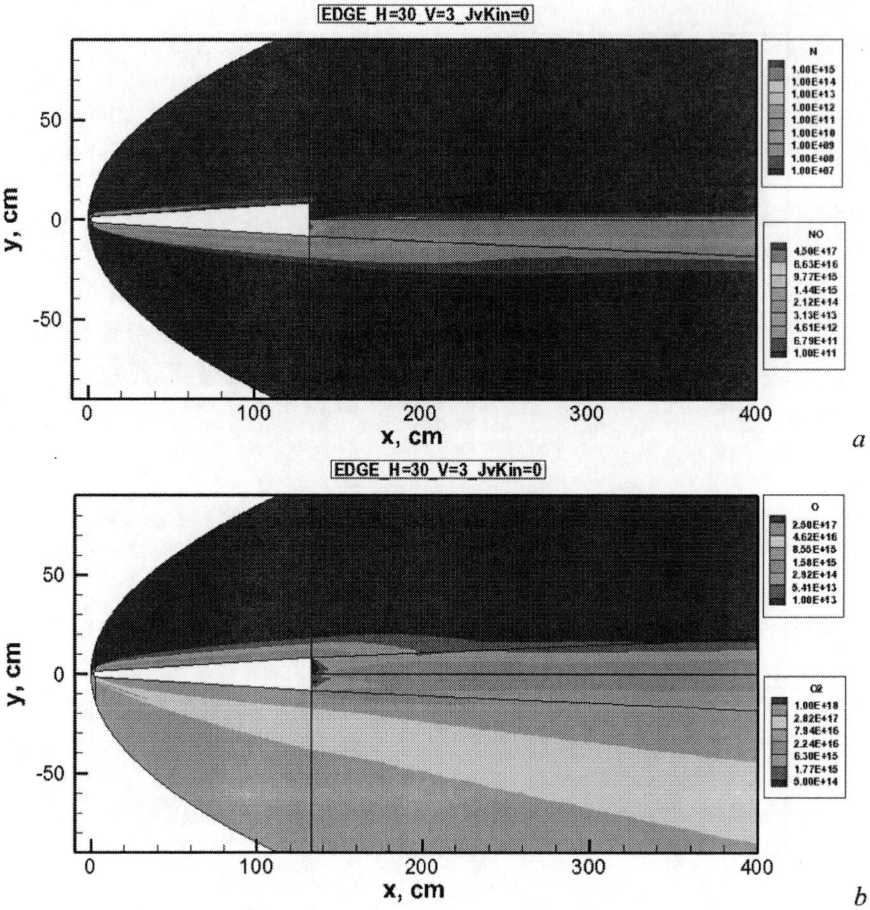

Figure 12. Fields of concentrations of N (a, above) and N_2 (a, below) molecules, atoms O (b, above) and molecules O_2 (b, below). Concentrations in cm^{-3}; $H = 30$ km, $V_\infty = 3$ km/s.

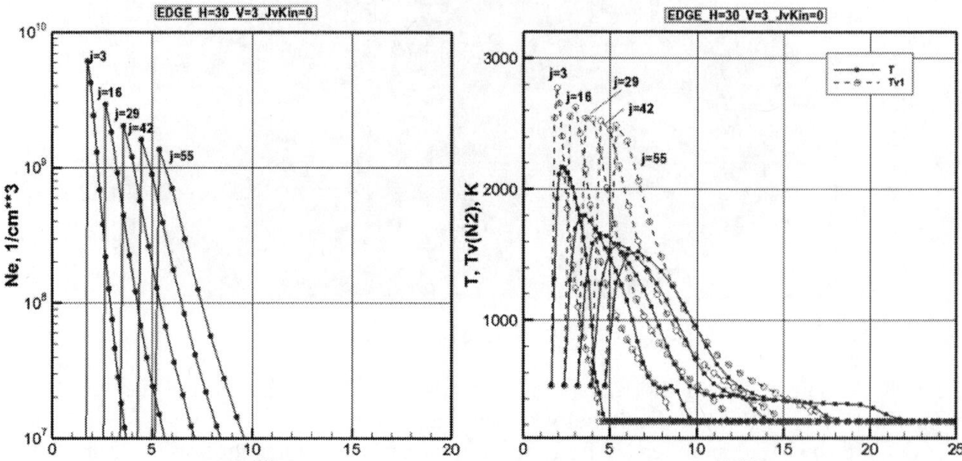

Figure 13. Axial distributions of the electron concentration (left), as well as translational and vibrational temperatures N_2 in 5 sections along the x axis. $H = 30$ km, $V_\infty = 3$ km/s.

The next series of drawings show the flow field above blunt wedges at still higher altitudes. At heights of $H = 40$ km, 50 km and 60 km with a successive decrease in speed $V_\infty = 5$ km/s, 4 km/s and 3 km/s (see figures 14–16, 17–19 and 20–22; figures 23–25, 26–28 and 29–31; figures 32–34, 35–37 and 38–40) there is a sharp decrease in concentrations n_e in the compressed layer. The flow becomes more non-equilibrium.

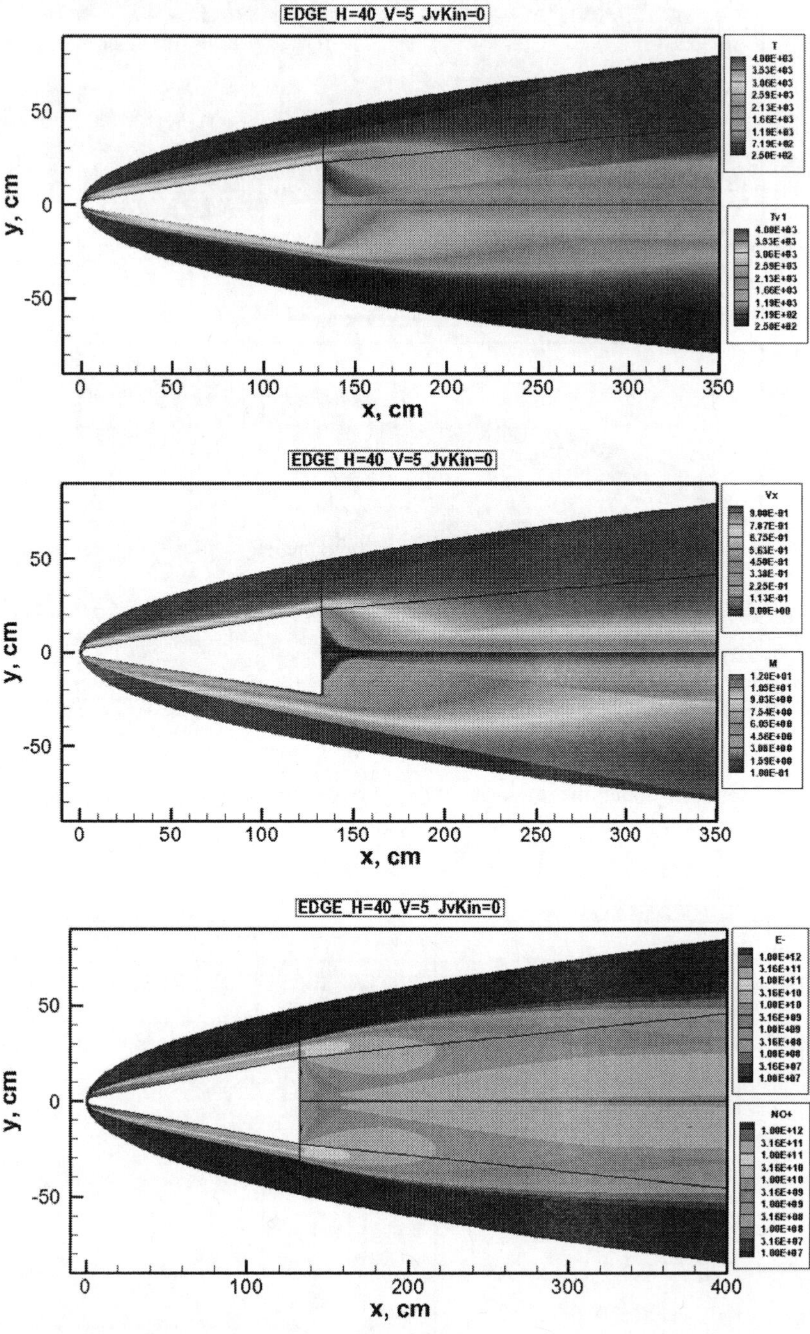

a

b

c

Figure 14. The fields of the translational temperature T (*a*, above) and the vibrational temperature T_V of N_2 (*a*, from below), the longitudinal velocity $Vx = u/V_\infty$ (*b*, from above) and the Mach number (*b*, from below), the electron concentration (*c*, from above) and NO^+ ions (*c*, from below). The temperature is in K, the concentration is in cm^{-3}; $H = 40$ km, $V_\infty = 5$ km/s.

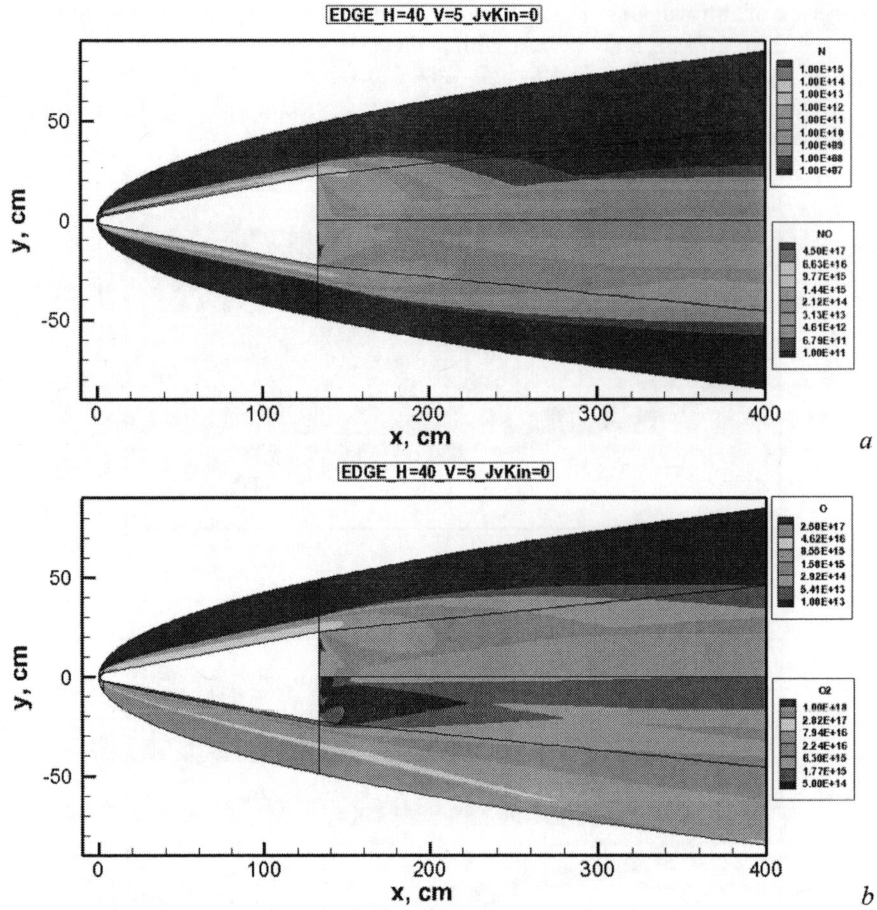

Figure 15. Fields of concentrations of N (a, above) and N_2 (a, below) molecules, atoms O (b, above) and molecules O_2 (b, below). Concentrations in cm^{-3}; $H = 40\,km$, $V_\infty = 5$ km/s.

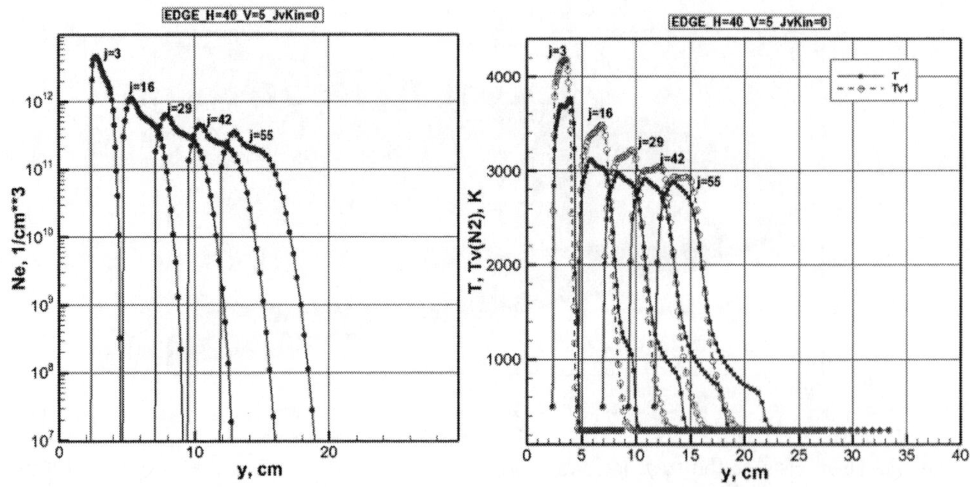

Figure 16. Axial distributions of the electron concentration (left), as well as translational and vibrational temperatures N_2 in 5 sections along the x axis. $H = 40\,km$, $V_\infty = 5\,km/s$.

IOP Conf. Series: Journal of Physics: Conf. Series **1009** (2018) 012022 doi:10.1088/1742-6596/1009/1/012022

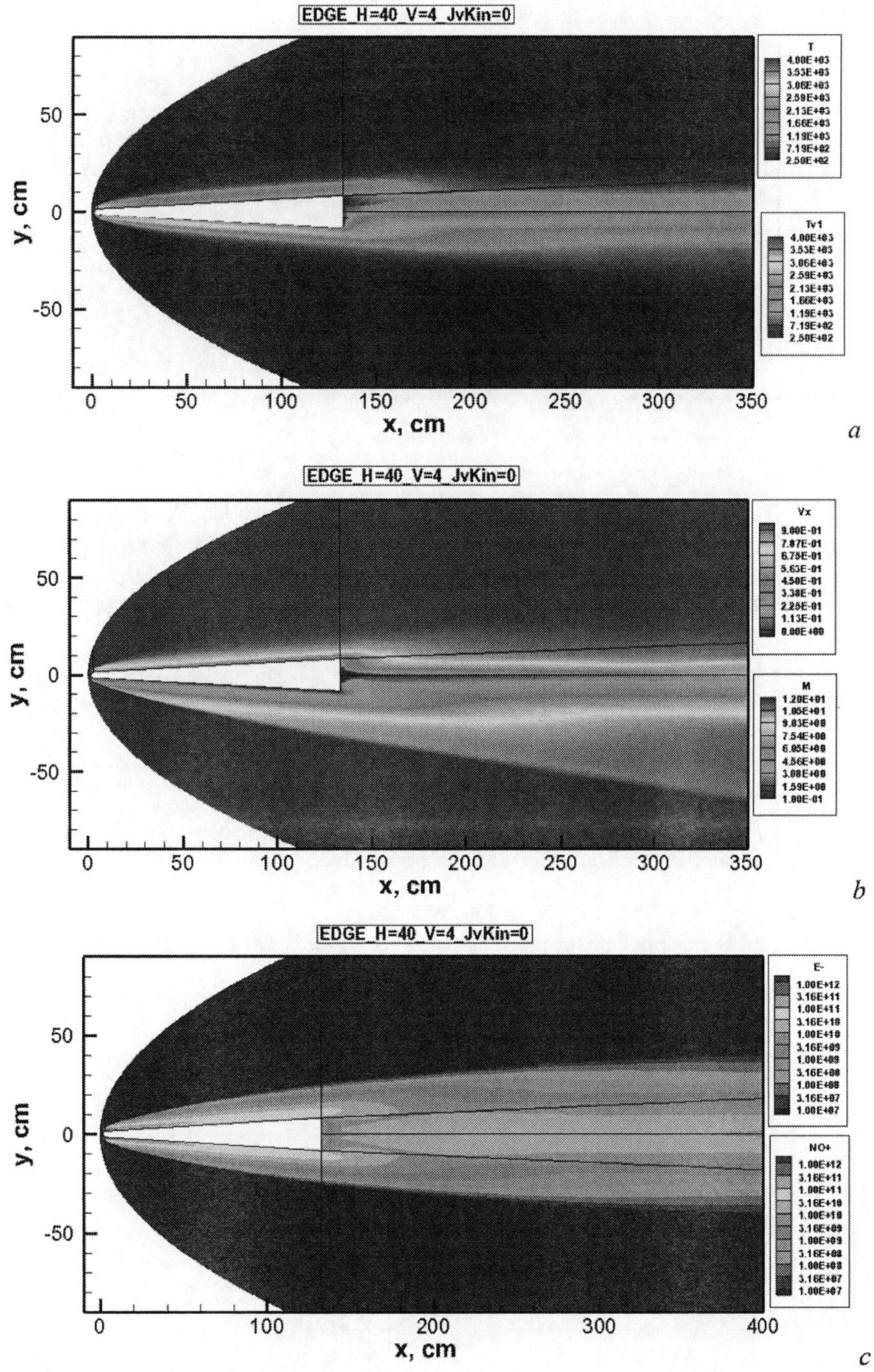

Figure 17. The fields of the translational temperature T (*a*, above) and the vibrational temperature T_V of N_2 (*a*, from below), the longitudinal velocity $Vx = u/V_\infty$ (*b*, from above) and the Mach number (*b*, from below), the electron concentration (*c*, from above) and NO^+ ions (*c*, from below). The temperature is in K, the concentration is in cm^{-3}; $H = 40\,km$, $V_\infty = 4\,km/s$.

IOP Conf. Series: Journal of Physics: Conf. Series **1009** (2018) 012022 doi:10.1088/1742-6596/1009/1/012022

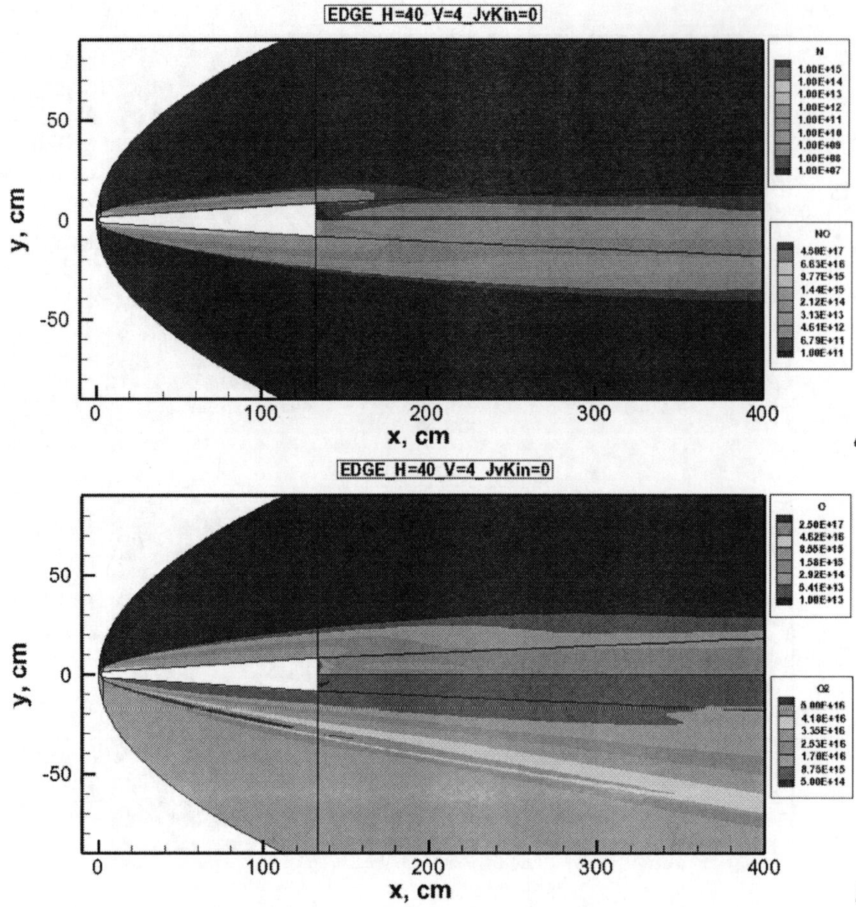

a

b

Figure 18. Fields of concentrations of N (a, above) and N_2 (a, below) molecules, atoms O (b, above) and molecules O_2 (b, below). Concentrations in cm^{-3}; $H = 40\,km$, $V_\infty = 4$ km/s.

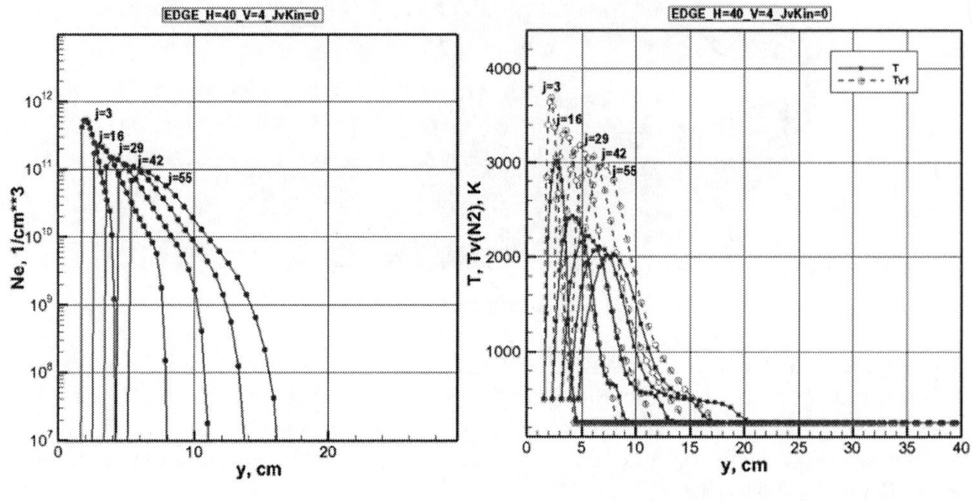

Figure 19. Axial distributions of the electron concentration (left), as well as translational and vibrational temperatures N_2 in 5 sections along the x axis. $H = 40\,km$, $V_\infty = 4\,km/s$.

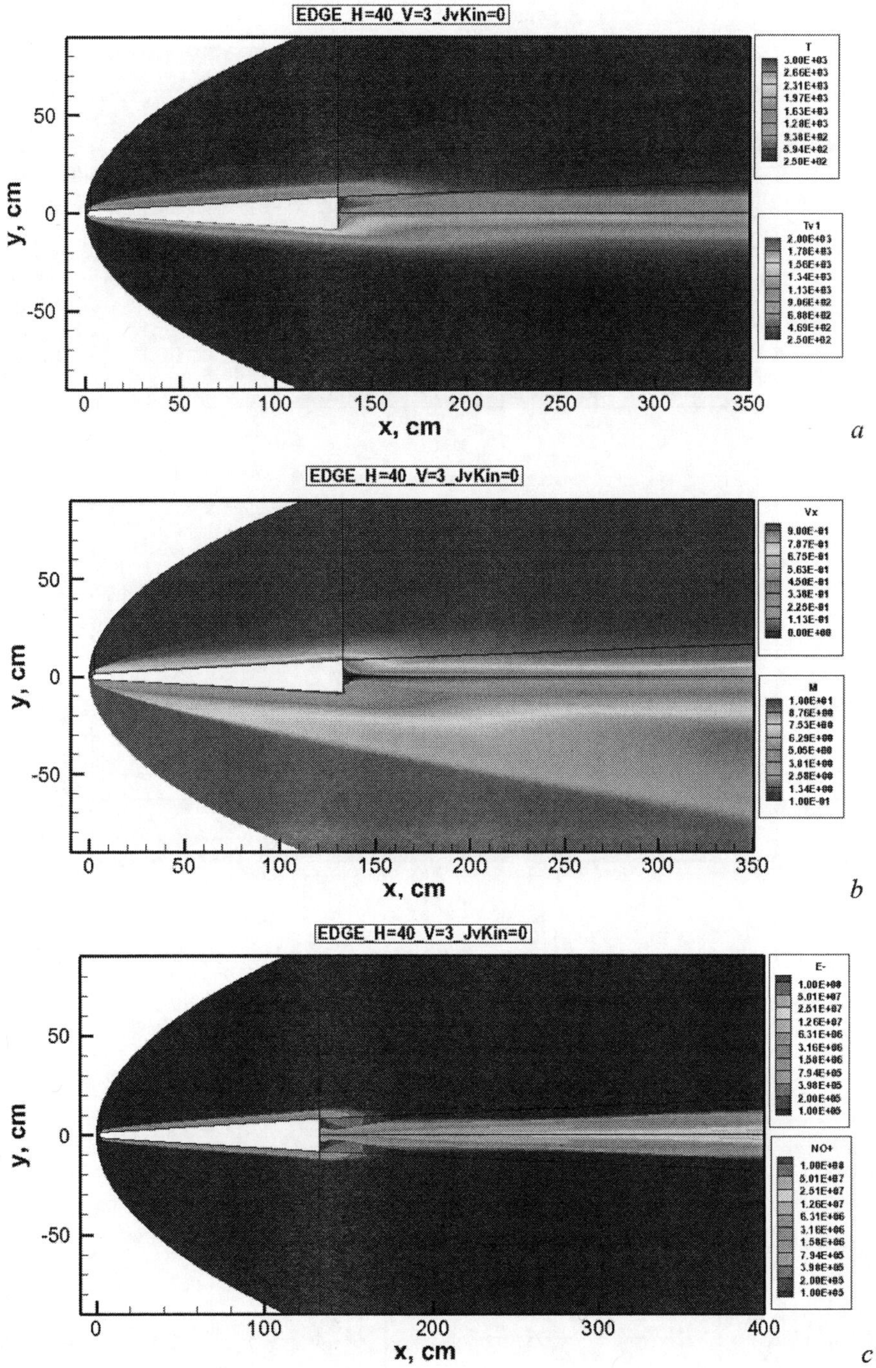

Figure 20. The fields of the translational temperature T (*a*, above) and the vibrational temperature T_V of N_2 (*a*, from below), the longitudinal velocity $Vx = u/V_\infty$ (*b*, from above) and the Mach number (*b*, from below), the electron concentration (*c*, from above) and NO^+ ions (*c*, from below). The temperature is in K, the concentration is in cm^{-3}; $H = 40\,km$, $V_\infty = 3\,km/s$.

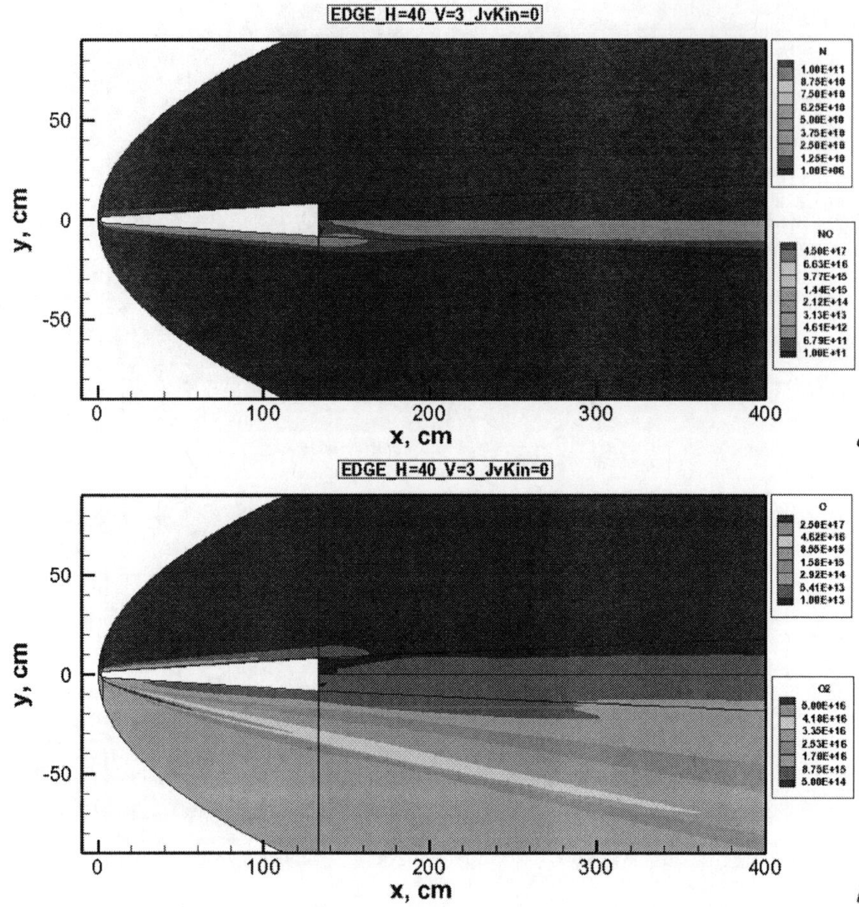

Figure 21. Fields of concentrations of N (a, above) and N_2 (a, below) molecules, atoms O (b, above) and molecules O_2 (b, below). Concentrations in cm^{-3}; $H = 40\,km$, $V_\infty = 3$ km/s.

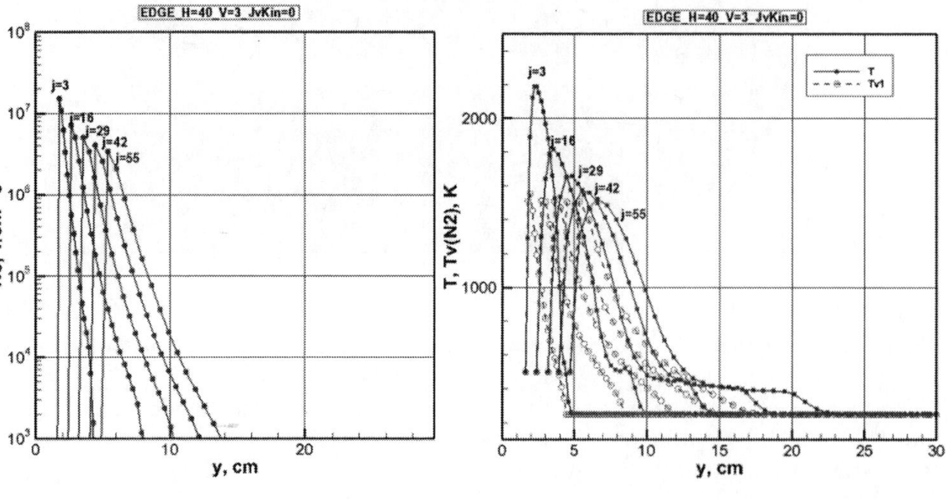

Figure 22. Axial distributions of the electron concentration (left), as well as translational and vibrational temperatures N_2 in 5 sections along the x axis. $H = 40\,km$, $V_\infty = 3\,km/s$.

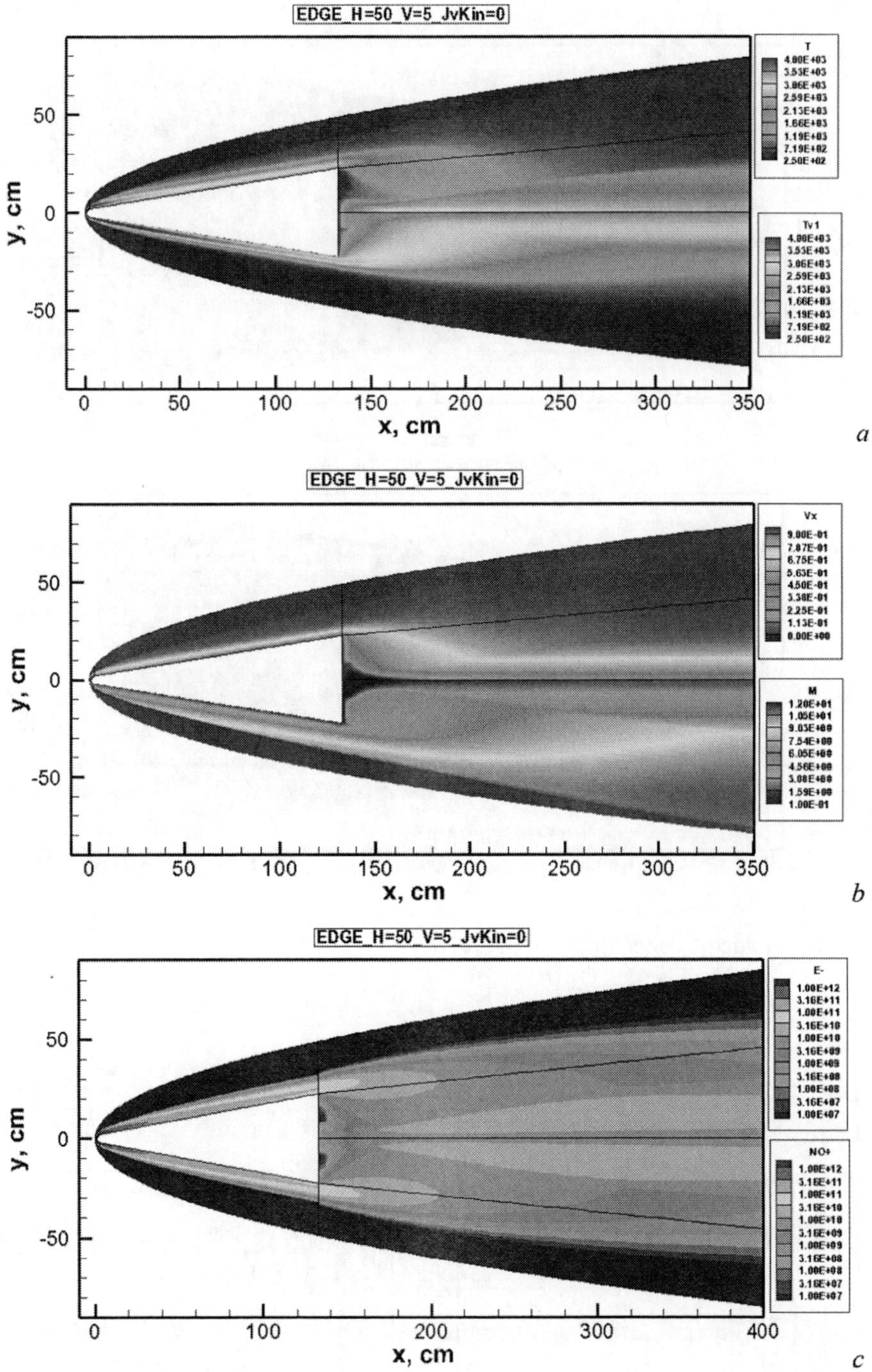

Figure 23. The fields of the translational temperature T (*a*, above) and the vibrational temperature T_V of N_2 (*a*, from below), the longitudinal velocity $Vx = u/V_\infty$ (*b*, from above) and the Mach number (*b*, from below), the electron concentration (*c*, from above) and NO^+ ions (*c*, from below). The temperature is in K, the concentration is in cm^{-3}; $H = 50$ km, $V_\infty = 5$ km/s.

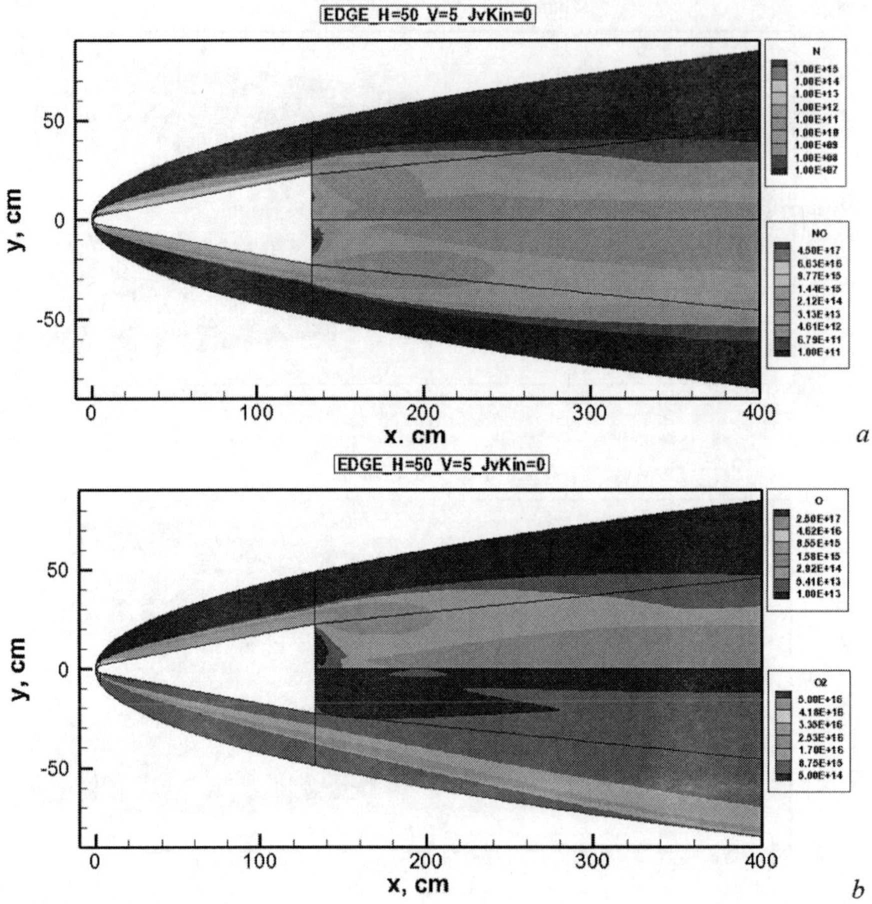

Figure 24. Fields of concentrations of N (*a*, above) and N_2 (*a*, below) molecules, atoms O (*b*, above) and molecules O_2 (*b*, below). Concentrations in cm^{-3}; $H = 50$ km, $V_\infty = 5$ km/s.

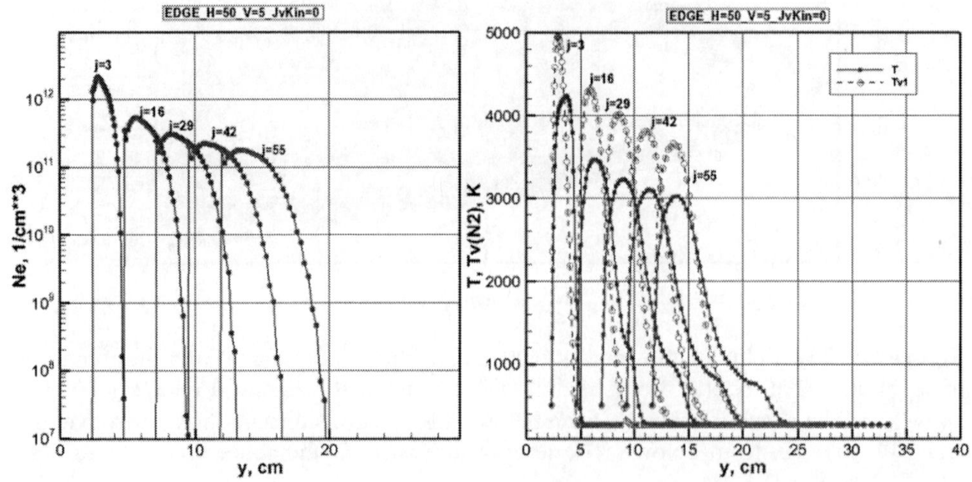

Figure 25. Axial distributions of the electron concentration (left), as well as translational and vibrational temperatures N_2 in 5 sections along the *x* axis. $H = 50$ km, $V_\infty = 5$ km/s.

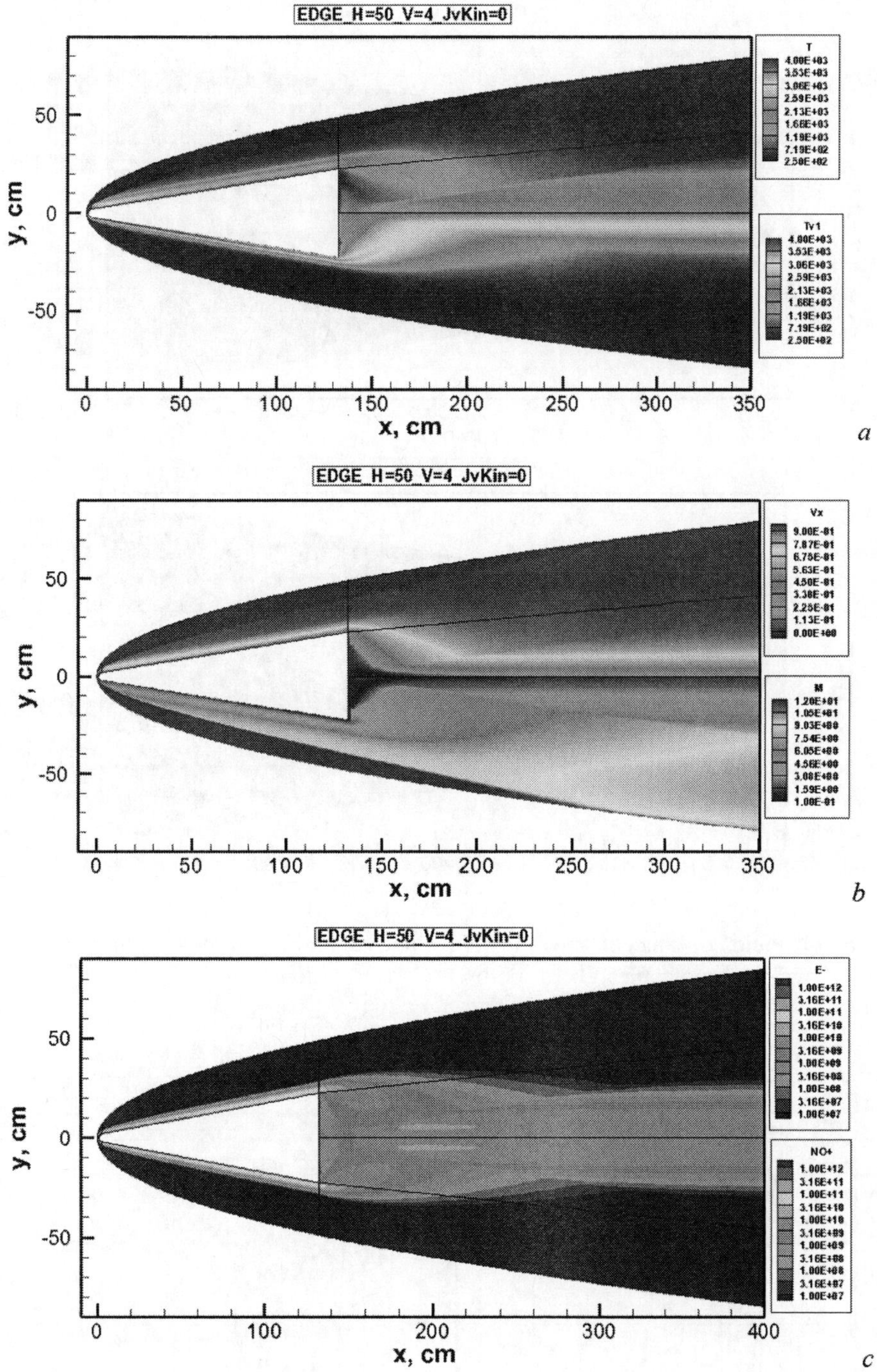

Figure 26. The fields of the translational temperature T (*a*, above) and the vibrational temperature T_V of N_2 (*a*, from below), the longitudinal velocity $Vx = u/V_\infty$ (*b*, from above) and the Mach number (*b*, from below), the electron concentration (*c*, from above) and NO^+ ions (*c*, from below). The temperature is in K, the concentration is in cm^{-3}; $H = 50\,km$, $V_\infty = 4\,km/s$.

IOP Conf. Series: Journal of Physics: Conf. Series **1009** (2018) 012022 doi:10.1088/1742-6596/1009/1/012022

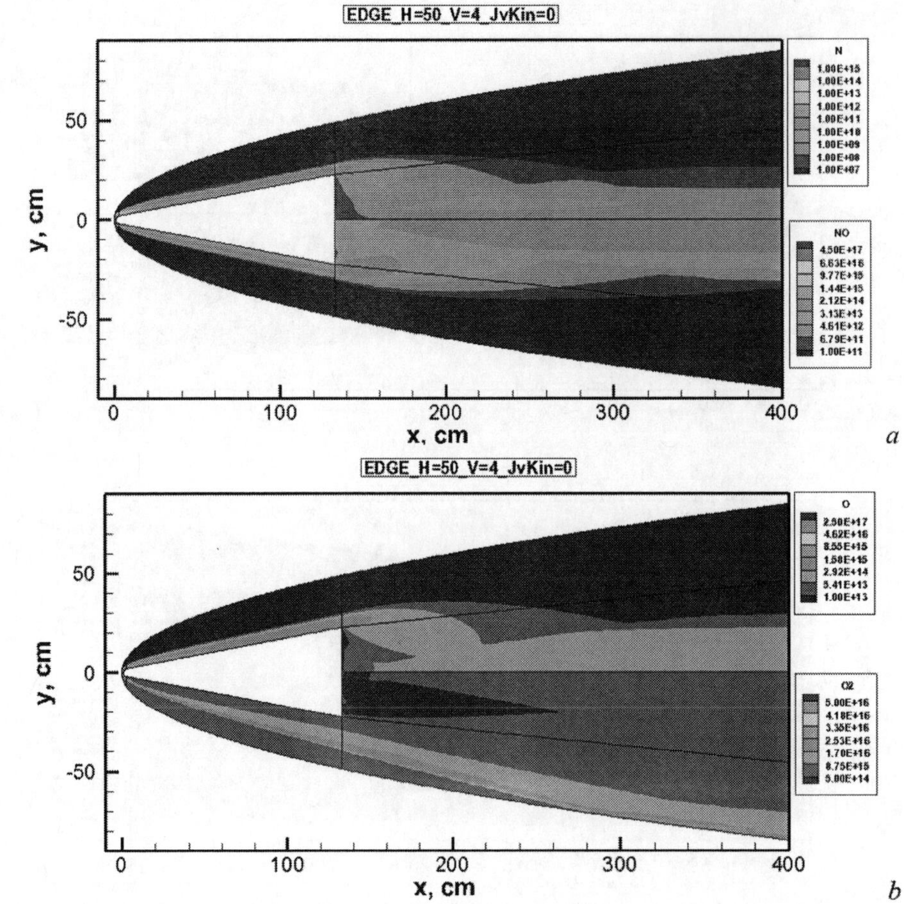

Figure 27. Fields of concentrations of N (a, above) and N_2 (a, below) molecules, atoms O (b, above) and molecules O_2 (b, below). Concentrations in cm^{-3}; $H = 50$ km, $V_\infty = 4$ km/s.

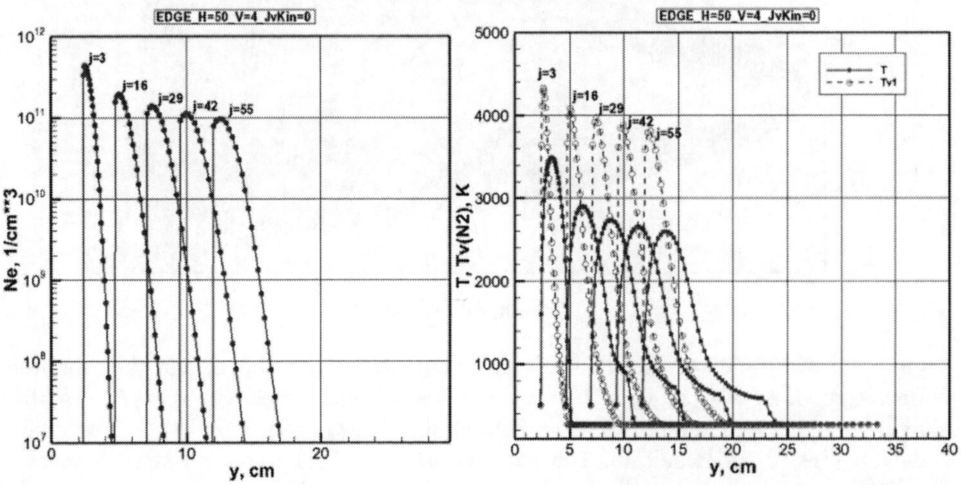

Figure 28. Axial distributions of the electron concentration (left), as well as translational and vibrational temperatures N_2 in 5 sections along the x axis. $H = 50$ km, $V_\infty = 4$ km/s.

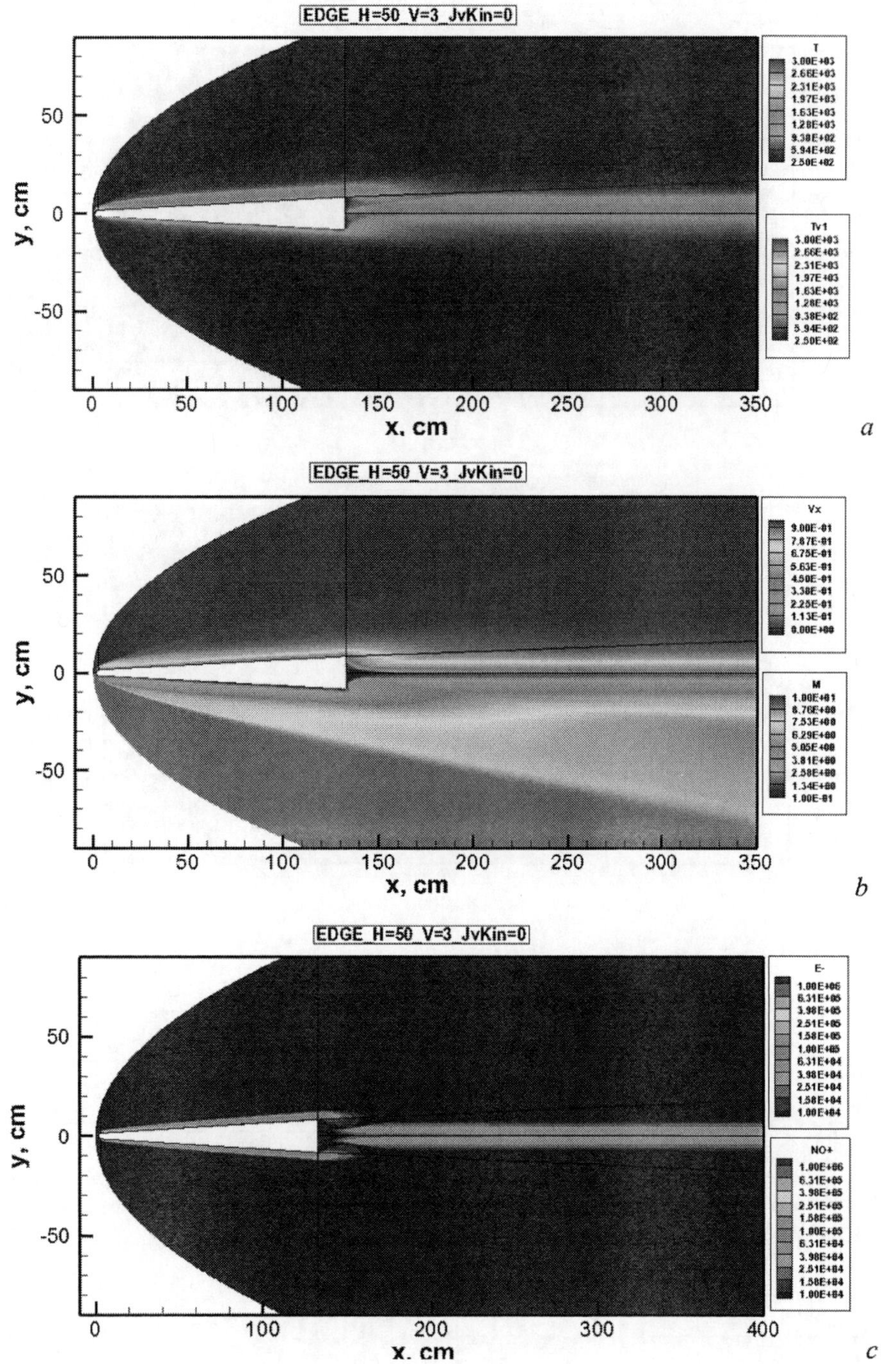

Figure 29. The fields of the translational temperature T (*a*, above) and the vibrational temperature T_V of N_2 (*a*, from below), the longitudinal velocity $Vx = u/V_\infty$ (*b*, from above) and the Mach number (*b*, from below), the electron concentration (*c*, from above) and NO^+ ions (*c*, from below). The temperature is in K, the concentration is in cm^{-3}; $H = 50$ km, $V_\infty = 3$ km/s.

IOP Conf. Series: Journal of Physics: Conf. Series **1009** (2018) 012022 doi:10.1088/1742-6596/1009/1/012022

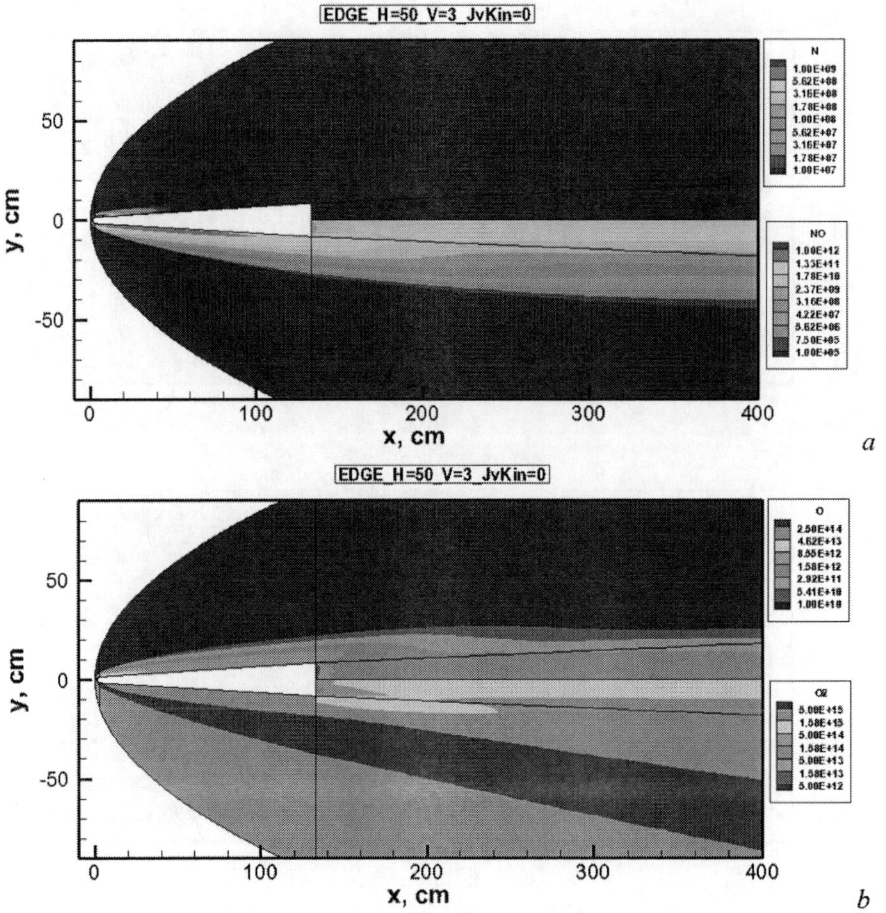

Figure 30. Fields of concentrations of N (a, above) and N_2 (a, below) molecules, atoms O (b, above) and molecules O_2 (b, below). Concentrations in cm^{-3}; $H = 50$ km, $V_\infty = 3$ km/s.

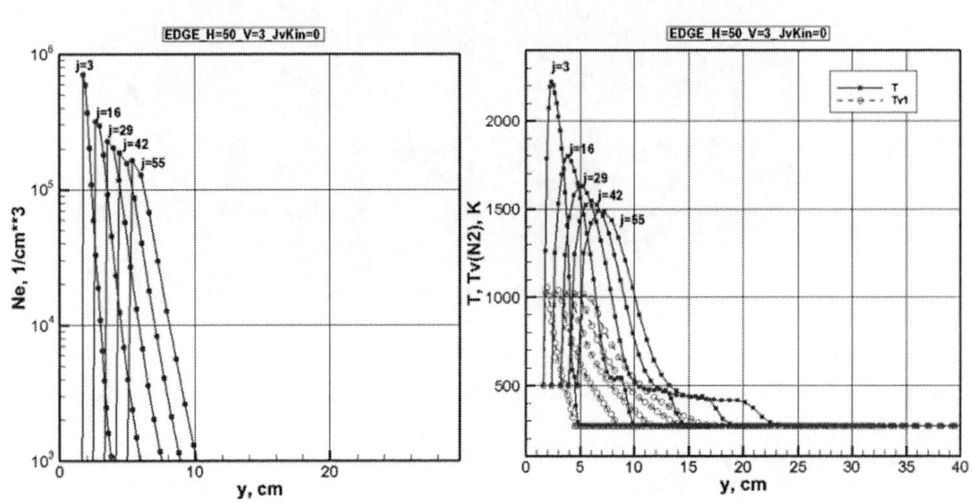

Figure 31. Axial distributions of the electron concentration (left), as well as translational and vibrational temperatures N_2 in 5 sections along the x axis. $H = 50$ km, $V_\infty = 3$ km/s.

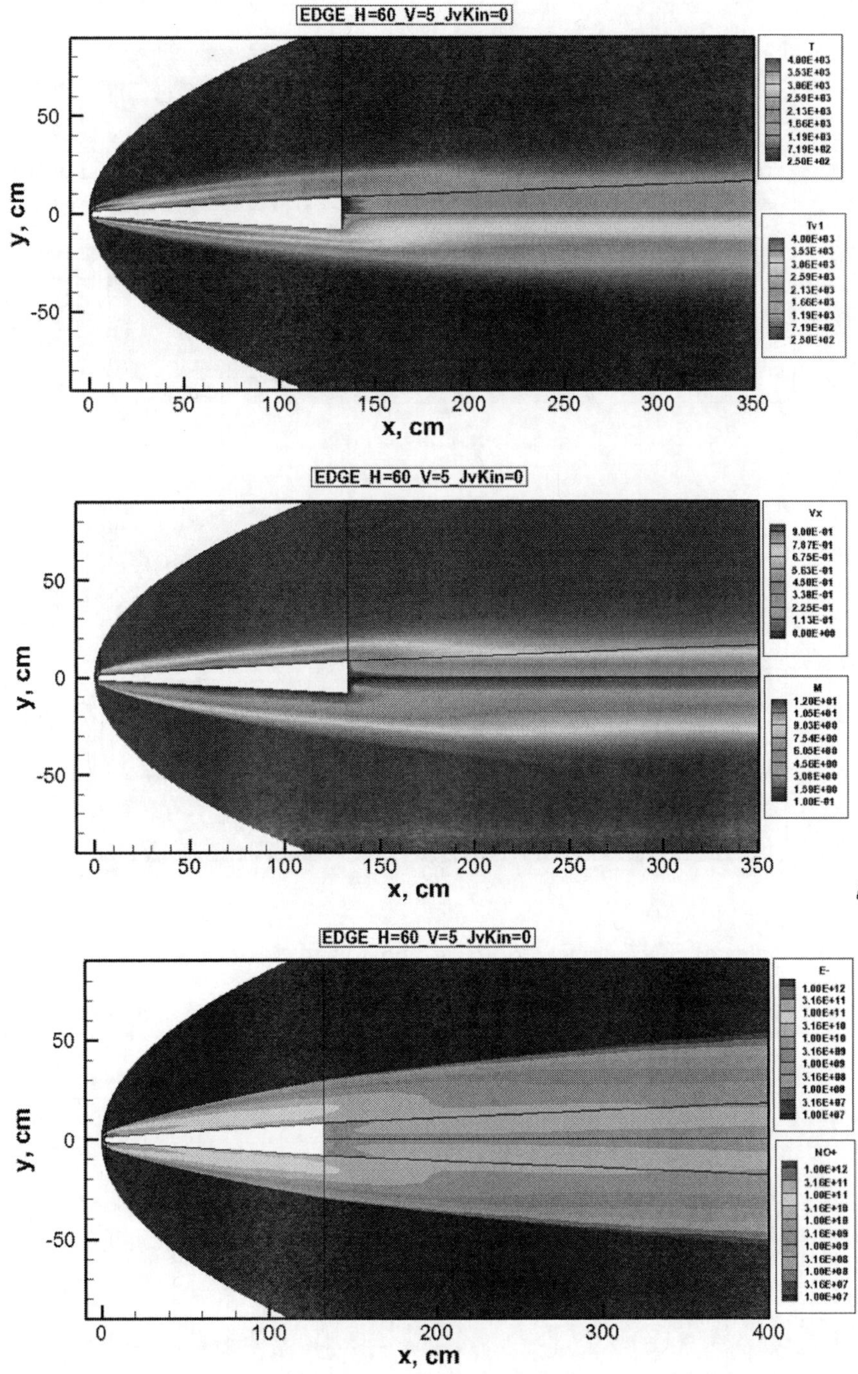

a, *b*, *c*

Figure 32. The fields of the translational temperature T (*a*, above) and the vibrational temperature T_V of N_2 (*a*, from below), the longitudinal velocity $Vx = u/V_\infty$ (*b*, from above) and the Mach number (*b*, from below), the electron concentration (*c*, from above) and NO^+ ions (*c*, from below). The temperature is in K, the concentration is in cm^{-3}; $H = 60$ km, $V_\infty = 5$ km/s.

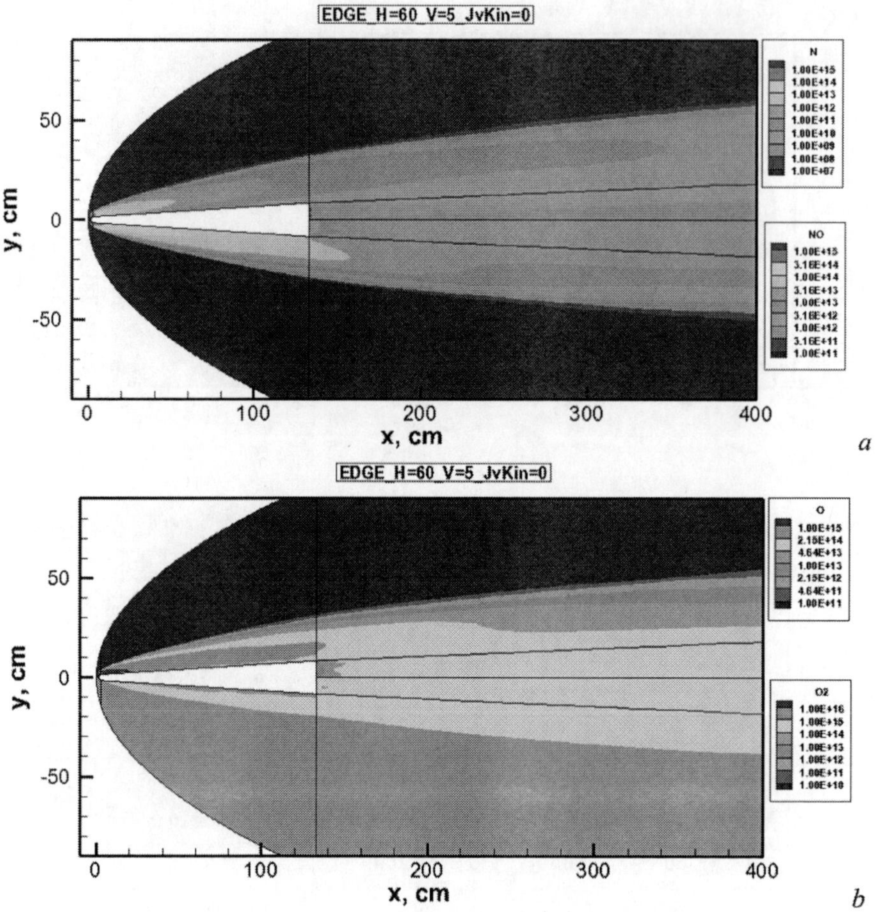

Figure 33. Fields of concentrations of N (a, above) and N_2 (a, below) molecules, atoms O (b, above) and molecules O_2 (b, below). Concentrations in cm^{-3}; $H = 60$ km, $V_\infty = 5$ km/s.

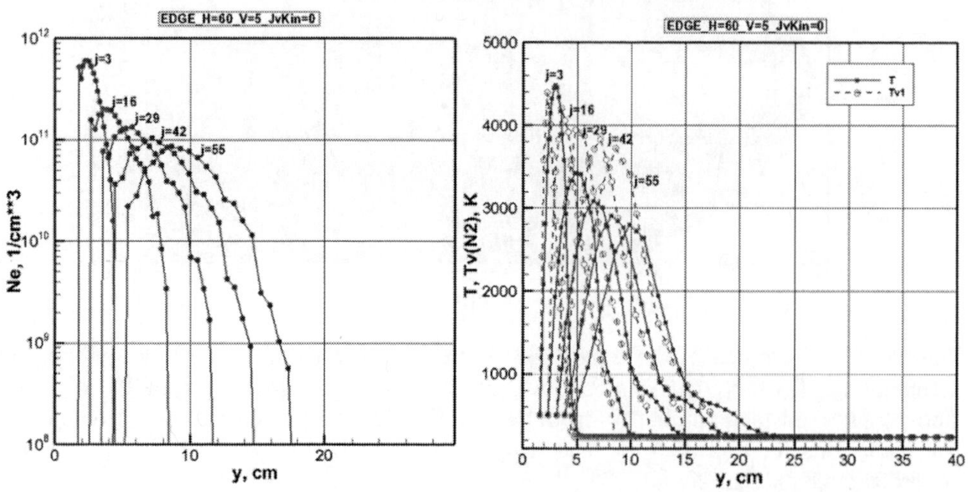

Figure 34. Axial distributions of the electron concentration (left), as well as translational and vibrational temperatures N_2 in 5 sections along the x axis. $H = 60$ km, $V_\infty = 5$ km/s.

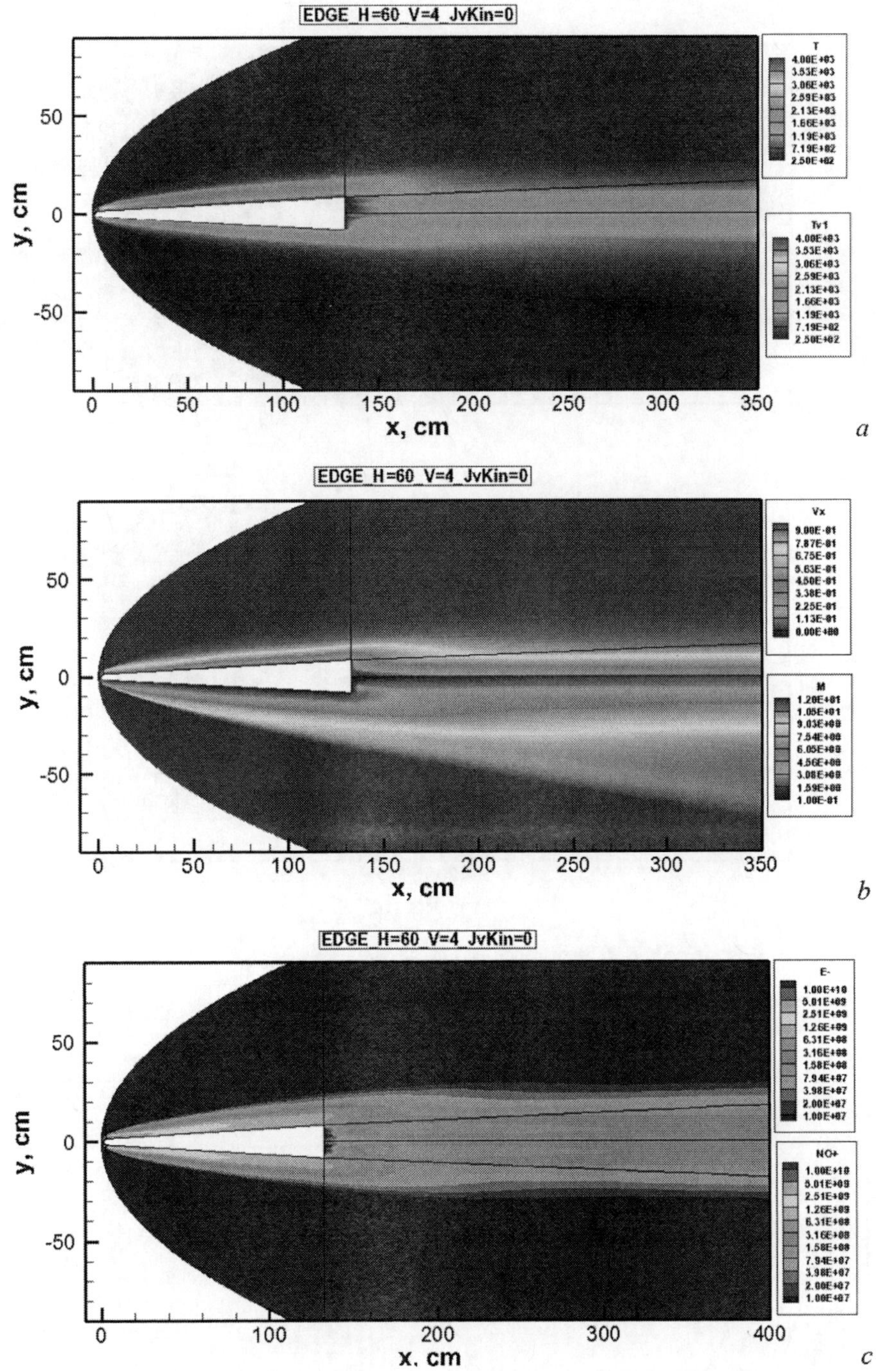

Figure 35. The fields of the translational temperature T (a, above) and the vibrational temperature T_V of N_2 (a, from below), the longitudinal velocity $Vx = u/V_\infty$ (b, from above) and the Mach number (b, from below), the electron concentration (c, from above) and NO^+ ions (c, from below). The temperature is in K, the concentration is in cm^{-3}; $H = 60$ km, $V_\infty = 4$ km/s.

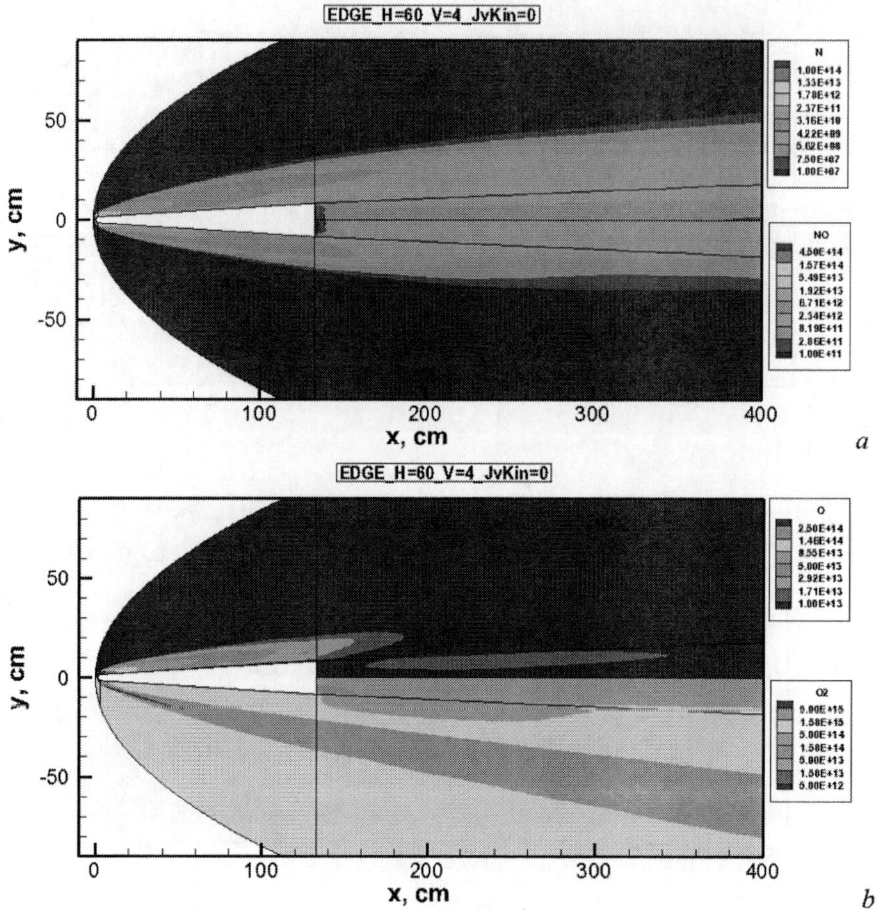

Figure 36. Fields of concentrations of N (*a*, above) and N_2 (*a*, below) molecules, atoms O (*b*, above) and molecules O_2 (*b*, below). Concentrations in cm^{-3}; $H = 60$ km, $V_\infty = 4$ km/s.

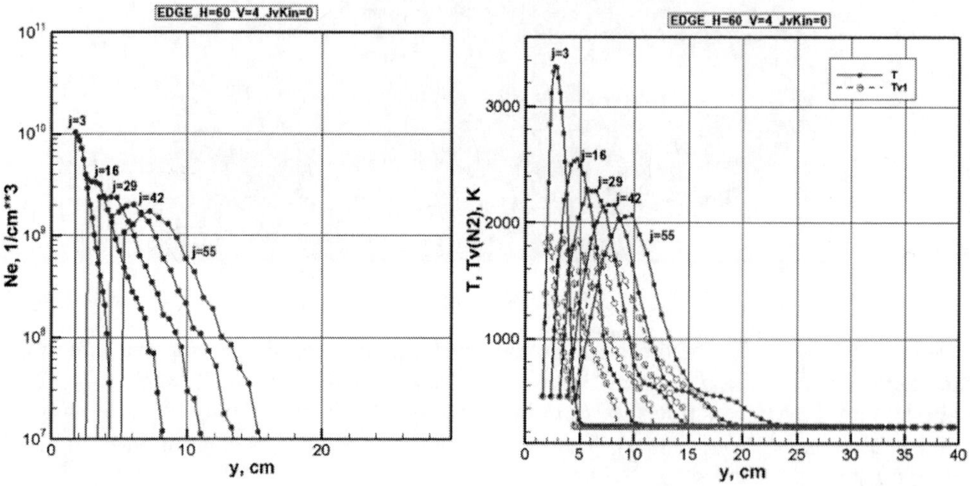

Figure 37. Axial distributions of the electron concentration (left), as well as translational and vibrational temperatures N_2 in 5 sections along the x axis. $H = 60$ km, $V_\infty = 4$ km/s.

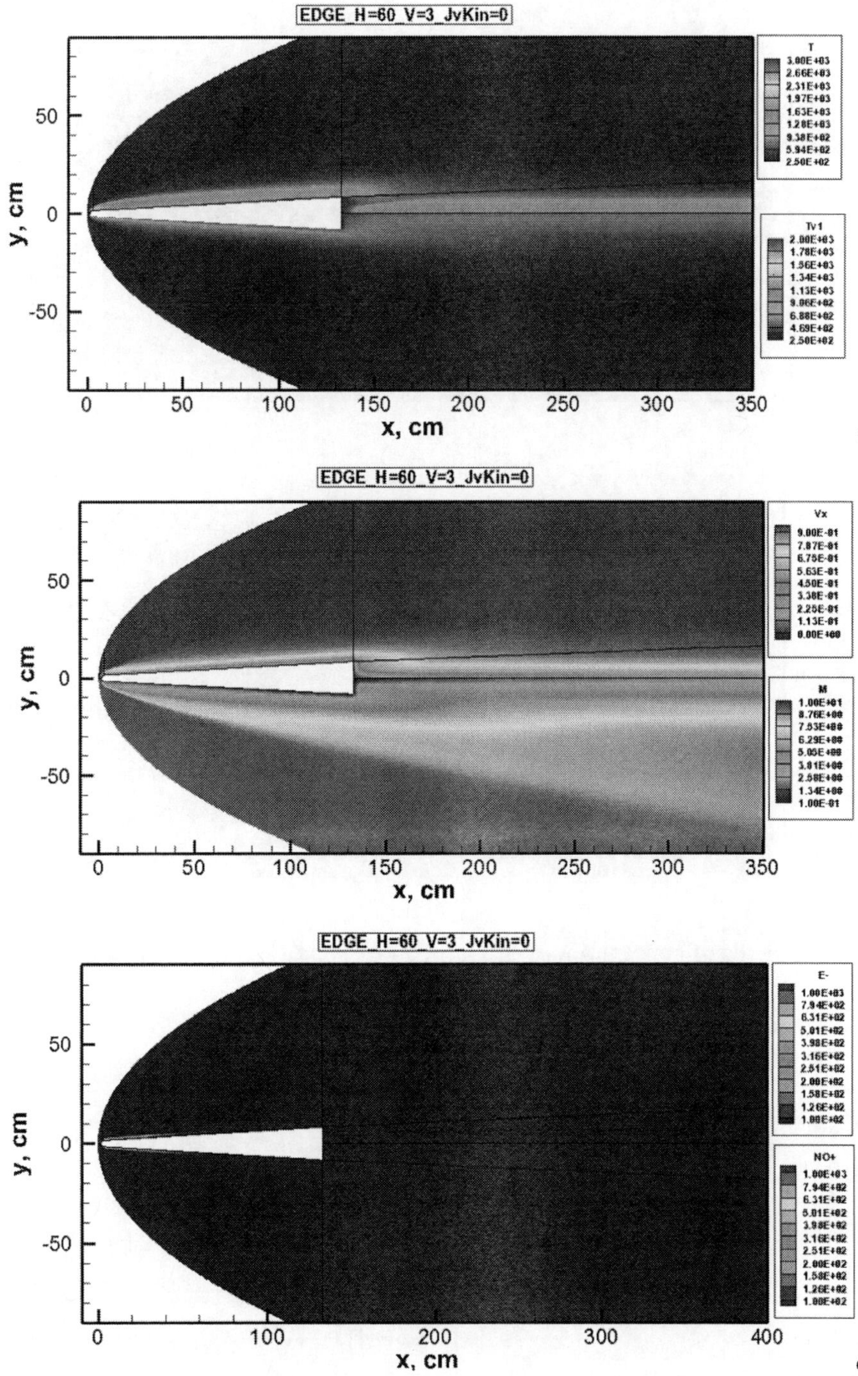

a, *b*, *c*

Figure 38. The fields of the translational temperature T (*a*, above) and the vibrational temperature T_V of N_2 (*a*, from below), the longitudinal velocity $Vx = u/V_\infty$ (*b*, from above) and the Mach number (*b*, from below), the electron concentration (*c*, from above) and NO^+ ions (*c*, from below). The temperature is in K, the concentration is in cm^{-3}; $H = 60$ km, $V_\infty = 3$ km/s.

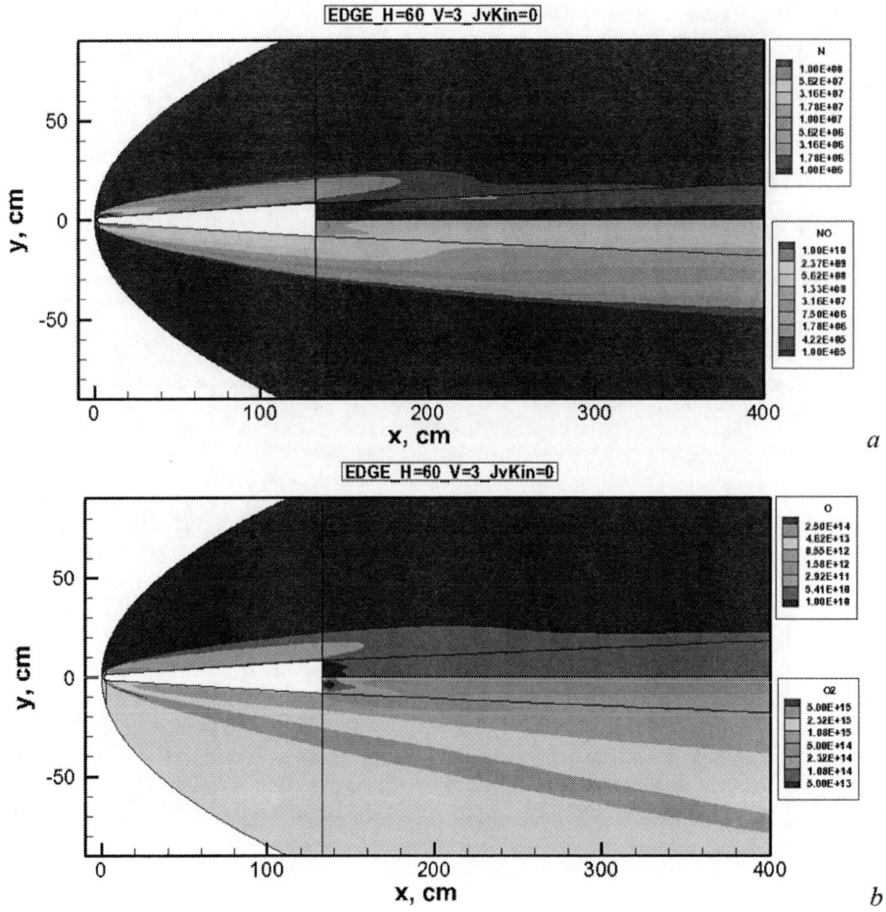

Figure 39. Fields of concentrations of N (a, above) and N_2 (a, below) molecules, atoms O (b, above) and molecules O_2 (b, below). Concentrations in cm^{-3}; $H = 60$ km, $V_\infty = 3$ km/s.

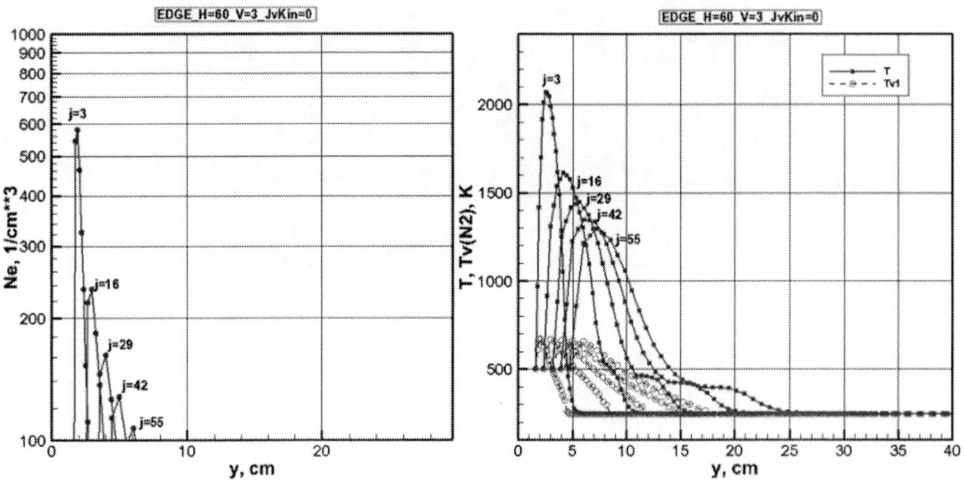

Figure 40. Axial distributions of the electron concentration (left), as well as translational and vibrational temperatures N_2 in 5 sections along the x axis. $H = 60$ km, $V_\infty = 3$ km/s.

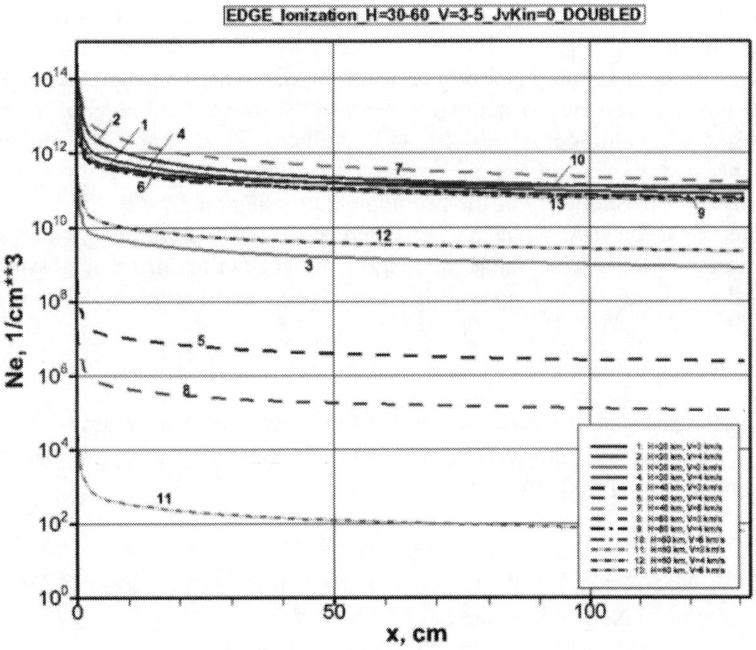

a

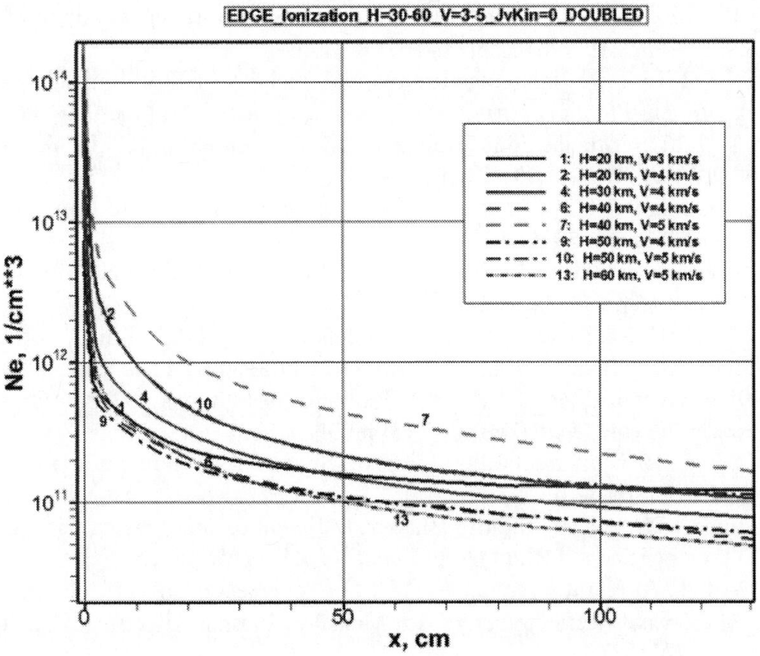

б

Figure 41. Distribution of electron concentrations along the surface of blunted wedges for all heights and velocities (*a*) and for relatively high velocities (*b*).

If at relatively high velocities the vibrational temperature T_V of N_2 molecules in the compressed layer at the lateral surface exceeds the translational temperature T, then at low velocities the reverse situation is observed: $T_V < T$. This is explained by the fact that at relatively low speeds at high altitudes the vibrational temperature T_V does not have time to reach the translational temperature T even near to blunted nose.

The distribution of the maximum electron concentrations along the generatrix of a blunt wedge in a wide range of heights and velocities is shown in figure 41. If the critical level of the electron density is $n_e^* \sim 10^{10}$ cm^{-3}, then at an altitude $H > 30$ km and a velocity $V_\infty < 3$ km/s, the ionization level is low. However, as mentioned above, the critical concentration depends not only on the parameters of the ionized air, but also on the frequency of the probing electromagnetic wave.

As the resume, it can be stated that the presented results of a preliminary analysis of aerophysics over a blunted wedge over a wide range of heights and velocities make it possible to determine a program for further studies of the nonequilibrium effects accompanying gas motion at moderate hypersonic speeds.

Acknowledgments
The work was performed within the framework of the Government program of basic research of the Russian academy of sciences (contract # AAAA– A17– 117021310372– 6) and partially under support of the RFBR grant # 016 – 01– 00379.

References

[1] Candler G V, MacCormack R W 1991 Computation of Weakly Ionized Hypersonic Flows in Thermochemical Nonequilibrium *J. of Thermophysics* **5** (3) pp 266–273
[2] Josyula E, Bailey W F 2003 Governing Equations for Weakly Ionized Plasma Flowfields of Aerospace Vehicles *J of Spacecraft and Rockets* **40** (6) pp 845–857
[3] Boyd I D 2007 Modeling of associative ionization reactions in hypersonic rarefied flows *The Phys. of Fluids* **19** 096102 pp 1–10
[4] Surzhikov S T 2015 Two-dimensional numerical analysis of flow ionization in the RAM-C-II flight experiment *Russian Journal of Chemical Physics B* **9** (1) pp 69– 86
[5] Surzhikov S T 2015 Spatial ionization effects of a shock layer in the RAM-C-II flight experiment *Doklady Physics* **60** (2) pp 89– 94
[6] Shang J S, Surzhikov S T 2012 Nonequilibrium radiative hypersonic flow simulation *Progress in Aerospace Sciences* **53** pp 46– 65
[7] Kozlov P V and Surzhikov S T 2017 Nonequilibrium radiation NO in shocked air *AIAA paper 2017–0157* 16 p
[8] Surzhikov S T 2012 Radiative-Collisional Models of Non-Equilibrium Aerothermodynamics of Entry Probes *J. of Heat Transfer – ASME Tr* **134**/031002-1 11 p
[9] Park C 1993 Review of Chemical-Kinetics Problems of Future NASA Missions, I: Earth Entries *J. of Thermophysics and Heat Transfer* **7** (3) pp 385– 398
[10] Surzhikov S T 2016 Numerical Rebuilding of Shock Layer Ionization for Two Flight Tests *AIAA Paper 2016-4305* 32 p
[11] Akey N D and Cross A E 1970 Radio Blackout Alleviation and Plasma Diagnostic Results from a 25000 Foot per Second Blunt Body Reentry *NASA TN D-5615*
[12] Grantham W L 1970 Flight Results of 25000 Foot per Second Blunt Body Reentry Experiment Using Microwave Reflectometers to Measure Plasma Electron Density and Standoff Distance *NASA TN D-6062*
[13] Jones W L Jr, Cross A E 1972 Electrostatic Probe Measurements of Plasma Parameters for Two Reentry Flight Experiments at 25000 Feet per Second *NASA TN D-6617*

IOP Conf. Series: Journal of Physics: Conf. Series **1009** (2018) 012023 doi:10.1088/1742-6596/1009/1/012023

Validation of Numerical Model of Penning Gas Discharge based on 2D/3V PIC-MCC Method

A S Dikalyuk[1,2,3]

[1]All-Russia Research Institute of Automatics, Moscow, 127055, Russia
[2]Ishlinsky Institute for Problems in Mechanics RAS, Moscow, 119526, Russia
[3]Moscow Institute of Physics and Technology, Dolgoprudny, 141701, Russia

aleks.dikalyuk@gmail.com

Abstract. The data obtained in the experiments conducted in University of California, Berkley are used to validate numerical model of Penning gas discharge. Numerical model is based on 2D/3V axisymmetric electrostatic particle-in-cell method implemented on unstructured grids. Elementary processes in the gas discharge plasma are simulated using Monte-Carlo collision method. Electrodynamic structure of the Penning gas discharge is obtained using mentioned numerical model under conditions relevant to experimental and analyzed. Distributions of the potential, electric field, electrons and ions number densities, temperatures, distribution functions are presented in the study. Validation of numerical model is performed by comparison of the dependence of discharge current on applied magnetic field.

1. Introduction

The plasma of Penning gas discharge is stable at relatively low pressure. This effect is achieved by placing plasma in quadrupole electric field and axial magnetic field. Electrons are trapped within these fields and produce substantial ionization in order the plasma to be selfsustained. Due to the fact that Penning gas discharge exists at low pressure, PIC-MCC method [1, 2] is a good approach for simulation of the considered physical system.

2D/3V axisymmetric electrostatic PIC-MCC method suitable for simulation of electrodynamic structure of the Penning gas discharge was implemented and analyzed in [3, 4]. Verification of the solver was performed on the basis of 2D Child-Langmuir problem. Results of verification were reported in [3]. In the present study numerical experiments were conducted in order to validate numerical method against available experimental data.

In the past ten years several studies has been performed dedicated to the experimental investigation of the Penning gas discharge [5–11], however not all of them is possible to use for the validation purposes. Kreindel and Ionov [5] conducted experiments with annular anode. They were able to record the dependence of discharge and extracted current on the magnetic field. Due to the usage of annular anode it was possible to make a photo of the discharge. However in the original study [5] several parameters (diameter of the annular anode and type of the gas) are missing which make impossible to use this work for the validation purposes.

In [6] volt-ampere characteristics of the Penning gas discharge in molecular hydrogen were obtained for the various geometric parameters of discharge chamber and H_2 pressure. Much data were gathered and presented in [6] however usage of anodes ("long" and "short") and anticathode made of wire mesh as well as annular magnets introduce uncertainties in the process of analysis of the discussed experimental results.

The studies [7, 8] were dedicated to the development of the compact Penning discharge chamber suitable for the usage as the ion source. Experimental investigation of the proposed design of the Penning discharge chamber was conducted for various gasses (molecular hydrogen, helium) in the wide range of pressures and anode voltages. Geometry of the proposed discharge chamber is given in details however lack of information regarding the permanent cylindrical magnet and substantial inhomogeneity of the magnetic field in the system does not allow using this work directly for the validation of the developed solver.

In the studies [9, 10] spectral radiation of the large volume Penning gas discharge in helium was investigated. Ring and double-ring anode configuration were considered. Parameters of the conducted experiments are given in details. For the purposes of validation of the computer code volt-ampere characteristics measured in the wide range of gas pressure can be used. However use of annular anode implies difficulties in the formulation of the boundary conditions in the process of numerical simulation.

In [11] classical Penning gas discharge has been studied. Cylindrical anode is placed between two circular cathodes with small gaps (relative to the size of anode) between cathode and anode. Experimental parameters relevant for the numerical simulation are given the study. External magnetic field is produced by electric coil. For the validation of the PIC-MCC solver dependence of the discharge current on the magnetic field is given in [11].

In the study analysis of electrodynamic structure of the Penning gas discharge relevant for the conditions of the experiments presented in [11] is performed. Distribution of potential, electric field, electrons and ions number densities, temperatures and distribution functions are given. Validation of the numerical model is performed by calculating dependence of the discharge current on the magnetic field and comparing it against available experimental data.

2. Brief description of the numerical model of Penning gas discharge

Particle-in-cell method introduces a concept of macroparticle [1, 2]. Macroparticle is a computational particle that simulates behaviour of a large number of real plasma particles (electrons or ions). Macroparticles follow the trajectory of real particles in the plasma since the Lorentz force depends only on the charge-to-mass ratio [12]. The movement of macroparticles changes the distribution of electromagnetic field in the system. It means that the whole simulation process is self-consistent. It can be shown that actually PIC method is a mean for solution of Vlasov-Maxvell system of equations [13].

For the numerical solution of the Penning gas discharge problem numerical model based on 2D/3V axisymmetric electrostatic PIC–MCC method was used. In order to briefly characterize numerical methods and approaches that have been used in the computer implementation of the PIC-MCC method mentioned above let us consider a computational cycle shown in figure 1.

The first stage of a computational cycle is the solution of the Poisson equation. It is formulated for the axisymmetric geometry:

$$\Delta\varphi = \frac{1}{r}\left(\frac{\partial}{\partial r}\left(r\frac{\partial\varphi}{\partial r}\right) + \frac{\partial}{\partial z}\left(r\frac{\partial\varphi}{\partial z}\right)\right) = \frac{\rho(r,z)}{\varepsilon_0} \tag{1}$$

In equation (1) φ is the electrostatic potential, $\rho(r,z)$ is the charge density and ε_0 is the electric constant. For the boundary value problem the boundary conditions can be formulated in the following form:

$$\left(\alpha_1\varphi + \alpha_2\frac{\partial\varphi}{\partial\vec{n}}\right)\bigg|_{\partial D} = g(r,z) \tag{2}$$

For the numerical solution of equation (1) the computational domain is subdivided into triangles. Usage of unstructured grids for the numerical simulation of gas discharges is necessary since discharge chamber may be of complex shape. Descritization of equation (1) is carried out using cell-centered finite volume method [14]. The descritization process leads to a problem of calculation of potential derivative along the edge of the triangle [3]. For this purpose the least square method was used.

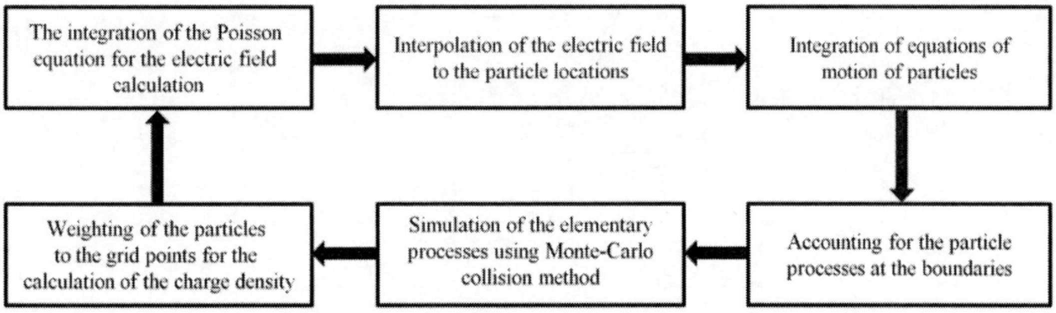

Figure 1. PIC-MCC computational cycle.

The corresponding template consists of all triangles that contain at least one point of the considered edge (figure 2). Geometrical multigrid method is used for the numerical solution of the system of linear algebraic equations resulted from the descritization process. Implementation of the multigrid procedure on the unstructured grid is performed using overset meshes approach [15]. Obtained potential distribution as the solution of the Poisson equation is used for the calculation of the electric field within the computational domain. This process is performed in two stages. At the first stage least square method and template shown in figure 3 are used in order to calculate the electric field in the cell centers. At the second stage values of electric field in the cell centers are interpolated to the vertices of the mesh [16, 17].

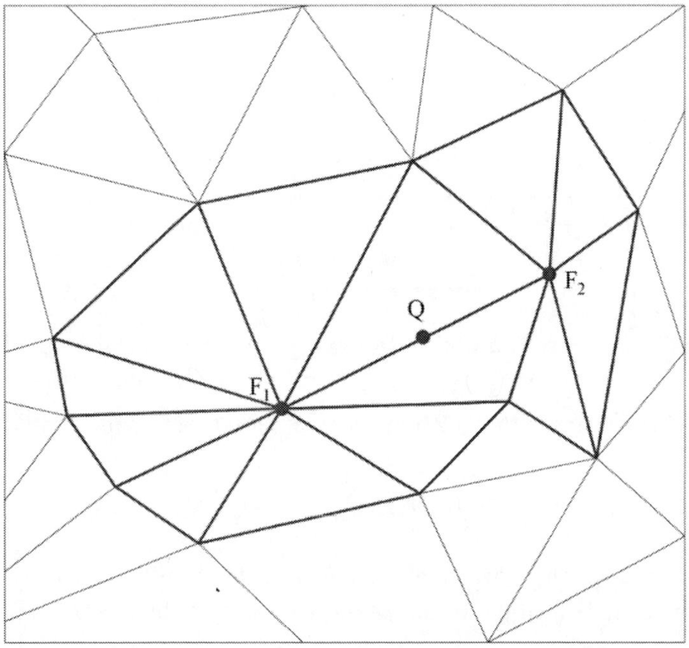

Figure 2. Template for calculation of potential derivative along the edge F_1F_2.

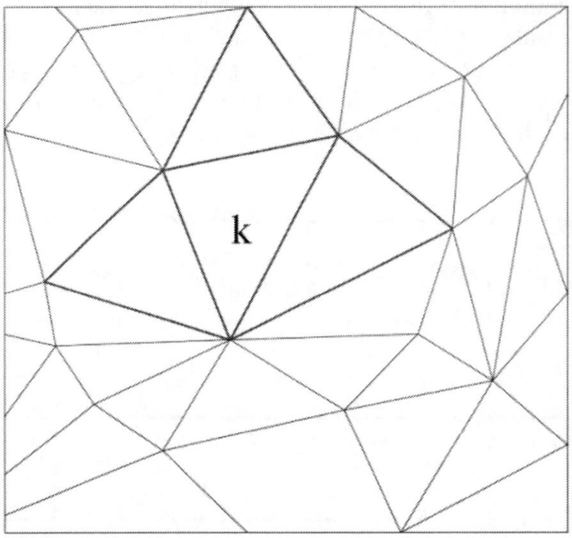

Figure 3. Template for the calculation of electric field in the cell centers.

Development of the particle-in-cell method on the unstructured grids requires special procedures of interpolation of the electric field to the particles position for the Lorentz force calculation and localization of macroparticles relative to the mesh elements (particle search algorithm) [13, 18]. These procedures constitute the second stage of the computational cycle. Both of these procedures can be easily implemented using local basis functions $N_n^k(z,r)$ constructed for the n-th node (local numbering of triangle nodes, varies from 1 to 3) of the k-th element (figure 4).

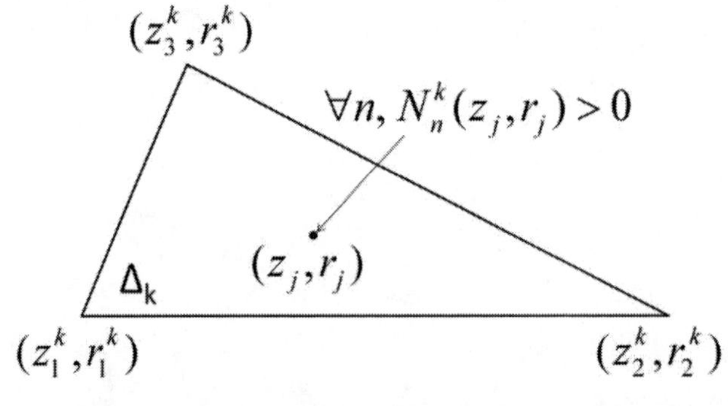

Figure 4. Local basis functions formalism used in the unstructured PIC method.

For the interpolation procedure the following property of the local basis function is important. If the values of interpolated function f are known in the k-th triangle nodes (z_m^k, r_m^k), then for every point with coordinates (z_j, r_j) that lies inside triangle:

$$f(z_j, r_j) = \sum_{n=1}^{3} N_n^k(z_j, r_j) f(z_n^k, r_n^k) \tag{3}$$

For the macroparticle search algorithm the following property is used [13, 18]. If the macroparticle lies within the k-th triangle then the values of all local basis functions $N_n^k(z_p, r_p)$ are nonnegative, where z_p and r_p are the axial and radial coordinates of the considered macroparticle.

In the third stage of the computational cycle one has to solve equations of particles motion. In the study only nonrelativistic case is considered. The resulting system of equations for the case of cylindrical coordinate system has singularity when particle with nonzero angular momentum approaches to the axis of symmetry [19]. If one is going to solve numerically this system it has to deal with the problem. Boris [12, 19] has proposed the numerical procedure which successfully overcomes mentioned above difficulty using Cartesian–cylindrical coordinate systems transformation.

Another problem is concerned with the Boris numerical method itself. Boris method is of the class of leap-frog methods, which means that velocities of the particles and their coordinates are not known in the same moments of time, however initial conditions are formulated in the same moment of time. In order to initialize calculation procedure correctly one has to obtain appropriate velocity of the particle. This problem was formulated and solved in [19].

On the next stage of computational cycle processes at the boundaries have to be considered. For example, plasma particles hitting the metal electrodes generate electric current. Dielectric boundaries are able to accumulate charge of plasma particles and impact on the distribution of the potential inside the discharge chamber [20]. Ions and electrons of plasma can hit surfaces of discharge chamber and generate fluxes of charge particles directed back to the gas discharge plasma due to the processes of secondary electron emission, ion-electron emission [21].

Accounting for the secondary emission processes at the boundaries can be implemented according to method described in [20]. Number of particles injected from the boundary into the computational area due to boundary-macroparticle collision is determined using the random number R uniformly distributed on the interval $[0,1]$. If $R<(\gamma-\text{int}(\gamma))$ then number of particles to be injected into the computational area is $N_{inj}=\text{int}(\gamma)+1$ otherwise $N_{inj}=\text{int}(\gamma)$. A model of the corresponding process has to be applied for the initialization of the new particles.

It is possible to account for elementary processes in the PIC algorithm using the Monte-Carlo collision approach based on the null collision method [22-24]. Monte-Carlo collision method consists of the following steps [23]. Calculation of the kinetic energy of the particle i of sort s:

$$\varepsilon_i = 0.5 m_s V_i^2 \tag{4}$$

Determining the magnitude of the total collision cross-section corresponding to this energy:

$$\sigma_{ttl}(\varepsilon_i) = \sigma_1(\varepsilon_i) + \ldots + \sigma_N(\varepsilon_i) \tag{5}$$

Here $\sigma_j(E_i)$ for $1 \le j \le N$ is the cross section of the j-th type of collision between the species s and the target species. The collision probability is calculated using the following formula:

$$P_i = 1 - \exp(-\Delta t V_i \sigma_{ttl}(\varepsilon_i) n_{trg}(\vec{x_i})) \tag{6}$$

In (6) V_i and $\vec{x_i}$ are the velocity and position of the i-th particle correspondingly, $n_{trg}(\vec{x_i})$ is the local density of the target species at the position of the i-th particle. A collision occur if a random number uniformly distributed on the interval $[0,1]$ is less then P_i. If the collision occurs then one has to determine the collision type using another random number. After that new particles have to be initialized based on the model [22-25] of the process corresponding to the given collision type.

Disadvantage of the approach presented above is due to the necessity of calculation of P_i for each particle in the simulation. In order to overcome this difficulty the null collision method was proposed. Detailed description of the method is given in [24]. The idea of the method is to introduce the constant collision frequency ν in the following manner:

$$\nu = \max_{x,E}\left(n_{trg}\sigma_{ttl}V\right) = \max_{x}\left(n_{trg}\right)\max_{E}\left(\sigma_{ttl}V\right) \tag{7}$$

IOP Conf. Series: Journal of Physics: Conf. Series **1009** (2018) 012023 doi:10.1088/1742-6596/1009/1/012023

Physically it means that additional collisional process is introduced with collision frequency of the special form. This collisional process is called null collision since no real interaction occurs. Then one can determine maximum fraction of the total number of particles of sort s which experience collision:

$$P_{null} = 1 - \exp(-\nu' \Delta t) \tag{8}$$

The colliding particles are chosen randomly among all the particles of the given sort s participating in the simulation (eliminating duplicates), and each particle is checked for the type of the collision.

The last stage of computational cycle is the calculation of the charge density of plasma particles. These fields are used for the construction of source term for the Poisson equation. In case of unstructured grid and cylindrical coordinate system one can utilize the formalism of local basis function [13, 18] in order to perform the calculations. The corresponding formula can be written as follows [13]:

$$q_n = \frac{1}{V_n} \sum_k \sum_j \frac{Q_j N_n^k (z_j, r_j)}{2\pi r_j} \tag{9}$$

In (9) summation is performed over all particles j that are in triangle k that contains the node n. z_j and r_j are the axial and radial coordinates of macroparticle, Q_j is the charge of macroparticle, in the two-dimensional case V_n is calculated according to the following formula:

$$V_n = \sum_k \frac{S_k}{3} \tag{10}$$

In (10) S_k is the area of the triangle k that contains node n.

3. Problem formulation for simulation of experiments [11]

Validation of the numerical model is performed on the basis of experiments, conducted in University of California, Berkley [11]. In the figure 5 schematic view of experimental setup is shown.

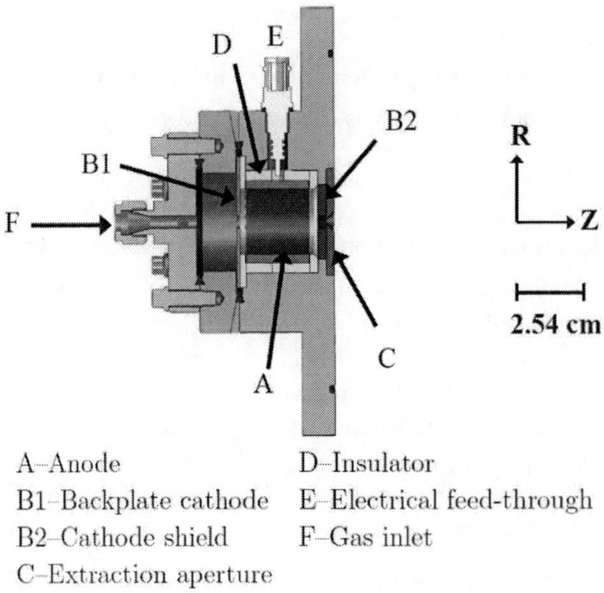

Figure 5. Schematic view of the experimental setup (figure is taken from [11]).

Cylindrical anode (A) with the length of 2.54 cm and diameter of 2.54 cm is placed inside the metallic chamber and isolated from it with insulator (D). There are two circular cathodes (B1, B2) on the both sides of anode. Cathodes are electrically isolated from the anode. Diameter of cathodes is 2.54 cm. The cathode B2 contains extraction orifice. Diameter of the extraction orifice is 0.2 cm. Electrodes of the discharge chamber are made of aluminium. External magnetic field in the system is created by an electromagnetic coil. This allowed varying the magnetic induction in the range 0.02÷0.05 T. In the experiments molecular hydrogen was used. The pressure of the gas was varied in the range 0.8÷1.0 mTorr. Anode potential was changed in the range 600÷800 V. In the experiments [11] dependence of the discharge current on the magnetic induction was measured (figure 6).

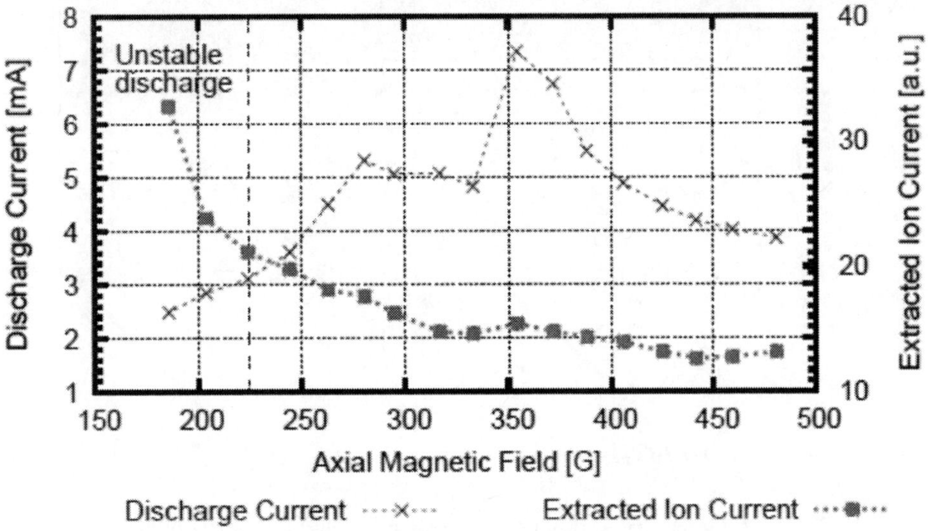

Figure 6. Experimental measurement of the dependence of the discharge current on the axial magnetic field (figure is borrowed from [11]).

Analysis of the described experiments was performed using the mesh presented in figure 7. Unstructured computational mesh consists of 2648 nodes and 5099 triangles. Computational mesh is uniform (no thickening). The boundaries of the computational domain coincide with the corresponding electrodes (see figure 5). Numerical simulation was performed at the anode voltage 800 V (cathodes are grounded), molecular hydrogen pressure 0.8 mTorr. Magnetic induction varied in the range 0.02÷0.033 T. For the numerical simulation time step is $4 \cdot 10^{-12}$ s.

Two kinds of macroparticles were taken into account: electrons and H_2^+ ions. At the beginning of the numerical simulation 120000 electrons and 120000 H_2^+ ions were placed in the cylindrical region of diameter of 1 cm and length of 1 cm in the center of computational area. Initial number density of charged particles is $1.5 \cdot 10^9$ cm^{-3}. Initial velocities of the macroparticles were sampled using Maxwell distribution at temperature 300 K.

Elementary processes accounted for in the study are elastic electron scattering on H_2 molecules ($e+H_2 \rightarrow e+H_2$) and ionization of H_2 molecules by electrons ($e+H_2 \rightarrow H_2^+ +e+e$). The data on the cross-section for these processes were taken from [26]. Also ion-electron emission from the cathode has been taken into account. The data on the dependence of the electron yield per ion versus energy was taken from [27].

In the following section results of the validation of the numerical model presented above on the basis of the experiments described in [11] for various magnetic induction are going to be presented and analyzed.

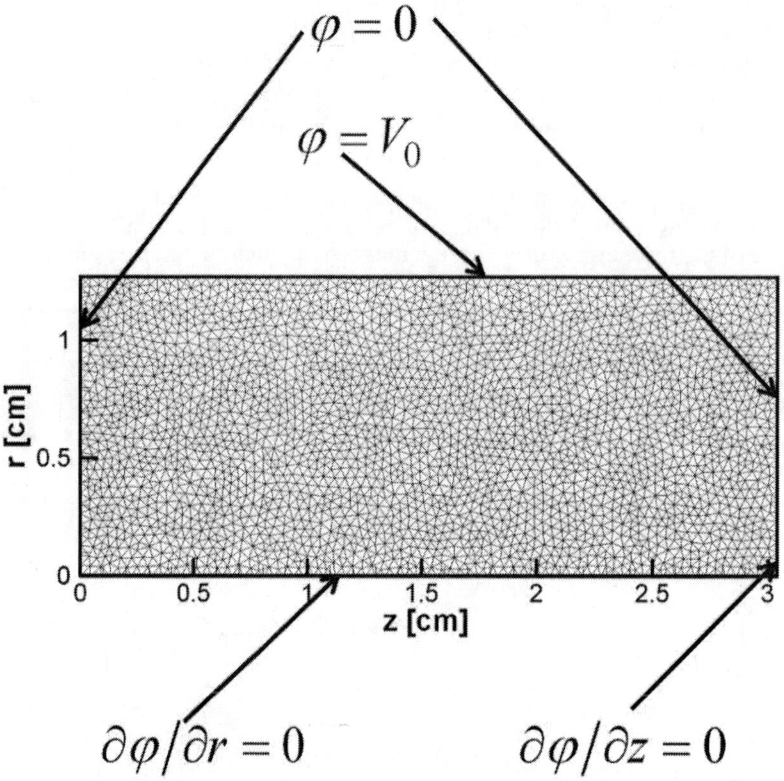

Figure 7. Geometry of the discharge chamber and computational mesh.

4. Numerical results of Penning gas discharge simulation

Results of numerical simulation of Penning gas discharge at conditions relevant to experiments [11] are presented in this section. In details we will show and analyze results obtained at $B_z = 0.033$ T.

In the figure 8 temporal evolution of number of macroparticles in computational area is presented. The plot indicates that plasma tends to stationary state. Number of macroelectrons in stationary state is about 1.16 millions, number of macroions in stationary state is about $7 \cdot 10^5$. Weight of each macroparticle is about 40000.

In the figures 9–12 distribution of the potential φ [V] and E_r [V/cm], E_z [V/cm] components of electric field in the stationary state are presented. In the figure 9 direction (length of the arrow does not reflect absolute value of the electric field) of the electric field is indicated with black arrows. It is easy to observe that the potential reaches minimum (~50 V) in the region r~0.45 cm. In this region electric field has only its z–component (figure 9). Potential rises to the anode which is obvious, however one can observe rise of the potential near the axis of symmetry (up to ~116 V).

E_z component of the electric field is equal ~10 V/cm in the region of the Penning gas discharge existence (center of the computational area, figure 10). The maximum absolute value of the E_z component of the electric field is about 4 kV/cm. This maximum value is reached near the edges of anode.

E_r component of the electric field reaches the local maximum (~500 V/cm) in the region r ~ 0.10 cm. The maximum absolute value of the E_r is reached near the anode and its edges.

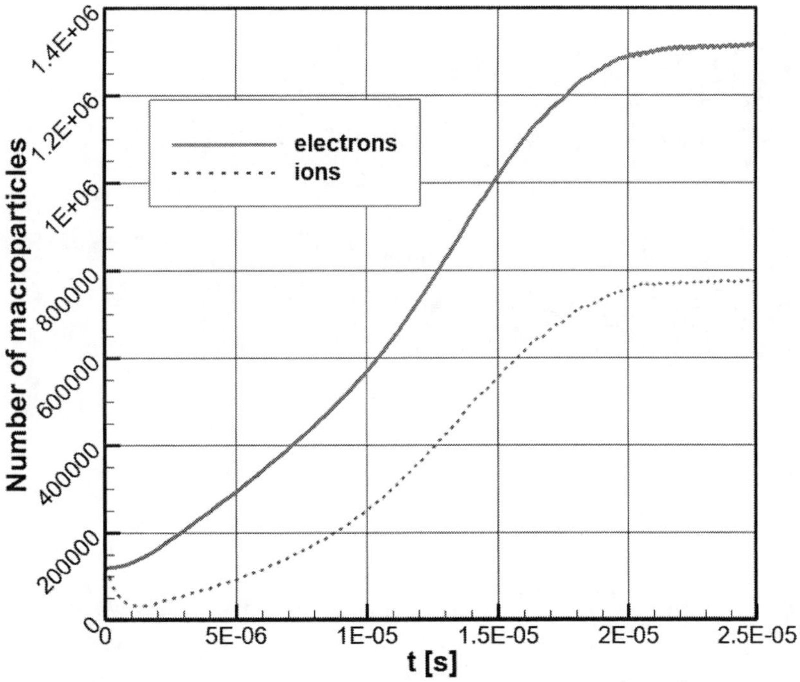

Figure 8. Temporal behaviour of number of macroparticles in the system, $B_z = 0.033$ T.

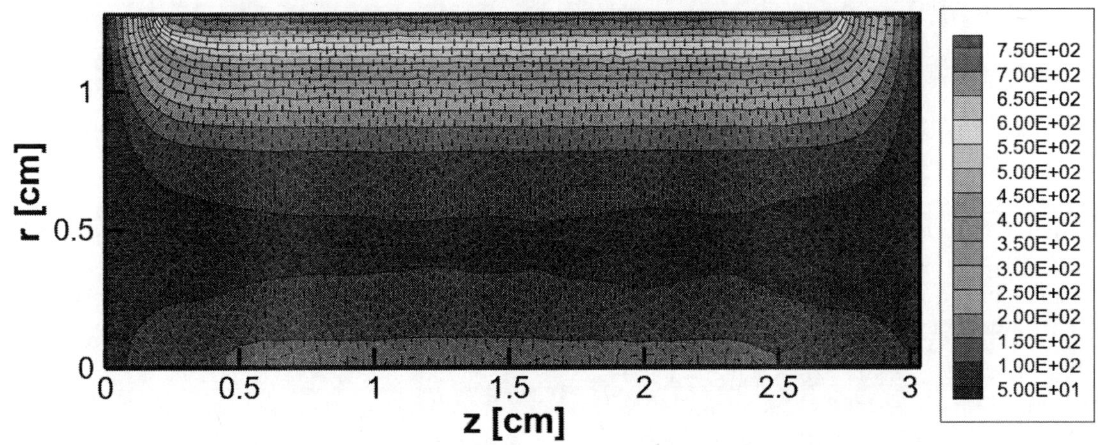

Figure 9. Spatial distribution of potential φ [V] in steady state, B_z=0.033 T.

In figures 13–15 distribution of number densities of electrons and H_2^+ ions in the computational area is presented. The distribution of electrons in computational area reaches the maximum near the axis of symmetry and then falls to the anode. Near the region r~1 cm one can observe the second local maximum in the distribution of the electron number density. The distribution of H_2^+ ions number density is similar to that of the electrons. The ions number density reaches the maximum near the axis of symmetry and then it falls down to the anode. The position of the global maximum in electrons and ions number density distributions are coincide. In the region r = 0.15 – 0.6 cm the number densities of electrons and ions coincide, which makes the plasma in that region quasineutral.

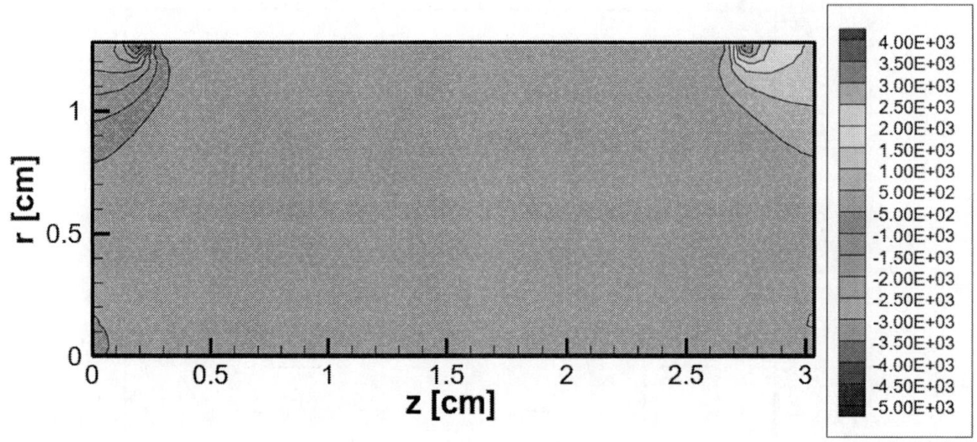

Figure 10. Spatial distribution of E_z [V/cm] component of electric field in steady state, B_z=0.033 T.

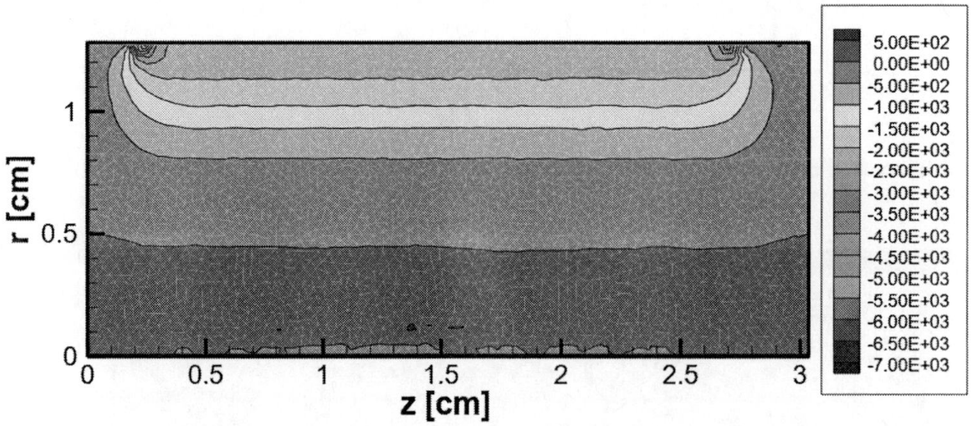

Figure 11. Spatial distribution of E_r [V/cm] component of electric field in steady state, B_z=0.033 T.

In figures 16–18 distribution of the electrons and H_2^+ ions temperatures in eV is presented. The temperatures of charged particles were calculated according to the following formula:

$$T(\vec{r},t)=\frac{m}{3n}\int f(\vec{r},\vec{\upsilon},t)\left(\vec{\upsilon}-\vec{u}\right)^2 d\vec{\upsilon}$$

There are two local maximums in the distribution of T_e. The first one is near the axis of symmetry, the second one is close to anode. The temperature of the electrons in the maximum reaches ~35 eV. The distribution of ions temperature has plateau in the region r ~ 0.2-0.8 cm. The temperature of ions in this region is about 150–160 eV. The temperature of electrons in this region is ~10–20 eV. Using presented data on the charged particles number densities and temperatures one can estimate Debye radius and plasma frequency in the system. Based on the corresponding maximum values: $\omega_p \sim 8\cdot 10^9$ s^{-1} and $\lambda_D \sim 0.03$ cm.

In the figure 19 energy distribution function (EDF) of charged particles is presented. The EDF can be calculated according to the following formula:

$$f(\varepsilon)=\frac{dN}{Nd\varepsilon}$$

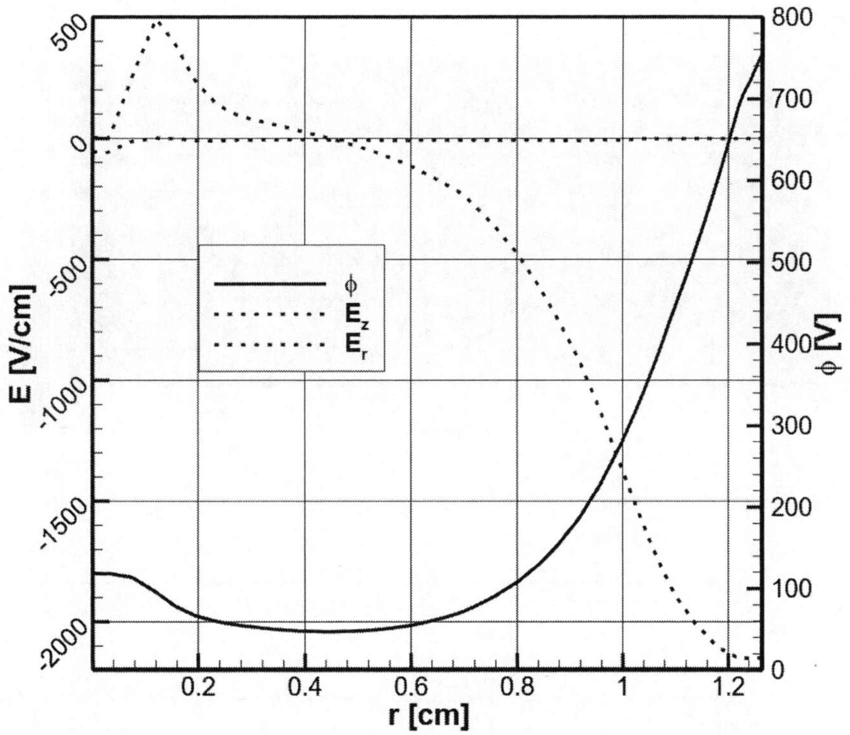

Figure 12. Distribution of the potential and component of electric field along line z=1.5 cm (from r = 0 cm– axis of symmetry to r = 1.27 cm - anode) in steady state, B_z=0.033 T.

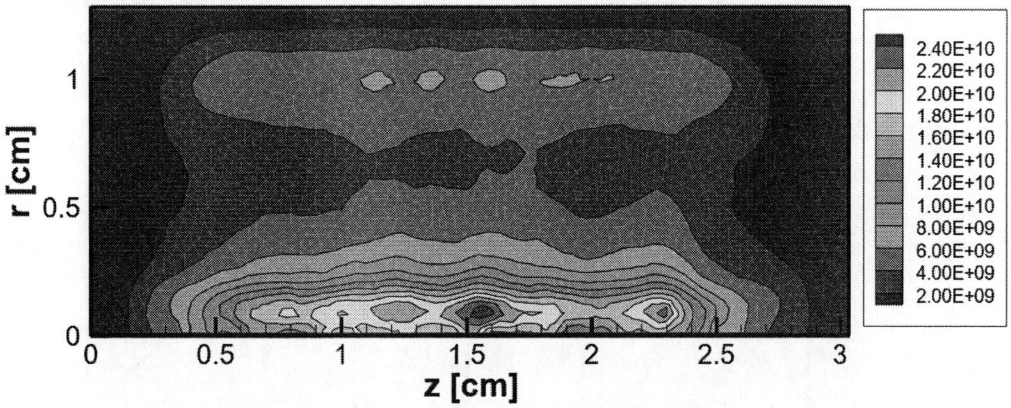

Figure 13. Spatial distribution of electrons number density n_e [cm^{-3}] in steady state, B_z=0.033 T.

One can see that the maximum energy of charged particles do not exceed 800 eV, which correspond to the applied anode voltage. One can see that the EDF of both charged species is deviate from the Maxwell distribution (in opposite case the EDF would be linear). The tail of electrons and ions EDF is overpopulated as compared to Maxwell distribution. In the EDF of ions one can observe behavior that is close to plateau in the region 200÷600 eV.

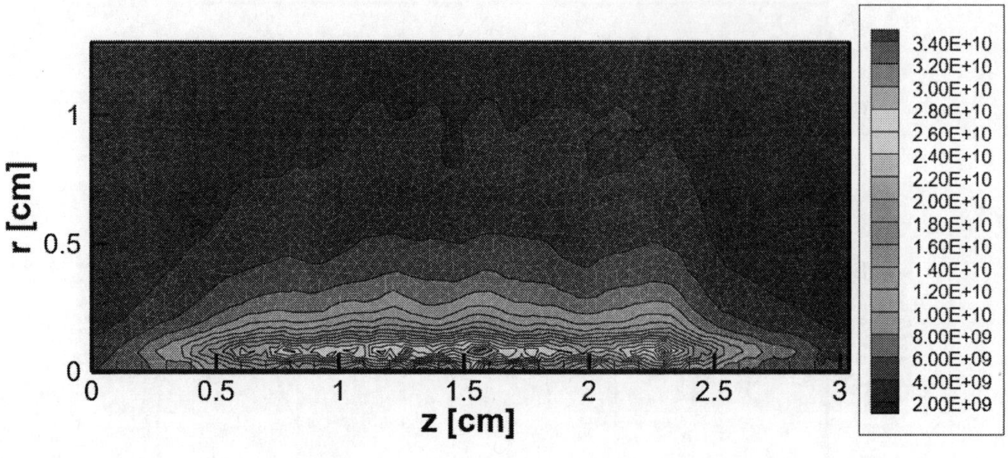

Figure 14. Spatial distribution of H_2^+ ions number density n_i [cm^{-3}] in steady state, B_z=0.033 T.

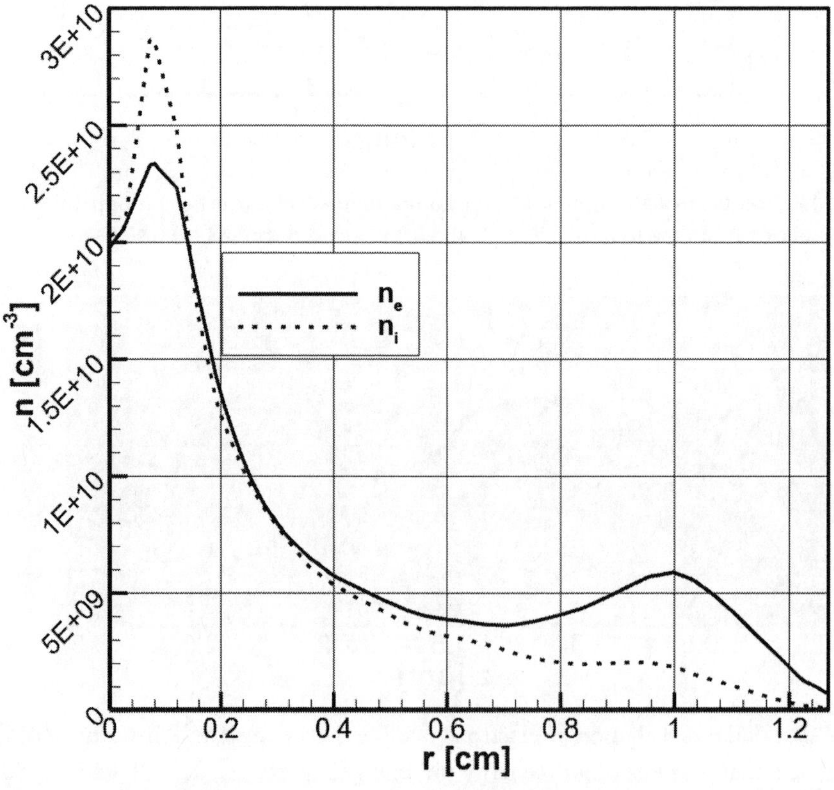

Figure 15. Distribution of the electrons and H_2^+ ions number density along line z=1.5 cm (from r = 0 cm– axis of symmetry to r = 1.27 cm - anode) in steady state, B_z=0.033 T.

In the figure 20 one can see the distribution of the ionization events versus energy of electrons. One can see that most electrons that participate in ionizing collisions have energies ~46.5 eV. Also according to the simulation it was found that most electrons that emerged in the process of ion-electron emission from the cathode acquire energy of 100 eV in approximately 1 ns. According to the numerical results most of the ions has lifetime ~10^{-7} s. For the electrons similar distributions shows two peaks at values of lifetime ~10^{-6} s and 10^{-10} s. However mentioned values are preliminary and require conduction of additional numerical experiments.

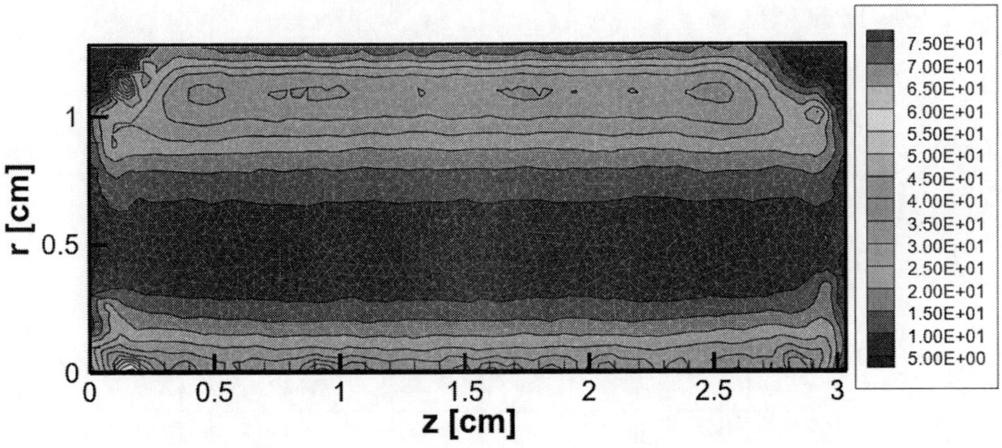

Figure 16. Spatial distribution of electron temperature T_e [eV] in steady state, B_z=0.033 T.

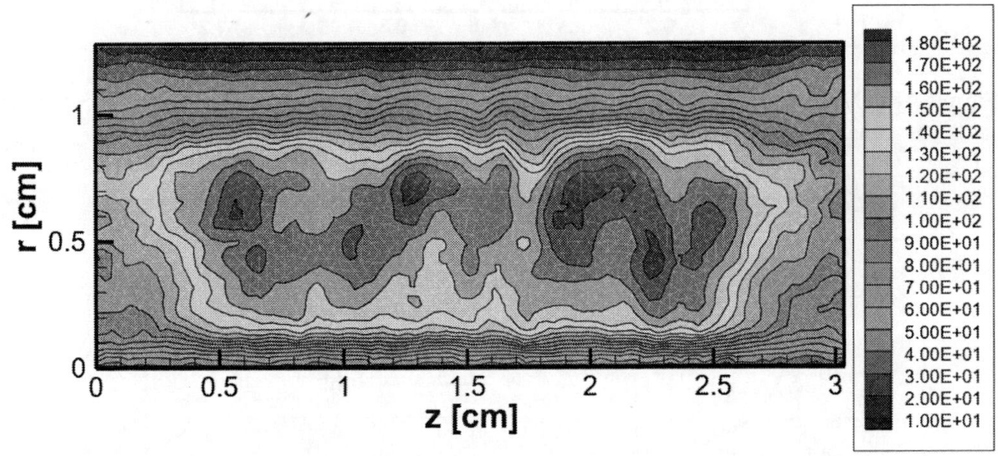

Figure 17. Spatial distribution of H_2^+ ions temperature T_i [eV] in steady state, B_z=0.033 T.

In figure 21 one can see comparison between experimental and numerical data. Dependence of the discharge current versus applied external magnetic field is shown. One can see reasonable agreement between measured and calculated data. However one can see (figure 6) that behavior of the current with increasing magnetic field is nonmonotonic. There exists I_{max} and corresponding B_{max} after which further increase in B leads to decrease in I. It is worth noting that similar behavior of the discharged current was observed in the experiments presented in [5] which might indicate that this is general behavior (not some feature of the considered experimental data) of current in dependence of the magnetic field. Further numerical investigation required in order to understand mentioned phenomena and check if the current version of the model is able to reproduce it.

Mass spectrum of the extracted current from the Penning ion source was measured in the experiments [11]. This measurement has shown the presence of hydrogen atoms in the extracted current. Validation of the model in that aspect is possible by introducing corresponding elementary

IOP Conf. Series: Journal of Physics: Conf. Series **1009** (2018) 012023 doi:10.1088/1742-6596/1009/1/012023

processes into it. In order to obtain estimate of the significance of the elementary processes that lead to the formation of hydrogen atoms studies [28-30] can be used.

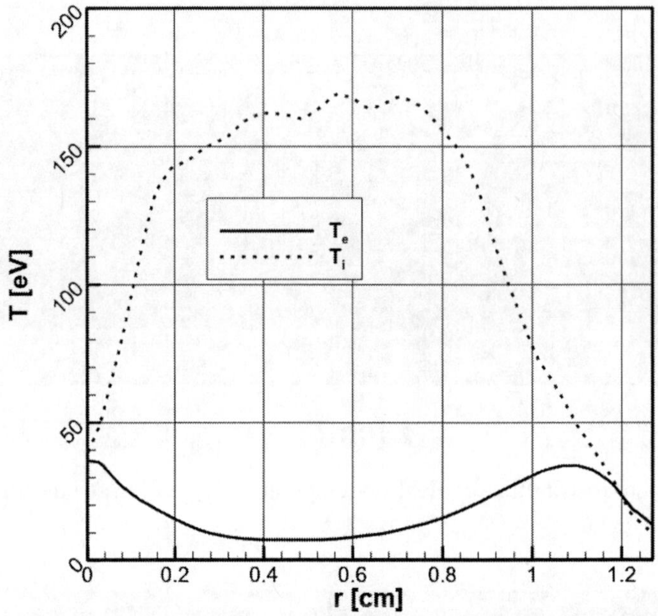

Figure 18. Distribution of the electrons and H_2^+ ions temperature along line z=1.5 cm (from r = 0 cm– axis of symmetry to r = 1.27 cm - anode) in steady state, B_z=0.033 T.

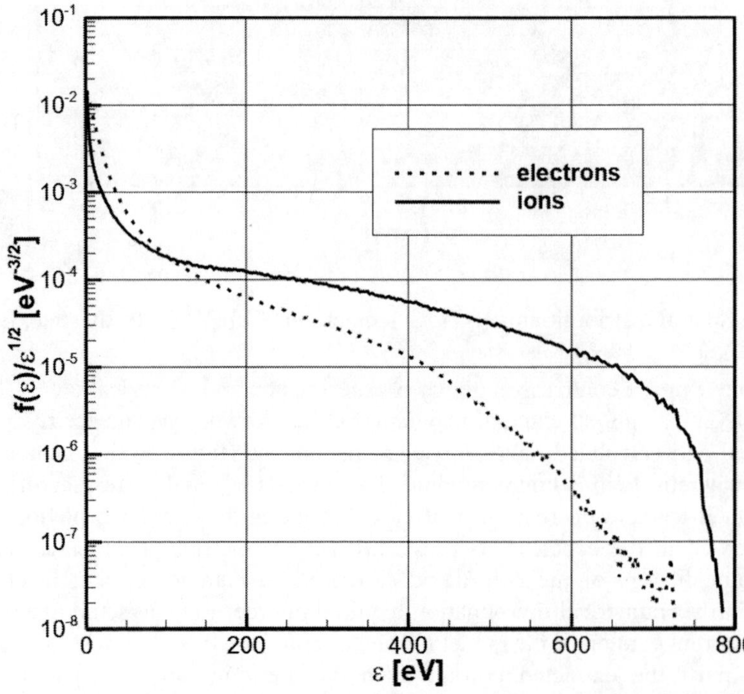

Figure 19. Energy distribution function of electrons and ions in steady state, B_z=0.033 T.

IOP Conf. Series: Journal of Physics: Conf. Series **1009** (2018) 012023 doi:10.1088/1742-6596/1009/1/012023

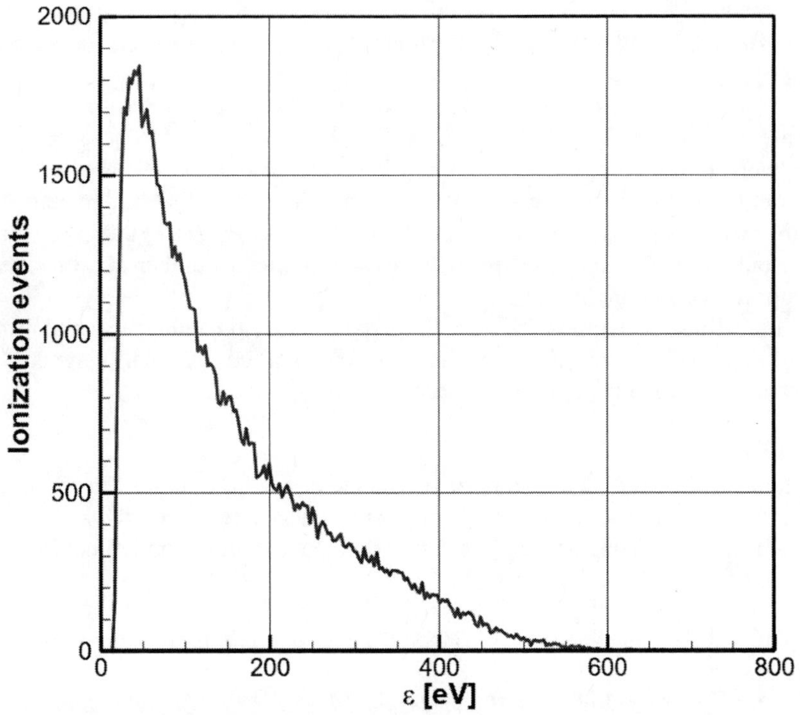

Figure 20. Distribution of ionization events versus energy of electrons, B_z=0.033 T.

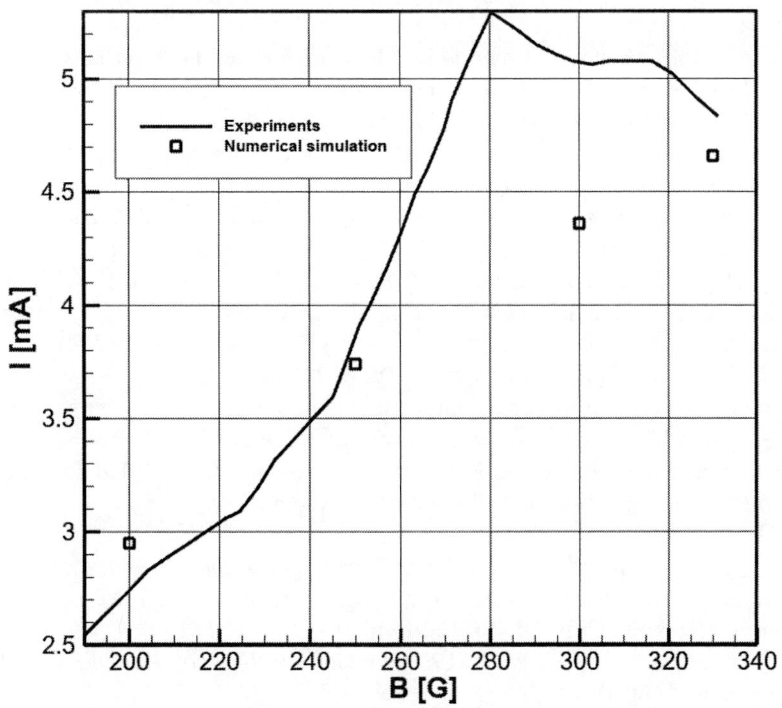

Figure 21. Comparison of experimental and numerical data.

Alternative method for the simulation of Penning gas discharge is modified drift-diffusion approach [31–34]. It seems that this model if properly set up might be used for the simulation of the Penning gas discharge.

5. Conclusion

In the study numerical model for the simulation of Penning gas discharge is applied for the analysis of experiments conducted in the University of California, Berkley. Electrodynamic structure of the Penning gas discharge at pressure $p = 0.8$ Torr, $V = 800$ V and $B = 0.02 \div 0.033$ T were obtained and analyzed. Distribution of potential, electric field, charged particles number densities, temperatures and energy distribution functions are presented.

Results of modelling are compared with the experimental data for the validation of numerical model. Dependences of the discharge current on the applied external magnetic field obtained experimentally and numerically are in reasonable agreement.

Acknowledgments

Author is grateful to the staff of Laboratory of Radiative Gas Dynamics (Institute for Problems in Mechanics RAS) for useful discussions on the problems considered in the study.

The presented study was supported by the Russian Science Foundation project № 16-11-10275.

References

[1] Hockney R W and Eastwood J W 1988 *Computer simulation using particles* (New York: CRC Press)
[2] Birdsall C K and Langdon A B 1985 *Plasma Physics Via Computer Simulation* (New York: McGraw-Hill, Inc.)
[3] Dikalyuk A S 2017 Development of Particle-In-Cell Solver for Numerical Simulation of Penning Discharge *Proc. of 55th AIAA Aerospace Meeting, AIAA SciTech Forum* AIAA 2017-0842
[4] Dikalyuk A S and Kuratov S E Numerical Investigation of Penning Discharge Characteristics using 2D/3V Particle-In-Cell Method *J. of Physics: Conference Series* **815** 012001
[5] Kreindel Yu E and Ionov A S 1964 *J. of Technical Physics* **34** 1199
[6] Schitov N N, Kanshin I A and Mamedov N V 2015 Investigation of the dependency of penning ion source operational characteristics on its geometric parameters *Physical-Chemical Kinetics in Gas Dynamics* **16** 4 http://chemphys.edu.ru/issues/2015-16-4/articles/590/
[7] Rovey J L, Ruzic B P and Houlahan T J 2007 Simple penning ion source for laboratory research and development applications *Review of Scientific Instruments* **78** 106101
[8] Rovey J L 2008 Design parameter investigation of a cold-cathode Penning ion source for general laboratory applications *Plasma Sources Sci. Technol.* **17** 035009
[9] Prakash R, Vyas G L, Jain J, Prajapati J, Pal U N, Chowdhuri M B and Manchanda R 2012 Development of large volume double ring penning plasma discharge source for efficient light emission *Review of Scientific Instrucments* **83** 123502
[10] Vyas G L, Prakash R, Pal U N, Manchanda R and Halder N 2016 Penning Plasma Based Simulutaneous Light Emission Source of Visible and VUV Lights *Plasma Physics Reports* **42** 601
[11] Sy A V 2013 *Advanced Penning-type source development and passive beam focusing techniques for an associated particle imaging neutron generator with enhanced spatial resolution* (Berkley: University of California)
[12] Berezin Yu A and Vshivkov V A 1980 *Particle Method in Dynamics of Rarified Plasma* (Novosibirsk: Nauka)
[13] Grigoriev Yu N, Vshivkov V A and Fedoruk M P 2004 *Numerical Simulation via Particle-In-Cell Methods* (Novosibirsk: Izdatelstvo SO RAN)
[14] Versteeg H K and Malalasekera W 2007 *An Introduction to Computational Fluid Dynamics: the*

IOP Conf. Series: Journal of Physics: Conf. Series **1009** (2018) 012023 doi:10.1088/1742-6596/1009/1/012023

Finite Volume Method (Harlow: Pearson Education)

[15] Mavriplis D J 1995 *Multigrid Techniques for Unstructured Meshes* (Hampton: Institute for Computer Applications in Science and Engineering)

[16] Holmes D G and Connell S D 1989 Solution of the 2-D Navier–Stokes Equations on Unstructured Adaptive Grids *AIAA Paper* 89-1392

[17] Rausch R D, Batina J T and Yang H T Y 1991 Spatial adaption procedures on unstructured meshes for accurate unsteady aerodynamic flow computation *AIAA paper* 1991–1106

[18] Spirkin A M 2006 *A Three-Dimensional Particle-In-Cell Methodology on Unstructured Voronoi Grids with Applications to Plasma Microdevices* (Worcester: Worcester Polytechnic Institute)

[19] Delzanno G L and Camporeale E 2013 On particle movers in cylindrical geometry for particle-in-cell simulations *J. of Comp. Phys.* **253** 259–277

[20] Bromley B P 2001 *Computational Modeling of the Axial-Cylindrical Inertial Electrostatic Confinement Fusion Neutron Generator* (Urbana-Champaign: University of Illinois at Urbana-Champaign)

[21] Raizer Yu P 1991 *Gas Discharge Physics* (Berlin: Springer)

[22] Birdsall C K 1991 Particle-in-cell charged-particle simulations plus monte carlo collisions with neutral atoms PIC-MCC *IEEE Trans. on Plasma Sci.* **19** 65–85

[23] Donko Z 2011 Particle simulation methods for studies of low-pressure plasma sources *Plasma Sources Sci. Technol.* **20** 024001

[24] Vahedi V and Surendra M 1995 A monte carlo collision model for the particle in cell method: applications to argon and oxygen discharges *Comp. Phys. Comm.* **87** 179–198

[25] Nanby K 2000 Probability theory of electron-molecules, ion-molecule, molecule-molecule, and Coulomb collisions for particle modeling of materials processing plasmas and gases *IEEE Trans. on Plasma Sci.* **28** 971–990

[26] Tawara H, Itikawa Y, Nishimura H and Yoshino M 1990 Cross sections and related data for electron collisions with hydrogen molecules and ions *J. Phys. Chem. Ref. Data.* **19** 617–636

[27] Phelps A V and Petrovic Z Lj 1999 Cold-cathode discharges and breakdown in argon: surface and gas phase production of secondary electrons *Plasma Sources Sci. Technol.* **8** R21–R44

[28] Storozhev D A, Surzhikov S T and Kuratov S E 2015 Numerical simulation of two-dimensional structure of glow discharge in molecular hydrogen *Proc. of 45th Thermophysics Conference* AIAA 2015–3108

[29] Storozhev D A, Surzhikov S T and Kuratov S E 2017 Numerical simulation of dissociation kinetics in the penning discharge plasma using 2D modified drift-diffusion model *Proc. of 55th AIAA Aerospace Meeting, AIAA SciTech Forum* AIAA 2017–1966.

[30] Storozhev D A and Kuratov S E 2017 Numerical simulation of the kinetics of dissociation and ionization of molecular hydrogen in the penning discharge plasma with the use of the reduced kinetic model *Journal of Physics: Conference Series* **815** 012002.

[31] Surzhikov S T 2017 Two-dimensional of the Penning discharge in a cylindrical chamber with the axial magnetic field *Technical Physics* **62** 1177

[32] Surzhikov S T 2017 Numerical simulating the two-dimensional structure of the Penning discharge using the modified drif-diffusion model *Journal of Physics: Conference Series* **815** 012004

[33] Surhzikov S T 2017 Two-dimensional structure of Penning discharge in a cylindrical chamber with axial magnetic field at pressure 1 Torr *Technical Physics Letter* **43** 169

[34] Surzhikov S T 2015 Application of the modified drift-diffusion theory to study of the two-dimensional structure of the penning discharge *AIAA paper* 2015–1832

Experimental study of the evolution of NO radiation in the air at the velocities up to second cosmic

P V Kozlov

Institute of Mechanics Lomonosov Moscow State University, Moscow, Russia

E-mail: kalevala@mail.ru

Abstract. The results of a study of the dynamics of radiation of nitric oxide, which is formed behind the front of a shock wave in air, are given. Results are presented for shock wave velocities of 8–11.4 km/s. The shock wave velocity at which N^+ radiation predominates in the range 210÷230 nm is determined experimentally.

1. Introduction

Assessment of the impact of heat and radiation fluxes to the surface lander in mixtures N_2– O_2 and CO_2– N_2 for the braking maneuver at present is of interest in connection with the design missions to Earth, Mars and Venus. When calculating the heat flux at the lander it is necessary to consider the radiation of the plasma of the shock layer. The magnitude of radiation fluxes increase with the size of the lander. The radiative processes have a significant influence on gas-dynamic flow if the size of the developed space vehicles over 3 meters [1]-[8]. Therefore it is necessary take into account the contribution of radiation processes significant in the development of thermal protection system of reentry vehicles. Testing computational models that describe radiation fluxes requires accurate experimental data in a wide range of speeds and pressures. One of the main sources of these data are the results of experiments made in shock tubes [2], [9]-[12].

The work continues to study the kinetics of formation and decomposition of NO behind the shock front, presented in [13], [14]. The results on the characteristics of radiation of nitric oxide with a high temporal resolution in the wavelength range 190-300 nm up to second cosmic velocity of the shock wave (SW) are presented.

2. Experimental Setup

The main elements of a double diaphragm shock tube (DDST) of the Institute of Mechanics of the Moscow State University and the registration system are described in [13]. The change in the conditions for the preparation of the driver gas in the high-pressure chamber made it possible to confidently obtain a second cosmic velocity at the initial pressure of the test gas of 0.25 Torr at the DDST facility.

The system of radiation registration has been changed figure 1.One channel is used that detects radiation in the spectral range of the spectrograph 190÷670 nm (UV) Horiba CP140 1824. This channel registers on the CCD detector an integral distribution of the spectral radiance during the passage of the shock wave through the measuring cross-section. Instead of a channel registering spectral radiation on a CCD detector in the range of 500-1100 nm, a spectrograph Horiba 1061 with a Hamamatsu R446 photomultiplier was installed. Now the measurement scheme allows one to record the temporal evolution of radiation at three wavelengths in one experiment. A low-noise Hamamatsu R4220 photomultiplier is installed in the channel with a high spectral resolution. This made it possible to improve the signal-to-noise ratio.

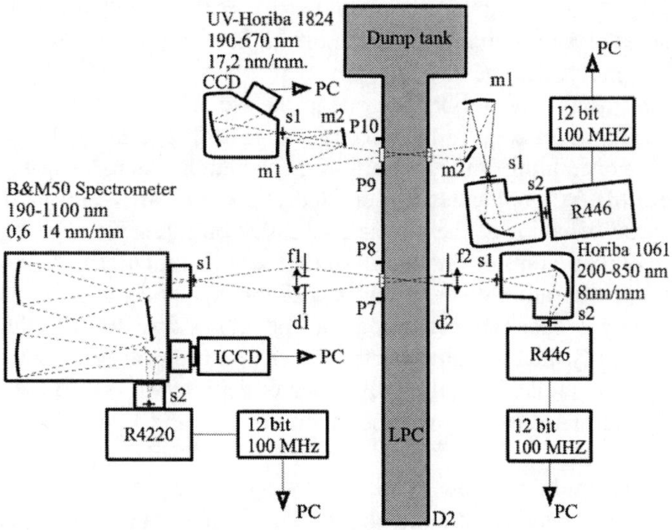

Figure 1. Experimental setup. System of registration.

3. Results of Experiment

As was shown in [8], at shock wave velocities of less than 8 km/s, radiation in the spectral range 190÷300 nm is basically a nonequilibrium emission of nitrogen oxide molecules. The main objective of the study was to study the evolution of the emission of an NO molecule in the UV spectral range for SW velocities of 8÷11.2 km/s in air. The experiments were carried out for the initial pressure P_1=0,25 Torr. Figure 2 shows the temporal behavior of the normalized radiation intensities at a wavelength of 213±2 nm for different velocities of the shock wave (V_{SW}).

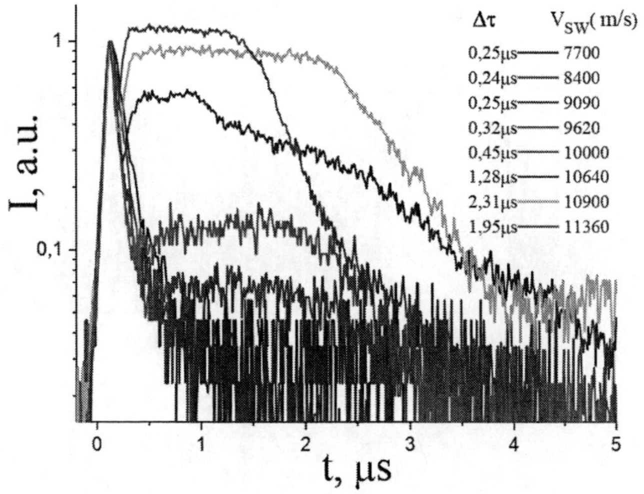

Figure 2. Experimental normalized profiles of the volumetric radiance of shock heated air on λ=213±2 nm for different V_{SW} (P_1=0.25 Torr).

The absolute intensity of the maximum of the volume radiation was practically independent of the velocity of the shock wave and was of the order of 10 W/(cm^3·sr·μm).

As can be seen from figure 2, for velocities exceeding 10 km/s, the intensity of radiation from the equilibrium zone becomes more than 10% of the intensity of the nonequilibrium peak. At a speed of 11 km/s, these intensities become equal. However nonequilibrium radiation peak due to formation and decomposition of NO molecules behind the front of the shock wave is still pronounced. The contribution of radiation from the equilibrium zone to the spectral radiance becomes comparable with radiation from a nonequilibrium zone in the spectral range 190÷290 nm for velocities above 9 km/s. Such a behavior of the nonequilibrium and equilibrium radiations is characteristic only for the region of the spectrum where the NO molecules are emitted. This is clearly seen in figure 3 and figure 4, which show the temporal behavior of the normalized radiation intensities at a wavelength of 213 nm (NO radiation) and on other regions of the spectrum (N_2^+ and N_2.radiation).

Due to the fact that the duration of radiation in different spectral ranges is not the same (figure 3 and figure 4), the integral spectrum obtained on the CCD detector should be divided by the corresponding duration. Only then will the spectral density values of the brightness in the spectrum be correct. The duration of the radiation in the required spectral interval is defined as the area under the function of the normalized intensity versus time. In figure 2, 3 and figure 4, Δτ corresponds to these durations.

It should be noted that in the previous work [13], when processing the spectral distribution of the integral intensity, the time Δτ was equal to the duration of the entire luminous shock layer. This time is much longer than the Δτ times indicated in figure 2, 3 and 4. Therefore, the spectral brightness density obtained by processing the integrated intensity data from the CCD detector will be much larger.

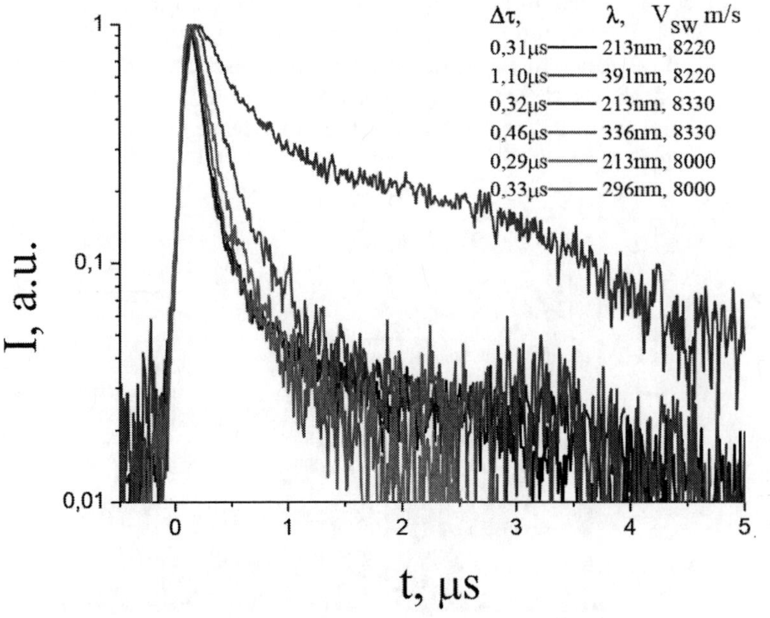

Figure 3. Experimental normalized profiles of the volumetric radiance of shock heated air on λ:213 nm(NO); 391 nm(N_2^+); 336 nm(N_2(2+)). The velocity of the shock wave is 8÷8.33 km/s; P_1=0.25 Torr.

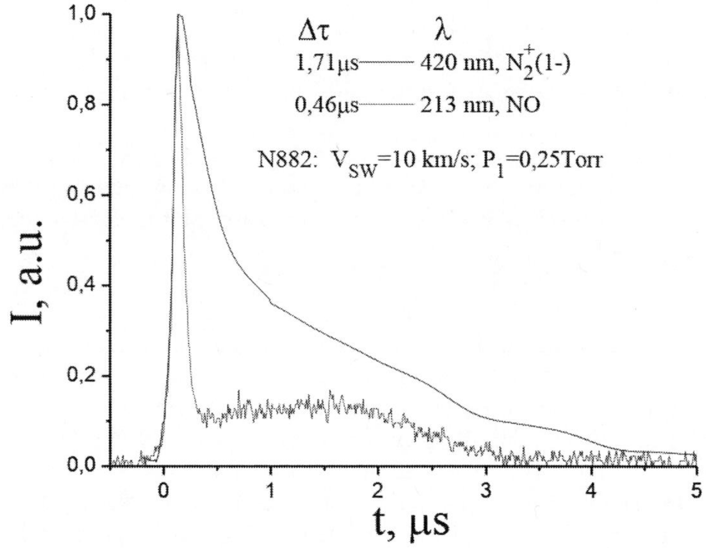

Figure 4. Experimental normalized profiles of the volumetric radiance of shock heated air on λ:213 nm(NO); 422 nm(N_2^+(1-)). P_1= 0.25 Torr.

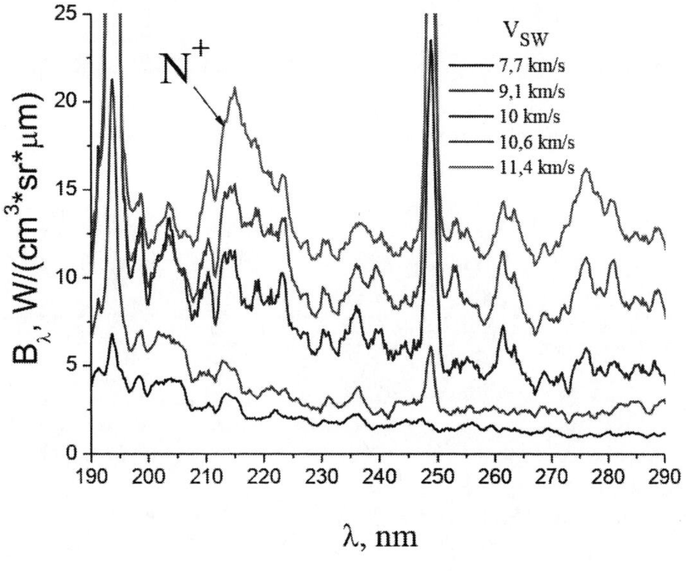

Figure 5. Spectral distribution of integral on time emissivity of air heated by shock wave P_1=0.25 Torr. V_{SW}=7.7 ÷ 11.36 km/s.

When processing the spectra recorded by the CCD of the detector, the integration times Δτ were determined as described above (shown in figure 2 for the corresponding V_{SW}). It is seen that as the shock wave velocity increases, the saw tooth form of the spectrum characteristic of NO changes. In the wavelength range 210÷240 nm, a characteristic hump appears at the velocity above 10 km/s (figure 5).

This hump is due to the equilibrium radiation of the shock layer after the nonequilibrium peak of radiation due to the formation and decomposition of the NO molecule. The source of radiation is the N^+ ion, which is produced in the process of photodissociation of nitrogen atoms. In [15], the absorption spectrum of air under conditions of local thermodynamic equilibrium for 15000 K is given (in [15], figure 7). This figure shows that in the range of 46500 cm^{-1}, intense N^+ absorption peaks are observed in the spectrum.

Calculations of the concentrations of the components behind the shock front also show that the concentration of the N^+ ion becomes significant and practically constant only after the concentrations of the excited electronic states of the NO molecule have reached their maximum and significantly decreased [14].

4. Conclusion

In this article, studies of the nature of the nonequilibrium radiation of nitric oxide started in [13], [14] are continued. Modernization of the recording system made it possible in one experiment to obtain temporal radiation profiles in three different regions of the spectrum with a high resolution (<100 ns). The measurements made showed that the nature of the radiation in different regions of the spectrum can differ substantially in shape and duration. This indicates various mechanisms for the decomposition and formation of molecules and atoms in the air behind the front of the shock wave.

A technique is proposed for processing the integrated spectra registered by CCD detectors in absolute units of spectral brightness.

The spectrum of the N^+ ion appearing in the range of 210÷240 nm in the equilibrium zone of the shock layer in the air at speeds above 10 km/s is identified.

In the future paper, the main attention will focus on the experimental study of the radiative contribution from the Schumann-Runge O_2 band in the spectral range 200÷300 nm for shock velocities of 5÷8 km/s in air.

Acknowledgments

This study was supported by RFBR grant No. 16-01-00379 and by the Program of Basic Researches of Russian Academy of Sciences. The author would like to thank Drs. Shatalov OP, Surzhikov ST and Babou Y for their insight during many fruitful discussions.

References

[1] Surzhikov S T, Shuvalov M P 2014 Estimation of Radiation-Convection Heating of Four Types of Reentry Spacecrafts *Physical-Chemical Kinetics in Gas Dynamics* **15**(4) (in Russian) http://chemphys.edu.ru/issues/2014-15-4/articles/237/

[2] Zalogin G N, Kozlov P V, Kuznetsova L A, Losev S A, Makarov V N, Romanenko Yu V and Surzhikov S T 2001 Radiation Excited by Shock Waves in a CO_2-N_2-Ar Mixture: Experiment and Theory*Technical Physics*, **46** 6 pp 654–661

[3] Surzhikov S T 2010 Spectral Emissivity of Shock Waves in Martian an Titan Atmospheres *10th AIAA/ASME Joint Thermophysics and Heat Transfer Conf. Fluid Dynamics and Co-located Conf. AIAA* 2010-4527 pp 1-32

[4] Sharikov I V, Surzhikov S T, Capitelli M and Colonna G 2006 Kinetic Models of Non-Equilibrium Radiation for Strong Air Shock Waves *44th AIAA Aerospace Sciences Meeting and Exhibit, Aerospace Sciences Meetings AIAA* 2006-0586, pp 1-11

[5] Gorelov V A, Kireev A Yu, Shilenkov S and Surzhikov S T 2004 Prediction of Nonequilibrium Ionization and Emission at Superorbital Flight in Air *37th AIAA Thermophysics Conf., Fluid Dynamics and Co-located Conf. AiAA* 2004-2380 pp 1-11

[6] Surzhikov S T 2009 Electronic Excitation in Air and Carbon Dioxide Gas *Lecture Series on "Non-Equilibrium Gas Dynamics - From Physical Models to Hypersonic Flights"* AVT-162 RTO AVT/VKI Lecture Series p 117

[7] Surzhikov S T 2009 Radiation Modeling in Shock-Tubes and Entry Flows *Lecture Series on*

Non-Equilibrium Gas Dynamics - From Physical Models to Hypersonic Flights AVT-162 RTO AVT/VKI Lecture Series p 98

[8] Surzhikov S T 2012 Radiative-Collisional Models in Non-Equilibrium Aerothermodynamics of Entry Probes *J. of Thermophysics and Heat Transfer* **134** 3 pp 031002-1 031002-11

[9] Kozlov P V, Romanenko Y V, Shatalov O P 2010 Radiation intensity measurement in simulated martian atmospheres on the double diaphragm shock tube *Proceedings of the 4th International Workshop on Radiation of High Temperature Gases in Atmospheric Entry* 2010 Lausanne. Swiss (ESA SP-689, on CD) pp 1-8

[10] Dikalyuk A S, Kozlov P V, Romanenko Y V, Shatalov O P and Surzhikov S T 2013 Nonequilibrium spectral radiation behind the shock waves in Martian and Earth atmospheres *44th AIAA Thermophysics Conf. AIAA 2013-2505* pp 1–10

[11] Dikalyuk A S, Kozlov P V, Romanenko Y V, Shatalov and O P Surzhikov S T 2012 Nonequilibrium radiation behind the strong shock waves in Martian and Titan atmospheres: Numerical rebuilding of experimental data *50th AIAA Aerospace Sciences Meeting including the New Horizons Forum and Aerospace Exposition, AIAA* 2012-0795 pp 1–10

[12] Gorelov V A, Gladyshev M K, Kireev A Yu, Yegorov I B, Plastinin Y A and Karabadzhak G G 1998 Experimental and Numerical Study of Nonequilibrium Ultraviolet NO andN_2^+ Emission in Shock Layer *J. of Thermophysics and Heat Transfer* **12** 2 pp 172–179

[13] Kozlov P V 2017 Experimental investigation of nitric oxide radiation in the shock-heated air *J. of Physics: Conf. Series* **815** 012033

[14] Kozlov P V, Surzhikov S T and 2017 Nonequilibrium radiation NO in shocked air *55th AIAA Aerospace Sciences Meeting, AIAA SciTech Forum AIAA* 2017-0157 pp 1-16

[15] Chauveau S, Deron C, Perrin M-Y, Rivière P and Soufiani A 2003 Radiative transfer in LTE air plasmas for temperatures up to 15000 K *J. Quant. Spectrosc. Radiat. Transfer* **77** pp 113-130

Calculation a Penning plasma discharge characteristics taking into account the magnetic field produced by annular magnets

V O Gumennov[1,2], **A S Dikalyuk**[1,2,3]

[1] Moscow Institute of Physics and Technology (State University), 9 Institutskiy pereulok, Dolgoprudny, Moscow Region, 141701, Russia
[2] All-Russia Research Institute of Automatics (VNIIA), 22 Sushchevskaya street, Moscow, 127055, Russia
[3] Institute for Problems in Mechanics of RAS (IPMech RAS), 101/1 Vernadskogo prospect, Moscow, 119526, Russia

Gumennov.vo@phystech.edu

Abstract. Penning discharge represent a gas discharge, that can exist only in presence of a magnetic field of permanent magnets or current solenoids. Based on this at VNIIA a series of experiments were performed where discharge currents were recorded for various cell geometries, voltage, and pressure. For the correct calculation of the current-voltage characteristics of the system it is necessary to calculate the distribution of the magnetic field inside the gas discharge chamber. Thus, a computer program was developed that calculates the magnetic field of a ring magnet with the specified parameters. The results obtained using the program was compared with the experimental and numerical data reported in literature. Using this approach two-dimensional distribution of magnetic field in the Penning cell was calculated and preliminary results on the electrodynamic structure of plasma of the Penning gas discharge corresponding to one studied in VNIIA experiments were obtained.

1. Introduction

Nowadays permanent magnets are used in many applications [1]: source of charged particles, manometers and ion pumps. One of the systems, in which permanent magnets are used, is Penning gas discharge [2]. It is an example of the system in which formation and sustaining of gas discharge plasma is determined by the presence of axial magnetic field created by permanent magnets or solenoids. The distribution of the magnetic field affects the motion of charged particles and thus the structure of the gas discharge plasma in the Penning cell.

Penning gas discharge plasma is rarefied. To model this kind of plasma particle-in-cell method can be used [3], however there are studies [4, 5] devoted to numerical investigation of Penning gas discharge in which other models are adopted.

In some cases Penning gas discharge is sustained by the non-uniform magnetic field. In order to study such systems one has to be able to calculate the distribution of it. An example of such system was studied in the experiments performed in VNIIA [6]. In mentioned experiments in the presence of external magnetic field created by permanent magnets Penning discharge currents were recorded for cells with various geometries, voltages and pressures,. Also visual observations of the glowing plasma were presented [6].

In the general case, the problem of finding the magnetic field distribution is solved by numerical methods, but for systems where the permanent magnets have a simple shape (ring topology) (e.g. [6]), it is possible to solve this problem analytically [7, 8].

In the paper based on the model described in [7, 8], computer program was developed for calculation of the distribution of magnetic field created by annular permanent magnet. The computer code was validated against experimental and numerical data presented in literature. Based on the magnetic field distribution calculated using the developed code preliminary results on the electrodynamics structure of Penning discharge studied in [6] were reported.

2. Analytical method

The calculation of the magnetic field generated by the annular permanent magnet can be obtained by the method that was presented in [7]. This method uses a Coulombian model and in the case of axial polarization, the formulas for calculation of the magnetic field are as follows:

$$\overrightarrow{H^+}(r,z) = \frac{\sigma^*}{4\pi\mu_0} \int_{\theta=0}^{\theta=2\pi} \int_{r=r_{in}}^{r=r_{out}} \frac{\overrightarrow{P_{1+}M}}{\left|\overrightarrow{P_{1+}M}\right|^3} r dr d\theta$$

$$\overrightarrow{H^+}(r,z) = \frac{-\sigma^*}{4\pi\mu_0} \int_{\theta=0}^{\theta=2\pi} \int_{r=r_{in}}^{r=r_{out}} \frac{\overrightarrow{P_{1-}M}}{\left|\overrightarrow{P_{1-}M}\right|^3} r dr d\theta \quad (1)$$

$$\vec{H}(r,z) = \overrightarrow{H^+}(r,z) + \overrightarrow{H^-}(r,z)$$

where

$$\overrightarrow{P_{1+}M} = (r - r_1\cos\theta)\vec{i_r} - r_1\sin\theta\vec{i_\theta} + (z-h)\vec{i_k}$$

$$\overrightarrow{P_{1-}M} = (r - r_1\cos\theta)\vec{i_r} - r_1\sin\theta\vec{i_\theta} + (z+h)\vec{i_k}$$

By substituting $\theta = \pi - 2\beta$, where $0 \le \beta \le \pi/2$ and integrating twice, one obtains 3 components of magnetic field (2) for point M. Assumption of axisymmetric geometry (which is valid for most of the Penning cells) leads to $H_\theta = 0$.

$$H_r(r,z,\theta) = H_r^+(r,z,\theta) + H_r^-(r,z,\theta)$$

$$H_z(r,z,\theta) = H_z^+(r,z,\theta) + H_z^-(r,z,\theta) \quad (2)$$

$$H_\theta(r,z,\theta) = H_\theta^+(r,z,\theta) + H_\theta^-(r,z,\theta) = 0$$

Expressions for only non-singular cases, where $r \ne r_{in}, r \ne r_{out}, z \ne h$ are:

$$H_r^+(r,z) = \frac{\sigma^*}{\pi\mu_0}\sum_{n=1}^{2}(-1)^{n-1}\frac{1}{k_n^+}\sqrt{\frac{r_n}{r}}\left[E(k_n^+) - (1 - \frac{k_n^{+2}}{2})K(k_n^+)\right]$$

$$H_r^-(r,z) = \frac{-\sigma^*}{\pi\mu_0}\sum_{n=1}^{2}(-1)^{n-1}\frac{1}{k_n^-}\sqrt{\frac{r_n}{r}}\left[E(k_n^-) - (1 - \frac{k_n^{-2}}{2})K(k_n^-)\right] \quad (3)$$

$$H_z^+(r,z) = \frac{\sigma^*}{4\pi\mu_0(z-h)}\sum_{n=1}^{2}(-1)^{n-1}\frac{k_n^+}{\sqrt{rr_n}}$$

$$\left\{\left[\sqrt{r^2+(z-h)^2} - r\right]\left[\sqrt{r^2+(z-h)^2} - r_n\right]\Pi(h_1^+, k_n^+)\right. \quad (4)$$

$$\left. + \left[\sqrt{r^2+(z-h)^2} + r\right]\left[\sqrt{r^2+(z-h)^2} + r_n\right]\Pi(h_2^+, k_n^+)\right\}$$

$$H_z^-(r,z) = \frac{\sigma^*}{4\pi\mu_0 z}\sum_{n=1}^{2}(-1)^{n-1}\frac{k_n^-}{\sqrt{rr_n}}$$
$$\left\{\left[\sqrt{r^2+z^2}-r\right]\left[\sqrt{r^2+z^2}-r_n\right]\Pi(h_1^-,k_n^-)\right.$$
$$\left.+\left[\sqrt{r^2+z^2}+r\right]\left[\sqrt{r^2+z^2}+r_n\right]\Pi(h_2^-,k_n^-)\right\}$$

And for the singular case (where $r = r_{in}, r = r_{out}$, $z = h$) we can use modification of (4), where the integrals of the third kind may be rewritten using the Heuman's lambda-function [9], which is the sum of elliptic integral of the first kind:

$$\Lambda_0(\varphi,k) = \frac{F(\varphi,(1-k^2)^{1/2})}{K((1-k^2)^{1/2})} + \frac{2}{\pi}K(k)Z(\varphi,(1-k^2)^{1/2}) \tag{5}$$

$$\Pi(n,k) = K(k) + \frac{\pi}{2}\delta[1-\Lambda_0(\varepsilon,k)];\ \delta = [n(1-n)^{-1}(n-\sin^2 k)^{-1}]^{1/2};$$
$$\varepsilon = \arcsin[(1-n)/\cos^2 k]^{1/2} \tag{6}$$

The modified expressions are as follows:

$$H_z^+(r,z) = \frac{\sigma^*}{2\pi\mu_0}\sum_{n=1}^{2}(-1)^{n-1}\left\{\frac{k_n^+(z-h)\sqrt{r^2+(z-h)^2}}{\sqrt{rr_n}(\sqrt{r^2+(z-h)^2}+r)}K(k_n^+)+\right.$$
$$\left.+\frac{\pi}{2}sign(z-h)sign(\sqrt{r^2+(z-h)^2}-r_n)[1-\Lambda(\theta_{1n}^+,k_n^+)]+\frac{\pi}{2}sign(z-h)[1-\Lambda(\theta_2^+,k_n^+)]\right\} \tag{7}$$

$$H_z^-(r,z) = \frac{-\sigma^*}{2\pi\mu_0}\sum_{n=1}^{2}(-1)^{n-1}\left\{\frac{k_n^- z\sqrt{r^2+z^2}}{\sqrt{rr_n}(\sqrt{r^2+z^2}+r)}K(k_n^-)+\right.$$
$$\left.+\frac{\pi}{2}sign(z)sign(\sqrt{r^2+z^2}-r_n)[1-\Lambda(\theta_{1n}^-,k_n^-)]+\frac{\pi}{2}sign(z)[1-\Lambda(\theta_2^-,k_n^-)]\right\} \tag{8}$$

where

$$k_n^+ = \left(\frac{4rr_n}{(r+r_n)^2+(z-h)^2}\right)^{1/2} \qquad k_n^- = \left(\frac{4rr_n}{(r+r_n)^2+z^2}\right)^{1/2}$$

$$h_1^+ = \frac{2r}{r+\left(r^2+(z-h)^2\right)^{1/2}} \qquad h_2^+ = \frac{2r}{r-\left(r^2+(z-h)^2\right)^{1/2}}$$

$$h_1^+ = \frac{2r}{r+\left(r^2+z^2\right)^{1/2}} \qquad h_2^+ = \frac{2r}{r-\left(r^2+z^2\right)^{1/2}}$$

$$r_1 = r_{in} \qquad r_2 = r_{out}$$

IOP Conf. Series: Journal of Physics: Conf. Series **1009** (2018) 012025 doi:10.1088/1742-6596/1009/1/012025

$$\theta_{1n}^{+} = \arcsin\left(\frac{1-h_1^{+}}{1-k_n^{+2}}\right)^{1/2} \qquad \theta_2^{+} = \arcsin\frac{|z-h|}{\left(r^2+z^2\right)^{1/2}+r}$$

$$\theta_{1n}^{-} = \arcsin\left(\frac{1-h_1^{-}}{1-k_n^{-2}}\right)^{1/2} \qquad \theta_2^{-} = \arcsin\frac{|z|}{\left(r^2+z^2\right)^{1/2}+r}$$

$K(k)$ – complete elliptic integral of the first kind.
$E(k)$ – complete elliptic integral of the second kind.
$\Pi(\varepsilon,k)$ – complete elliptic integral of the third kind.
$\Lambda_0(\varphi,k)$ – Heuman's lambda-function
$Z(\varphi,k)$ – Jacobi zeta-function

2.1 Solving elliptic integrals

In order to be able to use formulas reported in previous section one has to implement a procedure for evaluation of various special functions (e.g. elliptic integrals and so on). And in order to solve this problem there are different possibilities: approximation scheme presented in [9] or alternative forms of elliptic integrals. There were decided to use Carlson's integrals [10]. They are a modern alternative to Legendre forms. The Legendre forms may be expressed in terms of the Carlson forms. The Carlson's elliptic integrals are:

$$\begin{aligned} R_F(x,y,z) &= \frac{1}{2}\int_0^\infty \frac{dt}{((t+x)(t+y)(t+z))^{1/2}} \\ R_J(x,y,z,p) &= \frac{3}{2}\int_0^\infty \frac{dt}{[t+p]((t+x)(t+y)(t+z))^{1/2}} \\ R_C(x,y) = R_F(x,y,y) &= \frac{3}{2}\int_0^\infty \frac{dt}{[t+x](t+y)^{1/2}} \\ R_J(x,y,z,p) = R_J(x,y,z,z) &= \frac{3}{2}\int_0^\infty \frac{dt}{[t+z]((t+x)(t+y)(t+z))^{1/2}} \end{aligned} \tag{9}$$

The Carlson's elliptic integrals are calculated using an iterative scheme described in [10]. In the scheme we use the initial values to calculate the values at which the convergence condition is satisfied. Then, a series of expansions is calculated, which is the output value of the function. Incomplete elliptic integrals, which are used in calculation of Heuman's lambda-function and formulas (4-8), can be expressed by Carlson' elliptic integrals as:

$$\begin{aligned} F(\varphi,k) &= \sin\varphi R_F(\cos^2\varphi, 1-k^2\sin^2\varphi, 1) \\ E(\varphi,k) &= \sin\varphi R_F(\cos^2\varphi, 1-k^2\sin^2\varphi, 1) - \frac{1}{3}k^2\sin^3\varphi R_D(\cos^2\varphi, 1-k^2\sin^2\varphi, 1) \\ \Pi(\varphi,n,k) &= \sin\varphi R_F(\cos^2\varphi, 1-k^2\sin^2\varphi, 1) + \frac{1}{3}n\sin^3\varphi R_J(\cos^2\varphi, 1-k^2\sin^2\varphi, 1, 1-n\sin^2\varphi, 1) \end{aligned} \tag{10}$$

For the case of complete elliptic integrals we put $\varphi = \frac{\pi}{2}$. Also by Carlson's elliptic integrals we can obtain values for Heuman's lambda-function $\Lambda_0(\varphi,k)$:

IOP Conf. Series: Journal of Physics: Conf. Series **1009** (2018) 012025 doi:10.1088/1742-6596/1009/1/012025

$$\Lambda_0(\varphi,k) = \frac{2}{\pi}\sin(k)(R_F(0,\cos^2\varphi,1) - \frac{\sin^2\varphi R_d(0,\cos^2\varphi,1)}{3})R_F(\cos^2 k,1-\cos^2\varphi\sin^2 k,1)$$
$$-\frac{\cos^2\varphi\sin^3 k * R_F(0,\cos^2\varphi,1)}{3}R_D(\cos^2 k,1-\cos^2\varphi\sin^2 k,1) \quad (11)$$

3. Testing analytical method

For the checking the selected algorithm the computer program was developed, which computes complete and incomplete elliptic integrals as well as components for the magnetic field of permanent ring magnets. Further were made comparisons with known data.

3.1 *Comparison with numerical data*

As the example of numerical data was selected the data from [7]. Further was performed digitization of plots of distribution of the magnetic field components of the upper surface and have made a comparison between last one and our calculation. The parameters of the magnets were chosen according to [7]: r_{in} = 0.025 m, r_{out} = 0.028 m, h = 0.003 m, σ^*=1.0 T. The comparison is shown in figures 1–2.

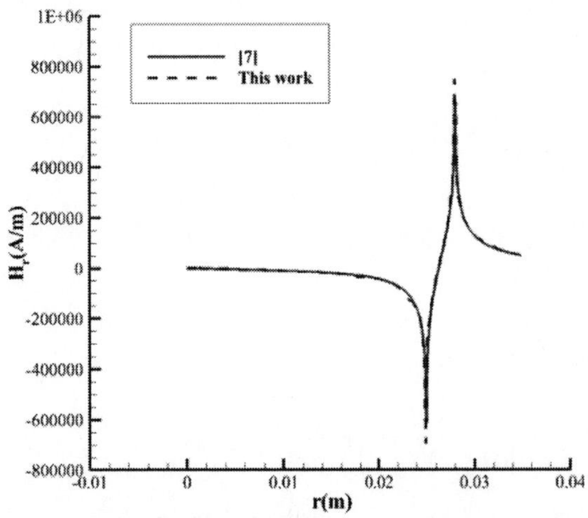

Figure 1. Comparison for radial component of magnetic field with z = 0.003 m.

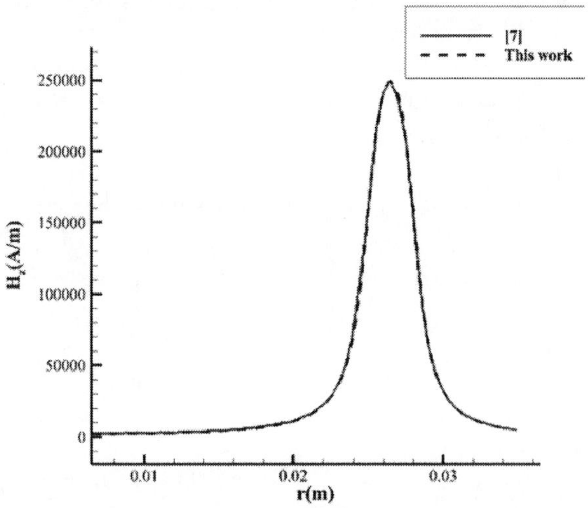

Figure 2. Comparison for axial component of magnetic field with z = 0.004 m.

Comparison of the analytical formula derived for the case $r=0$ and the calculated distribution of the magnetic field is presented in figure 3. For this formulas for k_n^+ and k_n^- were rewritten and linked with (4) and (6). Further comparison was performed with results obtained by the following formula:

$$B_z = \frac{\sigma}{2}\left[\left(\frac{z}{(r_{out}^2+z^2)^{1/2}} - \frac{z-h}{(r_{out}^2+(z-h)^2)^{1/2}}\right) - \left(\frac{z}{(r_{in}^2+z^2)^{1/2}} - \frac{z-h}{(r_{in}^2+(z-h)^2)^{1/2}}\right)\right] \quad (12)$$

In another case we have used the data reported in [11]. In the figure 4 one can see that 1D distribution obtained by the analytical method agrees well with numerical calculations obtained using finite element method (FEM) (r_{in} =0.0125 m, r_{out} =0.035 m, h =0.004 m, z =0.006 m, σ^* =1.2 T). As author of [11, 12] has noticed the problem of FEM is a relatively low speed of calculation. The 2D plane calculations with 140×140 points took about 300 minutes. Code based on analytical formulas performs similar calculations in 1–3 seconds.

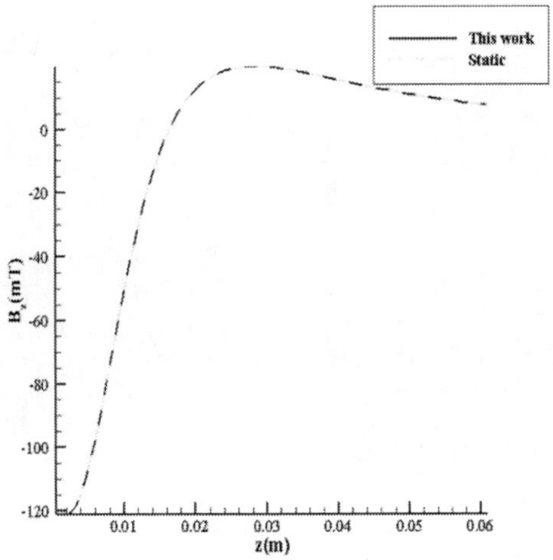

Figure 3. Difference between analytical methods. Preferences for magnets were r_{in} =0.0125 m, r_{out} =0.035 m, h =0.004 m, σ^* =1.2 T.

3.2 *Comparison with the experiment*

Further comparison of the magnetic field distribution was made using the experimental data presented in [11]. The permanent magnet parameters are as follows: r_{in} =0.0125 m, r_{out} =0.035 m, h=0.004 m, z =0.006 m, σ^* =1.2 T. Results of the comparison are given in figure 4.

In [8] the results of experiments performed in VNIIA are reported. In this research magnetic field is produced by 2 SmCo magnets with parameters: r_{in} = 0.010 m, r_{out} = 0.013 m, h = 0.010 m. Parameters of magnetic density were selected by SmCo magnetic characteristics, the left magnet had magnetic density σ = 1.02 T and right one σ = 0.89 T. The comparison (figure 5) showed that the computed data are close to the experimental results. Also a 2D distribution was modeled in order to see the full picture of axial and radial components of magnetic field (figures 6–7) that can be used in particle-in-cell calculations.

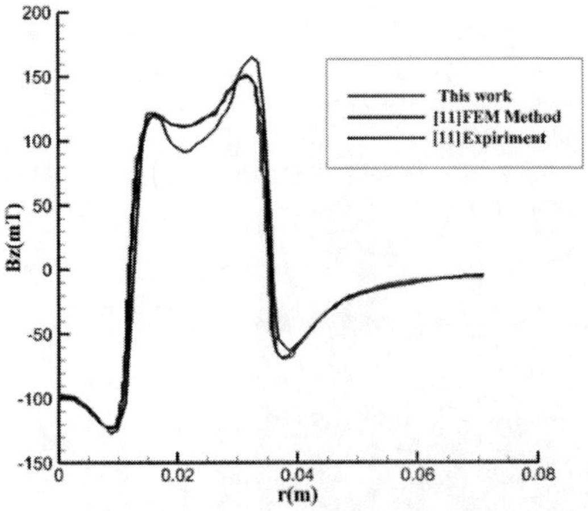

Figure 4.Comparisson with experimental data and numerical data from [11].

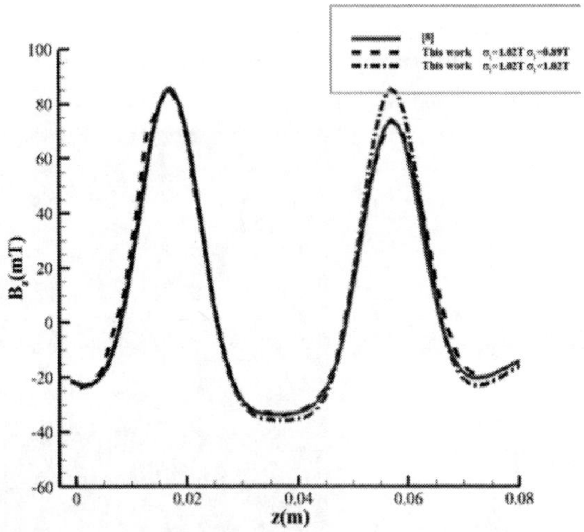

Figure 5.Comparisson with experimental data from [8].

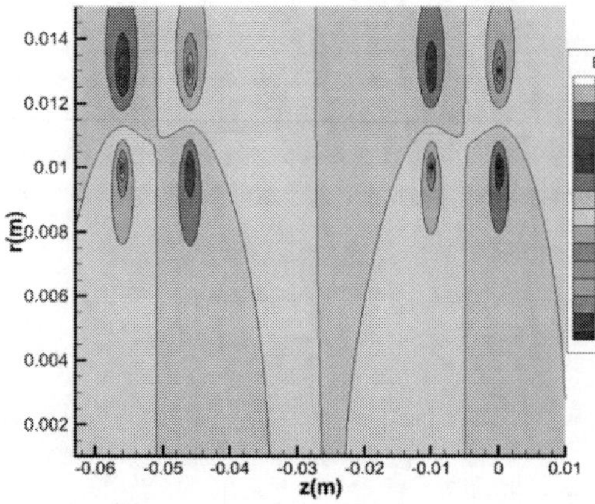

Figure 6. 2D distribution of B_r.

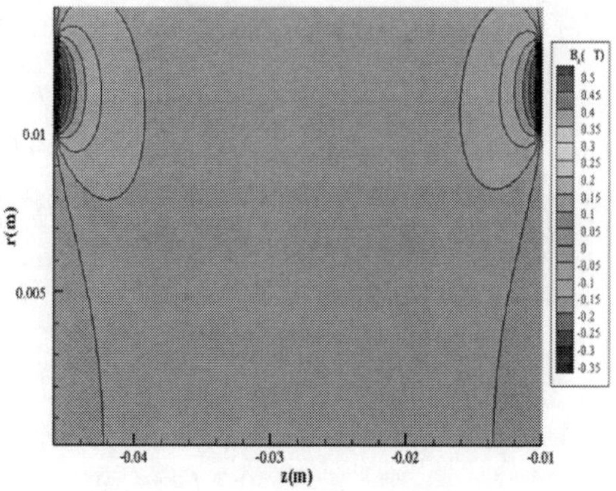

Figure 7. 2D distribution of B_z.

4. Electrodynamics structure of a plasma

After the calculation of magnetic field in the Penning discharge cell, it became possible to get preliminary results on the electrodynamics structure of plasma. Particle-in-cell method was chosen for the study. This method allows to obtain a lot of information about the structure of plasma in the Penning discharge cell.

For the numerical calculation parameters from Mamedov and Schitov experiment [8] was chosen: $D_a = L_a = 28$ mm, $L = 30$ mm, $p = 0.5$ mTorr, $U_a = 2$ kV, $I_{dis} = 1.187$ mA, $\sigma = 0.88$ T, $r_{in} = 10$ mm, $r_{out} = 13$ mm, $h = 5$ mm, $L_m = 36$ mm; here D_a – diameter of anode; L_a – length of anode; L – distance between anodes; p – pressure in discharge cell; U_a – voltage on anode; I_{dis} – discharge currents; r_{in} – inner radius of magnets; r_{out} – outer radius of magnets; h – height (length) of magnets; L_m – distance between two magnets. In the considered case modular ion source with Al anode and cathodes made of different materials was used. The permanent magnets were located behind the cathode and anticathode. The material of permanent magnets is SmCo.

As the result of numerical simulation 2D distributions (figures 8 – 12) of electric potential, electron density, H_2^+ ion density, temperature of electrons and H_2^+ ions were obtained. Also an energy distribution function of electrons and ions was calculated (figure 13).

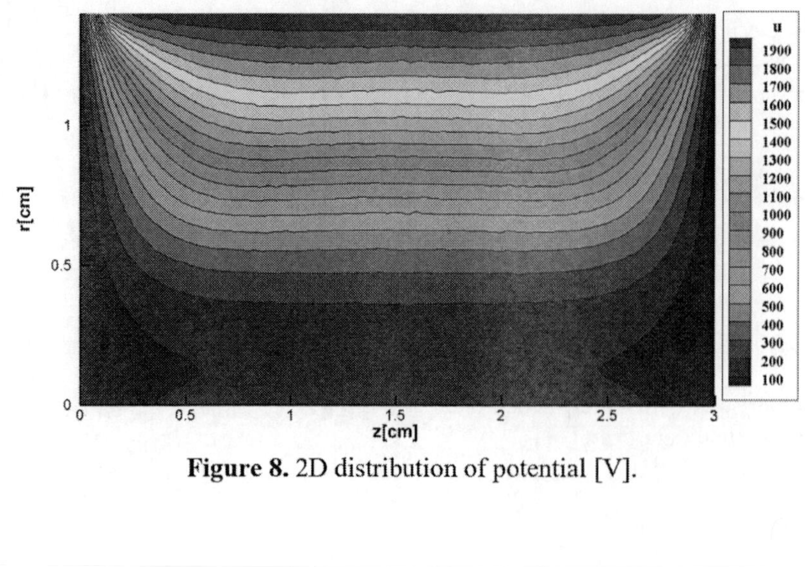

Figure 8. 2D distribution of potential [V].

IOP Conf. Series: Journal of Physics: Conf. Series **1009** (2018) 012025 doi:10.1088/1742-6596/1009/1/012025

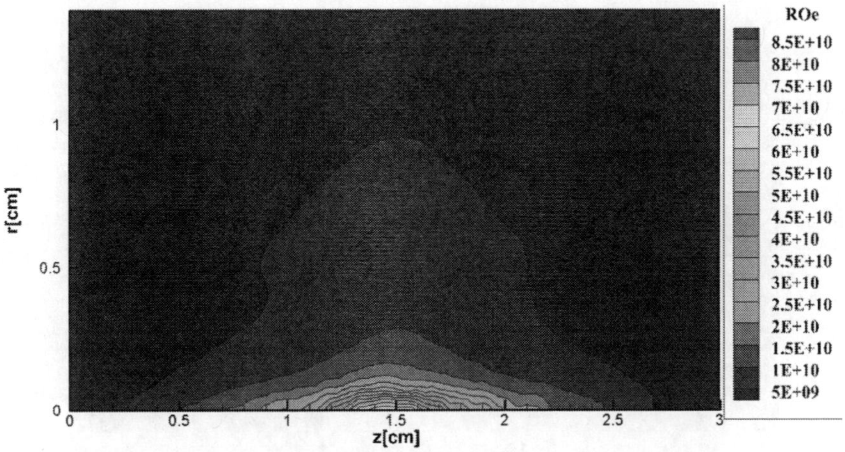

Figure 9. 2D distribution of electron density[cm^{-3}].

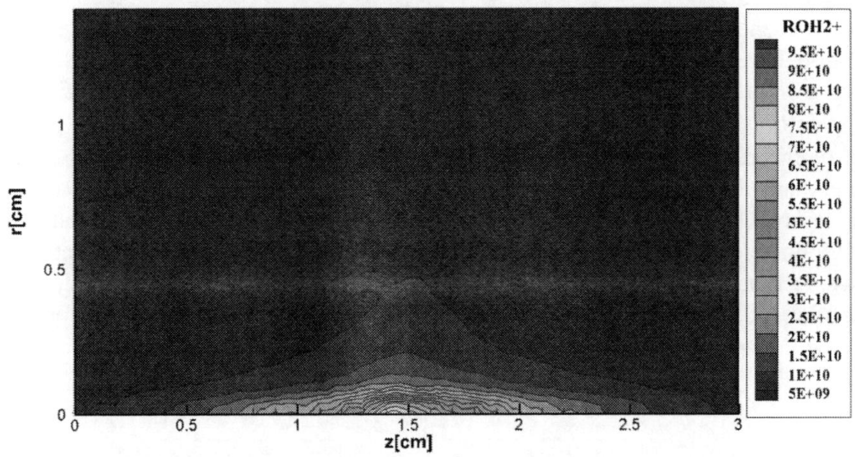

Figure 10. 2D distribution H_2 ion density[cm^{-3}].

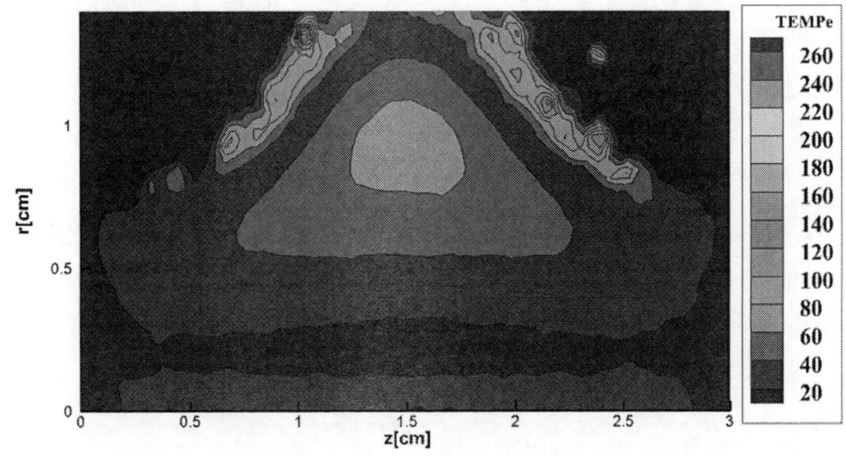

Figure 11. 2D distribution of electron's temperature[eV].

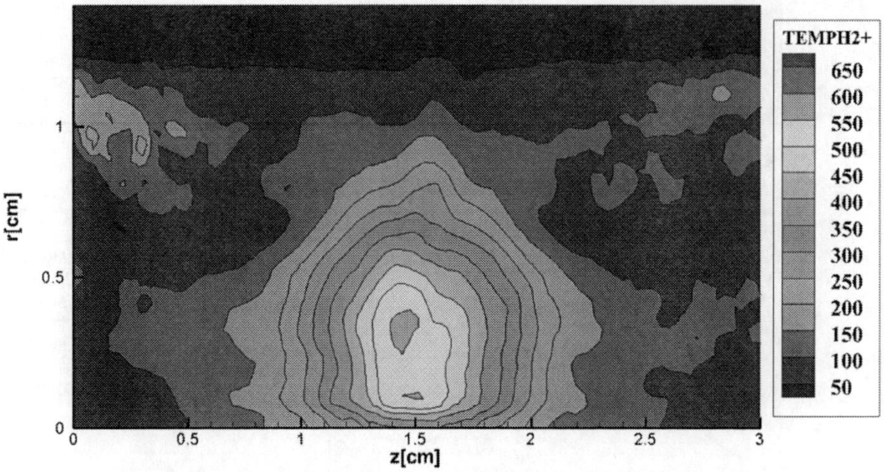

Figure 12. 2D distribution of H_2 ion's temperature[eV].

As we can see, the highest ion and electron density (see figure 9–10) are observed at the center of the computational area in the vicinity of the axis of symmetry. The electron density also decreases with distance along the r-coordinate. In this case, a large dependence on the position relative to the center of symmetry is observed.

Analysis of the distribution of ion and electron temperature (see figure 11–12) shows that the lowest temperatures are near the boundaries of the discharge cell. In the case of electron temperature distribution (figure 11), regions with the highest temperature are closer to the middle along the axial component. On the other hand, in the case of ion temperature distribution, we can note that the regions with the lowest temperature are located on the "upper" and "lower" side of the discharge cell with the temperature of about 50 eV. The highest temperature regions are located near the center of symmetry of discharge cell. Looking at the plot of the distribution function, we can say their energy does not exceed 2 keV.

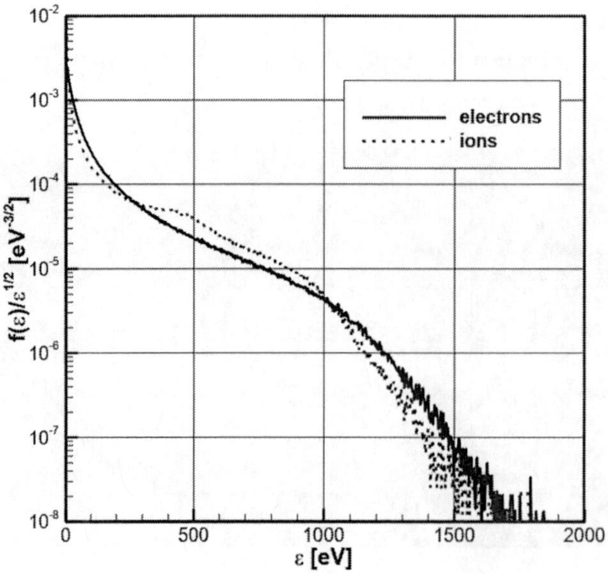

Figure 13. Function of distribution for electrons and H_2 ions by energy.

IOP Conf. Series: Journal of Physics: Conf. Series **1009** (2018) 012025 doi:10.1088/1742-6596/1009/1/012025

5. Conclusion

In conclusion computed code for evaluation of two-dimensional distribution of magnetic field created by annular magnets is developed based on the model presented in [7, 8]. The computed code was validated against numerical and experimental data available in literature. Preliminary results on the simulation of electrodynamic structure of Penning gas discharge plasma (at parameters given in [6]) in the presence of non-uniform external magnetic field created by annular permanent magnets were calculated using particle-in-cell method with the distribution of the magnetic field calculated using developed code.

Acknowledgments

Authors are grateful to prof. S.T. Surzhikov and staff of Radiative Gas Dynamics Laboratory of Institute for Problems in Mechanics for useful discussions on the problems considered in the study.

The presented study was supported by the Russian Science Foundation project №16–11–10275.

References

[1] Granovsky V L 1971 Electric current in gas. Steady current (Moscow: Nauka)

[2] Druyvesteyn M J and Penning F M 1940 The mechanism of electrical discharges in gases of low pressure *Reviews of Modern Physics* **12** 87

[3] Dikalyuk A S and Kuratov S E 2017 Numerical investigation of Penning discharge characteristics using 2D/3V Particle-In-Cell method *Journal of Physics: Conference Series* **815** 012001

[4] Surzhikov S T and Kuratov S E 2014 Diffusion-drift model of the Penning discharge at pressures of the order of 1 Torr *Physicochemical kinetics in gas dynamics* **15** 2 http://www.chemphys.edu.ru/pdf/2014-11-29-007.pdf

[5] Storozhev D A and Surzhikov S T 2013 Numerical Simulation of Glow Discharge in a Magnetic Field Through the Solution of the Boltzmann Equation *Journal of basic and applied physics* **2** 141–147

[6] Mamedov N V, Schitov N N and Kanshin I A 2015 Investigation of the Dependency of Penning Ion Source Operational Characteristics on its Geometric Parameters *Physical-Chemical Kinetics in Gas Dynamics* **16** 4 http://chemphys.edu.ru/issues/2015-16-4/articles/590/

[7] Ravaund R, Lemarquand G, Lemarquand V and Depollier C 2008 Analytical Calculation of the Magnetic Field Created by Permanent-Magnet Rings *IEEE Transactions on Magnetics* **44** 1982

[8] Babic S I and Akyel C 2008 Improvement in the analytical calculation of the magnetic field produced by permanent magnet rings *Progress In Electromagnetics Research C.* **5** 72

[9] Abramovitz M and Stegun I A 1972 Handbook of Mathematical Functions National Bureau of Standards Applied Mathematics Series **55** 590

[10] Carlson B C 1995 Numerical computation of real or complex elliptic integrals *Numerical Algorithms* **10** 13

[11] Mikolanda T, Kosec M and Richter A 2009 3D Magnetic Field Measurement, Visualisation and Modelling *Proc. of the 7th Intern. Conf. on Measurement* **306**

Application of polarization interferometry to plasma diagnostics

E Loktionov, V Telekh

Bauman Moscow State Technical University, Moscow, 105005, Russia

stcpe@bmstu.ru

Abstract. Polarization interferometer has some significant advantages over beam-splitting ones in a wide range of applications. Most of restrictions in interferometry regarding to optical elements quality and vibration protection can be neglected in Nomarski scheme, it is easy in alignment and allows studies in a broad spectral range without the replacement of mirrors. We present a description of two- and three-wavelength polarization interferometers and a technique for measurement of transparent media optical properties with high spatial (~ 10 μm) and temporal resolution (~ 10 ns) in UV-NIR (~ 380-900 nm) spectral range. As a result, atom and electron number densities have been evaluated in the case of about equal input of each of these components in refraction index change. Application of the proposed scheme to birefringent objects can be limited, but it also provides abilities to visualize strong electric and magnetic fields due to electro- and magneto-optical effects.

1. Introduction

Optical methods are widely used to study kinetics of chemical reactions and dynamics of inhomogeneous media. Widely used spectrophotometers have low temporal (10^{-3}–10^{-1} s) and no spatial resolution. For partial elimination of these shortcomings, polychromators can be used, incl. those with imaging diffraction gratings. Anyway, these devices are recording transmission (absorption) only; additional data on refractive index can significantly increase output of the study. Obtaining both optical constants is possible with partial overlapping of ordinary and extraordinary rays in a polarization interferometer, when the interference and absorption patterns of the region of interest are formed. The most common scheme of polarization interferometer is Nomarski differential interference-contrast microscope [1]. However, it was suggested to visualize phase shifts much smaller than wavelength and does not allow evaluating absorption and refractive indices for macroscopic objects.

To study the dynamics and macrostructure of laser-induced plasma flows, the following three methods are most often used: emission spectroscopy, interferometry and shadowgraphy (absorption photography). Each of these methods has its own advantages and disadvantages, and serves to record different, but partially overlapping, characteristics of such flows. Instability of impact radiation parameters and local properties of the target imposes a number of limitations on the processing of experimental results recorded under separate impacts, albeit under similar conditions. To eliminate these limitations, simultaneous application of several experimental techniques is desirable for the subsequent complex processing of the results obtained [2].

Emission spectroscopy, incl. LIBS, has been developing rapidly as an analytical technique, and led to release of commercial spectrometers for data recording with high temporal and spectral resolution. Interferometry, if obtained results are processed properly, is also a very informative technique [3]. But its high sensitivity is demanding to quality of the optical elements used and optical system vibration isolation. In some cases, additional alignment is required when buffer gas pressure changes in the experimental volume, especially if the mirrors (for example, due to lack of space) are located outside the gas cell or vacuum chamber. Proper (interferometric) quality wide-aperture windows for hermetic volumes, beam-splitters and mirrors are expensive. The quality requirements for optical elements can be significantly reduced for polarization interferometer [4-6]. Since probing beam is split into orthogonally polarized components close to the sensor, the distortion of the recorded interference pattern by optical elements is minimized.

The sensitivity of interferometers to vibrations can be significantly reduced by using short and ultrashort pulses of probing laser radiation [7]. An important problem in the implementation of beam-splitting interferometers with ultrashort probing pulses (especially for long paths) is matching spatial and temporal coherence, simultaneous pulses arrival, which requires high-precision linear motors for optical shoulders difference compensation. This problem is easily eliminated in polarization interferometer, since split beams optical paths difference is minimized and could be easily adjusted by the Wollaston prism inclination.

The disadvantage of the described scheme for randomly polarized sources is a relatively high radiation power (> 50 mW) required, since it is necessary to select only components that are linearly polarized at an angle of 45°. The advantages of this scheme are especially evident in multi-wavelength interferometer, an example of which is given below. For birefringent objects, the application of the proposed scheme can be limited, but also provides abilities to visualize strong electric and magnetic fields [8] due to corresponding electro- and magneto-optic effects.

2. Experimental layout

The principle of the interferometer (figure 1) is as follows. The probing radiation of DPSS lasers *1-3* (Lasever LSR 405NL-300, LSR 532NL-300, LSR 671NL-300), linearly polarized at 45° using half-wave plates, passes through the optical inhomogeneity in a single beam (unlike the classical Nomarski scheme). After that it is separated in a Wollaston prism *11* into two divergent beams (in our case, at an angle of ~10°, the angle is chosen according to the size of the region of interest). One of these beams is polarized vertically, and another one – horizontally. The distance between axes of the ordinary and extraordinary rays in the plane of the sensor can be changed (with an equal scale of the image) in proportion to the focal length of the lens *10*. The Wollaston prism *11* is located near the focus of the lens *10*, change in its position affects period of the interference pattern. This period is determined by probing wavelength and distance from the prism *11* to the sensor *13*, *14* divided by the product of the angle of ordinary and extraordinary beams divergence and the distance between focusing lens *10* and the prism *11* [9]. Easy period adjustment is very convenient for photoelectric mixing interferometers.

Using achromatic (Edmund Optics NT46 55x $\Delta\lambda$ ~ 200–300 nm) or tunable (zero order, e.g., Alphalas PO-TWP or liquid-crystal series, e.g., Thorlabs LCC series) half-wave plates *4-6*, tunable lasers could be used without the replacement of optical elements and adjustments similarly [8]. Interference pattern is formed in the area of ordinary and extraordinary beams overlap, after passing through the Wollaston prism. Since the orthogonally polarized components do not interfere with each other, it is necessary to align the polarization plane of both beams again at 45° by rotating the polarization filter *12* to approximately equal intensity in both beams, and so, maximum contrast of the interference pattern. A shadowgraph (absorption photo) of the investigated region is formed outside the overlapping region (figure 2).

Color CCD camera (Videoscan-205/C-USB) allows to use a single sensor, because blue (400 – 450 nm), green (530 –570 nm), and red (610 – 670 nm) colors can be easily extracted from the resulting image (Figure 3), providing the possibility of selecting resonant absorption bands in each of these ranges. In the case of insufficient sensitivity of usual CCD or in need to achieve high temporal resolution using continuous lasers, intensified cameras, e.g., based on microchannel plates (Nanoscan Nanogate-2) can be used. For single sensor registration, it is necessary to separate probing beams with an optical wedge located before the Wollaston prism. The advantages of such a solution are especially obvious when there is no need in 2D spatial resolution – using a broadband coherent radiation source (e.g., Fianium laser) a number of spectral intervals of interest can be cut out by slit diaphragms. The boundaries of 380–900 nm working spectral range are determined by the characteristics of the commercially available CCD's (190 – 1100 nm), optical elements of the scheme (350 –2200 nm) and commercial laser diodes or DPSS lasers (365 – 971 nm), and could be expanded.

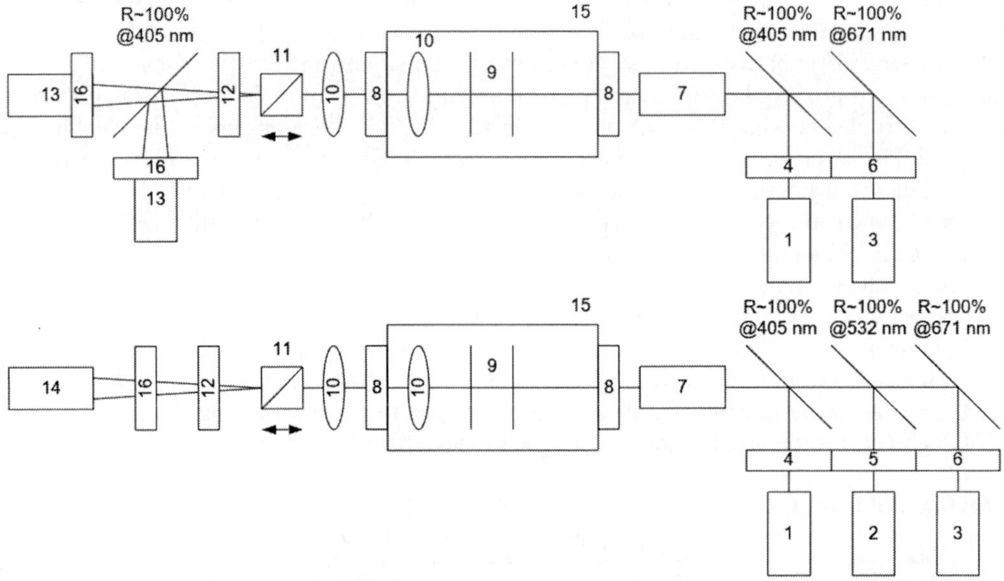

Figure 1. Optical schemes of experimental installations based on a polarization interferometer with monochrome and color sensors lasers: 1 – 405 nm, 2 – 532 nm, 3 – 671 nm; half-wave plates: 4 – 350 – 500 nm, 5 – 532 nm, 6 – 600 – 800 nm; 7 – telescope, 8 – vacuum chamber window, 9 – object, 10 – collecting lens, 11 – Wollaston prism, 12 – polarization filter, 13 – CCD camera with image intensifier, 14 – color CCD camera, 15 – vacuum chamber, 16 – color filter.

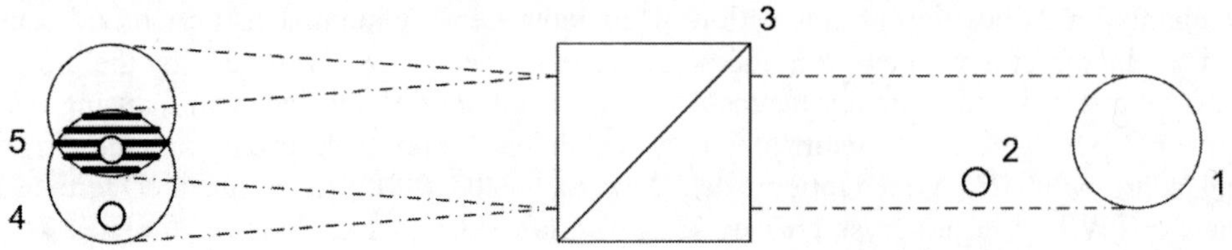

Figure 2. To the formation of shadowgraph (4) and interferograph (5) images: 1 – probing beam, 2 – optical inhomogeneity, 3 – Wollaston prism.

The refractive index resolution is determined not only by the features of the optical scheme, but also by the algorithm of the interferographs processing, and can reach $\delta\Delta nL \approx \lambda/200$ [7] (where 200 characterizes the discreteness of the phase shift detection), i.e., for the optical path $L \sim 10$ mm in the standard photometric cell it will be $\delta\Delta n \sim (1\text{–}4)\cdot 10^{-7}$. The upper limit of the refractive index value measured is limited only by the absolute deviation of the probing beam (it is important for large optical paths and small transverse dimensions of the sample); for automated processing, it is undesirable for the phase shift to exceed 2π [2]. Resolution on the absorption index is determined by the dynamic range of the sensor used (usually in the range from 8 to 14 bits), the instability of the spatial distribution of the radiation brightness along the profile of the probing beam and the optical path length; for the standard cell it is $\delta\alpha \sim (1–4)\cdot 10^{-3}$ cm^{-1}. It should be noted that these resolutions are mostly determined by random errors caused by fluctuations in the intensity of the probing radiation, CCD noise, convective currents, etc. These errors values and statistical processing capabilities should be evaluated for each particular installation. The limiting time resolution is determined by the duration of the probe radiation pulse ~ $10^{-14}–10^{-13}$ s or the exposure time of the camera (for usual CCD cameras ~ $10^{-5}–10^{-6}$ s, with MCP ~ $10^{-9}–10^{-8}$ s, streak-cameras ~ 10^{-13} s).

3. Results and discussion

Interferometry in plasma physics is used primarily to study the electron number density distribution by local refractive index analysis, which causes the phase shift of the probing radiation wave front. This approach is limited by the fact that the change in the refractive index depends on the contribution not only of electrons but also of atoms and ions [10], experimental establishing of each of these components influence is possible with multi-wavelength interferometry only [11].

As a result of processing using fast Fourier transform algorithm, the spectral refractive indices obtained in different color channels of interferographs of colliding laser-induced plasmas (Figure 4). The values obtained are substantially different for different wavelengths. The characteristic values of electron concentrations for 405 nm and 671 nm are determined respectively as: $2{,}4\cdot 10^{18}$ cm^{-3} and $4{,}5\cdot 10^{17}$ cm^{-3}. This difference occurs because the contribution of electrons to the refractive index formation not predominant $(n\text{-}1)_e/(n\text{-}1)_a \sim 1$ in the considered case. When the results obtained for both wavelengths are processed simultaneously, the characteristic value of n_e is ~1.1 10^{18} cm^{-3}, and the concentration of atoms is $n_a \sim 2{,}3\cdot 10^{19}$ cm^{-3}.

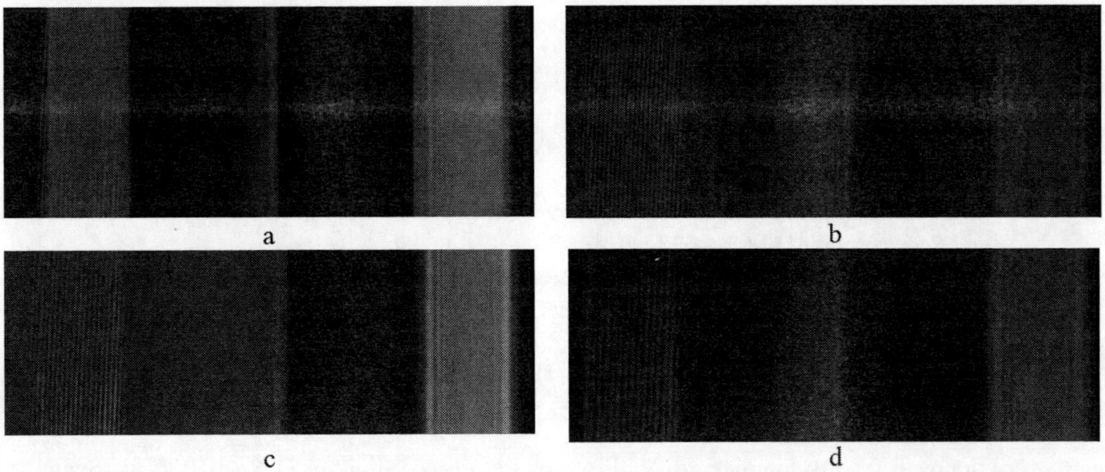

Figure 3. The use of a color CCD camera in a multi-wavelength interferometer: a – the recorded image; b, c, d – blue, green and red channels, respectively.

4. Conclusions

A multi-wavelength, tunable in a wide spectral range, polarization interferometer for simultaneous recording of spectral refractive and absorption indices of optically inhomogeneous media with high spatial and temporal resolution was developed and implemented for the first time. Features of the optical scheme described can significantly reduce its sensitivity to optical elements quality, vibrations, and simplify the alignment, especially when using ultrashort probing laser pulses.

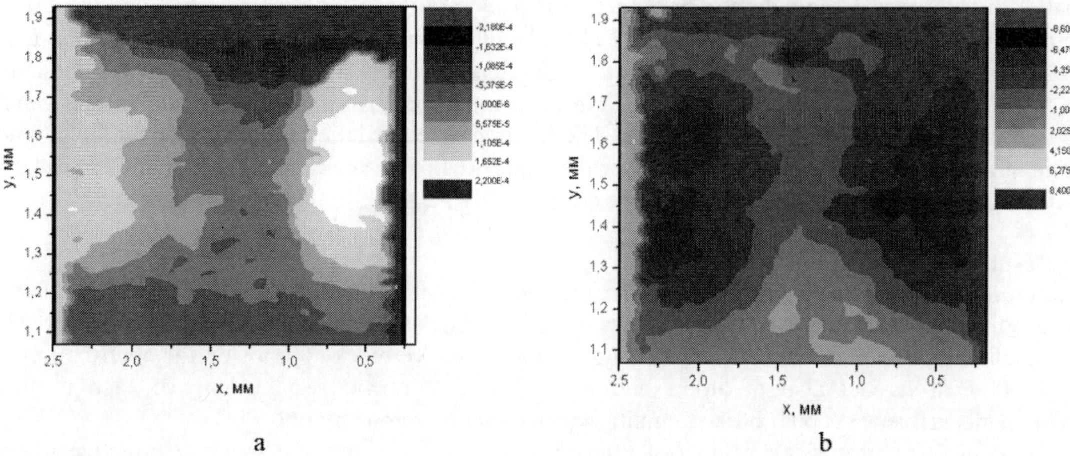

Figure 4. The distribution of the refractive index change relative to the initial for 405 nm (a) and 671 nm (b)

Acknowledgements
This research has been performed at "Beam-M" facility, supported by the Russian Ministry of Education and Science under the Federal Target Program 1.2 "Research and development in priority directions of science and technology in Russia 2014–2020" – contract No. 14.577.21.0247 (RFMEFI57717X0247).

References

[1] Murphy D B and Davidson M W 2012 *Fundamentals of Light Microscopy and Electronic Imaging*. (New York: Wiley-Blackwell)

[2] Loktionov E Y, Protasov Y Y, Telekh V D and Khaziev R R 2013 *Instruments and Experimental Techniques* **56** 46–54

[3] Loktionov E Y, Ovchinnikov A V, Protasov Y S, Protasov Y Y and Sitnikov D S 2012 *Optics and Spectroscopy* **112** 631–637

[4] Benattar R, Popovics C and Sigel R 1979 *Review of Scientific Instruments* **50** 1583–1586

[5] Hough P 2010 *Laser, optical and electrical diagnostics of colliding laser-produced plasmas* (Dublin: Ph. D. thesis)

[6] Borner M, Fils J, Frank A, Blazevic A, Hessling T, Pelka A, Schaumann G, Schokel A, Schumacher D, Basko M M, Maruhn J, Tauschwitz A and Roth M 2012 *Review of Scientific Instruments* **83** 043501

[7] Loktionov E Y, Ovchinnikov A V, Protasov Y Y and Sitnikov D S 2010 *Instruments and Experimental Techniques* **53** 416–422

[8] Pisarczyk T, Gus'kov S Y, Dudzak R, Chodukowski T, Dostal J, Demchenko N N, Korneev P, Kalinowska Z, Kalal M, Renner O, Smid M, Borodziuk S, Krousky E, Ullschmied J, Hrebicek J, Medrik T, Golasowski J, Pfeifer M, Skala J and Pisarczyk P 2015 *Physics of Plasmas* **22** 102706

[9] Stafe M, Marcu A and Puscas N N 2014 *Pulsed Laser Ablation of Solids*. (Berlin Heidelberg: Springer-Verlag)

[10] Zaidel A N and Ostrovskaya G V 1977 *Lazernye metody issledovania plazmy* [*Laser methods in plasma research*] (Leningrad.: Nauka) (in Russian)

[11] Kozin G I, Kuznetsov A P, Lebedinskii M O, Savel'ev A V, Sokolov A P and Savelov A S 2003 *Instruments and Experimental Techniques* **46** 188–193

Computation of inductively coupled air plasma flow in the torches

S A Vasilevskii, A F Kolesnikov, A I Bryzgalov, S E Yakush

Ishlinsky Institute for Problems in Mechanics RAS, Prospekt Vernadskogo 101-1, Moscow, 119526, Russia

vasil_ipmech@mail.ru

Abstract. Software was developed to calculate electric field amplitude in the induction plasmatron discharge channel in framework of 2D model with far field boundary conditions. This code was integrated with the solver for Navier-Stokes equations for plasma flow in the plasmatron channel. Numerical simulation of inductively coupled plasma flow was carried out for the channels of IPG-4 and IPG-3 induction plasmatrons operating in IPMech RAS. Computations were performed twice: first, the electric field amplitude was calculated with the more accurate 2D model, next - with the simplified 1D model to provide detailed compare of the accuracy of results obtained with use of 1D model. The gas-dynamic parameters (flow velocity, temperature) at the exit section of the channels obtained with 1D model are within the error from 0.4% to 4% for IPG-4 plasmatron; the corresponding error for IPG-3 plasmatron is 16% to 26%. Numerical simulation results for IPG-3 plasmatron are presented for the first time.

1. Introduction

Inductively coupled plasma (ICP) flow is one of the important objects for numerical simulation from the viewpoint of various applications and studying fundamental physical processes. A modern direction of induction plasma application is simulation of dissociated gas flow thermochemical action on a surface of thermal protection material and investigation of the materials surface catalycity [1]. Two induction plasmatrons, IPG-4 and IPG-3, are available in the Institute for Problems in Mechanics (IPMech RAS), they are used intensively for studying physical processes in high-enthalpy jets generated by induction plasmatrons and for material testing in such jets. The plasmatrons, experimental techniques and CFD codes developed for complex investigation of heat transfer from the jet to testing materials and for determination of material surface catalytic properties with respect to atoms recombination are described in [2–6].

Numerical simulation of ICP flow in a plasmatron discharge channel is a complicated problem even for the case of equilibrium plasma flow due to the large flow vortices that are formed after the inlet section. The room temperature gas inflows in the channel through a narrow (2-3 mm) inlet slot with a swirl angle β of about 45° to provide flow swirling, $\tan(\beta) = W/U$, U and W are the axial and tangential velocity components. The inductor coil with a high radio-frequency (RF) current encircles the channel and provides powerful electromagnetic field within the inductor region of the channel. The Joule heat production and Lorentz force lead to the strong plasma heating and so called "magnetic pressure" that affects the complicated flow structure with primary and secondary vortices.

Calculation of electromagnetic field within the channel is important part of the total problem of ICP numerical simulation. The more accurate 2D model for electromagnetic field was used in [7]; it requires the use of extended computational grid for electromagnetic field calculation to apply the far field boundary conditions - electromagnetic field turns to zero at far boundaries. The simplified 1D model was first proposed [8] and then used for ICP numerical simulation for IPG-4 [4-6]; it allows to save CPU time and programming efforts. The detailed comparison with the code developed in von Karman Institute for Fluid Dynamics confirmed the good accuracy of this 1D model for argon plasma flow in IPG-4 channel [9]. In this paper we provide detailed comparison of 2D and 1D models for electromagnetic field for ICP calculations in IPG-4 and IPG-3 plasmatrons for air plasma. The configuration of discharge channels for both plasmatrons is a simple cylinder; the air flow in the channels is always subsonic. Note that numerical simulation results for IPG-3 plasmatron are presented here for the first time.

2. Problem formulation

ICP flow in the plasmatron discharge channel is calculated on the basis of Navier-Stokes gas-dynamic equations coupled with the simplified Maxwell equation for the electric field amplitude. The following assumptions are used for the gas-dynamic part: the subsonic flow is stationary, laminar, equilibrium, and axisymmetric, with a swirl in azimuthal direction. High frequency electromagnetic field does not influence on transport properties of plasma, radiative processes are negligible. The air flow can be considered as equilibrium at sufficiently large pressure $P \geq 0.05$ atm, and radiative processes are not very essential for sufficiently low plasmatron power, i.e. sufficiently low flow temperature [10].

The governing equations are two-dimensional Navier-Stokes equations with three velocity components - axial, radial and tangential component due to the flow spinning. The equations are written in the cylindrical coordinate system, they include source terms corresponding to the electromagnetic field influence: F_Z, F_r are axial and radial components of the Lorentz force (the so called magnetic pressure); Q_J is Joule heat production [5]. These source terms are expressed by the electric field amplitude and will be written below.

Boundary conditions for the gas-dynamic equations are as follows. The necessary flow parameters including velocity tangential component are specified at the annular inlet slot at the channel entry section. Zero values for velocity components and certain values of temperature are prescribed at all solid surfaces (the channel wall, the face of the gas injector interface). Symmetry conditions are applied at the channel axis. The so-called "soft" boundary conditions are applied at the channel exit section, i.e. the unknown functions are assumed to be constant near this boundary.

RF electric current oscillating in the inductor coil produces a vortical electric field within the discharge channel. The inductor coil is treated as a number of separate turns with the same current in each of them. Inductor turns are represented as thin equidistant rings perpendicular to the symmetry axis. Navier-Stokes equations including the source terms are assumed to be time-averaged with respect to RF electric field oscillations. The electromagnetic field is treated as monochromatic one with use of complex amplitudes for electric and magnetic fields:

$$\vec{E}(t,z,r)=\vec{E}(z,r)\cdot exp(-i\omega t), \quad \vec{H}(t,z,r)=\vec{H}(z,r)\cdot exp(-i\omega t)$$

here z, r are the axial and radial coordinates, t is the time, $\omega = 2\pi f$ - circular frequency, f - frequency of the inductor current. For the experimental conditions in IPG-4, IPG-3 plasmatrons the following assumptions can be applied [11]: $\omega << \nu_e$ and $|\varepsilon| << 4\pi\sigma/\omega$, here ν_e is the electron collisions frequency, ε - dielectric permittivity of plasma. The last inequality means that the displacement current is negligible. Also, the following common assumptions are used: plasma is quasi-neutral; plasma magnetic permeance $\mu = 1$, and dielectric constant does not depend on electromagnetic field and so does not depend on z and r.

2.1. 2D formulation for the electric field amplitude equation

In frameworks of the above assumptions, Maxwell equations with use of Ohm's law and symmetry conditions lead to the following equations for tangential $E_\theta\,(z,r)$, axial E_z and radial E_r components of the electric field amplitude:

$$\frac{\partial^2 E_\theta}{\partial z^2}+\frac{\partial}{\partial r}\left(\frac{1}{r}\frac{\partial}{\partial r}rE_\theta\right)=-i\omega\mu_0\sigma E_\theta,\quad E_z=0,\ E_r=0 \qquad (1)$$

here μ_0 is vacuum magnetic permeance, σ is the plasma electric conductivity. The two-dimensional (2D) elliptic equation (1) for the tangential component of the electric field amplitude $E_\theta\,(z,r)$ is solved on far-field mesh, extended beyond the channel, to allow the use of the simple far-field boundary conditions:

$$E_\theta(z, r=\pm\infty)=0;\quad E_\theta(z=\pm\infty, r)=0 \qquad (2)$$

In addition, the symmetry condition $E_\theta(z, r=0) = 0$ is used also.

Following [7], to avoid singularities near the coil rings, the total field amplitude E_θ is split to a singular part $E_{\theta V}$ induced by the outer inductor rings and a nonsingular plasma-induced part $E_{P\theta}$:

$$E_\theta = E_{V\theta} + E_{P\theta} \qquad (3)$$

The equation (1) for E_θ then can be rewritten so:

$$\frac{\partial^2 E_{P\theta}}{\partial z^2}+\frac{\partial}{\partial r}\left(\frac{1}{r}\frac{\partial}{\partial r}rE_{P\theta}\right)=-i\omega\mu_0\sigma(E_{V\theta}+E_{P\theta}) \qquad (4)$$

Analytical formula for the electric field induced by a single thin current-carrying loop can be found in [12]. The vacuum-field amplitude $E_{V\theta}$ is then calculated by adding up contributions of the discrete coil rings of inductor. Singularities in $E_{V\theta}$ near the coil rings do not affect the equation (4) because the electrical conductivity σ is zero outside the channel.

The far-field boundary conditions can be rewritten so:

$$E_{P\theta}(z, r=\pm\infty)=0;\quad E_{P\theta}(z=\pm\infty, r)=0 \qquad (5)$$

2.2. 1D formulation for the electric field amplitude equation

Solution of 2D equation (4) requires additional computer time and efforts to solve elliptic equation on far-field mesh. To save time, the simplified 1D model for the electric field was proposed based on the assumption [8]: $\partial E_\theta/\partial z \ll \partial E_\theta/\partial r$. This assumption seems to be reasonable for the relatively thin plasmatron discharge channel (channel radius R_c is about 10 times smaller than its length). With use of this assumption, equation (1) can be simplified to the ordinary differential equation:

$$\frac{d}{dr}\left(\frac{1}{r}\frac{d}{dr}rE_\theta\right)=-i\omega\mu_0\sigma E_\theta \qquad (6)$$

Boundary condition for this equation is $E_\theta(z, r=0) = 0$ at the symmetry axis and the following approximate condition at the channel wall [5]:

$$r=R_c:\quad \frac{1}{r}\frac{d}{dr}(rE_\theta)=i\omega\mu_0 H_{Vz}(z) \qquad (7)$$

Here H_{Vz} is the vacuum-part of axial component of the magnetic field amplitude at the channel wall, produced only by the outer inductor rings. If we replace H_{Vz} by the total value $H_z = H_{Pz} + H_{Vz}$, then equation (7) will be accurate, but the value H_{Pz} is not known before the solution of the problem, so we have to assume that H_{Vz} could stand for H_z in the condition (7), e.g. for small plasmatron power regimes.

As a result, we obtain a quasi-one-dimensional formulation (6), (7) for the solution of electrodynamic part of the problem, here E_θ depends on z due to the boundary condition only. Note, that in 1D case the magnetic field amplitude has only axial non-zero component H_z, and F_r=0.

2.3. Source terms

Time-averaged source terms for the time-averaged Navier-Stokes equations are the axial and radial components of Lorentz force F_z, F_r (so called magnetic pressure) and Joule heat production Q_J. They are expressed via E_θ (here and further $E_\theta = E_{V\theta} + E_{P\theta}$):

$$F_z = -\frac{\sigma}{2\omega} Re\left\{ E_\theta \left(i\frac{\partial E_\theta}{\partial z} \right)^* \right\}, \quad F_r = -\frac{\sigma}{2\omega}\frac{1}{r} Re\left\{ E_\theta \left(i\frac{\partial r E_\theta}{\partial r} \right)^* \right\}, \quad Q_J = \frac{\sigma}{2} E_\theta E_\theta^*$$

here Re means the real part of a complex value, asterisk is the conjugation sign for a complex value.

3. Numerical solution techniques

Numerical solution of the 2D Navier-Stokes equations is achieved by the finite-volume method on the staggered grid within the channel region. The grid is non-uniform rectangular one both in axial (103 nodes) and radial (93 nodes) directions. The grid is refined in the radial direction near the torch wall; the grid is refined in the axial direction near the torch inlet section and within the inductor region.

Value of the inductor current amplitude I_0 is determined in process of iterations by the prescribed value of power input in plasma N_{pl}. As to a value of N_{pl} itself, it should be specified by both the measured value of anode power N_{ap} and the plasmatron efficiency η_{eff} obtained from special experimental measurements: $N_{pl} = \eta_{eff} \cdot N_{ap}$. So, to determine the I_0 at each iteration we first calculate the electric field assuming that the inductor current amplitude equals to unit (I_1=1), and then integrate Joule heat production inputs from all the computation cells to obtain the total computational Q_{Jcomp} and the relation of the specified N_{pl} and Q_{Jcomp}: $\alpha = N_{pl} / Q_{Jcomp}$, and then we rescale the inductor current amplitude: $I_0 = \sqrt{\alpha}$. The necessary transport coefficients (including electrical conductivity) and thermodymamic properties are determined simultaneously in process of numerical simulation for each flow point by a rather accurate interpolation across the previously calculated tables. Such tables for equilibrium air mixture have been calculated as functions of pressure and temperature in advance in the temperature and pressure regions $300 \le T \le 20000$ K, $0.001 \le P \le 10$ atm. The calculation of transport coefficients for the tables were made with use of accurate formulas of Chapman-Enskog method [13-15] for the second and third non-zero approximation numbers, i.e. the numbers of terms in Sonine polynomials expansions of Boltzmann's equation solution that provide a good convergence. The reliable data on the interaction potentials and collision integrals were used for all pairs of particles [16]. The 22 components were accounted for the dissociated partially ionized air mixture including some second ions.

3.1. Numerical solution of Navier-Stokes equations

Navier-Stokes equations solution technique is based on the control volumes method and SIMPLE algorithm of Patankar and Spalding with use of the staggered grid [17]. Convective terms are approximated by the finite differences with the first order accuracy to provide better monotonicity property. Modifications of the Patankar & Spalding method were made to provide the convergence of iterations for a complicated ICP flow with large vortices. In particular, the unknown functions are determined with use of additional under-relaxation procedure to increase stability of the solution technique.

3.2. Computation of electric field amplitude with 1D model

Thomas algorithm ("sweep") for 3-diagonal matrix equations with use of complex variables is applied to solve the boundary value problem (6)-(7) to obtain the complex value of electric field amplitude tangential component E_θ and magnetic field amplitude axial component H_z.

3.3. Computation of electric field amplitude with 2D model

The two-dimensional boundary-value problem described by the equations (4)–(5) was solved numerically in a domain which was extended beyond the induction channel, so that the far-field boundary conditions (5) could be applied. The electric field was decomposed into two components described by (3), and the actual approximation according to (4) was performed for the electric field induced by plasma, $E_{P\theta}$.

The elliptic operator on the left-hand side of (4) was approximated by the standard second-order scheme applied to the real and imaginary parts of the electric field. In (4), the same operator applies to both the real-valued and imaginary parts, coupling between these occurs on the right-hand side. The resulting system of algebraic equations with sparse matrix was solved iteratively by an efficient ILUT-preconditioned GMRES method [18] implemented in SPARSKIT numerical package [19].

4. Results and discussion

Computations were made for the experimental regimes of the two IPMech RAS facilities - induction plasma generators IPG-4 and IPG-3 at the single pressure value P=100 mbar.

IPG-4 plasmatron parameters are as follows: the inductor current frequency f=1.76 MHz, the discharge channel length L_c=400 mm, the discharge channel diameter D_c= 80 mm. IPG-3 parameters are as follows: f= 0.44 MHz, L_c= 870 mm, D_c=192 mm. For each plasmatron the computations were made for three values of power. Each computation was made twice, with use of 1D and 2D models for calculation of electric field. The same gas-dynamic solver was used for all cases. The main goal of this work was detailed comparison of 1D and 2D models for electric field calculation for the two plasmatrons for different power regimes. The basic results of comparison of the 1D and 2D calculations are presented in the tables 1 and 2 for IPG-4 and IPG-3 plasmatrons.

Table 1. Comparison of 1D and 2D calculations for IPG-4.

N_{ap}	N_{pl}	U_c, m/s			h_c, MJ/kg			T_c, K			I_0, A		
kW	kW	1D	2D	%	1D	2D	%	1D	2D	%	1D	2D	%
70	44	220	228	*4*	49.7	51.3	*3*	9539	9735	*2*	373	367	*2*
45	29	155	157	*1*	36.5	37.8	*3*	7032	7223	*3*	317	320	*1*
20	12	69.0	69.3	*0.4*	17.6	17.9	*2*	5672	5691	*0.3*	312	310	*0.6*

The notations in the tables 1 and 2 are as follows: N_{ap} and N_{pl} are the plasmatron anode power and power input in plasma; U_c, h_c, T_c are plasma velocity, enthalpy and temperature at the symmetry axis at the channel exit section; I_0 is the amplitude of the inductor current. The sub-columns 1D and 2D correspond to the parameters calculated with use of 1D and 2D models for electric field amplitude. The sub-column % corresponds to the percent difference of the parameters calculated with use of 1D and 2D models.

Table 2. Comparison of 1D and 2D calculations for IPG-3.

N_{ap}	N_{pl}	U_c, m/s			h_c, MJ/kg			T_c, K			I_0, A		
kW	kW	1D	2D	%	1D	2D	%	1D	2D	%	1D	2D	%
400	240	219	261	*16*	55.3	67.1	*18*	10148	11014	*8*	899	1126	*20*
300	180	170	213	*20*	51.4	62.0	*17*	9745	10674	*9*	775	1010	*23*
200	120	107	145	*26*	44.0	52.7	*17*	8531	9896	*14*	637	856	*26*

It is clear from tables 1, 2 that for IPG-4 plasmatron the use of 1D model to calculate the electric field amplitude provides good (for low power) or satisfactory (for high power) results for the gas-dynamic parameters at the channel exit section. But for the IPG-3 facility the use of 1D model leads to essential errors even in the gas-dynamic parameters at the channel exit section.

IOP Conf. Series: Journal of Physics: Conf. Series **1009** (2018) 012027 doi:10.1088/1742-6596/1009/1/012027

The detailed comparison of the results obtained with use of 1D and 2D models is presented in the Figures below. The contours of dimensionless stream function (f), isotherms (T), contours of the electric and magnetic field amplitudes (E_θ, H_z), calculated with use of 1D and 2D models, are presented in figures 1–4 for IPG–4 for P=100 mbar, N_{ap}= 45 kW, N_{pl}=29 kW.

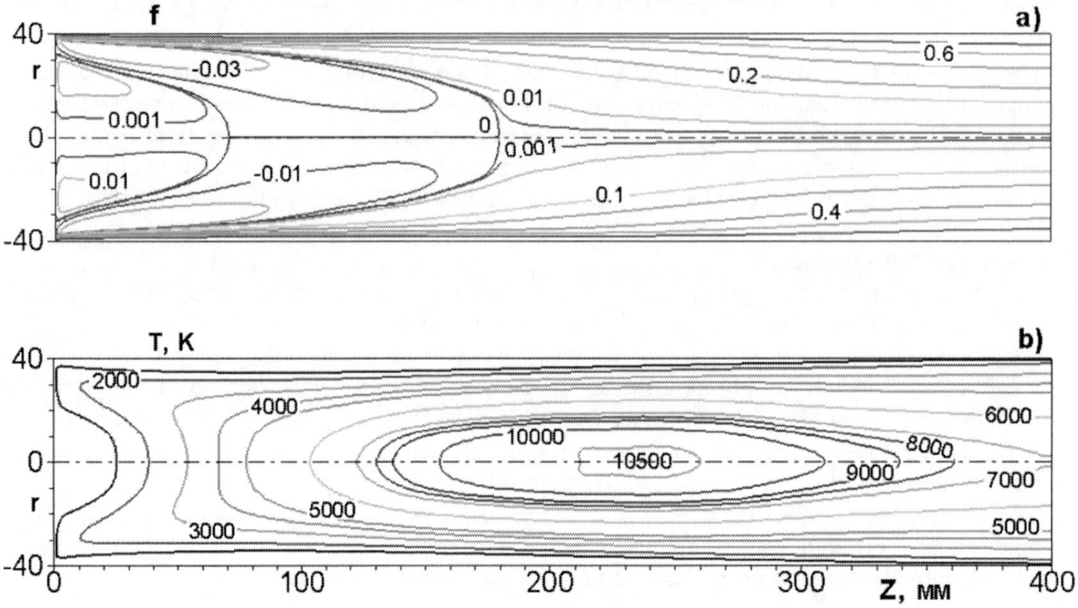

Figure 1. Contours of dimensionless stream function (a) and isotherms (b) in IPG–4 discharge channel calculated with use of 1D model for the electric field amplitude.

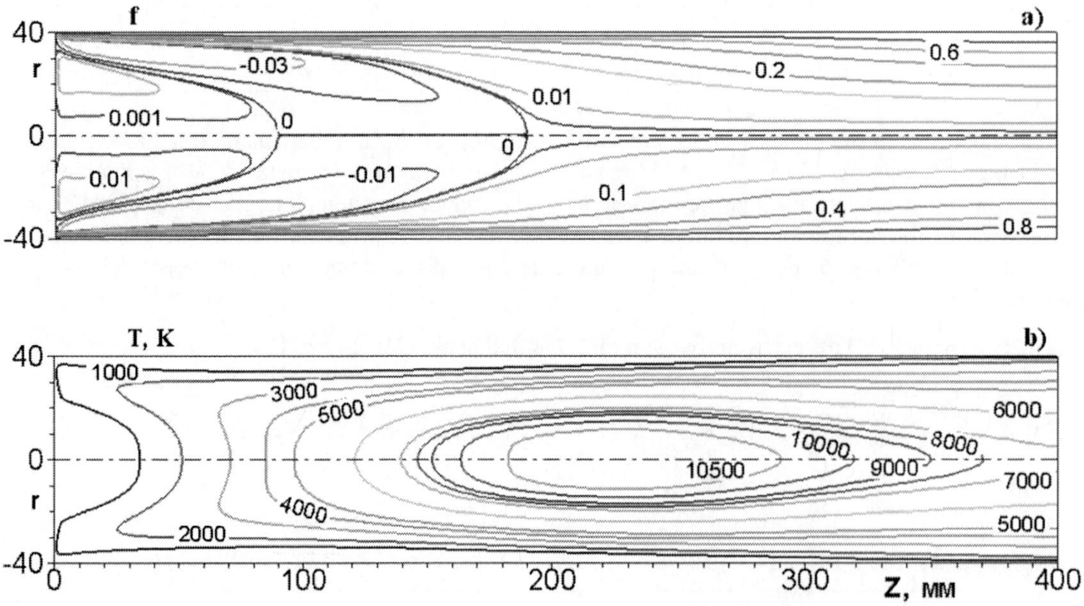

Figure 2. Contours of dimensionless stream function (a) and isotherms (b) in IPG–4 discharge channel calculated with use of 2D model for the electric field amplitude.

IOP Conf. Series: Journal of Physics: Conf. Series **1009** (2018) 012027 doi:10.1088/1742-6596/1009/1/012027

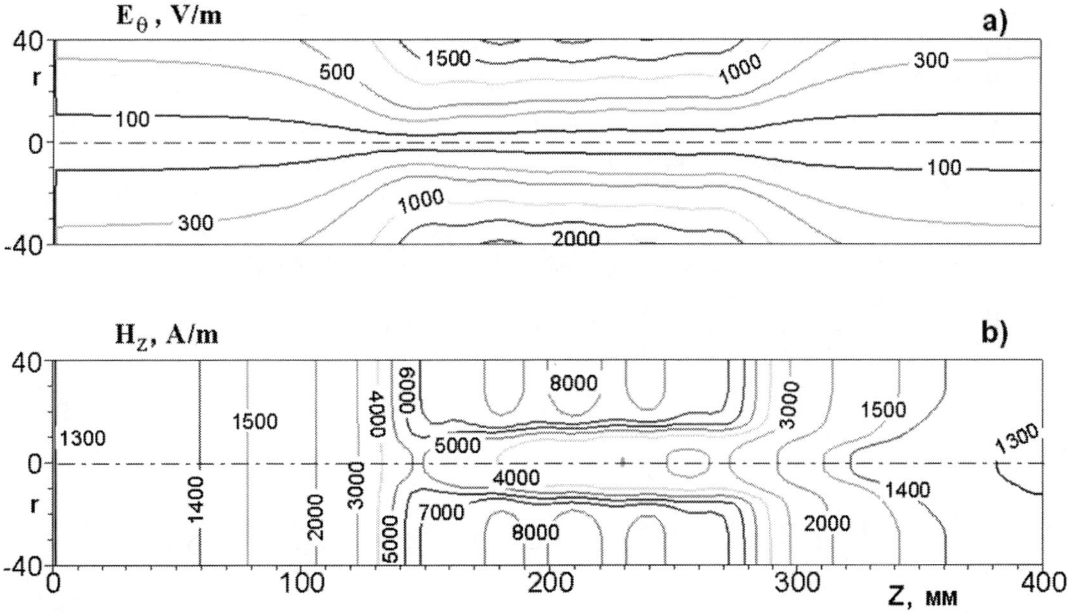

Figure 3. Contours of the amplitudes of electric field (a) and magnetic field (b) in IPG-4 discharge channel calculated with use of 1D model.

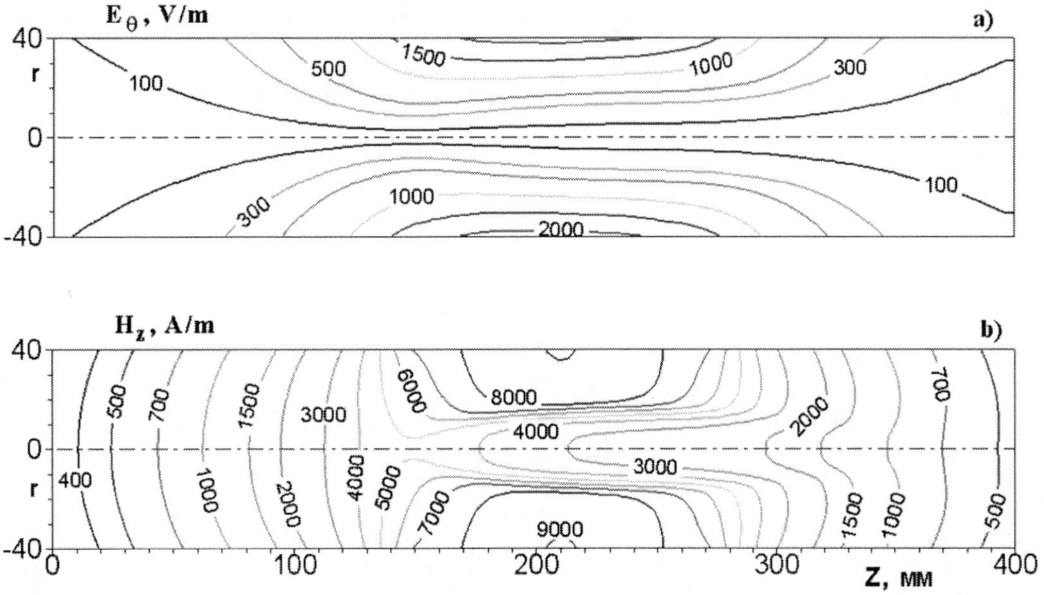

Figure 4. Contours of the amplitudes of electric field (a) and magnetic field (b) in IPG-4 discharge channel calculated with use of 2D model.

It is clear from the comparison of Figures 1 and 2, that the flow structures calculated with use of 1D and 2D models are similar, but slightly different, e.g. the length of secondary vortex near the inflow interface is greater for 2D case; the temperature maximums are nearly the same for 1D and 2D cases, but the geometry of isotherms is slightly different. The comparison of figures 3 and 4 shows that the maximum values of electric and magnetic field amplitudes within the inductor region are nearly the same for 1D and 2D cases, but the geometry of the contours is different.

IOP Conf. Series: Journal of Physics: Conf. Series **1009** (2018) 012027 doi:10.1088/1742-6596/1009/1/012027

The contours of dimensionless stream function, isotherms, electric and magnetic field amplitudes, calculated with use of 1D and 2D models, are presented in the figures 5–8 for IPG-3 for $P = 100$ mbar, N_{ap}=300 kW, N_{pl}=180 kW.

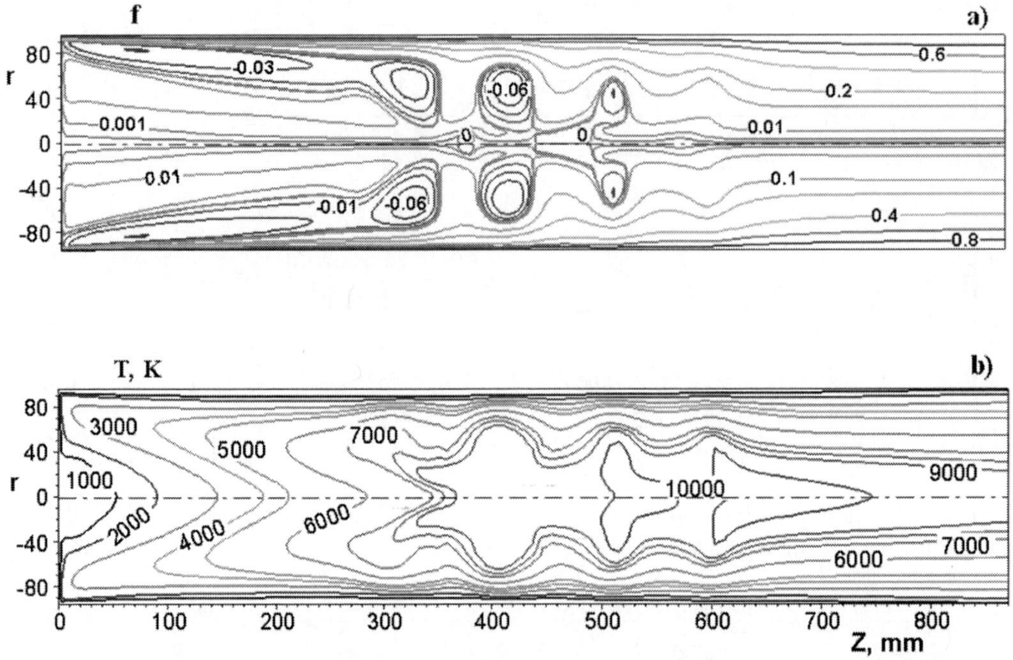

Figure 5. Contours of dimensionless stream function (a) and isotherms (b) in IPG-3 discharge channel calculated with use of 1D model for the electric field amplitude.

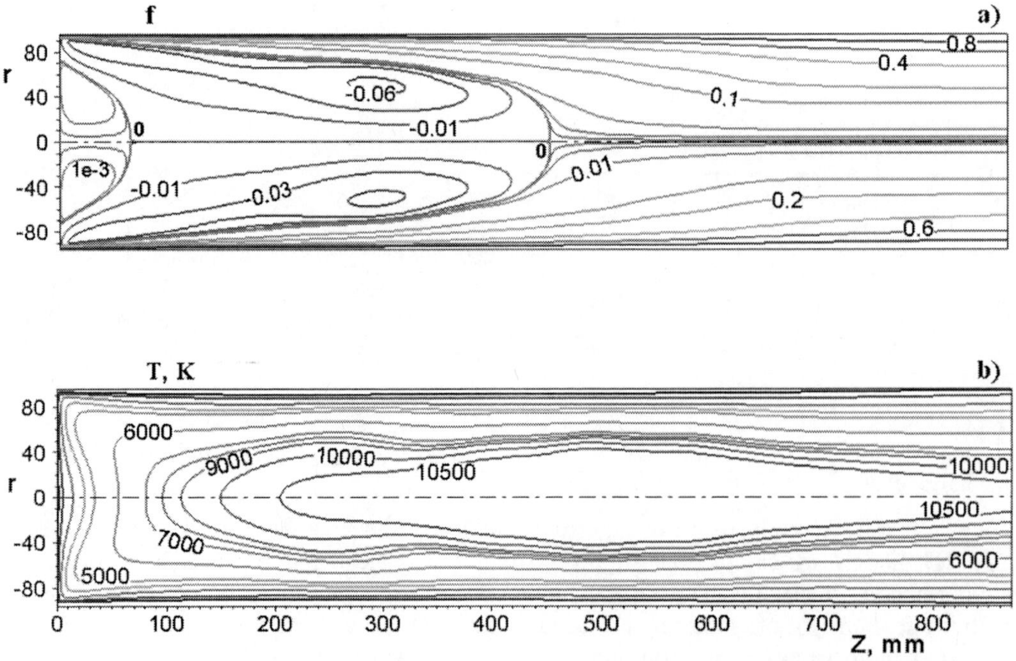

Figure 6. Contours of dimensionless stream function (a) and isotherms (b) in IPG-3 discharge channel calculated with use of 2D model for the electric field amplitude.

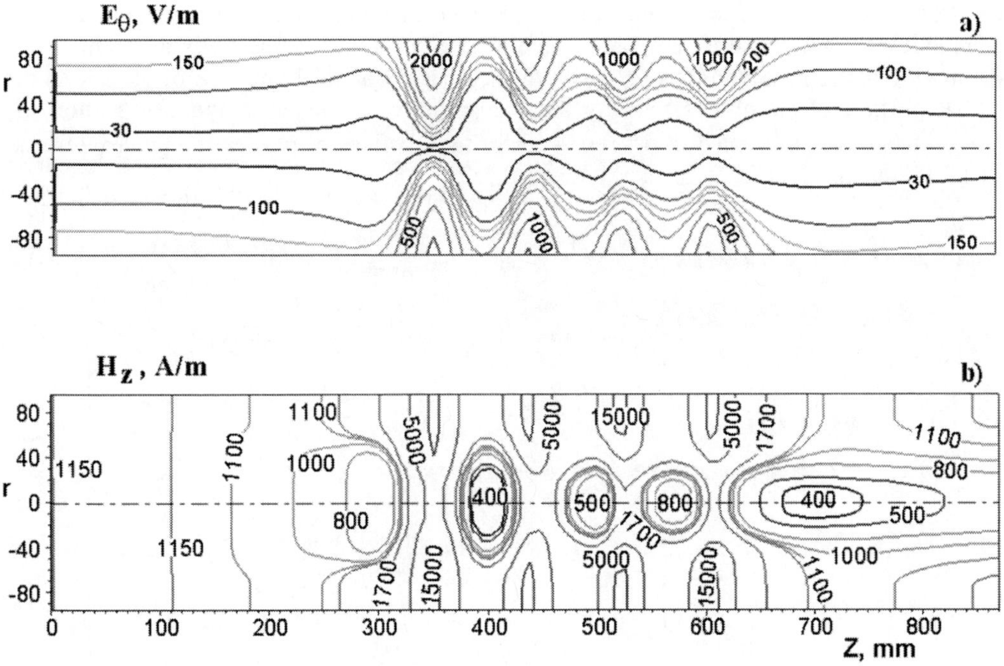

Figure 7. Contours of electric field (a) and magnetic field (b) amplitudes in IPG-3 channel, 1D model.

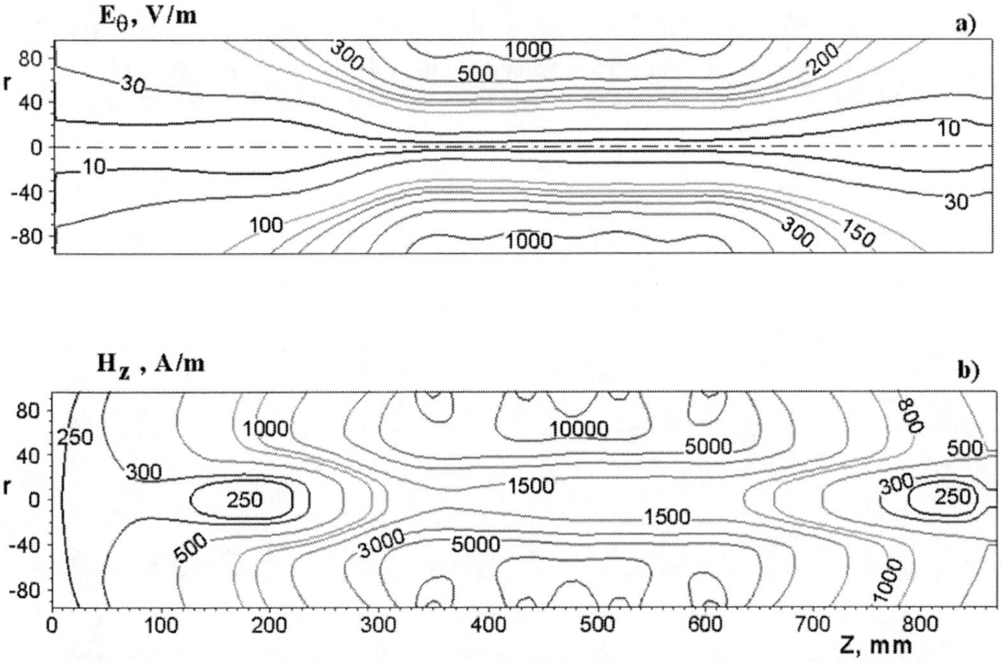

Figure 8. Contours of electric field (a) and magnetic field (b) amplitudes in IPG-3 channel, 2D model.

It is clear from the comparison of the corresponding figures 5 and 6, figures 7 and 8, that for IPG–3 plasmatron the difference between 1D and 2D results is much greater than for IPG-4 one. The

IOP Conf. Series: Journal of Physics: Conf. Series **1009** (2018) 012027 doi:10.1088/1742-6596/1009/1/012027

temperature maximums, the maximums of electric and magnetic field amplitudes obtained with 1D and 2D models are not very different, but the geometry of contours is essentially different.

To provide better quantitative comparison of the results obtained with use of 1D and 2D models, the radial profiles of air plasma velocity and temperature calculated at the exit section of IPG-4 discharge channel are presented in figures 9–10 for the 3 values of plasmatron power, obtained with use of 1D (blue circles) and 2D (black squares) models.

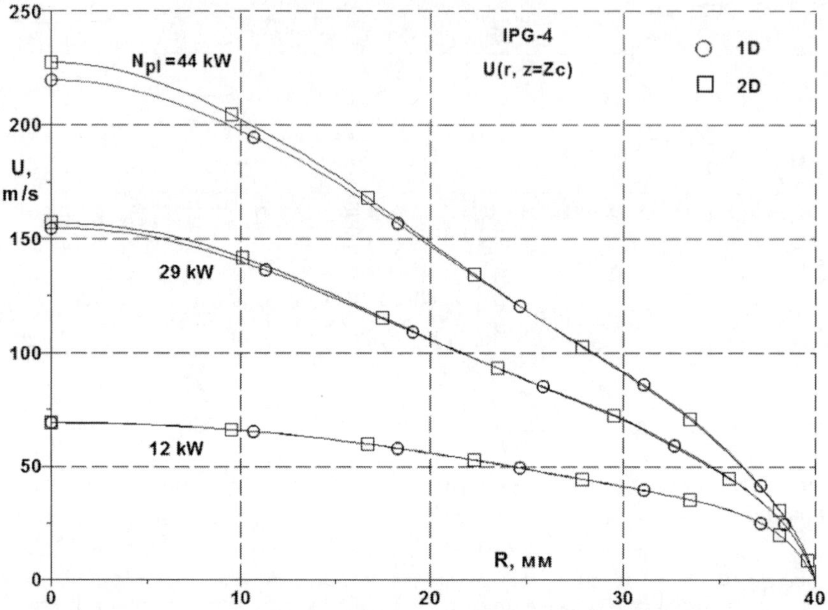

Figure 9. Radial profiles of plasma velocity at the IPG-4 channel exit section calculated with use of 1D and 2D models for 3 values of plasmatron power N_{pl}.

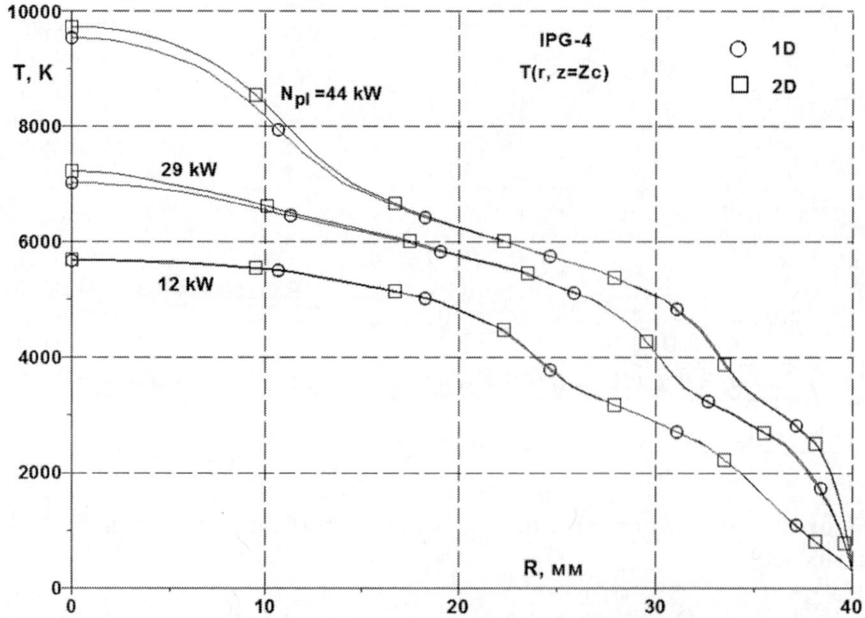

Figure 10. Radial profiles of plasma temperature at the IPG-4 channel exit section calculated with use of 1D and 2D models for 3 values of plasmatron power N_{pl}.

Similar radial profiles at the IPG-3 channel exit section are presented in figures 11-12.

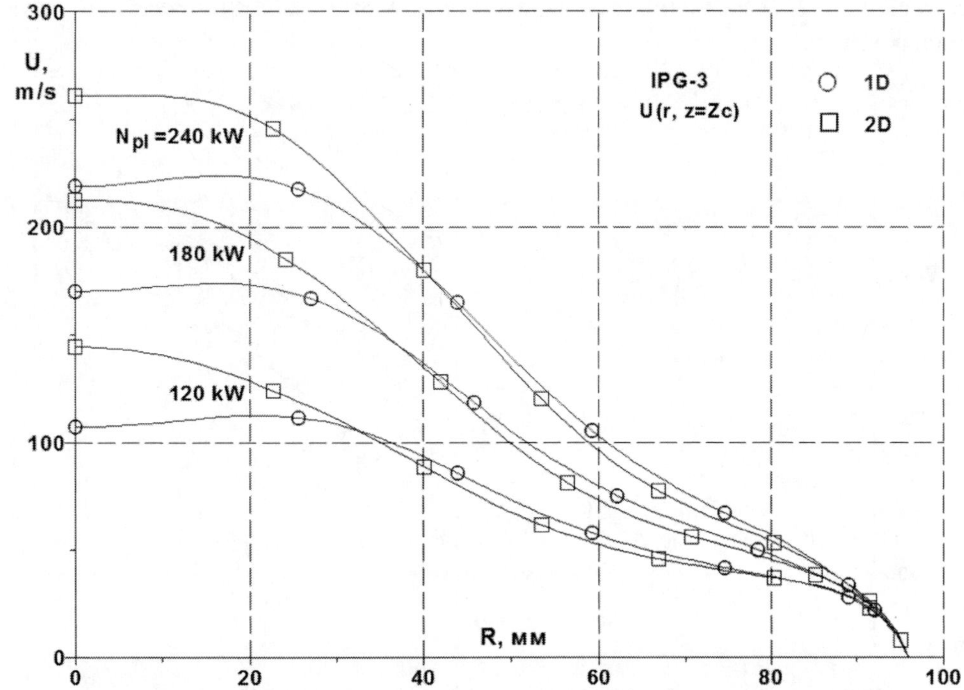

Figure 11. Radial profiles of plasma velocity at the IPG-3 channel exit section calculated with use of 1D and 2D models for 3 values of plasmatron power N_{pl}.

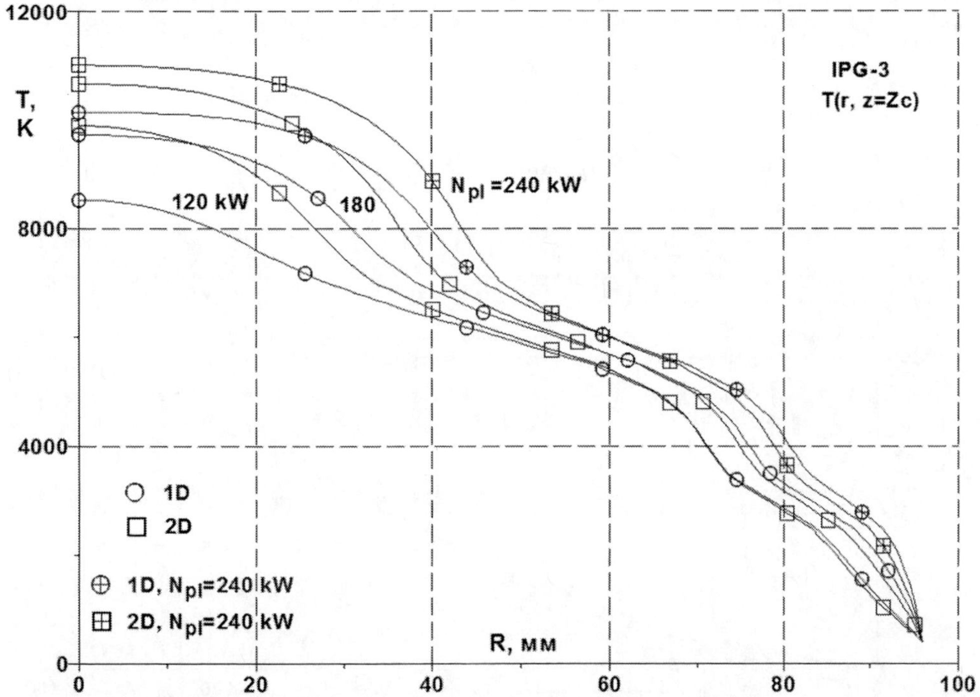

Figure 12. Radial profiles of plasma temperature at the IPG-3 channel exit section calculated with use of 1D and 2D models for 3 values of plasmatron power N_{pl}.

IOP Conf. Series: Journal of Physics: Conf. Series **1009** (2018) 012027 doi:10.1088/1742-6596/1009/1/012027

Figures 13–14 show the comparison of radial profiles of electric field amplitude E_θ at the middle of inductor region for IPG-4 and IPG-3, obtained with use of 1D and 2D models for 3 values of plasmatron power.

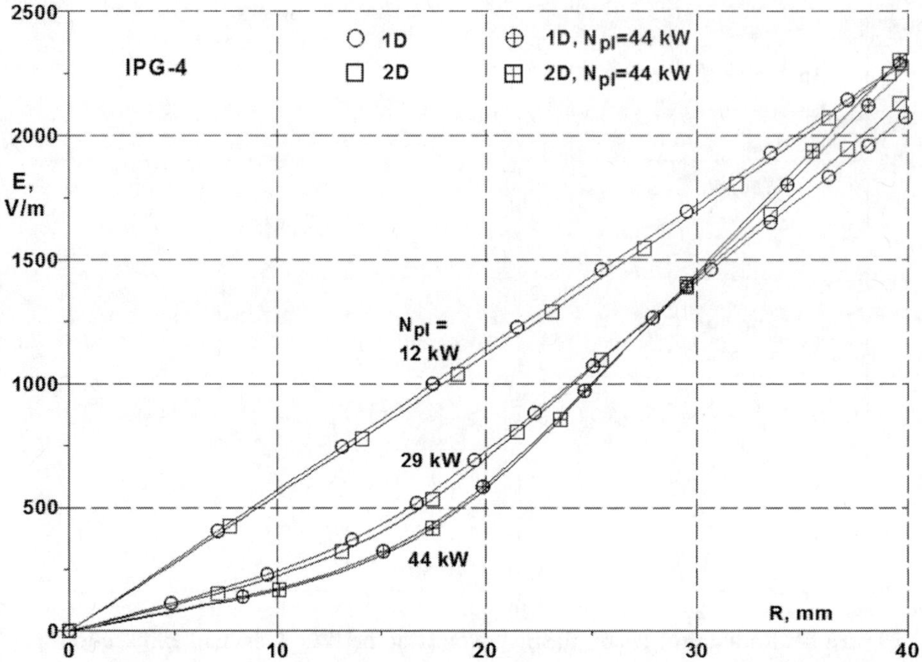

Figure 13. Radial profiles of electric field amplitude E_θ at the middle of inductor region for IPG-4, obtained with use of 1D and 2D models for 3 values of plasmatron power.

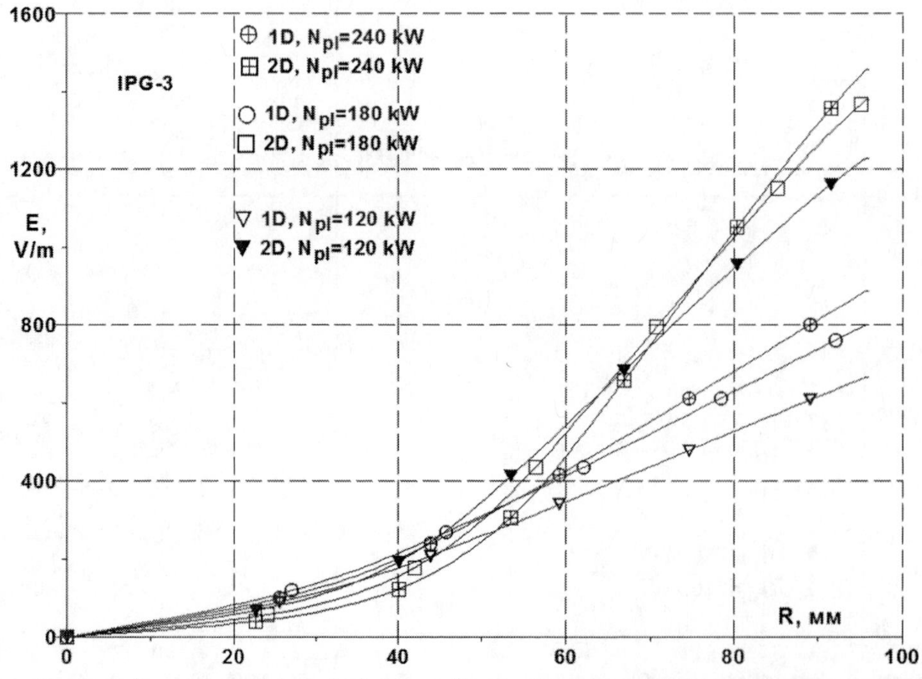

Figure 14. Radial profiles of electric field amplitude E_θ at the middle of inductor region for IPG-3, obtained with use of 1D and 2D models for 3 values of plasmatron power

Significant differences of 1D and 2D results are clear from figures 11, 12 and 14 for IPG-3 plasmatron, while for IPG-4 the corresponding differences are relatively small (figures 9, 10, 13).

Comparison of axial distributions of plasma velocity and temperature along the symmetry axis obtained by 1D and 2D models is presented in figures 15, 16 for IPG-4 plasmatron for the three values of plasmatron power. The similar results for IPG-3 plasmatron are presented in figures 17, 18.

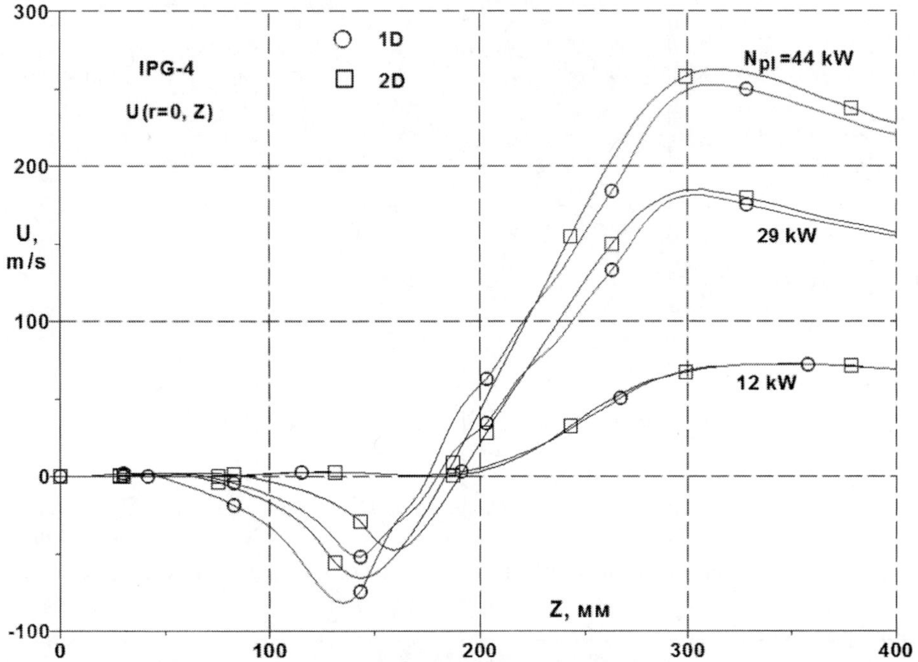

Figure 15. Axial distributions of plasma velocity along IPG-4 channel symmetry axis.

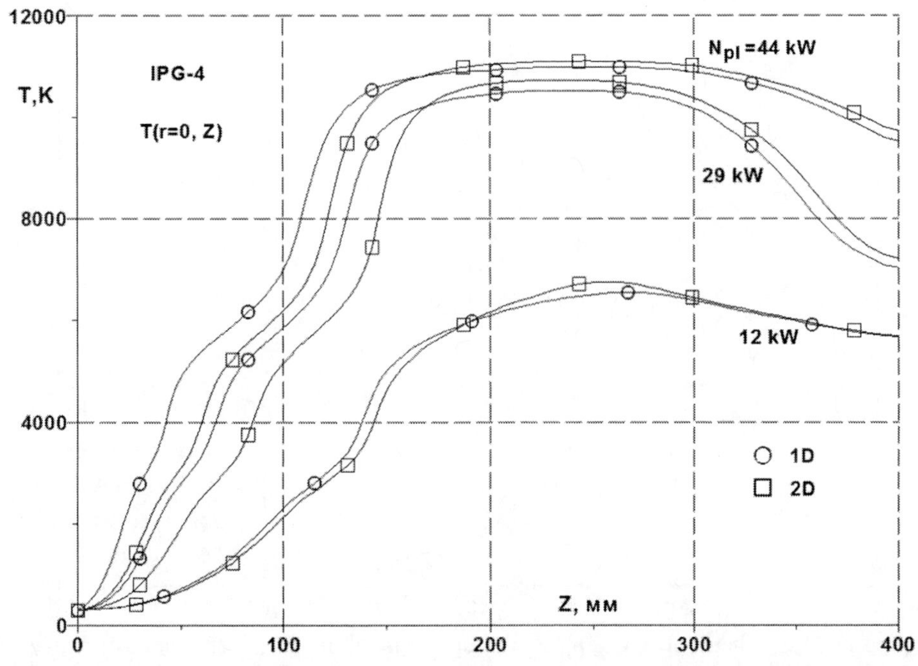

Figure 16. Axial distributions of plasma temperature along IPG-4 channel symmetry axis.

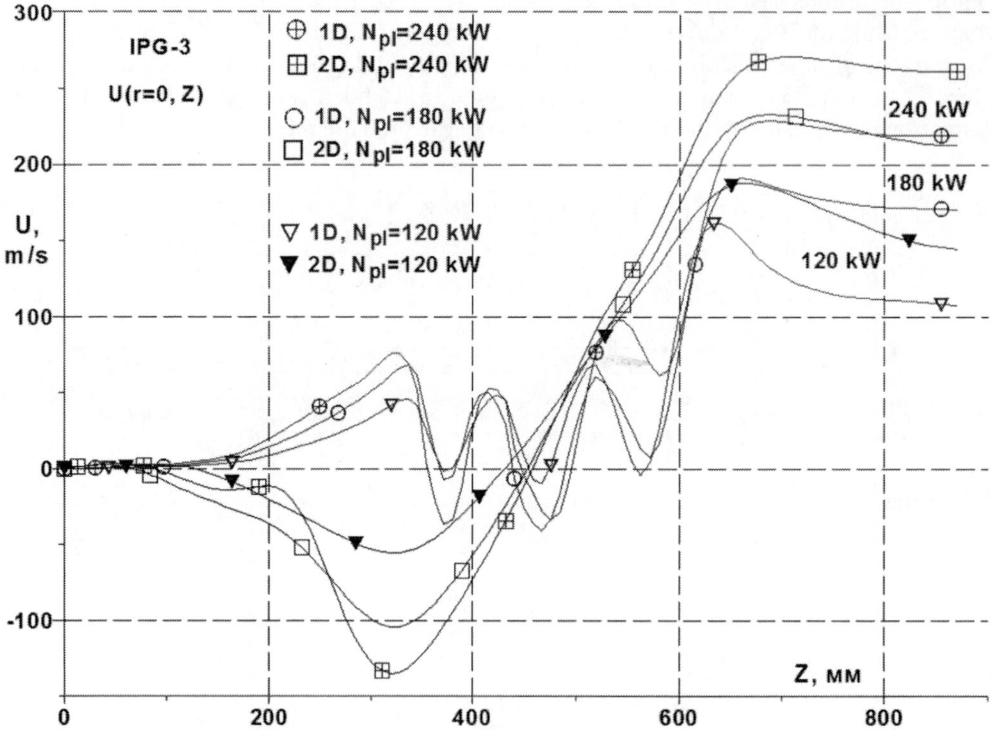

Figure 17. Axial distributions of plasma velocity along IPG-3 channel symmetry axis.

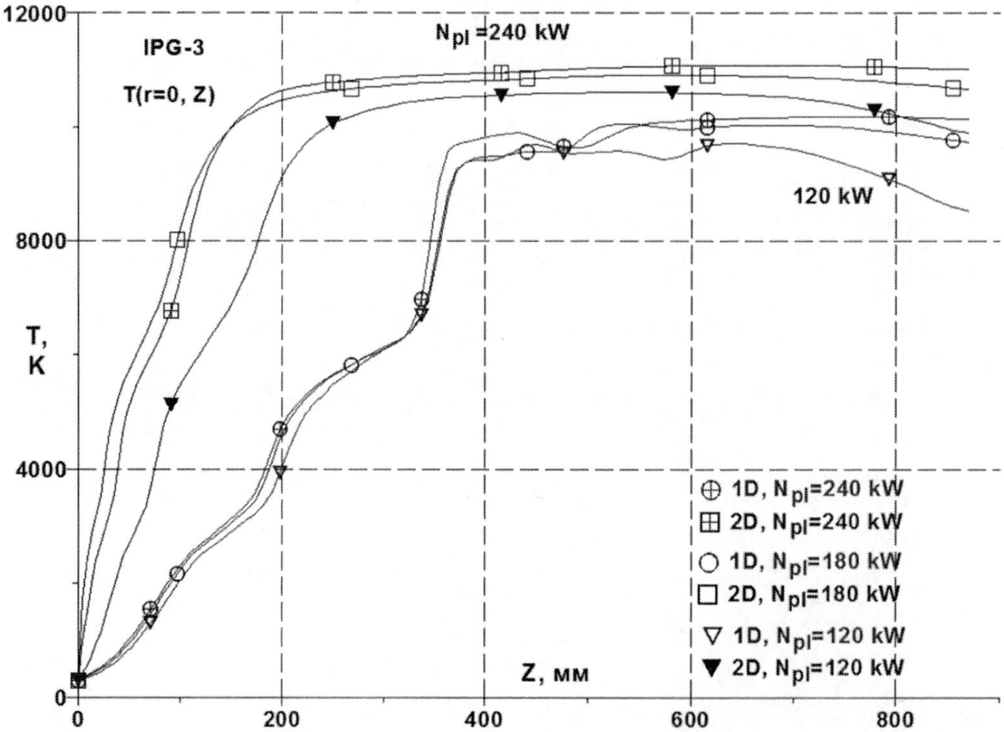

Figure 18. Axial distributions of plasma temperature along IPG-3 channel symmetry axis.

Again, essential differences of 1D and 2D results is clear from figures 17, 18 for IPG-3 plasmatron, while for IPG-4 the corresponding differences are relatively small (figures 15, 16).

Summing up the differences due to the use of 1D and 2D electromagnetic field models in flow parameters at the discharge channel exit section, important from practical viewpoint:

for IPG-4 plasmatron the differences in U_c and T_c between 1D and 2D results are very small for low power regime (0.4% and 0.3% for U_c and T_c) and are relatively small for large power regime (4% and 2% for U_c, T_c), while for h_c the difference is 2% - 3% for all regimes;

for IPG-3, 1D and 2D results are essentially different (up to 26%), and the differences in U_c and T_c are greater for low power regime (26% and 14% for U_c and T_c), while for h_c the difference is 17%÷18% for all regimes.

Acknowledgements
This work was partially supported by RFBR Grant No. 17-01-00054-a.

References

[1] Vlasov A V, Zalogin G N, Zemlyanskii B A and Knot'ko V B 2003 *Fluid Dynamics* **38** 815
[2] Gordeev A N, Kolesnikov A F and Yakushin M I 1992 *SAMPE Journal* **28** 29
[3] Vasil'evskii S A, Kolesnikov A F and Yakushin M I 1991 *High Temperature* **29** 411
[4] Vasil'evskii S A, Kolesnikov A F and Yakushin M I 1996 Mathematical models for plasma and gas flows in induction plasmatrons *Molecular Physics and Hypersonic Flows (NATO ASI Series* C vol 482) Ed M Capitely (Dordrecht: Kluwer) p 495
[5] Vasil'evskii S A and Kolesnikov A F 2000 *Fluid Dynamics* **35** 769
[6] Kolesnikov A F, Pershin I S, Vasil'evskii S A and Yakushin M I 2000 *The Journal of Spacecraft and Rockets* **37** 573
[7] Vanden Abeele D and Degrez G 2000 *AIAA Journal* **38** 234
[8] Boulos M I 1976 *IEEE Trans. on Plasma Sc.* **4** 28
[9] Vanden Abeele D, Vasil'evskii S A, Kolesnikov A F, Degrez G and Bottin B 1999 Code-to-code validation of inductive plasma computations *Progress in Plasma Processing of Materials* Ed P Fauchais and J Amouroux (New York: Begell House) p 245
[10] Bykova N G, Vasilevskii S A and Kolesnikov A F 2004 *High Temperature* **42** 12
[11] Ginzburg V L 1970 *The Propagation of Electromagnetic Waves in Plasmas* (Oxford, New York: Pergamon Press)
[12] Landau L D and Lifshitz E M 1963 *Electrodynamics of Continuous Media* (Oxford: Pergamon Press)
[13] Hirschfelder J O, Curtiss C F and Bird R B 1966 *Molecular Theory of Gases and Liquids* (New York: Wiley)
[14] Devoto R S 1966 *Physics of Fluids* **9** 1230
[15] Tirskii G A 1999 *Journal of Applied Mathematics and Mechanics* **63** 841
[16] Sokolova I A and Tirskiy G A 1997 Transport properties of gases and plasma mixtures for gas-dynamics simulation *AIAA Paper* 97-2584
[17] Patankar S 1980 *Numerical Heat Transfer and Fluid Flow* (New York: Hemisphere Publishing)
[18] Saad Y 2003 *Iterative Methods for Sparse Linear Systems (2nd edition)* (SIAM, Society for Industrial and Applied Mathematics)
[19] http://www-users.cs.umn.edu/~saad/software/SPARSKIT/

IOP Conf. Series: Journal of Physics: Conf. Series **1009** (2018) 012028 doi:10.1088/1742-6596/1009/1/012028

History of scramjet propulsion development

R K Seleznev [1,2,3]

[1] Dukhov Research Institute of Automatics (VNIIA), Suschevskaya Street 22, Moscow 127055, Russia
[2] Moscow Institute of Physics and Technology (State University), 9 Institutskiy per., Dolgoprudny, Moscow Region, 141701, Russia
[3] Ishlinsky Institute for Problems in Mechanics RAS, Vernadsky Street 101(1), Moscow, 119526, Russia

rkseleznev@gmail.com

Abstract. In the paper a historical overview of scramjet propulsion development is presented. Three main stages of development of hypersonic summer programs are distinguish. The conclusion about the importance of the development of three-dimensional computer models describing the process in the scramjet is made.

1. Introduction

This historical review will primarily be useful to engineers, who are engaged in the development and implementation of promising hypersonic technologies. This paper describes the path that the programs on the creation of scramjet for the last 60 years have passed. Hypersonic ramjet air-jet engines (scramjet) have a basic scheme of operation is much simpler than other types of aircraft engines.

However, the practical implementation of the scramjet is very complex due to a number of principal reasons. To overcome these problems will have to spend a considerable amount of time. One of the basic problems of creating a scramjet is the organization of combustion of fuel in combustion chambers, organization of effective ignition of fuel components and the instability of the processes of ignition and combustion with respect to various gasdynamic perturbations. An important problem in the operation of the ramjet is the maintenance of the thermal regime of its design throughout the time of operation.

In this paper, it is proposed to consider how the development of experimental and computer technology influenced the trends in the development of hypersonic technologies. For this purpose, a well-known book by W. Heiser and D. Pratt [1], the book of E. Kuran and S. Marti [2] and the book of S. Segal [3] were used. The article [4], devoted to the history of the development of supersonic and hypersonic topics in the laboratory of the US Navy, was also used [4]. It should also be noted the work of Sabelnikov and Penzin [5,6], which provides an overview of the research on the development of a scramjet in Russia.

2. Advantages and disadvantages of ramjet and scramjet

The efficiency of jet engines is usually estimated by the specific impulse - the ratio of thrust to the second fuel consumption. This indicator is also a measure of the efficiency of the engine. The specific impulse of various engines is shown in figure 1.

It follows from the diagram that according to the specific impulse rocket engines are much inferior to air-jet engines of all types. This is due to the fact that the oxidant is also included in the fuel consumption of the rocket engine, which the air-jet engine takes from the atmosphere. In connection with this, possible applications of ramjet and scramjet are the creation of hypersonic aircraft (cruisers), which will make it possible to make large flights quickly enough. (from Moscow to Tokyo for 2-3 hours) at a speed of 4÷8 M, as well as the creation of reusable accelerators for launching missiles into space, which will reduce the cost of bringing cargo into space.

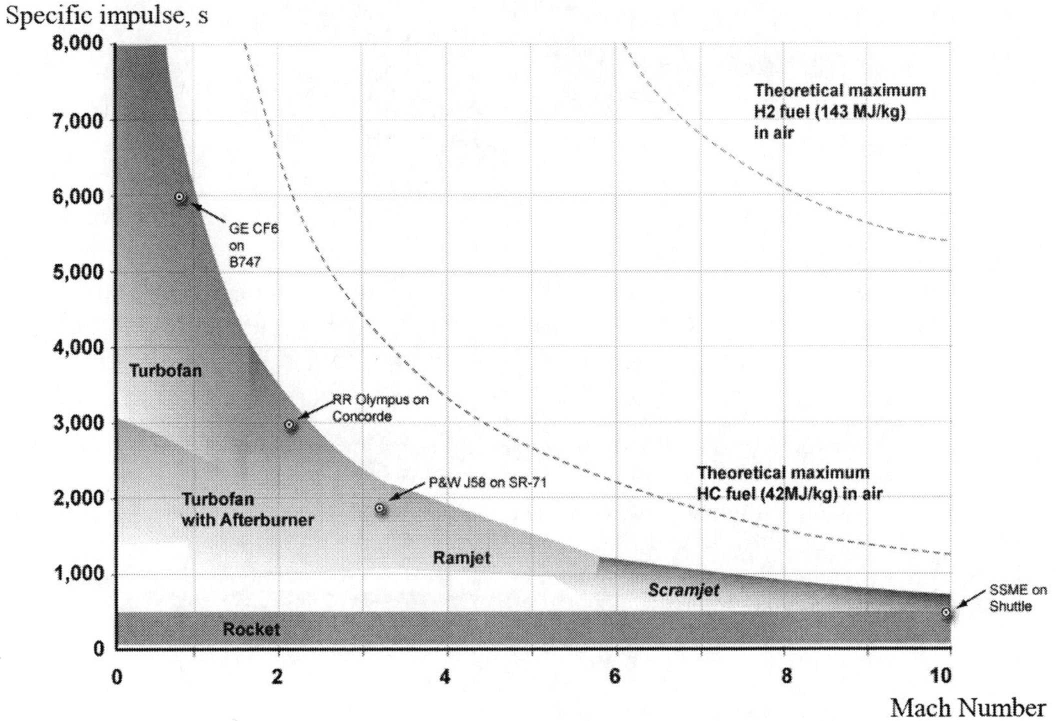

Figure 1. The specific impulse of various engines (https://en.wikipedia.org/wiki/Scramjet).

However, the ramjet and scramjet have several limitations. First, the drag of ramjet and scramjet should be more than that of a rocket due to the fact that it is necessary to take air from the atmosphere. This leads to more heating of the ramjet and scramjet than for rocket engines. In connection with this, the problem of creating heat-resistant strong materials is especially acute.

Secondly, high flight speeds lead to the fact that the characteristic time of the gas passage through the combustion chamber is very small (1 ms) and comparable to the time of self-ignition. In connection with this burning of fuel can take place in the nozzle. Therefore, it is necessary to use tricks (caverns, throttle valves, pylons, etc.) for efficient mixing of fuel and air. It should also be noted that the combustion in the duct can sometimes cause chocking. Therefore, the following restrictions exist. First, the scramjet can not operate at a speed of less than 4 M (another engine is needed to accelerate to the operating interval). Secondly, it is necessary to correctly integrate the engine and the "bird" to implement flight mode in a wide range of Mach numbers. In this paper, three stages of development of the ramjet are highlighted. Below you will find detailed information about each stage.

3. The first stage. Podded engine

At the first stage of the creation of the scramjet, an external accelerator was investigated. At this stage, the engine was created separately from the aircraft. Several experimental installations were created, operating according to the scheme of the connected air duct. The best example of such

IOP Conf. Series: Journal of Physics: Conf. Series **1009** (2018) 012028 doi:10.1088/1742-6596/1009/1/012028

engines is NASA HRE (Hypersonic Research Engine) [1] (figure 2), as well as the Russian device created by the group E.S. Shchetinkov in NII-1 [2] (figure 3). These engines use a large axisymmetric ledge for external compression. And the appearance of the engine is very similar to a gas turbine. As a fuel that cooled the walls of the combustion chamber in NASA HRE hydrogen was used. Russian engine uses kerosene.

Figure 2. Appearance of NASA HRE [7].

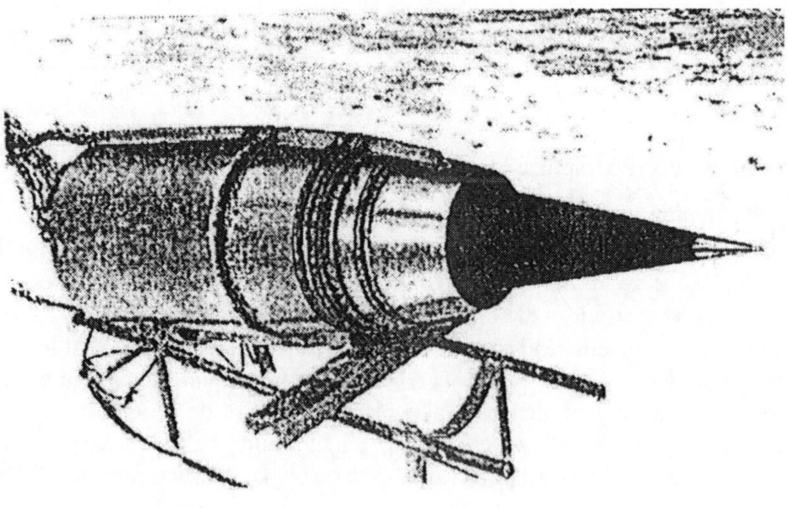

Figure 3. The appearance of the first Russian scramjet tested at NII–1 in 1964 by Kulibyakin and Penzin [6].

In 1954, it was decided to create an experimental rocket-propelled aircraft equipped with rocket engines. After the X-15 plane crash, the aircraft was damaged, but it was decided to use the restored X-15A-2, with the podded NASA HRE engine [3]. The main purpose of the research was to create a jet engine, which will achieve a speed of 3 to 8 Mach, and uses hydrogen as a fuel. In 1968, it was decided to close the NASA HRE program. By that time, two axisymmetric models of different geometry were created. To cool the walls of one model, hydrogen was used, which takes heat from the wall, heats up and is more easily ignited. To cool the walls of another model located on the experimental platform, water was used. Measurements of heat fluxes were made. Strength properties of the structure were calculated in such a way as to withstand more than 50 cycles of starting the propulsion system.

Unfortunately, in the last flight of the X-15A-2 due to the high flight speed of 6.7 M, the outer shell of the aircraft was heated to 1480 °C. That led to the burning of the airplane shell and the appearance of holes in the metal parts (figure 4). Despite such damage, the plane landed successfully, but never flew again. A huge number of scientific results were obtained as a result of ground tests of this engine. As a result, it was found that the thrust of this type of engine does not exceed the external drag of the aircraft. Therefore, it was decided that it is necessary to integrate the engine into the aircraft.

Figure 4. Burning of the airplane shell and the appearance of holes in the metal parts of the X–15A–2 aircraft, with podded NASA HRE engine. Photos after landing.

4. The second stage. 2D engine integrated in aircraft

In the second generation of scramjets, the engines of fairly simple geometries (2D) were investigated, which were integrated into the aircraft. It should be especially noted that in these types of aircraft the engine itself is located after the front of the head shock wave. This means that the necessary compression for the engine is created by the aircraft itself. Also in this configuration, the engine drag is much less than the drag of hypersonic aircraft. One example of such integration is the spacecraft NASP (National Aero-Space Plane) [4], which was created by the USA in 1985÷1994. The main idea of NASP was to create a hypersonic aircraft of a new generation with a horizontal launch and landing, developed by the United States to create a reliable and simple means of mass lifting people and cargo into space. The expected form of the device is shown in figure 5. For this purpose, it was planned to create a power plant, which would consist of a gas turbine engine, a ramjet engine, and a rocket engine. The program proved to be very expensive and technologically unfeasible. Despite the fact that because of this program was canceled, the results obtained in basic research, found application in subsequent programs (Hyper–X). For example, the Jachimowski kinetic schemes [10] for describing the combustion of hydrogen fuel in air are used today.

Figure 5. Prospective view of the apparatus a NASP (drawing (left) and photo (right) taken from the site https://ntrs.nasa.gov/).

In the new NASA program "Hyper–X" [11], which is a continuation of the NASP program, it was decided to abandon the creation of a spacecraft. Instead, it was proposed to create a small hypersonic aircraft with hydrogen-fueled scramjet. NASA's Hyper–X program was designed to conduct flight tests of the unmanned experimental hypersonic X–43A. Appearance of the device is shown in figure 6.

Figure 6. Appearance of the X-43A (drawing).

To accelerate (output to the required speed and altitude), the upper stage of the Pegasus rocket was used. The first version of the X–43A was designed to achieve a speed above the Mach number of 7 - about 8000 km/h (2.24 km/s) at an altitude of 30000 m or more. It was developed as a single-use system. Only three models have been built. The first flight - June 2001 (failure, destroyed 11 seconds after the release by command of a security officer, fell in the Pacific) because of a control system error. The other two successfully completed the program - a ram jet engine worked for 10 seconds, followed by a 10-minute glider drop. Both models were drowned in the Pacific Ocean. The second flight X–43A - March 27, 2004, was in normal mode. The third flight set a speed record at 11.850 km/h (Mach 9.6 = 3.2 km / s) on November 16, 2004.

In addition to the X–43A, the X–43B, X–43C, X–43D are developed in the Hyper-X program. The appearance of the X–43B is shown in figure 7. In this hypersonic aircraft, it is planned to use the ISTAR engine (Integrated Systems Test of an Air-Breathing Rocket) [5]. ISTAR uses a liquid-rocket hydrocarbon engine for initial acceleration, a ramjet engine at speeds higher than 2.5 M, and a scramjet at speeds higher than 5 M. The device must reach speeds above 7 M. When an X–43B exits into space, a rocket engine is again launched.

Figure 7. Appearance of the X-43B (drawing).

The X–43C is an X–43A version in which the scramjet engine on hydrocarbon fuel HyTECH [6] (figure 8). While most ramjet projects use hydrogen fuel today, HyTech operates on conventional kerosene-type hydrocarbon fuels. Currently, a full-scale engine is being built.

In 2002, HyShot hypersonic aircraft was successfully launched [14,15], and demonstrated successful combustion in the combustion chamber at a speed of 7.8 M and in the altitude range from 36 km to 25 km. At the first stage of the experiment, the apparatus was lifted by the Terrier-Orion Mk70 to a height of 300 km and dropped from this height. As soon as the device reaches the speed of 7.5 M, the process of measuring the operation of the scramjet starts. After that, the engine lasted 6 seconds [14]. The apparatus consisted of a wedge, and two rectangular scramjet chambers arranged symmetrically relative to the plane of symmetry of the apparatus. During the flight one of the engines worked, while the injection of fuel into the second was not carried out. Appearance of HyShot–2 is shown in figure 9.

From 2009 to 2012, flight tests were carried out under the HIFiRE program [16,17]. As in the case of the HyShot experiment [14,15], in the first stage of the flight the apparatus is lifted with a Terrier-Orion Mk70 missile and dropped. The appearance of the HIFiRE-2 is shown in figure 10. The apparatus consisted of an air inlet (two wedges) and scramjet chamber with two cavities arranged symmetrically relative to the symmetry plane of the apparatus. As a fuel, a surrogate mixture of ethylene, methane and heptane was used. The engine is dual-mode, therefore both the ramjet mode and the scramjet are implemented. In the experiment, the pressure distribution on the chamber wall and the

IOP Conf. Series: Journal of Physics: Conf. Series **1009** (2018) 012028 doi:10.1088/1742-6596/1009/1/012028

Mach number were measured. Before the flight, ground tests were conducted on the HIFiRE Direct-Connect Rig (HDCR) experimental setup located at NASA Langley [7].

Figure 8. Appearance of the scramjet engine on hydrocarbon fuel HyTECH (https://www.globalsecurity.org/military/systems/munitions/hytech.htm).

Figure 9. Appearance of HyShot–2 [14].

IOP Conf. Series: Journal of Physics: Conf. Series **1009** (2018) 012028 doi:10.1088/1742-6596/1009/1/012028

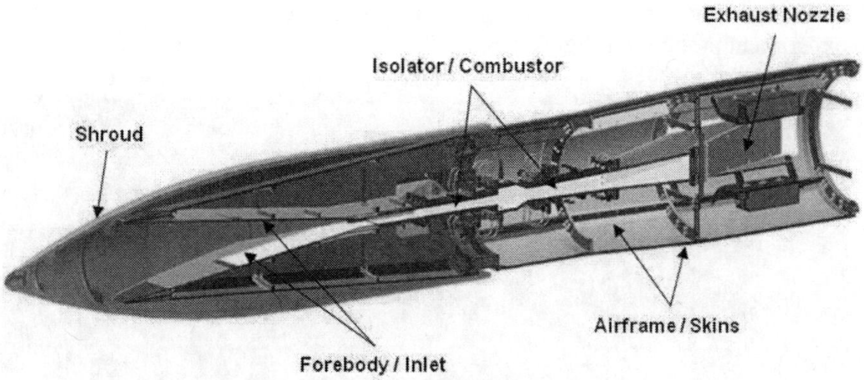

Figure 10. Appearance of HIFiRE–2 [8].

It should be noted about Russian works on this topic. In CIAM, the development of the hypersonic aircraft "Kholod" was carried out [9,10]. The hypersonic flying machine "Kholod" (figure 11) was created on the basis of the SA–5 surface-to-air missile 5B28, developed at Khimki KB "Fakel" under the direction of the general designer P.D. Grushin [2]. The choice of this missile was due to the fact that the parameters of its flight trajectory were close to those required for flight tests of the scramjet. It was also considered important that this missile was removed from armament, and its cost was low. The combat part of the missile was replaced by the head compartments of the hypersonic aircraft "Kholod", which housed the flight control system, a container for liquid hydrogen with a displacement system, a system for regulating the flow of hydrogen with measuring devices, and, finally, an experimental scramjet E–57 of an axisymmetric configuration. By 1999, a total of seven flights were conducted. The first two flights with overall-weight mock-ups of the head compartments according to the flight design test program made it possible to debug a new missile control system to provide the required trajectory. In five flights, a real ramjet was used with detailed preparation of the combustion chamber's flow path. In three flights, liquid hydrogen was fed into the combustion chamber of the scramjet. The scramjet time operation in flight increased from one test to another and in the latter was 77 seconds, corresponding to the maximum flight time of the S–200 missile. The maximum achieved speed of the flight "Kholod" was 1855 m/s, which corresponds to the Mach number M = 6.49. It was established that the combustion chamber was still operable after it was turned off.

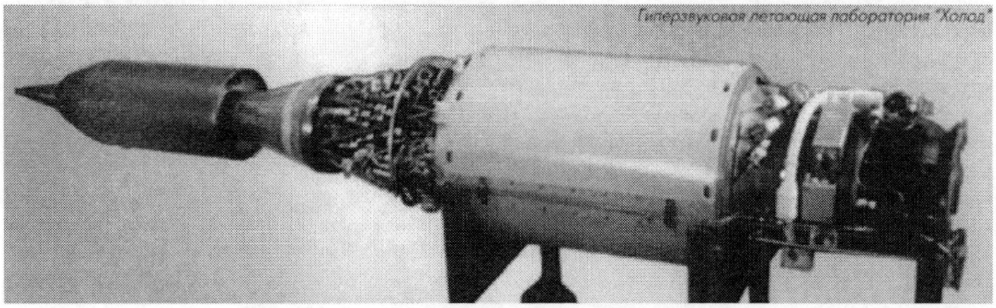

Figure 11. Appearance of Kholod [19,20].

Another of the well-known hypersonic projects is the Hypersonics Flight Demonstration program (HyFly). The hypersonic aircraft is launched from an F–15E aircraft and accelerated to operating speed using a solid-propellant rocket booster. The HyFly engine is a dual-mode Dual-Combustion Ramjet (DCR). In this engine, two different air intake systems are implemented, which allow the engine to work either as ramjet or scramjet. In addition, the engine operates on a conventional liquid hydrocarbon fuel (JP–10), which is much easier to operate than liquid hydrogen. The DCR concept was developed by the Applied Physics Laboratory (APL) of Johns Hopkins University and in May

2002. APL successfully tested the HyFly engine in a wind tunnel at a speed of 6.5 M. The photograph of the experiment is shown in figure 12.

Figure 12. Appearance of HyFly. Photo of the Johns Hopkins University.

Another of the experiments that gave hope for the soonest implementation of the scramjet was X–51 [21]. The X–51A was launched from Edwards Air Force Base in California. The device was raised to a height of 50 000 feet under the left wing of the B–52 Stratofortress airborne test center, as shown in figure 13. Then the missile accelerated the X–51A to a speed of 4.8 M.

Figure 13. Hypersonic aircraft X-51 before launch from the aircraft B-52.

Once the X–51A disconnected from the accelerating rocket engine, its own engine SJY61, which is shown in figure 14, has started. In the first seconds of operation, a mixture of ethylene and jet fuel JP–7, and then exclusively jet fuel JP–7, was used as fuel. As a result, the device reached an altitude of about 70 000 feet and a maximum speed of 5 M.

Figure 14. The appearance of the SJX61–2 engine, installed on the X–51. (Photo from the site https://www.nasa.gov).

5. The third stage. 3D engine integrated in aircraft

In connection with the development of new computer methods in the 90's, it became possible to explore different types of 3D virtual models of scramjets. Therefore, in the third generation of ramjets, engines with complex three-dimensional geometry are used. This decision was made due to the fact that the second generation of ramjets had limitations. These include the strength of the entire structure; as well as insufficient system efficiency. Three-dimensional scramjets include engines with rounded or elliptical forms. The main advantages of elliptical motors are that for the same mass flow, a smaller surface area of the chamber is required, which leads to a reduction in drag and cooling requirements. The design weight also decreases due to the inherent strength of the rounded structures.

An example of such a ramjet with complex geometry is a rectangular channel that goes into an elliptical (Rectangular-to-Elliptical Shape Transition - REST) [22–25]. This configuration was investigated by NASA Langley in 1997 [22]. This scramjet consists of a rectangular inlet, which allows it to effectively integrate it into the engine and elliptical engine. A photograph of this engine is shown in figure 15.

Figure 15. Photo of NASA Langley REST [22].

IOP Conf. Series: Journal of Physics: Conf. Series **1009** (2018) 012028 doi:10.1088/1742-6596/1009/1/012028

There are various types of integration of this type of engine in a hypersonic aircraft. Figure 16 shows the proposed integration scheme of NASA's Langley REST engine, proposed in [22]. In this scheme, it is planned to place several engines on the lower surface of the apparatus. In the HIFiRE 7 project [24, 26], the REST engines are located as shown in figure 17.

Figure 16. The proposed integration scheme of NASA Langley REST engine [22].

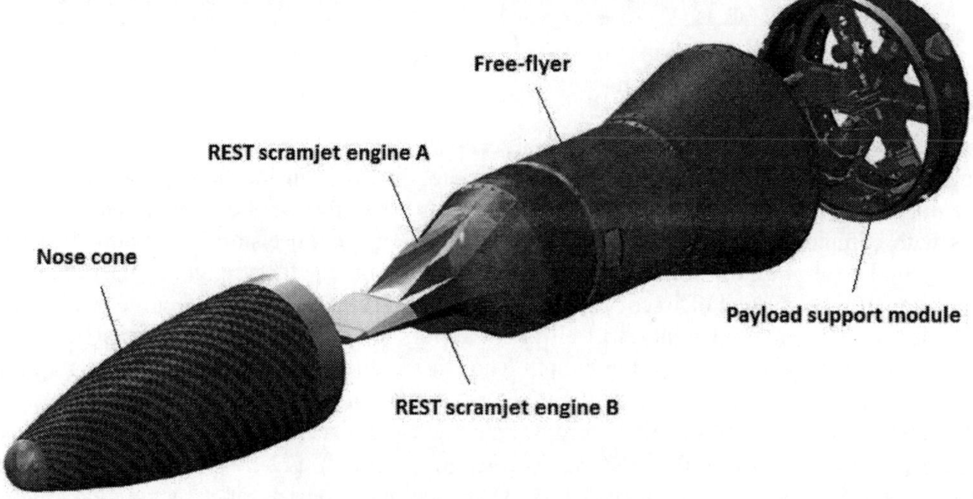

Figure 17. Scheme HIFiRE 7 REST [24,26].

The REST ideas are also implemented in the hypersonic aircraft LAPCAT–MR2 [27] (HEXAFLY–INT - High-Speed Experimental Fly Vehicles - International). In this hypersonic aircraft, the engine is integrated into the aircraft, as shown in figure 18.

Another example of using the three-dimensional geometry is scramjet Hypersonic Collaborative Australia / United States Experiment (HyCAUSE) [28]. The appearance of this hypersonic aircraft is shown in figure 19. As a result of wind tunnel experiments at the University of Queensland, this engine shape was found to be optimal and was used in further flight tests [28].

6. Conclusion

This paper presents a historical overview of the three stages of development of the ramjet. In the third generation of ramjets, engines with complex three-dimensional geometry are used. Therefore, in addition to creating quasi-one-dimensional and two-dimensional computer codes designed for modeling processes in ramjet [29 –35], it is especially important to develop three-dimensional codes.

Figure 18. Appearance of the hypersonic aircraft LAPCAT-MR2 [27].

Figure 19. Appearance of the hypersonic aircraft HyCAUSE [28].

Acknowledgments
This work was preformed within the framework of the Government program of basic research of the Russian academy of sciences (contract #AAAA-A17-117021310372-6) and partially under support of RFBR grant #016-01-00379.

References

[1] Heiser W, Pratt D, Daley D and Mehta U 1994 *Hypersonic Airbreathing Propulsion* vol 26(Washington, DC: American Institute of Aeronautics and Astronautics, Inc.)

[2] Murthy S N B and Curran E T 2001 *Scramjet Propulsion* (Reston ,VA: American Institute of Aeronautics and Astronautics)

[3] Segal C 2009 *The scramjet engine: Processes and characteristics* vol 9780521838(Cambridge: Cambridge University Press)

[4] Waltrup P J, White M E, Zarlingo F and Gravlin E S 2002 History of U.S. Navy Ramjet, Scramjet, and Mixed-Cycle Propulsion Development *J. Propuls. Power* **18** 14–27

[5] Walther R 1993 New partnerships to meet hypersonic propulsion challenges - Joint MTU-TSAGI cooperation in scramjet technology development *5th International Aerospace Planes*

and Hypersonics Technologies Conference (Reston, Virigina: American Institute of Aeronautics and Astronautics)

[6] Sabel'nikov V A and Penzin V I 2001 Scramjet Research and Development in Russia *Scramjet Propulsion* vol 189(Reston ,VA: American Institute of Aeronautics and Astronautics)pp 223–367

[7] Andrews E H and Mackley E A 1994 NASA's Hypersonic Research Engine Project: A review *Nasa Tech. Memo.*

[8] Andrews E H and Mackley E A 1993 Review of NASA's Hypersonic Research Engine Project *29th Joint Propulsion Conference and Exhibit* (Reston, Virigina: American Institute of Aeronautics and Astronautics)

[9] Augenstein B, Harris E, Aroesty J, Blumenthal I, Brooks N, Frelinger D R, Garber T B, Gottemoeller R E, Hiland J R, Liu S, Pace S, Rosen J H, Rowell L N and Stucker J P 1993 The National Aerospace Plane (NASP): Development Issues for the Follow-on Vehicle 1–56

[10] Jachimowski C J 1988 An analytical study of the hydrogen-air reaction mechanism with application to scramjet combustion *NASA Tech. Pap.* 16

[11] McClinton C, Rausch V, Sitz J and Reukauf P 2001 Hyper-X program status *39th Aerospace Sciences Meeting and Exhibit* (Reston, Virigina: American Institute of Aeronautics and Astronautics)

[12] Faulkner R 2001 Integrated system test of an airbreathing rocket (ISTAR) *10th AIAA/NAL-NASDA-ISAS International Space Planes and Hypersonic Systems and Technologies Conference* (Reston, Virigina: American Institute of Aeronautics and Astronautics) pp 1–12

[13] Powell O A, Edwards J T, Norris R B, Numbers K E and Pearce J A 2001 Development of Hydrocarbon-Fueled Scramjet Engines: The Hypersonic Technology (HyTech) Program *J. Propuls. Power* **17** 1170–6

[14] Hass N, Smart M and Paull A 2005 Flight Data Analysis of the HYSHOT 2 *AIAA/CIRA 13th International Space Planes and Hypersonics Systems and Technologies Conference* (Reston, Virigina: American Institute of Aeronautics and Astronautics) p 17

[15] Boyce R, Gerard S and Paull A 2003 The HyShot Scramjet Flight Experiment - Flight Data and CFD Calculations Compared *12th AIAA International Space Planes and Hypersonic Systems and Technologies* (Reston, Virigina: American Institute of Aeronautics and Astronautics) p 8

[16] Storch A, Bynum M, Liu J and Gruber M 2011 Combustor Operability and Performance Verification for HIFiRE Flight 2 *17th AIAA International Space Planes and Hypersonic Systems and Technologies Conference* (Reston, Virigina: American Institute of Aeronautics and Astronautics) p 13

[17] Jackson K, Gruber M and Barhorst T 2009 The HIFiRE Flight 2 Experiment: An Overview and Status Update *45th AIAA/ASME/SAE/ASEE Joint Propulsion Conference & Exhibit* (Reston, Virigina: American Institute of Aeronautics and Astronautics) p 19

[18] Hass N, Cabell K, Storch A and Gruber M 2011 HIFiRE Direct-Connect Rig (HDCR) Phase I Scramjet Test Results from the NASA Langley Arc-Heated Scramjet Test Facility *17th AIAA International Space Planes and Hypersonic Systems and Technologies Conference* (Reston, Virigina: American Institute of Aeronautics and Astronautics) p 18

[19] Roudakov A, Semenov V and Hicks J 1998 Recent flight test results of the joint CIAM-NASA Mach 6.5 scramjet flight program *8th AIAA International Space Planes and Hypersonic Systems and Technologies Conference* (Reston, Virigina: American Institute of Aeronautics and Astronautics) p 15

[20] Voland R, Auslender A, Smart M, Roudakov A, Semenov V and Kopchenov V 1999 CIAM/NASA Mach 6.5 scramjet flight and ground test *9th International Space Planes and Hypersonic Systems and Technologies Conference* (Reston, Virigina: American Institute of Aeronautics and Astronautics) p 9

[21] Hank J, Murphy J and Mutzman R 2008 The X-51A Scramjet Engine Flight Demonstration Program *15th AIAA International Space Planes and Hypersonic Systems and Technologies*

Conference (Reston, Virigina: American Institute of Aeronautics and Astronautics) pp 1–13

[22] Smart M and Ruf E 2006 Free-Jet Testing of a REST Scramjet at Off-Design Conditions *25th AIAA Aerodynamic Measurement Technology and Ground Testing Conference* (Reston, Virigina: American Institute of Aeronautics and Astronautics) pp 190–201

[23] Barth J 2014 *Mixing and combustion enhancement in a Mach 12 shape-transitioning scramjet engine* (The University of Queensland)

[24] Wilson Chan Y K, Razzaqi S A, Wise D J and Smart M 2014 Freejet Testing of the 75%-scale HIFiRE 7 REST Scramjet Engine *19th AIAA International Space Planes and Hypersonic Systems and Technologies Conference* (Reston, Virginia: American Institute of Aeronautics and Astronautics) pp 1–20

[25] Suraweera M V. and Smart M K 2009 Shock-Tunnel Experiments with a Mach 12 Rectangular-to-Elliptical Shape-Transition Scramjet at Offdesign Conditions *J. Propuls. Power* **25** 555–64

[26] Roberts M, Smart M and Frost M 2012 HIFiRE 7: Design to Achieve Scientific Goals *18th AIAA/3AF International Space Planes and Hypersonic Systems and Technologies Conference* (Reston, Virigina: American Institute of Aeronautics and Astronautics) pp 1–8

[27] Walker S H, Rodgers F, Paull A, Van Wie D M, S Walker, F Rodgers, A Paull and Wie D Van 2008 HyCAUSE Flight Test Program *15th AIAA International Space Planes and Hypersonic Systems and Technologies Conference* vol 32(Reston, Virigina: American Institute of Aeronautics and Astronautics)pp 1–14

[28] Seleznev R and Surzhikov S 2013 Generalized Newton Method For Solving Differential Equations of Chemical Kinetics *44th AIAA Thermophysics Conference* (Reston, Virginia: American Institute of Aeronautics and Astronautics) pp 1–17

[29] Seleznev R and Surzhikov S 2015 Quasi-One-Dimensional and Two-Dimensional Numerical Simulation of Scramjet Combustors *51st AIAA/SAE/ASEE Joint Propulsion Conference* (Reston, Virginia: American Institute of Aeronautics and Astronautics) p 28

[30] Polezhaev Y V. and Seleznev R K 2014 Numerical study of the processes of resonance emergence in the experimental set-up of a pulse detonation engine *High Temp.* **52** 225–9

[31] Seleznev R and Surzhikov S 2016 A Quasi-One-Dimensional Analysis of Hydrogen-Fueled Scramjet Combustors *52nd AIAA/SAE/ASEE Joint Propulsion Conference* (Reston, Virginia: American Institute of Aeronautics and Astronautics) pp 1–27

[32] Seleznev R K 2017 Comparison of two-dimensional and quasi-one-dimensional scramjet models by the example of VAG experiment *J. Phys. Conf. Ser.* **815** 12007

[33] Shang J and Surzhikov S 2013 Numerical Prediction of Convective and Radiative Heating of Scramjet Combustion Chamber with Hydrocarbon Fuels *51st AIAA Aerospace Sciences Meeting including the New Horizons Forum and Aerospace Exposition* (Reston, Virigina: American Institute of Aeronautics and Astronautics) pp 1–16

[34] Shang J and Surzhikov S 2013 Radiative Heat Exchange in a Hydrogen-Fueled Scramjet Combustion Chambers *51st AIAA Aerospace Sciences Meeting including the New Horizons Forum and Aerospace Exposition* (Reston, Virigina: American Institute of Aeronautics and Astronautics)

[35] Kotov M A, Ruleva L B, Solodovnikov S I and Surzhikov S T 2017 Experimental and numerical study of supersonic flow over two blunted wedges *J. Phys. Conf. Ser.* **815** 12025

Boron particle agglomeration and formation of solid residues in combustion of boron-containing solid propellants

S A Rashkovskiy

Ishlinsky Institute for Problems in Mechanics RAS. Vernadskogo Ave., 101/1, Moscow, 119526, Russia

rash@ipmnet.ru

Abstract. The model of the boron particles agglomeration in combustion of boron-containing solid propellants is suggested. The process of the boron particles agglomeration is considered as a result of competition between two main processes: the formation of adhesive bonds between contacting boron particles and rupture of these bonds under the action of gaseous combustion products. The criterion of solid residues formation and dependence of their relative mass on pressure are obtained. It is shown that depending on the propellant burning rate law, two different types of propellants with different behavior of the agglomeration process at pressure change can exist.

1. Introduction

Boron–containing solid propellants which have an oxidizer deficit and a high content (up to 30% by weight and more) of boron are considered as the promising propellants for the ducted rockets. Such solid propellants are capable to self-sustained combustion and their combustion occurs in a special gas generator. Due to low content of oxidizer, such propellants are characterized by incomplete conversion of fuel components (primarily, binder and boron) which leads to residues (slag) formation. Moreover, the temperature of the combustion products in the chamber of the gas generator is typically less than 2000K. At this temperature, the boron particles are in solid state and weakly react with the oxidizing components of the solid propellant. Combustion products of the solid propellant flow out of the gas generator through the nozzle into the secondary combustor in which air is fed from the environment. In the secondary combustor, combustion of the boron particles and other condensed products occurs in oxygen of air. Most part of the energy which is used to creation of the thrust is released during the combustion of condensed products (primarily, boron particles) in air flow in the secondary combustor. For this reason, the efficiency of the ducted rocket is largely determined by the completeness of the boron particles combustion in the secondary combustor. Taking this in mind, the design of the ducted rocket is directed to obtain a maximum efficiency of boron combustion [1-3].

For theoretical estimation of the completeness of boron combustion in secondary combustor, the data obtained in the experiments on ignition and combustion of the single boron particles are usually used. However, there are direct and indirect data indicating that combustion of the boron-containing propellants is accompanied by an intense agglomeration of the boron particles near the burning surface. As a result, the coarse conglomerates (aggregates) containing a lot of primary boron particles leave the propellant burning surface [4]. We use the term "conglomerate" with respect to boron particles instead of the term "agglomerate" which is adopted in the literature when describing the process of agglomeration of aluminum particles on the burning surface of solid propellant. This highlights a fundamental difference in the structure of the boron particles conglomerates from aluminum particles agglomerates [5].

The interferometric measurements [6] showed that the particles with sizes of 20÷200 μm leave the burning surface of the boron-containing propellant, although the initial boron particles in propellant have usually the sizes of 1÷2 μm.

In [7], it was observed an irregular combustion of propellant with high boron content (40–50%) with a periodic variation in burning rate level with a detachment of unburned layers. An increase in the size of the conglomerates, detached from the burning surface, was observed at pressures above 50 atm.

Experiments [1,2] showed that boron combustion efficiency in the same secondary combustor can vary from 0.6 up to 0.95, depending on the organization of the air flow and flow of solid propellant combustion products in the secondary combustor. It is significantly less than the estimations obtained from the models of combustion of single boron particles which are in propellant. The experimental data [1, 2] indirectly indicate the presence of the boron particles conglomerates in combustion products of the boron-containing solid propellant. The combustion time of the boron particles conglomerate can significantly exceed the combustion time of single primary boron particle due to the difference in their sizes and to the essentially difference in the ignition and combustion behaviour of the conglomerates compared to single boron particle. The experiments [1,2] demonstrate an ability to control the combustion time of the boron particles conglomerates due to optimization of a design of the secondary combustor.

An ignition of the boron particles conglomerates in air for different sizes of the primary boron particles and different diameters and densities of the conglomerate was investigated experimentally in [8]. It was found out that the ignition temperature of a conglomerate substantially less than the ignition temperature of single boron particles (~1000 K vs ~1900 K). Experimental data [8] showed that the conglomerate ignition temperature varies with their density monotonically and at a certain conglomerate density, the ignition temperature reaches a minimum. Decrease in the ignition temperature of the conglomerate compared to single boron particle is a positive factor contributing to the completeness of the boron combustion in the secondary combustor of the limited length. At the same time, the combustion time of the conglomerate is essentially long compared to that of single boron particles, thus the total time required for ignition and complete combustion of the conglomerate is significantly longer (at the same conditions) than the time required for complete combustion of the single particles entering the conglomerate.

Note that the agglomeration of the boron particles in ducted rockets plays more important role than the agglomeration of aluminum particles in conventional solid rocket motors, because, in ducted rockets, the incompleteness of boron combustion, connected with a formation of conglomerates, leads to significant additional (along with the two-phase losses) decrease in the specific impulse (down to 20%).

The main way of reducing the size of the conglomerates entering the secondary combustor from the gas generator is a reducing the intensity of the boron particles agglomeration on the burning surface of solid propellant. This can be achieved by optimizing the composition and structure of solid propellant.

For effective reducing the agglomeration of the boron particles on the burning surface, we need a clear understanding of the mechanisms of this complicated process and identification of the controlling factors, by changing of which one can control the intensity of the agglomeration.

Moreover, for the correct calculation of the ignition and combustion behaviour of the conglomerate, one needs to know not only its size but also its structure and shape that can significantly affect these processes.

In contrast to the detailed studies of the agglomeration of aluminium particles, in the literature, there is very little of the experimental data on the boron particles agglomeration in combustion of the boron-containing solid propellants.

Some experimental study of the agglomeration of refractory powders was done in the work [9].

In the works [10,11], the condensed combustion products of boron-containing propellant was investigated experimentally. The propellant formulation was 19.0% hydroxyl-terminated polybutadiene (HTPB), 4.0% dioctyl sebacate (DOS), 4.0% catocene, 34.0% ammonium perchlorate (AP), 26.0% boron powder, 7.5% magnesium powder, 4.0% aluminium powder and 4.5% toluene-2.4–diisocyanate (TDI). Initial boron particles in propellant have the sizes about 1μm. It was shown that the condensed-phase products mainly consist of B, C, B_4C (or $B_{12}C_2$), BN, Mg, MgO, $MgAl_2O_4$, Al, Al_2O_3, $AlCl_3$, $NH_4[Mg(H_2O)_6]Cl_3$, NH_4Cl and Fe_3O_4; there were large amounts of boron oxide and boron carbide in the condensed-phase products, which indicates that elemental boron is highly active during the combustion process in gas generator. The content of elemental boron decreases significantly with an increase in chamber pressure and this is mainly due to the production of boron carbide, but the content of boron oxide does not increase simultaneously.

Considerable quantities of CO are produced at low chamber pressures and the content of CO decreases significantly with an increase in chamber pressure. The composition of the combustion products shows that carbon exists mainly in the form of boron carbide, carbon and carbon monoxide at low chamber pressures, while it exists mainly as boron carbide and carbon at high chamber pressures. Boron carbide is produced by the reactions of elemental boron with binder's hydrocarbons (CH_4, C_2H_4, and C_4H_6), so a decrease in CO may be the result of the higher reactivity of elemental boron and binder's hydrocarbons.

The content of unreacted boron in the condensed-phase products decreased exponentially while the content of boron carbide increased with the increase in chamber pressure [11].

The burning rate of propellant plays an important role in the consumption of elemental boron and production of boron carbide. The reaction of elemental boron cannot reach the chemical equilibrium because of a short reaction time; therefore, the composition of boride was strongly affected by the burning rate of the propellants: the higher the burning rate, the lower the reaction percentage of boron when the residence time was the same.

Experimental data [10,11] showed that a decrease in the unreacted boron correlated with an increase in boron carbide.

Illustrative result is the dependence of the mean size of conglomerates leaving the burning surface on pressure: for some compositions, the mean size of conglomerates increases with increasing pressure, while for others it decreases [11]. This indicates the existence of several competing mechanisms that control the process of formation of condensed-phase products in combustion of boron-containing solid propellants. For the last case it was found that most of the boride in the condensed-phase combustion products is either elemental boron or boron carbide particles produced with a longer time; therefore, the large particle size may result from the agglomerated boron or boron carbide at a higher temperature.

The goal of this work is to develop a model of the residues formation (primarily, the boron particles agglomeration) in combustion of the boron-containing propellants, and to identify the factors that govern the process.

2. Mechanism of solid residues formation

The mechanism of boron particles agglomeration is largely similar to the mechanism of aluminum particles agglomeration in combustion of composite solid propellants [12], but it has its own features [5]. As in the case of aluminium, the boron particles agglomeration is connected with the presence of contacting particles in the condensed phase of the propellant. At passing of a heat wave in the propellant, a conversion of its components (binder, oxidizer particles etc.) and simultaneously a heating of the boron particles occur. In contrast to aluminium particles which melting point is less than the temperature of the propellant burning surface, boron has a melting point above the temperature of the propellant burning surface, and even higher than the temperature of the combustion products in the gas generator and therefore boron is always present at solid state on the propellant burning surface. The boron particles, located in propellant are covered with a thin film of boron oxide B_2O_3. The boron oxide melting point ($T_{m B_2O_3} = 720K$) is commensurable with a temperature of the propellant burning surface ($T_s = 700 - 1000\ K$) and considerably less than a temperature of combustion products of the propellant in a gas generator ($T_b = 1500 - 2000\ K$). This means that the boron particles on the propellant burning surface have the liquid oxide film and contacting particles are bound by the liquid bridge of B_2O_3. If the contacting boron particles did not have the oxide film or it was insufficiently thick to create the liquid bridge between the particles after its melting, the sintering of the particles can occur due to their rapid heating in the propellant burning wave. Furthermore in the vicinity of the contact of the particles, the various chemical compounds can form due to the chemical reactions of the species of the products of decomposition of solid propellant both between themselves and with boron. In particular, the sufficiently strong boron carbide B_4C and boron nitride BN can be formed. In [10,11,13] have shown that for HTPB-based solid propellants, a significant increase in B_4C content with increasing pressure is observed, at that at low pressures, there is a marked deviation from thermodynamic equilibrium, while at high pressures, the B_4C concentration approaching to equilibrium. The GAP-based propellants demonstrate the opposite behaviour [13]: for these propellants, the content of boron carbide in the combustion products decreases with increasing pressure. Results of [10,11,13] showed that the boron carbide can be predominant boron compound on the burning surface of the propellant and directly above the surface. The strong chemical compounds which are formed in the vicinity of the contact points of the boron particles can create the mechanical (adhesive) bonds between the particles.

Due to oxidizer deficit in the boron-containing solid propellants, the incomplete conversion of the binder occurs on the burning surface which leads to formation of condensed-phase products consisting, primarily, of carbon. This slag forms the skeleton layer on the propellant burning surface which creates the additional bonds holding the boron particles on the burning surface and promoting their agglomeration.

As a result of all these processes, the condensed-phase products form on the propellant burning surface the coral-like structures, which have the different shapes, sizes and porosity [7]. Gaseous combustion products of decomposition of easily degradable components of propellant, affect these coral-like structures, creating an aerodynamic detached force. This force depends on the shape, size and porosity of the coral-like structure on the propellant burning surface and increases in average with increasing the size of these structures. When the detached aerodynamic force becomes greater than the adhesive force which binds the skeleton layer with the burning surface, the breaking of the bonds occurs and conglomerate is detached from the burning surface.

As a result, the conglomerates consisting of a plurality of interconnected primary boron particles being in solid state come into the propellant combustion products. Note that unlike the aluminium particle agglomerates which are almost spherical drops [9] consisting of molten aluminium and solid aluminium oxide, the boron particles conglomerates are the porous structures consisting, essentially of the solid boron, boron carbide and carbon [11]. The structure of the propellant is a determining factor in formation of the boron particles conglomerates [5].

The mechanism of agglomeration, described above, imposes certain limitations on the methods of simulation of the boron particles agglomeration. For example, the pockets model [14-18] can be used in some cases for calculation of the aluminium particles agglomeration, but it quite inapplicable in order to calculate the boron particles agglomeration because the conglomerate can unite the boron particles belonging to different, sometimes separated by a considerable distance, “pockets”.

With some modifications, the statistical model [19-21] developed for the aluminium particles agglomeration and considering the process in dynamics can be used for calculation of the boron particles agglomeration. But the method of the direct numerical simulation of the agglomeration [12] is more appropriate for this purpose. This method consists in modelling the solid propellant structure [22, 23] and the subsequent calculations of the behaviour of each boron particle near the propellant burning surface, taking into account changes in its thermal and chemical states and accounting its mechanical interaction with the gaseous combustion products and with other boron particles. The method [12] was applied to simulation of boron particle agglomeration in combustion of boron-containing solid propellants [5].

IOP Conf. Series: Journal of Physics: Conf. Series **1009** (2018) 012029 doi:10.1088/1742-6596/1009/1/012029

3. Model of the boron particles agglomeration

In general the propellant burning rate u as well as the boron particles conglomerates size depend on the propellant structure. This means that changes in the composition and structure of the propellant lead to a simultaneous change in the burning rate and in the intensity of the agglomeration process. Obviously, a complete model of combustion of boron-containing solid propellant should take into account both the effect of burning rate on agglomeration process and the effect of agglomeration process on the burning rate. In [5], a simplified approach was used in which the propellant burning rate is considered to be given and only the calculations of the boron particles agglomeration are performed. This is due to the fact that the propellant burning rate can be easily determined experimentally, while the measurement of sizes of the boron particles conglomerates which are formed during the propellant combustion is much more complicated problem. For this reason, it is appropriate to estimate the sizes, shape and structure of the conglomerates by calculation using a model of the agglomeration process and the experimentally determined propellant burning rate.

Boron particle agglomeration is considered as a result of competition between two main processes: the formation of adhesive bonds between contacting boron particles and rupture of these bonds under the action of gaseous combustion products.

At low Reynolds numbers, which corresponds to the flow above the burning surface, the force $\mathbf{f}_i$ acting on i–th particle from the side of gaseous combustion products flowing from the burning surface can be written in the form

$$\mathbf{f}_i = \mu(\mathbf{U}_i - \mathbf{V}_i) \quad (1)$$

where $\mathbf{V}_i$ is the velocity of i-th boron particle; $\mathbf{U}_i = (U_i, 0, 0)$; U_i is the velocity of the gaseous combustion products, flowing from the burning surface under i-th boron particle; μ is some coefficient ("aerodynamic drag" coefficient of the particle) which is considered as a matching coefficient of the model.

Due to heterogeneity of the propellant, the velocity U_i will be different in different points of burning surface. As an approximation, one can consider the mean velocity of the gaseous combustion products: $U_i = U$,

$$U = (1 - m_r)\frac{\gamma}{\rho}u \quad (2)$$

where m_r is the mass fraction of condensed-phase products of decomposition of propellant on the burning surface; γ is the propellant density, $\rho = \frac{p}{RT_{as}}$ is the density of gaseous combustion products, p is pressure; T_{as} is a characteristic temperature of gaseous combustion products above the burning surface, e.g. $T_{as} = 0.5(T_s + T_b)$.

The adhesive force between contacting particles i and k prevents their detachment can be written in the form [5]

$$\mathbf{F}_{ik} = K_{ik}\frac{\mathbf{r}_k - \mathbf{r}_i}{|\mathbf{r}_k - \mathbf{r}_i|}\left(|\mathbf{r}_k - \mathbf{r}_i| - \frac{1}{2}(D_i + D_k)\right) \quad (3)$$

where K_{ik} is the factor characterizing the rigidity of the bonds between contacting particles i and k; D_i is the diameter of i-th particle. Further, the factor K is considered to be the same for all pairs of interacting particles, i.e. $K_{ik} = K$ and is considered to be a matching coefficient of the model.

In this case, the equations that describe the behavior of the boron particles above the burning surface have the form [5]

$$\frac{d\mathbf{r}_i}{dt}=\frac{K}{\mu}\sum_k\frac{\mathbf{r}_k-\mathbf{r}_i}{|\mathbf{r}_k-\mathbf{r}_i|}\left(|\mathbf{r}_k-\mathbf{r}_i|-\frac{1}{2}(D_i+D_k)\right)+\mathbf{U}_i \quad (4)$$

Each bond has a tensile strength F_{ik}^0: if the tearing force (3) exceeds the tensile strength, the breaking of the bond between the particles i and k occurs, and this particles will cease to interact. The tensile strength of the particles bonds is a random variable that depends on many factors: the nature of the particles connection, the temperatures of the particles and ambient gas, the ambient gas composition, the duration of the particle stay in the flow of gaseous combustion products (time elapsed since the release of these particles on the burning surface), etc. All these factors are not currently studied, so for all pairs of particles, we take the same value of F_{ik}^0 equal to F^0 which will be considered to be a mean strength of the bonds between the contacting particles.

Condition of the breaking of the bonds is as follows [5]

$$|\mathbf{r}_k-\mathbf{r}_i|>(D_i+D_k)/2+F^0/K \quad (5)$$

For analysis of the model, one can introduce the nondimensional variables, keeping for $\mathbf{U}$, $\mathbf{r}_k$ and t the previous notations. We choose the characteristic size of the boron particles D_0 as a characteristic spatial scale, the solid propellant burning rate u as a characteristic velocity scale and the ratio D_0/u, as the characteristic time scale. Then nondimensional diameter of the boron particle is $d=D/D_0$. As follows from (2), the nondimensional velocity of the gaseous combustion products near the burning surface

$$\mathbf{U}=\left((1-m_r)\gamma/\rho,0,0\right) \quad (6)$$

The system of equations which describes in nondimensional variables the behavior of the boron particles above the propellant burning surface, takes the form

$$\frac{d\mathbf{r}_i}{dt}=\mathbf{U}_i+\varepsilon\sum_k\xi_{ik}\frac{\mathbf{r}_k-\mathbf{r}_i}{|\mathbf{r}_k-\mathbf{r}_i|}\left(|\mathbf{r}_k-\mathbf{r}_i|-\frac{1}{2}(d_i+d_k)\right) \quad (7)$$

where $\xi_{ik}=0$ if the condition

$$|\mathbf{r}_k-\mathbf{r}_i|>\frac{1}{2}(d_i+d_k)+\delta \quad (8)$$

was satisfied at any preceding time for particles i and k, i.e. they never contacted or the breaking of the bond already occurred; otherwise ξ_{ik}= 1; if at some point in time, there was a collision of the particles i and k which did not contact previously, i.e. the condition $|\mathbf{r}_k-\mathbf{r}_i|\le(d_i+d_k)/2$ was satisfied, we assume ξ_{ik}=1 until the condition (8) satisfies; this allows taking into account the emergence of new adhesive bonds between boron particles in the course of their movement;

$$\varepsilon=\frac{KD_0}{\mu u} \quad (9)$$

$$\delta=\frac{F^0}{KD_0} \quad (10)$$

Besides the parameters characterizing the propellant structure (content and dispersity of the components), the model (7)-(10) contains three nondimensional parameters: δ,ε and U which can vary independently of each other when changing the propellant composition and combustion conditions. Nondimensional velocity of the gaseous combustion products near the burning surface U

IOP Conf. Series: Journal of Physics: Conf. Series **1009** (2018) 012029 doi:10.1088/1742-6596/1009/1/012029

(2) is considered to be known, because the density of the gaseous combustion products near the burning surface can be calculated with sufficient accuracy, either thermodynamically or using the temperature near the burning surface measured in experiments. However, the parameters K and μ, and therefore, the parameters ε and δ cannot be calculated at present from the first principles and should be considered as the matching coefficients. However, some information about the nature of changes of the parameters ε and δ can be obtained from their definitions.

Let the propellant burning rate is described by the power law

$$u = u_0 \left(p / p_0 \right)^{\nu} \tag{11}$$

where u_0 and ν are the constant parameters of the propellant which are considered to be known.

Then one can write

$$\varepsilon = \varepsilon_0 \left(p / p_0 \right)^{-\nu} \tag{12}$$

$$U = U_0 \left(p / p_0 \right)^{-1} \tag{13}$$

Where

$$\varepsilon = \frac{KD_0}{\mu u_0} \tag{14}$$

$$U_0 = \left(1 - m_r\right) \frac{\gamma}{\rho_0} u \tag{15}$$

$$\rho_0 = \frac{p_0}{RT_{as}} \tag{16}$$

It can be expected that the strength of the interparticle bonds δ will depend on the residence time of particles in the heated layer of the condense phase of propellant $\tau = \kappa / u^2$ (κ is the thermal diffusivity of the condensed phase of propellant): the longer the boron particles are in heated layer of the propellant condensed phase, the stronger the bonds are established between them due to forming the liquid boron oxide bridges and strong chemical compounds formed at chemical reactions between the propellant species. We assume that the dependence of the strength of adhesive bonds between the boron particles on the residence time of the particles in the heated layer of the propellant condense phase can be described by the power function: $\delta \sim \tau^{\alpha}$. In this case one obtains

$$\delta = Au^{-2\alpha} \tag{17}$$

where A and $\alpha > 0$ are the constants.

Then one can write

$$\delta = \delta_0 \left(p / p_0 \right)^{-2\alpha\nu} \tag{18}$$

Where

$$\delta_0 = Au_0^{-2\alpha} \tag{19}$$

is the constant.

If the dependencies of the parameters K, μ, F^0 and T_{as} on pressure are weak, then the expressions (12), (13) and (18) completely determine the dependence of the boron particles agglomeration process on pressure.

Although, the parameters ε_0, U_0 and δ_0 do not depend on the pressure, they can be varied due to changing the composition of the propellant (the content and dispersion of the components, the types of the binder and oxidizer, etc.). Furthermore, it is obvious that any surface coating of the boron particles, which leads to a decrease in their mutual adhesion (for example, easily removed coating) will contribute to a reduction of the coefficient δ_0 and hence to a reduction of the intensity of the agglomeration process. For example, in [24], combustion of the micron-size boron particles (average particles diameter ~51 μm) covered with glycidyl azide polymer (GAP) was investigated in a hot gas flow of a flat-flame burner and in composite solid propellant. The results [24] show that propellants containing the amorphous boron particles coated with GAP, demonstrate a higher burning rate and a lower intensity of agglomeration compared to the base composition containing the uncoated boron particles.

Obviously, an intense agglomeration is possible only if the coefficient ε satisfies the condition

$$\varepsilon\delta > U \tag{20}$$

Otherwise the single boron particles will leave the burning surface and agglomeration will not occur.

Using (12), (13) and (18), one can rewrite the condition (20) in the form

$$\varepsilon_0\delta_0\left(p/p_0\right)^{-(1+2\alpha)\nu} > U_0\left(p/p_0\right)^{-1} \tag{21}$$

From this inequality, it follows that two different kinds of propellants are possible:

A. The propellants which have

$$(1+2\alpha)\nu < 1 \tag{22}$$

For example, for $\alpha = 1$ this corresponds to $\nu < 0.33$.

In this case the condition (21) takes the form

$$\left(p/p_0\right)^{1-(1+2\alpha)\nu} > \frac{U_0}{\varepsilon_0\delta_0} \tag{23}$$

It follows that there is a threshold pressure p_a:

$$p_a/p_0 = \left(\frac{U_0}{\varepsilon_0\delta_0}\right)^{\frac{1}{1-(1+2\alpha)\nu}} \tag{24}$$

such that there is no agglomeration at $p < p_a$ while the agglomeration occurs when $p > p_a$ and it is amplified with increasing pressure.

B. The propellants which have

$$(1+2\alpha)\nu > 1 \tag{25}$$

For example, for $\alpha = 1$ this corresponds to $\nu > 0.33$.

In this case the condition (21) takes the form

$$\left(p/p_0\right)^{(1+2\alpha)\nu-1} < \varepsilon_0\delta_0/U_0 \tag{26}$$

It follows that there is a threshold pressure p_b:

$$p_b/p_0 = \left(\varepsilon_0\delta_0/U_0\right)^{\frac{1}{(1+2\alpha)\nu-1}} \tag{27}$$

such that there is no agglomeration at $p > p_b$ while the agglomeration occurs when $p < p_b$ and it is amplified with decreasing pressure.

Thus, theoretically, there may exist two different types of boron-containing propellants, depending on their composition which are different by a behavior of the agglomeration at pressure change: for some propellants (with low ν) agglomeration increases with increasing pressure, while for the others (with high ν) it increases with decreasing pressure. This conclusion is supported by the available experimental data [10,11].

Results of numerical simulation of boron particle agglomeration are given in [5].

4. Critical conditions for the slag formation

It has been experimentally established that a layer of condensed-phase residues (slag) forms on the burning surface of boron-containing propellants, under certain conditions. In some cases, these residues represent a porous skeleton layer on the burning surface, but in some cases, the residues can completely fill the combustion chamber of the gas generator.

Experimental data show that for some propellants, the residues are formed at high pressures (above some critical pressure p_{cr}, which is a characteristic of propellant), but no residues are formed at low pressures ($p < p_{cr}$). For other propellants, on the contrary, the residues are formed at low pressures ($p < p_{cr}$), while at high pressures ($p > p_{cr}$), the combustion products of the propellant are completely take away from the combustion chamber of the gas generator.

This result completely corresponds to the analysis carried out above, in particular, to the relation (21).

To determine the critical pressure p_{cr}, we can use equation (21), writing it in dimensional form.

The same result can be obtained by considering the detachment of the skeleton layer (slag) formed on the burning surface.

Assuming that this layer is porous and the gaseous products of decomposition of the propellant components are filtered through it, for detachment force we obtain the expression

$$F_r = 3\pi\beta N\mu_g VdhS$$

where $N = \dfrac{6}{\pi d^3}(1-\varepsilon)$ is the number of particles of condensed phase per unit volume of slag; ε is the porosity of the slag; h is the thickness of the slag layer; d is the characteristic diameter of the particles of the particles in the slag (it is of order of the pore sizes); S is the area of the slag layer; μ_g is gas viscosity, depending on temperature; β is a coefficient of order unity.

The mass of gaseous products filtered through the slag per unit time $\varepsilon\rho VS$ is equal to the mass input of the gaseous combustion products of propellant $(1-m_r)\gamma uS$; u is the burning rate of propellant; γ is the density of propellant; ρ is the density of the gas.

Thus

$$V = \frac{(1-m_r)}{\varepsilon}\frac{\gamma}{\rho}u$$

Taking into account that $p = \rho RT$, one obtains

$$F_r = \frac{18\beta(1-m_r)}{d^2}\frac{(1-\varepsilon)}{\varepsilon}\frac{\gamma\mu_g RT}{p}uhS$$

The strength of adhesion of the slag layer to the burning surface (and slag layers among themselves)

$$F_c = \sigma_s \frac{\pi d^2}{4} NSd$$

where σ_s is the strength of the slag (the stress at which the bond between the particles in the slag breaks down).

We take

$$\sigma_s = B\tau^{\alpha} T^{-\beta}$$

where α, β, B are the factors depending on the propellant properties.

The dependence of σ_s on the residence time of particles in the heated layer of the condense phase of propellant $\tau = \kappa/u^2$ has been discussed above (see (17)). The dependence σ_s on temperature reflects the fact that the strength of the slag layer decreases with temperature increase.

Then

$$F_c = B\tau^{\alpha} T^{-\beta} \frac{3(1-\varepsilon)}{2} S$$

The condition for detachment of slag from the burning surface has the form $F_r > F_c$ or

$$12\beta(1-m_r)\frac{\gamma\mu_g RT}{pd^2} uh > B\varepsilon\tau^{\alpha} T^{-\beta}$$

This allows finding the maximum thickness of the slag layer on the burning surface

$$\frac{h_{\max}}{d} = \frac{\varepsilon B\kappa^{\alpha} d}{12\beta\gamma R} \frac{p}{u^{1+2\alpha}} \frac{T^{-(1+\beta)}}{(1-m_r)\mu_g}$$

If $h/d \sim 1$, then we can believe that the slag is not formed. Then the condition for the absence of slag will have the form

$$(1-m_r)\frac{u^{1+2\alpha}}{p}\mu_g T^{(1+\beta)} > A_0 \qquad (28)$$

where $A_0 = \dfrac{\varepsilon B\kappa^{\alpha} d}{12\beta\gamma R}$ is the parameter that can be considered a constant characteristic of propellant.

If we assume that $\mu_g T^{(1+\beta)} \approx const$ (i.e., we do not take into account the temperature changes at the burning surface with the pressure), and take the burning rate law in the form (11), then the condition (A1) can be written in the form

$$(p/p_0)^{(1+2\alpha)\nu-1} > \frac{A_0 p_0}{(1-m_r)\mu_g T^{1+\beta} u_0^{1+2\alpha}}$$

which completely corresponds to the relation (21), which determines the conditions of intensive agglomeration.

We again come to the conclusion that for propellants A that satisfy the condition (22), the residues are formed at pressures greater than p_{cr} and grow with increasing pressure, while for $p < p_{cr}$ the residues are not formed for these propellants, where

IOP Conf. Series: Journal of Physics: Conf. Series **1009** (2018) 012029 doi:10.1088/1742-6596/1009/1/012029

$$p_{cr}/p_0 = \left[\frac{(1-m_r)\mu_g T^{1+\beta} u_0^{1+2\alpha}}{A_0 p_0}\right]^{\frac{1}{1-(1+2\alpha)\nu}} \tag{29}$$

In particular, for these propellants, p_{cr} increases with increasing u_0, which can be regulated by additives with practically no changes in the composition and structure of propellant.

On the other hand, for propellants B that satisfy the condition (25), the residues are formed at pressures less than p_{cr} and grow with decreasing pressure, while for $p > p_{cr}$ there is no residues for these propellants, where

$$p_{cr}/p_0 = \left[\frac{A_0 p_0}{(1-m_r)\mu_g T^{1+\beta} u_0^{1+2\alpha}}\right]^{\frac{1}{(1+2\alpha)\nu-1}} \tag{30}$$

For these propellants, p_{cr} decreases with increasing u_0.

Comparison with experimental data shows that $\alpha = 1$.

5. Dependence of the mass of residues on mean pressure in gas generator

Experiments show that the relative mass of slag remaining in the combustion chamber after the end of the gas generator operation depends on the mean pressure during the operation of the gas generator.

We assume that the pressure in the combustion chamber increases with time (progressive diagram) according to a linear law: $p = p_1 + at$, where p_1 is the initial pressure; $a > 0$. Some experimental pressure-time diagram is shown in figure 1.

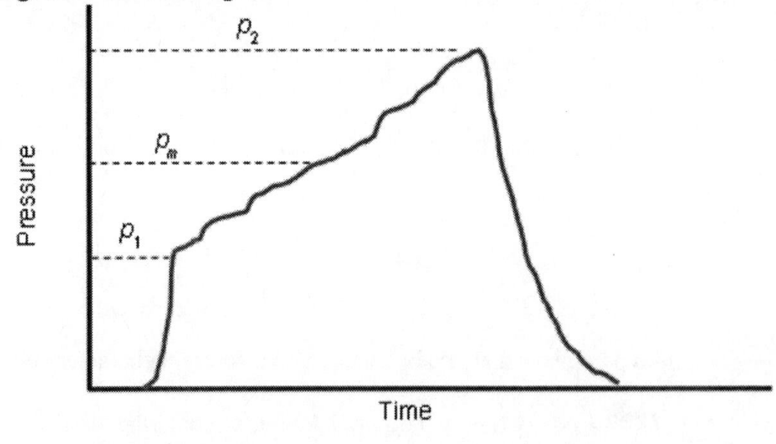

Figure 1. Experimental progressive pressure-time diagram.

Let us consider the case of propellants A that satisfy the condition (22), when the slag is formed if the pressure in the combustion chamber exceeds the critical pressure (29).

Then the slag is not formed at $p \le p_{cr}$. The instant of operation of the gas generator, up to which the slag is not formed, is determined by the condition: $p_{cr} = p_1 + at_{cr}$. Hence $t_{cr} = (p_{cr} - p_1)/a$. Then the mass of propellant burned at the range $[0, t_{cr}]$ is

$$M_1 = A\sigma \int_0^{t_{cr}} p\,dt = A\sigma(p_1 + 0.5at_{cr})t_{cr}$$

The mass of whole propellant (which was burned during the total time of operation)

IOP Conf. Series: Journal of Physics: Conf. Series **1009** (2018) 012029 doi:10.1088/1742-6596/1009/1/012029

$$M_p = A\sigma\int_0^{t_m} pdt = A\sigma(p_1 + 0.5at_m)t_m$$

Then the relative mass of propellant which was burned at the range when the slag was formed (i.e., at $t > t_{cr}$), is

$$m = 1 - \int_0^{t_{cr}} pdt \bigg/ \int_0^{t_m} pdt$$

or

$$m = 1 - \frac{(p_1 + 0.5at_{cr})t_{cr}}{(p_1 + 0.5at_m)t_m}$$

where $t_m = (p_2 - p_1)/a$ is the total operating time of the gas generator; p_2 is the pressure at the end of the operation of the gas generator.

Substituting t_{cr} and t_m, one obtains

$$m = 1 - \frac{p_{cr}^2 - p_1^2}{p_2^2 - p_1^2}$$

This expression can be rewritten in the form

$$m = 1 - \frac{(p_{cr}/p_1)^2 - 1}{z^2 - 1} \quad (31)$$

where $z = p_2/p_1$ is the characteristic of the gas generator.

The mean pressure in the gas generator $p_m = (p_2 + p_1)/2$. Thus $p_m/p_1 = (z+1)/2$.

Substituting these expressions in (31), one obtains

$$m = 1 - \frac{(p_{cr}/p_m)^2((z+1)/2)^2 - 1}{z^2 - 1} \quad (32)$$

The calculation according to equation (32) corresponds only to those parameters for which $0 \le m \le 1$. In the case where the calculation by formula (32) gives negative values, it is necessary to put $m = 0$.

The relative mass of the slag

$$m_{slag} = m_r m \quad (33)$$

Figure 2 shows the dependence of m_{slag}/m_r on p_m/p_{cr} at different $z = p_2/p_1$ for propellants A.

Let us consider the case of propellants B that satisfy the condition (25), when the slag is formed if the pressure in the combustion chamber less than the critical pressure (30).

In this case we similarly obtain

$$m = \frac{(p_{cr}/p_m)^2((z+1)/2)^2 - 1}{z^2 - 1} \quad (34)$$

Figure 3 shows the dependence of m_{slag}/m_r on p_m/p_{cr} at different $z = p_2/p_1$ for propellants B.

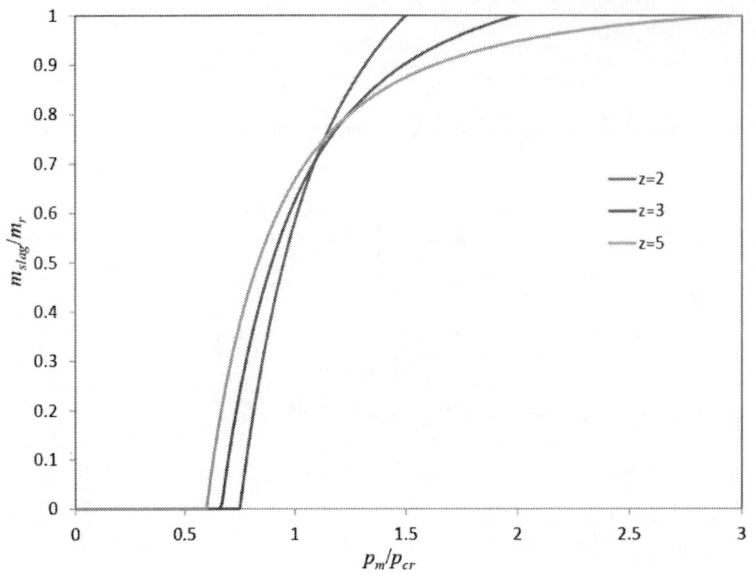

Figure 2. Dependence of m_{slag}/m_r on p_m/p_{cr} at different $z = p_2/p_1$ for propellants A.

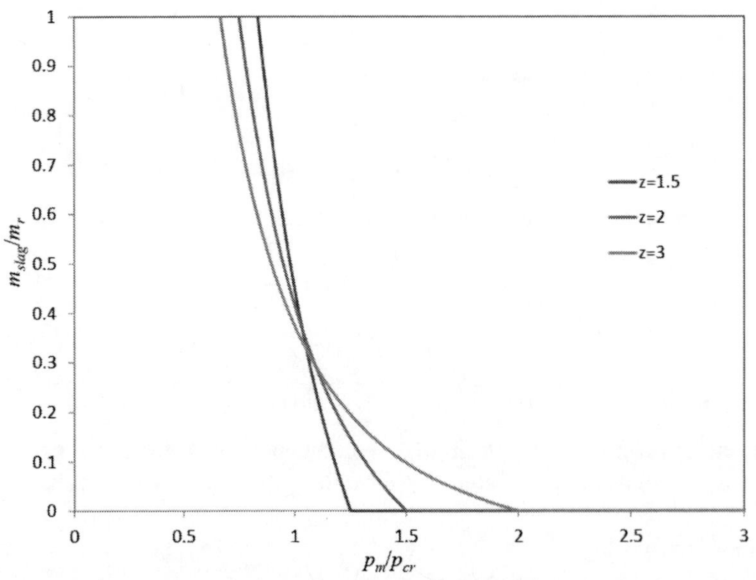

Figure 3. Dependence of m_{slag}/m_r on p_m/p_{cr} at different $z = p_1/p_2$ for propellants B.

It is necessary to note that the critical pressure, at which the formation of slag on the burning surface begins, is a property of only the propellant itself. However, the mean pressure in the gas generator, at which the formation of propellant residues in the gas generator begins, depends also on the ratio z. For this reason, in experiments, the mean pressure at which the accumulation of residues in the gas generator chamber begins, differs from p_{cr} and has a scatter related to the scatter of the diagram $p(t)$.

Note that when processing the real experimental data, we should use not a linear but a real experimental pressure diagram.

6. Conclusion

The model of the formation of residues in combustion of boron-containing propellants, developed in this work, allows explaining some regularities observed in experiments.
In this work, we do not take into account the inverse effect of the conglomerates and slag, which form on the burning surface, on the burning rate of solid propellant, which can lead to a change in the nature of the agglomeration and slag formation.

The presented study was supported by Russian Science Foundation through project No 18-19-00081.

References

[1] Vigot C, Bardelle L and Nadaud L 1986 Proc. *AIAA/ASME/SAE/ASEE 22nd Joint Propulsion Conference* (June 16-18, 1986/Huntsville Alabama) pp 1–9
[2] Vigot C, Cochet A and Guin C 1991 *Proc. Second International Symposium on Special Topics in Chemical Propulsion "Combustion of Boron-Based Solid Propellants and Fuels"* (Lampoldshausen (FRG) March 4-6) pp 1–18
[3] Rashkovskii S A, Milekhin Y M and Fedorychev A V 2016 *Doklady Physical Chemistry* **471** 203
[4] Gany A 1993 Combustion of Boron-Containing Fuels in Solid Fuel Ramjets. *Combustion of Boron-Based Solid Propellants and Solid Fuels* eds K K Kuo and R Pein (CRC Press Boca Raton FL)
[5] Rashkovskiy S A 2017 *Combust. Sci. Technol.* **189** 1277
[6] King M K 1984 *Proc. of 21st JANNAF Combustion Meeting* (The Johns Hopkins University Applied Physics Laboratory Laurel MD October 1–4)
[7] Laredo D and Gany A 1983 *Acta Astronautica* **10** 437
[8] Shevchuk V G, Zolotko A N and Polishchuk D I 1975 *Combust. Expl. Shock Waves* **11** 189
[9] DeLuca L T, Marchesi E, Spreafico M et al 2010 *Int. J. Energ. Mater. Chem. Propuls.* **9** 91
[10] Liu L L, He G Q, Wang Y H and Hu S Q 2015 *RSC Advances* **5** 101416
[11] Liu L L, He G Q, Wang Y H, Hu S Q and Liu Y M 2017 *J. Propulsion and Power* **33** 333
[12] Rashkovskiy S A 2009 *Int. J. Energ. Mater. Chem. Propul.* **8** 1
[13] Pein R and Anders S 2000 Proc. *The Fifth International Symposium on Special Topics in Chemical Propulsion, Combustion of Energetic Materials* (Stresa (Lake Maggiore) Italy June 18–22) p 33
[14] Grigor'ev V G, Kutsenogii K P and Zarko V E 1981 *Combust. Expl. Shock Waves* **17** 356
[15] Cohen N S 1983 *AIAA J.* **21** 720
[16] Kovalev O B, Petrov A P and Fol'ts A V 1987 *Combust. Expl. Shock Waves* **23** 133
[17] Gallier S and Hiernard F 2008 *J. Propulsion and Power* **24** 147
[18] DeLuca L T, Galfetti L, Colombo G, Maggi F, Bandera A, Babuk V A and Sindistkii V P 2010 *J. Propulsion and Power* **26** 724
[19] Rashkovsky S A 1998 *Combust. Sci. Technol.* **139** 125
[20] Rashkovsky S A 1998 *Combust. Sci. Technol.* **139** 149
[21] Rashkovskii S A 2005 *Combust. Expl. Shock Waves* **41** 174
[22] Rashkovskii S A 1999 *Combust. Expl. Shock Waves*, **35** 523
[23] Rashkovskii S A 2002 *Combust. Expl. Shock Waves* **38** 435
[24] Shyu I M and Liu T K 1995 *Combust. Flame* **100** 634

Validation of two-dimensional model by the example of a supersonic inlet-isolator

R K Seleznev [1,2,3]

[1] Dukhov Research Institute of Automatics (VNIIA), Suschevskaya Street 22, Moscow 127055, Russia
[2] Moscow Institute of Physics and Technology (State University), 9 Institutskiy per., Dolgoprudny, Moscow Region, 141701, Russia
[3] Ishlinsky Institute for Problems in Mechanics RAS, Vernadsky Street 101(1), Moscow, 119526, Russia

rkseleznev@gmail.com

Abstract. In the paper two-dimensional model for hypersonic flow are described. Comparing the result of calculation with wall pressure data, obtained in the inlet-isolator model mounted to the floor of a Mach 5 wind tunnel at the University of Texas, presented in this article. The present results show that two-dimensional simulations are able to predict the shock wave structures in a scramjet inlet/isolator.

1. Introduction

The inlet is an essential element of the ramjet and scramjet, and their design greatly affects the overall performance of the engine. Therefore, at the initial stage of creating a gas-dynamic code for describing processes in ramjet and scramjet it is very important to correctly describe the inlet and the shock-wave structure in it. Inlets are with external compression, mixed compression and internal compression [1], but they have the same function in common. In the inlet, due to the appearance of the shock-wave structure, the incoming air is compressed and the Mach number decreases to the desired value at the entrance to the engine. The isolator is also an essential part of the ramjet and scramjet. The isolator is a constant cross-section passage in order to prevent chocking. There are many experiments with inlets [2][3][4–6]. In this paper we consider the model of inlet-isolator mounted to the floor of a Mach 5 wind tunnel at the University of Texas [4,6,7]. In this experiment, it is important to consider the thickness of the boundary layer at the inlet. In this paper, numerical experiments are carried out that take into account the thickness of the boundary layer, as well as experiments with a uniform flow at the inlet. The results of the calculation are compared with the results obtained by other authors [8–10], as well as with the results of the experiment.

2. Description of University of Texas inlet-isolator experimental setup

Wagner [4,6–8] conducted a series of experiments on the study of shock-wave structure in the supersonic inlet-isolator mounted to the floor of a Mach 5 wind tunnel (figure 1). The experimental setup consists of an input ramp with an inclination angle of 6°, and a constant-section isolator (figure 2). At the end of the isolator a special flap is installed, raising it can lead to nonstationary formation with subsequent choking of the flow. With the flap is completely lowered, a small part of it protrudes from the bottom wall of the chamber, as shown in figure 2. However, in this paper, as in the works of other authors[9–13], it is considered that the floor is completely smooth. The following incoming air flow parameters were used:

Incoming flow pressure: $P = 0.0538$ atm;
Incoming flow temperature: $T = 57.4$ K;
Incoming flow Much Number: $M = 4.9$;
Incoming flow gas mixture: Air.

Figure 1. Schematic of the model of the inlet/isolator model mounted on the wind tunnel floor [4].

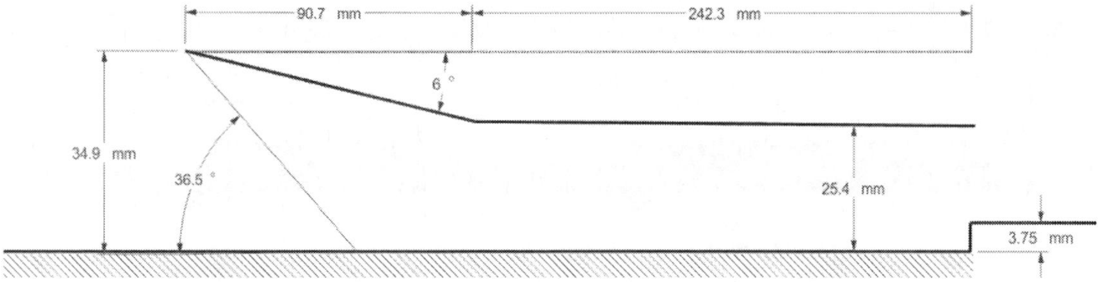

Figure 2. Inlet/isolator model mounted on the wind tunnel floor [4].

3. Two-dimensional computational fluid dynamic model

For two-dimensional calculations we used the NERAT-2D computer code [13]. NERAT-2D realizes the time-relaxation method. At each time step the following groups of governing equations were integrated successively: the Navier – Stokes and continuity equations, the equations of mass conservation of chemical species, the equation of energy conservation. These equations are formulated in the following form:

$$\frac{\partial \rho}{\partial t} + \operatorname{div}(\rho \mathbf{V}) = 0 \tag{1}$$

$$\frac{\partial \rho u}{\partial t} + \operatorname{div}(\rho u \mathbf{V}) = -\frac{\partial p}{\partial x} - \frac{2}{3}\frac{\partial}{\partial x}(\mu \operatorname{div}\mathbf{V}) + \frac{1}{r}\frac{\partial}{\partial r}\left[r\mu\left(\frac{\partial \mathrm{v}}{\partial x} + \frac{\partial u}{\partial r}\right)\right] + 2\frac{\partial}{\partial x}\left(\mu\frac{\partial u}{\partial x}\right), \tag{2}$$

$$\frac{\partial \rho \mathrm{v}}{\partial t} + \operatorname{div}(\rho \mathrm{v} \mathbf{V}) = -\frac{\partial p}{\partial r} - \frac{2}{3}\frac{\partial}{\partial r}(\mu \operatorname{div}\mathbf{V}) + \frac{\partial}{\partial x}\left[\mu\left(\frac{\partial \mathrm{v}}{\partial x} + \frac{\partial u}{\partial r}\right)\right] + 2\frac{\partial}{\partial r}\left(\mu\frac{\partial \mathrm{v}}{\partial r}\right) + 2\mu\frac{\partial}{\partial r}\left(\frac{\mathrm{v}}{r}\right), \tag{3}$$

$$\frac{\partial \rho_i}{\partial t} + \operatorname{div}\rho_i \mathbf{V} = -\operatorname{div}\mathbf{J}_i + \dot{w}_i, \quad i = 1,2,\ldots,N_s, \tag{4}$$

$$\rho c_p \frac{\partial T}{\partial t} + \rho c_p \mathbf{V} \operatorname{grad} T = \operatorname{div}(\lambda \operatorname{grad} T) + \frac{\partial p}{\partial t} + \mathbf{V} \operatorname{grad} p + \\ + \Phi_\mu - \sum_{i=1}^{N_s} h_i \dot{w}_i + \sum_{i=1}^{N_s} \rho c_{p,i} D_i (\operatorname{grad} Y_i \cdot \operatorname{grad} T), \quad (5)$$

where $\Phi_\mu = \mu\left[2\left(\frac{v}{r}\right)^2 + 2\left(\frac{\partial v}{\partial r}\right)^2 + 2\left(\frac{\partial u}{\partial x}\right)^2 + \left(\frac{\partial v}{\partial x} + \frac{\partial u}{\partial r}\right)^2 - \frac{2}{3}\left(\frac{\partial u}{\partial x} + \frac{\partial v}{\partial r} + \frac{v}{r}\right)^2\right]$ is the dissipative function; ; μ, λ are the viscosity and heat conductivity coefficients, c_p is the specific heat capacity of gas mixture; $c_p = \sum_i^{N_s} Y_i c_{p,i}$; Y_i is the mass fraction of species i; $c_{p,i}, h_i$ are the specific heat capacity at constant pressure and specific enthalpy of species i; $\dot{w}_i$ is the reaction rate for species i; D_i is the effective diffusion coefficient of species i; $\rho_i, \mathbf{J}_i$ are the density and mass diffusion flux for species i; $\mathbf{J}_i = -\rho D_i \operatorname{grad} Y_i$; N_s is the number of species. The calculations use a structured mesh of 500x100 dimensions, if not stated other. Chemical reactions are neglected.

4. Results

At the initial stage of calculations, at the entrance to the inlet a uniform flow is set. Figure 3 shows the results of calculated pressure distribution within the experimental setup on structured mesh of 1000 x1000.

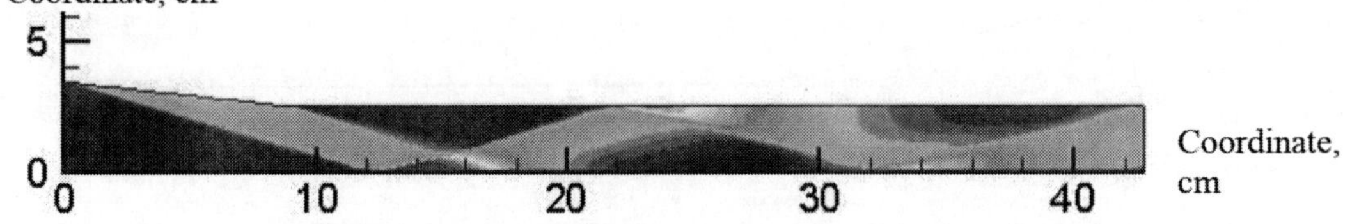

Figure 3. Pressure distribution within the experimental setup.

Figure 4 shows the results of calculated pressure distribution on the bottom wall obtained in this work with the results by Lutsky [9] and Zhukov [8].

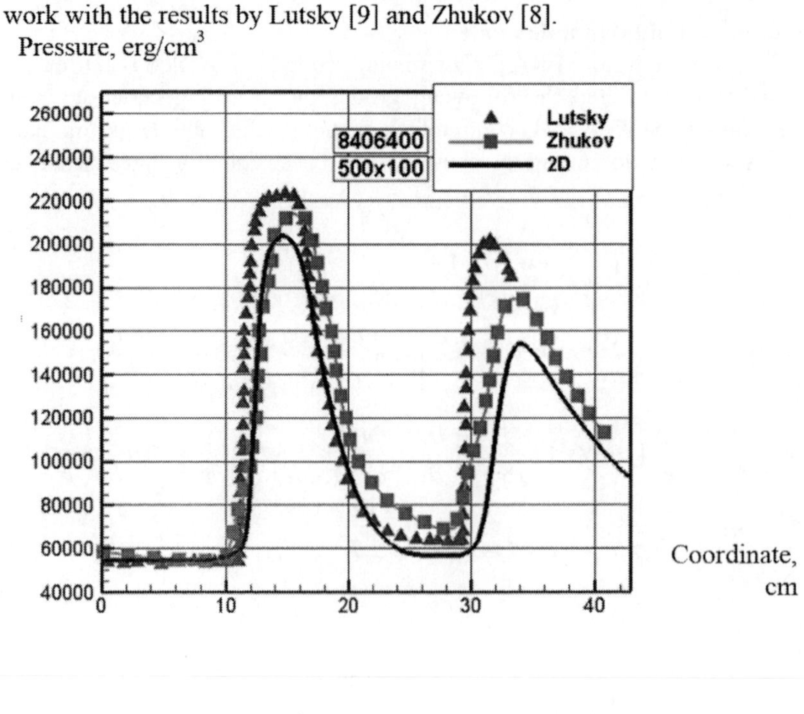

Figure 4. Comparison of calculated pressure distribution on the bottom wall obtained in this work (black line) and the results obtained by Lutsky [9](red tangles) and Zhukov[8] (green squares).

IOP Conf. Series: Journal of Physics: Conf. Series **1009** (2018) 012030 doi:10.1088/1742-6596/1009/1/012030

It can be seen that the flow pattern is described qualitatively. However, in the experiment, the input flow was uneven with a thick boundary layer. The incoming boundary layer profiles were extracted from a method, based on well-known formulas for the boundary layer on a thick plate together with a modification from the work [9,14]. However, it should be noted that in the case of using this model without using the turbulence model, a vortex zone appears before the first shock wave, which changes the flow in the channel. As the results show [10–13,15], the use of the turbulence model avoids the appearance of a vortex. In this work, it was decided to artificially increase the viscosity of a factor of 1000 in the region ahead of the shock wave, as shown in figure 5. This led to the fact that the vortex disappeared. Figure 6 shows a comparison of the pressure distribution on the lower chamber wall obtained in this work (black line) with the experimental results [4,6] (green squares), and the results obtained by Lutsky [9] (red squares)and Koo[10,15] (blue delta). The pressure amplitude obtained in my calculations is just below the amplitude obtained by other authors. However, it should be noted that the position of the shock wave peaks is consistent, which allows us to conclude that the results are qualitatively identical.

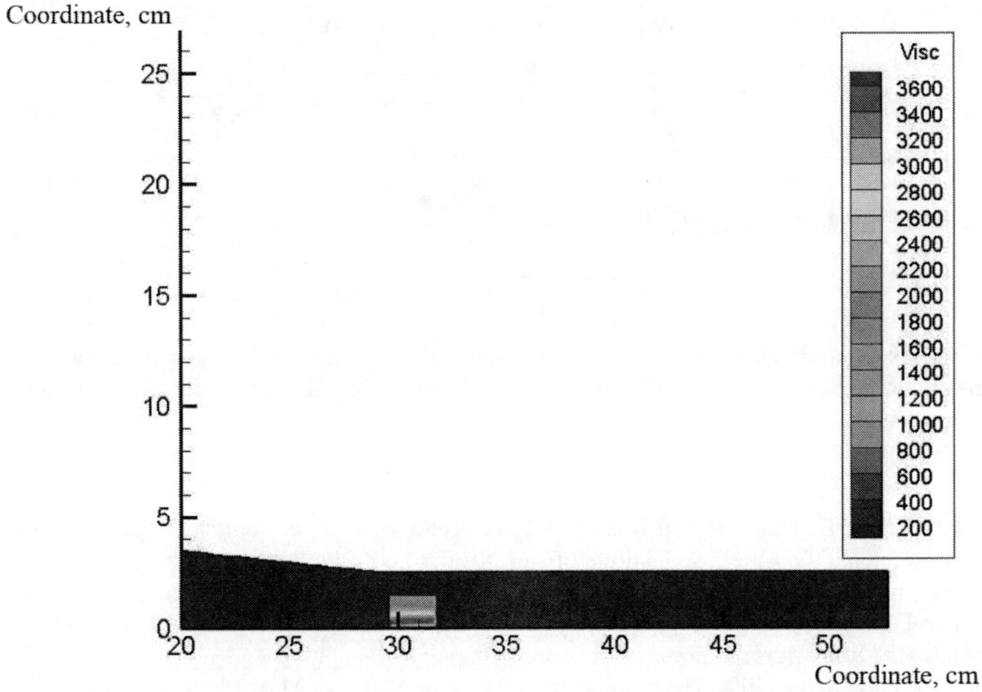

Figure 5. Viscosity distribution within the experimental setup.

5. Conclusion

Comparison of the pressure distribution calculated by two-dimensional model, the results obtained by other authors and pressure distribution in the inlet-isolator model mounted to the floor of a Mach 5 wind tunnel at the University of Texas is presented. The present results show that two-dimensional simulations are able to predict the shock wave structures in a scramjet inlet/isolator. This work presents a continuation of our efforts on the verification and validation of numerical methods and computational codes for calculation of various hypersonic vehicles and energetic devices [16–23]. The results obtained in this study in the study of the shock wave structure in a supersonic air inlet can be used later to study the effect of a surface discharge on a hypersonic flow as it was numerically studied in [24,25].

Acknowledgments

The presented study was supported by the Russian Science Foundation project № 16-11-10275.

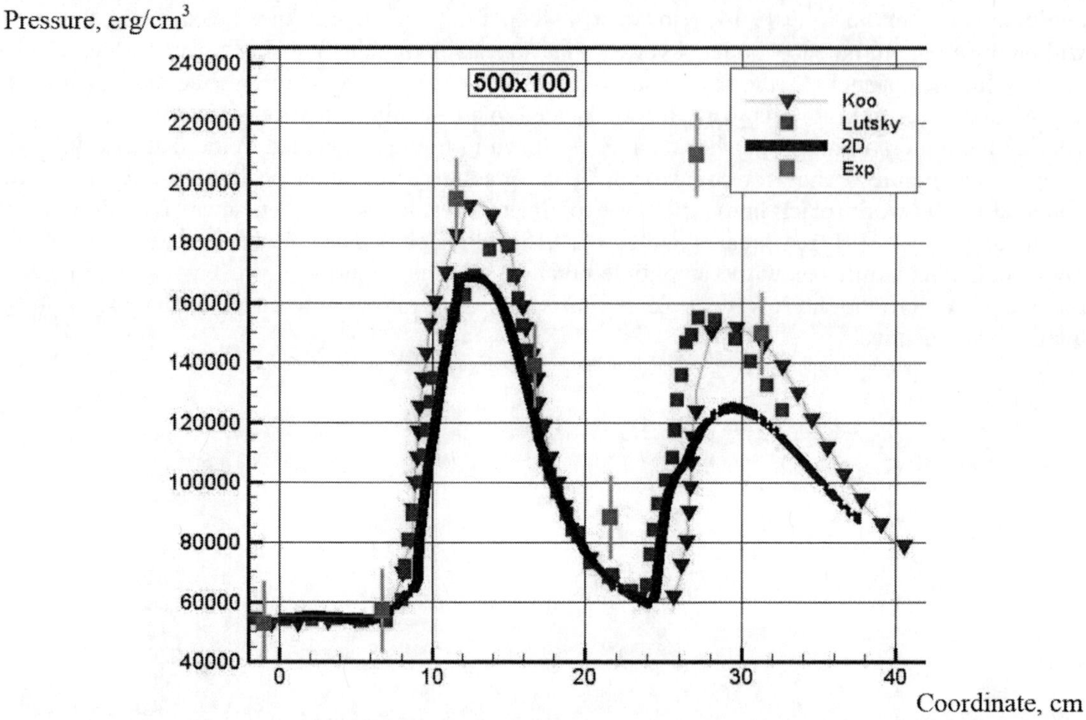

Figure 6. Comparison of calculated pressure distribution on the bottom wall obtained in this work (black line) with the experimental results [4,6] (green squares), and the results obtained by Lutsky [9] (red squares)and Koo[10,15] (blue delta).

References

[1] Heiser W, Pratt D, Daley D and Mehta U 1994 *Hypersonic Airbreathing Propulsion* vol 26 (Washington, DC: American Institute of Aeronautics and Astronautics, Inc.)

[2] Xing F, Zhang S and Yao Y 2012 Numerical Simulation of Shock-Induced-Combustion in Three-Dimensional HyShot Scramjet Model *50th AIAA Aerosp. Sci. Meet. Incl. New Horizons Forum Aerosp. Expo.* 1–12

[3] Boon S and Hillier R 2006 Hypersonic Inlet Flow Analysis at Mach 5, 6 and 7 *44th AIAA Aerospace Sciences Meeting and Exhibit* (Reston, Virigina: American Institute of Aeronautics and Astronautics) pp 1–13

[4] Wagner J L 2009 *Experimental Studies of Unstart Dynamics in Inlet / Isolator Configurations in a Mach 5 Flow*

[5] Gamba M, Miller V, Mungal M G and Hanson R 2011 Ignition and Flame Structure in a Compact Inlet/Scramjet Combustor Model *17th AIAA Int. Sp. Planes Hypersonic Syst. Technol. Conf.*

[6] Wagner J L, Yuceil K B, Valdivia A, Clemens N T and Dolling D S 2009 Experimental Investigation of Unstart in an Inlet/Isolator Model in Mach 5 Flow *AIAA J.* **47** 1528–42

[7] Wagner J L, Yuceil K B and Clemens N T 2010 Velocimetry Measurements of Unstart of an Inlet-Isolator Model in Mach 5 Flow *AIAA J.* **48** 1875–88

[8] Zhukov V T, Manukovskii K V., Novikova N D, Rykov Y G and Feodoritova O B 2015 *Numerical simulation of the pseudo-shock region formation in the channel*

[9] Borisov V E, Kudryashov I Y and Lutsky A E 2016 Numerical simulation of the pseudo-shock

IOP Conf. Series: Journal of Physics: Conf. Series **1009** (2018) 012030 doi:10.1088/1742-6596/1009/1/012030

region formation in the channel *Keldysh Inst. Prepr.* **2898** 1–24
[10] Koo H 2010 Large-Eddy Simulations of Scramjet Engines *Dr. Diss.*
[11] Jang I, Pečnik R and Moin D P 2010 A numerical study of the unstart event in an inlet/isolator model *Cent. Turbul. Res. Annu. Res. Briefs* 93–103
[12] Boles J, Edwards J, Choi J-I and Baurle R 2009 Simulations of High-Speed Internal Flows Using LES/RANS Models *47th AIAA Aerosp. Sci. Meet. Incl. New Horizons Forum Aerosp. Expo.* 1–22
[13] Surzhikov S, Seleznev R, Tretjakov P and Zabaykin V 2014 Unsteady Thermo-Gasdynamic Processes in Scramjet Combustion Chamber with Periodical Input of Cold Air *50th AIAA/ASME/SAE/ASEE Joint Propulsion Conference* (Reston, Virginia: American Institute of Aeronautics and Astronautics) p 25
[14] Xu S and Martin M P 2004 Assessment of inflow boundary conditions for compressible turbulent boundary layers *Phys. Fluids* **16** 2623–39
[15] Koo H and Raman V 2012 Large-Eddy Simulation of a Supersonic Inlet-Isolator *AIAA J.* **50** 1596–613
[16] Seleznev R and Surzhikov S 2016 A Quasi-One-Dimensional Analysis of Hydrogen-Fueled Scramjet Combustors *52nd AIAA/SAE/ASEE Joint Propulsion Conference* (Reston, Virginia: American Institute of Aeronautics and Astronautics) pp 1–27
[17] Seleznev R K 2017 Comparison of two-dimensional and quasi-one-dimensional scramjet models by the example of VAG experiment *J. Phys. Conf. Ser.* **815** 12007
[18] Kotov M A, Ruleva L B, Solodovnikov S I and Surzhikov S T 2017 Experimental and numerical study of supersonic flow over two blunted wedges *J. Phys. Conf. Ser.* **815** 12025
[19] Shang J and Surzhikov S 2013 Radiative Heat Exchange in a Hydrogen-Fueled Scramjet Combustion Chambers *51st AIAA Aerospace Sciences Meeting including the New Horizons Forum and Aerospace Exposition* (Reston, Virigina: American Institute of Aeronautics and Astronautics)
[20] Shang J and Surzhikov S 2013 Numerical Prediction of Convective and Radiative Heating of Scramjet Combustion Chamber with Hydrocarbon Fuels *51st AIAA Aerospace Sciences Meeting including the New Horizons Forum and Aerospace Exposition* (Reston, Virigina: American Institute of Aeronautics and Astronautics) pp 1–16
[21] Polezhaev Y V. and Seleznev R K 2014 Numerical study of the processes of resonance emergence in the experimental set-up of a pulse detonation engine *High Temp.* **52** 225–9
[22] Seleznev R and Surzhikov S 2015 Quasi-One-Dimensional and Two-Dimensional Numerical Simulation of Scramjet Combustors *51st AIAA/SAE/ASEE Joint Propulsion Conference* (Reston, Virginia: American Institute of Aeronautics and Astronautics) p 28
[23] Seleznev R and Surzhikov S 2013 Generalized Newton Method For Solving Differential Equations of Chemical Kinetics *44th AIAA Thermophysics Conference* (Reston, Virginia: American Institute of Aeronautics and Astronautics) pp 1–17
[24] Surzhikov S T 2017 Surface electromagnetic actuator in rarefied hypersonic flow *J. Phys. Conf. Ser.* **815** 12005
[25] Shang J S, Kimmel R L, Menart J A and Surzhikov S T 2008 Hypersonic Flow Control Using Surface Plasma Actuator *J. Propuls. Power* **24** 923–34

IOP Conf. Series: Journal of Physics: Conf. Series **1009** (2018) 012031 doi:10.1088/1742-6596/1009/1/012031

Validation of 3D model by the example of a supersonic inlet-isolator

R K Seleznev [1,2,3]

[1] Dukhov Research Institute of Automatics (VNIIA), Suschevskaya Street 22, Moscow 127055, Russia
[2] Moscow Institute of Physics and Technology (State University), 9 Institutskiy per., Dolgoprudny, Moscow Region, 141701, Russia
[3] Ishlinsky Institute for Problems in Mechanics RAS, Vernadsky Street 101(1), Moscow, 119526, Russia

rkseleznev@gmail.com

Abstract. In the paper three-dimensional model for hypersonic flow are described. Comparing the result of calculation with wall pressure data, obtained in the inlet-isolator model mounted to the floor of a Mach 2.75 wind tunnel at the Stanford University, presented in this article. The present results show that three-dimensional model is able to predict the shock wave structures in a scramjet inlet/isolator.

1. Introduction

The inlet is an essential element of the ramjet and scramjet, and their design greatly affects the overall performance of the engine. Therefore, at the initial stage of creating a gas-dynamic code for describing processes in ramjet and scramjet it is very important to correctly describe the inlet and the shock-wave structure in it. Inlets are with external compression, mixed compression and internal compression [1], but they have the same function in common. In the inlet, due to the appearance of the shock-wave structure, the incoming air is compressed and the Mach number decreases to the desired value at the entrance to the engine. The isolator is also an essential part of the ramjet and scramjet. The possibility of a correct description of the shock wave structure is very important for accurate calculations of the combustion chamber. In other words, by setting the parameters at the inlet to the air intake, it becomes possible to obtain the parameters at the inlet to the combustion chamber. The model described allows us to calculate the flow in the combustion chamber, which will be demonstrated in subsequent work. Recently, experiments have appeared with a complex shape of the air intake, so it is especially important to create three-dimensional codes. For example, Rectangular-to-Elliptical Shape Transition (REST) [2–4] scramjet become common in recent years.

There are many experiments with inlets [2–12]. In this paper we consider the model of inlet-isolator mounted to the floor of a Mach 2.75 wind tunnel at the Stanford University[9]. This type of experimental setup is called jet in supersonic cross-flow (JISCF). Absolutely similar experiments were conducted in China [5,6]. In this experiment, in addition to visual observations, pressure measurements were made on the upper and lower walls. In this paper, the validation is performed on the pressure distribution in this experiment on the top wall. Besides this verification case involves the computation of the supersonic flowfield past a wedge with a half-angle of 15 degrees.

IOP Conf. Series: Journal of Physics: Conf. Series **1009** (2018) 012031 doi:10.1088/1742-6596/1009/1/012031

2. Description of Stanford University inlet-isolator experimental setup

Stanford University conducted a series of experiments to study the injection of fuel into the supersonic flow, as well as to study of shock-wave structure in the supersonic inlet-isolator mounted to the floor of a Mach 2.75 at the Expansion Tube Facility of the High Temperature Gasdynamics Laboratory at Stanford University. (figure 1). The experimental setup consists of an input ramp with an inclination angle of 10°, and a constant-section isolator. The height at the entrance to the air intake is 23 mm. The height of the constant part of the isolator is 15mm. The width of the chamber is constant and equal to 75 mm. From the bottom wall of the chamber perpendicular to the flow, fuel is injected through a nozzle 2 mm in diameter. The injection site is located in the center of the bottom wall of the chamber at a distance of 70 mm down from the leading edge of the inlet. In the experiment, in addition to visual observations, the pressure distribution on the lower and upper walls was investigated, as well as the interaction of the blown jet with the main air flow. In this paper, injection of fuel was not considered. The following incoming air flow parameters were used:

Incoming flow pressure:	$P = 0.40$ atm;
Incoming flow temperature:	$T = 1250$ K;
Incoming flow Much Number:	$M = 2.75$;
Incoming flow gas mixture:	Air.

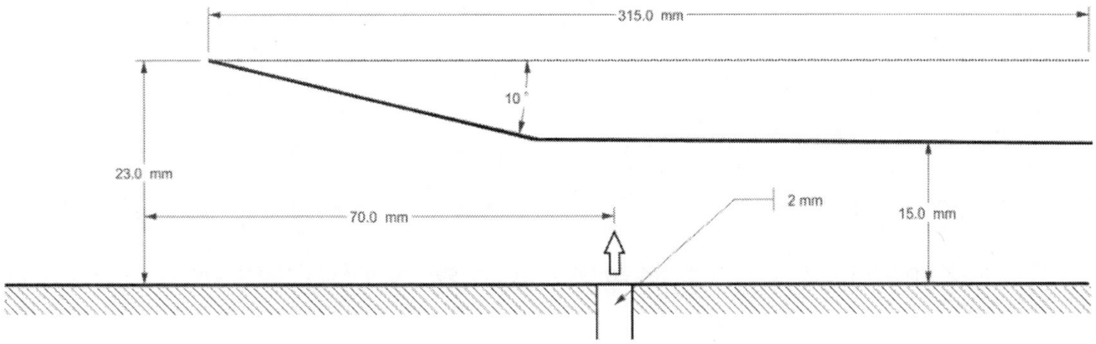

Figure 1. The side-view of inlet/isolator model [9].

3. Three-dimensional computational fluid dynamic model

For three-dimensional calculations we used the NERAT-3D computer code [13]. NERAT-3D realizes the time-relaxation method. At each time step the following groups of governing equations were integrated successively: the Navier – Stokes and continuity equations, the equations of mass conservation of chemical species, the equation of energy conservation. These equations are formulated in the following form:

$$\frac{\partial \rho}{\partial t} + \mathrm{div}(\rho \mathbf{V}) = 0 , \tag{1}$$

$$\frac{\partial \rho \mathbf{V}}{\partial t} + div\left(\rho \mathbf{V}\mathbf{V} + \hat{\mathbf{\Pi}}\right) = 0 \tag{2}$$

$$\rho c_p \frac{\partial T}{\partial t} + \rho c_p \mathbf{V}\,\mathrm{grad}\,T = \mathrm{div}(\lambda\,\mathrm{grad}\,T) + \sum_{i=1}^{N_s} \rho c_{p,i} D_i (\mathrm{grad}\,Y_i \cdot \mathrm{grad}\,T) - \\ - \sum_{i=1}^{N_s} h_i \dot{w}_i + \frac{\partial p}{\partial t} + \mathbf{V}\,\mathrm{grad}\,p + \Phi_\mu \tag{3}$$

$$\frac{\partial \rho_i}{\partial t} + \mathrm{div}\,\rho_i \mathbf{V} = -\mathrm{div}\,\mathbf{J}_i + \dot{w}_i, \quad i = 1,2,\ldots,N_s \tag{4}$$

where: t – time; $\mathbf{V} = \mathbf{i}u + \mathbf{j}v + \mathbf{k}w$ – velocity vector with projections on the axis of the Cartesian coordinate system x, y, z; p, ρ – pressure and density; T – temperature of translational movement of particles; μ, λ – dynamic coefficient of viscosity and coefficient of thermal conductivity; c_p – specific heat of the mixture at constant pressure; $c_p = \sum_i^{N_s} Y_i c_{p,i}$; N_s – number of gas mixture components; Y_i – mass fraction of the i-th component of the mixture; $c_{p,i}, h_i$ – the specific heat at constant pressure associated with translational and rotational degrees of freedom, and the enthalpy of the i-th component of the mixture; $\dot{w}_i$ – the mass velocity of chemical transformations for the i-th component of the mixture; D_i – the effective diffusion coefficient of the i-th component of the mixture; $\mathbf{J}_i$ – the diffusion flux density of the i-th component; $\mathbf{J}_i = -\rho D_i \operatorname{grad} Y_i$; N_s number of chemical components of the gas mixture. The components of the viscous stress tensor and the dissipative function were calculated by the formulas:

$$\Pi_{i,j,k} = -p\delta_{i,j,k} + \mu\left[\left(\frac{\partial u_i}{\partial x_j} + \frac{\partial u_j}{\partial x_i}\right) - \frac{2}{3}\delta_{i,j}\frac{\partial u_k}{\partial x_k}\right],\ i,j,k = 1,2,3$$

$$\Phi_\mu = \mu\left[2\left(\frac{\partial u}{\partial x}\right)^2 + 2\left(\frac{\partial v}{\partial y}\right)^2 + 2\left(\frac{\partial w}{\partial z}\right)^2 + \right.$$

$$\left. + \left(\frac{\partial v}{\partial x} + \frac{\partial u}{\partial y}\right)^2 + \left(\frac{\partial w}{\partial y} + \frac{\partial v}{\partial z}\right)^2 + \left(\frac{\partial u}{\partial z} + \frac{\partial w}{\partial x}\right)^2 - \frac{2}{3}\left(\frac{\partial u}{\partial x} + \frac{\partial v}{\partial y} + \frac{\partial w}{\partial z}\right)^2\right].$$

The closing relations for the system of equations to be solved include the thermal equation of state of an ideal gas:

$$\frac{p}{\rho} = \frac{R_0}{M_\Sigma}T,\ \frac{1}{M_\Sigma} = \sum_i^{N_s}\frac{Y_i}{M_i} \qquad (5)$$

The system of equations (1)–(5) was integrated numerically by the establishment method using computer code NERAT-3D. Equations (1)–(2) were integrated by an explicit finite-difference method according to the AUSM scheme [14]. The diffusion equations (4) and energy conservation were solved using an implicit finite-difference scheme of the second order of the Crank-Nicholson approximation. Numerical solution of finite-difference equations was carried out using multi-block technology. The number of blocks used in this technology is determined by the complexity of the geometry of the calculation area. In this case, 8 blocks of the grid were used. The calculations use a structured mesh of 60×50×14 dimensions in each block, if not stated other. Chemical reactions are neglected.

4. Results

The first verification case for 3D code involves the computation of the supersonic flowfield past a wedge with a half-angle of 15 degrees. Figures 2– 3 shows the main features of the flow field. The freestream consists of air. Mach number is of 2.5. The inlet temperature is equal to 280 K. The inlet pressure is equal to 1 atm. As the flow meets the leading edge of the wedge, an oblique shock is formed as the flow turns to become tangent with the wedge surface. The parameters of the flowfield (Mach number, pressure, temperature and angle of oblique shock) past the shock is uniform and presented in table 1. The calculations use a structured mesh of 40×40×30 dimensions in each of eight blocks. Table 1 presents a comparison of the calculation results behind the oblique shock for the

NERAT 3D and WIND 2D [15] with an analytical solution [16]. It can be seen that the relative error does not exceed 4 percent.

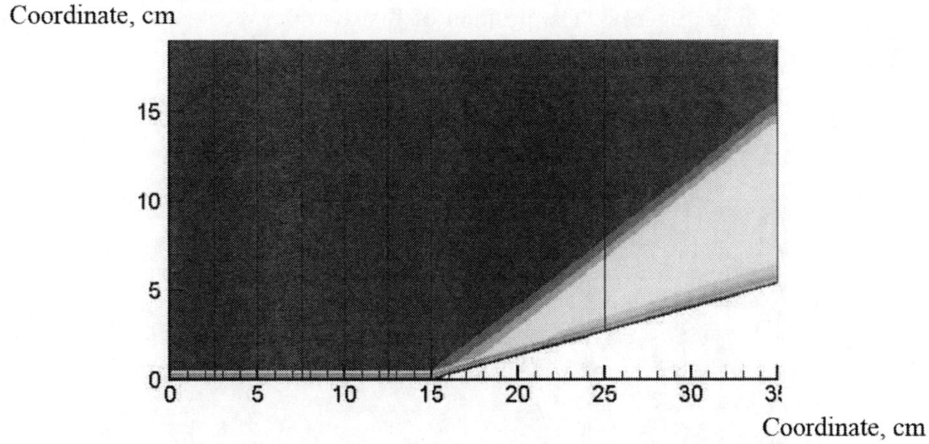

Figure 2. Mach number distribution within the experimental setup.

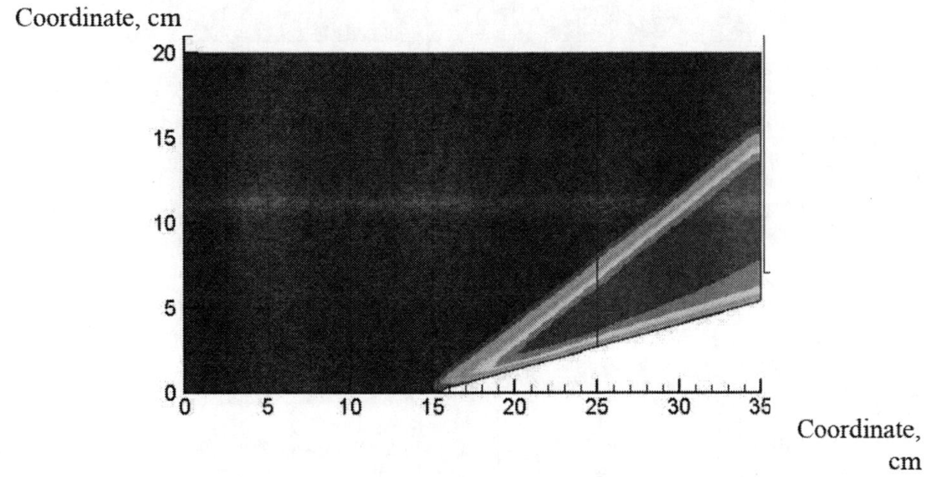

Figure 3. Pressure distribution within the experimental setup.

Table 1. Comparison of the results of calculating a flat wedge with an angle of 15° with analytical and calculated data.

	Mach	Pres, *atm*	T, *K*	angle
NERAT 3D (this work)	1.860	2.560	373.0	37.14
Analytical solution [16]	1.870	2.467	370.1	36.95
Relative error,%	0.540	3.610	0.780	0.530
WIND 2D [15]	1.868	2.469	372.5	-

The second validation case is Stanford University experimental setup. At the calculations the entrance to the inlet is set uniform flow. Figure 4 shows the results of calculated pressure distribution within the experimental setup. It should be noted the occurrence of a shock-wave structure in the

IOP Conf. Series: Journal of Physics: Conf. Series **1009** (2018) 012031 doi:10.1088/1742-6596/1009/1/012031

channel. However, the intensity of the shock waves attenuates. This is particularly noticeable on the results of calculated pressure distribution on the top wall. Comparison of calculated pressure distribution on the top wall obtained in this work (red line) and the experimental results (green squares) [9] presented in the figure 5. The position of first two the shock wave peaks is consistent, which allows us to conclude that the results are qualitatively identical.

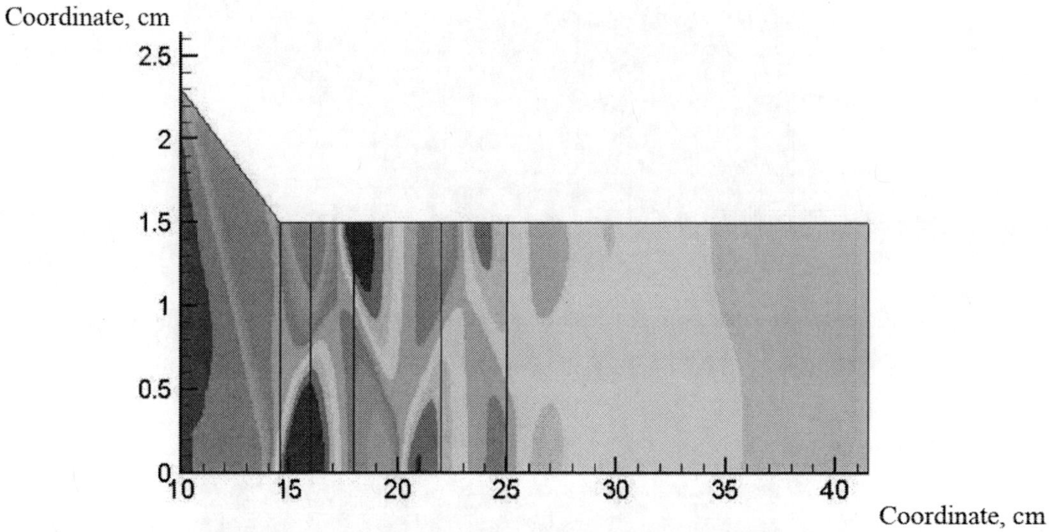

Figure 4. Pressure distribution within the experimental setup.

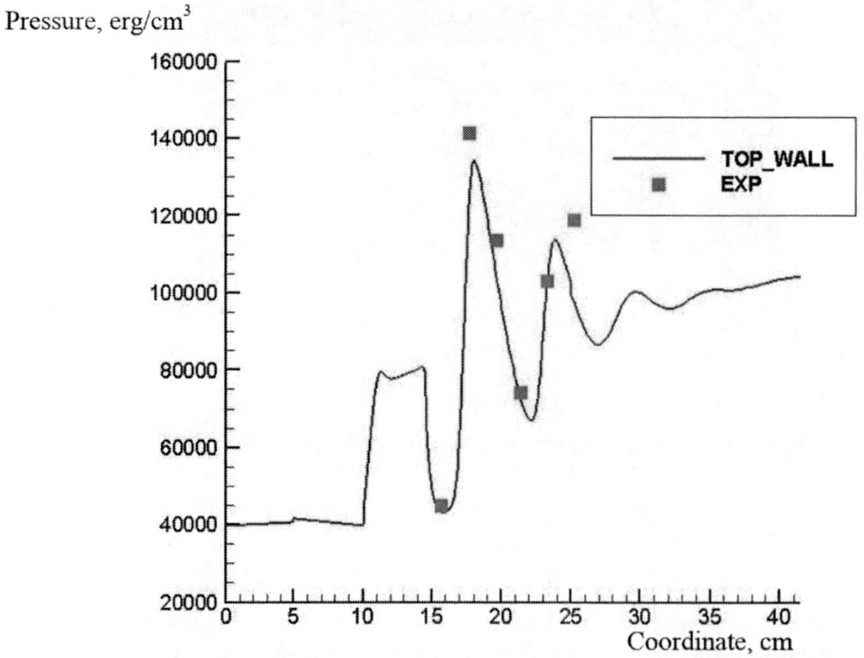

Figure 5. Comparison of calculated pressure distribution on the top wall obtained in this work (red line) and the experimental results (green squares) [9].

5. Conclusion

Comparison of the calculation results behind the oblique shock for the NERAT 3D and WIND 2D and an analytical solution are presented. Comparison of the pressure distribution calculated by three-dimensional model and pressure top wall distribution in the inlet-isolator model mounted to the floor of a Mach 2.75 wind tunnel at the Stanford University are presented. The present results show that our three-dimensional simulations are able to predict the shock wave structures in a scramjet inlet/isolator. This work presents a continuation of our efforts on the verification and validation of numerical methods and computational codes for calculation of various hypersonic vehicles and energetic devices [18–23].

Acknowledgments
This work was preformed within the framework of the Government program of basic research of the Russian academy of sciences (contract #AAAA-A17-117021310372-6) and partially under support of RFBR grant #016-01-00379.

References

[1] Heiser W, Pratt D, Daley D and Mehta U 1994 *Hypersonic Airbreathing Propulsion* vol 26(Washington, DC: American Institute of Aeronautics and Astronautics, Inc.)

[2] Roberts M, Smart M and Frost M 2012 HIFiRE 7: Design to Achieve Scientific Goals *18th AIAA/3AF International Space Planes and Hypersonic Systems and Technologies Conference* (Reston, Virigina: American Institute of Aeronautics and Astronautics) pp 1–8

[3] Smart M and Ruf E 2006 Free-Jet Testing of a REST Scramjet at Off-Design Conditions *25th AIAA Aerodynamic Measurement Technology and Ground Testing Conference* (Reston, Virigina: American Institute of Aeronautics and Astronautics) pp 190–201

[4] Zhao M, Zhou T, Ye T, Zhu M and Zhang H 2017 Large eddy simulation of reacting flow in a hydrogen jet into supersonic cross-flow combustor with an inlet compression ramp *Int. J. Hydrogen Energy* **42** 16782–92

[5] Zhao M, Ye T, Cao C, Zhou T and Zhu M 2016 Study of sonic injection from circular injector into a supersonic cross-flow using large eddy simulation *Int. J. Hydrogen Energy* **41** 17657–69

[6] Wagner J L 2009 *Experimental Studies of Unstart Dynamics in Inlet / Isolator Configurations in a Mach 5 Flow*

[7] Wagner J L, Yuceil K B, Valdivia A, Clemens N T and Dolling D S 2009 Experimental Investigation of Unstart in an Inlet/Isolator Model in Mach 5 Flow *AIAA J.* **47** 1528–42

[8] Gamba M, Miller V, Mungal M G and Hanson R 2011 Ignition and Flame Structure in a Compact Inlet/Scramjet Combustor Model *17th AIAA Int. Sp. Planes Hypersonic Syst. Technol. Conf.*

[9] Wilson Chan Y K, Razzaqi S A, Wise D J and Smart M 2014 Freejet Testing of the 75%-scale HIFiRE 7 REST Scramjet Engine *19th AIAA International Space Planes and Hypersonic Systems and Technologies Conference* (Reston, Virginia: American Institute of Aeronautics and Astronautics) pp 1–20

[10] Barth J 2014 *Mixing and combustion enhancement in a Mach 12 shape-transitioning scramjet engine* (The University of Queensland)

[11] Suraweera M V. and Smart M K 2009 Shock-Tunnel Experiments with a Mach 12 Rectangular-to-Elliptical Shape-Transition Scramjet at Offdesign Conditions *J. Propuls. Power* **25** 555–64

[12] Surzhikov S, Seleznev R, Tretjakov P and Zabaykin V 2014 Unsteady Thermo-Gasdynamic Processes in Scramjet Combustion Chamber with Periodical Input of Cold Air *50th AIAA/ASME/SAE/ASEE Joint Propulsion Conference* (Reston, Virginia: American Institute of Aeronautics and Astronautics) p 25

[13] Shang J and Surzhikov S 2013 Numerical Prediction of Convective and Radiative Heating of Scramjet Combustion Chamber with Hydrocarbon Fuels *51st AIAA Aerospace Sciences Meeting including the New Horizons Forum and Aerospace Exposition* (Reston, Virigina: American Institute of Aeronautics and Astronautics) pp 1–16

[14] Shang J and Surzhikov S 2013 Radiative Heat Exchange in a Hydrogen-Fueled Scramjet Combustion Chambers *51st AIAA Aerospace Sciences Meeting including the New Horizons Forum and Aerospace Exposition* (Reston, Virigina: American Institute of Aeronautics and Astronautics)

[15] Edwards J, Liou M-S, Edwards J and Liou M-S 1997 Low-diffusion flux-splitting methods for flows at all speeds *13th Computational Fluid Dynamics Conference* vol 36 (Reston, Virigina: American Institute of Aeronautics and Astronautics)

[16] Slater J W https://www.grc.nasa.gov/www/wind/valid/wedge/wedge01/wedge01.html

[17] Marsh H 1983 Modern compressible flow *Int. J. Heat Fluid Flow* **4** pp 59–60

[18] Seleznev R and Surzhikov S 2013 Generalized Newton Method For Solving Differential Equations of Chemical Kinetics *44th AIAA Thermophysics Conference* (Reston, Virginia: American Institute of Aeronautics and Astronautics) pp 1–17

[19] Seleznev R and Surzhikov S 2015 Quasi-One-Dimensional and Two-Dimensional Numerical Simulation of Scramjet Combustors *51st AIAA/SAE/ASEE Joint Propulsion Conference* (Reston, Virginia: American Institute of Aeronautics and Astronautics) p 28

[20] Polezhaev Y V. and Seleznev R K 2014 Numerical study of the processes of resonance emergence in the experimental set-up of a pulse detonation engine *High Temp.* **52** 225–9

[21] Seleznev R and Surzhikov S 2016 A Quasi-One-Dimensional Analysis of Hydrogen-Fueled Scramjet Combustors *52nd AIAA/SAE/ASEE Joint Propulsion Conference* (Reston, Virginia: American Institute of Aeronautics and Astronautics) pp 1–27

[22] Seleznev R K 2017 Comparison of two-dimensional and quasi-one-dimensional scramjet models by the example of VAG experiment *J. Phys. Conf. Ser.* **815** 12007

[23] Kotov M A, Ruleva L B, Solodovnikov S I and Surzhikov S T 2017 Experimental and numerical study of supersonic flow over two blunted wedges *J. Phys. Conf. Ser.* **815** 12025

IOP Conf. Series: Journal of Physics: Conf. Series **1009** (2018) 012032 doi:10.1088/1742-6596/1009/1/012032

Combustion stability in a solid-fuel ramjet engine

S A Rashkovskiy[1,2], **S E Yakush**[1,2] **and A A Baranov**[1,2]

[1] Ishlinsky Institute for Problems in Mechanics RAS, Prospekt Vernadskogo 101 (1), Moscow, 119526, Russia
[2] Bauman Moscow State Technical University, Baumanskaya 2-ya 5, Moscow, 105005, Russia

E-mail: yakush@ipmnet.ru

Abstract. A computational model is presented for a solid-fuel ramjet engine with an axisymmetric combustion channel in a fuel grain containing a wider section near the air inlet. The latter part of the channel serves as a flame holder where a recirculation zone develops providing necessary combustion stabilization. The model is based on RANS equations, with finite-rate chemistry in the turbulent gas phase flame taken into account. Reacting multicomponent gas equations are solved by a low-dissipation numerical scheme suitable for a wide range of Mach numbers. Solid fuel is gasified at the surface, and gasification products react in the core flow. Combustion of PMMA is studied numerically in the subsonic flow regime for different pressures, inlet velocities, and flame holder geometry parameters. It is shown that under certain conditions combustion in the engine becomes unstable, with flame blown off either in the main channel, or in the flame holder as well. Different combustion regimes are plotted as stability maps in the parameters space.

1. Introduction

Solid-fuelled ramjet (SFRJ) engines are quite promising for unmanned aerial propulsion vehicles, and interest to their development and application has been increasing over the past few decades [1-3]. In SFRJs, thrust is created by reacting gas flow in the grain channel, with fuel supplied into the reacting zone by decomposition of grain material, and oxygen taken from the atmospheric air. Unlike the rocket engines, the fuel in SFRJ contains only a small fraction of oxidizer and is not capable of self-sustained combustion. SFRJs can also be used as a part of hybrid propulsion systems where the hot gases in the grain channel are created by burning a rocket fuel [4].

An important issue in the design of solid-fuel ramjet engines is to make sure that combustion in the fuel grain channel is stable in a wide range of flight conditions. One approach to the solution of this problem relies on the specially profiled channel geometry: a ledge (wider section) is formed in the inlet part of the grain channel, which acts as a flame holder due to development of recirculation zone. In the hot recirculation zone, the incoming air is mixed continuously with the gas fuel to burn and maintain the high temperature necessary for steady ignition of the main grain channel.

Combustion in SFRJs was studied experimentally and numerically in a number of papers [1-6]. In particular, numerical study on the effects of the shape and sizes of the flame holder on the flow was carried out in [5]. The influence of the ratio of flame holder length to its depth and the slope angle of the wall at the exit of the flame holder was analysed by a steady-state model, which does not allow simulation of SFRJ operation instability. In [6], self-ignition of solid fuel in a ramjet was investigated, but stability of combustion of solid fuels and connection between the stability and geometry of the flame holder was not considered.

In a recent paper [7], a non-stationary numerical model for simulation of combustion processes in solid-fuel ramjet engines was developed, and some screening results of numerical simulation for several inlet parameters of a solid fuel ramjet combustor were presented. Then, the model was further enhanced to include finite-rate turbulent combustion in order to provide the capability of simulating ignition and extinction of fuel-air mixture [8, 9]. Here, we present extensive numerical simulations of combustion in SFRE carried out for subsonic flow regimes, with emphasis on the stability of combustion and its dependence on the inlet flow velocity, pressure, and flame holder geometry.

2. Mathematical model

2.1. Gas flow

Multicomponent reacting mixture consisting of five components is considered: $C_5H_8O_2$ (fuel), O_2, CO_2, H_2O, and N_2. Each component is described by the ideal gas equation of state, the thermal properties (specific heat, internal energy, enthalpy) as functions of temperature are described by polynomials [10]. The temperature of gas mixture is determined from its internal energy and composition by solving a corresponding nonlinear equation.

The gaseous fuel is methylmethacrylate (MMA) resulting from decomposition of the solid fuel, polymethylmethacrylate (PMMA). Being a complex polymer, PMMA produces, upon gasification, a whole spectrum of gas components, however, in this study we take a simple single-product model of its gasification (see review, e.g., in [4]). More detailed kinetics schemes will be taken upon in the future work.

In the cylindrical coordinates (r,z), gas flow is described by the Reynolds-averaged Navier-Stokes equations (RANS) in the fully compressible formulation:

$$\frac{\partial \mathbf{Q}}{\partial t}+\frac{\partial \mathbf{F}}{\partial r}+\frac{\partial \mathbf{G}}{\partial z}+\frac{\mathbf{S}}{r}=\frac{\partial \mathbf{F}^D}{\partial r}+\frac{\partial \mathbf{G}^D}{\partial z}+\frac{\mathbf{S}^D}{r}+\mathbf{R} \tag{1}$$

where $\mathbf{Q}$ is the vector of state variables, $\mathbf{F}$ and $\mathbf{G}$ are inviscid fluxes in the $r-$ and $z-$directions, and $\mathbf{S}$ is the source term, while $\mathbf{F}^D$, $\mathbf{G}^D$, and $\mathbf{S}^D$ are corresponding dissipation terms; $\mathbf{R}$ is the chemical reaction source term, subscript i denotes a chemical component ($i=1,\ldots,N_c$).

$$\mathbf{Q}=\begin{pmatrix}\rho\\ \rho u\\ \rho v\\ E\\ \rho Y_i\end{pmatrix},\quad \mathbf{F}=\begin{pmatrix}\rho u\\ \rho u^2+p\\ \rho uv\\ (E+p)u\\ \rho Y_i u\end{pmatrix},\quad \mathbf{G}=\begin{pmatrix}\rho u\\ \rho uv\\ \rho v^2+p\\ (E+p)v\\ \rho Y_i v\end{pmatrix},\quad \mathbf{S}=\begin{pmatrix}\rho u\\ \rho u^2\\ \rho uv\\ (E+p)u\\ \rho Y_i u\end{pmatrix} \tag{2}$$

$$\mathbf{F}^D=\begin{pmatrix}0\\ \tau_{rr}\\ \tau_{rz}\\ u\tau_{rr}+v\tau_{rr}+\lambda T_r\\ \rho DY_{i,r}\end{pmatrix},\ \mathbf{G}^D=\begin{pmatrix}0\\ \tau_{rz}\\ \tau_{zz}\\ u\tau_{rz}+v\tau_{zz}+\lambda T_z\\ \rho Y_{i,z}\end{pmatrix},\ \mathbf{S}^D=\begin{pmatrix}0\\ \tau_{rr}-\tau_{\theta\theta}\\ \tau_{rz}\\ u\tau_{rr}+v\tau_{rr}+\lambda T_r\\ \rho Y_{i,r}\end{pmatrix},\ \mathbf{R}=\begin{pmatrix}0\\ 0\\ 0\\ \dot{w}\Delta H_c\\ \dot{w}_i\end{pmatrix} \tag{3}$$

where

$$\tau_{rr}=\mu\left(2u_r-\frac{2}{3}\mathrm{div}\,\mathbf{U}\right),\ \tau_{rz}=\tau_{zr}=\mu(u_z+v_r),\ \tau_{zz}=\mu\left(2u_z-\frac{2}{3}\mathrm{div}\,\mathbf{U}\right),\ \tau_{\theta\theta}=\mu\left(2\frac{u}{r}-\frac{2}{3}\mathrm{div}\,\mathbf{U}\right) \tag{4}$$

are components of the stress tensor with $\mathrm{div}\,\mathbf{U} = u_r + v_z + u/r$ (here, for brevity, u_r, u_z, v_r, v_z, T_r, T_z, $Y_{i,r}$, and $Y_{i,z}$ denote the partial derivatives of velocity components, temperature and mass fractions of components with respect to the corresponding coordinate). The effective viscosity $\mu = \mu_l + \mu_t$, heat conductivity $\lambda = \lambda_l + \lambda_t$, and diffusivity $D = D_l + D_t$, are the sums of laminar and turbulent values, the latter obtained from the standard $k-\varepsilon$ turbulence model.

2.2. Reactions and combustion model

The mass rate of gasification corresponding to the incident heat flux q_s is described by a simple model taking into account the heating of solid material from its initial temperature T_0 to the boiling temperature T_s on the surface of decomposing material and further gasification:

$$\dot{m} = \frac{q_s}{c_s(T_s - T_0) + \Delta H_G} \tag{5}$$

Here, c_s is the specific heat capacity of solid fuel. The heat flux q_s can generally include the radiation part, but here it was not taken into account due to small sizes of the combustor considered. The properties of PMMA are: density $\rho_s = 1.18$ g/cm^3, boiling temperature $T_s = 200$ °C, specific heat capacity $c_s = 1500$ J/(kg K), heat of gasification $\Delta H_G = 1591$ kJ/kg.

Combustion of MMA in the gas phase is described by a single irreversible gross reaction [11]

$$C_5H_8O_2 + 6O_2 \rightarrow 5CO_2 + 4H_2O + \Delta H_c \tag{6}$$

with the heat of combustion per unit mass of fuel $\Delta H_c = 25.6$ MJ/kg and kinetics

$$\dot{\omega}_k = B \cdot T \cdot \frac{s_k}{W_{O_2}} \rho^2 Y_F Y_{O_2} \exp\left(-\frac{E_c}{RT}\right) \tag{7}$$

Here, $B = 6.6 \cdot 10^6$ [m^3/mol s K] is the pre-exponential factor, $E_c = 144$ kJ/mole is the activation energy, s_k is the mass stoichiometric coefficient of *k*-th species (for fuel $s_F = -1$), W_{O_2} is the molar mass of oxygen.

The turbulent combustion rates for all species on the right-hand side of conservation equations (3) are evaluated as $\dot{w}_k = s_k \dot{w}$, where the reaction rate $\dot{w}$ is obtained from the corresponding turbulent combustion model. In the previous work [7], the Eddy Break-up (EBU) model [12] was used in which assumes that infinitely fast chemistry, so that the reaction rate is controlled by the rate of turbulent mixing of fuel and oxidizer. In the current work, this turbulent combustion model was substituted by the Eddy Dissipation Concept (EDC) [13] which takes into account the finite-rate kinetics in turbulent flame. EDC considers combustion as occurring in fine structures occupying a fraction of volume, each fine structure described as a perfectly stirred reactor with the flowrate and residence time related to the turbulent quantities, and reaction rate described by the kinetical formula (7). In this way, turbulence-kinetics interaction is taken into account, which, importantly, allows us to model flame ignition/extinction without any additional assumptions on the critical temperature or strain rate.

2.3. Initial and boundary conditions

The inlet conditions corresponded to air with normal oxygen contents (21% vol.) supplied at a given linear velocity V_{in}, temperature T_{in}, and pressure P_{in}.

The mass flux of fuel gas (MMA) at the channel boundaries is described by equation (5) which relates the decomposition rate of PMMA with the heat flux incident on the wall. The temperature of gaseous fuel at the wall was taken equal to the boiling temperature T_s; the density was then obtained from the equation of state using the local pressure near the wall (possible incomplete gasification due to charring is not taken into account). The normal velocity component at the wall was obtained from the mass flux and density, for the tangential velocity component the wall functions implemented in the $k-\varepsilon$ turbulence model were used, implying the universal logarithmic wall law; thermal wall functions were used to evaluate the incident heat flux q_s.

In a working ramjet engine, combustion is maintained due to feedback between the hot burning zone and solid fuel gasification at the channel walls. In order to establish the working regime, some ignition procedure (external heating or hot gas flow from a gas generator) is necessary. Studying the ignition stage was outside the scope of current paper, which focuses on the simulation of reacting flows in a working ramjet engine and evaluation of combustion stability. Therefore, an "artificial" ignition procedure was followed. Simulations were started with gas flow parameters corresponding to the inlet conditions throughout the domain; during short initial time (1 ms), the heat flux in the gasification formula (5) was set to $q_s = 1$ kW/m^2, which was sufficient to initialize the gasification, but did not lead to development of high temperature in the combustor. Then, simulations were run for another 4 ms with infinite reaction rate in the EDC model (equivalent to EBU model), which was sufficient for obtaining steady combustion. At time 5 ms, the combustion model was switched to finite-rate EDC (imitating switching off the igniter), and further evolution of gas flow and combustion zone was simulated.

3. Numerical implementation

The governing equations (1) – (4) were solved in the axisymmetric coordinates by an explicit finite-volume high-resolution scheme on a uniform Cartesian grid. The "inviscid" fluxes are approximated by a scheme HR-SLAU2 [14] belonging to the AUSM family of numerical schemes. HR-SLAU2 scheme is an "all-speed" numerical scheme applicable to wide range of Mach number flows. Viscosity, diffusion, and heat conduction terms are approximated by the second-order central difference scheme; chemical reactions are solved on a separate substep, prior to the solution of gas dynamics equations.

The solid fuel surface is described on the Cartesian grid by the "embedded" sharp interface approach [15] in which the internal boundary is the zero level set of a signed distance function. In the near-boundary cells, fluxes and finite-volume operators are modified to take into account that cells are cut by the surface. The wall functions of $k-\varepsilon$ turbulence model were implemented for the embedded boundaries by extending a normal from the surface into the flow, setting the wall laws along this normal, and subsequent interpolation of the boundary layer variables back onto the Cartesian grid [16].

4. Results

4.1. Geometry and inlet conditions

Numerical simulations were carried out for the geometry similar to that used in the experiments [1, 2]. The geometry is shown in figure 1, with the main geometrical parameters listed in table 1. The same geometry was studied in our previous works [7–9]; here, a wider range of geometrical parameters and operating conditions is covered.

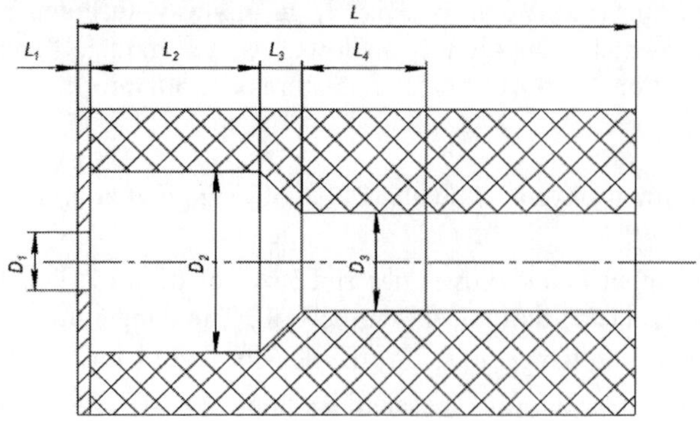

Figure 1. Geometry of solid fuel combustor: air in injected through the inlet of diameter D_1 into the main channel (of diameter D_3) via a flow expansion chamber (of diameter D_2) serving as flame holder.

Table 1. Combustion chamber dimensions.

Parameter	Value (mm)
D_1	10
D_2	26 – 40
D_3	15
L	100
L_1	5
L_2	20 – 60
L_3	5

Simulations were carried out for two values of operating pressure, $P = 1$ and 2 atm, the inlet velocity was varied in the range of $V_{in} = 100 - 400$ m/s, the inlet temperature was fixed at $T_{in} = 500$ K. It was found that five main operating regimes are possible: 1) stable combustion, 2) short-duration combustion instability with eventual reignition; 3) oscillatory regime; 4) flame extinction in the channel; 5) total flame extinction. In the following sections, we demonstrate the flow and combustion zone structure corresponding to these regimes, after which maps of possible regimes in the parameter space will be shown.

4.2. Stable combustion

In figure 2, steady-state distributions corresponding to $P_{in} = 2$ atm and $V_{in} = 300$ m/s are presented; the flame holder dimensions are $D_2 = 30$ mm, and $L_2 = 45$ mm. On the temperature field, streamlines demonstrating recirculation in the flame holder are shown. It can be seen that hot gases recirculating in the flame holder cause fuel gasification in the absence of oxygen near the surface (see plates c) and d)). Gaseous fuel produced in the wide section mixes near the inlet with the incoming air, however, temperature there is low, and ignition occurs later on, as the fuel-air mixture travels towards the main channel. As a result, a "hanging" combustion zone is formed in the flame holder (see plate b)). Hot gases from the burning flame holder ignite the channel, and a near-surface combustion zone is stabilized there along the channel walls. These distributions are typical of all cases where steady-state combustion was observed.

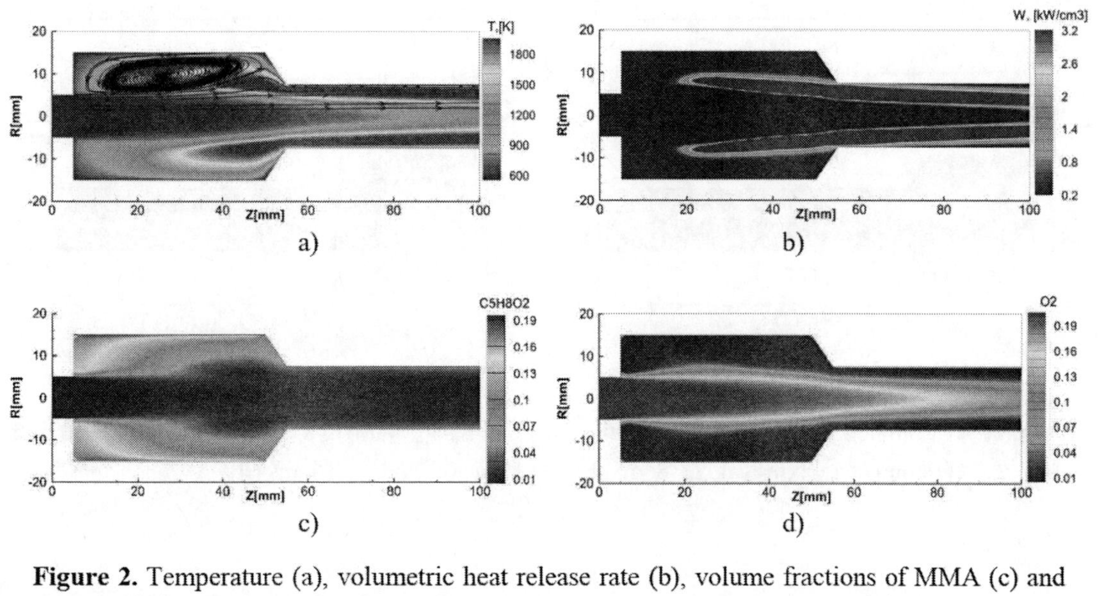

Figure 2. Temperature (a), volumetric heat release rate (b), volume fractions of MMA (c) and oxygen (d) for steady combustion regime: $P_{in} = 2$ atm, $v_{in} = 300$ m/s, $T_{in} = 500$ K.

4.3. One-time extinction and reignition in main channel

The sequence of events in this case is shown in figures 3 and 4. The flame is blown down the channel by the time 8 ms, however, after that the main channel is reignited by the hot gases from the flame holder, and steady combustion is established by 16 ms.

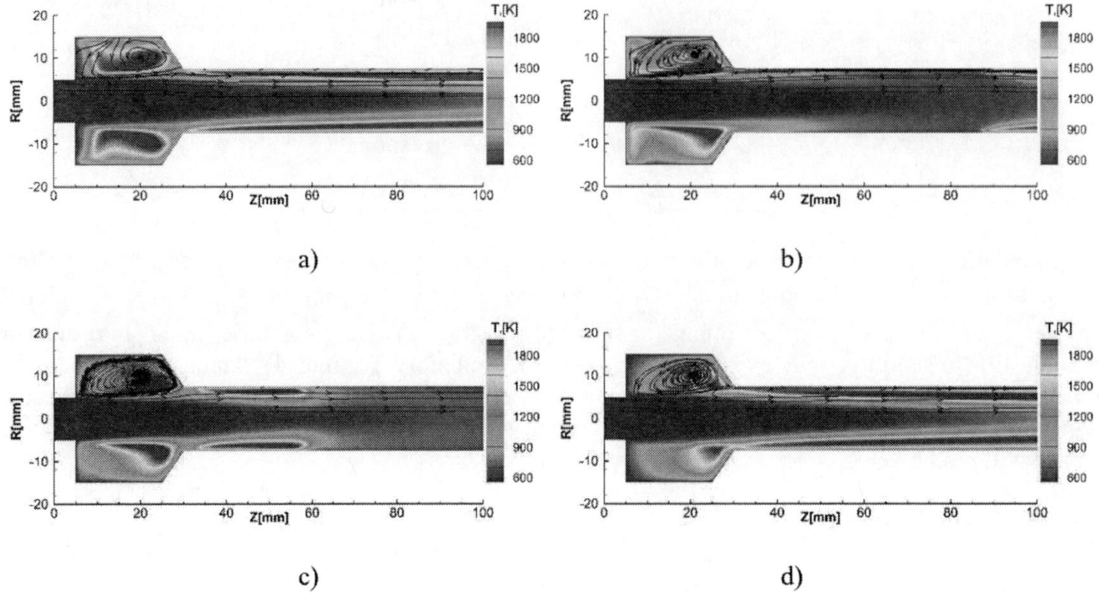

Figure 3. Temperature for $P_{in} = 1$ atm, $v_{in} = 100$ m/s, $T_{in} = 500$ K, $D_2 = 30$ mm and $L_2 = 20$ mm at $t = 5$ ms (a), $t = 8$ ms (b), $t = 11$ ms (c) and $t = 16$ ms (d).

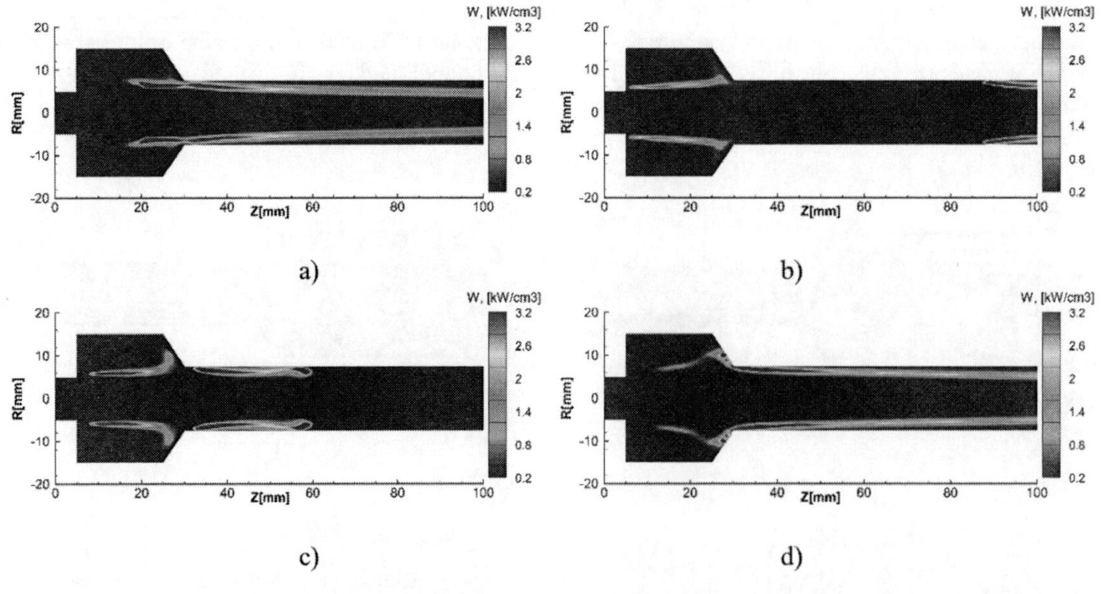

Figure 4. Volumetric heat release rate for $P_{in} = 1$ atm, $v_{in} = 100$ m/s, $T_{in} = 500$ K, $D_2 = 30$ mm and $L_2 = 20$ mm at $t = 5$ ms (a), $t = 8$ ms (b), $t = 11$ ms (c) and $t = 16$ ms (d).

IOP Conf. Series: Journal of Physics: Conf. Series **1009** (2018) 012032 doi:10.1088/1742-6596/1009/1/012032

4.4. *Periodic extinction and reignition in main channel*

In figures 5 and 6, the case where partial extinction of combustion in the main channel occurs, however, the flame is not blown out of the channel because of fast fuel reignition. This process is periodic, as will be demonstrated below on time histories of the total power (see figure 8).

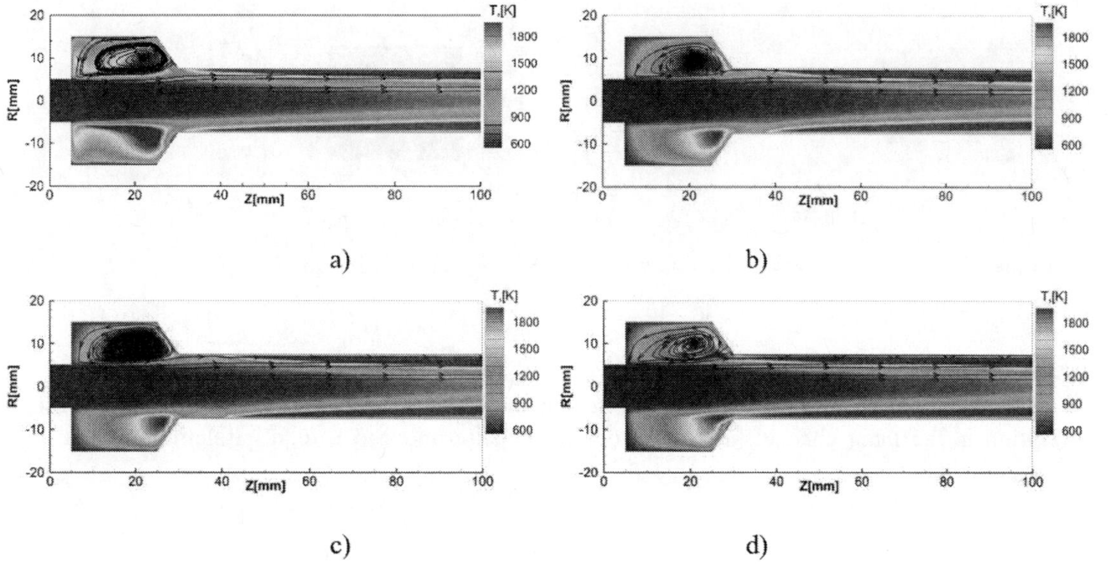

Figure 5. Temperature for $P_{in} = 2$ atm, $v_{in} = 200$ m/s, $T_{in} = 500$ K, $D_2 = 30$ mm and $L_2 = 20$ mm at $t = 5$ ms (a), $t = 12$ ms (b), $t = 17$ ms (c) and $t = 25$ ms (d).

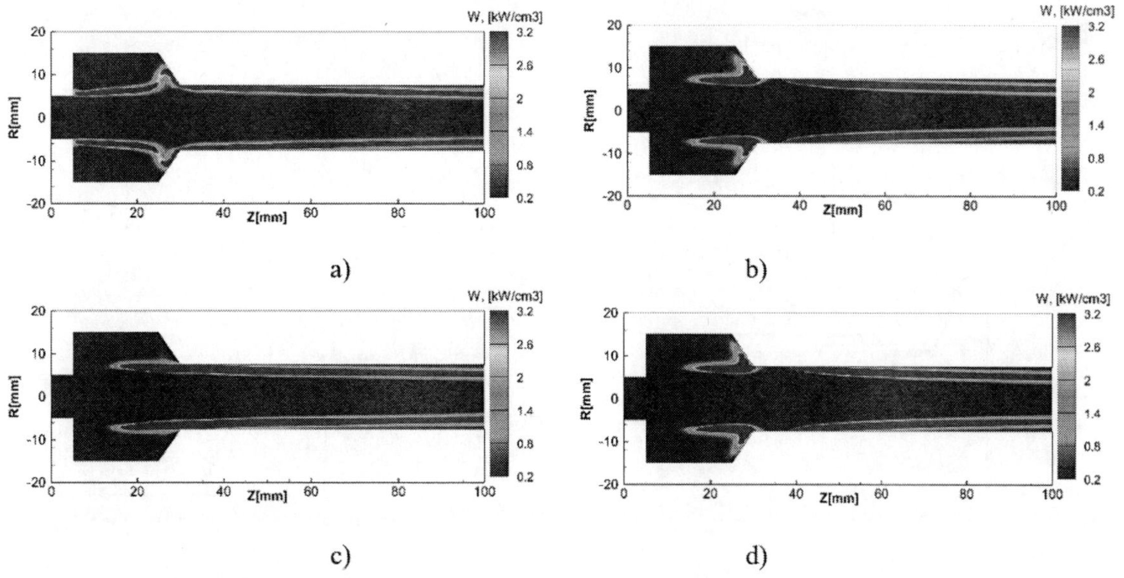

Figure 6. Volumetric heat release rate for $P_{in} = 2$ atm, $v_{in} = 200$ m/s, $T_{in} = 500$ K, $D_2 = 30$ mm and $L_2 = 20$ mm at $t = 5$ ms (a), $t = 12$ ms (b), $t = 17$ ms (c) and $t = 25$ ms (d).

IOP Conf. Series: Journal of Physics: Conf. Series **1009** (2018) 012032 doi:10.1088/1742-6596/1009/1/012032

4.5. Total extinction in main channel

In some cases extinction in the main channel occurred without subsequent reignition, however, combustion in the flame holder was stable. The steady temperature and volumetric heat release rate distributions corresponding to such a case are shown in figure 7.

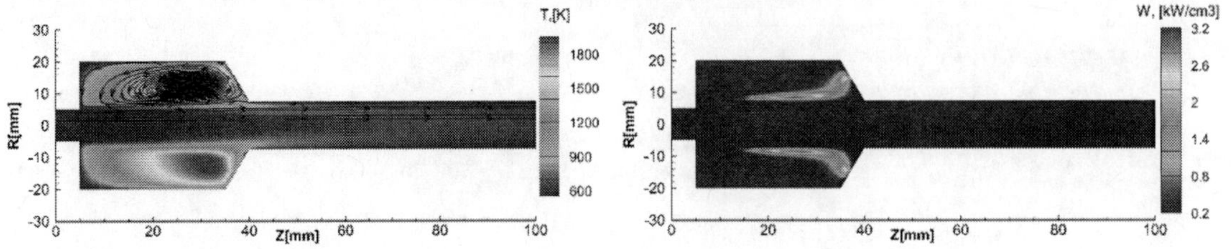

a) b)

Figure 7. Temperature (a) and volumetric heat release for $P_{in} = 1$ atm, $v_{in} = 300$ m/s, $D_2 = 40$ mm, and $L_2 = 30$ mm.

4.6. Total combustion power

In figure 8, typical time histories are plotted, demonstrating the regimes of flame blow-off and reignition in the main channel (a), oscillatory combustion mode (b), total extinction (c), and stable combustion (d).

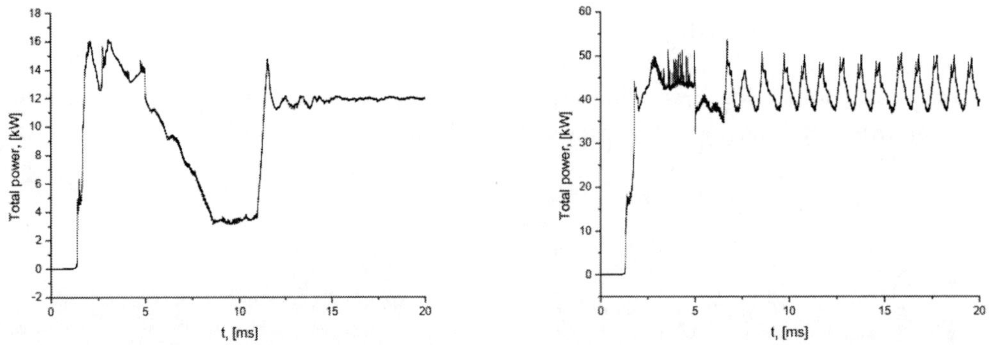

a) b)

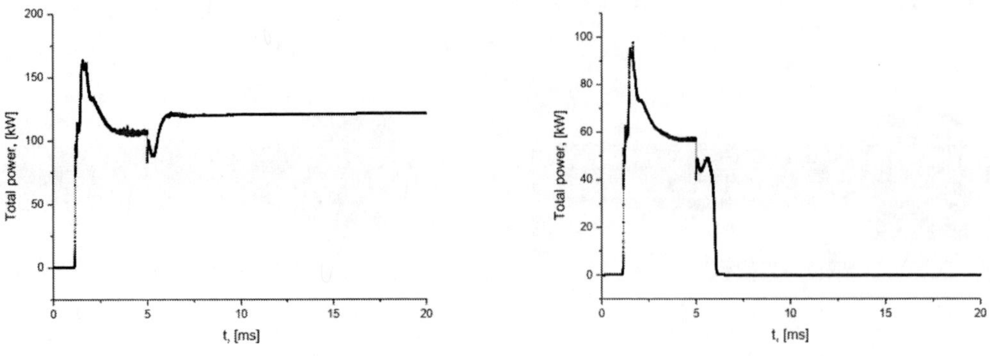

c) d)

Figure 8. Time histories of total power for $P_{in} = 1$ atm, $V_{in} = 100$ m/s (a) and $P_{in} = 2$ atm, $v_{in} = 200$ m/s (b) with the geometry $D_2 = 30$ mm, $L_2 = 20$ mm; $P_{in} = 1$ atm, $V_{in} = 400$ m/s (c) and $P_{in} = 2$ atm, $v_{in} = 400$ m/s (d) with the geometry $D_2 = 30$ mm, $L_2 = 45$ mm.

IOP Conf. Series: Journal of Physics: Conf. Series **1009** (2018) 012032 doi:10.1088/1742-6596/1009/1/012032

4.7. Map of combustion regimes

A number of simulations were performed in which the flow parameters and flame holder geometry were varied. It was found that combustion regimes observed in the whole parameter space covered by the simulations fall into the categories demonstrated in the previous sections. An overview of the results is given in figures 9 and 10, where color maps of combustion regimes are given for flame holder of the diameter $D_2 = 30$ mm. On each map, rows correspond to different flame holder lengths L_2, whereas columns correspond to different inlet velocities. Figure 10 shows the map obtained for pressure $P_{in} = 1$ atm, while figure 11 corresponds to a higher pressure $P_{in} = 2$ atm. Different flow regimes obtained for each combination of flame holder length and inlet velocity are shown by colours, with the following colour key:

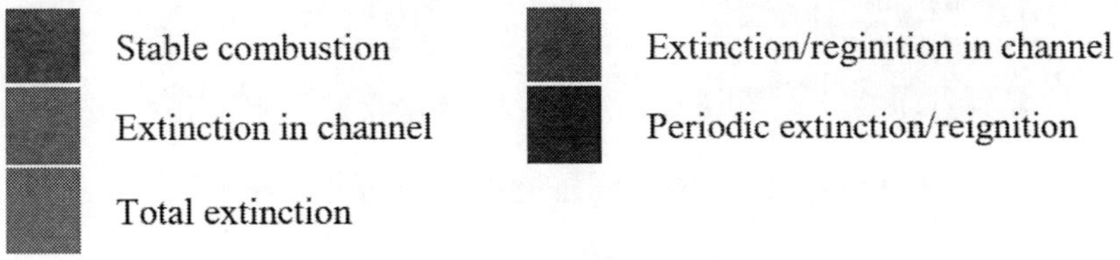

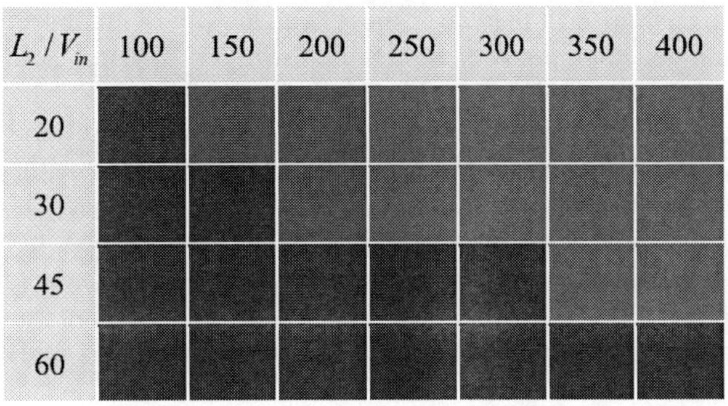

Figure 9. Combustion regime map for pressure $P_{in} = 1$ atm.

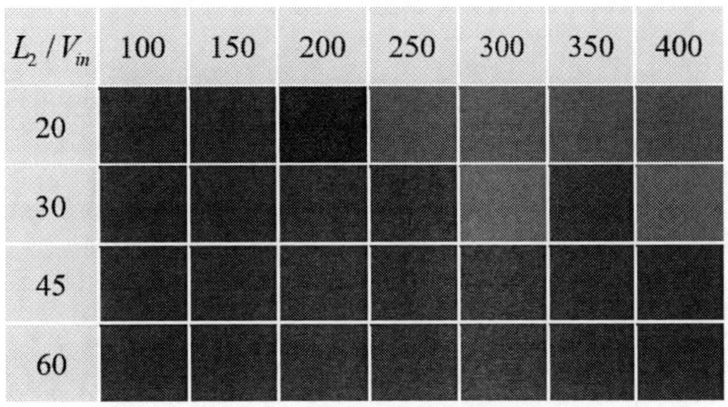

Figure 10. Combustion regime map for pressure $P_{in} = 2$ atm.

One can see that combustion stability deteriorates with the increase in the flow velocity. Short flame holders are incapable of providing necessary combustion stabilization. Also, with the increase in pressure combustion becomes more stable, which can be attributed to the increase in the burning rate proportional to the square of density (second-order reaction, see (7)).

Effect of flame holder geometry is also demonstrated in figure 11 corresponding to a single value of inlet velocity $V_{in} = 300$ m/s, $P_{in} = 1$ atm, with the colour keys given in the legend.

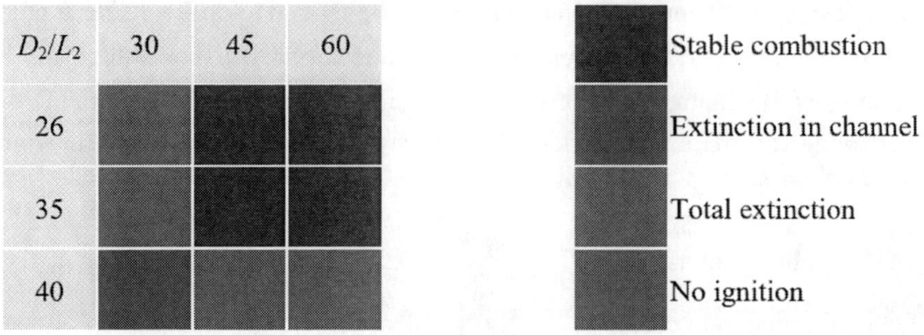

Figure 11. Effect of flame holder geometry.

5. Conclusions

Numerical simulations performed in this work show that stability of combustion in a solid fuel ramjet engine depend on a number of parameters, so that optimization of flame holder geometry is vital for achievement of stable operation of SFRJ in wide range of conditions.

In the subsonic regime considered here, at higher velocities combustion in the fuel channel becomes unstable, the instability manifests itself first via oscillations with periodic extinction and reignition near the boundary between the flame holder and main channel; further increase in the velocity may result in complete flame extinguishment in the channel, possibly followed by secondary ignition. Combustion in the flame holder is more stable due to the presence of recirculation zone; the "hanging" combustion zone is located at the periphery of the recirculation vortex providing necessary ignition source for the main channel. However, at high velocities complete extinguishment was observed in the whole SFRJ.

Further work will be focused on the extension of the model to supersonic flow conditions, typical of scramjets. Other planned improvements include implementation of a more detailed model for solid fuel gasification. Validation of the computational model against the data on supersonic tests will also be performed.

Acknowledgment

This work was supported by Russian Foundation for Basic Research, Grant RFFI 16-29-01084 ofi_m.

References

[1] Zvuloni R, Gany A and Levy Y 1989 Geometric effects on the combustion in solid fuel ramjets *J. Propulsion* **5** 32–7

[2] Ben-Yakar A, Natan B and Gany A 1998 Investigation of a solid fuel scramjet combustor *J. Propulsion and Power* **14** 447–55

[3] Wang L, Wu Z, Chi H, Liu C, Tao H and Wang Q 2014 Numerical and experimental study on the solid-fuel scramjet combustor *Journal of Propulsion and Power* **31** 685–93

[4] Novozhilov V, Joseph P, Ishiko K, Shimada T, Wang H and Liu J 2011 Polymer combustion as a basis for hybrid propulsion: A comprehensive review and new numerical approaches *Energies* **4** 1779–839

[5] Pei X and Hou L 2014 Numerical investigation on cavity structure of solid-fuel scramjet combustor *Acta Astronautica* **105** 463–75

[6] Chi H, Wei Z, Wang L, Li B and Wu Z 2015 Numerical investigation of self-ignition characteristics of solid-fuel scramjet combustor *Journal of Propulsion and Power* **31** 1019–32

[7] Rashkovskiy S A , Yakush S E and Baranov A A 2017 Stabilization of solid fuel combustion in a ramjet engine *J. of Physics: Conference Series* **815**(1) 012008 DOI: 10.1088/1742–6596/815/1/012008

[8] Rashkovskiy S A , Yakush S E and Baranov A A 2017 Combustion Stabilization in a Solid-fuel Ramjet *8th European Combustion Meeting, 18-21 April 2017, Dubrovnik, Croatia* 2067–2072, ISBN 978-953-59504-1-7

[9] Rashkovskiy S A , Yakush S E and Baranov A A 2017 Numerical Simulation of Solid-fuel Ramjet Combustor with a Flame Holder *7th European Conference for Aeronautics and Space Sciences (EUCASS), Italy, Milan, 3-6 Sept 2017* 1-11 DOI: 10.13009/EUCASS2017-106

[10] Goos E, Burcat A and Ruscic B 2017 Extended Third Millennium Ideal Gas and Condensed Phase Thermochemical Database for Combustion with Updates from Active Thermochemical Tables http://burcat.technion.ac.il/dir/

[11] Bedir H and T'ien J S 1998 A computational study of flame radiation in PMMA diffusion flames including fuel vapor participation *27th Symp. Combust.* 2821–2828

[12] Magnussen B F and Hjertager B H 1976 On the mathematical modelling of turbulent combustion with special emphasis on soot formation and combustion *Sixteenth Symp. (Int.) on Combustion* The Combustion Institute Pittsburgh PA 711–729

[13] Gran I R and Magnussen B F 1996 A numerical study of a bluff-body stabilized diffusion flame Part 2. Influence of combustion modeling and finite-rate chemistry *Combust. Sci. Technol.* **119** 191–217

[14] Kitamura K and Hashimoto A 2016 Reduced dissipation AUSM-family fluxes: HR-SLAU2 and HR-AUSM+-up for high resolution unsteady flow simulations *Computers and Fluids* **126** 41–57

[15] Hu X Y, Khoo B C, Adams N A and Huang F L 2006 A conservative interface method for compressible flows *Journal of Computational Physics* **219** 553–78

[16] Berger M and Aftosmis M J 2012 Progress towards a Cartesian cut-cell method for viscous compressible flow *50th AIAA Aerospace Sciences Meeting (Nashville)* Jan. 9-12 1–17

IOP Conf. Series: Journal of Physics: Conf. Series **1009** (2018) 012033 doi:10.1088/1742-6596/1009/1/012033

Numerical modeling of nozzle gas flow using continuum approach in transition regime

M Yu Timokhin[1,2], I E Ivanov[1,2] and I A Kryukov[2,3]

[1]Faculty of Physics, Moscow State University, 119992, Moscow, Russia
[2]Moscow Aviation Institute, 125993, Moscow, Russia
[3]Ishlinsky Institute for Problems in Mechanics RAS, prosp. Vernadskogo 101-1, Moscow, 119526, Russia

timokhin@physics.msu.ru

Abstract. This study is devoted to the numerical modelling of two-dimensional micro nozzle gas flow in continuum and transition regime with continuum mathematical models. Regularized 13-moment (R13) set of equations and Navier-Stokes-Fourier (NSF) equations are used as continuum approaches here. The numerical results for two-dimensional nozzle flow of the R13 equations and NSF equations have been compared for two cases of input and output pressures.

1. Introduction

It is known that the assumption of a continuum medium is valid in case where the mean free path of the molecules is much smaller than the characteristic linear size of the flow. In this case the Navier-Stokes-Fourier (NSF) equations are correct. If this condition is violated, the fluid will not be in thermodynamic equilibrium, then the result - occurrence a variety rarefied effects. The criterion of a flow rarefaction is a Knudsen number (Kn). Kn is defined as the ratio of the average mean free path to the typical length scale of the flow. Numerous analytical, experimental and numerical investigations have helped to make the classification of flow regimes of gas into the rarefied effects, based on the local Knudsen number. The range of Knudsen number $Kn<10^{-3}$ corresponds to the continuum flow, $10^{-3}<Kn<10^{-1}$ is slip flow, $10^{-1}<Kn<10$ is flow in the transitional regime, for $Kn>10$ is the free-molecular regime. So the formal area of the applicability of NSF equations with kinetic boundary conditions is $Kn<10^{-1}$ [1].

In micro-nozzles a continuum flow occurs in the combustor, transonic section, and the flow core of the supersonic section, while the slip and transition regimes are observed near the walls, at the nozzle exit, in high-gradient zones of shock-wave structures, and in the initial section of the jet. In the far-field region of the jet, we can observe a free-molecular flow. Due to the variety of flow regimes, flows in such devices are difficult to simulate in sense of mathematical approach and numerical methods. Accordingly, different medium models have to be used in different zones of the computational domain and the resulting solutions have to be matched at the boundaries of the zones or we need to use a universal approach to the entire problem, which is not always an optimal way out. Examples of universal approaches, in which all the flow regimes are computed using unified kinetic algorithms, include nonequilibrium flow simulation based on the direct solution of the Boltzmann equation [2,3] or model equations [4-7], the direct simulation Monte Carlo (DSMC) method [8-10]. However, for low-velocity gas flows in three dimensions, an exact solution is difficult to compute by applying kinetic methods, and this task often lies beyond the capabilities of modern computers (e.g., the DSMC method yields a wide statistical scatter under such conditions, while the direct solution of the Boltzmann equation and molecular dynamics methods lead to high computational costs at low Knudsen numbers).

For processes in the transition regime, where, approximately, 0.01<Kn<10, extensions of the hydrodynamic equations are based either on the Chapman-Enskog method [1,11,12], or on Grad's moment method [1,11,13,14]. In the given work we have stopped on the question of the applicability of 13 moment system of Grad with its proposed modification [14,15] to two-dimensional supersonic flows. The obtained system is here called the «regularized Grad's set» (or R13). The choice of the system of thirteen moments indicates that all variables in this case have clear physical meaning (density, components of a vector of speed, pressure tensor and a vector of a heat flux).

2. Mathematical model

2.1. Bulk equations

The regularization of the Grad's original 13-moment system [13] was derived in 2003 [15] by a Chapman-Enskog expansion [12] of higher moment equations only, based on the assumption of faster relaxation times for higher moments. Later derivations of the R13 equations were developed explicitly without this assumption (Order of Magnitude Method) [11]. The resulting equations contain higher order terms which behave well in the description of shocks [16-20] and some two dimensional supersonic flow simulations [21-23]. The tensor form of the regularized 13-moment system (R13) can be written as

$$\frac{\partial \rho}{\partial t}+\frac{\partial \rho \upsilon_k}{\partial x_k}=0, \tag{1}$$

$$\rho\frac{\partial \upsilon_i}{\partial t}+\rho\upsilon_k\frac{\partial \upsilon_i}{\partial x_k}+\frac{\partial p}{\partial x_i}+\frac{\partial \sigma_{ik}}{\partial x_k}=0, \tag{2}$$

$$\frac{3}{2}\rho\frac{\partial \theta}{\partial t}+\frac{3}{2}\rho\upsilon_k\frac{\partial \theta}{\partial x_k}+\frac{\partial q_k}{\partial x_k}+p\frac{\partial \upsilon_k}{\partial x_k}+\sigma_{ij}\frac{\partial \upsilon_i}{\partial x_j}=0, \tag{3}$$

$$\frac{\partial \sigma_{ij}}{\partial t}+\frac{\partial \sigma_{ij}\upsilon_k}{\partial x_k}+\frac{4}{5}\frac{\partial q_{\langle i}}{\partial x_{j\rangle}}+2p\frac{\partial \upsilon_{\langle i}}{\partial x_{j\rangle}}+2\sigma_{k\langle i}\frac{\partial \upsilon_{j\rangle}}{\partial x_k}+\frac{\partial m_{ijk}}{\partial x_k}=-\frac{\sigma_{ij}}{\tau}, \tag{4}$$

$$\begin{aligned}&\frac{\partial q_i}{\partial t}+\frac{\partial q_i\upsilon_k}{\partial x_k}+\frac{5}{2}p\frac{\partial \theta}{\partial x_i}+\frac{5}{2}\sigma_{ik}\frac{\partial \theta}{\partial x_k}+\theta\frac{\partial \sigma_{ik}}{\partial x_k}-\sigma_{ik}\theta\frac{\partial \rho}{\partial x_k}-\frac{\sigma_{ij}}{\rho}\frac{\partial \sigma_{jk}}{\partial x_k}+\\ &+\frac{7}{5}q_k\frac{\partial \upsilon_i}{\partial x_k}+\frac{2}{5}q_k\frac{\partial \upsilon_k}{\partial x_i}+\frac{2}{5}q_i\frac{\partial \upsilon_k}{\partial x_k}+\frac{1}{2}\frac{\partial R_{ik}}{\partial x_k}+\frac{1}{6}\frac{\partial \Delta}{\partial x_i}+m_{ijk}\frac{\partial \upsilon_j}{\partial x_k}=-\frac{2}{3}\frac{q_i}{\tau},\end{aligned} \tag{5}$$

where the mass density ρ, velocity υ_i, temperature in energy units $\theta=\frac{k}{m}T$ (k is the Boltzmann constant and m the particle mass), trace-free viscous stress tensor σ_{ij} (with σ_{ii} = 0), and heat flux $q_i\,(i=x,y,z)$ form 13 primitive variables. The pressure is given by the ideal gas law $p=\rho\theta$, and $\tau=\mu/p$ is the relaxation time, with the viscosity coefficient μ. The viscosity in this study is defined by combined law [24]

$$\mu=\begin{cases}\mu(T_*)\left(\dfrac{T}{T_*}\right)^a, T<T_*\\[2ex] \mu(T_*)\left(\dfrac{T}{T_*}\right)^{3/2}\dfrac{T_*+S}{T+S}, T\geq T_*\end{cases}, \tag{6}$$

where T_* is the critical temperature ($T_* = 150K$, $S = 128.35K$ and $a = 0.945$ for argon) [24]. The angular brackets in the subscripts indicate the trace-free and symmetric part of the tensor [11]. Equations (1) – (3) are the conservation laws for mass, momentum and energy; equations (4) and (5) are the moment equations for stress tensor and heat flux vector, respectively. These 13 equations must be closed by constitutive relations for the higher moments R_{ij}, Δ, m_{ijk}, and these differ based on the method of (regularizing) closure [19,20]. For Grad's original 13 moment equations [13], $R_{ij} = \Delta = m_{ijk} = 0$. Higher-order moments have the following analytical form [15]:

$$m_{ijk} = -2\tau\left[\frac{k}{m}T\frac{\partial\sigma_{\langle ij}}{\partial x_{k\rangle}} - \frac{k}{m}T\sigma_{\langle ij}\frac{\partial \ln\rho}{\partial x_{k\rangle}} + \frac{4}{5}q_{\langle i}\frac{\partial \upsilon_j}{\partial x_{k\rangle}} - \frac{\sigma_{\langle ij}}{\rho}\frac{\partial\sigma_{k\rangle l}}{\partial x_l}\right], \tag{7}$$

$$\begin{aligned} R_{ij} = -\frac{24}{5}\tau\Bigg[&\frac{k}{m}T\frac{\partial q_{\langle i}}{\partial x_{j\rangle}} + \frac{k}{m}q_{\langle i}\frac{\partial T}{\partial x_{j\rangle}} + \frac{5}{7}\frac{k}{m}T\left(\sigma_{k\langle i}\frac{\partial \upsilon_{j\rangle}}{\partial x_k} + \sigma_{ki}\frac{\partial \upsilon_k}{\partial x_j} - \frac{2}{3}\sigma_{ij}\frac{\partial \upsilon_k}{\partial x_k}\right) - \\ &-\frac{k}{m}Tq_{\langle i}\frac{\partial \ln\rho}{\partial x_{j\rangle}} - \frac{1}{\rho}q_{\langle i}\frac{\partial\sigma_{j\rangle k}}{\partial x_k} - \frac{5}{6}\frac{\sigma_{ij}}{\rho}\frac{\partial q_k}{\partial x_k} - \frac{5}{6}\frac{\sigma_{ij}}{\rho}\sigma_{kl}\frac{\partial \upsilon_k}{\partial x_l}\Bigg], \end{aligned} \tag{8}$$

$$\Delta = -12\tau\left[\frac{k}{m}T\frac{\partial q_k}{\partial x_k} + \frac{5}{2}\frac{k}{m}q_k\frac{\partial T}{\partial x_k} - \frac{k}{m}Tq_k\frac{\partial \ln\rho}{\partial x_k} + \frac{1}{\rho}q_j\frac{\partial\sigma_{jk}}{\partial x_k} + \frac{k}{m}T\sigma_{ij}\frac{\partial \upsilon_i}{\partial x_j}\right]. \tag{9}$$

In this paper we studied the behavior of the linear version [21] of this system. In the case, the calculation of high-order moments involves only terms that are responsible for the so-called gradient transport mechanism (GTM) [25] for the stress tensor and heat flux. It is these terms that ensure stabilization of the original Grad's system of equations. The remaining terms omitted in the linear case form the so-called non-gradient transport mechanism (NGTM) [25].

In the linear case (GTM), higher-order moments have the following form:

$$m_{ijk} = -2\tau\frac{k}{m}T\frac{\partial\sigma_{\langle ij}}{\partial x_{k\rangle}},\ R_{ij} = -\frac{24}{5}\tau\frac{k}{m}T\frac{\partial q_{\langle i}}{\partial x_{j\rangle}},\ \Delta = -12\tau\frac{k}{m}T\frac{\partial q_l}{\partial x_l}. \tag{10}$$

The NSF equations of classical hydrodynamics can be obtained from the above by Chapman-Enskog expansion, which yields the Navier-Stokes and Fourier laws as

$$\sigma_{ij}^{NSF} = -2\mu\frac{\partial v_{\langle i}}{\partial x_{j\rangle}},\ q_i^{NSF} = -\frac{15}{4}\mu\frac{\partial\theta}{\partial x_i}. \tag{11}$$

In 2006 M. Torrilhon in work [21] has offered the divergent form of the R13 equations for the two-dimensional case. For the two-dimensional case (the number of the equations then decreases to nine) this system of the equations can be written down as follows:

$$\frac{\partial U(W)}{\partial t} + divF(W) = P(W),\ W = \left\{\rho, \upsilon_x, \upsilon_y, p, p_x, p_y, \sigma, q_x, q_y\right\}^T, \tag{12}$$

where W is the vector of primitive variables, $\sigma = \sigma_{xy}$, $p = \left(p_x + p_y + p_z\right)/3$. Vectors $U(W)$, $F(W)$ and $P(W)$ are vectors of conservative variables, fluxes and relaxation terms accordingly. The form of these vectors is presented in [21].

2.2. Boundary conditions

The first attempt to derive the boundary conditions on a solid wall for R13 system was done by Gu and Emerson in [26]. Further a similar method was used by Struchtrup and Torrilhon to derive their version of the boundary conditions [27] on the solid wall. In both cases, Maxwell's accommodative model [28] was the base for derivation these boundary conditions. The idea of such formulation of the boundary conditions is that some part (1–α) of the molecules reflects specularly from the wall. The rest α part of the molecules diffusely reflects with a Maxwellian distribution f_M^w at the temperature of the wall. As result the compound «wall» distribution function is as follows:

$$f^{w}(\upsilon_x,\upsilon_y,\upsilon_z)=\begin{cases}\alpha f_M^{w}+(1-\alpha)f(\upsilon_x,-\upsilon_y,\upsilon_z), & \upsilon_y\geq 0\\ f(\upsilon_x,\upsilon_y,\upsilon_z), & \upsilon_y\leq 0\end{cases}, \tag{13}$$

Then, in order to obtain conditions for each moment on the wall, it is necessary to make the integration of the equations for the corresponding weight of each moment. After integration the following boundary conditions for the solid isothermal wall were obtained [27]:

$$u=u_w-\frac{1}{p_\alpha}\left(n_y\sqrt{\frac{\pi}{2}\theta}\,\frac{2-\alpha}{\alpha}\sigma_{xy}+\frac{m_{xyy}}{2}+\frac{q_x}{5}\right), \tag{14}$$

$$\theta=\theta_w-\frac{1}{p_\alpha}\left(n_y\sqrt{\frac{\pi}{2}\theta}\,\frac{2-\alpha}{\alpha}\frac{q_y}{2}+\frac{\theta\sigma_{yy}}{4}+\frac{5R_{yy}}{56}+\frac{\Delta}{30}\right)+\frac{(u-u_w)^2}{4}, \tag{15}$$

$$R_{xy}=n_y\sqrt{\frac{2}{\pi\theta}}\frac{\alpha}{2-\alpha}\left[p_\alpha\theta u_s-\frac{1}{2}\theta m_{xyy}-\frac{11}{5}\theta q_x-p_\alpha(u-u_w)^3+6p_\alpha(\theta-\theta_w)(u-u_w)\right], \tag{16}$$

$$m_{xxy}=-n_y\sqrt{\frac{2}{\pi\theta}}\frac{\alpha}{2-\alpha}\left[\frac{p_\alpha}{5}(\theta-\theta_w)-\frac{4p_\alpha}{5}(u-u_w)^2+\frac{R_{xx}}{14}+\theta\left(\sigma_{xx}-\frac{\sigma_{yy}}{5}\right)+\frac{\Delta}{150}\right], \tag{17}$$

$$m_{yyy}=n_y\sqrt{\frac{2}{\pi\theta}}\frac{\alpha}{2-\alpha}\left[\frac{2p_\alpha}{5}(\theta-\theta_w)-\frac{R_{yy}}{14}+\frac{\Delta}{75}-\frac{7}{5}\theta\sigma_{yy}-\frac{3p_\alpha}{5}(u-u_w)^2\right], \tag{18}$$

where n_y is the normal to wall surface directed into gas, α is the factor of accommodation of a surface, u_w and θ_w are the tangential velocity and wall temperature respectively.

These equations represent the five so-called kinetic boundary conditions. The integrals for the moments, which weight is proportional to the weight of even degree of the normal component of particle velocity are equal to zero [27]. The first equation determines the slip velocity on the wall, the second boundary condition - the temperature jump. The remaining equations are expressions for the moments of higher order (the new moments in comparison with Grad's 13-moment system). Missing relations for the determination of the remaining variables can be obtained from the main flow equations. The question of formulating proper boundary conditions for the R13 equations is under ongoing investigation [29].

The variant of kinetic boundary conditions [30] for NSF equations is used for numerical simulation of solid wall [30]

$$u=u_w+\frac{2-\alpha}{\alpha}\lambda\frac{\partial u}{\partial y}+\frac{3}{4}\frac{\mu}{\rho T}\frac{\partial T_w}{\partial x} \tag{19}$$

$$T=T_w+\frac{2-\alpha}{\alpha}\frac{2\gamma}{\gamma+1}\frac{\lambda}{\mathrm{Pr}}\frac{\partial T}{\partial y} \tag{20}$$

where λ is the mean of the average free path, γ is the ratio of specific heats and Pr is Prandtl number.

3. Numerical scheme

Implemented numerical method for R13 and NSF equations is a version of high resolution Godunov method with a linear reconstruction of the flow parameters [31]. To approximate both systems of equations we introduce a regular computational grid consisting of convex quadrilateral cells.

The second-order accuracy in space for smooth solutions is achieved using essentially two-dimensional reconstruction procedures [32,33] for the primitive variables within each computational cell. For a discretization "elliptical" part of the fluxes (diffusion terms) the standard central difference approximation is used.

A modified second order Runge-Kutta method [32-34] is used to approximate the equation system in time. So the divergent term is approximated explicitly, and the relaxation term - implicitly. Therefore, in each cell at each time stage of the method a system of nonlinear equations has to be

IOP Conf. Series: Journal of Physics: Conf. Series **1009** (2018) 012033 doi:10.1088/1742-6596/1009/1/012033

solved. But it doesn't increase computational time too much, because this nonlinear system can be easily solved due to structure of the relaxation terms.
Numerical scheme based on Newton's method is used for solid wall boundary conditions modeling for R13 equations [32-34].

4. Numerical results

4.1. Formulation of the problem
The two dimensional argon nozzle flow is considered. The half-angle of the solution of the nozzle confuser part is 30^0 and the half-angle of the diffuse one is 45^0. The width of critical cross section is 2 mm. The input temperature is 300 K. The ratio of input and output pressure is equal to 10^3. The numerical results for both methods have been obtained for three variants of input pressure ($p_{inp} = 10$ atm., $p_{inp} = 1$ atm. and $p_{inp} = 10^{-1}$ atm.)

4.2. Continuum and slip flow regime
The numerical results for the variant of deep continuum regime (with $p_{inp} = 10$ atm.) are almost the same for R13 and NSF approaches. This fact just proves that R13 equations give the same physical result for supersonic nozzle flow in continuum regime as NSF equations. With the increase of input pressure in ten times the small difference in Mach number distribution fields can be seen (see figure 1). The flow inside of the nozzle is same and the difference grows with the distance from the nozzle edge.

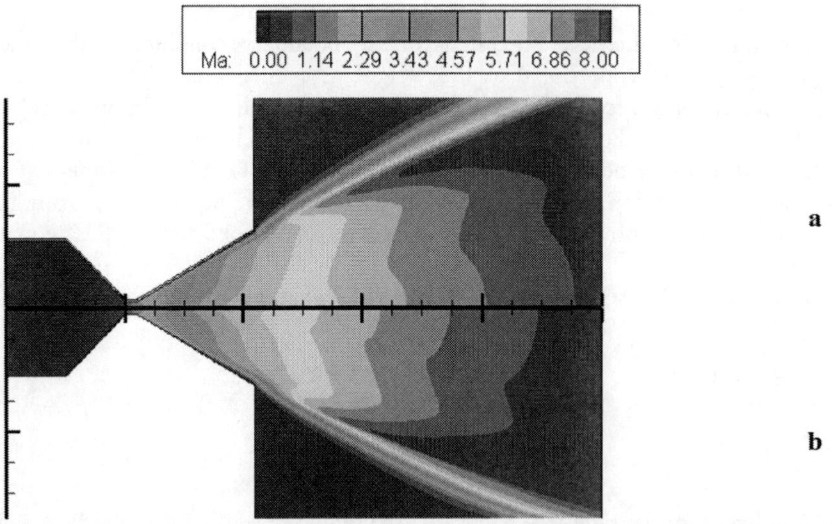

Figure 1. Mach number distribution for NSF (a) and R13 (b) equations for $p_{inp} = 1$ atm.

The figure 2 presents the field of local Knudsen number for this case counted by the average free path for Maxwell molecules

$$Kn = \frac{\lambda}{h} = \sqrt{\frac{\pi}{2}} \frac{\mu(\theta)\sqrt{\theta}}{ph} \tag{21}$$

where h is the critical cross section size. As it can be seen, Knudsen number is moderate for the applicability of NSF approach here. The area of moderately rarefied gas ($Kn > 0.1$) is outside of the interest for the nozzle flow problem here.

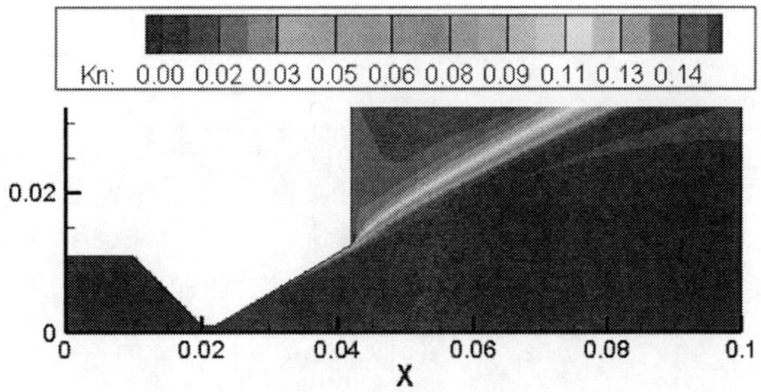

Figure 2. The distribution of Knudsen number for p_{inp} = 1 atm.

4.3. Transition regime

The further increasing of pressure ($p_{inp} = 10^{-1}$ atm.) has obvious influence to the flow in the low pressure chamber (figure 3). The difference between numerical results can be seen inside the nozzle. The width of boundary layers is different. So we have the difference of Mach number function along the symmetry plane. The field of local Knudsen number (figure 4) shows that rarefaction nonequilibrium effect can be observed inside the nozzle geometry. The level of $Kn = 0.1$ (formal boundary of transitional regime) is overstepped for significant part of considered flow.

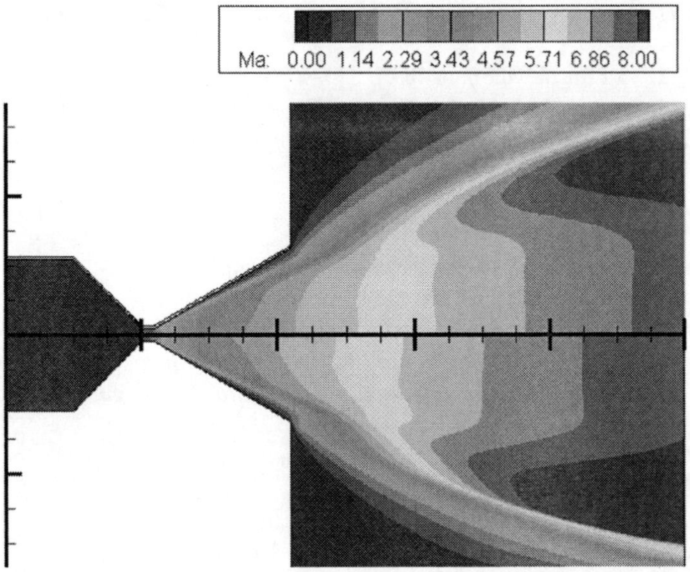

a

b

Figure 3. Mach number distribution for NSF (a) and R13 (b) equations for p_{inp} = 0.1 atm.

5. Conclusion

A numerical method for regularized 13-moment equations has been tested for supersonic nozzle flow. The numerical results presented here show the capability of R13 equations to extend the area of continuum approach applicability. In contrast with the kinetic methods, this approach is applicable for as an efficient numerical tool for simulation of transitional-regime gas flows in micro scale devices.

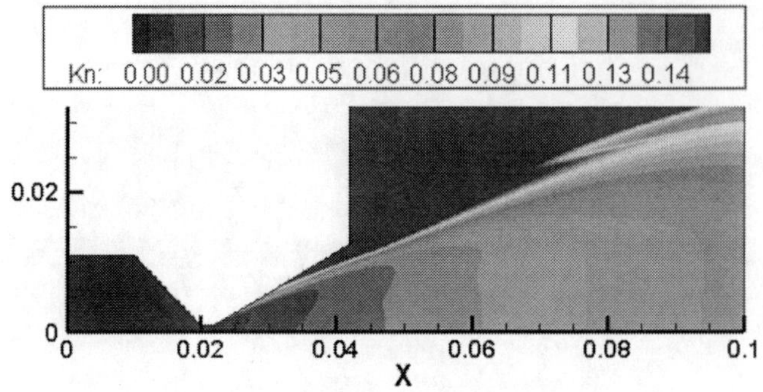

Figure 4. The distribution of Knudsen number for $p_{inp} = 0.1$ atm.

Acknowledgments
This work was supported by the Russian Foundation for Basic Research (projects No. 16-31-60034 and 16-01-00444).

References

[1] Kogan M N 1969 *Rarefied gas dynamics* (Plenum, New York)
[2] Cercignani C 1988 *The Boltzmann equation and its applications* (New York: Springer-Verlag)
[3] Aristov V V and Cheremisin FG 1980 *USSR Comp. Math. Math. Phys.* **20** pp 208–225
[4] Bhatnagar P L, Gross E P and Krook M A 1954 *Phys. Rev.* **94** pp 511–525
[5] Shakhov E M 1974 *The Method of Investigation of Motion of a Rarefied Gas* (Moscow: Nauka) (in Russian)
[6] Larina I N and Rykov V A 2013 Fluid Dynamics **48** pp 837–849
[7] Kudryavtsev A N and Shershnev A A 2013 *J. Sci. Comput* **57** pp 42–73
[8] Bird G A 1994 *Molecular Gas Dynamics and The Direct Simulation of Gas Flows* (Oxford: Oxford University Press)
[9] Ivanov M S, Kashkovsky A V, Vashchenkov P V and Bondar Ye A 2011 *AIP Conf. Proc.* **1333** pp 211–218
[10] Malkov E A, Bondar Ye A, Kokhanchik A A, Poleshkin S O and Ivanov M S 2015 *Shock Waves* **25** pp 387-397
[11] Struchtrup H 2005 *Macroscopic transport equations for rarefied gas flows* (Springer)
[12] Chapman S and Cowling T G 1991 *The mathematical theory of non-uniform gases: an account of the kinetic theory of viscosity, thermal conduction and diffusion in gases* (Cambridge mathematical library)
[13] Grad H 1949 On the kinetic theory of rarefied gases *Comm. Pure Appl. Math.* **2** pp 331–407
[14] Torrilhon M 2016 *Annu. Rev. Fluid Mech.* **48** pp 429–458
[15] Struchtrup H and Torrilhon M 2003 *Phys. Fluids* **15** pp 2668–2680
[16] Torrilhon M and Struchtrup H 2004 *J. Fluid Mech* **513** pp 171–198
[17] Ivanov I E, Kryukov I A, Timokhin M Yu, Bondar Ye A, Kokhanchik A A and Ivanov M S 2012 *AIP Conf. Proc.* **1501** pp 215–222
[18] Timokhin M Yu, Bondar Ye A, Kokhanchik A A, Ivanov M S, Ivanov I E and Kryukov I A 2015 *Phys. Fluids* **27** 037101
[19] Timokhin M Yu, Struchtrup H, Kokhanchik A A and Bondar Ye A 2016 *AIP Conf. Proc.* **1786** 140006
[20] Timokhin M Yu, Struchtrup H, Kokhanchik A A and Bondar Ye A 2017 *Phys. Fluids* **29** 037105
[21] Torrilhon M 2006 *Multiscale Model. Simul.* **5** pp 695–728
[22] Znamenskaya I A, Ivanov I E, Kryukov I A, Mursenkova I V and Timokhin M Yu 2014 *Tech. Phys. Lett.* **40** pp 533–536

IOP Conf. Series: Journal of Physics: Conf. Series **1009** (2018) 012033 doi:10.1088/1742-6596/1009/1/012033

[23] Shoev G, Kokhanchik A A, Timokhin M Yu and Bondar Ye A 2017 *30th International Symposium on Shock Waves* **1** pp 685–689
[24] Dubov D Yu 2013 *Poceedings of All-Russian conference on rarefied gas dynamic problems* pp 85–87 (in Russian)
[25] Gu X J and Emerson D R 2009 *J. Fluid Mech.* **636** pp 177–216
[26] Gu X J and Emerson D R 2007 *J. Comput. Phys.* **225** pp 263–283
[27] Struchtrup H and Torrilhon M 2008 *J. Comput. Phys.* **227** pp 1982–2011
[28] Maxwell J. C. 1879 *Phil. Trans. R. Soc. (London)* **170** pp 231–256
[29] Rana A S and Struchtrup H 2016 *Phys. Fluids* **28** 027105
[30] Roy S, Raju R, Chuang H F, Cruden B A and Meyyappan M 2003 *J. Applied Phys.* **93** pp 4870–4879
[31] Glushko G S, Ivanov I E and Kryukov I A 2010 *Math. Model. Comput. Simul.* **2** pp 407–422
[32] Timokhin M Yu, Ivanov I E and Kryukov I A 2012 *AIP Conf. Proc.* **1501** pp 843–848
[33] Ivanov I E, Kryukov I A and Timokhin M Yu 2013 *Comput. Math. Math. Phys.* **53** pp 1534–1550
[34] Timokhin M Yu, Ivanov I E and Kryukov I A 2014 *AIP Conf. Proc.* **1628** pp 748–755

Numerical study of the flow structure in the supersonic inlet-isolator

R K Seleznev [1,2,3]

[1] Dukhov Research Institute of Automatics (VNIIA), Suschevskaya Street 22, Moscow 127055, Russia
[2] Moscow Institute of Physics and Technology (State University), 9 Institutskiy per., Dolgoprudny, Moscow Region, 141701, Russia
[3] Ishlinsky Institute for Problems in Mechanics RAS, Vernadsky Street 101(1), Moscow, 119526, Russia

rkseleznev@gmail.com

Abstract. In the paper three-dimensional model for hypersonic flow with turbulence model are described. Comparing the result of calculation with wall pressure experimental data, obtained in the inlet-isolator model mounted to the floor of a Mach 5 wind tunnel at the University of Texas, presented in this article. The influence of flow parameters such as the thickness of the boundary layer and turbulence models on the flow character is investigated in this paper. It was shown that a correct description of turbulence significantly affects the structure of the flow, and its consideration is critically important under comparing the calculation results with the experimental data.

1. Introduction

Development of high-speed aircraft has led to the need to create experimental installations and numerical models [1–5] for studying the process in the ramjet and scramjet. The inlet is an important part of ramjet and scramjet, and their design greatly affects the overall performance of the engine. The isolator is also an essential part of the ramjet and scramjet. The isolator is a constant cross-section passage in order to prevent chocking. There are many experiments with inlets [6][7][8–10]. In this paper we consider the model of inlet-isolator mounted to the floor of a Mach 5 wind tunnel at the University of Texas [8,10,11]. In this experiment, it is important to consider the thickness of the boundary layer at the inlet. In this paper, numerical experiments are carried out that take into account the thickness of the boundary layer, as well as experiments with a uniform flow at the inlet. In this paper, we study the influence of various factors on the flow structure. The following factors are chosen as the investigated factors: the dimension of the problem, the wall temperature, the size of the computational grid, and also the influence of the turbulence model. The results of the calculation are compared with the results obtained by other authors [12–16], as well as with the results of the experiment.

2. Description of University of Texas inlet-isolator experimental setup

Wagner [8,10,11] conducted a series of experiments on the study of shock-wave structure in the supersonic inlet-isolator mounted to the floor of a Mach 5 wind tunnel (figure 1). The experimental setup consists of an input ramp with an inclination angle of 6°, and a constant-section isolator (figure 2). At the end of the isolator a special flap is installed, raising it can lead to nonstationary formation with subsequent choking of the flow. With the flap is completely lowered, a small part of it protrudes from the bottom wall of the chamber, as shown in figure 2. However, in this paper, as in the works of other authors[9–13], it is considered that the floor is completely smooth. The following incoming air flow parameters were used:

Incoming flow pressure: $P = 0.0538$ atm;
Incoming flow temperature: $T = 57.4$ K;
Incoming flow Much Number: $M = 4.9$;
Incoming flow gas mixture: Air.

Figure 1. Schematic of the model of the inlet/isolator model mounted on the wind tunnel floor [8].

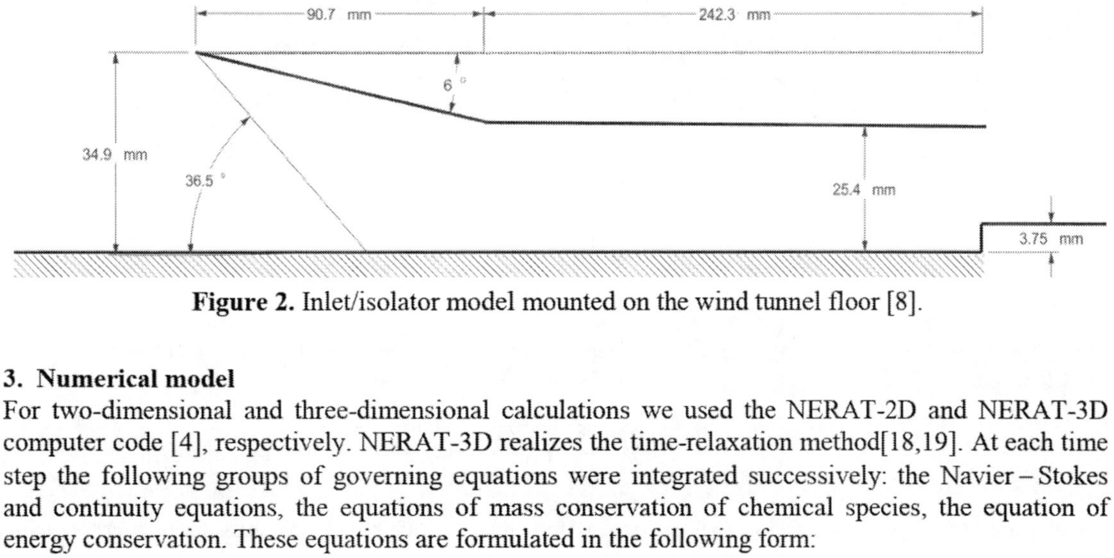

Figure 2. Inlet/isolator model mounted on the wind tunnel floor [8].

3. Numerical model

For two-dimensional and three-dimensional calculations we used the NERAT-2D and NERAT-3D computer code [4], respectively. NERAT-3D realizes the time-relaxation method[18,19]. At each time step the following groups of governing equations were integrated successively: the Navier–Stokes and continuity equations, the equations of mass conservation of chemical species, the equation of energy conservation. These equations are formulated in the following form:

$$\frac{\partial \rho}{\partial t} + \mathrm{div}(\rho \mathbf{V}) = 0 \,, \tag{1}$$

$$\frac{\partial \rho \mathbf{V}}{\partial t} + div\left(\rho \mathbf{V}\mathbf{V} + \hat{\mathbf{\Pi}}\right) = 0 \,, \tag{2}$$

$$\rho c_p \frac{\partial T}{\partial t} + \rho c_p \mathbf{V}\,\mathrm{grad}\,T = \mathrm{div}\left(\lambda\,\mathrm{grad}\,T\right) + \sum_{i=1}^{N_s} \rho c_{p,i} D_i \left(\mathrm{grad}\,Y_i \cdot \mathrm{grad}\,T\right) - \\ - \sum_{i=1}^{N_s} h_i \dot{w}_i + \frac{\partial p}{\partial t} + \mathbf{V}\,\mathrm{grad}\,p + \Phi_\mu \tag{3}$$

$$\frac{\partial \rho_i}{\partial t} + \mathrm{div}\,\rho_i \mathbf{V} = -\mathrm{div}\,\mathbf{J}_i + \dot{w}_i, \quad i = 1,2,\ldots,N_s \tag{4}$$

where: t – time; $\mathbf{V} = \mathbf{i}u + \mathbf{j}v + \mathbf{k}w$ – velocity vector with projections on the axis of the Cartesian coordinate system x, y, z; p, ρ – pressure and density; T – temperature of translational movement of particles; μ, λ – dynamic coefficient of viscosity and coefficient of thermal conductivity; c_p – specific heat of the mixture at constant pressure; $c_p = \sum_i^{N_s} Y_i c_{p,i}$; N_s – number of gas mixture components; Y_i – mass fraction of the i-th component of the mixture; $c_{p,i}, h_i$ – the specific heat at constant pressure associated with translational and rotational degrees of freedom, and the enthalpy of the i-th component of the mixture; $\dot{w}_i$ – the mass velocity of chemical transformations for the i-th component of the mixture; D_i – the effective diffusion coefficient of the i-th component of the mixture; $\mathbf{J}_i$ – the diffusion flux density of the i-th component; $\mathbf{J}_i = -\rho D_i\,\mathrm{grad}\,Y_i$; N_s number of chemical components of the gas mixture. The components of the viscous stress tensor and the dissipative function were calculated by the formulas:

$$\Pi_{i,j,k} = -p\delta_{i,j,k} + \mu\left[\left(\frac{\partial u_i}{\partial x_j} + \frac{\partial u_j}{\partial x_i}\right) - \frac{2}{3}\delta_{i,j}\frac{\partial u_k}{\partial x_k}\right], \; i,j,k = 1,2,3$$

$$\Phi_\mu = \mu\left[2\left(\frac{\partial u}{\partial x}\right)^2 + 2\left(\frac{\partial v}{\partial y}\right)^2 + 2\left(\frac{\partial w}{\partial z}\right)^2 + \right. \\ \left. + \left(\frac{\partial v}{\partial x} + \frac{\partial u}{\partial y}\right)^2 + \left(\frac{\partial w}{\partial y} + \frac{\partial v}{\partial z}\right)^2 + \left(\frac{\partial u}{\partial z} + \frac{\partial w}{\partial x}\right)^2 - \frac{2}{3}\left(\frac{\partial u}{\partial x} + \frac{\partial v}{\partial y} + \frac{\partial w}{\partial z}\right)^2\right].$$

The closing relations for the system of equations to be solved include the thermal equation of state of an ideal gas:

$$\frac{p}{\rho} = \frac{R_0}{M_\Sigma}T, \; \frac{1}{M_\Sigma} = \sum_i^{N_s} \frac{Y_i}{M_i} \tag{5}$$

The system of equations (1)–(5) was integrated numerically by the establishment method using computer code NERAT-3D. Equations (1)–(2) were integrated by an explicit finite-difference method according to the AUSM scheme [20]. The diffusion equations (4) and energy conservation were solved using an implicit finite-difference scheme of the second order of the Crank-Nicholson approximation. Numerical solution of finite-difference equations was carried out using multi-block technology. The number of blocks used in this technology is determined by the complexity of the geometry of the calculation area. In this case, 8 blocks structured of the grid were used. Chemical reactions are neglected.

IOP Conf. Series: Journal of Physics: Conf. Series **1009** (2018) 012034 doi:10.1088/1742-6596/1009/1/012034

4. Thickness of boundary layer

In the Wagner experiment observed boundary layer entering the air inlet has a thickness $\delta = 19.3$ mm, so for the correct description of the experiment is necessary to consider this fact. In this works [12–16] the simulation was run until the flow became turbulent and a target thickness for the boundary layer at the exit was reached. However, there is another approach proposed in [14] in which for forming the profile of the boundary layer of a given thickness in the cross-section at the inlet the well-known formulas for the boundary layer on a thick plate used together with the modification from the work [21]. Let x be the longitudinal coordinate, and y - the distance to the plate. We assume that the pressure is constant along the cross-section x = const, the longitudinal component of the velocity has a power-law dependence, the temperature is found according to the formula proposed by Walz [22], where r – is the flux recovery constant is expressed in terms of the wall temperature T_w [21]. Then the density is recalculated from the equation of state, i.e.

$$p(y) = p_\infty, \tag{6}$$

$$u(y,\delta) = u_\infty \left(\frac{y}{\delta}\right)^{\frac{1}{7}}, \tag{7}$$

$$T(y,\delta) = T_\infty \left[r\frac{\gamma-1}{2}M_\infty^2 \left(1 - \left(\frac{u(y,\delta)}{u_\infty}\right)^2\right) + 1\right], \tag{8}$$

$$r = \frac{T_w - T_\infty}{T_\infty M_\infty^2 (\gamma - 1)}, \tag{9}$$

$$\rho(y,\delta) = \frac{p(y)}{RT(y,\delta)}. \tag{10}$$

here δ – is the given thickness of the boundary layer. It is assumed that the transversal component of the velocity w is zero (that is, an effectively two-dimensional problem is considered). Then, for a boundary layer on a thick plate:

$$\frac{\delta(x)}{x} = 0.382\,\mathrm{Re}_x^{-\frac{1}{5}},\ \mathrm{Re}_x = \frac{\rho_\infty u_\infty x}{\mu_\infty}. \tag{11}$$

$$\delta(x) = 0.382\left(\frac{\mu_\infty}{\rho_\infty u_\infty}\right)^{\frac{1}{5}} x^{\frac{4}{5}}, \tag{12}$$

Substituting (12) into (7), we obtain:

$$u(y,x) = u_\infty \left(y \Big/ 0.382\left(\frac{\mu_\infty}{\rho_\infty u_\infty}\right)^{\frac{1}{5}} x^{\frac{4}{5}}\right)^{\frac{1}{7}}, \tag{13}$$

$$T(y,x) = T_\infty \left[r\frac{\gamma-1}{2}M_\infty^2 \left(1 - \left(y \Big/ 0.382\left(\frac{\mu_\infty}{\rho_\infty u_\infty}\right)^{\frac{1}{5}} x^{\frac{4}{5}}\right)^{\frac{2}{7}}\right) + 1\right], \tag{14}$$

$$\rho(y,x)=\frac{p_\infty}{RT_\infty\left[r\frac{\gamma-1}{2}M_\infty^2\left(1-\left(y\Big/0.382\left(\frac{\mu_\infty}{\rho_\infty u_\infty}\right)^{\frac{1}{5}}x^{\frac{4}{5}}\right)^{\frac{2}{7}}\right)+1\right]}. \quad (15)$$

The space derivative can be presented in the following form:

$$\left.\frac{\partial\rho u}{\partial x}\right|_{(x,y)}=\frac{\rho(x+h_x,y)u(x+h_x,y)-\rho(x,y)u(x,y)}{h_x}. \quad (16)$$

Then v can be found by numerical integration from the continuity equation:

$$-\left.\frac{\partial\rho u}{\partial x}\right|_{(x,y)}=\frac{\rho(x+h_y,y)v(x,y+h_y)-\rho(x,y)v(x,y)}{h_y}, \quad (17)$$

$$v(x,y+h_y)=-\frac{h_y}{h_x}\frac{\rho(x+h_x,y)u(x+h_x,y)-\rho(x,y)u(x,y)+\rho(x,y)v(x,y)}{\rho(x+h_y,y)}. \quad (18)$$

5. Turbulence model

In this paper, the algebraic turbulence model of Cebeci–Smith is used [23]. In a two-layer model, the boundary layer is considered to comprise two layers: inner (close to the surface) and outer. The kinematic eddy viscosity is calculated separately for each layer and combined using:

$$\nu_t=\min(\nu_{in},\nu_{out}). \quad (19)$$

The inner-region eddy viscosity is given by:

$$\nu_{in}=l_m^2\left[\left(\frac{\delta u}{\delta y}\right)^2+\left(\frac{\delta v}{\delta x}\right)^2\right]^{1/2}, \quad (20)$$

$$l_m=ky\left[1-\exp\left(-y^+\big/A^+\right)\right], \quad (21)$$

where: $A^+=26$, $k=0.4$, $y^+=\frac{y}{\nu}\sqrt{\frac{\tau_w}{\rho}}=\frac{y}{\nu}\sqrt{\nu\left(\frac{\partial u}{\partial y}\right)_w}$.

The eddy viscosity in the outer region is given by:

$$\nu_{out}=\alpha u_\infty\delta_v^*F_{Kleb}(y,\delta), \quad (22)$$

where: $\alpha=0.0168$, $F_{Kleb}(y,\delta)=\left[1+5.5\left(\frac{y}{\delta}\right)^6\right]^{-1}$. δ_v^* - is the displacement thickness, given by

$\delta_v^*=\int_0^\delta\left(1-\frac{u(y)}{u_\infty}\right)dy$. It should be noted that in this experiment $\delta_v^*=8.89$ mm [8,10,11].

6. Results

At the initial stage of calculations two-dimensional laminar code was used. At the entrance to the inlet a uniform flow is set. Figures 3–5 show the results of calculated density, transverse velocity and

IOP Conf. Series: Journal of Physics: Conf. Series **1009** (2018) 012034 doi:10.1088/1742-6596/1009/1/012034

pressure distribution within the experimental setup on structured mesh of 800 x142, respectively. A distinct shock-wave structure is seen that occurs when an incoming supersonic airflow interacts with an inlet. Figure 6 shows the results of calculated pressure distribution on the bottom wall obtained in this work with the results by Lutsky [14] and Zhukov[17]. It can be seen that the flow pattern is described qualitatively. A comparison of the results of the two-dimensional and three-dimensional code was also carried out. The results of such a comparison are shown in the figure 7. For three-dimensional laminar calculation, an eight-block structured grid of dimension 120x160x8 is used in each block. it can be seen that Zhukov's laminar calculations [17]. are closer to the results of two-dimensional calculations with the resolution of the first peak of pressure, and to the results of three-dimensional calculations at the resolution of the second peak of pressure. Only the results of the three-dimensional calculation will be presented below..

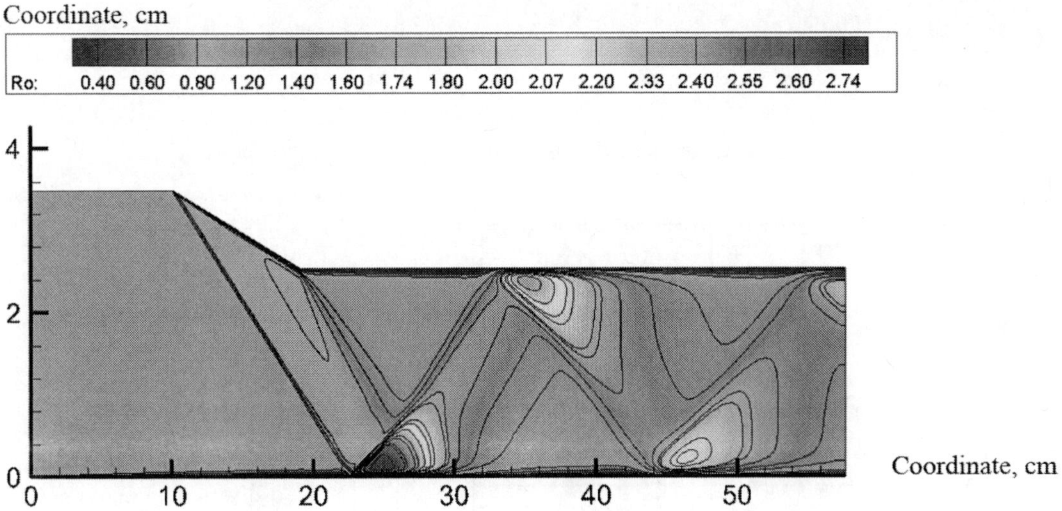

Figure 3. Density distribution within the experimental setup.

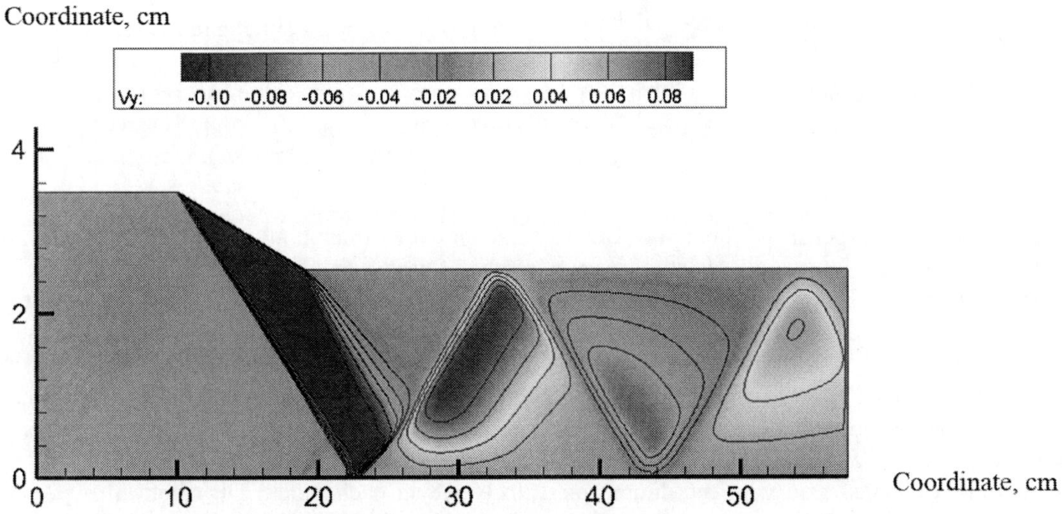

Figure 4. Transverse velocity distribution within the experimental setup.

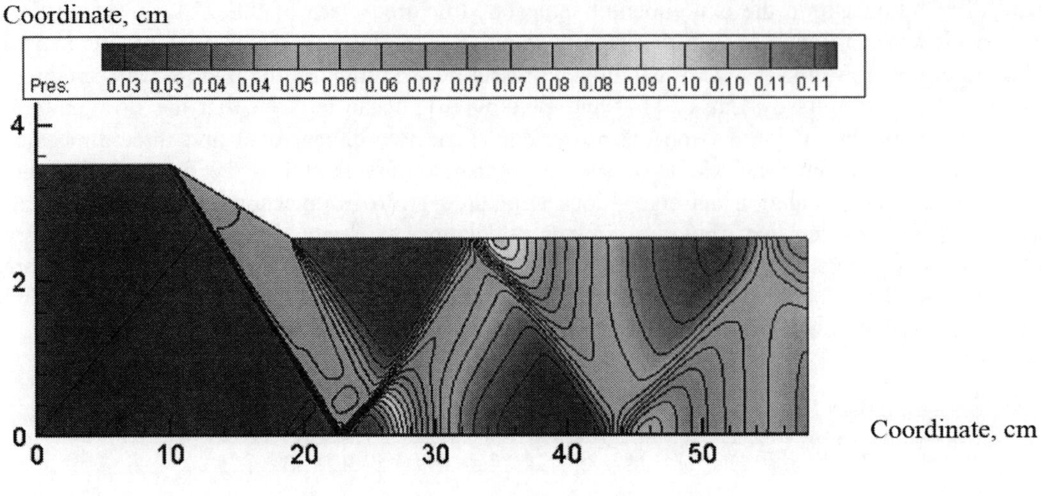

Figure 5. Pressure distribution within the experimental setup.

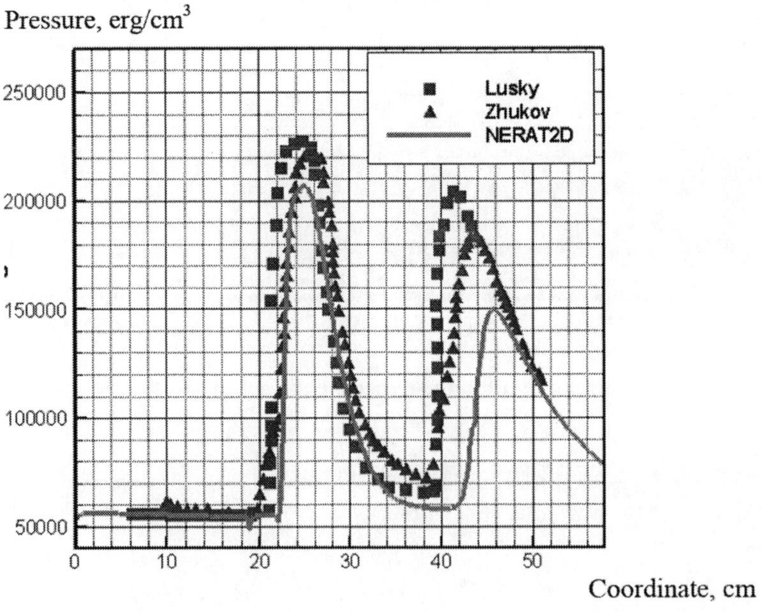

Figure 6. Comparison of calculated pressure distribution on the bottom wall obtained in this work (line) and the results obtained by Lutsky [14] (blue squares) and Zhukov[17] (green tangles).

However, in the experiment, the input flow was uneven with a thick boundary layer. The incoming boundary layer profiles were extracted from a method, based on well-known formulas for the boundary layer on a thick plate together with a modification from the work [14, 21]. At the first stage, a coarse calculation grid was used (30x40x8). The pressure distribution within the experimental setup is shown in figure 8. Comparison of calculated pressure distribution on the bottom wall and experimental results [8] is shown in figure 9. It should be noted that the calculated pressure amplitude is less than the experimental data. Most likely this is due to the fact that on a coarse grid numerical viscosity plays an essential role. Therefore, in the subsequent calculations we will use in the calculations a detailed grid with the dimension 120x160x8 in each block. Distribution of pressure within the experimental setup. Distribution of pressure within the experimental setup in this case is shown in figure 10. It should be noted that in the case of using this model without using the turbulence model, a vortex zone (figure 11) appears before the first shock wave, which changes the flow in the channel. The change in the structure of the flow in the channel is well evidenced by the distribution of pressure on the bottom wall, which is shown in figure 12. Therefore, in subsequent calculations, the Cebeci–Smith turbulence model will be used. In this case, the vortex before the first shock wave disappears, as demonstrated in figure 13. It may be noted that in this case, the bottom wall of the pressure increases, which is demonstrated in figure 14. This fact can be explained by the increased viscosity (the distribution of turbulent viscosity in this case is shown in figure 15). Increased viscosity slows down the main air flow, as shown in figure 16. In this paper, we also investigated the influence of the wall temperature on the flow structure. The calculation was carried out at a wall temperature equal to room temperature (Tw=300K) and equal to the flow temperature (Tw=57K). A comparison of these two cases is shown in figure 14. It can be seen that small differences appear in the second peak of pressure.

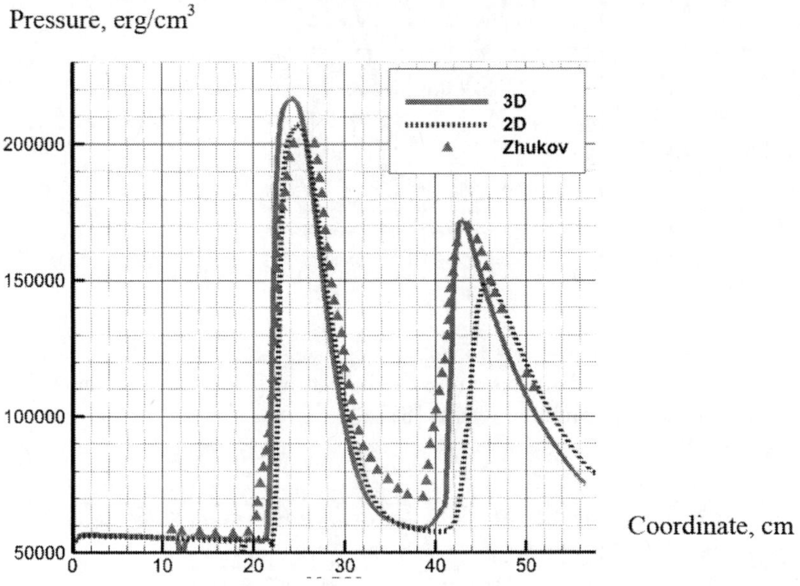

Figure 7. Comparison of calculated pressure distribution on the bottom wall obtained in this work by 2D (blue dots), 3D (line) and the results obtained by Zhukov [17] (green tangles).

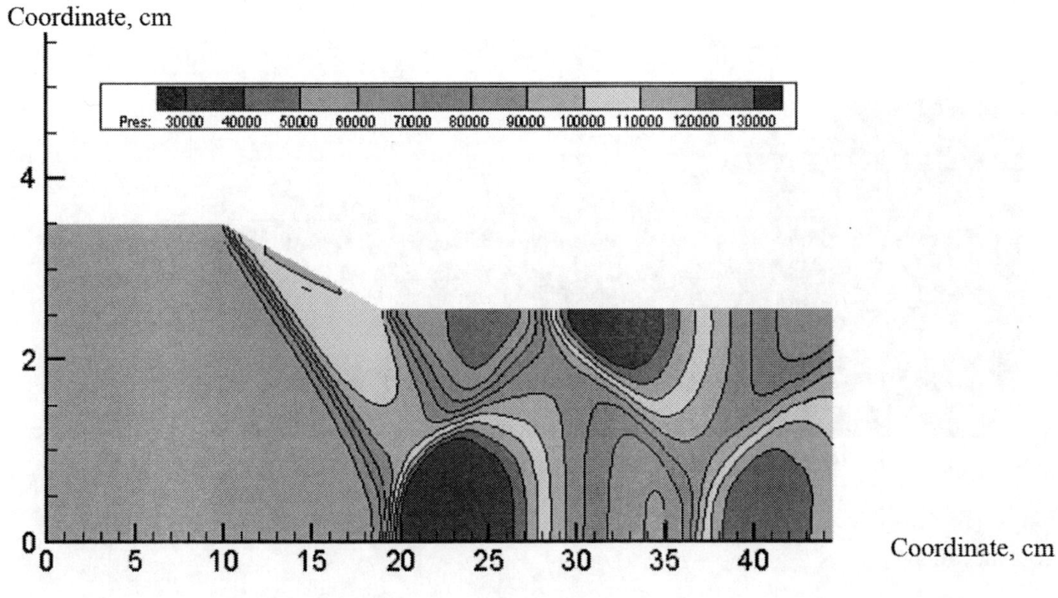

Figure 8. Pressure distribution within the experimental setup on coarse calculation grid.

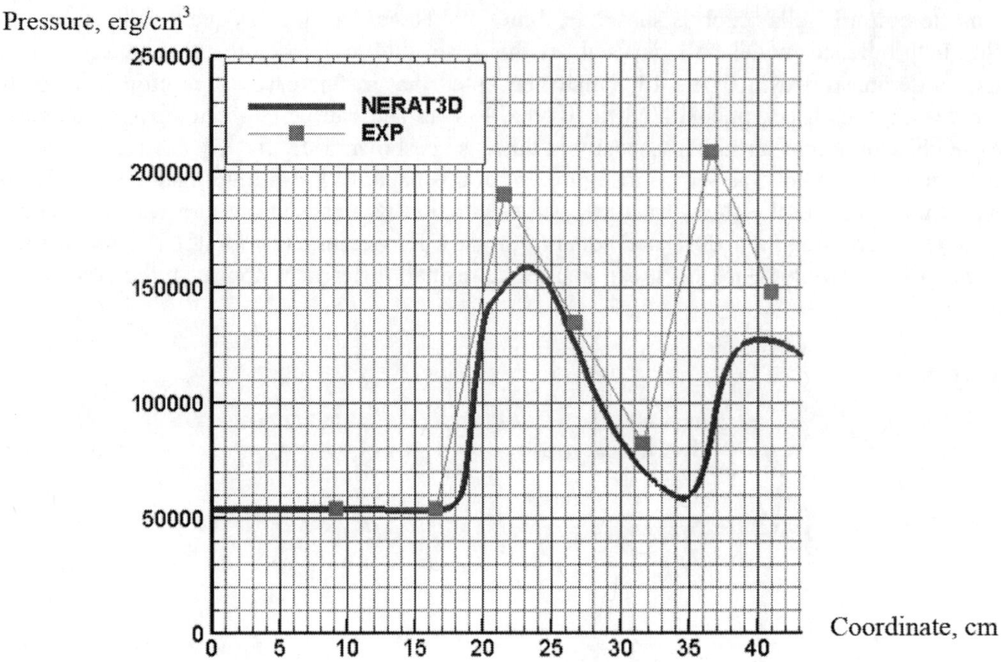

Figure 9. Comparison of calculated pressure distribution on the bottom wall obtained in this work (red line) on coarse calculation grid and experimental results [8].

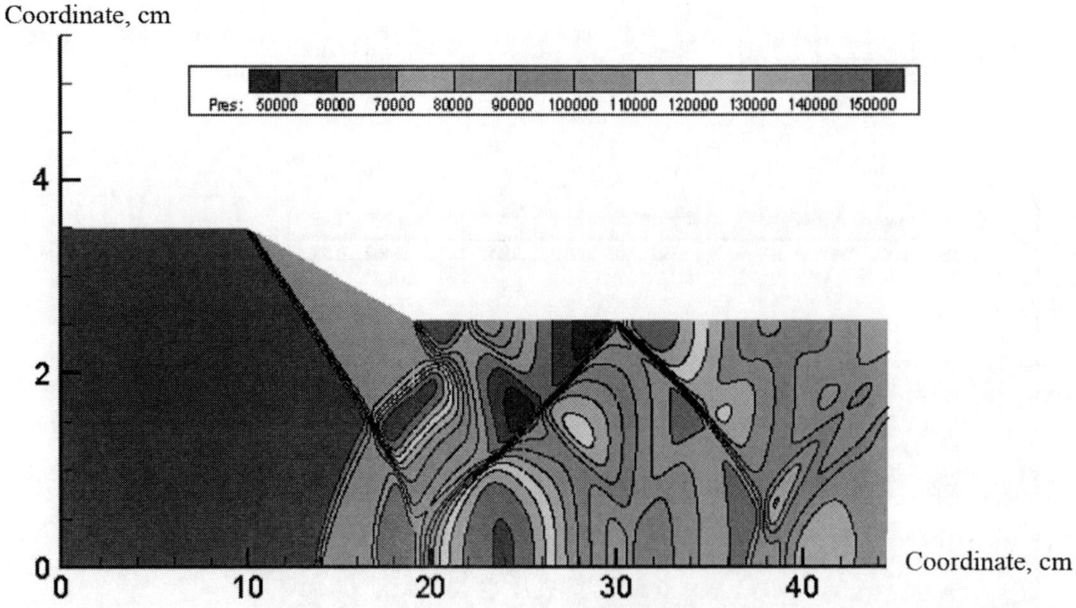

Figure 10. Pressure distribution within the experimental setup on a detailed grid without a turbulence model.

IOP Conf. Series: Journal of Physics: Conf. Series **1009** (2018) 012034 doi:10.1088/1742-6596/1009/1/012034

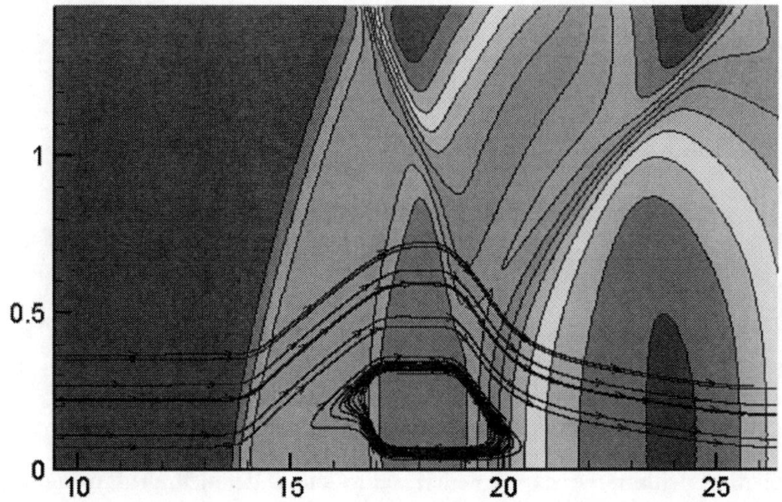

Figure 11. Vortex generation within the experimental setup on a detailed grid without a turbulence model.

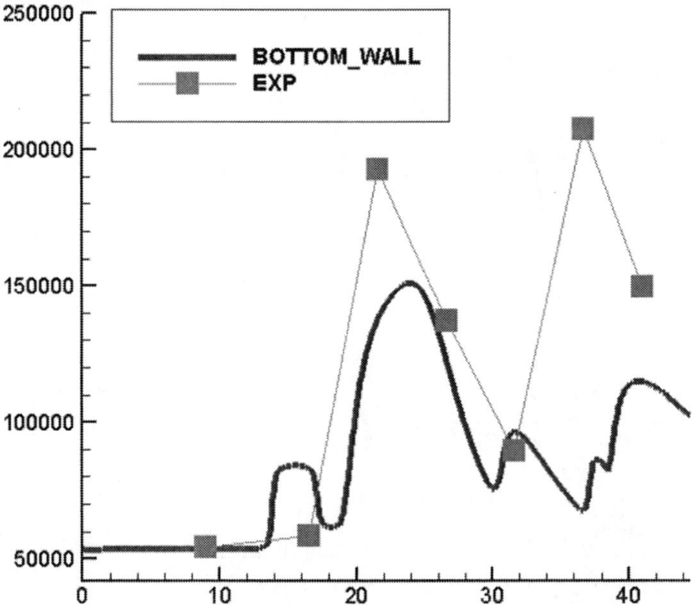

Figure 12. Comparison of calculated pressure distribution on the bottom wall obtained in this work on a detailed grid without a turbulence model. (red line) and experimental results [8].

Also a comparison with the results obtained by other authors was carried out. Figure 17 shows a comparison of the pressure distribution on the chamber wall obtained in this work (black line) with the experimental results [8,10] (green squares), and the results obtained by Lutsky [14] (red squares) and Koo[13,15] (blue delta), Edwards[16] and Jang [12].

IOP Conf. Series: Journal of Physics: Conf. Series **1009** (2018) 012034 doi:10.1088/1742-6596/1009/1/012034

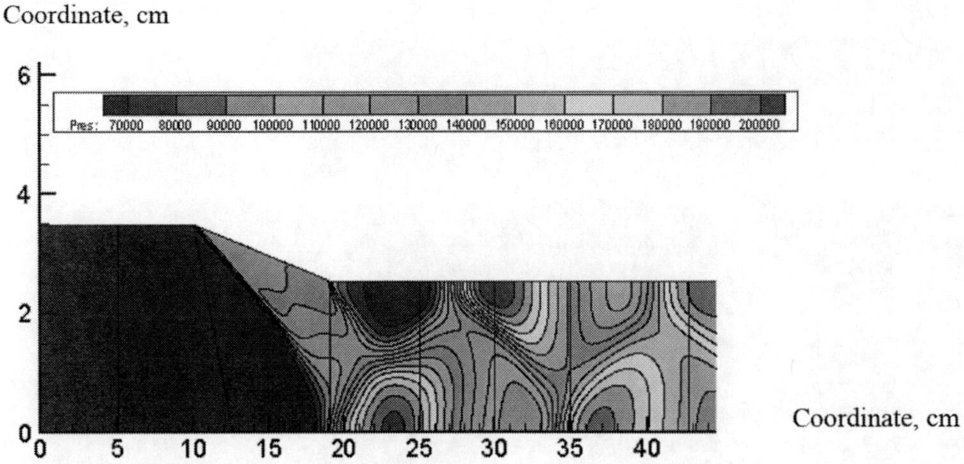

Figure 13. Pressure distribution within the experimental setup on a detailed grid using the turbulence model.

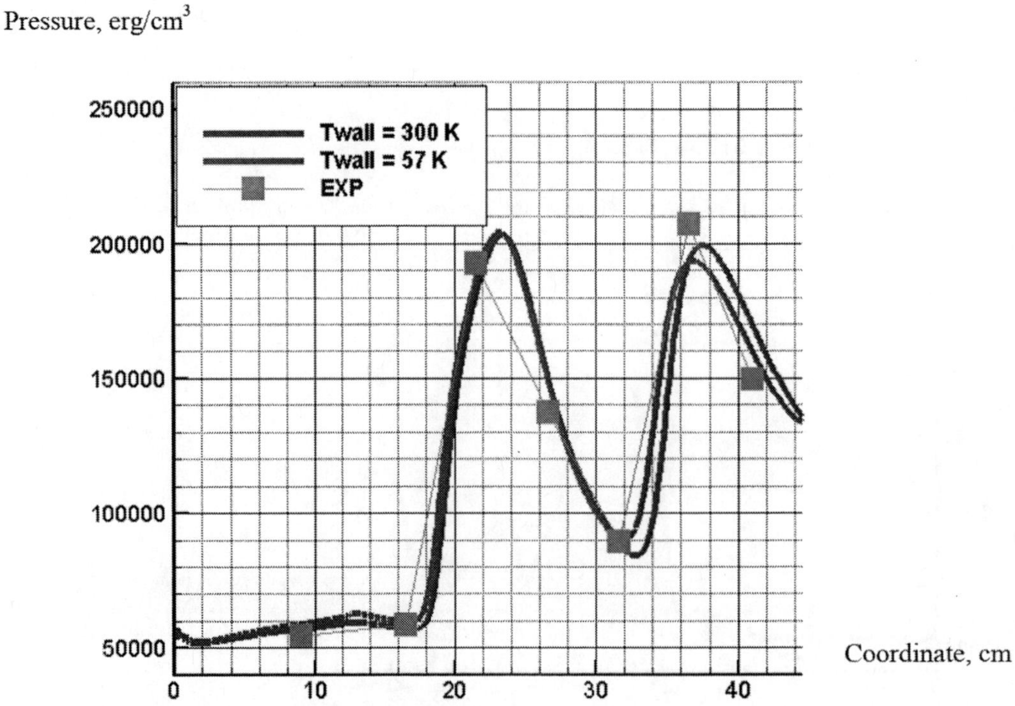

Figure 14. Comparison of calculated pressure distribution on the bottom wall obtained in this work at a wall temperature equal to room temperature (Tw=300K) and equal to the flow temperature (Tw=57K) with experimental results [8]

IOP Conf. Series: Journal of Physics: Conf. Series **1009** (2018) 012034 doi:10.1088/1742-6596/1009/1/012034

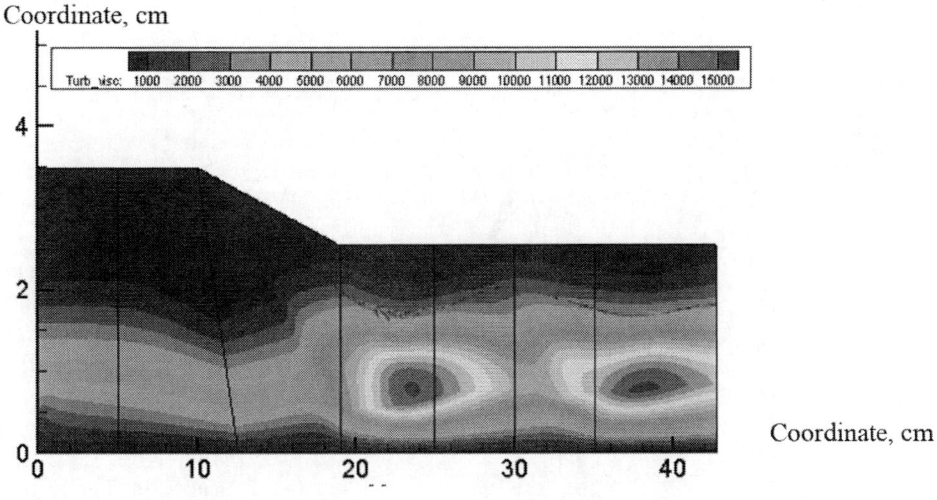

Figure 15. Turbulent viscosity distribution within the experimental setup.

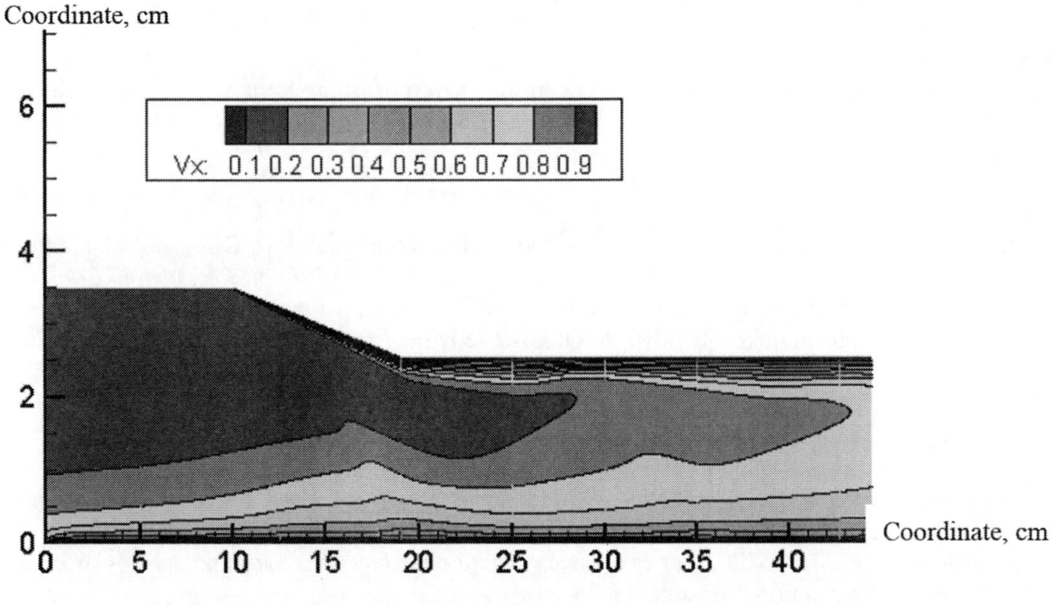

Figure 16. Longitudinal speed distribution within the experimental setup.

7. Conclusion

Results of the pressure distribution calculated by two-dimensional and three-dimensional model is presented. The present results show that two-dimensional simulations are able to predict the shock wave structures in a scramjet inlet/isolator. The influence of flow parameters such as the thickness of the boundary layer and turbulence models on the flow character is investigated in this paper. It was shown that a correct description of turbulence significantly affects the structure of the flow, and its consideration is critically important under comparing the calculation results with the experimental data.

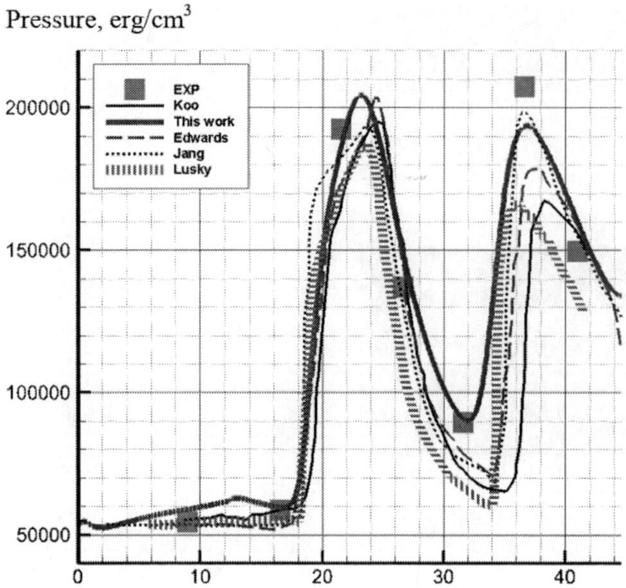

Figure 17. Comparison of calculated pressure distribution on the bottom wall obtained in this work (red line) and the results by Lutsky [14], Koo [13,15], Edwards [16] and Jang [12] with experimental results [8].

Acknowledgments
This work was preformed within the framework of the Government program of basic research of the Russian academy of sciences (contract #AAAA-A17-117021310372-6) and partially under support of RFBR grant #016-01-00379.

References

[1] Seleznev R and Surzhikov S 2015 Quasi-One-Dimensional and Two-Dimensional Numerical Simulation of Scramjet Combustors *51st AIAA/SAE/ASEE Joint Propulsion Conference* (Reston, Virginia: American Institute of Aeronautics and Astronautics) p 28

[2] Seleznev R and Surzhikov S 2016 A Quasi-One-Dimensional Analysis of Hydrogen-Fueled Scramjet Combustors *52nd AIAA/SAE/ASEE Joint Propulsion Conference* (Reston, Virginia: American Institute of Aeronautics and Astronautics) pp 1–27

[3] Seleznev R K 2016 Comparison of two-dimensional and quasi-one- dimensional scramjet models by the example of VAG experiment *J. Phys. Conf. Ser.* **755** 11001

[4] Surzhikov S, Seleznev R, Tretjakov P and Zabaykin V 2014 Unsteady Thermo-Gasdynamic Processes in Scramjet Combustion Chamber with Periodical Input of Cold Air *50th AIAA/ASME/SAE/ASEE Joint Propulsion Conference* (Reston, Virginia: American Institute of Aeronautics and Astronautics) p 25

[5] Seleznev R and Surzhikov S 2013 Generalized Newton Method For Solving Differential Equations of Chemical Kinetics *44th AIAA Thermophysics Conference* (Reston, Virginia: American Institute of Aeronautics and Astronautics) pp 1–17

[6] Xing F, Zhang S and Yao Y 2012 Numerical Simulation of Shock-Induced-Combustion in Three-Dimensional HyShot Scramjet Model *50th AIAA Aerosp. Sci. Meet. Incl. New Horizons Forum Aerosp. Expo.* 1–12

[7] Boon S and Hillier R 2006 Hypersonic Inlet Flow Analysis at Mach 5, 6 and 7 *44th AIAA Aerospace Sciences Meeting and Exhibit* (Reston, Virigina: American Institute of Aeronautics and Astronautics) pp 1–13

[8] Wagner J L 2009 *Experimental Studies of Unstart Dynamics in Inlet / Isolator Configurations in a Mach 5 Flow*

[9] Gamba M, Miller V, Mungal M G and Hanson R 2011 Ignition and Flame Structure in a Compact Inlet/Scramjet Combustor Model *17th AIAA Int. Sp. Planes Hypersonic Syst.*

Technol. Conf.

[10] Wagner J L, Yuceil K B, Valdivia A, Clemens N T and Dolling D S 2009 Experimental Investigation of Unstart in an Inlet/Isolator Model in Mach 5 Flow *AIAA J.* **47** 1528–42

[11] Wagner J L, Yuceil K B and Clemens N T 2010 Velocimetry Measurements of Unstart of an Inlet-Isolator Model in Mach 5 Flow *AIAA J.* **48** 1875–88

[12] Jang I, Pečnik R and Moin D P 2010 A numerical study of the unstart event in an inlet/isolator model *Cent. Turbul. Res. Annu. Res. Briefs* 93–103

[13] Koo H and Raman V 2012 Large-Eddy Simulation of a Supersonic Inlet-Isolator *AIAA J.* **50** 1596–613

[14] Borisov V E, Kudryashov I Y and Lutsky A E 2016 Numerical simulation of the pseudo-shock region formation in the channel *Keldysh Inst. Prepr.* **2898** 1–24

[15] Koo H 2010 Large-Eddy Simulations of Scramjet Engines *Doctoral dissertation*

[16] Boles J, Edwards J, Choi J-I and Baurle R 2009 Simulations of High-Speed Internal Flows Using LES/RANS Models *47th AIAA Aerosp. Sci. Meet. Incl. New Horizons Forum Aerosp. Expo.* 1–22

[17] Zhukov V T, Manukovskii K V., Novikova N D, Rykov Y G and Feodoritova O B 2015 *Numerical simulation of the pseudo-shock region formation in the channel*

[18] Shang J and Surzhikov S 2013 Numerical Prediction of Convective and Radiative Heating of Scramjet Combustion Chamber with Hydrocarbon Fuels *51st AIAA Aerospace Sciences Meeting including the New Horizons Forum and Aerospace Exposition* (Reston, Virigina: American Institute of Aeronautics and Astronautics) pp 1–16

[19] Shang J and Surzhikov S 2013 Radiative Heat Exchange in a Hydrogen-Fueled Scramjet Combustion Chambers *51st AIAA Aerospace Sciences Meeting including the New Horizons Forum and Aerospace Exposition* (Reston, Virigina: American Institute of Aeronautics and Astronautics)

[20] Edwards J, Liou M-S, Edwards J and Liou M-S 1997 Low-diffusion flux-splitting methods for flows at all speeds *13th Computational Fluid Dynamics Conference* vol 36(Reston, Virigina: American Institute of Aeronautics and Astronautics)

[21] Xu S and Martin M P 2004 Assessment of inflow boundary conditions for compressible turbulent boundary layers *Phys. Fluids* **16** 2623–39

[22] Walz A 1959 Compressible Turbulent Boundary Layers With Heat Transfer and Pressure Gradient In Flow *Journal of Research of the National Bureau of Standards-B. Mathematics and Mathematical Physics* **63** 53–72

[23] Smith A. M. and Cebeci T. 1967 *Numerical solution of the turbulent boundary layer equations* (Long Beach, Calif.: Douglas Aircraft Co., Douglas Aircraft Division)

IOP Conf. Series: Journal of Physics: Conf. Series **1009** (2018) 012035 doi:10.1088/1742-6596/1009/1/012035

Development of a visualization module for aerogasdynamic computations

P Silvestrov[1] and O Bessonov[2]

[1] Tactical Missiles Corporation
ul. Ilyicha 7, Korolev, Moscow district, 141080, Russia
[2] Ishlinsky Institute for Problems in Mechanics RAS
prosp. Vernadskogo 101-1, Moscow, 119526, Russia

E-mail: pavelsilvestrov@yandex.ru

Abstract. In this paper we present a concept and implementation details of a new visualization subsystem for presentation of the results of scientific computations. The subsystem is built into the interactive information and simulation system for gas dynamics and multiphysics problems as a comprehensive visualization module. This module is based on the open-source cross-platform application frameworks Qt and VTK. The paper describes basic functions of the visualization subsystem and presents examples of its graphical capabilities for drawing spatial images and two-dimensional plots, as well as examples of dialogue boxes for controlling the process of visualization. The subsystem supports main geometric and grid formats. The new subsystem has the unified implementation for Windows and Linux operation systems.

1. Introduction

In developing new flying vehicles (aircrafts, spacecrafts, landers) it is necessary to perform numerical simulations of their aerothermodynamic characteristics [1-5]. These simulations allow significantly decrease the number of ground and flight tests, and, consequently, greatly reduce the cost of developing the future models. In the previous work [6] an interactive information and simulation system (IISS) was presented. This system combines computational modules for performing aerogasdynamic simulations (both outside a vehicle and within its engine), interactive environment for controlling and monitoring computations, databases of physical and chemical properties, and different information facilities.

When working with the information and simulation system, it is necessary to have capabilities for visual presentation and interactive analysis of obtained results. At the moment, most computational modules in the IISS produce results in the formats suitable for visual processing by the commercial software Tecplot [7]. This software has become widespread in computational fluid dynamics and aerodynamics. In fact, it is now one of de facto standards in the field of scientific visualization.

In order to achieve independency on commercial software and to avoid problems of a licensing nature, we've set a goal to create a domestic visualization subsystem for aerogasdynamic computations. By this goal, such subsystem should possess some functionality of Tecplot and be partially compatible with it.

This paper presents the first version of the visualization module implemented within the framework of the interactive information and simulation system. The module can be deployed on a computer running a Windows or Linux operating system.

2. Functional structure of the visualization module

Interface of the visualization module was implemented similar to the interface of the commercial visualizer Tecplot. This similarity allows users familiar with the Tecplot to easily switch to the new module.

The new visualizer can be used at all stages of the computational session. The main function of the visualizer is the graphical presentation of the following sorts of objects:

- electronic geometry of the surface of a vehicle;
- computational grid of the domain surrounding a vehicle (for controlling grid quality), in particular spatial grid and surface grid;
- spatial distributions of physical quantities around a vehicle, on its surface and in selected cross-sections by different ways for intermediate and final results of computations;
- combination of the above spatial objects;
- two-dimensional plots (linear diagrams) and distributions of quantities.

Figure 1 presents an example of the interface of the visualizer, along with the distribution of some physical quantities in the vicinity of a space vehicle. It can be seen that the visualization screen consists of several fields – interface field, control field and window for graphical presentation.

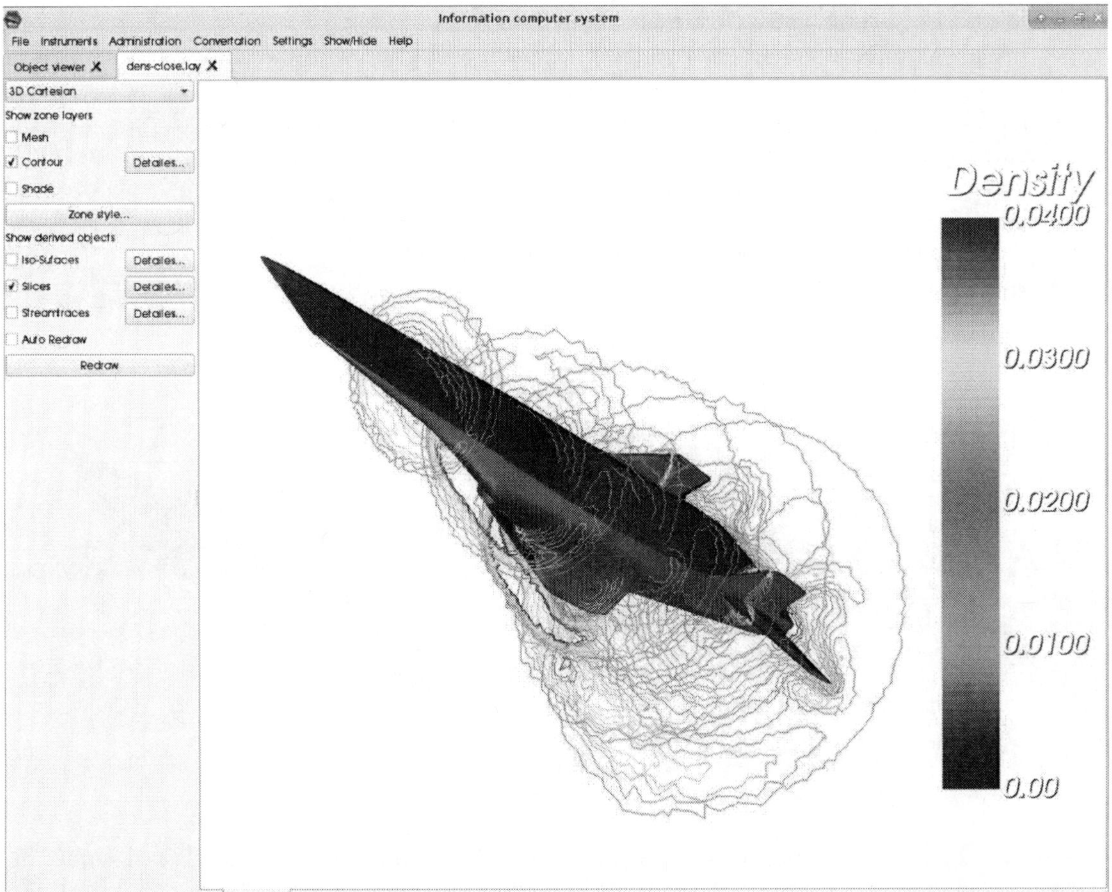

Figure 1. Example of the interface and graphical window showing the spatial distribution of flow density and temperature around a vehicle.

The visualization module supports the following input data formats:

- standard formats for geometric presentation of objects (* .stl in binary and text forms);
- standard formats for presentations of computational grids (*. neu);
- graphical formats of common visualization systems used for scientific computations – Tecplot (* .plt, * .dat) and ParaView (* .vtk, * .vtu).

The module supports some formats of control files used by the Tecplot (*.lay). Up to the moment, a number of configuration and tuning option are supported.

The visualizer supports two main regimes of graphical presentation and analysis of data – spatial mode and two-dimensional mode.

2.1. Spatial visualization mode

The spatial visualization mode is intended for representing images of three-dimensional objects and their fragments, including pictures of two-dimensional cross-sections, with spatial distributions of physical quantities on them. The visualizer supports several ways of graphical presentation of computational results (physical quantities), as well as the combination of thereof. Below is a list of main types of spatial images, with examples of visualization for all image types:

- contours (isolines) and color distributions of physical quantities on the surfaces of an object and in cross-sections around it (figure 1);
- images of surface and spatial (volume) computational grids with the ability to hide invisible lines (figure 2, *a*);
- halftone images of the surfaces of objects and their structural elements (figure 2, *b*);
- smooth and discrete color distributions of physical quantities in cross-sections of a computational domain with the ability of combining with other spatial images (figure 3);
- isosurfaces (surfaces of equal value) of physical quantities with the ability to depict color distributions of another quantities on these surfaces (figure 4);
- spatial trajectories and streamlines.

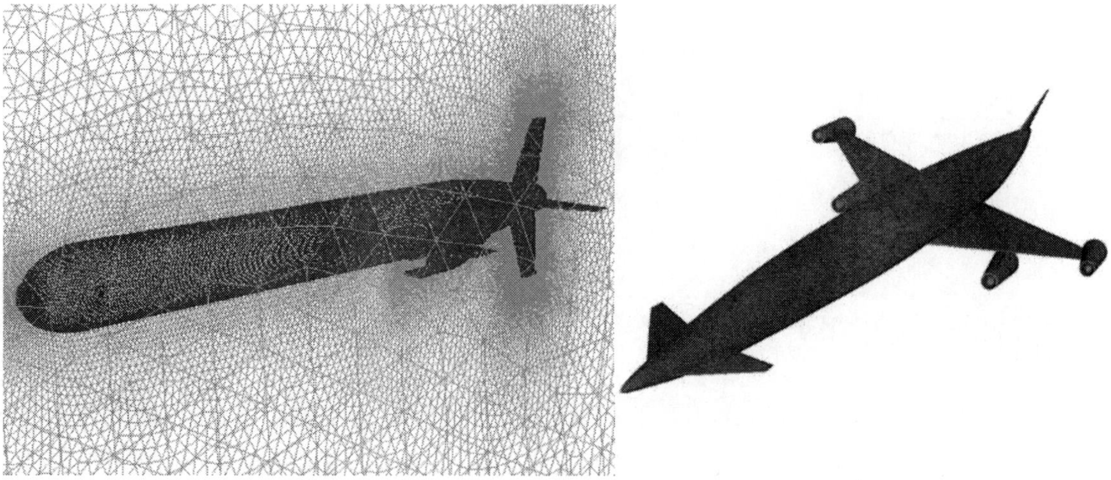

a *b*

Figure 2. Example of the presentation of a computational grid; zone that corresponds to a surface of a vehicle is highlighted (*a*). Halftone visualization of the surface of an aircraft (*b*).

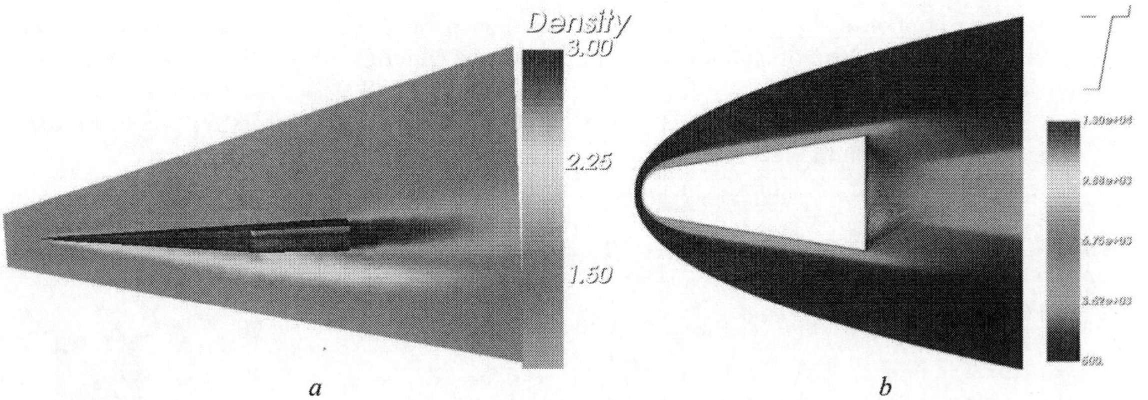

Figure 3. Examples of smooth color distribution of physical quantities in the cross-section of a computational domain. Distribution of flow density around a vehicle and on its surface (*a*). Distribution of flow temperature around a vehicle and streamlines behind it (*b*).

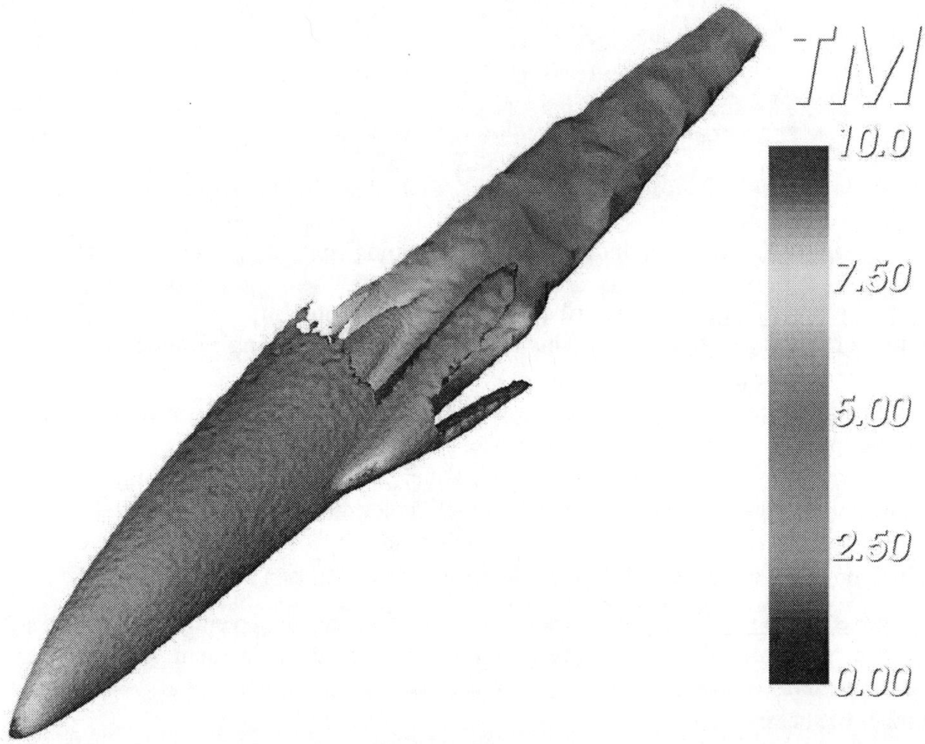

Figure 4. Example of the surface of equal flow density (isosurface) around a vehicle, with the color distribution of flow temperature depicted on this surface.

For all sorts of spatial images, the visualizer support different graphical manipulations – moving around the screen, scaling and rotation. These manipulations can be performed through the control fields of the screen and by the computer mouse.

The overall control on the graphical presentation, its configuration and tuning are performed in dialog boxes. To control pictures in the above example, several types of dialog boxes are used, describing correspondingly contours, zones, slices (cross-sections), isosurfaces and streamlines.

Contour description dialog box describes the way and parameters of representation of physical quantities at a plane in form of isolines or color distributions (figure 5, *a*). There are two main sets of parameters in this dialog box:

- ranges and contour levels for presenting quantities as isolines, color grades for presenting them as a color distribution;
- legends of contour or color presentations of quantities.

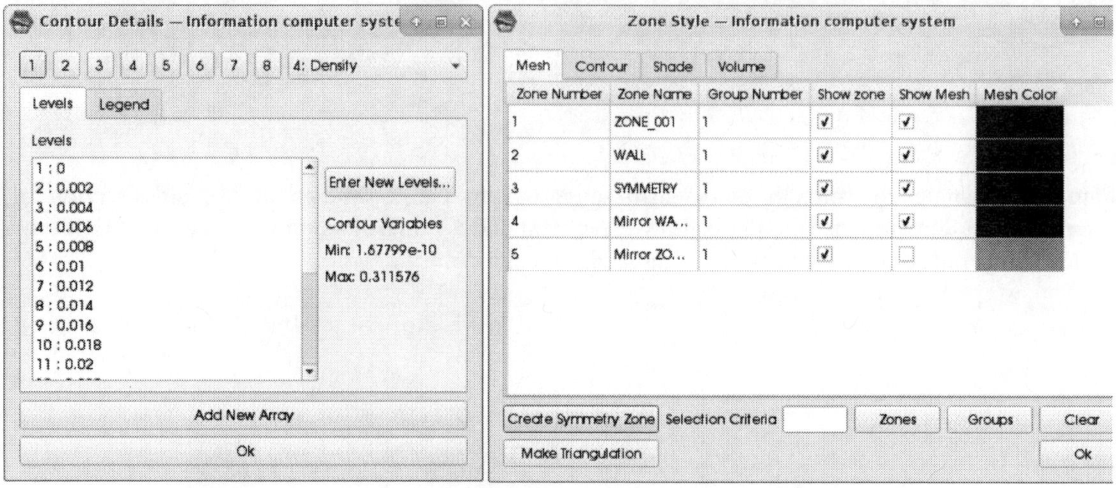

a *b*

Figure 5. Examples of dialog boxes describing contours (*a*) and zones (*b*).

Zone description dialog box describes a picture that presents several zones, with separate configuration of each zone (figure 5, *b*). This box specifies the following parameters:

- list of zones to be presented on the picture;
- specification of a type of graphical presentation for each zone – grid, surface, contour or color distribution, slice (cross-section), isosurface, streamline;
- particular way of the presentation in each zone, parameters of presenting the distribution of physical quantities (contours, discrete or smooth color distributions);
- colors for grids, surfaces and other line in each zone;
- indication of the symmetry of a zone with respect to the chosen plane.

Slices (cross-sections) dialog box describes configurations of several parallel slices and their mutual placement (figure 6) and the way of presentation of data on them. This box specifies the following parameters:

- number of slices;
- parameter of a plane to which the slices will be parallel;
- points on the axis perpendicular to this plane on which the slices will be places;
- indication whether a slice will be colored (for each slice), slice color;
- indication whether the surface will be colored, surface color;
- indication whether the grid will be projected on a slice (for each slice), grid color;
- parameters of presenting the distribution of physical quantities for each slice (contours, discrete or smooth color distributions).

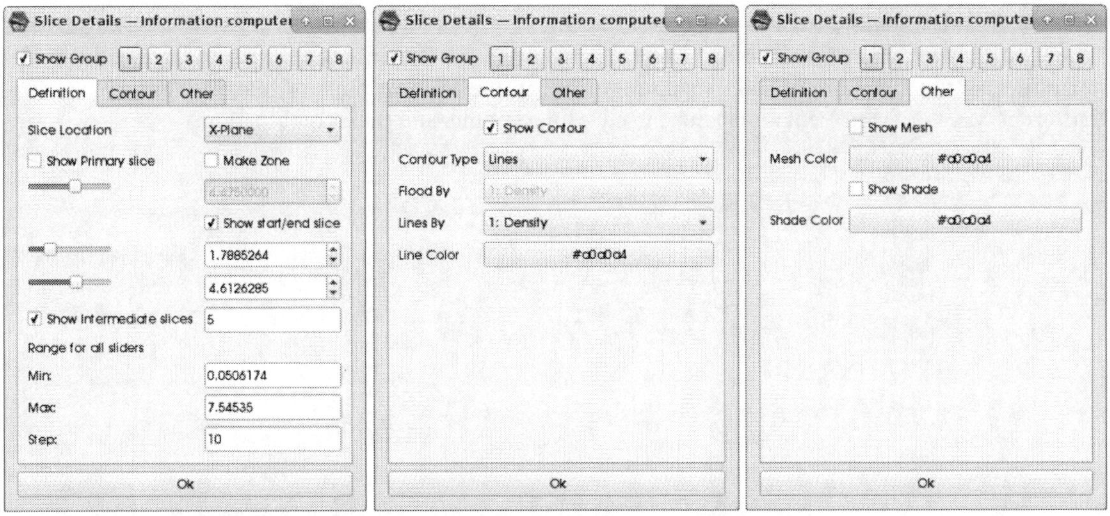

Figure 6. Example of dialog boxes describing a group of several slices.

Isosurface dialog box describes surfaces of equal value and distributions of a physical quantity on them (figure 7). This box specifies the following parameters:

- physical quantity represented as isosurfaces;
- values for which isosurfaces will be presented;
- indication whether the isosurfaces will be colored, isosurface color;
- indication whether the grid will be projected on isosurfaces, grid color;
- parameters of presenting the distribution of physical quantities on isosurfaces (contours, color distributions).

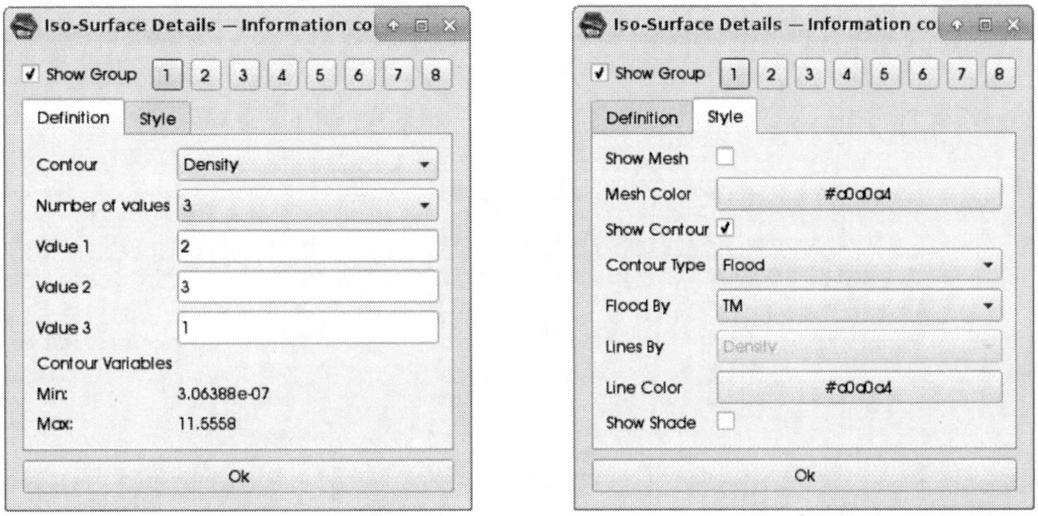

a *b*

Figure 7. Example of dialog boxes describing the isosurface (*a*) and the color distribution of a physical quantity on it (*b*).

Streamline description dialog box describes the characteristics of trajectories (streamlines) for a given velocity field. It is possible to manipulate with trajectories – add new trajectories by indication of their starting points, delete existing trajectories. This facility is convenient for analyzing flow fields interactively.

IOP Conf. Series: Journal of Physics: Conf. Series **1009** (2018) 012035 doi:10.1088/1742-6596/1009/1/012035

2.2. Two-dimensional visualization mode

The two-dimensional visualization mode is intended for representing distributions of physical quantities as functions of variables on an argument in the form of plots or linear diagrams (figure 8). Different ways of the presentation can be used – lines, points and their combinations.

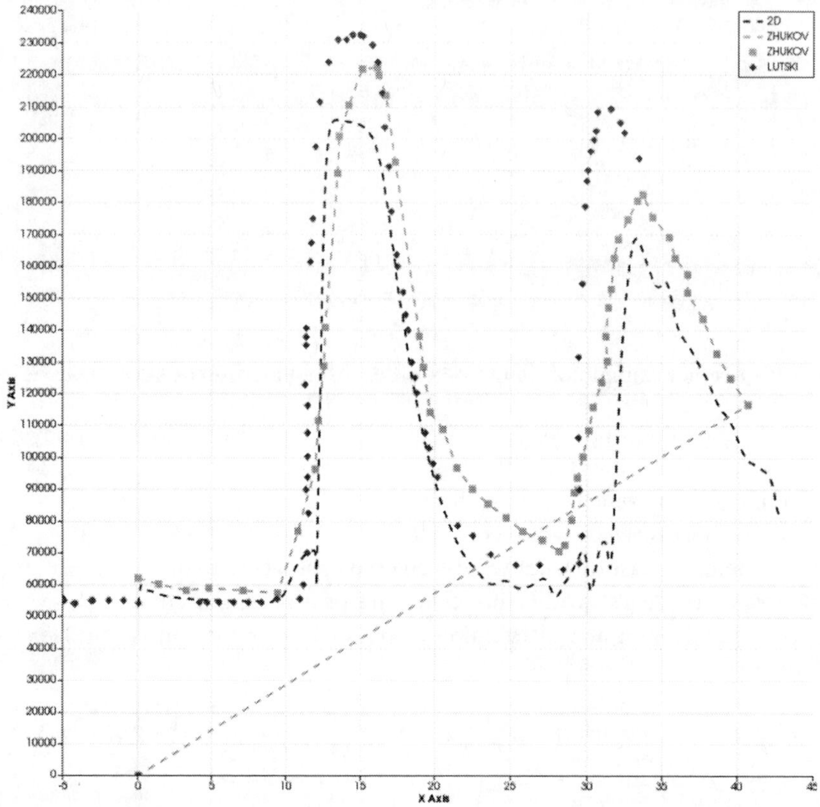

Figure 8. Two-dimensional mode. Example of a comparison of calculated and experimental data.

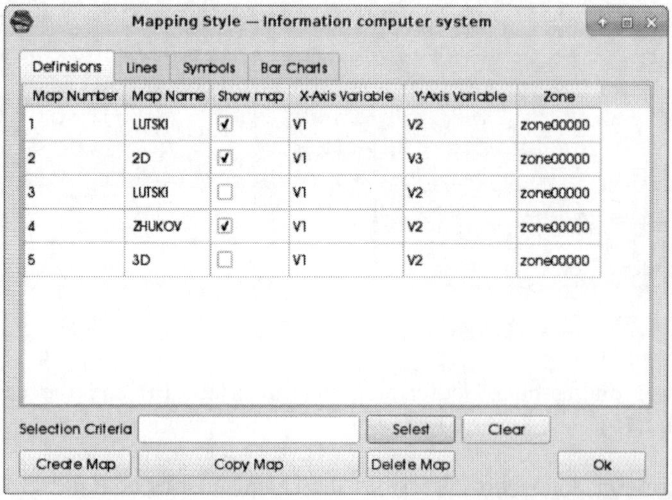

Figure 9. Example of the dialog describing two-dimensional visualization.

Two-dimensional visualization if controlled by the corresponding dialog box (figure 9). This box specifies the following parameters:

- dependent variables and their argument;
- zones from which this variables will be extracted;
- way of the representation of values (lines, points), the appearance and color of lines and points;
- way of the representation of legends, texts of legends.

Thus, the developed visualization module contains all necessary elements and methods of graphical representation of data, as well as tools for interactive control of the viewing process.

3. Implementation of the visualization module

The visualization module is implemented as a constituent part of the interactive information and simulation system (IISS). It is written on C++ language, using the open-source cross-platform software package Qt [8, 9] version 5.5 for implementing the interactive window interface. The graphical output and manipulation functions are built upon the open-source visualization toolkit VTK [10, 11] version 7.1.

The new module was tested and validated by several authors when performing aerogasdynamic computation. It was successfully used for spatial visualization and analysis of external flows [12-15], two-dimensional visualization [16] and simulation of an engine [17-19].

4. Conclusion

The developed visualization module provides a wide range of possibilities for graphical representation of the results of numerical simulations, as well as for viewing the geometries of modelled object and the computational grids. The visualizer support different manipulations with the resulting images – moving around the screen, scaling and rotation. The overall control on the graphical presentation, its configuration and tuning are performed in dialog boxes

The visualization module supports the geometric and grid formats and provides partial compatibility on data formats and control methods with the widespread commercial visualizer Tecplot. Throughout the development of the module, the amount of functional capabilities and the number of supported data formats will be increased.

The new visualizer is integrated into the interactive information and simulation system (IISS) and is a constituent part of it. The developed platform will become an effective and convenient tool both for performing aerogasdynamic calculations and for visual analysis of obtained results.

Acknowledgements
The authors wish to thank Dr. A. Dikalyuk for his help in implementing some functional element of the module, Dr. A. Rybakov for the assistance and support in our work.

References

[1] Surzhikov S T 2011 Radiative-Convective Heat Transfer of a Spherically Shaped Space Vehicle in Carbon Dioxide *High Temperature* **49** (1) pp 92–107
[2] Surzhikov S T 2012 Coupled radiative gasdynamic interaction and non-equilibrium dissociation for large-scale returned space vehicles *Chemical Physics* **398** 56–63
[3] Shang J S and Surzhikov S T 2012 Nonequilibrium radiative hypersonic flow simulation *Progress in Aerospace Sciences* **53** 46–65
[4] Surzhikov S T 2012 Radiative-Collisional Models in Non-Equilibrium Aerothemodynamics of Entry Probes *Journal of Heat Transfer* **134** 031002
[5] Surzhikov S T 2015 Two-dimensional numerical analysis of flow ionization in the RAM-C-II flight experiment *Russian J. Physical Chemistry* B **9** 69-86
[6] Bessonov O and Silvestrov P 2017 On the concept of the interactive information and simulation system for gas dynamics and multiphysics problems *Journal of Physics: Conference Series*

815 012019

[7] *Tecplot Home – CFD visualization & analysis post-processing software* https://tecplot.com/

[8] *Qt | Cross-platform software development for embedded & desktop* https://www.qt.io/

[9] Dalheimer M K 2002 *Programming with Qt: Writing Portable GUI Applications on Unix and Win32* (O'Reilly Media)

[10] *VTK – The Visualization Toolkit* https://www.vtk.org/

[11] Schroeder W, Martin K and Lorensen B 2006 *The visualization toolkit* (Kitware)

[12] Ermakov M and Kryukov I 2017 Supercomputer modeling of flow past hypersonic flight vehicles *Journal of Physics: Conference Series* **815** 012016

[13] Surzhikov S 2017 Validation of computational code UST3D by the example of experimental aerodynamic data *Journal of Physics: Conference Series* **815** 012023

[14] Kotov M, Ruleva L, Solodovnikov S and Surzhikov S 2017 Experimental and numerical study of supersonic flow over two blunted wedges *Journal of Physics: Conference Series* **815** 012025

[15] Yatsukhno D 2017 Numerical simulation of the flow over a hypersonic waverider using the method for splitting into physical processes *Journal of Physics: Conference Series* **815** 012022

[16] Surzhikov S 2017 Surface electromagnetic actuator in rarefied hypersonic flow *Journal of Physics: Conference Series* **815** 012005

[17] Seleznev R K 2017 Comparison of two-dimensional and quasi-one-dimensional scramjet models by the example of VAG experiment *J. Phys.: Conf. Ser.* **815** 12007

[18] Seleznev R K, Surzhikov S T, Shang J A 2016 Quasi-One-Dimensional Analysis of Hydrogen-Fueled Scramjet Combustors *AIAA 2016-4569*, 52nd AIAA/ASME/SAE/ASEE Joint Propulsion Conf., *AIAA Paper*-2016-4569 pp 1–27

[19] Surzhikov S, Seleznev R, Tretjakov P and Zabaykin V 2014 Unsteady Thermo-Gasdynamic Processes in Scramjet Combustion Chamber with Periodical Input of Cold Air *AIAA Paper*-2014-3917 p 25

IOP Conf. Series: Journal of Physics: Conf. Series **1009** (2018) 012036 doi:10.1088/1742-6596/1009/1/012036

Preliminary experimental results of heat flux surface field registration at the hypersonic aerodynamic shock tube using temperature sensitive paint

V E Mosharov[1], V N Radchenko[1], I V Senuev[1], M A Kotov[2], L B Ruleva[2], S I Solodovnikov[2] and S T Surzhikov[2]

[1] Central Aerohydrodynamic Institute, Zhukovsky Street, TsAGI, Zhukovsky, Moscow Region, 140180, Russia
[2] Ishlinsky Institute for Problems in Mechanics RAS, Vernadsky prospekt 101(1), Moscow, 119526, Russia

E-mail: vladimir.mosharov@tsagi.ru, ruleva@ipmnet.ru

Abstract. The temperature gradient fields of a wedge with double ramp were explored by using applied method temperature sensitive paint [1-4]. The model was studied experimentally in the HAST facility [5-8] at incoming airflow Mach numbers of M = 5...7. With the technique, perturbations of luminescent elements on the model surface during stationary hypersonic flow were recorded. The results of experimental data processing of surface temperature gradient fields are presented.

1. Introduction

Temperature sensitive paint (TSP) is a luminescent paint that provides the measurement of global temperature distribution. TSP consists of polymer binder and organic luminophore molecules and can be thin enough for measurements in short duration wind tunnels. Temperature Sensitive Paint application techniques for heat flux measurements in Ludwieg wind tunnel were described in [1–4].

In present tests the binary (two color) temperature sensitive paint was used. Binary paint contains two types of organic luminophore molecules. One luminophore (europium complex) is temperature sensitive. It's luminescence decreases with temperature growth. Other luminophore (Coumarin) is temperature insensitive. Its luminescence is used as a reference for pixel-by-pixel correction of excitation light variation from flash to flash. Both luminophores are excited simultaneously by UV flash lamp (280–390 nm), but emit light in different spectral ranges. Temperature sensitive luminescence is red (580–630 nm) and temperature insensitive luminescence is blue (420–500 nm) and thus they can be separated spectrally.

Intensity of emitted red light decreases at temperature increase with the rate of 3–5% °C and is absolutely insensitive to pressure. TSP is optimized for the temperature range of 10–80°C. TSP degrades at the temperature above 120°C.

TSP is applied on the model surfaces by airbrushing like an ordinary paint. The thickness of the dry TSP layer is about 3÷5 micrometers. After TSP application a set of markers are placed on the model surface. Markers are contrasted (black) points on TSP surface and are used for correction of the model displacement in the flow in respect with the position without the flow.

Six images (three by each CCD camera) are acquired for each facility run: dark images (without excitation light), wind-off (at known constant temperature without air flow) and wind-on (with air flow) images.

Data processing included the next steps:

- Dark signal correction (dark image subtraction);
- Flat field correction;
- Alignment of wind-off and wind-on images using the markers on the model images;
- Pixel-by-pixel correction of excitation light intensity variation: temperature sensitive (red) images are divided on corresponding reference (blue) images;
- Image normalization: wind-on image is divided on corresponding wind-off image. The ratio of 'wind-on' to 'wind-off' images depends only on temperature. It does not depend on paint layer thickness;
- Temperature field calculation using TSP calibration characteristics. TSP calibration is performed in laboratory calibration setup on TSP sample prepared simultaneously with the model covering (a priori temperature calibration);
- Heat flux calculation;
- Distortion correction using an image of etalon body;
- Projection of resulting 2D Stanton number field on the 3D mesh describing the model geometry.

Global heat flux image is computed from temperature field using exact solution of one-dimensional heat transfer equation:

$$\vartheta = 1 - exp(\beta^2)\cdot erfc(\beta),$$
$$\vartheta = (T_m - T_{in})/(T_r - T_{in}),$$

where T_{in} is initial model temperature before wind tunnel start, measured with thermocouple;

T_m is the model surface temperature at the moment t after flow initialization, measured by TSP;

T_r is recovery temperature and is assumed $T_r \approx T_0$;

$\beta = h\sqrt{t}/\sqrt{\lambda c\rho}$ is non-dimensional heat flux (h is heat output coefficient and $\sqrt{\lambda c\rho}$ is thermal product of model material);

erfc is a complementary error function. Results are presented as Stanton number fields ($St = h/\rho_\infty V_\infty C_p$). The relation to maximum intensity presented in figure 1.

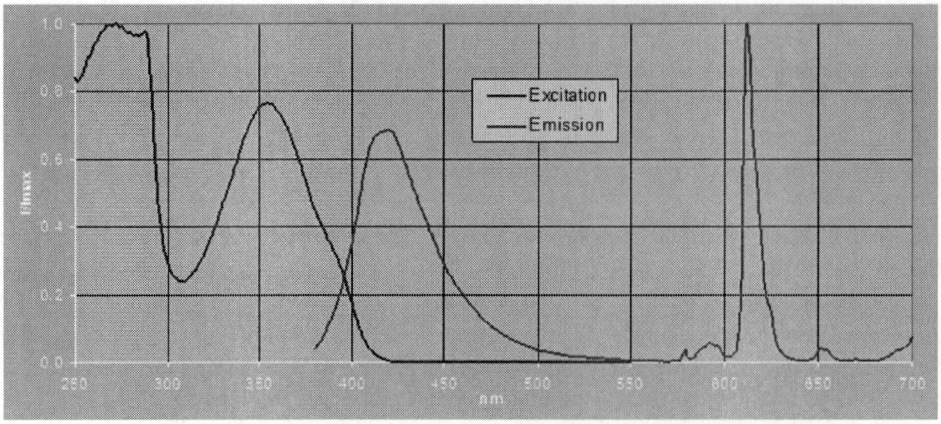

Figure 1. The spectrum of excitation and emission characteristics.

These procedures were tested on experimental facilities in TsAGI corporation and first used in experiments in the HAST [5-8] at the Institute for Problems in Mechanics RAS.

2. Operation of the measuring technique in the HAST

The observation windows were modified to accommodate two cameras and flash lights as shown in figure 2.

Figure 2. New observation windows in HAST for providing the measuring technique.

The model was made of acrylic and painted with TSP in black colour, as shown in figure 3. The experiments of temperature field visualization on the model surface in the HAST were made under Mach numbers $M = 5.3$ and $M = 5.5$ with estimated time delay of the flash and the photo frames.

Figure 3. Investigated model.

3. The results of experiments on heat flux field visualisation

In experimental studies of temperature gradient on the model surface by using fluorescent temperature transducer method the measured quantity is the intensity of luminescence, which was chosen by the maximum range of the CCD camera VS–STT–285 charge-coupled in different spectral regions. Processing of the results was carried by analysing data received in two stages: before the experiment and during the experiment. The time of measurement was chosen so that the signal did not come from the selected range of time.

The results of the measurements obtained with the help of a luminescent coating were processed into the Stanton numbers (the Stanton number is proportional the measured heat flow and inversely proportional to the density, gas velocity in the undisturbed flow and the increment of temperature). Maximum values were located at different distances from the sharp edge. Figure 4 shows the processed results of the measurement of increment of the model surface temperature.

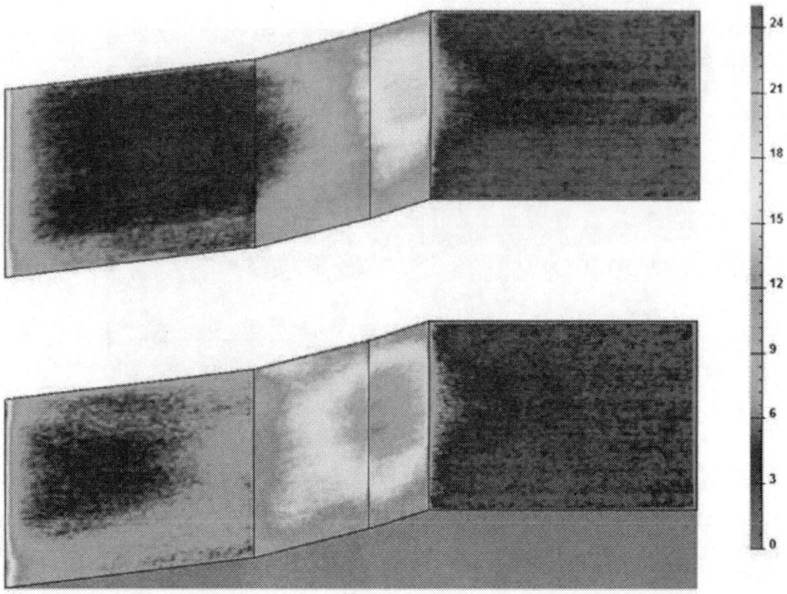

Figure 4. The processed results of the measurement of increment of the model surface temperature in two runs of the HAST: $M = 5.3$ (top) and $M = 5.5$ (bottom).

After experiments to determine the temperature field on the model surface the visualisation of shock-wave structures was carried out with a shooting video speed of 2000 frames/s at the oncoming airflow at $M = 5.3$, 5.5. Figure 5 contains pictures of the received shock interaction.

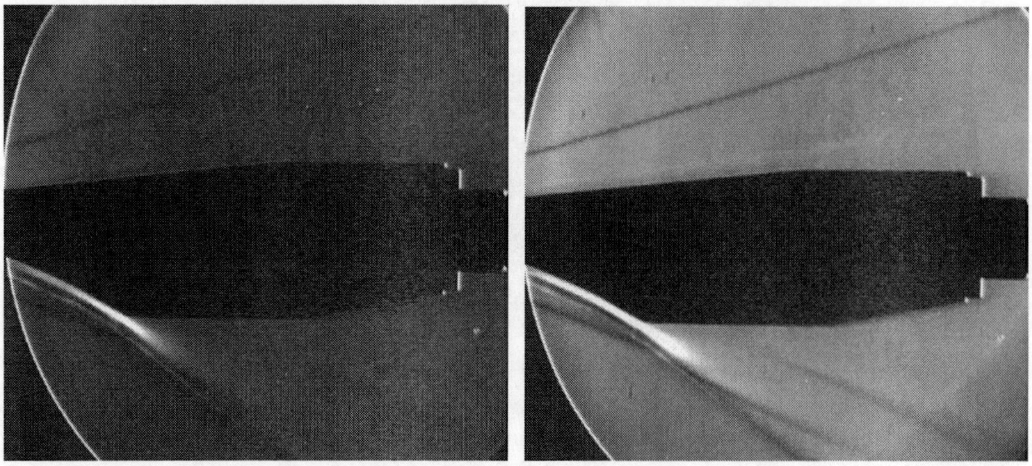

Figure 5. Shadow pictures of airflow shock structure at $M = 5.3$ (left) and $M = 5.5$ (right).

4. Conclusion

Preliminary experiments have been prepared and carried out to visualize the temperature field on the surface of models at the GUAT. The temperature parameters of the model system were measured during experiment. Preliminary thermal data on the temperature gradient of the model surface are obtained.

Acknowledgement

The presented study was partially supported by Government program of basic research of Russian academy of science (contract #AAAA-A17-117021310372-6) and partially supported by Russian Foundation for Basic Research grant # 16-01-00379.

References

[1] Mosharov V E 2009 Luminescent methods for investigating surface gas flows *Instruments and Experimental Techniques* **52**
[2] Troyanovsky I, Sadovskii N, Kuzmin M, Mosharov V, Orlov A, Radchenko V and Phonov S 1993 Set of luminescence pressure sensors for aerospace research *Sensors and Actuators B: Chemical* **11** pp 1–3
[3] Borovoy V Y, Mosharov V N, Noev A Y and Radchenko V N 2012 Temperature sensitive paint application for investigation of boundary layer transition in short-duration wind tunnels *Progress in Flight Physics* **3**
[4] Borovoy V Y, Mosharov V N, Noev A Y and Radchenko V N 2009 Laminar-turbulent flow over wedges mounted on sharp and blunt plates *Fluid Dynamics* **44**
[5] Kotov M A, Kryukov I A, Ruleva L B, Solodovnikov S I and Surzhikov S T 2015 Supersonic Air Flows Around Some Geometrical Primitives 33rd *AIAA Applied Aerodynamics Conference AIAA paper* 2015–3012
[6] Kotov M A, Kryukov I A, Ruleva L B, Solodovnikov S I and Surzhikov S T 2014 Multiple Flow Regimes in a Single Hypersonic Shock Tube Experiment 30th *AIAA Aerodynamic Measurement Tchnology and Ground Testing Coferece AIAA paper* 2014–2657
[7] Kotov M A, Kryukov I A, Ruleva L B and Solodovnikov S I 2017 The Investigation of Shock-Wave Interaction with Aerodynamic Models 55th *AIAA Aerospace Sciences Meeting AIAA paper* 2017–0262
[8] Kotov M A, Ruleva L B, S I Solodovnikov S I and Surzhikov S T 2017 Experimental and numerical study of supersonic flow over two blunted wedges *J. Physysics: Conference Series* **815** 012025

IOP Conf. Series: Journal of Physics: Conf. Series **1009** (2018) 012037 doi:10.1088/1742-6596/1009/1/012037

Investigation of gas dynamic parameters of the conical nozzle block functioning in the Hypersonic Aerodynamic Shock Tube

I A Ivanov[1], M A Kotov[2], I A Kryukov[2], L B Ruleva[2] and S I Solodovnikov[2]

[1] Lomonosov Moscow State University, GSP–1, Leninskie Gory, Moscow, 119991, Russia
[2] Ishlinsky Institute for Problems in Mechanics RAS, Vernadsky prospekt 101(1), Moscow, 119526, Russia

E-mail:ivanovmai@mail.ru, ruleva@ipmnet.ru

Abstract The results of investigations of the conical nozzle block of the hypersonic shock aerodynamic tube (HAST) [1–4] of the Laboratory of Radiative Gas Dynamics of the Institute for Problems in Mechanics RAS are presented. The efficiency of short conical nozzle is calculated. The gas dynamic flow parameters were measured experimentally at different distances from the nozzle exit. The design pressure values are compared with the measurements at the nozzle outlet. Comparison of numerical calculation and experimental results shows that the nozzle is started, successfully worked and it is possible to obtain a uniform flow at Mach numbers $M = 5...7$.

1. Introduction

Experimental studies in the Hypersonic Aerodynamic Shock Tube (HAST) of the Laboratory of Radiative Gas Dynamics at the Institute for Problems in Mechanics RAS are continuing. The investigated gas flows are formatting by using the short conic nozzle and oncoming onto models of high speed aircraft primitives [1–4]. The flow quality should be investigated at several distances from the nozzle to ensure the flow uniformity. A direct numerical study of gas-dynamic processes in the complete configuration of HAST involves overcoming a number of problems.

The main problem is connected with the necessity of non stationary calculation of the large number of interacting gas-dynamic discontinuities taking into account viscous phenomena on very detailed computational grids. The latter is due to the large length of the impact tube (15 m) [1], limited by the possibility of using 1D calculations, the presence of an additional diaphragm in the nozzle throat and the need to resolve the boundary layers in the tube sections and in the nozzle block.

In this paper, at first, a simplified mathematical modeling was carried out, in which numerical calculations can be divided into two stages. At the first stage non stationary processes in the shock tube sections (driver and driven tubes) after opening the diaphragm are considered. At the second, the quasi-stationary gas flow through the nozzle block and the formation of the near-to-nozzle field of gas-dynamic parameters in the jet incoming to the vacuum chamber are calculated. The questions of the first stage were considered in the [2, 3], and the parameters are experimentally refined, so let's use them.

2. Conical nozzle block scheme in the HAST facility

The HAST experimental setup, 15 m of length, is shown in diagram (figure.1).

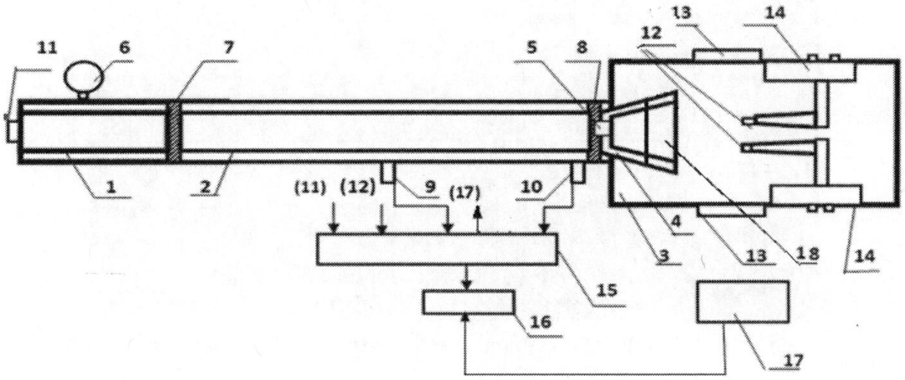

Figure 1. The HAST scheme.

The HAST scheme shows the following: high-pressure chamber (1), cylindrical channel / driven tube (2), vacuum chamber (3) and installed in it hypersonic nozzle (4) with a critical section (5), pressure gauge (6), high-speed solenoid valve (7), second membrane unit (8), high-frequency dynamic pressure sensors (9), (10), (11), (12), observation windows (13), model holders (14) with models, analog-to-digital converters (15), PC (16), high-speed video camera (17). High-vacuum pumps (pumping stations), gas mixing and filling systems are not shown. View of the HAST part with the nozzle block inlet and the vacuum camera is shown in the photo of figure 2.

Figure 2. Photo of the shock tube section and aerodynamic part of the HAST facility.

The short conical nozzle in the aerodynamic block of the HAST, in front of which the model is mounted, is shown in figure 3.

The nozzle block inlet with critical section is located in the shock tube driven section. It has a built-in part for mounting the membrane before entering the nozzle. A sketch of the simplified conical nozzle for the calculations below is shown in figure 4. Here: r^* - the radius of the nozzle critical cut, R – the radius of the nozzle output section, L – the length of the nozzle, θ – the half angle of the nozzle supersonic part.

Figure 3. Photo of the short conical nozzle output in the vacuum chamber.

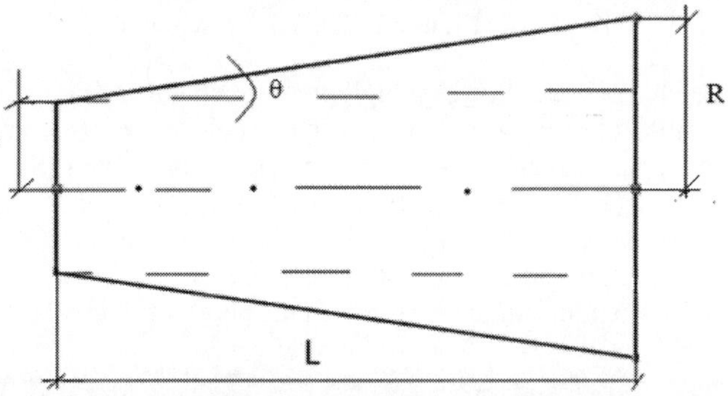

Figure 4. The sketch of the simplified short conical nozzle.

3. Quasi-stationary calculation of the HAST nozzle block

The base of the nozzle block used in HAST experiments equipped with a conical nozzle with the radius of the critical cross-section $r^* = 4$ mm and the radius of output section $R = 43.5$ mm, the half-angle of the supersonic cone part $\theta = 13.4°$, the length of the supersonic part $L = 168$ mm. the geometric degree of the nozzle expansion is:

$$\left(\frac{R}{r^*}\right)^2 = 118.266$$

The pressure at the nozzle inlet, according to preliminary estimates and experimental data, was set at three levels: $Pk = 0.5$, 1.9, 2.3 atm. The gas inlet temperature $Tk = 297$ K. The pressure in the space surrounding nozzle (in the vacuum chamber) $Ph = 0.1$ Pa. The ratio of specific heat capacities for the gas environment - air is equal to:

$$\gamma = \frac{c_p}{c_v} = 1.4$$

The temperature dependence of the viscous transfer coefficient is described by the Sutherland formula:

$$\mu = \mu_0 \frac{T_0 + C}{T + C}\left(\frac{T}{T_0}\right)^{\frac{3}{2}},$$

here:

μ – dynamic viscosity in (Pa•s) at a given temperature T,

IOP Conf. Series: Journal of Physics: Conf. Series **1009** (2018) 012037 doi:10.1088/1742-6596/1009/1/012037

μ_0 – control viscosity in (Pa•s) at control temperature T_0,
T – set temperature in Kelvin,
T_0 – control temperature in Kelvin,
C – Sutherland constant.

The calculation area is the space inside the nozzle, including the subsonic part, the area of the vacuum chamber above the nozzle, the area adjacent to the nozzle section and covering the near field of the jet flow. A regular calculation grid adapted to the boundaries of the computational domain is generated in the region. The thickening of the grid was carried out near the solid walls of the nozzle, in the area of the critical section and near of the output section of the nozzle.

In this numerical study, the steady flow of viscous gas in axisymmetric nozzle and jet was calculated. The system of two-dimensional (axisymmetric) nonstationary Navier-Stokes equations written in divergent form and supplemented by thermal and caloric equations of ideal gas state and boundary conditions at the boundaries of the computational domain is solved numerically.

Only molecular (laminar) viscosity was taken into account. On the boundaries corresponding to solid walls, the adhesion conditions were set. At the nozzle inlet, the enthalpy and entropy calculated for the total pressure and the stagnation temperature on the axis of symmetry were preserved - the condition for the absence of motion. At the remote boundary, the conditions for the absence of reflection were set. The terms in the system of equations describing the convective transfer were approximated by means of a modified high-order Godunov scheme [5-8]; the terms of the equations describing the diffusion (viscous) transfer were approximated by a finite volume scheme. The calculation process developing in time was carried out using the Runge-Kutta method of the second order [6].

The results of calculations for the stationary flow regimes in the conical nozzle in the coordinates normalized to the critical radius are presented in figures 5-10. Here one can see the Mach number field and the distribution of Mach numbers along the axis of the nozzle symmetry. The results of calculations for the inlet pressure values are presented: P_k = 0.5 atm in figures 5 and 6; P_k = 1.9 atm in figures 7 and 8, P_k = 2.3 atm in figures 9 and 10.

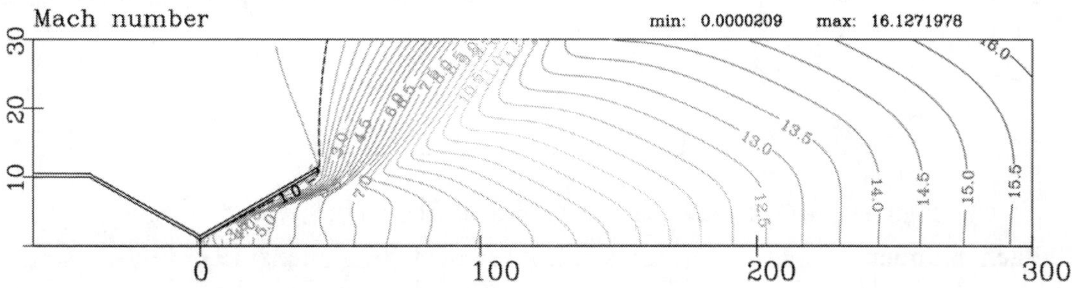

Figure 5. Field of Mach numbers. Conical nozzle, P_k = 0.5 atm, r^* = 4 mm.

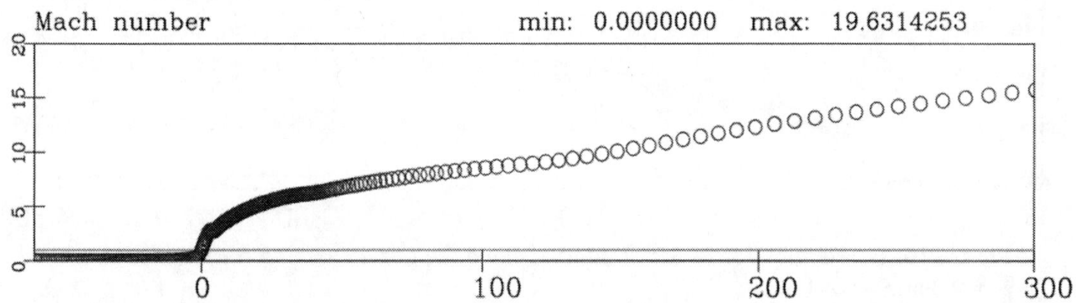

Figure 6. The distribution of Mach number along the nozzle axis. Conical nozzle, P_k = 0.5 atm, r^* = 4 mm.

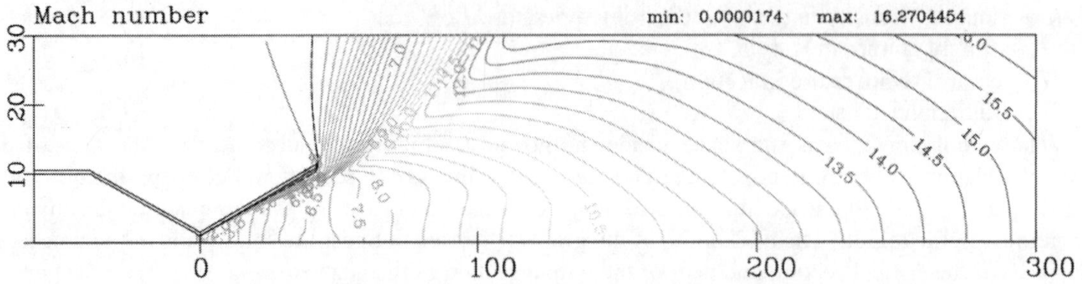

Figure 7. Field of Mach numbers. Conical nozzle, P_k= 1.9 atm, $r*$ = 4 mm.

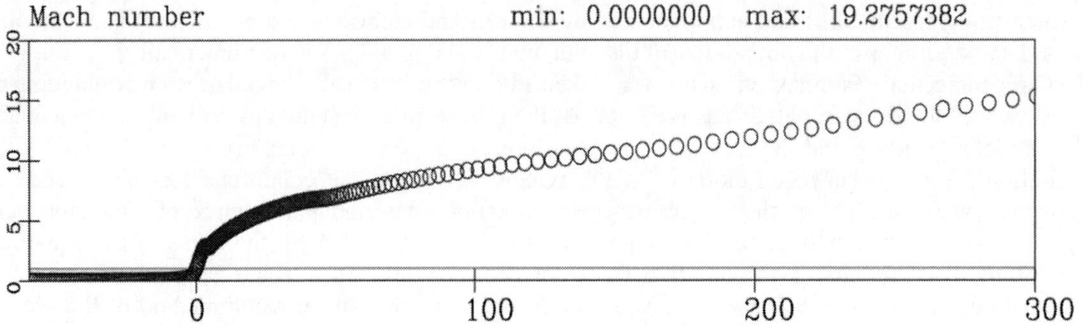

Figure 8. The distribution of Mach number along the nozzle axis. Conical nozzle, P_k = 1.9 atm, $r*$ = 4 mm.

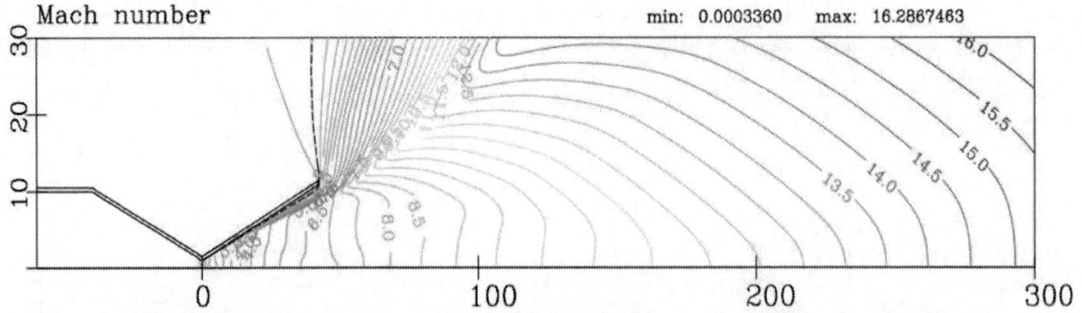

Figure 9. Field of Mach numbers. Conical nozzle, P_k= 2.3 atm, $r*$ = 4 mm.

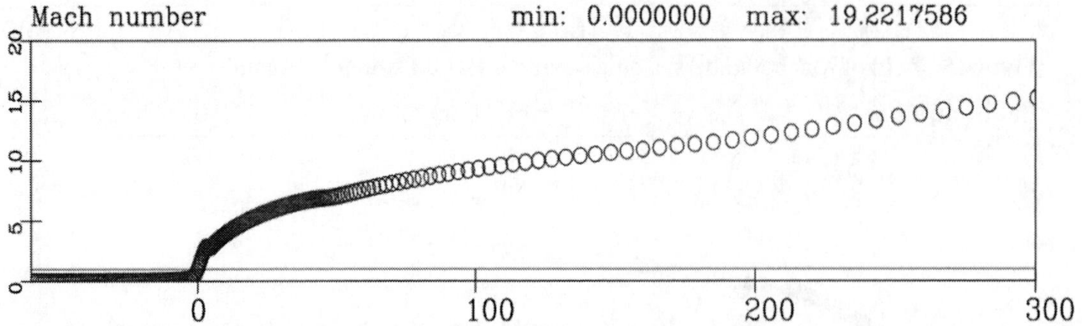

Figure 10. The distribution of Mach number along the nozzle axis. Conical nozzle, P_k= 2.3 atm, $r*$ = 4 mm.

The velocity field at a pressure of P_k = 1.9 atm was constructed for the flow regime studied in the experiment (figure 11).

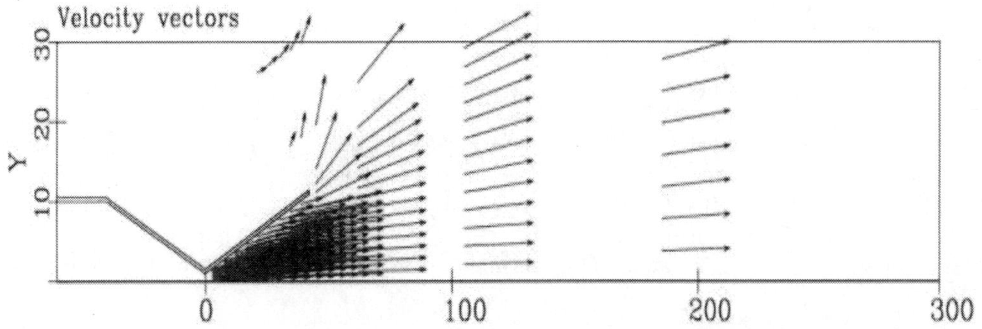

Figure 11. The velocity field. Short conical nozzle, P_k= 1.9 atm., r^* = 4 mm.

Increasing the inlet pressure leads to an increase in static pressure and flow density inside the nozzle, that is, an increase in the Reynolds numbers and decrease in the thickness of the boundary layer on the wall.

It can be seen from the graphs that the field of flow parameters is rather uneven both along the x-axis and the y-axis (figure 5, 7, 9), with increasing the thickness of the boundary layer in the nozzle, there is a tendency both to a certain increase in the "one-dimensional" flow and to a slight decrease in the average cross-section value of the Mach number. So for P_k = 0.5 atm the flow at the nozzle section has a Mach number M = 6.5, P_k= 1.9 atm – M = 6.8. The current lines behind the nozzle section have smaller slopes than in the conical section, the Mach number grows more slowly with increasing spatial coordinates along the nozzle axis, for example, in figure 5 and figure 10. Thus, the growth of the boundary layer leads to the fact that the flow becomes more uniform in the area of influence of the nozzle cut parameters.

The flow parameters along the nozzle axis were evaluated. The calculated data of stagnation pressure at different distances from the nozzle exit are obtained for comparison with experimental data obtained at the HAST.

4. Experimental study of the HAST nozzle block

On the axis of the short conical nozzle, instead of the model (figure 3) certified dynamic pressure sensor PCB 113B28 was installed in the position of the Pitot tube. Sensor readings and calculated values of stagnation pressure at different distances from the nozzle exit are given in the table 1.

Table 1. Experimental and calculated values of stagnation pressure at distances from the nozzle exit.

Distance from the nozzle exit, mm	P_{pitot} - stagnation pressure (Pitot), Pa. Calculation	P_t - stagnation pressure (Pitot), mV. Experiment
0	0.35642E+04	54
49	0.26669 E+04	41.4
100	0.18369E+04	27.7
144	0.13498E+04	21
193	0.10452E+04	15.5

IOP Conf. Series: Journal of Physics: Conf. Series **1009** (2018) 012037 doi:10.1088/1742-6596/1009/1/012037

The stagnation pressure (Pitot) in the numerical study was calculated by the ratio:

$$p_{\text{pitot}} = p\left[\frac{2\gamma}{\gamma+1}M^2 - \frac{\gamma-1}{\gamma+1}\right]\left[\frac{4\gamma}{(\gamma+1)^2} - \frac{2(\gamma-1)}{(\gamma+1)^2 M^2}\right]^{-\frac{\gamma}{\gamma-1}},$$

here: P – is the static pressure at a given point.

Taking into account the transmission coefficient C = 15 mV/kPa of dynamic pressure sensor PCB 113B28, comparison of the calculated and experimental values of the stagnation pressure is shown in the graph (figure 12).

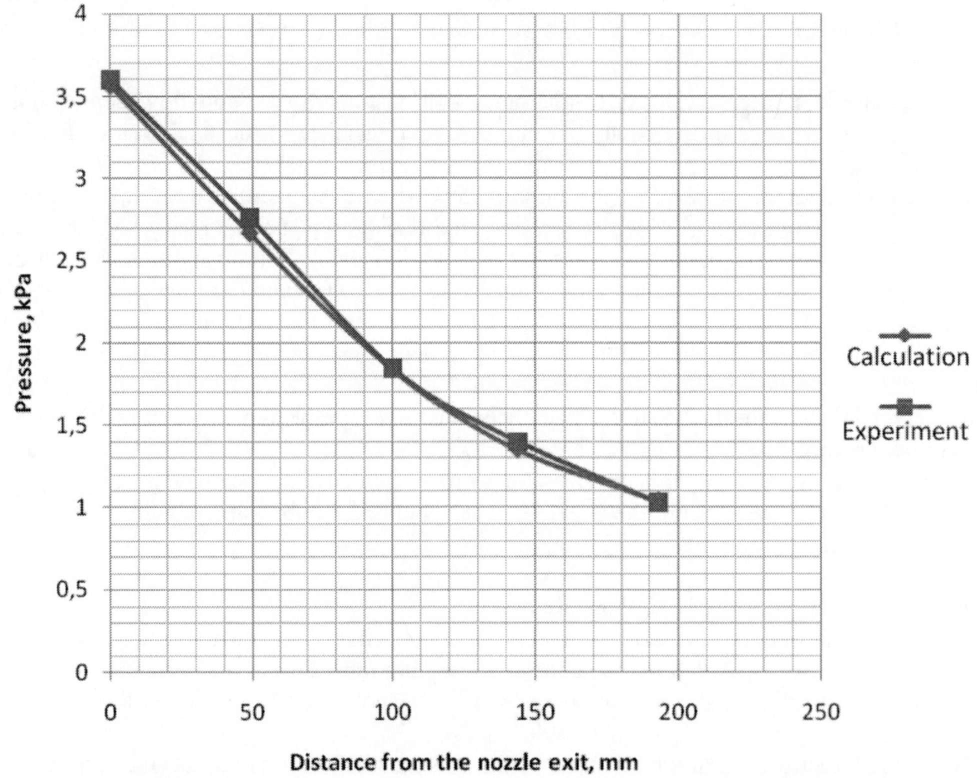

Figure 12. Flow stagnation pressure. Comparison of calculation and experiment.

It can be seen that the calculated data are satisfactory reproduce the experimental results (error less than 4%).

Conclusion

Calculated-experimental researches of the HAST nozzle block at the Laboratory of Radiative Gas Dynamics at the IPMech RAS showed that the short conical nozzle suitable for supersonic flow for Mach number range M = 5..7. The calculations of the real conical nozzle of the experimental facility and of the formation of gas flows are made for real parameters in experiments.

The structure of the gas-dynamic flow near the conical nozzle section of the experimental unit is studied numerically. The calculated flow fields are compared with the nozzle outlet pressures measured experimentally at Mach numbers M = 5...7, which showed a satisfactory coincidence.

Acknowledgement
The presented study was partially supported by Government program of basic research of Russian academy of science (contract #AAAA-A17-117021310372-6) and partially supported by Russian Foundation for Basic Research grant #16-01-00379.

References

[1] Kotov M A, Kryukov I A, Ruleva L B, Solodovnikov S I and Surzhikov S T 2015 Supersonic Air Flows Around Some Geometrical Primitives 33rd *AIAA Applied Aerodynamics Conference AIAA paper* 2015–3012
[2] Kotov M A, Kryukov I A, Ruleva L B and Solodovnikov S I 2017 The Investigation of Shock-Wave Interaction with Aerodynamic Models 55th *AIAA Aerospace Sciences Meeting AIAA paper* 2017–0262
[3] Kotov M A, Ruleva L B, S I Solodovnikov S I and Surzhikov S T 2017 Experimental and numerical study of supersonic flow over two blunted wedges *J. Physysics: Conference Series* **815** 012025
[4] Kotov M A, Ruleva L B, Solodovnikov S I and Surzhikov S T 2014 Experimental Researches of Two Flat Halfwedges *Physical-Chemical Kinetics in Gas Dynamics* **16** (4) http://chemphys.edu.ru/issues/2015-16-4/articles/589/ (in Russian)
[5] Godunov S K 1959 A difference method for numerical calculation of discontinuous solutions of the equations of hydrodynamics *Mat. Sb.* **47** 271–306
[6] Ivanov I E, Kryukov I A and Timokhin M Y 2013 Application of moment equations to the mathematical simulation of gas microflows *Computational Mathematics and Mathematical Physic* **53**
[7] Glushko G S, Ivanov I E and Kryukov I A 2010 Computational method for turbulent supersonic flows *Mathematical Models and Computer Simulations* **2**
[8] Ivanov I E, Kryukov I A and Semenov V V 2009 Numerical simulation of separated flow in nozzle with slots 26th *International Symposium on Shock Waves. Springer Berlin Heidelberg* **2** 973–978

Preliminary experimental assessment of supersonic airflow behavior over ExoMars and X–43 inlet models using multiple flow regime shock tube

M A Kotov, L B Ruleva, S I Solodovnikov and S T Surzhikov

Ishlinsky Institute for Problems in Mechanics RAS, Vernadsky prospekt 101(1), Moscow, 119526, Russia

E-mail:ma_kotov@mail.ru, ruleva@ipmnet.ru

Abstract. Models of the ExoMars spacecraft [1-3] and variants of the X-43 air intake [4-9] were first tested in the hypersonic aerodynamic shock tube (HAST) [10-13] facility. The shock-wave structures of the gas flow under Mach numbers M=7…4.5 around the model surfaces were filmed by high-speed video camera. The obtained data were analyzed using image recognition software systems. Preliminary estimates of the flow behavior over the investigated models in the test section of the installation during the whole test are given.

1. Introduction

The interaction of shock waves with the surface of a spacecraft and a high-speed aircraft at supersonic and hypersonic flight speeds is one of the fundamental problems of modern aerodynamics. The relevance of this issue is related to the practical needs that arise when developing high-speed aircraft of complex shape. Of particular importance are aspects of this type of flow around the elements of thermal protection of the spacecraft, aerodynamic controls, elements of the feathering of a high-speed aircraft, wings, elements of the scramjet air intake. The experimental data obtained are the basis for further computer aerothermodynamic studies.

“ExoMars” — a joint program of the European space Agency (ESA) and the Russian state corporation Roscosmos [1–3]. The launch of interplanetary probe "ExoMars-2016" implemented in 2016 was unsuccessful. "Schiaparelli" module has not carried out a soft landing. New launch "ExoMars-2020" it is planned in 2020. In the framework of this paper experiments on the “ExoMars” spacecraft model were conducted with the aim of obtaining general patterns governing the formation of shock-wave interactions near the apparatus surface arising from the incoming high-speed air flow with different Mach numbers during one experimental run in the facility.

If one distinguish the characteristic features that can be inherent in perspective aerospace systems, then such aspects as the possibility of repeated use of the aerospace apparatus and close aerodynamic integration of the glider with control elements, propulsion system and launch vehicle are very important. A considerable amount of data in open sources of literature contains the hypersonic aircraft models (in particular, such as X-43) in terms of the geometry characteristics of the scramjet air intake [10-13]. This paper presents the results of preliminary experimental modeling of some air intake configurations of hypersonic aircraft.

Experimental studies were conducted in the hypersonic aerodynamic shock tube (HAST, figure 1). Detailed description of the HAST installation and its technical capabilities are presented in [12, 13].

Figure 1. Photo of the HAST facility.

2. ExoMars model experiments. Multiple flow regimes

ExoMars model was mounted in front of the hypersonic nozzle in the HAST test section (figure 2). The PCB sensors registered pressures at different distances from the nozzle exhaust. Schlieren photos of experiments with the structure detached shock wave over the model are shown in figure 3.

Figure 2. ExoMars model mounted near the nozzle in the HAST.

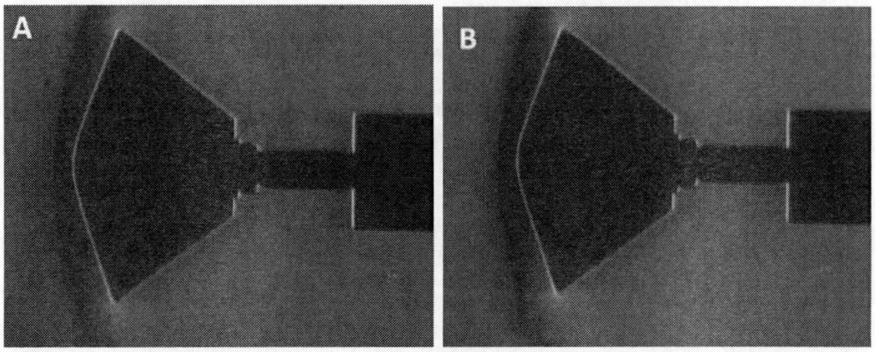

Figure 3. Schlieren images of M=7 (A) and M=4.5 (B) air flow near the ExoMars model in different phases of experiment.

When considering the entire period of one test in shock-tube type installations with an aerodynamic section working under reflected shock wave regime, special attention is paid to the so-called periods of the quasi-stationary flow process around the model under study [11]. When using the same gases as a driver and driven, during these periods the gas dynamic parameters of the flow at the nozzle block inlet undergo small changes. This is directly related to the character of the shock-wave interactions occurring in the combined volume of the driver and driven tubes during the experiment [11].

In the framework of the paper, a preliminary analysis of the model's motion during the experiment was carried out using software image recognition systems [14]. The files of high-speed video material of shadow images of shock-wave interaction near the model surface were analyzed. These data were recorded during the course of one experiment, i.e. the so-called periods of the quasistationary flow process were taken into account. Figure 4 shows the location of the coordinate axis on the video frames for recording the motion of the model. Two-dimensional graphs of the dependence of the spatial displacement on time are shown in figure 5. The center of the coordinate axis was specified at the deflection of the model's rear surface.

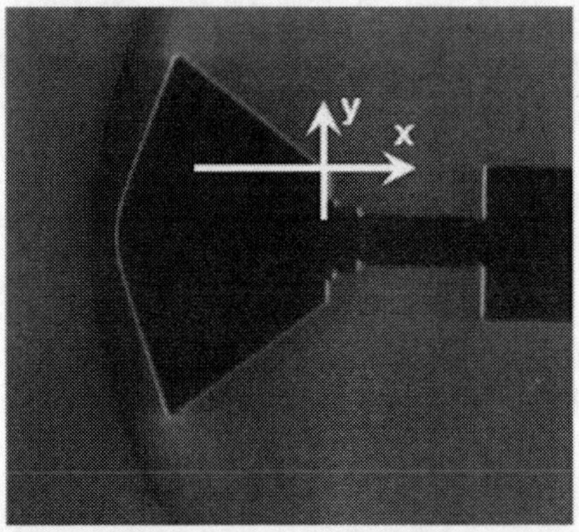

Figure 4. Registration the model movement. Coordinate axes.

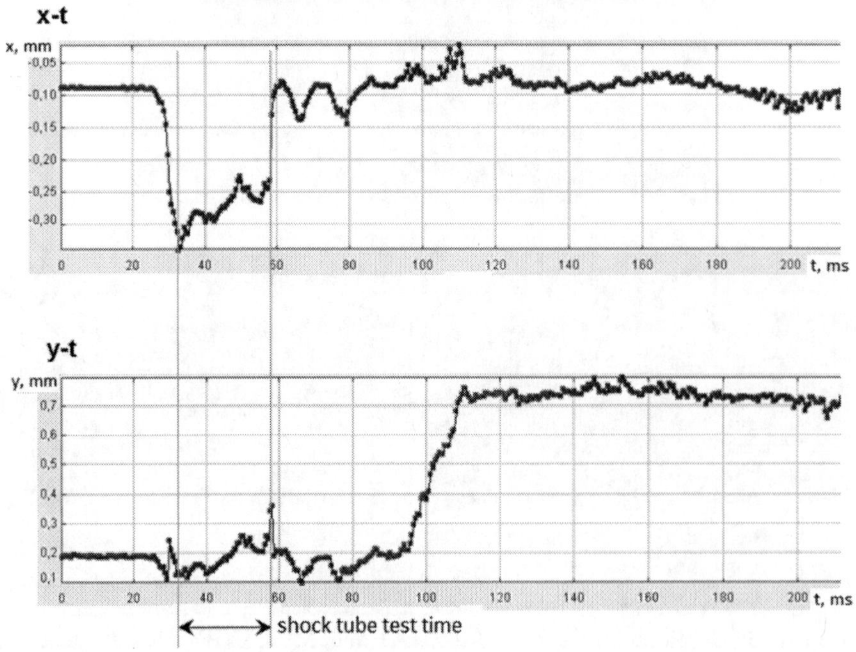

Figure 5. x-t and y-t diagrams of model movement during the experiment.

In the graphs of figure 5, one can select several phases of the experiment. The start of the test is determined by the static position of the model (0-30 ms). Then comes the period of a supersonic flow of the gas under investigation with stationary parameters. The first drop in the x-t graph in figure 5 identifies the onset of the impact of the incoming flow from the nozzle on the model (backward motion). This period (30-55 ms on figure 5) shows rather small movements of the model in comparison with the general nature of displacements. For this consideration it can be neglected because of their fairly clear time comparison with the processes occurring at the entrance to the nozzle block (reflected shock tube test time, figure 3A).

The following jumps in the x-t and y-t graphs (figure 5, 55 ms) show the attenuation of the oncoming flow, the termination of the stationary flow process. This is followed by the onset of shock wave perturbations, which is due to the variable parameters of the gas-dynamic characteristics at the nozzle entrance.

The next significant change shows y-t graph of 100 ms. The change in the position of the model by 0.5 mm is due to the quasistationary flow process beginning (figure 3B). Diagnostics of gas-dynamic parameters of this period is of particular interest, since under the conditions of one test it is possible to investigate the influence of oncoming flows with different gas dynamic characteristics onto the model (under certain assumptions). The flow pattern shown in figure 3B lasts several tens of milliseconds on a high-speed video file. A more detailed picture of these periods can be obtained, in particular, by comparing the experimental data with the results of numerical simulation [10,13]. The further minor fluctuations in the x-t and y-t graphs are evidence of the static position of the model and are not of interest for consideration in the framework of this preliminary assessment.

With a more detailed study of the vibration nature of the investigated models during the entire time of the experiment, it is necessary to take into account not only the shock-wave interaction before entering the nozzle and changes in the parameters of the incoming flow at the outlet from it. In particular, additional factors attract attention, such as: the design features of the model holder, its attachment, its own resonant frequencies, acoustic oscillations of the facility sections during the experiment.

3. Variations of hypersonic air inlet model

The shock wave profile of the flow in the nose part of the hypersonic aircraft and directly near the air intake of the propulsion system plays an important role, since it directly affects the processes of the engine operation (such as local flow compaction, supersonic combustion, thrust, heat loads of the scramjet and etc.). Of particular importance is the adherence to these aspects in all phases of aircraft flight with a scramjet working (at various points of flight trajectory, at the accelerator separation period, etc.).

The procedure for aerothermodynamic modeling of hypersonic aircraft propulsion system contains several items (figure 6):

- Aerothermodynamic analysis of the air inlet configuration;
- Modelling of gas-dynamics and physico-chemical combustion aspects in the combustion chamber;
- Nozzle flow and system analysis.

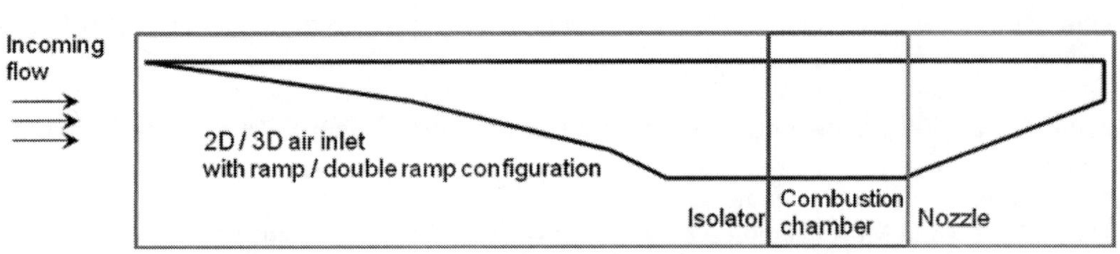

Figure 6. Design of a scramjet propulsion system [5].

Figure 7 shows shock-wave interactions for ramp/double ramp configuration of hypersonic aircraft inlet. The separation of the boundary layer (figure 7A) is one of the critical moments in the design of the hypersonic air intake geometry as this can cause a local change in the flow pattern, pressure

IOP Conf. Series: Journal of Physics: Conf. Series **1009** (2018) 012038 doi:10.1088/1742-6596/1009/1/012038

parameters, heat transfer increment. This can lead to locking the engine path, causing unevenness and irregularity in its operation. According to Heiser-Pratt [6,15], there are certain values of the critical angles of inclination for laminar and turbulent flow cases to avoid the boundary layer separation and the appearance of flow recirculation zones. For example, the hypersonic aircraft 14-X B model has a second inclination angle of 14.5° and for some experiments with the Mach number M = 7, it was possible to observe these phenomena [6].

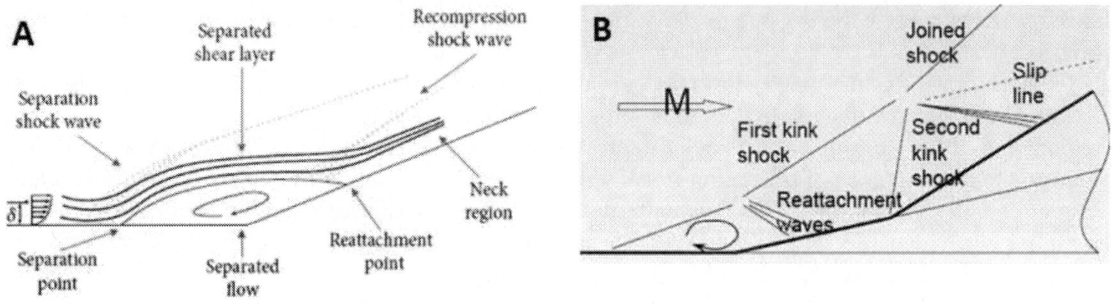

Figure 7. Configuration scheme with ramp (A) / double ramp (B) [5, 6].

Also one of the biggest problems in the design of the hypersonic air intake is to enable the engine to operate in a wide range of Mach numbers without using elements of variable geometry (rotating edge, moving lower cowl, etc.). With a fixed inlet geometry (ramp configuration angles), the angle of impact (the angle formed by the flow direction and the leader shock) becomes smaller for a larger Mach numbers. This changes the structure of the flow field along the inlet (figure 8).

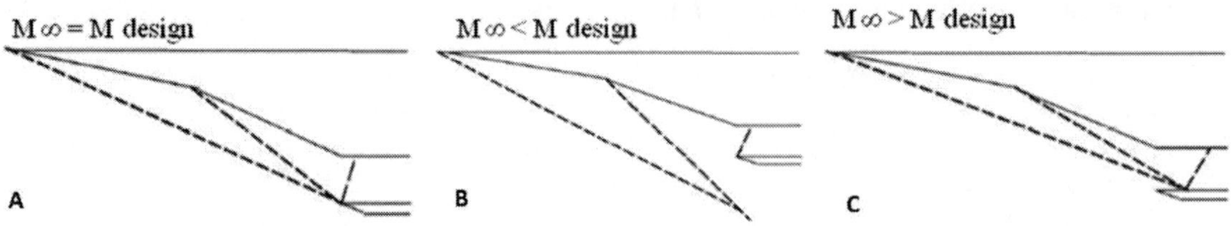

Figure 8. Schemes of shock-wave pictures for different Mach numbers with a fixed geometry of the air inlet [7].

When the engine is operating in the design mode all external inclined shock waves should be focused on the tip of the lower edge. The Mach number corresponding to this test case is called the design Mach number (figure 8A). When working at the Mach number below the calculated, the deviation angles increase and a leakage occurs (figure 8B). If the flight Mach number is higher than the design Mach number, the inclined shock waves deflect into the entrance throat (figure 8C). In this operating state, the interaction of the shock-boundary layer causes zones of detachment and re-attachment of the flow, high-temperature spots inside the entrance throat. This creates huge heat loads on the intake structure [7].

As a geometric source of compression wave concentration at the tip of the lower edge of the apparatus, the Prandtl Meyer compression fan can also be used [8]. Example of this geometry is shown in figure 9.

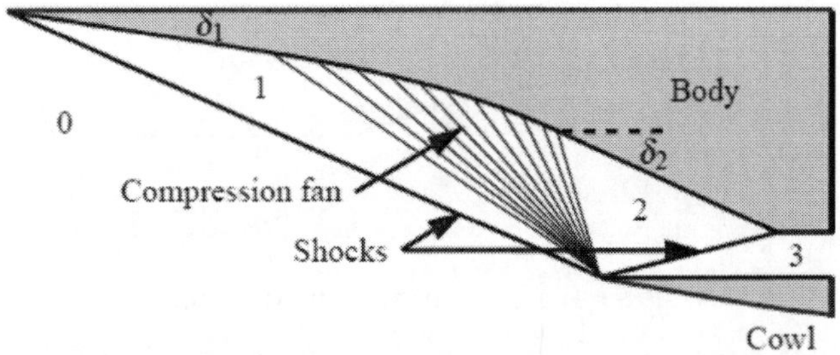

Figure 9. Inlet configuration with Prandtl Meyer compression fan [8].

By means of analyzing of literature open sources of literature [4-9] the three-dimensional CAD model of the X-43 (figure 10) was created at the Laboratory of Radiative Gas Dynamics at the Institute for Problems in Mechanics of the Russian Academy of Sciences. This model is used for numerical simulation of aerothermodynamic processes [16].

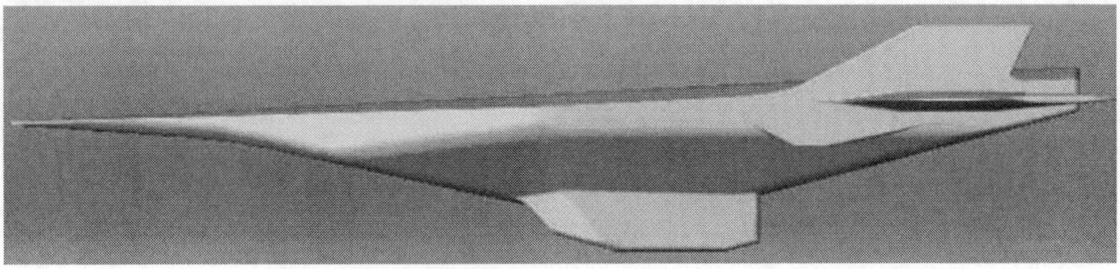

Figure 10. Hypersonic aircraft X-43 CAD model [16].

4. Models of X-43 air inlet configuration

Taking into account the above-mentioned aspects of the hypersonic aircraft inlet design, a variant of the model for carrying out the experiments was created (figures 11, 12).

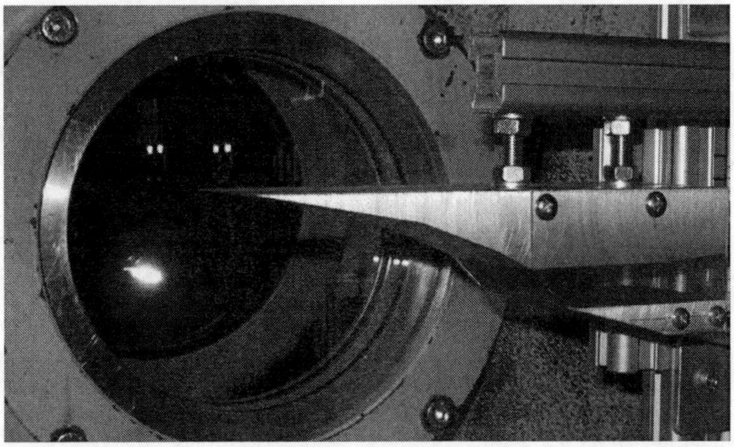

Figure 11. The experimental X-43 air inlet model in the HAST test section.

IOP Conf. Series: Journal of Physics: Conf. Series **1009** (2018) 012038 doi:10.1088/1742-6596/1009/1/012038

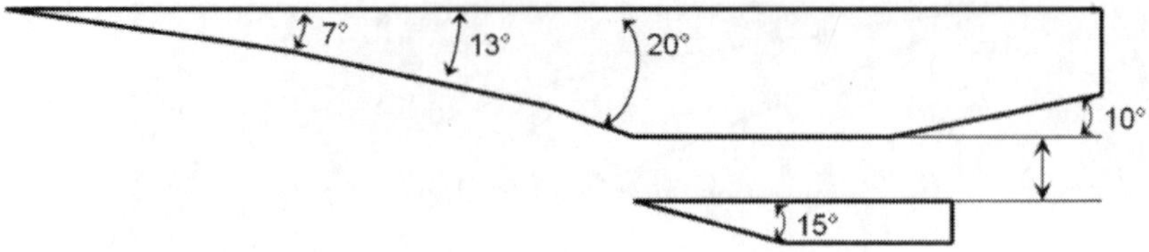

Figure 12. X-43 air inlet model scheme.

In the case, the ramp angles were 6° and 7° so there was no boundary layer separation. Figure 13 shows shadow picture results combinated from two experimental runs with M=7 incoming airflow. Features of flow motion are shown in figures 14, 15.

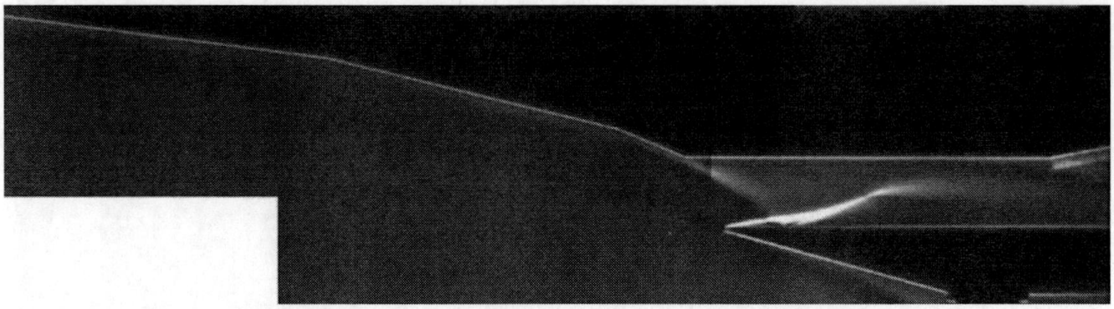

Figure 13. Shadow picture combination of M=7 incoming airflow onto X-43 air inlet model.

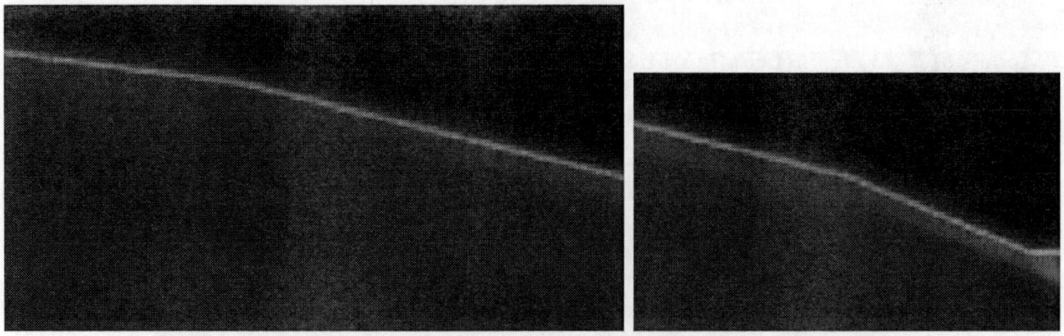

Figure 14. Formation of the second and third oblique shock waves.

The formation of the second and third oblique shock waves occurs immediately at the first ramp surface, which indicates that there is no boundary separation (figure 14). Figure 15 shows the enlarged flow picture in the flow path.

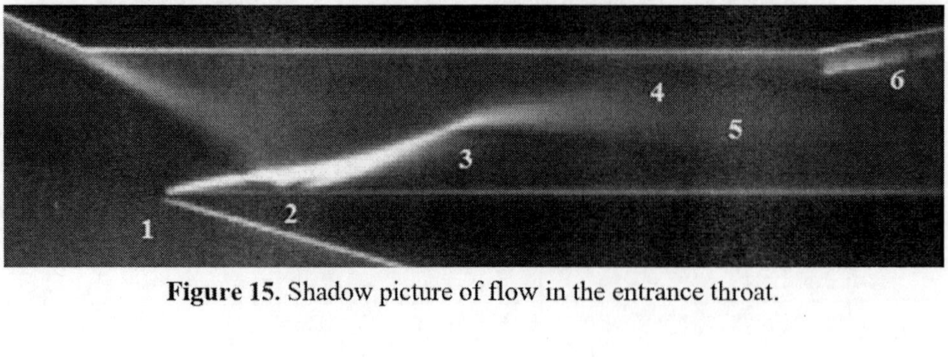

Figure 15. Shadow picture of flow in the entrance throat.

The lower edge in the experiment was located in such a way that the place of encounter of shock waves from the first and second deflections (figures.13,14) of the upper edge occurred at its tip (figure 15, 1). This was done to enhance the picture of the shock-wave interaction in the flow path. An oblique shock wave from the second ramp (figure 14) occurs with the bottom edge slightly further downstream (figure 15, 2). Particular attention is drawn to the separation of shock waves at the point of reflection and their subsequent curvilinear structure (figure 15, 3). At the location of the shock waves (figure 15, 3), reflected from the head shocks of the leading edge (figure 15, 1 and 2), attention is drawn to the presence of a detached flow from the upper edge of the air intake. The places of encounter of shock wave re-reflections and separated flows are shown in figure 15, 4-6.

To assess the influence of the side walls on the flow structure in the throat section, experiments with lateral edges simulating the sidewalls of the propulsion system of the apparatus were performed (figures 16, 17). The edges were made of plexiglass 2 mm thick with a blunt radius of 1 mm.

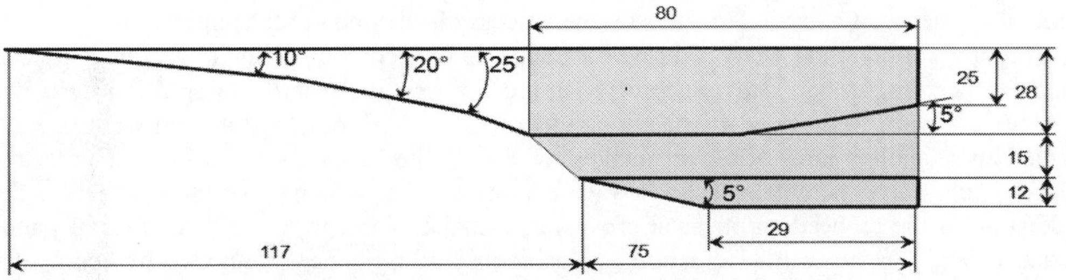

Figure 16. Schematic of the model with side edges.

Figure 17. Comparison of experimentally obtained flow patterns for configurations with side edges and without. M = 7 airflow.

As can be seen from figure 17, the main features of the flow pattern are traced in both model configurations. However, aspects of the influence of the side edge characteristics on the shock-wave processes inside the flow path require additional experimental studies. In particular, for more complete and detailed implementation of these studies, it is expedient to use the so-called panoramic recording methods [17].

5. Conclusion

At the HAST facility, preliminary studies of the ExoMars model were carried out under multiregime flow conditions. Shock patterns are obtained near the model for Mach numbers M = 7...4.5. Using the pattern recognition system, the results of model oscillations were analyzed throughout the entire experiment. This allowed to distinguish several periods of shock-wave interactions in the shock tube sections and to compare them with the behavior of the high-speed flow oncoming to the model. A more detailed study of the model movement during the test is expedient to carry out taking into account the influence of additional factors on the process of carrying out the experiment.

Open sources of literature were analyzed for modeling the geometry of the hypersonic aircraft inlet. The main aspects of design and issues of ensuring the operation of the propulsion system under various conditions of high-speed flight are singled out. At the HAST installation experiments were conducted on several variants of the air intake model with the Mach number M = 7. The results of a preliminary analysis of the obtained shock-wave interaction pictures are presented. For a more detailed consideration of the proceeding physical processes, it makes sense to use the methods of panoramic diagnostics and carry out validation calculations using the methods of computational aerothermodynamics which is planned for implementation in the future.

Acknowledgement

The presented study was partially supported by Government program of basic research of Russian academy of science (contract #AAAA-A17-117021310372-6) and partially supported by Russian Foundation for Basic Research grant # 16-01-00379.

References

[1] Surzhikov S T and Omaly P 2011 Radiative Gas Dynamics of Martian Space Vehicles 49th *AIAA Aerospace Sciences Meeting AIAA paper* 2011–0452 p 28

[2] Surzhikov S T 2016 Radiative Gas Dynamics of MSL at angle of attack 54th *AIAA Aerospace Sciences Meeting* AIAA paper 2016–0742 p 25

[3] Yatsukhno D, Surzhikov S, Bessonov O, Andrienko D, Annaloro J, Hebert J P and Omaly P 2018 Radiative Gas Dynamics of MSL and Exomars at Angle of Attack in view of Turbulent Heating *AIAA Aerospace Sciences Meeting AIAA paper* 2018–0982

[4] Woods W C, Holland S D and DiFulvio M 2001 Hyper-X Stage Separation Wind-Tunnel Test Program *J. of Spacecraft and Rockets* **38** pp 811–819

[5] Weigand B 2009 Aero-Thermodynamic Design of a Scramjet Propulsion System: The graduate school GRK 1095/2 *Deutsche Forschungsgemeinschaft (DFG)*

[6] Martos J F A, Rêgo I S, Laiton S N P, Lima B C, Costa F J and Paula Toro P G 2017 Experimental Investigation of Brazilian 14-X B Hypersonic Scramjet Aerospace Vehicle *International J. of Aerospace Engineering* **2017** p 10

[7] Quan L H, Hung N P, Quang L D and Long V N 2016 Analysis and Design of a Scramjet Engine Inlet Operating from Mach 5 to Mach 10 *Int.l J. of Mechanical Engineering and Applications* **4** pp 11–23

[8] Chan J, Sislian J and Alexander D 2010 Numerically Simulated Comparative Performance of a Scramjet and Shcramjet at Mach 11 *J. of Propulsion and Power* **26** pp 1125–1134

[9] Seleznev R K, Surzhikov S T and Shang J A 2016 Quasi-One-Dimensional Analysis of Hydrogen-Fueled Scramjet Combustors 52nd *AIAA Joint Propulsion Conf. AIAA paper* 2016–4569 p 27

IOP Conf. Series: Journal of Physics: Conf. Series **1009** (2018) 012038 doi:10.1088/1742-6596/1009/1/012038

[10] Kotov M A, Kryukov I A, Ruleva L B, Solodovnikov S I and Surzhikov S T 2015 Supersonic Air Flows Around Some Geometrical Primitives 33rd *AIAA Applied Aerodynamics Conf. AIAA paper* 2015–3012

[11] Kotov M A, Kryukov I A, Ruleva L B, Solodovnikov S I and Surzhikov S T 2014 Multiple Flow Regimes in a Single Hypersonic Shock Tube Experiment 30th *AIAA Aerodynamic Measurement Tchnology and Ground Testing Coferece AIAA paper* 2014–2657

[12] Kotov M A, Kryukov I A, Ruleva L B and Solodovnikov S I 2017 The Investigation of Shock-Wave Interaction with Aerodynamic Models 55th *AIAA Aerospace Sciences Meeting AIAA paper* 2017–0262

[13] Kotov M A, Ruleva L B, S I Solodovnikov S I and Surzhikov S T 2017 Experimental and numerical study of supersonic flow over two blunted wedges *J. Physysics: Conference Series* **815** 012025

[14] http://physlets.org/tracker/

[15] Heiser W and Pratt D 1994 Hypersonic Airbreathing Propulsion *AIAA Education Series* p 594

[16] Kotov M A and Kuzenov V V 2012 Geometric modeling of complex surfaces of hypersonic aircraft in CAD *Physical-Chemical Kinetics in Gas Dynamics* **13** (4) http://chemphys.edu.ru/issues/2012-13-4/articles/372/

[17] Borovoy V, Egorov I, Mosharov V, Radchenko V, Skuratov A, Struminskaya I 2016 Entropy-Layer Influence on Single-Fin and Double-Fin/Boundary-Layer Interactions *AIAA Journal* **54** pp 443–457

IOP Conf. Series: Journal of Physics: Conf. Series **1009** (2018) 012039 doi:10.1088/1742-6596/1009/1/012039

Numerical simulation of the underexpanded plume spectral radiance using Monte-Carlo method

D A Storozhev [1,2,3]

[1]N.L. Dukhov All-Russian Research Institute of Automatics (State Corporation «ROSATOM» company), Suschevskaya Street 22, Moscow 127055, Russia
[2]Moscow Institute of Physics and Technology (State University), 9 Institutskiy per., Dolgoprudny, Moscow Region, 141701, Russia
[3]Ishlinsky Institute for Problems in Mechanics RAS, Vernadsky Prospect 101(1), Moscow, 119526, Russia

dmitry.stor@gmail.com

Abstract. Numerical simulation of the tactical rocket plumes spectral radiance is performed. An influence of the chemical composition and temperature of the plume exhausts as well as the angle of sight and scattering on the solid particles is discussed. Numerical simulation results for homogeneous plume of tactical rocket are in a good agreement with published data.

1. Introduction

Most of the modern subsonic and supersonic vehicles produces plumes which radiate in the infrared region of the electromagnetic spectrum. Due to partial atmospheric opacity in this spectral region the radiance reaches the detector with some distortion. Particular shape of the spectrum depends on the type of the engine (solid or liquid), chemical composition of the engine fuel, radius of the nozzle exit, presence of solid particles, its phase and radius, altitude of the flight.

For this reason, numerical simulation of the plume spectral radiance is complex problem which requires the knowledge of large amount of data such as: spatial distribution of temperature and pressure in the plume, chemical composition, absorbance coefficients for different chemical species, complex refractive index of solid particles as well as physical and mathematical models[1–5] included different nonequilibrium processes proceeded in the plume. It is well known that the most universal algorithm to calculate spectral radiance is the Monte-Carlo method [1, 6, 7]. The main purpose of this work is to calculate spectral signature of the underexpanded plumes with the use of Monte-Carlo method and investigate its dependence on the plume characteristics: temperature, emitted volume, number density of solid particles and its size, imaginary part of the refractive index.

2. Spectral radiance of the plume with Al_2O_3 particles

The plume flowfield was modelled as a homogeneous cylinder. The cylinder size, temperature and pressure, as well as chemical composition was assumed to be the same as in [8]. It's length is equal to $L = 600\,\mathrm{cm}$ and radius $r = 10\,\mathrm{cm}$. The cylinder is isothermal with temperature $T = 1000\,\mathrm{K}$ and pressure $p = 1\,\mathrm{atm}$. Gas and solid particle concentrations were uniform. All the particles are spherical and have the same radius. Chemical composition of the plume and its particle properties are given in table 1. The aspect angle was assumed to be equal to 90°.

Table 1. Chemical composition and solid particle properties of the plume.

Species	Mole fraction
HCL	–
HF	–
H_2	(0.40)
H_2O	0.30
CO	0.02
CO_2	–
Al_2O_3	(0.5)

Mole fraction	Density, $g * cm^{-3}$	Concentration, cm^{-3}		
		R = 0.1 mcm	R = 1 mcm	R = 5 mcm
0.001	1.25E-06	8.08E+07	8.08E+04	6.46E+02
0.01	1.26E-05	8.15E+08	8.15E+05	6.52E+03
0.1	1.39E-04	8.96E+09	8.96E+06	7.17E+04
0.5	1.25E-03	8.07E+10	8.07E+07	6.45E+05

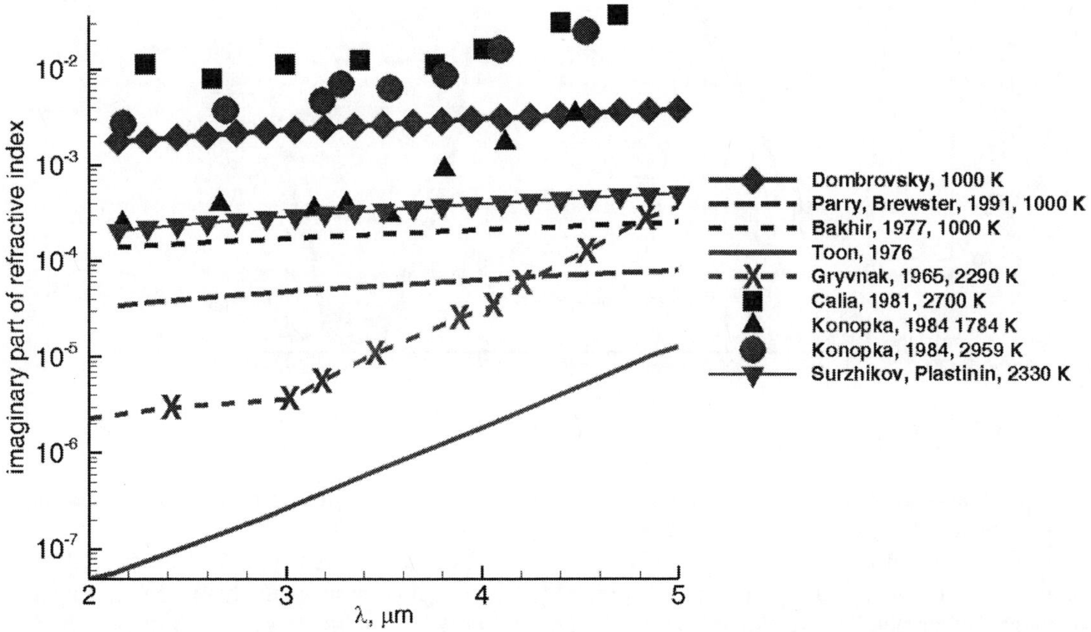

Figure 1. Comparison of the imaginary part of the refractive indexes provided by different authors [9 –15].

The absorption and scattering coefficients of the solid particles are calculated with the use of the Mie theory. So the spectral signature of the plume with scattering particles is significantly depends on the imaginary part of the refractive index which in its turn depends on the wavelength, temperature and phase of solid particles. Comparison of the values of imaginary part of the refractive index for Al_2O_3 particles given by different authors [9 –15] is presented in figure 1. It can be seen that its values differ by several orders of magnitude for different temperature and phase of particles.

The influence of the imaginary part of the refractive index on the spectral radiance of the plume is given in figure 2. Numerical simulation of the signature performed for the particle radius $r = 1$ mcm and different concentrations of solid Al_2O_3. Two approximations of the imaginary part of the refractive index were used provided by Dombrovskii [9] and Toon [12]. With the increase of the solid particle concentration difference in the spectral radiance of the plume for data [12] and data [9] increases. The main change in spectrum is observed in the spectral range $\omega = 2400 \div 3000\,\text{cm}^{-1}$ and $\omega = 4100 \div 5000\,\text{cm}^{-1}$ where optical depth is small. At the same time in the range $\omega = 2000 \div 2400\,\text{cm}^{-1}$ and $\omega = 3000 \div 4100\,\text{cm}^{-1}$ where the absorption coefficient for CO_2 and H_2O the differences in spectral radiance are insignificant.

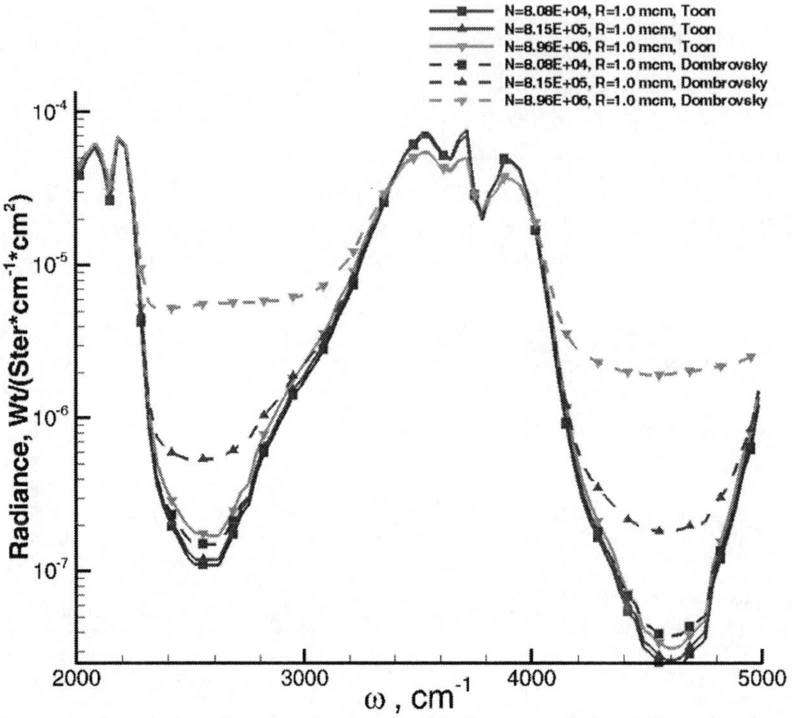

Figure 2. Spectral radiance calculated with the use of the different values of the imaginary part of the refractive index.

Numerical simulation results for the matrix of solid particle concentrations and radiuses and its comparison to the data [8] are given in figure 3.

The are two sets of numerical simulation results. The first set (left column) is the total volume emission of the homogeneous plume and the second set (right column) is the emission only for a path through the diameter of the radiative cylinder. The numerical simulation results with the use the values of the refraction index provided by Toon [12] give satisfactory agreement with the data [8].

3. Spectral radiance of the plume with B_2O_3 particles

As in previous section the plume flowfield was modelled as a homogeneous cylinder. The cylinder size, temperature and pressure, as well as chemical composition was assumed to be the same as in [16]. It's length is equal to $L = 1500\,\text{cm}$ and radius $r = 75\,\text{cm}$. The cylinder is isothermal with temperature and pressure given in table 2. The temperature is assumed to be equal to $T = 1000\,\text{K}$ (variant 1) and $T = 1600\,\text{K}$ (variant 2) and pressure $p = 1\,\text{atm}$ (both cases). Its gas and solid particle concentrations were uniform. All the particles are spherical and have the same radius. It was assumed that B_2O_3 are the scattering solid particles in this case. Values of the complex refractive index for B_2O_3 particles were taken from [16]. Chemical composition of the plume and its particle properties are given in table 2. The aspect angle was assumed to be equal to 90°.

IOP Conf. Series: Journal of Physics: Conf. Series **1009** (2018) 012039 doi:10.1088/1742-6596/1009/1/012039

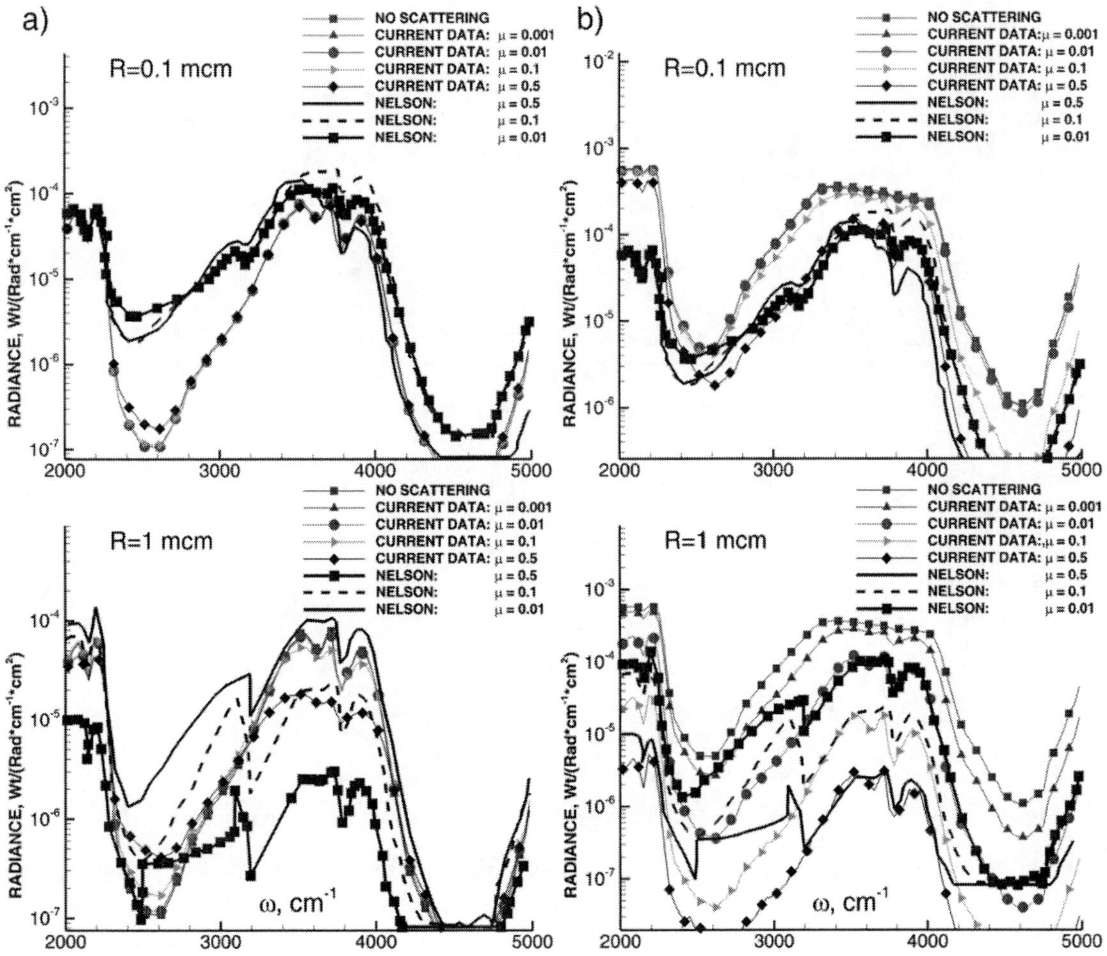

Figure 3. Total volume emission of the plume (a) and the emission for a path through the diameter of the cylinder (b) and its comparison to data [8].

Numerical simulation results for the matrix of solid particle concentrations and radiuses and its comparison to the data [16] are given in figure 4.

There are two sets (variant 1 and variant 2) of numerical simulation results. The conditions for any case are given in table 2. A good agreement with the data [16] is obtained for all considered cases.

Table 2. Chemical composition and solid particle properties of the plume with B_2O_3 particles.

Mole fraction	Density, $g * cm^{-3}$	Concentration, cm^{-3}		CO_2	H_2O	CO	HBO_2	N_2
		R = 0.1 mcm	R = 1 mcm					
			Variant 1					
0.00011	9.53E-08	1.26E+07	1.26E+04					
0.00025	2.14E-07	2.84E+07	2.84E+04	0.140	0.060	0.030	0.001	0.579
0.00100	8.58E-07	1.14E+08	1.14E+05					
			Variant 2					
0.00011	5.95E-08	7.90E+06	7.90E+03					
0.00025	1.34E-07	1.78E+07	1.78E+04	0.30	0.020	0.020	0.020	0.790
0.00100	5.36E-07	7.11E+07	7.11E+04					

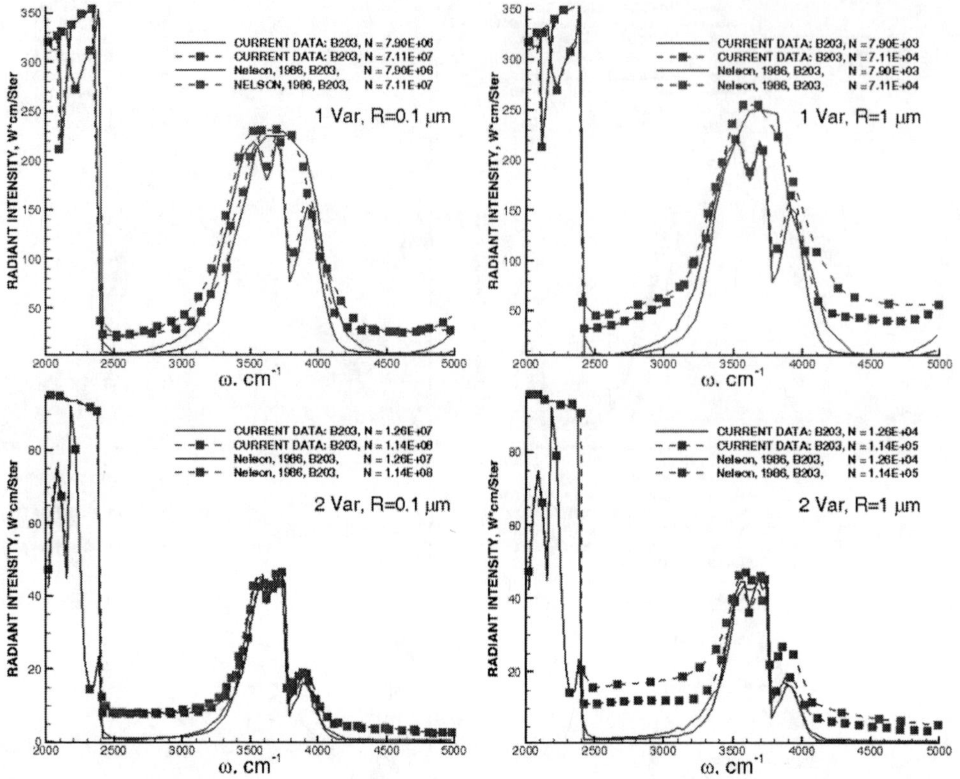

Figure 4. Spectral radiance of the plume with B_2O_3 particles and its comparison with data [16].

In figure 5 comparison of the spectral signature for different angles of observation is given ($\theta = 90^0$ and $\theta = 45^0$). Numerical simulation results are presented for the Variant 2 (see table 2) for concentration of solid particles $N = 7.11E + 04$ and its radius $r = 1\,mcm$. With the decrease of the angle between the axis of the cylinder and the line of sight from $\theta = 90^0$ to $\theta = 45^0$ the local maximum in the spectral radiance in the range $\omega = 3300 - 4000\,cm^{-1}$ (which corresponds to H_2O emission) is decreased.

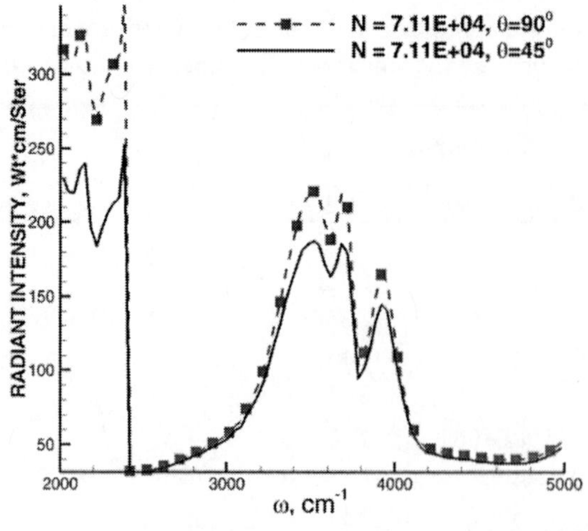

Figure 5. Spectral radiance of the plume for different angles of observation.aa.

IOP Conf. Series: Journal of Physics: Conf. Series **1009** (2018) 012039 doi:10.1088/1742-6596/1009/1/012039

4. Spectral radiance of the nonhomegeneous plume of tactical rocket

The plume flowfield in this section is calculated from a system of a Navier-Stokes equations for compressible chemically reacting gas [2,17,18]. The system of equations is formulated in two-dimensional axissimmetric geometry:

$$\frac{\partial \boldsymbol{f}}{\partial t}+\frac{\partial \boldsymbol{E}}{\partial r}+\frac{\partial \boldsymbol{G}}{\partial z}=\frac{\partial \boldsymbol{E}_\mu}{\partial r}+\frac{\partial \boldsymbol{G}_\mu}{\partial z}+\boldsymbol{H}$$

$$\boldsymbol{f}=\begin{bmatrix}\rho\\ \rho U^r\\ \rho U^z\end{bmatrix};\ \boldsymbol{E}=\begin{bmatrix}\rho U^r\\ \rho U^r U^r+P\\ \rho U^z U^r\end{bmatrix};\ \boldsymbol{G}=\begin{bmatrix}\rho U^z\\ \rho U^r U^z\\ \rho U^z U^z+P\end{bmatrix};$$

$$\boldsymbol{E}_\mu=\begin{bmatrix}0\\ 2\mu\dfrac{\partial U^r}{\partial r}-\dfrac{2}{3}\mu div(\boldsymbol{U})\\ \mu\left(\dfrac{\partial U^z}{\partial r}+\dfrac{\partial U^r}{\partial z}\right)\end{bmatrix};\ \boldsymbol{G}_\mu=\begin{bmatrix}0\\ \mu\left(\dfrac{\partial U^z}{\partial r}+\dfrac{\partial U^r}{\partial z}\right)\\ 2\mu\left(\dfrac{\partial U^z}{\partial z}\right)-\dfrac{2}{3}\mu div(\boldsymbol{U})\end{bmatrix};$$

$$\boldsymbol{H}=\begin{bmatrix}-\dfrac{\rho U^r}{r}\\ 2\mu\left(\dfrac{\partial}{\partial r}\left(\dfrac{U^r}{r}\right)\right)-\dfrac{\rho\left(U^r\right)^2}{r}\\ \dfrac{\mu}{r}\left(\dfrac{\partial U^z}{\partial r}+\dfrac{\partial U^r}{\partial z}\right)-\dfrac{\rho U^z U^r}{r}\end{bmatrix};$$

Diffusion equation for chemical species:

$$\frac{\partial \rho_i}{\partial t}+\mathrm{div}\rho_i\boldsymbol{U}=\text{-}\mathrm{div}\boldsymbol{J}_i+\dot{w}_i$$

Heat equation:

$$\rho c_p\frac{\partial T}{\partial t}+\rho c_p\boldsymbol{U}\mathrm{grad}T=\mathrm{div}\left(\lambda\mathrm{grad}T\right)+\frac{\partial p}{\partial t}+\mathrm{grad}p+\Phi_\mu-\sum_{i=1}^{N_s}h_i\dot{w}_i$$

$$\Phi_\mu=\mu\left(2\left(\frac{U^r}{r}\right)^2+2\left(\frac{\partial U^r}{\partial r}\right)^2+2\left(\frac{\partial U^z}{\partial z}\right)^2+\left(\frac{\partial U^r}{\partial z}+\frac{\partial U^z}{\partial r}\right)^2-\frac{2}{3}\left(\frac{\partial U^z}{\partial z}+\frac{\partial U^r}{\partial r}+\frac{U^r}{r}\right)^2\right)$$

where Φ_μ is dissipative function; t – is time; z, r – orthogonal cylindrical coordinates; U^r, U^z are the projections of the velocity vector $\boldsymbol{U}$ on the axis r and z; p, ρ are pressure and density; T is temperature of the gas; μ,λ are viscosity and conductivity coefficients; c_p is the specific heat capacity; h_i, ρ_i are the enthalpy and density of i-th component; $\dot{w}_i$ is the mass production of the i-th component; $\boldsymbol{J}_i$ is the density vector of the diffusion flow for i–th component; D_i is the diffusion coefficient for the i-th component; Y_i is mass fraction for i–th component; N_s is the number of chemical species.

The algebraic Penner-Haselman-Edwards model of turbulent transfer [19] (PHE), were used to consider the effects of turbulent mixing in the model:

$$\mu_t = \frac{1}{2}\rho l^2 \left(\left(\frac{\partial U^z}{\partial r} + \frac{\partial U^r}{\partial z} \right)^2 + 2\left[\left(\frac{\partial U^r}{\partial r} \right)^2 + \left(\frac{\partial U^z}{\partial z} \right)^2 + \left(\frac{U^r}{r} \right)^2 \right] \right)^{1/2}$$

$$l = K\left(\frac{\left(\left(U^z\right)^2 + \left(U^r\right)^2 \right)^{1/2}}{B^{1/2}} + B^{1/2}\frac{1}{\left(\left(\frac{\partial^2 U^r}{\partial r^2} + \frac{1}{r}\frac{\partial U^r}{\partial r} \right)^2 + \left(\frac{U^r}{r^2} \right)^2 + \left(\frac{\partial^2 U^r}{\partial z^2} \right)^2 + \left(\frac{\partial^2 U^z}{\partial r^2} + \frac{1}{r}\frac{\partial U^z}{\partial r} \right)^2 + \left(\frac{\partial^2 U^z}{\partial z^2} \right)^2 \right)^{1/2}} \right)$$

$$B = \left(\frac{\partial U^z}{\partial r} \right)^2 + \left(\frac{\partial U^z}{\partial z} \right)^2 + \left(\frac{\partial U^r}{\partial r} \right)^2 + \left(\frac{\partial U^r}{\partial x} \right)^2 + \left(\frac{U^r}{r} \right)^2$$

where K is an empirical constant with recommended value of 0.125 [19]. Its value is varied in calculations.

For the numerical simulation of the afterburning processes in the plume the following kinetic scheme was used:

1)* $OH+H_2 \leftrightarrow H_2O+H$
2)* $OH+O \leftrightarrow H+O_2$
3)* $H+H+M \leftrightarrow H_2+M$
4)* $H+OH+M \leftrightarrow H_2O+M$
5)* $OH+OH \leftrightarrow H_2O+O$
6)* $O+H_2 \leftrightarrow OH+H$
7)* $H+O+M \leftrightarrow OH+M$
8)* $O+O+M \leftrightarrow O_2+M$
9) $CO+OH \leftrightarrow CO_2+H$
10) $CO+O+M \leftrightarrow CO_2+M$

Rate coefficients for the processes (1)−(8) were taken from the [20]. Rate coefficients for the reaction (9) – (10) were taken from the [21].

The initial data for the numerical simulation were assumed to be the same as in [22]. The temperature is equal to $T = 2070\,K$ and pressure is equal to $p = 1.069\,K$. Chemical composition is given in table 3.

Table 3. Chemical composition.

Chemical component:	Mass fraction
CO_2	0.103
H_2O	0.0124
CO	0.169

The rocket radius is equal to $r = 8\,cm$. The nozzle exit radius is assumed to be equal to $r = 3.81\,cm$. The nozzle exit velocity isn't given in the paper [22] so it has been varied. Numerical simulation results for the plume flowfield are presented in figure 6. Spatial two-dimensional distribution in the plane trough the axis of the plume for the density, temperature, and mass fraction of chemical species are presented in this figure. This results were obtained for the nozzle exit velocity $V = 1476$ m/s. Comparison of the axial distributions of temperature and mass fraction of chemical species for different nozzle exit velocities ($V = 1476$ m/s and $V = 3000$ m/s) is given in figure 7.

IOP Conf. Series: Journal of Physics: Conf. Series **1009** (2018) 012039 doi:10.1088/1742-6596/1009/1/012039

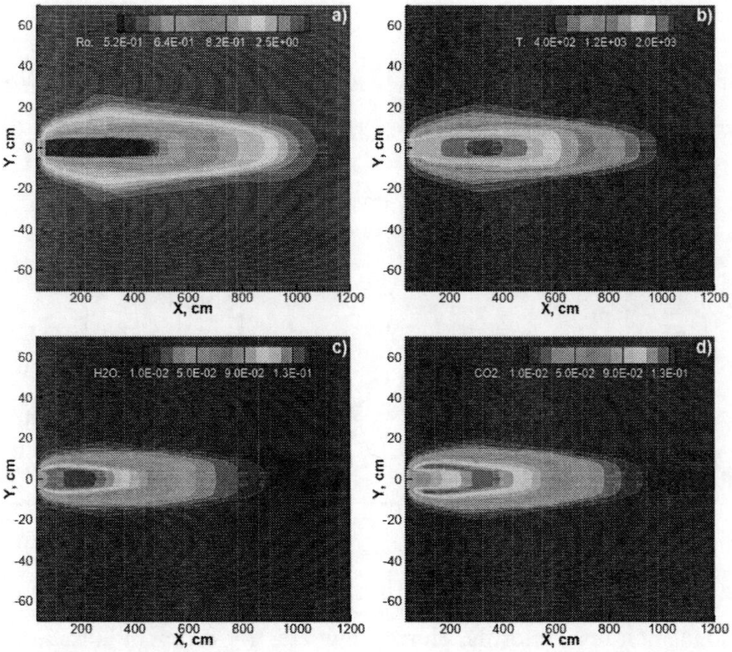

Figure 6. Spatial distribution of the density (a), temperature(b), H_2O (c) and CO_2 (d) mass fractions.

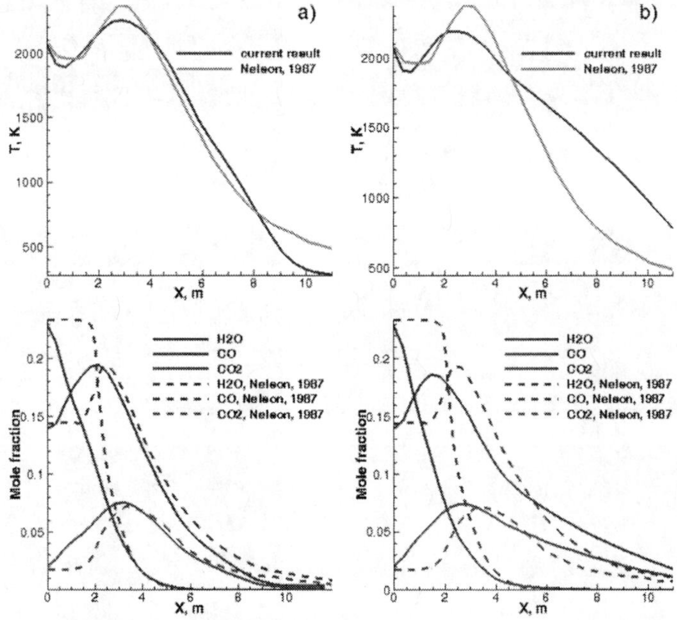

Figure 7. Axial distributions of the temperature (top) and mass fractions of chemical species (bottom) for different values of the nozzle exit velocities ($V = 1476$ m/s (a) and $V = 3000$ m/s (b)).

Spectral signature of the plume for the different angles of observation (broadside and nose on) are given in figure 8. It is considered that in the case of the nose on signature ($\Theta = 0^{\circ}$) the rocket body partially obscure the radiative plume.

IOP Conf. Series: Journal of Physics: Conf. Series **1009** (2018) 012039 doi:10.1088/1742-6596/1009/1/012039

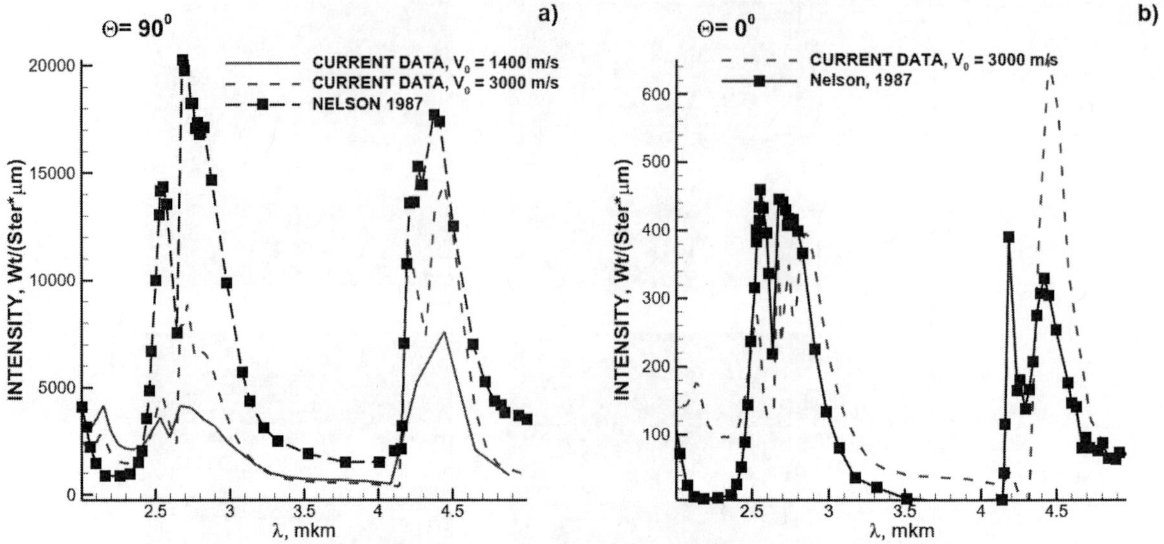

Figure 8. Broadside (a) and nose on (b) spectral radiance of the tactical rocket plume.

Numerical simulation results for the plume flowfields for different altitudes are given in figure 9. With the increase of the altitude plume significantly expands and the absolute value of temperature on the plume axis is decreased. Comparison of the plume radiance for different altitudes are presented in figure 10.

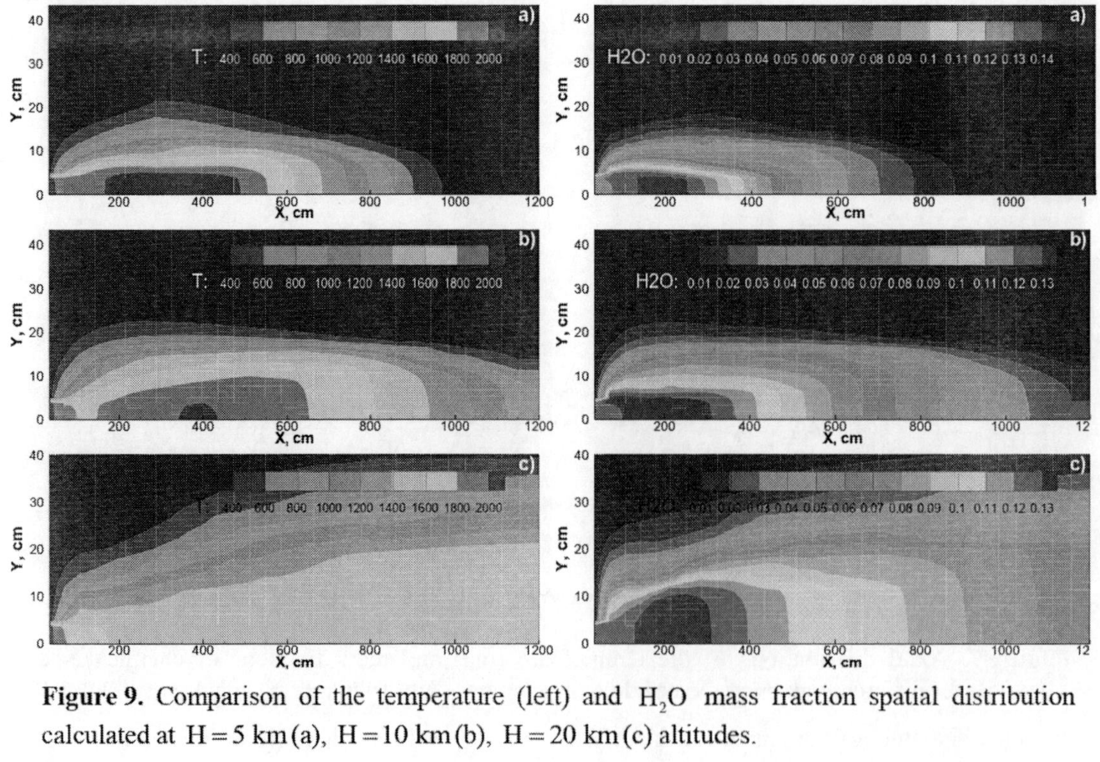

Figure 9. Comparison of the temperature (left) and H_2O mass fraction spatial distribution calculated at H = 5 km (a), H = 10 km (b), H = 20 km (c) altitudes.

IOP Conf. Series: Journal of Physics: Conf. Series **1009** (2018) 012039 doi:10.1088/1742-6596/1009/1/012039

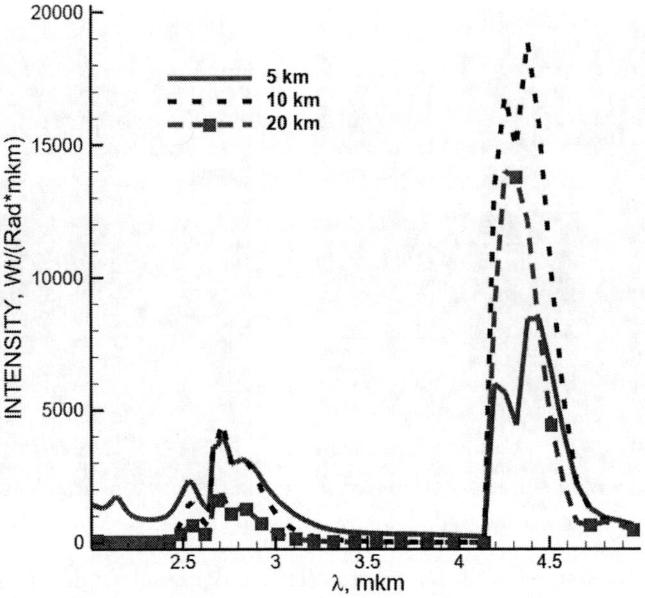

Figure 10. Spectral radiance of the plume with B_2O_3 particles and its comparison with data [5].

5. Spectral radiance of the Atlas plume

Numerical simulation of the Atlas rocket plume flowfield was performed with the use of the technique described in the previous section. The boundary conditions for the freestream and nozzle exit regions were taken from the [23]. The specified boundary conditions are given in the table 4.

Table 4. Freestream and nozzle exit properties.

	Freestream	Nozzle exit
Velocity (m/s)	1476	2960
Pressure (Pa)	278	68850
Temperature (K)	251	2230
Mass Fractions:		
CO_2	0.000458	0.302262
H_2O	0.000000	0.272443
CO	0.000000	0.412135
N_2	0.766388	0.000000
O_2	0.233154	0.000028
H_2	0.000000	0.012194
OH	0.000000	0.000797
O	0.000000	0.000014
H	0.000000	0.000127

Spatial distributions of temperature and density in the Atlas plume are presented in figure 11. Comparison of the axial and spatial distributions calculated with those obtained in the [23]. for perfect and chemically reacting gas is given in figure 12. Numerical simulation results presented in this figure were obtained for the laminar case. There is a satisfactory agreement for both (perfect gas and chemically reacting gas) cases.

Comparison of the spectral signature with those obtained by [24] is given in figure 13.

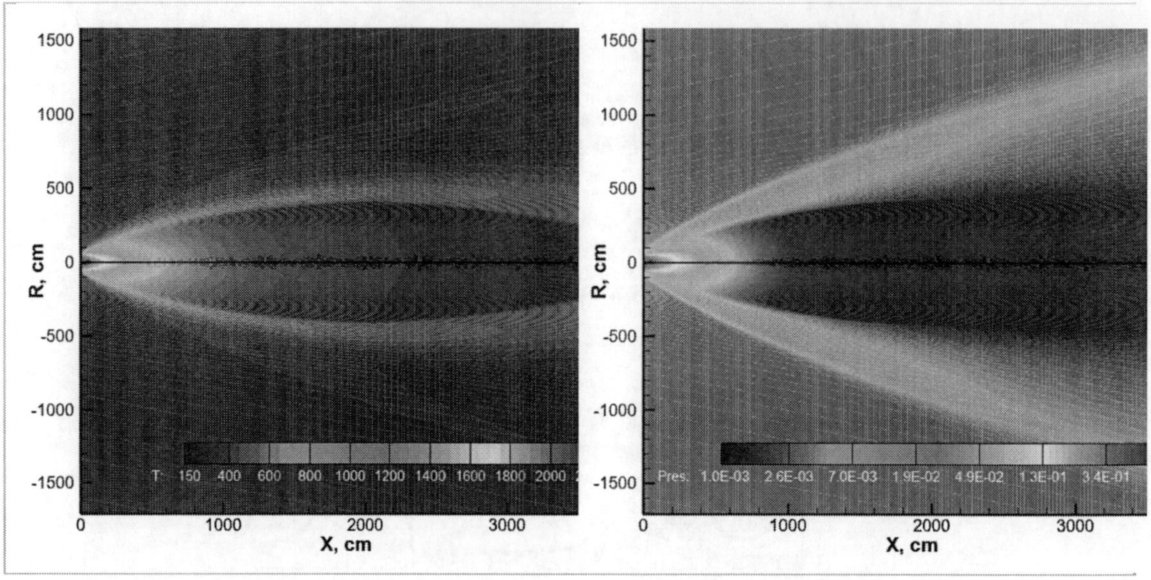

Figure 11. Spatial distribution of the temperature(left) and pressure(right) in the plume.

6. Conclusion

Numerical simulation of the plume flowfields and spectral signatures for different altitudes, chemical composition, altitudes and solid particle properties were performed. A satisfactory agreement with published results are obtained. An influence of the altitude and imaginary part of the refractive index on the plume spectral signature was shown.

The present study was supported by the Russian Science Foundation project № 16–11–10275.

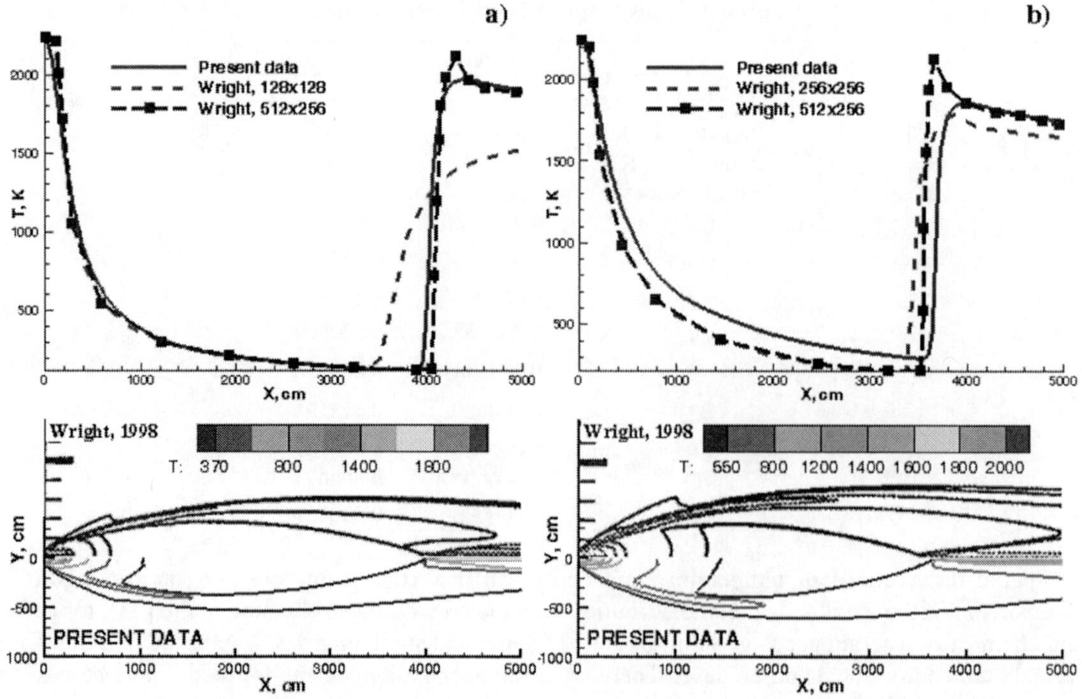

Figure 12. Comparison of the axial (top) and spatial (bottom) distributions with the data [23] for perfect(a) and chemically reacting gas(b).

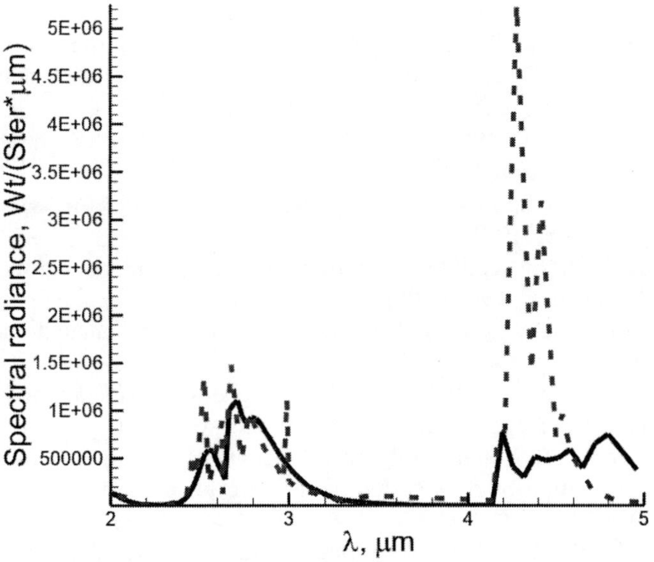

Figure 13. Spectral signature of the Atlas plume and its comparison to data [24].

References

[1] Surzhikov S 2002 Spectral and Narrow Band Directional Emissivity of Light-Scattering and Non-Scattering Volumes *8th AIAA/ASME Joint Thermophysics and Heat Transfer Conf.*

[2] Surzhikov S T 2011 Radiative-convective heat transfer of a spherically shaped space vehicle in carbon dioxide *High Temp.* **49** 92–107

[3] Surzhikov S and Storozhev D 2017 Numerical simulation of dissociation kinetics in the Penning discharge plasma using 2D modified drift-diffusion model *55th AIAA Aerospace Sciences Meeting*

[4] Seleznev R, Surzhikov S and Shang J 2016 Quasi-one-dimensional analysis of hydrogen-fueled scramjet combustors *Proc. 52nd AIAA/ASME/SAE/ASEE Joint Propulsion Conf. AIAA 2016*–4569

[5] Dikalyuk A S 2017 Development of Particle-in-Cell Solver for Numerical Simulation of Penning Discharge *AIAA 2017* **842** 22

[6] Surzhikov S 2004 Numerical Simulation of Heat Radiation Generated by Entering Space Vehicle *37th AIAA Thermophysics Conf.*

[7] Surzhikov S 2003 Monte Carlo Simulation of Plumes Spectral Emission *36th AIAA Thermophysics Conf.*

[8] Nelson H F 1984 Influence of particulates on infrared emission from tactical rocket exhausts *J. Spacecr. Rockets* **21** 425–32

[9] Dombrovsky L A 1996 *Radiation heat transfer in disperse systems* (Begell House: New York)

[10] Parry D L and Brewster M Q 1991 Optical constants of Al2O3 smoke in propellant flames *J. Thermophys. Heat Transf.* **5** 142–9

[11] Bakhir L P, Levashenko G I and Tamanovich V V 1977 Refinement of the imaginary part of the complex refractive index of liquid aluminum oxide *J. Appl. Spectrosc.* **26** 378–83

[12] Toon O B, Pollack J B and Khare B N 1976 The optical constants of several atmospheric aerosol species: Ammonium sulfate, aluminum oxide, and sodium chloride *J. Geophys. Res.* **81** 5733–48

[13] Gryvnak D A and Burch D E 1965 Optical and Infrared Properties of Al_2O_3 at Elevated Temperatures* *JOSA* **55** 625–9

[14] Ludwig C B, Malkmus W, Walker J, Slack M and Reed R 1981 The standard infrared radiation model *AIAA 16th Thermophysics Conf., Palo Alto, CA, Jun* 23–25
[15] Konopka W L, Reed R A and Calia V S 1984 Measurements of infrared optical properties of Al2O3 rocket particles *Prog. Astronaut. Aeronaut. Spacecr. Contam. Sources Prev.* **91** 180–196
[16] Nelson H F and Tucker E O 1986 Boron slurry-fueled jet engine exhaust plume infrared signatures *J. Spacecr. Rockets* **23** 527–33
[17] Surzhikov S T and Shang J S 2004 Two-component plasma model for two-dimensional glow discharge in magnetic field *J. Comput. Phys.* **199** 437–64
[18] Surzhikov S T 2012 Radiative-collisional models in non-equilibrium aerothermodynamics of entry probes *J. Heat Transf.* **134** 031002
[19] Penner J E, Haselman L C and Edwards L L 1985 Buoyant plume calculations *23rd Aerospace Sciences Meeting*
[20] Evans J S and Schexnayder C J 1980 Influence of chemical kinetics and unmixedness on burning in supersonic hydrogen flames *AIAA J.* **18** 188–193
[21] Warnatz J, Maas U, Dibble R W and Warnatz J 1996 *Combustion* vol 3 (Springer)
[22] Nelson H F 1987 Evaluation of rocket plume signature uncertainties *J. Spacecr. Rockets* **24** 546–51
[23] Wright M, Rao R, Candler G, Hong J, Schilling T and Levin D 1998 Modeling issues in the computation of plume radiation signatures *34th AIAA/ASME/SAE/ASEE Joint Propulsion Conf. and Exhibit*
[24] Ozawa T, Garrison M B and Levin D A 2007 Accurate Molecular and Soot Infrared Radiation Model for High-Temperature Flows *J. Thermophys. Heat Transf.* **21** 19–27

IOP Conf. Series: Journal of Physics: Conf. Series **1009** (2018) 012040 doi:10.1088/1742-6596/1009/1/012040

Model of thermal radiation using heat absorption by CO2 in submicron pores with application to magnesium-zinc ferrite fine disperse particles synthesis via combustion

A A Markov[1] and I A Filimonov[2]

[1]Ishlinsky Institute for Problems in Mechanics RAS, Moscow 119526, Russia
[2]ISMAN RAS Chernogolovka, Moscow Region, 142432 Russia

E-mail: markov.ipm@yandex.ru

Abstract. Microscopic and macroscopic heat fluxes have been calculated on the basis of the self-consistent analysis of micro and macro scales taking into account absorption of thermal radiation by the CO2 molecules in micro pores. The macroscopic fluxes have been considered with the help of A. Shack modification to the Stefan-Boltzmann law at the given degree of blackness of a surface of a solid phase. These fluxes are received by averaging of the microscopic fluxes taking into account the influence of Knudsen's layer and absorption of heat during combustion of carbon. In the case of a submicron scale the models of slipping of a gas phase and jump of temperature of gas inside Knudsen's layers in the vicinity of pore borders are used. The intensities of macroscopic fluxes are presented through the parameters of similarity and depend on the emissivity of pore surfaces, on the coefficients of molecule reflection and thermal accommodation. The governing system of equations on a macroscopic scale includes heat exchange between the gaseous and solid phases. The macro fluxes are calculated by means of averaging of micro fluxes taking into account the effect of Knudsen's layer and the heat absorption during the carbon combustion. The synthesis of magnesium-zinc ferrite submicron particles is modelled numerically. The results of thermal front numerical simulation allow us to evaluate the impact of CO2 absorption, emissivity along with the slippage on the heat and mass transfer. It was found that an increase in the intensity of radiation corresponds to the decrease in the rate of submicron particles growth.

1. Introduction

At submicron scales when the size of a particle or of a pore filled with gas is less than the wavelength of radiation, the properties of absorption and emission significantly depend on the surface molecular structure.

Results and references concerning thermal radiation at submicron scales are presented in [1-6]. Measurements of thermal radiation in a nanometric gap between two golden plates [1,2] have detected a heat flux which greatly exceeds the radiation limit of the absolutely black body. The obtained result was originally explained by the coupling of surface phonon modes inside the film, which creates additional channels for the photon tunneling [6].

The absorption and emission properties of clusters as systems of bound atoms are demonstrated to be the sources of thermal radiation that like the soot radiation in flames. The cluster properties are reviewed in [4]. Peculiarities of absorption of thermal radiation by gas molecules are formulated in [7]. The approximate formulas for the heat absorption by molecules CO2 and H2O suggested by Schack with the Hottel's and Edgbert's modification are in good agreement with experimental data. The formulas are based on black-body radiation with given emissivity for gas and solid surface.

The mathematical models of self-propagating high temperature synthesis (SHS) of oxides or Combustion Synthesis (international name) are considered in numerous papers: [8-30]. The economical and energy efficient synthesis of nanostructured complex oxides from inexpensive reactant mixtures, referred to as Carbon Combustion Synthesis of Oxides (CCSO) was recently developed [8-11]. The CCSO is a modified form of the Self-propagating High-temperature Synthesis (SHS) or of so called internationally: Combustion Synthesis [16-18]. The experimental and numerical studies of CCSO have revealed that the gas transport behavior is rather complex with respect to traditional SHS systems. It involves at least two gaseous components: oxygen (oxygenator) as the main and practically unique gaseous reagent in SHS for gas-solid reactions and carbon dioxide which is released in the post combustion zone during CCSO. No carbon was used previously in original SHS systems [16]. Correspondingly, formation and release of carbon dioxide (the 2nd gas phase in CCSO) was not considered in models [18-20]. They were simpler in comparison with the multiphase gas dynamics taking place in CCSO [8-14, 26-30].

The effects of complex CCSO kinetics on dispersion of the CCSO condensed product are reported in [12-15, 24-30]. The models take into account in detail the gas dynamics as well as the mass and thermal transport accompanying CCSO. The result of theoretical predictions is in good agreement with experiments [28-30]. In SHS of materials, the thermal radiation is one of the fundamental mechanism of heat transfer. The characteristics of thermal radiation depends on various properties of the surface, including its temperature, its spectral absorptivity and spectral emissivity power, as is expressed by Kirchhoff's law. The simple approximation for the emitted power of radiation from particle surface into micro pore can be found using the Stefan-Boltzmann law for a gray body with a given emissivity [13, 31-33].

The estimates of radiation in combustion systems are given in [12, 21, 31-35]. The radiation can be neglected if the combustion temperature T_b doesn't exceed the critical value $T_b < T_{crit}^{rad}$. In the case of slow propagating front of carbon combustion synthesis of oxides (CCSO) [31-33] by smoldering combustion, the value of was estimated to be: $T_b << T_{crit}^{rad}$. Thus for slow CCSO the radiation impact is not important. The impact of radiation into the integral heat flux of slow CCSO [11] was estimated to be less than $6 \div 14\%$ while for fast CCSO for which the velocity of propagating combustion front $v_c \cong 4mm/s$ the impact of thermal radiation rises up to $\cong 43\%$ and radiation should be taken into account.

During synthesis of submicron particles via CCSO the formation of Knudsen layer in gas close to pore surface becomes important [13, 14]. The models of slipping of a gas phase and jump of temperature of gas inside Knudsen's layers in the vicinity of pore borders are used.

The aim of the present research is to analyze the thermal radiation taking into account the CO2 absorption in submicron pores. The Shack's formula with the Hottel's and Edgbert's modification is applied [7]. According to [1, 2] the black-body model can be applied for length scales that are about 0.5 micron and more. For these scales the measured heat fluxes are not significantly differed from those predicted by the black-body radiation law. These length scales are considered in the present paper.

The approach applied is the generalization of the slip-jump technique [14]. The various intensity of macro fluxes due to gas velocity slippage on the surface of nanopores with allowance for a thermal radiation impact are simulated numerically. The temperature jump depends on the thermal reflection coefficient and on the thermal diffusivity. The radiation intensity, velocity slippage and temperature jump fluxes are formulated using the appropriate similarity parameters. The synthesis of magnesium-zinc ferrite submicron particles is modeled numerically. The results of thermal front numerical simulation allow us to evaluate the impact of CO2 absorption, emissivity along with the slippage on the heat and mass transfer.

2. Simplified radiation equations at micro scale for isolated pore

In contrast to solids, gases have usually just lines, series of lines or bands as the spectra of absorption and radiation, i.e., the appreciable infrared (IR) absorption or radiation takes place only in certain local parts localized ranges of wavelength in of their spectrum (in so called absorption-emission bands), while solids have a continuous absorption-emission spectrum, since they absorb rays of any wavelength. Monatomic and diatomic gases practically do not absorb IR radiation. They are almost transparent in relation to IR radiation since a greater flux of energy (for example the high energetic UV radiation as in the case of nitrogen or oxygen) is needed for their excitement For a simplified calculation of radiation, a gray body model is used in which the area under the radiation intensity distribution curve via the wavelength is equal to the sum of the areas of the emission bands of the real gas. Because of the combustion acceleration due to gas slippage in submicron pores in the synthesis of particles behind the combustion front of carbon, it is necessary to take into account the IR absorption-emission of CO2. The emissivity of CO2 depends on the thickness of the gas layer and is proportional to the temperature $T^{3.5}$ while the flux density of the IR radiation from the solid phase is proportional to the temperature T^4.

The CO2 emission density per unit area of pore surface using Schack's formula [7] reads $E_{CO2} = 3.5\times\sqrt[3]{l\cdot p_{CO2}}\left(T_g/100\right)^{3.5}\ \left(W/m^2\right)$. Here p_{CO2} is the partial pressure (bar), l is the average thickness of the layer (m), and T_g is the gas temperature (K). The thickness of the layer depends on the shape of the pores and is proportional to the ratio of the pore volume to its area. The intensity of radiation and absorption of the pore surface according to the Stefan-Boltzmann law is equal to $E_S = \varepsilon_S c_0\left(T_g/100\right)^4\ \left(W/m^2\right)$, where $c_0 = 5.67$ is the Stefan-Boltzman constant, ε_S is the emissivity of the pore surface, T_g and T_S is the temperature of gas and solid phase.

Shack's formula with the Hottel's and Edgbert's modification, including the emissivity of the pore surface [7] is as follows $q_{CO2} = 0.01\cdot\varepsilon_S\cdot\left(p_{CO2}\cdot l\right)^{0.4}\left[\left(T_g/100\right)^{3.2} - \left(T_S/100\right)^{3.2}\times\left(T_g/T_S\right)^{0.65}\right]\left(W/m^2\right)$, where p_{CO2} – the partial CO2 pressure and l – the layer thickness.

2.1. *Micro flow caused by thermal radiation on isolated pore surface*

Our computations of slipping flow in an isolated tube showed the fast formation of steady state that depends only on radial (transverse) coordinate and doesn't depend on axial coordinate. The generalized Poiseuille flow using slippage boundary condition on the pore surface was obtained.
We begin with dimensional variables using cylindrical coordinates. Neglecting the advection transport, the steady state energy conservation equation reads

$$-\frac{1}{r}\frac{\partial}{\partial r}\left(\lambda_g r\frac{\partial T}{\partial r}\right) = Q_{rad}^{micro},\quad 0 < r < r_1 \tag{1}$$

Here Q_{rad}^{micro} denotes the micro heat flux due to thermal radiation. The equation (1) is solved using the symmetry and radiation boundary condition on the wall of a given pore

$$T'(0) = 0 \text{ and } -\lambda_g T'(r_1) = q_{CO2}$$

$$q_{CO2} = c_{02} \cdot \varepsilon_S \cdot \left(p_{CO2} \cdot r_1\right)^{0.4} \left[-\left(\frac{T(r_1)}{100}\right)^{3.2} + \left(\frac{T_W}{100}\right)^{3.2} \times \left(\frac{T(r_1)}{T_w}\right)^{0.65} \right] \left(\frac{W}{m^2}\right)$$

and $c_{02} = 10.427 \left(WBa^{-0.4}m^{-2.4}K^{-3.2}\right)$. For the region close to the tube surface, we obtain

$$Q_{rad}^{micro} = c_{02} \times r_1^{-1} \cdot \varepsilon_S \cdot \left(p_{CO2} \cdot r_1\right)^{0.4} \left[-\left(\frac{T(r_1)}{100}\right)^{3.2} + \left(\frac{T_W}{100}\right)^{3.2} \times \left(\frac{T(r_1)}{T_W}\right)^{0.65} \right]$$

2.2. *Average fluxes for computation at macro step*

Using the averaging over all pores inside the elementary volume V, we find the average flux as

$$\sum_l \int_{V_l} c_{02}\varepsilon_S r_1^{-1} \left[-\left(\frac{T(r_1)}{100}\right)^{3.2} + \left(\frac{T_W}{100}\right)^{3.2} \times \left(\frac{T(r_1)}{T_W}\right)^{0.65} \right]_l dV = V_{pore}^{\Sigma} c_{02}\varepsilon_S \hat{r}_1^{-1} \left[-\left(\frac{\hat{T}_g}{100}\right)^{3.2} + \left(\frac{\hat{T}_S}{100}\right)^{3.2} \times \left(\frac{\hat{T}_g}{\hat{T}_S}\right)^{0.65} \right]$$

The summation is assumed over all pores inside the given volume. Bar refers to the averaged values. Thus, the averaged dimensional radiation flux is:

$$Q_{rad}^{\dim} = \frac{\chi}{\hat{r}_1} c_{02} \cdot \varepsilon_S \cdot \left(\hat{p}_{CO2} \cdot \hat{r}_1\right)^{0.4} \left[-\left(\frac{\hat{T}_g}{100}\right)^{3.2} + \left(\frac{\hat{T}_S}{100}\right)^{3.2} \times \left(\frac{\hat{T}_g}{\hat{T}_S}\right)^{0.65} \right] \left(\frac{W}{m^3}\right).$$

3. Frank-Kamenetskii dimensionless variables

We apply the time and length scales for combustion proposed by Frank-Kamenetskii [38]. The reference time scale and length scales are as follows:

$$t_0 = k^{-1} \exp\left(E/RT_0\right) \text{ and } l_0 = \sqrt{\frac{\lambda_0 t_0}{c_V \rho_0}}.$$

We use below the transformed non-dimensional variables marked with tilde above the letters. For Cartesian coordinates the variables read

$$\begin{aligned} &\tilde{x}_k = x_k/l_0,\ \tilde{t} = t/t_0, \tilde{u}_k = u_k/u_0, k = 1,2,3,\ u_0 = l_0/t_0 \\ &\tilde{\rho}_{jg} = \bar{\rho}_{jg}/\rho_0, j = 1,2,\ \tilde{\rho}_S = \bar{\rho}_S/\rho_0 \end{aligned} \quad (2)$$

Here $p_0 = R\rho_0 T_0/M_0$ is the reference gas pressure. The effective density of gas and solid is marked with above bar (the mass of a given specie per unit volume occupied by this specie). $\bar{\rho}_g = \sum_j \bar{\rho}_{jg}$, $\rho_S = (1-\chi)\bar{\rho}_S$, $\rho_{jg} = \chi\bar{\rho}_{jg}$ where χ is a porosity. Using the similarity parameters $\beta = RT/E$, $\gamma = c_P T_0 \beta/Q$ the non-dimensional temperature is expressed as $\tilde{T} = (T - T_0)E/(RT_0^2)$; $T/T_0 = 1 + \beta\tilde{T}$. The dimensionless average macro heat flux due to thermal radiation can be presented in the form

$$\tilde{Q}_{g,rad}^{macro} = \frac{t_0 Q_{g,rad}^{\dim}}{\rho_0 c_g T_0} \text{ and } \tilde{Q}_{g,rad}^{macro} = \frac{A_R^0}{\beta} \tilde{p}_{CO2}^{0.4} \left(\frac{\left(1+\beta\tilde{T}_G\right)^{0.65}}{\left(1+\beta\tilde{T}_S\right)^{0.65}} \left(1+\beta\tilde{T}_S\right)^{3.2} - \left(1+\beta\tilde{T}_g\right)^{3.2} \right)$$

where the dimensionless intensity of thermal radiation is

$$A_R^0 = \chi \frac{c_{02}}{l_0 r_1} p_0^{0.4} \left(l_0 r_{pore}\right)^{0.4} \left(\frac{T_0}{10^2}\right)^{3.2} \times \frac{t_0}{\rho_0 c_g T_0}$$

and r_{pore} is dimensionless pore radius. For submicron pores, we take into account the formulas for the intensity of fluxes due to temperature jump and gas velocity slippage [14]. Thus, the heat flux due to temperature jump on the pores boundary reads

$$Q_{jump}^{macro} = \frac{2\chi}{r_1} \frac{\lambda_g \sqrt{\bar{\rho}_g P_g}}{\mu_g} \left(\frac{T_g}{T_W}\right)^{\alpha} \left(b_T\left(T_g - T_S\right) + b_W T_S\right)\left(\frac{W}{m^3}\right),\ b_T = \frac{\alpha'}{2-\alpha'}\sqrt{\frac{2}{\pi}} \frac{\gamma'+1}{2\gamma'},\ \gamma' = c_p/c_V$$

b_W characterizes the rate of a lateral pore surface cooling ($0 \le b_W \le 1$) and α' is the coefficient of thermal accommodation [14]. For the non-dimensional variables (2), we write down the macro heat flux due to temperature jump in the form

$$\tilde{Q}_{jump}^{macro} = A_T \sqrt{\tilde{\rho}_g \tilde{P}_g} \left(\left(-\tilde{T}_g + \tilde{T}_S\right) + \frac{b_W}{b_T \beta}\left(1 + \beta \tilde{T}_S\right)\right)\left(\frac{1+\beta \tilde{T}_g}{1+\beta \tilde{T}_S}\right)^{\alpha},\ A_T = \frac{2\chi b_T}{r_{pore}} \frac{l_0^2 \sqrt{\rho_0 P_0}}{\mu_0}$$

where A_T is the dimensionless temperature jump intensity.

The velocity slippage flux is given as $\vec{J}_{slip}^{macro} = 2\chi/\hat{r}_1\, b_u \sqrt{\bar{\rho}_g P_g}\,\mathbf{u}$ and $b_u = \theta\sqrt{2}/\left(\left(2-\theta\right)\sqrt{\pi}\right)$, where θ is the reflection coefficient of gas molecules from the pores boundary, $0 \le \theta \le 1$. Applying the variables (2), we obtain $\tilde{J}_{slip}^{macro} = A_u \tilde{\mathbf{u}} \sqrt{\tilde{\rho}_g \tilde{P}_g}$. Here A_u is the dimensionless intensity of slippage

$$A_u = \frac{2\chi b_u}{r_{pore}} \frac{t_0 \sqrt{\rho_0 P_0}}{\rho_0} \quad \text{or} \quad A_u = \frac{2\chi b_u}{r_{pore}} \frac{l_0^2 \sqrt{\rho_0 P_0}}{\mathrm{Re}_0\, \mu_0}.$$

Using the local Mach and Reynolds numbers $Ma_0 = u_0\sqrt{\rho_0}/\sqrt{\gamma P_0}$ and $\mathrm{Re}_0 = l_0 u_0 \rho_0/\mu_0$ the intensities of fluxes can be expressed as follows

$$A_u = \frac{2\chi b_u l_0}{r_{pore}\sqrt{\gamma}} \frac{1}{Ma_0} \quad \text{and} \quad A_T = \frac{2\chi l_0 b_T}{r_{pore}\sqrt{\gamma}} \frac{\mathrm{Re}_0}{Ma_0 Pe_0}.$$

Estimates

The radiation intensity is evaluated as $A_R^0 = c_{02} p_0^{0.4}\left(l_0 r_{pore}\right)^{-0.6} \left(T_0/100\right)^{3.2} t_0 \left(\rho_0 c_g T_0\right)^{-1}$, using $t_0 = l_0^2 \rho_0 c_v/\lambda_0$, it follows $A_R^0 = \chi c_{02} p_0^{0.4}\left(l_0 r_{pore}\right)^{-0.6}\left(T_0/100\right)^{3.2} l_0^2 \left(T_0 \lambda_0 \gamma\right)^{-1}$. For reference values $l_0 \approx 10^{-5}$, $T_0 \approx 10^3$, $\mu_0 = 5 \cdot 10^{-5}$ we estimate $A_R^0 \approx \chi\left(r_{pore}\right)^{-0.6} l_0^{1.4} c_R$, where $c_R = c_{02} p_0^{0.4} T_0^{2.2} 10^{-6.2} \lambda_0^{-1}$ and $c_R \approx 10^5$. For pores of submicron size [13] $A_u \approx 200\chi b_u$ and $A_T \approx 200\chi b_T$.

4. The governing equations

The magnesium -zinc ferrite synthesis using the carbon combustion (CCSO) is considered. The kinetics [11] is applied:

$$\begin{aligned} &0.25MnCO_3(s) + 0.75ZnO(s) + Fe_2O_3(s) + C(s) + O_2(g) \rightarrow \\ &0.25CO_2(g) + Mn_{0.25}Zn_{0.75}Fe_2O_4(s) \end{aligned} \tag{3}$$

The densities of gas species O2 and CO2 are $\rho_{1g} = \rho_{O2}$ and $\rho_{2g} = \rho_{CO2}$. The densities of solid species C, MnCO3, ZnO, Fe2O3, and magnesium- zinc ferrite are as follows

$\rho_{1S} = \rho_C$, $\rho_{2S} = \rho_{MnCO3}$, $\rho_{3S} = \rho_{ZnO}$, $\rho_{4S} = \rho_{Fe2O3}$, and $\rho_{5S} = \rho_{Fer}$

We use below the dimensionless variables (2), tilde is omitted. The mass conservation equation for total gas density reads

$$\frac{\partial \chi \rho_g}{\partial t} + \nabla \cdot \left(\chi \rho_g \mathbf{u}\right) = (1-\chi) J^{macro}_{S\to g}, \text{ where mass flux } J^{macro}_{S\to g} = \rho_{1S}\rho_{1g} k_1 \exp\left(\frac{T_g}{\beta T_g + 1}\right)$$

The mass conservation equations for gas species are $C_2 = 1 - C_1$

$$\chi \rho_g \frac{\partial C_1}{\partial t} + \chi \rho_g \mathbf{u} \nabla \cdot (C_1) = \nabla \cdot \left(\frac{\chi}{Pe_1} \rho_g \nabla C_1\right) - \left(1 + \frac{M_{1g}}{M_{1S}}\right) J^{macro}_{S\to g}, \quad 1 + \frac{M_{1g}}{M_{1S}} = \frac{M_{2g}}{M_{1S}}$$

The mass conservation equations for solid species are as follows

$$\frac{\partial \rho_{1S}}{\partial t} = -\frac{M_{1S}}{M_{4S}} J^{macro}_{S\to S}, \quad \frac{\partial \rho_{2S}}{\partial t} = -0.25 \frac{M_{2S}}{M_{4S}} J^{macro}_{S\to S}, \quad \frac{\partial \rho_{3S}}{\partial t} = -0.75 \frac{M_{3S}}{M_{4S}} J^{macro}_{S\to S}, \quad \frac{\partial \rho_{4S}}{\partial t} = -J^{macro}_{S\to S}$$

$$\frac{\partial \rho_{5S}}{\partial t} = \frac{M_{5S}}{M_{4S}} J^{macro}_{S\to S}, \quad J^{macro}_{S\to S} = \rho_{2S}\rho_{3S}\rho_{4S} k_2 \exp\left(\frac{T_S}{\beta T_S + 1}\right) \text{ and } \rho_S = \sum_j \rho_{jS}$$

The momentum conservation includes the distributed porous resistance and the flux caused by gas slippage on porous surface

$$\frac{\partial \chi \rho_g \mathbf{u}}{\partial t} = \mathbf{S}_V + \vec{J}^{macro}_{slip} - \mathrm{Ma}^{-2} \nabla p_g + \mathrm{Re}^{-1} \nabla \cdot \boldsymbol{\tau}; \quad \boldsymbol{\tau} = \mu \left[\nabla \mathbf{u} + (\nabla \mathbf{u})^T - \frac{2}{3} (\nabla \cdot \mathbf{u}) \mathbf{I}\right]$$

The porous resistance and slippage fluxes are as follows

$$(\mathbf{S}_V)_i = -u_i \kappa_i, \quad \kappa_i = \alpha_i |u| + \varsigma_i \;, i = 1,2,3.$$

The thermal balance for gas phase reads

$$\chi \rho_g c_g \left(\frac{\partial T_g}{\partial t} + \mathbf{u} \cdot \nabla T_g\right) = \nabla \cdot \left(\frac{\lambda_g}{Pe} \nabla T_g\right) - \kappa (T_g - T_S) + Q^{macro}_{g,rad} + Q^{macro}_{g} + Q^{macro}_{jump} \text{ and } Q^{macro}_{g} = Q J^{macro}_{S\to g}$$

The thermal fluxes caused by temperature jump and thermal radiation are

$$\vec{J}^{macro}_{slip} = A_u \mathbf{u} \sqrt{p_g \rho_g}, \; Q^{macro}_{jump} = A_T \sqrt{\rho_g p_g} \left(\left(T_g - T_S\right) + \frac{b_W}{b_T \beta}(1 + \beta T_S)\right) \left(\frac{1 + \beta T_g}{1 + \beta T_S}\right)^{\alpha} \tag{4}$$

$$Q^{macro}_{g,rad} = \frac{A^0_R}{\beta} (p_{CO2})^{0.4} \left(\frac{(1 + \beta T_g)^{0.65}}{(1 + \beta T_S)^{0.65}} (1 + \beta T_S)^{3.2} - (1 + \beta T_g)^{3.2}\right) \tag{5}$$

The heat balance equation for solid phase reads

$$(1-\chi) \sum_j \rho_{jS} c_{jS} \frac{\partial T_S}{\partial t} = \nabla \cdot \left(\frac{\lambda_S}{Pe_T} \nabla T_S\right) + \kappa (T_g - T_S) + Q^{macro}_{gS} - Q^{macro}_{g,rad}$$

$$Q^{macro}_{gS} = Q \rho_{1S} \rho_{1g} k_1 \exp\left(\frac{T_S}{\beta T_S + 1}\right)$$

The heat exchange between solid and gas phases is described using the Leveque formula [38] that depends on Peclet and Reynolds numbers as follows $\kappa = \kappa_0 \left(1 + \mathrm{Re}_0^{0.3} Pe_T^{0.3}\right)$. The pressure satisfies the perfect gas equation $P_g = \rho_g \left(1 + \beta T_g\right)$.

5. Results of modelling and discussions

We report the carbon combustion simulation in a sample with submicron pores. The effect of thermal radiation, temperature jump as well as the velocity slippage on the pore surface has been analyzed. The results of computation of combustion presented below refer to the values of parameters as follows. $\beta \approx 0.1, \gamma \approx 0.288, \chi = 0.05$, where $\beta = RT_0/E$ and $\gamma = c_P T_0 \beta / Q$. The dimensionless resistance of porosity is given as $\alpha_i = 0$ and $\varsigma_i = 50$. The Mach and Reynolds number are $Ma_0 = 0.2$ and $\mathrm{Re}_0 = 10^{-2}$ The intensities of slippage, jump (4) and radiation (5) are given using the values: $A_u = 10^3$, $A_T = 10^2$, $\alpha = 0$, and $A_R^0 = 0,10,30,400$. The thermal heat in the ignition is given as $q_f(t) = \begin{cases} \mathrm{Q_f}, 0 < t < t_{init} \\ 0, t \geq t_{init} \end{cases}$, where $\mathrm{Q_f} = 1200$. The ignition time is $t_{init} = 0.25 \div 0.5$. The cylindrical sample was placed in vertical position. The ignition initiated from the bottom at $x = 0$. In the figures below $0 \leq r \leq 1, 0 \leq x \leq 5$. The results of submicron particles is presented for radiation intensities: $A_R^0 = 0$, $A_R^0 = 10$, $A_R^0 = 30$, $A_R^0 = 400$, emissivity $\varepsilon_s = 0.85$, slippage intensity $A_u = 100$, and temperature jump $A_T = 10$. The data in figures 4÷8 refers to variable emissivity (5), which depends on CO2 partial pressure.

The influence of the heat CO2 absorption is demonstrated in figure 1 via gas temperature versus time distribution for various positions on the symmetry axis $(0,0),(0,0.3),(0,1),(0,1.5),(0,2)$, and $(0,3)$ lines 1, 2, 3, 4, 5, and 6.

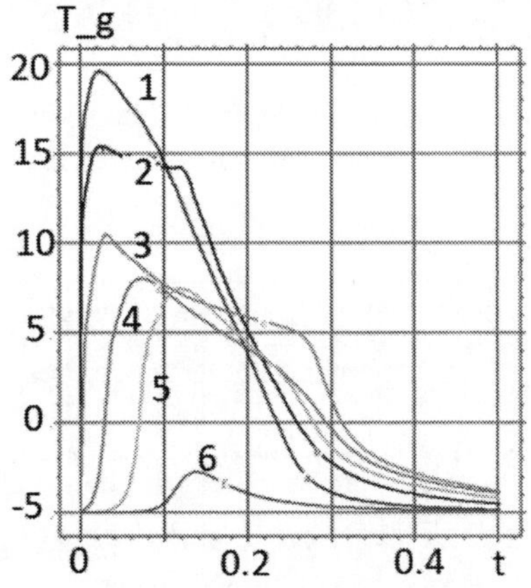

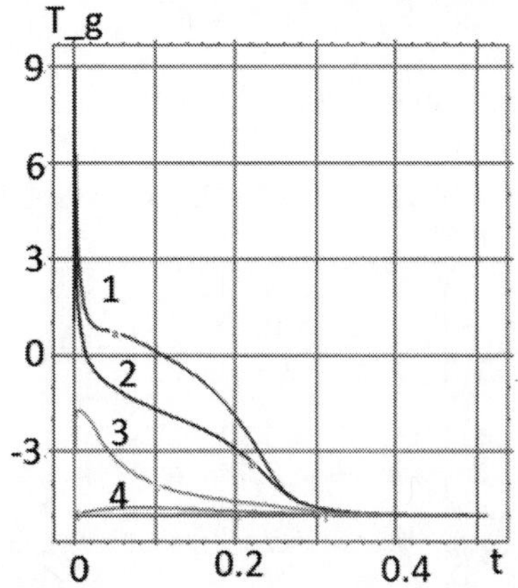

Figure 1. The CO2 heat absorption effect on the gas temperature is presented for radiation intensity $A_R^0 = 400$ -Temperature of the gas phase versus time on the axis of symmetry for positions with coordinates $(0,0),(0,0.3),(0,1),(0,1.5),(0,2)$, and $(0,3)$ lines 1÷6 respectively is shown. Left and right column refers to the computation for constant emissivity of gas and variable one (5).

The radiation heat flux caused by CO2 molecules absorption results in remarkable (about double decrease) in the gas temperature (compare left and right columns in figure 1). The gas specie CO2 is generated just behind the combustion front. The heat absorbed in this region decreases both temperature and the velocity of the propagating combustion waves. Thus, neglecting the radiation

CO2 heat absorption gives the essential higher temperature maximum in combustion wave compared to the model that includes absorption (5).

Figure 2 shows the rate of propagating combustion front that is characterized by gas temperature versus time distribution for various positions $(0,0)$, $(0,0.3)$, $(0,1)$, $(0,1.5)$, $(0,2)$ and $(0,3)$ on the axis of symmetry of the sample. We can see the gas temperature decrease due to thermal radiation. The gas heat loss due to radiation absorption by CO2 molecules becomes higher with increasing values of radiation intensity. Using $A_R^0 \approx \chi\left(r_{pore}\right)^{-0.6} l_0^{1.4} c_R$ we conclude the remarkable lower gas temperature for smaller pore size.

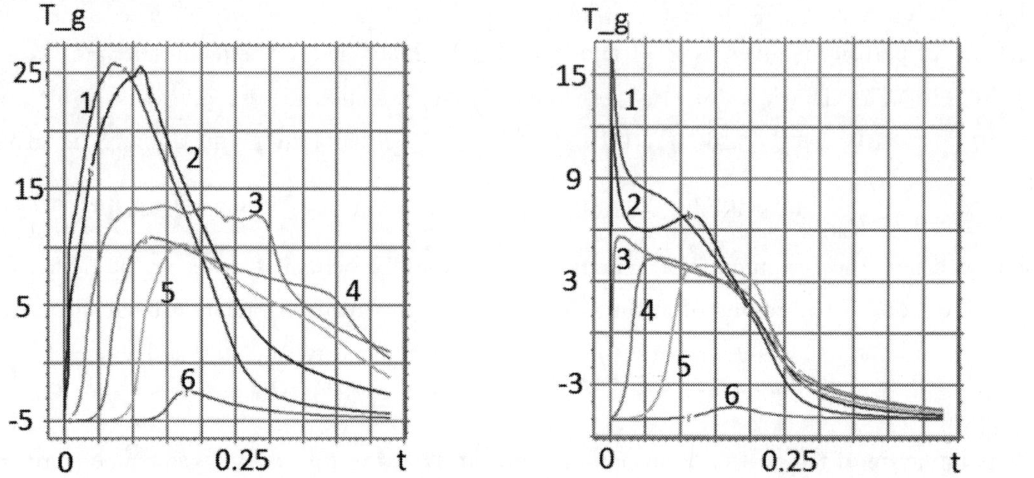

Figure 2. The temperature of gas phase versus time on the axis of symmetry for positions with coordinates $(0,0)$, $(0,0.3)$, $(0,1)$, $(0,1.5)$, $(0,2)$ and $(0,3)$ lines $1 \div 6$ respectively is shown. Left and right column refers to $A_R^0 = 0$ (no radiation) and to $A_R^0 = 10$ (variable emissivity (5)).

The solid temperature versus time on the symmetry axis is presented in figure 3. The radiation heat flux from gas to solid using (5) results in the increase of solid temperature. Radiation heat flux is provided by the sufficiently high temperatures of the solid.

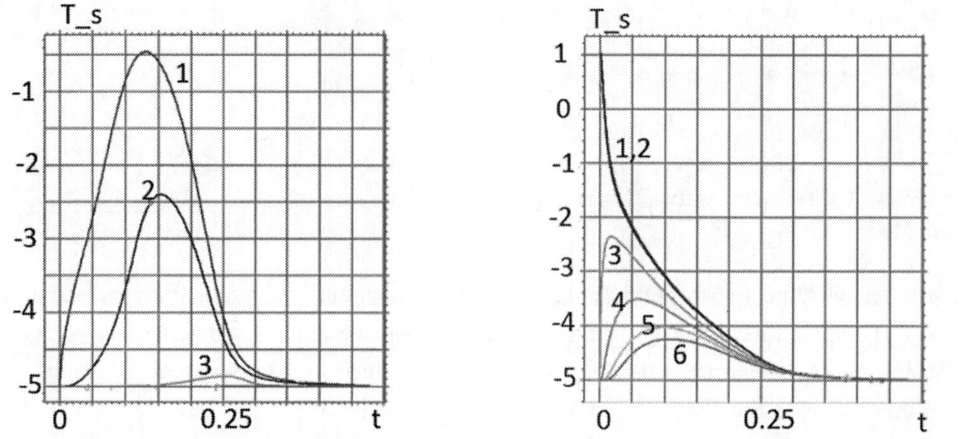

Figure 3. Temperature of solid phase versus time on the axis of symmetry for positions with coordinates $(0,0)$, $(0,0.3)$, $(0,1)$, $(0,1.5)$, $(0,2)$ and $(0,3)$ lines $1 \div 6$ respectively is shown. Left and right column refers to $A_R^0 = 0$ (no radiation) and to $A_R^0 = 10$ (the emissivity of gas depends on CO2 partial pressure).

IOP Conf. Series: Journal of Physics: Conf. Series **1009** (2018) 012040 doi:10.1088/1742-6596/1009/1/012040

The comparison of gas temperature for sufficiently big particle synthesis of $124\mu m$ and $1.7\mu m$ is shown in figure 4. Note the higher temperature and more intensive combustion takes place for bigger particles (see left column in figure 4).

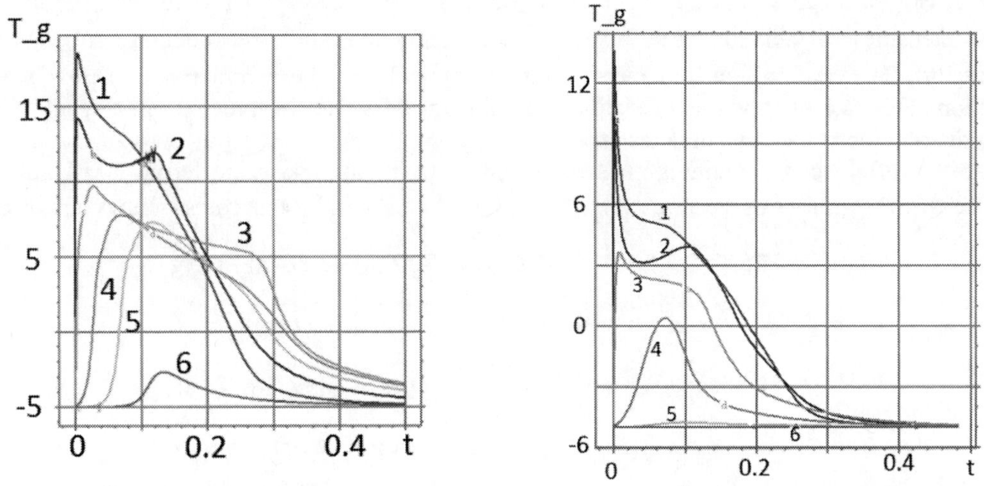

$A_R^0 = 2.3,\ r_{pore} = 124\mu m$ $A_R^0 = 30, r_{pore} = 1.7\mu m$

Figure 4. The temperature of gas phase versus time on the axis of symmetry for positions with coordinates $(0,0)$, $(0,0.3)$, $(0,1)$, $(0,1.5)$, $(0,2)$ and $(0,3)$ lines $1 \div 6$ respectively is shown. Left and right column refers to $A_R^0 = 2.3,\ r_{pore} = 124\mu m$ and $A_R^0 = 30, r_{pore} = 1.7\mu m$.

We estimated the radiation heat fluxes using the constant emissivity radiation for slipping model and the similar computation for which the emissivity of CO2 was varying with partial CO2 pressure (5). The comparison (see figure 1) demonstrates the remarkable increase of radiation heat fluxes for the constant emissivity computation. The rate of carbon oxide generation behind the propagating combustion front is illustrated in figure 5.

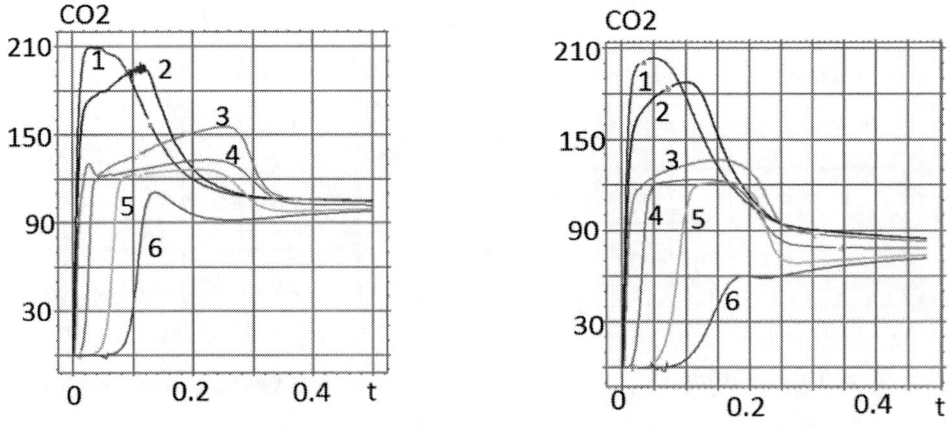

$A_R^0 = 2.3,\ r_{pore} = 124\mu m$ $A_R^0 = 10,\ r_{pore} = 10\mu m$

Figure 5. The CO_2 density versus time on the axis of symmetry for positions with coordinates $(0,0)$, $(0,0.3)$, $(0,1)$, $(0,1.5)$, $(0,2)$ and $(0,3)$ lines $1 \div 6$ respectively is shown. Left and right column refers to $A_R^0 = 2.3,\ r_{pore} = 124\mu m$ and $A_R^0 = 10,\ r_{pore} = 10\mu m$.

IOP Conf. Series: Journal of Physics: Conf. Series **1009** (2018) 012040 doi:10.1088/1742-6596/1009/1/012040

The significant higher gas pressure due to thermal radiation is found in pores of bigger size, i.e. for smaller radiation intensity, compare $A_R^0 = 10,\ r_{pore} = 10\mu m$ and $A_R^0 = 400, r_{pore} = 2nm$ in figure 6.

The density of magnesium-zinc ferrite at the same time instance $t = 0.5$ is shown in figure 7 for various intensities of radiation heat flux $A_R^0 = 2.3, 10, 30$, and 400 (see columns A,B,C, and D respectively). The heat lost due to thermal radiation results in the submicron particles synthesis retardation. This rate of slowness of synthesis is illustrated as well in figure 8 via reagent MnCO3 density distribution at the same time instance.

Note the difference of synthesis magnesium-zinc ferrite and nickel-zinc ferrite via combustion [11]. The CCSO synthesis of the $Mn_{0.25}Zn_{0.75}Fe_2O_4$ and $Ni_{0.35}Zn_{0.65}Fe_2O_4$ ferrites follows the reactions

$$0.25MnCO_3(s) + 0.75ZnO(s) + Fe_2O_3(s) + \alpha\left(C(s) + O_2(g)\right)$$

$$\rightarrow \left(\alpha + 0.25\right)CO_2(g) + Mn_{0.25}Zn_{0.75}Fe_2O_4(s) \text{ and}$$

$$0.35NiO(s) + 0.65ZnO(s) + Fe_2O_3(s) + \alpha\left(C(s) + O_2(g)\right)$$

$$\rightarrow Ni_{0.35}Zn_{0.65}Fe_2O_4(s) + \left(\alpha\right)CO_2(g)$$

The main difference between reactions of magnesium-zinc and nickel-zinc ferrite formation is that at the same content of carbon in the green mixture (that contains (α) moles of carbon) the greater amount of gaseous carbon dioxide is formed during the synthesis of magnesium-zinc ferrite. As one can see from the kinetic equations the magnesium-zinc ferrite synthesis is more promising than the synthesis of nickel-zinc ferrite in terms of gaseous carbon oxide generation since it produces $1 + 0.25/\alpha$ moles of carbon oxide per each mole of solid carbon contained initially in the green charge. While during the synthesis of nickel-zinc ferrite just about one mole of only is released per each mole of the carbon contained initially in the green mixture

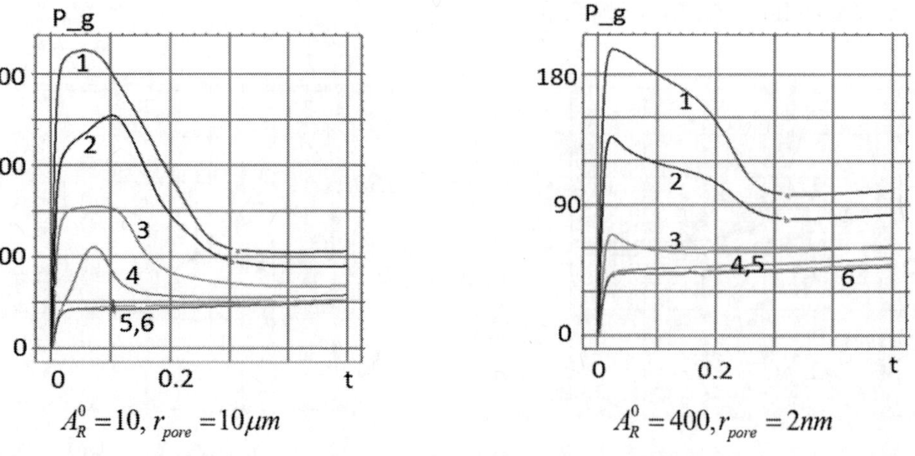

Figure 6. Pressure of gas phase versus time on the axis of symmetry for positions with coordinates (0,0), (0,0.3), (0,1), (0,1.5), (0,2) and (0,3) lines $1 \div 6$ respectively is shown. The emissivity of gas depends on CO2 partial pressure. Left and right column refers to radiation intensity $A_R^0 = 10$ and $A_R^0 = 400$.

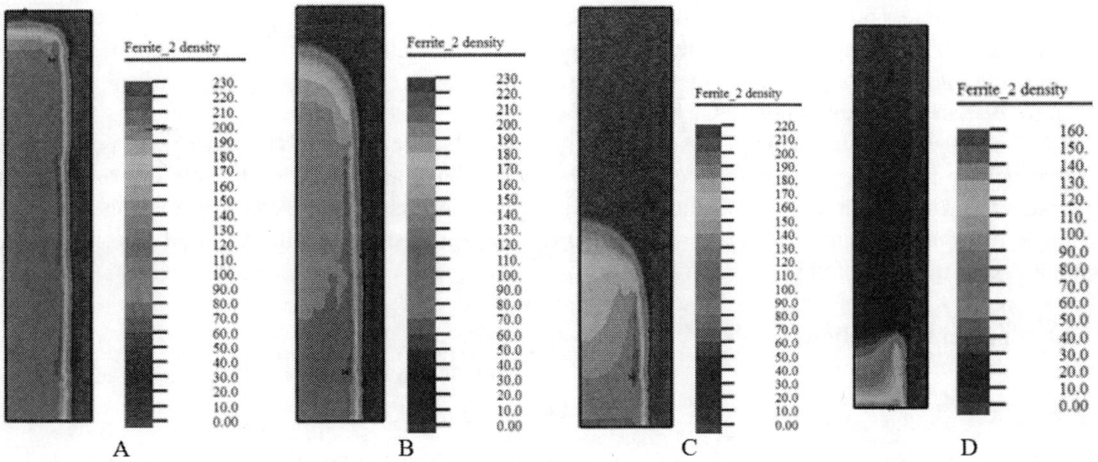

Figure 7. The product of synthesis $Mn_{0.25}Zn_{0.75}Fe_2O_4$ magnesium-zinc ferrite density in the reactor region $0<r<1;\ 0<z<5$ at the time instant $t=0.5$ is demonstrated. The columns A, B, C and D refer to increasing values of radiation intensity $A_R^0 = 2.3, 10, 30$ and 400. The corresponding pore size is as follows $24\mu m, 10\mu m, 1.7\mu m$ and $2nm$.

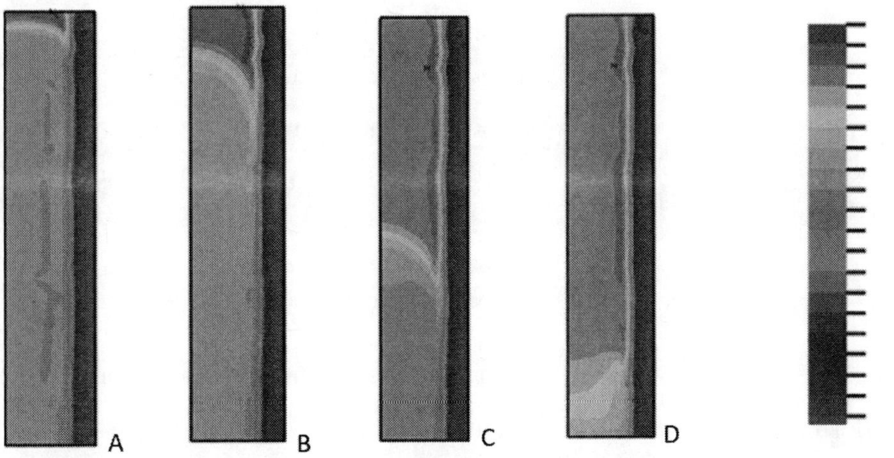

Figure 8. The reagent magnesium-zinc oxide $MnCO_3$ density when the emissivity of gas depends on CO2 partial pressure in the reactor region $0<r<1;\ 0<z<5$ at the time instant $t=0.5$ are shown for increasing values of radiation intensity $A_R^0 = 2.3, 10, 30$ and 400. Column A, B, C, and D refers to $A_R^0 = 2.3, 10, 30$ and 400 respectively.

6. Conclusion

The fluxes induced by thermal radiation and velocity slipping fluxes in submicron pores as well as the fluxes of gas induced by temperature jumps are formulated using similarity parameters and applied in numerical simulation of the thermal front motion in a porous channel. The results allow us to predict characteristic features of the heat and mass transfer behind the thermal front in the sample with submicron pores. The significant impact of radiation as well as of velocity slipping for large Knudsen Number is reported. The results of thermal front numerical simulation allow us to evaluate and predict the impact of the CO2 emissivity and absorbtivity on the heat mass transfer including the effect of gas velocity slipping and temperature jump at submicron pore surface. An intense gas flow from the combustion zone and the sample as a whole provides the excessive heat loss and decelerates propagation of the combustion wave during slipping.

We obtained that an increase in the intensity of radiation corresponds to the decrease of the submicron particles growth rate in CCSO wave.

The stronger slipping promotes the more intensive advection. The CO2 gas specie is generated intensively in the wave region close to a propagating combustion front from the side of products. The radiation heat flux caused by CO2 molecules absorption results in remarkable decreasing of the gas temperature and pressure. The smaller size of pores is considered the smaller value of gas temperature and pressure in pore is found.

List of principal notations

T'	Dimensional temperature
$T_0 = 1000^0 C$	Referred temperature
$T = T'/T_0$	Dimensionless temperature
$\tilde{T} = (T-1)/\beta$	Dimensionless normalized temperature
T_g , T_{jS}	Gas temperature and solid phase temperature
T_{init}	Ignition temperature
$T_{initial}$, $T_{initial} = -1/\gamma$	Initial temperature of the sample
c_{Pk}	Specific capacity at constant pressure
ρ_{lg}, $l = 1,2$	Densities of gas species
$\rho_g = \rho_{1g} + \rho_{2g}$	Density of gas mixture
$\bar{\rho}_g$, $\bar{\rho}_S$, $\rho_g = \chi\bar{\rho}_g$, $\rho_S = (1-\chi)\bar{\rho}_S$	Effective values of gas density ρ_g and solid density ρ_S
$\beta = RT_0/E$	Dimensionless parameter
$\gamma = c_P T_0 \beta / Q$	Dimensionless parameter (Zeldovich number)
A_R^0	Dimensionless radiation heat parameter
$E_S = \varepsilon_S c_0 (T_g/100)^4 \; (W/m^2)$	Radiative flux
$c_0 = 5.67$	Stefan-Boltzmann constant
ε_S, $(0 \le \varepsilon_S \le 1)$	emissivity
$Pe_l = u_0 l_0 / D_{l,0}, l = 1,2$	Peclet numbers
$Pe = u_0 l_0 / \lambda_0$	Thermal Peclet number
$u_0 = l_0 / t_0$	Referred velocity
$\mathrm{Re}_0 = u_0 l_0 \rho_0 / \mu_0$	Reynolds number
χ	Porosity coefficient
P_0	Referred pressure
μ_0	Referred gas viscosity
λ_0	Referred thermal conductivity
Ma_0	Mach number
Q_f	Thermal flux
Q	Combustion heat

List of principal notations

$Q^{macro}_{g,rad}, Q^{macro}_{g}, Q^{macro}_{jump}$	Heat fluxes due to radiation, chemical reaction, and temperature jump
b_u; $b_u = \theta\sqrt{2}/\left((2-\theta)\sqrt{\pi}\right)$	Slip parameter
θ, $(0 \le \theta \le 1)$	Reflection coefficient
A_T $b_T = \sqrt{2/\pi}\,\alpha'\left(\gamma'+1\right)/\left((2-\alpha')2\gamma'\right), \gamma' = c_p/c_V$	Temperature jump parameters
α', $(0 \le \alpha' \le 1)$	Thermal accommodation coefficient
D_T	Thermal diffusivity coefficient
R	Universal gas constant
$t_0 = k^{-1}\exp\left(E/RT_0\right)$, $l_0 = \sqrt{\lambda_0 t_0/(c_V\rho_0)}$	Referred time and length scales for the Frank-Kamenetskii variables
t_{init}	Ignition time

References

[1] Sheng S, Mavrokefalos A, Sambegoro P and Gang Chen G 2012 *Appl. Phys Let.* **100** (23) http://dx.doi.org/10.1063/1.4723713
[2] Sheng S Narayanaswamy A and Chen G 2009 *Nano Lett.* **9** (8) pp 2909–13 DOI: 10.1021/nl901208v http://pubs.acs.org/doi/abs/10.1021/nl901208v
[3] Martinenko Y and Ognev L 2005 *J. Tech. Phys*. **V. 75** (11) pp 130–32
[4] Smirnov B 1997 *Physics-Uspekhi* **40** (11):1117. (In Russian)
[5] Xi Liu X and Zhang Zh 2015 *ACS Photonics* Publication Date (Web): 17 Aug
[6] Xianglei Liu and Zhuomin Zhang 2015 Near-field thermal radiation between metasurfaces *ACS Photonics* ACS Photonics **2** (9) pp 1320–26 DOI: 10.1021/acsphotonics.5b00298
[7] Schack Alfred 1965 *Industrial Heat Transfer Practical and theoretical with basic numerical examples*. Translated from the 6th German ed. by I. Gutman Publisher: Springer US; Copyright Holder: Springer Science+Business Media New York. eBook ISBN: 978–1–5041–2079–1
[8] Martirosyan K S and Luss D 2005 Carbon Combustion Synthesis of Oxides: Process Demonstration and Features *AIChE.* **51** (10) pp 2801–10
[9] Martirosyan K Iliev M and Luss D 2007 *Internat. J. SHS* **16** (1) pp 36–45
[10] Martirosyan K S and Luss D 2009 *Chem. Eng. Technol.* **32** (9) pp 1376–83
[11] Martirosyan K S and Luss D 2007 Carbon Combustion Synthesis of Ferrites: Synthesis and Characterization *Ind. Eng. Chem. Res.* **46** pp 1492–99
[12] Markov A, Filimonov I and Martirosyan K 2012 *J. Comput. Phys*. **231** (20) pp 6714–24
[13] Markov A A 2015 Heat and Mass transfer in submicron pores at the movement of thermal front in a channel *Institute for Problems in Mechanics RAS Moscow (Preprint №1108*) p 45 (in Russian).
[14] Markov A 2014 *Computers and Fluids* **99** pp 83–92
[15] Markov A 2010 Micro and Macro Scale Technique for Particle Growth Simulation *Computational Fluid Dynamics Review* 2010 Chapter 24 pp 583–600
[16] Markov A, Barinov V, Umarov L and Filimonov I 2016 (in Russian) *Proc. NPNJ'2016* 25–31 May Alushta pp 97-99 ISBN 978-5-4316-0300-6
[17] Filimonov I and Poletaev A 2016 *Current Opinion in Chemical Engineering* **11** pp 42–45
[18] Rogachev A and Mukasyan A 2012 Combustion for synthesis of materials: Introduction into Structural Macrokinetics *Physmatlit* 398 p (in Russian)
[19] Merzhanov A 1997 *Reports of RAS* **503** (7) pp 353–55

[20] Merzhanov A, Peregudov A and Gontkovskaya V 1997 *Reports of RAS* **360** pp 217–19
[21] Filimonov I 1998 Combust. Explos. and Shock Waves **343** pp 69–76
[22] Zeldovich Y and Frank-Kamenetskii D 1938 *J Phys. Chem.* **12** pp 93–99
[23] Zeldovich Y 1942 *J of Experiment. and Theoret. Phys.* **12** (11/12) pp 498–524
[24] Markov A, Filimonov I and Martirosyan K 2011 in: Kuzmin A (Ed.) *Comput. Fluid Dyn.* (Springer London) pp 929–936
[25] Markov A 2009 *Computers & Fluids* **38** pp 1435–44
[26] Markov A, Filimonov I and Martirosyan K 2013 *Internat. J. of SHS* **22** (1) pp 11–17
[27] MarkovA A, Hobosyan M A and Martirosyan K S 2016 Slip-Jump Model for Carbon Combustion Synthesis of Complex Oxide Nanoparticles *Eurasian Chemico-Technological Journal* **18** pp 223–32
[28] Markov A A, Hobosyan M A and Martirosyan K S 2015 Simulation of heat and mass transfer in pores as applied to synthesis of magnesium–zinc and nickel–zinc ferrite nanoparticles *Nanomechanics Science and Technology: An International Journal* DOI: 10.1615 Nanomechanics Sci.Technol.Int.J. **v6**. (i3.40) pp 209–22
[29] Markov A A, Hobosyan M A and Martirosyan K S 2016 Ferrite Synthesis Simulation via Carbon Combustion using Slip Temperature and Concentration Gas Species Jump at Pore Surface. *Physical-Chemical Kinetics in Gas Dynamics* **V16** (1) (in Russian)
[30] Markov A, Filimonov I, Hobosyan M and Martirosyan K 2016 *Wschodnioeuropejskie Czasopismo Naukowe (East European Scientific J.)* **9**(13) pp 58–63 http://eesa-journal.com/ru/o-zhurnale
[31] Filimonov I and Luss D 2005 *AIChE J.* **51** pp 1521–31
[32] Filimonov I 1998 *27th Intern. Symp on Combust.* (Boulder) pp 2441–50
[33] Filimonov I, Kidin N and Mukasyan A 2000 *Proc. of the Combust. Inst.* **28** pp 1421-29
[34] Seplyarskii B 2017 Abstracts of Internat. Conf Internat. SHS-50 dated for 50-year anniversary of SHS Chernogolovka (november 20-21st) p 192
[35] Seplyarskii B G et al 2014 *Mendeleev Communic.* **4** pp 242–44
[36] Karnidakis G, Beskok A and Aluru N 2005 *Microflows and nanoflows* Springer Science+Business Media Inc. p 817
[37] Kogan M 1967 Dynamics of rarefied gas *Moscow Nauka* p 440 (in Russian)
[38] Frank-Kamenetskii D A 1969 *Diffusion and Heat Transfer in Chemical Kinetics* (Second Enlarged and Revised Edition Translation Editor: J.P. Appleton Plenum Press) p 491
[39] Williams F A 1985 *Combustion theory* Westview Press p 704
[40] Koutateladze S 1970 *Basics of the Theory of Heat Exchange* Nauka Siberian Branch Novosibirsk p 658